Enhanced Oil Recovery
Field Case Studies

Enhanced Oil Recovery Field Case Studies

James J. Sheng

Bob L. Herd Department of Petroleum Engineering,
Texas Tech University,
Lubbock, TX 79409-3111
USA

AMSTERDAM • BOSTON • HEIDELBERG • LONDON
NEW YORK • OXFORD • PARIS • SAN DIEGO
SAN FRANCISCO • SINGAPORE • SYDNEY • TOKYO
Gulf Professional Publishing is an imprint of Elsevier

Gulf Professional Publishing is an imprint of Elsevier
225 Wyman Street, Waltham, MA 02451, USA
The Boulevard, Langford Lane, Kidlington, Oxford OX5 1GB, UK

Notice
Knowledge and best practice in this field are constantly changing. As new research and experience broaden our understanding, changes in research methods or professional practices, or medical treatment may become necessary.

Practitioners and researchers must always rely on their own experience and knowledge in evaluating and using any information, methods, compounds, or experiments described herein. In using such information or methods they should be mindful of their own safety and the safety of others, including parties for whom they have a professional responsibility.

To the fullest extent of the law, neither the Publisher nor the authors, contributors, or editors, assume any liability for any injury and/or damage to persons or property as a matter of products liability, negligence or otherwise, or from any use or operation of any methods, products, instructions, or ideas contained in the material herein.

Library of Congress Cataloging-in-Publication Data
A catalog record for this book is available from the Library of Congress

British Library Cataloguing-in-Publication Data
A catalogue record for this book is available from the British Library

ISBN: 978-0-12-386545-8

For information on all Gulf Professional Publishing publications visit our website at http://store.elsevier.com

Printed and bound by CPI Group (UK) Ltd, Croydon, CR0 4YY

Contents

Preface xix
Contributors xxi
Acknowledgments xxiii

1. Gas Flooding **1**

Russell T. Johns and Birol Dindoruk

1.1 What Is Gas Flooding? **1**
1.2 Gas Flood Design **2**
1.3 Technical and Economic Screening Process **3**
1.4 Gas Injection Design and WAG **5**
1.5 Phase Behavior **9**
1.5.1 Standard (or Basic) PVT Data 9
1.5.2 Swelling Test 9
1.5.3 Slim-Tube Test 10
1.5.4 Multicontact Test 11
1.5.5 Fluid Characterization Using an Equation-of-State 12
1.6 MMP and Displacement Mechanisms **12**
1.6.1 Simplified Ternary Representation of Displacement Mechanisms 13
1.6.2 Displacement Mechanisms for Field Gas Floods 15
1.6.3 Determination of MMP 15
1.7 Field Cases **16**
1.7.1 Slaughter Estate Unit CO_2 Flood 16
1.7.2 Immiscible Weeks Island Gravity Stable CO_2 Flood 17
1.7.3 Jay Little Escambia Creek Nitrogen Flood 19
1.7.4 Overview of Field Experience 20
1.8 Concluding Remarks **21**
Abbreviations **21**
References **22**

2. Enhanced Oil Recovery by Using CO_2 Foams: Fundamentals and Field Applications **23**

S. Lee and S.I. Kam

2.1 Foam Fundamentals **23**
2.1.1 Why CO_2 Is so Popular in Recent Years? 23
2.1.2 Why CO_2 Is of Interest Compared to Other Gases? 24
2.1.3 Why CO_2 Is Injected as Foams? 24

2.1.4 Foam in Porous Media: Creation and Coalescence Mechanisms 25
2.1.5 Foam in Porous Media: Three Foam States and Foam Generation 25
2.1.6 Foam in Porous Media: Two Strong-Foam Regimes—High-Quality and Low-Quality Regimes 27
2.1.7 Modeling Foams in Porous Media 28
2.1.8 Foam Injection Methods and Gravity Segregation 30
2.1.9 CO_2-Foam Coreflood Experiments 31
2.1.10 Effect of Subsurface Heterogeneity—Limiting Capillary Pressure and Limiting Water Saturation 32
2.1.11 Foam–Oil Interactions 34
2.2 Foam Field Applications **34**
2.2.1 The First Foam Field Applications, Siggins Field, Illinois 34
2.2.2 Steam Foam EOR, Midway Sunset Field, California 35
2.2.3 CO_2/N_2 Foam Injection in Wilmington, California (1984) 37
2.2.4 CO_2-Foam Injection in Rock Creek, Virginia (1984–1985) 38
2.2.5 CO_2-Foam Injection in Rangely Weber Sand Unit, Colorado (1988–1990) 39
2.2.6 CO_2-Foam Injection in North Ward-Estes, Texas (1990–1991) 40
2.2.7 CO_2-Foam Injection in the East Vacuum Grayburg/San Andres Unit, New Mexico (1991–1993) 42
2.2.8 CO_2-Foam Injection in East Mallet Unit, Texas, and McElmo Creek Unit, Utah (1991–1994) 43
2.3 Typical Field Responses During CO_2-Foam Applications **45**
2.3.1 Diversion from High- to Low-Permeability Layers 45
2.3.2 Typical Responses from Successful SAG Processes 46
2.3.3 Typical Responses from Successful Surfactant–Gas Coinjection Processes 51
2.4 Conclusions **52**
Acknowledgment **53**
Appendix—Expression of Gas-Mobility Reduction in the Presence of Foams **53**
References **59**

3. Polymer Flooding—Fundamentals and Field Cases **63**
James J. Sheng

3.1 Polymers Classification **63**
3.2 Polymer Solution Viscosity **64**
3.2.1 Salinity and Concentration Effects 64
3.2.2 Shear Effect 65
3.2.3 pH Effect 65
3.3 Polymer Flow Behavior in Porous Media **65**
3.3.1 Polymer Viscosity in Porous Media 65
3.3.2 Polymer Retention 67

3.3.3 Inaccessible Pore Volume 68
3.3.4 Permeability Reduction 69
3.3.5 Relative Permeabilities in Polymer Flooding 70
3.4 Mechanisms of Polymer Flooding 70
3.5 Polymer Mixing 72
3.6 Screening Criteria 72
3.7 Field Performance and Field Cases 73
3.7.1 Overall Field Performance 73
3.7.2 Polymer Flooding in a Very Heterogeneous Reservoir 74
3.7.3 Polymer Flooding Using High MW and High Concentration Polymer 75
3.7.4 Polymer Flooding in Heavy Oil Reservoirs 76
3.7.5 Polymer Flooding in the Marmul Field, Oman 77
3.7.6 Polymer Flooding in a Carbonate Reservoir—Vacuum Field, New Mexico 78
3.8 Post-Polymer Conformance Control Using Movable Gel 78
References 80

4. Polymer Flooding Practice in Daqing 83

Dongmei Wang

4.1 Mechanism 83
4.1.1 Mobility Control 83
4.1.2 Profile Modification 84
4.1.3 Microscopic Mechanism 86
4.2 Reservoir Screening 87
4.2.1 Reservoir Type 87
4.2.2 Reservoir Temperature 88
4.2.3 Reservoir Permeability 88
4.2.4 Reservoir Heterogeneity 89
4.2.5 Oil Viscosity 90
4.2.6 Formation Water Salinity 90
4.3 Key Points of Polymer Flood Design 91
4.3.1 Well Pattern Design and Combination of Oil Strata 92
4.3.2 Injection Sequence Options 94
4.3.3 Injection Formulation 95
4.3.4 Individual Production and Injection Rate Allocation 101
4.4 Polymer Flooding Dynamic Performance 102
4.4.1 Stages and Dynamic Behavior of Polymer Flooding Process 102
4.4.2 Problems and Treatments During Different Phases 104
4.5 Surface Facilities 104
4.5.1 Mixing and Injection 105
4.5.2 Produced Water Treatment 106
4.6 A Field Case 107
4.6.1 Well Pattern and Oil Strata Combination 107
4.6.2 Polymer Injection Case Design 108

4.6.3 Polymer Performance Prediction 109
4.6.4 Polymer Performance Evaluation 111
4.7 Conclusions **111**
Nomenclature **112**
References **114**

5. Surfactant–Polymer Flooding **117**

James J. Sheng

5.1 Introduction **117**
5.2 Surfactants **117**
5.2.1 Parameters to Characterize Surfactants 118
5.3 Types of Microemulsions **119**
5.4 Phase Behavior Tests **120**
5.5 Interfacial Tension **121**
5.6 Viscosity of Microemulsion **122**
5.7 Capillary Number **122**
5.8 Capillary Desaturation Curve **123**
5.9 Relative Permeability **123**
5.10 Surfactant Retention **124**
5.11 SP Interactions **125**
5.12 Displacement Mechanisms **126**
5.13 Screening Criteria **126**
5.14 Field Performance Data **126**
5.15 Field Cases **127**
5.15.1 Loma Novia Field Low-Tension Waterflooding 127
5.15.2 Wichita County Regular Field Low-Tension Waterflooding 128
5.15.3 El Dorado M/P Pilot 130
5.15.4 Sloss M/P Pilot 132
5.15.5 Torchlight M/P Pilot 134
5.15.6 Delaware-Childers M/P Project 136
5.15.7 Minas SP Project Preparation 136
5.15.8 SP Flooding in the Gudong Field, China 139
References **141**

6. Alkaline Flooding **143**

James J. Sheng

6.1 Introduction **143**
6.2 Comparison of Alkalis Used in Alkaline Flooding **143**
6.3 Alkaline Reactions **144**
6.3.1 Alkaline Reaction with Crude Oil 144
6.3.2 Alkaline Interaction with Rock 145
6.3.3 Alkaline–Reactions with water 146
6.4 Recovery Mechanisms **146**

6.5 Field Injection Data 147
6.6 Application Conditions of Alkaline Flooding 149
6.7 Field Cases 151
6.7.1 Russian Трехозephое Field (Abbreviated as Field T) 151
6.7.2 Russian Шагцр-Гоаеан Field (Abbreviated as Field W) 154
6.7.3 Hungarian H Field 155
6.7.4 North Gujarat Oil Field, India 156
6.7.5 Whittier Field in California 157
6.7.6 Torrance Field in California 158
6.7.7 Wilmington Field in California 159
6.7.8 Court Bakken Heavy Oil Reservoir in Saskatchewan, Canada 164
6.8 Conclusions 165
References 165

7. Alkaline-Polymer Flooding 169
James J. Sheng

7.1 Introduction 169
7.2 Interactions Between Alkali and Polymer 169
7.3 Synergy Between Alkali and Polymer 169
7.4 Field AP Applications 171
7.4.1 Almy Sands (Isenhour Unit) in Wyoming, USA 171
7.4.2 Moorcroft West in Wyoming, USA 172
7.4.3 Thompson Creek Field in Wyoming, USA 174
7.4.4 David Lloydminster "A" Pool in Canada 174
7.4.5 Etzikom Field in Alberta, Canada 176
7.4.6 Xing-28 Block, Liaohe Field, China 176
7.4.7 Yangsanmu in China 177
7.5 Concluding Remarks 178
References 178

8. Alkaline-Surfactant Flooding 179
James J. Sheng

8.1 Introduction 179
8.2 Interactions and Synergies Between Alkali and Surfactant 179
8.2.1 Alkaline Salt Effect 179
8.2.2 Effect on Optimum Salinity and Solubilization Ratio 179
8.2.3 Synergy Between Soap and Surfactant to Improve Phase Behavior 180
8.2.4 Effect on IFT 183
8.2.5 Effect on Surfactant Adsorption 183
8.3 Simulated Results of an Alkaline-Surfactant System 184
8.4 Field Cases 185
8.4.1 Big Sinking Field in East Kentucky 186
8.4.2 White Castle Field in Louisiana 186
References 188

9. ASP Fundamentals and Field Cases Outside China 189
James J. Sheng

9.1 Introduction **189**
9.2 Synergies and Interactions of ASP **189**
9.3 Practical Issues of ASP Flooding **190**
9.3.1 Produced Emulsions 190
9.3.2 Chromatographic Separation of Alkali, Surfactant, and Polymer 191
9.3.3 Precipitation and Scale Problems 192
9.4 Amounts of Chemicals Injected in Chinese Field ASP Projects **192**
9.5 Overall ASP Field Performance **194**
9.6 ASP Examples of Field Pilots and Applications **194**
9.6.1 Lawrence Field in Illinois 194
9.6.2 Cambridge Minnelusa Field in Wyoming 196
9.6.3 West Kiehl Field in Wyoming 198
9.6.4 Tanner Field in Wyoming 199
9.6.5 Lagomar LVA-6/9/21 Area in Venezuela 199
References **200**

10. ASP Process and Field Results 203
Harry L. Chang

10.1 Introduction **203**
10.2 Background **204**
10.3 Laboratory Studies and Mechanistic Modeling **207**
10.3.1 Laboratory Studies 207
10.3.2 Mechanistic Modeling 212
10.3.3 Other Laboratory Studies and Field Experiments 215
10.4 The Screening Process **216**
10.5 Field Applications and Results **218**
10.5.1 ASP Flooding in the Daqing Oil Field 221
10.5.2 ASP Flooding in the Shengli Oil Field 225
10.5.3 ASP Flooding in the Karamay Oil Field 225
10.5.4 Other Field Test Results 226
10.6 Interpretation of Field Test Results **227**
10.6.1 Assessment of Oil Recovery Efficiency 227
10.6.2 Interpretation of Recovery Mechanisms 229
10.6.3 Process Application 229
10.7 Lessons Learned **230**
10.8 Future Outlook and Focus **232**
10.9 Conclusions **235**
10.10 Recommendation on Field Project Designs **235**
Nomenclature and Abbreviations **239**
References **240**

11. Foams and Their Applications in Enhancing Oil Recovery 251

James J. Sheng

11.1 Introduction **251**
11.2 Characteristics of Foam **251**
11.3 Foam Stability **252**
11.4 Mechanisms of Foam Flooding to Enhance Oil Recovery **257**
11.4.1 Foam Formation and Decay 258
11.4.2 Foam Flooding Mechanisms 260
11.5 Foam Flow Behavior **260**
11.5.1 Foam Viscosity 260
11.5.2 Relative Permeabilities 261
11.5.3 Mobility Reduction 261
11.5.4 Flow Resistance Factor 262
11.6 Foam Application Modes **262**
11.6.1 CO_2 Foam 262
11.6.2 Steam-Foam 263
11.6.3 Foam Injection in Gas Miscible Flooding 264
11.6.4 Gas Coning Blocking Foam 264
11.6.5 Enhanced Foam Flooding 264
11.6.6 Foams for Well Stimulation 264
11.7 Factors That Need to Be Considered in Designing Foam Flooding Applications **265**
11.7.1 Screening Criteria 265
11.7.2 Surfactants 265
11.7.3 Injection Mode 266
11.8 Results of Field Application Survey **267**
11.8.1 Locations of Conducted Foam Projects 267
11.8.2 Applicable Reservoir and Process Parameters 267
11.8.3 Injection Mode 268
11.8.4 Gas Used in Foam 268
11.9 Individual Field Applications **268**
11.9.1 Single Well Polymer-Enhanced Foam Flooding Test 268
11.9.2 Nitrogen Foam Flooding in a Heavy Oil Reservoir After Steam and Waterflooding 271
11.9.3 Snorre Foam-Assisted-Water-Alternating-Gas Project 273
References **276**

12. Surfactant Enhanced Oil Recovery in Carbonate Reservoirs 281

James J. Sheng

12.1 Introduction **281**
12.2 Problems in Carbonate Reservoirs **282**

12.3 Models of Wettability Alteration Using Surfactants **283**
12.4 Upscaling **286**
12.5 Oil Recovery Mechanisms in Carbonates Using Chemicals **289**
12.6 Chemicals Used in Carbonate EOR **291**
12.7 Chemical EOR Projects in Carbonate Reservoirs **292**
12.7.1 The Mauddud Carbonate in Bahrain 292
12.7.2 The Yates Field in Texas 293
12.7.3 The Cottonwood Creek Field in Wyoming 294
12.7.4 The Baturaja Formation in the Semoga Field in Indonesia 294
12.7.5 Cretaceous Upper Edwards Reservoir (Central Texas) 295
12.8 Concluding Remarks **296**
Nomenclature **296**
References **297**

13. Water-Based EOR in Carbonates and Sandstones: New Chemical Understanding of the EOR Potential Using "Smart Water" 301

Tor Austad

13.1 Introduction **301**
13.1.1 Wetting in Carbonates 302
13.1.2 Wetting in Sandstones 304
13.1.3 Smart Water Flooding 304
13.2 "Smart Water" in Carbonates **306**
13.2.1 Introduction 306
13.2.2 Reactive Potential Determining Ions 307
13.2.3 Suggested Mechanism for Wettability Modification 312
13.2.4 Optimization of Injected Water 312
13.2.5 Viscous Flood Versus Spontaneous Imbibitions 315
13.2.6 Environmental Effects 315
13.2.7 Smart Water in Limestone 316
13.2.8 Condition for Low Salinity EOR Effects in Limestone 317
13.3 "Smart Water" in Sandstones **320**
13.3.1 Introduction 320
13.3.2 Conditions for Low Salinity Effects 320
13.3.3 Suggested Low Salinity Mechanisms 320
13.3.4 Improved Chemical Understanding of the Mechanism 321
13.3.5 Chemical Verification of the Low Salinity Mechanism 321
13.4 Field Examples and EOR Possibilities **326**
13.4.1 Carbonates 326
13.4.2 Sandstones 328
13.4.3 Statoil Snorre Pilot 330
13.5 Conclusion **332**
Acknowledgments **332**
References **332**

14. Facility Requirements for Implementing a Chemical EOR Project 337

John M. Putnam

14.1 Introduction 337
14.2 Overall Project Requirements 339
14.3 Modes of Chemical EOR Injection 343
14.3.1 Polymer Flooding 344
14.3.2 Surfactant-Polymer Flooding 345
14.3.3 Alkaline-Polymer Flooding 345
14.3.4 Alkaline-Surfactant-Polymer 347
14.4 Water Treatment and Conditioning 347
14.5 Handling and Processing EOR Chemicals On-site 350
14.5.1 Polymer Handling, Processing, and Metering 350
14.5.2 Surfactant Handling and Metering 354
14.5.3 Alkaline Agent Handling, Processing and Metering 355
14.6 Injection Schemes and Strategies 358
14.7 Materials of Construction 359
14.8 Conclusion 360
References 360

15. Steam Flooding 361

James J. Sheng

15.1 Thermal Properties and Energy Concepts 361
15.1.1 Heat Capacity (C) 361
15.1.2 Latent Heat (L_v) 361
15.1.3 Sensible Heat 362
15.1.4 Total Volumetric Heat Capacity 362
15.1.5 Thermal Diffusivity (α) 363
15.1.6 Enthalpy (H, h) 363
15.1.7 Vapor Pressure, Saturation Pressure, and Saturation Temperature 363
15.1.8 Steam Quality 363
15.1.9 Temperature-Dependent Oil Viscosity 363
15.1.10 Gravitational Potential Energy 364
15.1.11 Kinetic Energy 364
15.1.12 Total Energy 364
15.2 Modes of Heat Transfer 364
15.2.1 Heat Conduction 365
15.2.2 Heat Convection 365
15.2.3 Thermal Radiation 365
15.3 Heat Losses 366
15.3.1 Heat Loss from Surface Pipes 366
15.3.2 Heat Loss from a Wellbore 366
15.3.3 Heat Loss to Over- and Underburden Rocks 366
15.3.4 Heat Loss from Produced Fluids 367

15.4 Estimation of the Heated Area 367
15.5 Estimation of Oil Recovery Performance 370
15.6 Mechanisms 371
15.7 Screening Criteria 371
15.8 Practice in Steam Flooding Projects 373
15.8.1 Formation 373
15.8.2 Injection Pattern and Well Spacing 374
15.8.3 Injection and Production Rates 375
15.8.4 Injection Schemes 376
15.8.5 Time to Convert Steam Soak to Steam Flood 376
15.8.6 Oil Recovery and OSR 377
15.8.7 Completion Interval 377
15.8.8 Production Facilities 378
15.8.9 Water Treatment 378
15.8.10 Monitoring and Surveillance 379
15.9 Field Cases 379
15.9.1 Kern River in California 379
15.9.2 Duri Steam Flood (DSF) Project in Indonesia 381
15.9.3 WASP in West Coalinga Field, CA 382
15.9.4 Karamay Field, China 382
15.9.5 Qi-40 Block in Laohe, China 383
References 386

16. Cyclic Steam Stimulation 389
James J. Sheng

16.1 Introduction 389
16.2 Mechanisms 389
16.3 Estimating Production Response from CSS—Boberg and Lantz Model 391
16.4 Screening Criteria 395
16.5 Practice in CSS Projects 396
16.5.1 General Producing Methods 396
16.5.2 Injection and Production Parameters 397
16.5.3 Completion Interval 400
16.5.4 Wellbore Heat Insulation 400
16.5.5 Incremental Oil Recovery and OSR 400
16.5.6 Monitoring and Surveillance 400
16.6 Field Cases 401
16.6.1 Cold Lake in Alberta, Canada 401
16.6.2 Midway Sunset in California 402
16.6.3 Du 66 Block in the Liao Shuguang Field, China 404
16.6.4 Jin 45 Block in Liaohe Huanxiling Field, China 406
16.6.5 Gudao Field, China 407
16.6.6 Blocks 97 and 98 in Karamay Field, China 408
16.6.7 Gaosheng Field, China 411
References 412

17. SAGD for Heavy Oil Recovery 413

Chonghui Shen

17.1 Introduction 413
17.2 Evaluation of SAGD Resource 416
17.2.1 Importance of Resource Quality 416
17.2.2 Focus of Delineation 419
17.3 Start-Up 420
17.3.1 Circulation Heating and Inter-Well Communication Initialization 420
17.3.2 Well Separation and Start-Up Period 423
17.3.3 Wellbore Effects 423
17.4 Well Completion and Work-Over 424
17.4.1 Steam Circulation for Start-Up 424
17.4.2 Thermal Wellbore Insulation 424
17.4.3 Sand Control Liner 425
17.4.4 Liner Plugging Issue and Treatment 426
17.4.5 Recompletion to Fix Local Steam Breakthrough 428
17.4.6 Intelligent Well Completion 429
17.5 Production Control 431
17.5.1 Steam Trap 431
17.5.2 Wellbore Lift 432
17.5.3 Geysering Phenomenon Under Natural Lift 433
17.6 Well, Reservoir, and Facility Management 434
17.6.1 Wellbore Pressure and Temperature 435
17.6.2 Reservoir Monitoring 435
17.6.3 Rock Deformation Evaluation and Surface Monitoring 436
17.7 SAGD Wind-Down 438
17.8 Integration of Subsurface and Surface 440
17.9 Solvent-Enhanced SAGD 440
References 442

18. *In Situ* Combustion 447

Alex Turta

18.1 Fundamentals 447
18.1.1 Introduction and Qualitative Description of *In Situ* Combustion Techniques 447
18.1.2 Design, Operation, and Evaluation of an ISC Field Project 454
18.2 Field Applications 467
18.2.1 Screening Guide 467
18.2.2 Monitoring and Evaluation of an ISC Pilot/Project 469
18.2.3 ISC Pilots 473
18.2.4 Commercial ISC Projects in Heavy Oil Reservoirs 493
18.2.5 Wet ISC Projects 497
18.3 ISC Projects in Light Oil Reservoirs 512
18.3.1 Commercial HPAI Projects in Very Light, Deep, Williston Basin Oil Reservoirs 512

18.3.2 ISC Projects in Waterflooded Reservoirs Containing Very Light Oil 516
18.3.3 ISC Failures in Reservoirs with Light-Medium Oils 519
18.4 CISC Applications 520
18.4.1 CISC Application for Heavy Oil Production Stimulation 521
18.4.2 Increase of Injectivity for Water Injection Wells 524
18.4.3 Sand Consolidation by Hot Air Injection ("Controlled Coking") 524
18.5 New Approaches to Apply ISC in Combination with Horizontal Wells 525
18.5.1 Horizontal Wells Drilled in Old Conventional ISC Projects 525
18.5.2 Long-Distance Versus Short-Distance Displacement 526
18.5.3 THAI Process 528
18.5.4 Other ISC Approaches (COSH and Top-Down ISC) 531
18.6 Operation Problems and Their Remedies 532
18.6.1 Critical Problems 533
18.7 Noncritical Problems 534
References 536

19. Introduction to MEOR and Its Field Applications in China 543

James J. Sheng

19.1 Introduction 543
19.2 MEOR Mechanisms 544
19.3 Microbes and Nutrients Used in MEOR 548
19.4 Screening Criteria 549
19.5 Field Applications 550
19.5.1 Single-Well Microbial Huff-and-Puff 551
19.5.2 Microbial Waterflooding 552
19.5.3 Well Stimulation to Remove Wellbore or Formation Damage 554
19.5.4 MEOR Using Indigenous Microbes 555
Acknowledgments 558
References 558

20. The Use of Microorganisms to Enhance Oil Recovery 561

Lewis Brown

20.1 Origin of the MEOR Concept 561
20.2 Early Work on MEOR 562
20.3 Patents on MEOR 563
20.4 Our Projects on MEOR 567
20.5 Future Studies 576
References 577

21. Field Applications of Organic Oil Recovery—A New MEOR Method 581

Bradley Govreau, Brian Marcotte, Alan Sheehy, Krista Town, Bob Zahner, Shane Tapper and Folami Akintunji

21.1 Introduction 581
21.2 Oil Release Mechanism 582
21.3 Discussion of Applications 584
21.3.1 Screening Reservoirs Is Critical to Success 584
21.3.2 Organic Oil Recovery Can Be Applied to a Wide Range of Oil Gravities 585
21.3.3 Reservoir Plugging or Formation Damage Is No Longer a Risk 587
21.3.4 Microbes Reside in Extreme Conditions and Can Be Manipulated to Perform Valuable *In Situ* "Work" 588
21.3.5 Organic Oil Recovery Can Be Successfully Applied in Dual-Porosity Reservoirs 589
21.3.6 Applying Organic Oil Recovery Can Reduce Reservoir Souring 590
21.3.7 Organic Oil Recovery Can Be Used in Tight Reservoirs 591
21.3.8 An Oil Response Is Not Always Seen When Treating Producing Wells 591
21.4 Case Study 1—Trial Field, Saskatchewan 595
21.4.1 Background 595
21.4.2 Reservoir Screening and Laboratory Work 595
21.4.3 Field Application Process 596
21.4.4 Nutrient Test in Producer 596
21.4.5 Pilot 597
21.4.6 Additional Producer Applications 600
21.4.7 Expanding the Pilot 601
21.4.8 Discussion 604
21.5 Case Study 2—Beverly Hills Field, California 604
21.5.1 Background 604
21.5.2 Nutrient Test in Producer 605
21.5.3 Injection Well Treatments 606
21.5.4 Additional Producer Treatments 608
21.5.5 OS-8 609
21.5.6 BH-15 610
21.5.7 Discussion of Results 612
21.6 Conclusion 613
References 613

22. Cold Production of Heavy Oil 615

Bernard Tremblay

22.1 Introduction 616
22.2 Mechanisms 618

22.2.1 Solution-Gas Drive 618
22.2.2 Sand Production 627
22.3 Field Case **645**
22.3.1 Heterogeneity of Reservoirs 645
22.3.2 History Matching Cold Production Wells 651
22.3.3 Predicting CHOPS Production 652
22.3.4 Predicting Post-CHOPS Production 656
22.4 Conclusions **660**
Acknowledgments **662**
References **662**

Index 667

Preface

Conducting an enhanced oil recovery (EOR) project requires experts and their experience from different disciplines. Realistically, a group of experts may not be readily available within an organization. Even though such a group of experts are available, they may have limited experience in meeting special challenges in a particular project. Any EOR project costs millions of dollars. We cannot afford to make mistakes in designing a project. And the experience required cannot be obtained from text books. It must be collectively from many actual field projects. This motivated us to publish this book.

The book collects experience from all sources of information. In particular, the personal field experience and the extensive research knowledge from the experts in different EOR areas are summarized here, in addition to their collection from literature and other sources. Based on the recent developments, this book outlines the most recent techniques in EOR. The fundamentals of EOR methods are briefly presented as well. This strikes an ideal balance between project design, theory, and EOR field practices.

The book covers all EOR areas: gas injection, chemical flooding, thermal recovery, and microbial EOR. However, it is not feasible to include all the field cases in a single book. Efforts have been made to discuss the main lessons and experience from important field projects. Overall field experience and practices are summarized from published or authors' own survey work.

The book is designed for professionals working on practical EOR projects. The fundamentals included are also useful to students and those who plan to learn EOR technology.

The following chapters are included:

Chapter 1: Gas Flooding
Chapter 2: Enhanced Oil Recovery by Using CO_2 Foams: Fundamentals and Field Applications
Chapter 3: Polymer Flooding—Fundamentals and Field Cases
Chapter 4: Polymer Flooding Practice in Daqing
Chapter 5: Surfactant–Polymer Flooding
Chapter 6: Alkaline Flooding
Chapter 7: Alkaline-Polymer Flooding
Chapter 8: Alkaline-Surfactant Flooding
Chapter 9: ASP Fundamentals and Field Cases Outside China
Chapter 10: ASP Process and Field Results
Chapter 11: Foams and Their Applications in Enhancing Oil Recovery

Chapter 12: Surfactant Enhanced Oil Recovery in Carbonate Reservoirs
Chapter 13: Water-Based EOR in Carbonates and Sandstones: New Chemical Understanding of the EOR Potential Using "Smart Water"
Chapter 14: Facility Requirements for Implementing a Chemical EOR Project
Chapter 15: Steam Flooding
Chapter 16: Cyclic Steam Stimulation
Chapter 17: SAGD for Heavy Oil Recovery
Chapter 18: *In Situ* Combustion
Chapter 19: Introduction to MEOR and Its Field Applications in China
Chapter 20: The Use of Microorganisms to Enhance Oil Recovery
Chapter 21: Field Applications of Organic Oil Recovery—A New MEOR Method
Chapter 22: Cold Production of Heavy Oil

Contributors

Folami Akintunji Atinum E&P, 333 Clay Street, Suite 700 Houston, TX 77002, USA

Tor Austad University of Stavanger, 4036 Stavanger, Norway

Lewis Brown Mississippi State University, Biological Sciences, 449 Hardy Road, Room 131 Etheredge Hall, P.O. Box GY, Mississippi State, MS 39762, USA

Harry L. Chang Chemor Tech International, LLC, 4105 W. Spring Creek Parkway, #606, Plano, TX 75024, USA

Birol Dindoruk Shell Exploration and Production Inc. Houston, TX, USA

Bradley Govreau Titan Oil Recovery, Inc., 9595 Wilshire Blvd., Suite 303 Beverly Hills, CA 90212, USA

Russell T. Johns The Pennsylvania State University, Department of Energy and Mineral Engineering, School of Earth Sciences Energy Institute, University Park, PA 16802, USA

S.I. Kam Craft and Hawkins Department of Petroleum Engineering, Louisiana State University, Patrick F. Taylor Hall, Baton Rouge, Louisiana 70803, USA

S. Lee Craft and Hawkins Department of Petroleum Engineering, Louisiana State University, Patrick F. Taylor Hall, Baton Rouge, Louisiana 70803, USA

Brian Marcotte Titan Oil Recovery, Inc.

John M. Putnam SNF Holding Company, FLOQUIP Engineering, P.O. Box 250, Riceboro, GA 31323, USA

Alan Sheehy Titan Oil Recovery, Inc.

Chonghui Shen Shell Canada Limited, 400, 4th Ave., SW Calgary, Alberta T2P 2H5, Canada

James J. Sheng Bob L. Herd Department of Petroleum Engineering, Texas Tech University, Lubbock, TX 79409, USA

Shane Tapper Pengrowth Energy Corporation, 222 Third Avenue SW, Suite 2100, Calgary, Alberta T2P 0B4, Canada

Krista Town Pengrowth Energy Corporation, 222 Third Avenue SW, Suite 2100, Calgary, Alberta T2P 0B4, Canada

Bernard Tremblay Saskatchewan Research Council, EOR Field Development, Energy Division, Regina, Saskatchewan, Canada

Alex Turta Alberta Innovates Technology Futures, Calgary , Canada

Dongmei Wang Harold Hamm School of Geology and Geological Engineering, University of North Dakota, Grand Forks, ND 58202, USA

Bob Zahner Venoco, Inc., 6267 Carpinteria Avenue, Suite 100 Carpinteria, CA 93013, USA

Acknowledgments

First of all, I would like to express my appreciation to the contributing authors. Without their contribution, this book could not be written. Second, I am grateful to Kenneth P. McCombs, Acquisition Editor at Elsevier, for his trust that I would be able to get the book written. My thanks also go to Kattie Washington, Renata Corbani, Jill Leonard, and the other staff at Elsevier for their support which made my editing work more enjoyable.

I am most grateful to my family for their sacrifice, support, and love. They are waiting patiently for my return to a normal life of a husband, a father, a son, and a brother.

Chapter 1

Gas Flooding

Russell T. Johns[1] and Birol Dindoruk[2]
[1]The Pennsylvania State University, Department of Energy and Mineral Engineering, School of Earth Sciences Energy Institute, University Park, PA 16802, USA, [2]Shell Exploration and Production Inc., Houston, TX, USA

This chapter first defines what gas flooding is, and explains how recovery is enhanced by increasing both sweep and displacement efficiencies. The basic steps in gas flood design are described followed by the important technical parameters and scoping economics used in screening the best reservoirs for a gas flood. It is shown how gas is injected in wells through slug, continuous, or water-alternating-gas schemes. The importance of phase behavior on miscibility and equation-of-state (EOS) tuning is stressed, along with experiments needed for proper fluid characterization. It is also discussed how miscibility is developed through a multicontact process either by a vaporizing, condensing, or a combined condensing and vaporizing mechanism. The best techniques to estimate the minimum miscibility pressure (MMP) are given. Finally, three case studies and an overall summary of field experience are presented. Field displacements considered are CO_2 flooding, nitrogen flooding, and an immiscible gravity stable CO_2 flood.

1.1 WHAT IS GAS FLOODING?

Gas flooding is the injection of hydrocarbon or nonhydrocarbon components into oil reservoirs that are typically waterflooded to residual oil (and perhaps in some cases as a primary or secondary method). Injected components are usually vapors (gas phase) at atmospheric temperature and pressure and may include mixtures of hydrocarbons from methane to propane, and nonhydrocarbon components such as carbon dioxide, nitrogen, and even hydrogen sulfide or other exotic gases such as SO_2. Although these components are usually vapors at atmospheric temperature and pressure, they may be supercritical fluids at reservoir temperature and pressure in that some of their properties may be more liquid-like. Carbon dioxide, for example, has a density similar to that of oil, but a viscosity more like vapor at most reservoir conditions. Gas injection today often means CO_2 or rich hydrocarbon gas injection to recovery residual oil, and in some cases to also store or sequester CO_2 from the atmosphere.

Enhanced Oil Recovery Field Case Studies.

The primary mechanism for oil recovery by high pressure gas flooding is through mass transfer of components in the oil between the flowing gas and oil phases, which increases when the gas and oil become more miscible. Secondary recovery mechanisms include swelling and viscosity reduction of oil as intermediate components in the gas condense into the oil.

The key to gas flooding is to contact as much of the reservoir with the gas as possible and to recover most of the oil once contacted. Injection gases are designed to be miscible with the oil so that oil previously trapped by capillary forces mixes with the injected gas. The injected gas or hydrocarbon phase then drives the oil components to the production well.

Ideally, miscible flow is piston-like in that whatever gas volume is injected displaces an approximately equal volume of reservoir hydrocarbon fluid. Unfortunately, in real field applications such piston-like behavior does not occur because reservoir heterogeneities and gravity override cause gas to cycle through one or more high-permeability layers, bypassing some oil and leading to poor sweep efficiency. Mixing of oil and gas components within a single phase will also lead to nonpiston-like behavior even without geological heterogeneities.

A proper gas flood design will consider both the microscopic displacement efficiency and sweep efficiency. The profitability of that process is a function of the overall recovery, which is expressed by $E_R = E_V E_D$. E_V is the volumetric sweep efficiency, which is the fraction of the reservoir that is contacted by the gas, while the displacement efficiency E_D is the fraction of contacted oil that is displaced. Displacement efficiencies for miscible floods at field scale are often on the order of 70–90%, while sweep efficiencies can be much worse, leading to typical incremental recoveries above waterflood recovery of only between 10–20% OOIP.

Gas flooding designs are limited by both economics and physics of displacement so that there is often a trade-off between the sweep and displacement efficiencies. Because it is not possible to give exact values for these efficiencies in the field, they are useful only to qualitatively explain how key parameters such as injection fluid viscosity, phase behavior, heterogeneities, and other fluid and rock properties affect recovery and the design of gas flooding processes. Each reservoir is unique so that an engineer must have a good understanding of the fundamental processes.

1.2 GAS FLOOD DESIGN

The engineering steps in gas flood design depend on whether a flood is a small or large project. For a large project there is more risk involved so that the process involves three basic steps; screening, design, and implementation. The basic design steps for a large flood are the following:

1. Technical and economic screening to eliminate reservoirs under consideration before a more detailed study is done;

2. Reservoir/geologic study, including 2-D and 3-D reservoir simulation to make performance predictions;
3. Wells and surface facility design based on forecasted fluid volumes, compositions and reservoir continuity;
4. Economic studies where key input variables are varied to understand associated risks;
5. Management approval (or disapproval) of the gas flood based on uncertainties and economic considerations; and
6. Implementation of the gas flood design by making wellbore modifications as needed, installing field facilities and any required recycle plant (if one is not already nearby), and injecting initial gas.

These steps often require iteration as more is learned about the field when new wells are drilled, and laboratory data is obtained. Iterations in the design may also be required to maximum present value profit, for example, by changing the volume of gas injected.

Small projects require fewer steps than larger projects as detailed reservoir studies, simulations, and associated predictions and economics may not be done to reduce costs. The screening process (step 1) is typically used to provide required predictions and economics for small gas floods.

1.3 TECHNICAL AND ECONOMIC SCREENING PROCESS

The primary objectives of the screening process are to:

1. Rank potential candidate reservoirs for gas flooding;
2. Identify potential injection fluids;
3. Identify analogue fields;
4. Make some preliminary production rate estimates and scoping economic calculations; and
5. Identify which reservoirs should be examined in a later more detailed analysis, especially if the gas flood is a large project.

A good screening process will consider several key technical factors in addition to investment and operating costs. Typical reservoir screening considerations include

1. Residual oil saturation to waterflooding;
2. Average reservoir pressure (and temperature);
3. Oil viscosity and minimum pressure for miscibility;
4. Available miscible gas source and cost;
5. Reservoir heterogeneity and conformance issues at injection well and well pattern scale;
6. Reservoir permeability and ability to inject and produce fluids at economic rates; and
7. Reservoir geometry and flow: gravity effects and vertical permeability.

Most fields undergo waterflooding prior to gas injection. This is typically done to increase reservoir pressure and reduce risks associated with potential gas flood projects. Risk is reduced if a secondary waterflood is done first because much is learned about well connections within the reservoir during water injection, and facility costs associated with water injection are already built. Thus, one of the most important initial screening factors is the residual oil saturation to waterflooding. If residual oil saturation is small (say less than 0.15), then there is little oil left to recover by gas flooding.

Other important key technical factors are the average reservoir pressure, minimum pressure for miscibility, and the oil viscosity. The reservoir pressure must usually be near or above the minimum pressure for miscibility to achieve good displacement efficiency. The MMP is typically smaller for low viscosity oils. Rough "rules of thumb" for oils with bubble-point viscosities less than about 10 cp and an API oil gravity of 25 or greater are that CO_2 or enriched gases become miscible with the oil when the reservoir pressure is above 1000 psia, while methane can become miscible with light oils at pressures greater than about 3000 psia, and nitrogen at pressures greater than about 5000 psia. Of course, reservoir temperature and oil composition play an important role in this assessment as well. The miscible fluid chosen should be available and less costly than other alternatives.

Reservoir heterogeneity and conformance also plays an important role in screening. Conformance is defined as injecting fluids where you want them to go, usually into the pay zone. If a waterflood had poor conformance, either at injection wells or at the pattern scale, the sweep from a gas flood is likely to be worse. For example, a reservoir where permeability varies greatly is not likely a good candidate for gas flooding, i.e., a Dykstra–Parsons coefficient greater than 0.7. A reservoir with many high-permeability fractures is also typically a poor candidate, especially when these fractures are aligned from injectors to producers. Injection well completions can also be faulty so that fluids do not go into the pay zones. For example, fluids can travel behind the casing if the cementing job is not good. Wells with poor conformance issues typically require significant upfront costs to redo their completion, which should be considered in the gas flood economics.

Water and gas must also be able to be injected in sufficient quantities that oil can be produced at economic levels. The ability of fluids to be injected or produced can be poor if the formation permeability is low or the fluid viscosities are large. In most cases if water was able to be injected at economic rates then gas can be injected at the same or higher rates. Well injectivity tests should be done in the field if the gas flood is to be done as a secondary recovery method, instead of as a tertiary flood following a waterflood.

One other consideration in the screening process is the effect of gravity on where the fluids go. Injection fluids with a higher density than the oil (such as water) can move downward through the reservoir bypassing oil, while injection fluids with a smaller density (such as gas) can move upward through the reservoir. Low density gases may require new perforations at the bottom of the pay zone to lessen the effect of gravity. Gravity effects are lessened in fields with low vertical permeability, such as the CO_2 floods in West Texas. Gravity effects can also be used to their advantage as has been demonstrated for many Gulf-Coast reservoirs with a significant dip. For example, gas can be injected up-dip and oil produced down-dip in a gravity stable process. As oil is produced the injected gas can move down-dip contacting more oil. Such a gravity stable process can overcome large heterogeneities within the reservoir and yield high oil recoveries.

Finally, the screening process also involves scoping economics. Both investment (capital) and operating costs should be considered at an early stage. Capital and operating costs depend on rates and volumes predicted from a scoping model or detailed simulation. Such rates are typically calculated from simplified simulations or by analogue gas floods that are similar to the current reservoir under consideration. The cumulative injection and production rates must be scaled up by considering the relative size of the proposed gas flood compared to the analogue flood. The procedure for generating rates for the scoping model is well described by Jarrell et al. (2002).

The investment costs include initial gas and water purchasing costs, injection gas recovery plants, compressors for reinjection of produced gas, gas injection facilities, pipelines for injected gas transmission, modifications to wells and production facilities for handling increased amounts of produced gas, water treatment facilities (if not already present), new injection and production wells, and separators and gas gathering surface facilities. Gas recovery plants can be the most expensive part of a gas flood if one is not already nearby.

Operating costs include chemicals used for corrosion-, scale-, and paraffin-inhibition, labor, well servicing and workovers, power, water disposal, and injection and production facility maintenance costs. Workovers are typically a large percentage of field operating costs.

Implementation of gas floods for offshore and deep water fields becomes even more riskier, mainly due to well costs and reach. In such economic environments, understanding reservoir heterogeneities and well connectivity become extremely important.

1.4 GAS INJECTION DESIGN AND WAG

Gas flooding can be implemented in a variety of ways either as continuous gas injection, continuous gas injection chased with water (or possibly another gas), conventional water-alternating-gas (WAG), or tapered WAG (TWAG) as illustrated in Figure 1.1. For WAG, the total volume of gas to be injected must be determined along with the frequency (number of gas−water cycles) and the water−gas ratio (volume of water divided by the volume of gas injected in each cycle). In West Texas, gas injection is sometimes tapered so

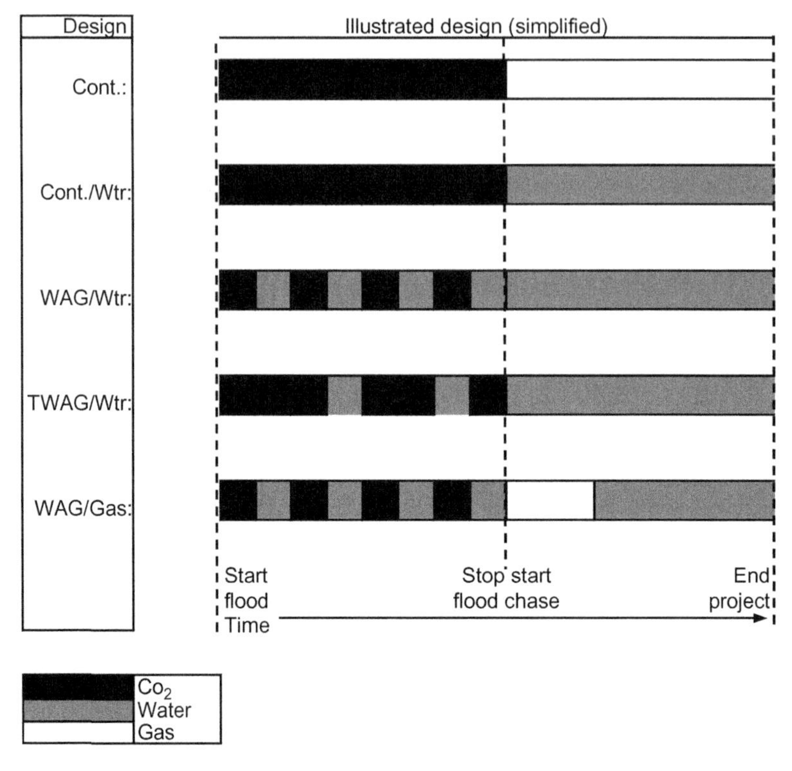

FIGURE 1.1 Water-alternating-gas floods can take on many forms (illustration for CO_2 from Jarrell et al. 2002). "Cont." = continuous gas injection; "Cont./Wtr" = continuous gas injected chased with water; "WAG/Wtr" = conventional water alternating gas (WAG) flood chased with water; "TWAG/Wtr" = tapered alternating gas and water chased with water; and "WAG/Gas" conventional WAG chased with gas.

that more gas is delivered up front, and gas is changed to water injection once gas breaks through a production well. Simultaneous water–gas injection (SWAG) has also been attempted, although this requires significant monitoring owing to injection of multiple phases. The specific injection scheme to be used must be examined using compositional simulation for each reservoir under consideration since each reservoir and oil is unique.

Figure 1.2 illustrates a typical gas flooding process where CO_2 or another gas component is injected in several slugs alternating with water (WAG). WAG can significantly improve sweep efficiency since water is less mobile (greater viscosity) than CO_2 and hence improves the average mobility ratio of the flood while exhibiting better coverage at the deeper sections of the reservoir due to underriding.

One definition of the mobility ratio is the displacing fluid mobility (injection gas permeability over its viscosity) divided by the displaced fluid mobility (reservoir fluid permeability over its viscosity). A large gas mobility

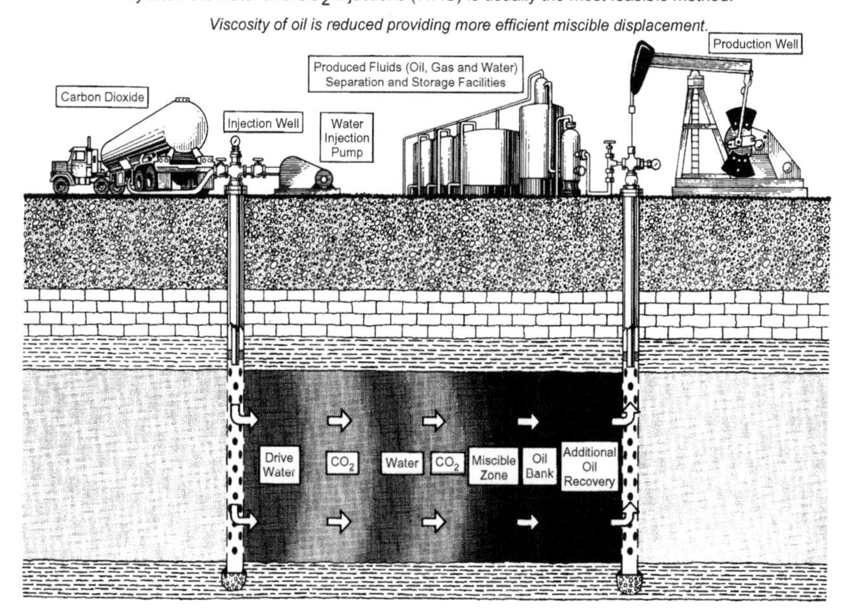

FIGURE 1.2 Illustration of gas flooding process showing a typical water-alternating-gas process where carbon dioxide is injected in several slugs followed by water. Source: *Data from Lake (1989), drawing by Joe Lindley, U.S. Department of Energy, Bartlesville, OK.*

primarily results from the low viscosity of the gas. A large mobility ratio is unfavorable, because a smaller and more favorable mobility ratio means delayed breakthrough of the injected gas and less gas cycling through the high-permeability layers. One advantage of WAG outside of mobility control is that less gas (per total pore volume injected) is injected in favor of cheaper water. WAG or TWAG has proven to be very effective as a mobility control measure and is nearly always used in practice to improve sweep efficiency.

Although WAG does increase recoveries, gas can still tongue upward in the formation away from the wells during the gas injection cycle, while water can move downward in the water cycle (Figure 1.3). This segregation of fluids will occur when there is a sufficient vertical permeability and density difference between the gas and reservoir fluids. Channeling of gas and water through high-permeability layers usually dominates over gravity tonguing and becomes more significant as heterogeneities increase, permeability and density differences decrease, and fluid velocities become larger.

It is important to adjust the volume of water and gas injected during WAG so that a maximum in the recovery efficiency is achieved. Too much water or too much gas can result in poorer vertical sweep efficiency.

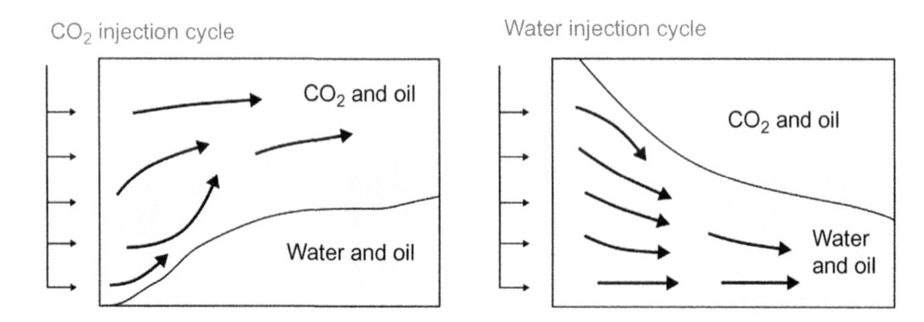

FIGURE 1.3 During WAG, gas can move upward owing to its low density, while injected water can move downwards. Source: *Data from Jarrell et al. (2002).*

Reservoirs that have smaller permeability at the top generally perform better than those with larger permeability at the top since low density gas wants to gravitate to the top of the reservoir increasing channeling through that top layer.

Gas flooding can also be implemented in a single well. In a standard gas flood, gas is transported through the formation from one well to another. In a cyclic single-well treatment, gas is injected into a single well and then shut in. That same well is then produced after the gas is allowed to soak. This process may be repeated several times. This single-well method is generally not as good in recovering oil as multiwell gas flooding but is increasingly being used in an immiscible process to reduce viscosity and swell heavier oils (Jarrell et al., 2002). However, as a miscible injectant the single-well cyclic-soak option could be one of the few methods that can be applied to heavy oils, sometimes in combination with reservoir heating. Such a process would be analogous to cyclic steam injection. Miscible gases in various forms, like foams, can also be beneficial for tight unconventional reservoirs (i.e. liquid rich shales) in terms of delivering proppants to hydraulic fractures and/or stimulating the wells by removing liquids. The wells can also be operated in a huff-and-puff mode to recover more hydrocarbons.

Other mobility control methods include foams, which are injected primarily into the large permeability layers. Unlike WAG, foam is still in the research stage. The idea is to create stable CO_2 or N_2 foam by injection of a small amount of surfactant in either the gas or more typically the water phase. If a stable foam can be generated *in situ* within the high-permeability layers it can restrict flow through them causing the gas to be diverted to other less permeable layers. Other potential conformance methods include gels as outlined in detail by Green and Willhite (1998).

All of these conformance measures (and gas injection in general) can suffer from injectivity problems in that water or gas may not be able to be injected in sufficient quantities. Injectivity tests or pilot tests are often performed in the field to reduce the risk associated with these methods.

In recent years, steam–gas foam was also introduced as a technique; however, there is not enough field data for the success of this method. This application is normally studied under thermal methods, and is not discussed further here.

1.5 PHASE BEHAVIOR

Phase behavior controls whether an oil and gas mixture will be miscible at reservoir temperature and pressure. Thus, an important step in any gas flood design is to properly characterize the fluid system over a wide range of reservoir pressures (and temperatures to mimic the surface facilities). This is also a key for optimizing production as the fluid properties change from the reservoir to the surface facilities.

1.5.1 Standard (or Basic) PVT Data

PVT experiments are nearly always performed to determine the relationships between pressure, volume, and temperature for the fluid of interest. There are essentially four key experiments that are done on most fluids, independent of whether gas is injected or not. These are constant-mass expansion tests, constant-volume depletion tests, differential liberation tests, and separator tests. In addition, viscosity of the *in situ* fluid is measured for the entire range of pressures, from reservoir pressure to atmospheric pressure and is usually reported along with the differential liberation data. All of these tests attempt to duplicate in some limited way the primary and secondary oil recovery process that exist in a reservoir. A complete description of these tests can be found in Amyx et al. (1960), Dake (1994), Danesh (1998), McCain (1990), Standing (1977), and Pedersen and Christensen (2007).

1.5.2 Swelling Test

When gas injection is planned, a swelling test is done where the selected injection gas is mixed with the oil at various proportions at constant reservoir temperature. A swelling test is used to determine the pressure required to dissolve a given amount of gas (or how much gas dissolves in the oil at a given pressure), how much the oil will swell as intermediate components in the gas are dissolved by the oil, and the resulting saturation pressures as injection gas is progressively added. Beyond calibration of PVT data, a swelling test can help to quantify the amount of the oil recovered due to an increase in oil volume that results from transfer of gas components into oil. In many cases, it is possible to identify the likely retrograde condensation behavior of the oil.

A swelling test begins with reservoir oil at its bubble-point pressure. At the reservoir temperature, a fixed amount of gas is mixed with the oil. The pressure is then increased maintaining the same reservoir temperature until all of the gas goes back into the oil. When the last gas bubble is dissolved, the oil–gas mixture is at its new bubble-point pressure. The pressure and volume of the mixture is then measured. The new volume is called the "swollen" volume. Although swelling is of secondary importance in miscible gas flooding, some oils can swell by factors of 1.7 when contacted by CO_2. Swelling causes a fraction of trapped oil to flow owing to mass transfer-based volume expansion within the reservoir. This process of adding gas and measuring the new bubble-point pressure is repeated several times at different percentages of gas injected. At large gas/oil ratios the gas–oil mixture can exhibit dew-point behavior rather than bubble-point behavior. This portion of the swelling experiments around the dew points is important since it is an analogue of the fluid properties for gas-rich zones in the reservoir.

Gas and oil viscosities are often measured in swelling tests as these are very important in determining fluid mobilities and mobility ratios. Swelling tests rigorously represent miscible gas injection processes for first-contact miscibility (FCM), which are explained in detail in the next section.

1.5.3 Slim-Tube Test

Slim-tube experiments are very useful to define the minimum pressure for miscibility using crude oil obtained from the field. The slim tube consists of a very small internal diameter coiled tube filled with crushed core, sand, or glass bead material. The tube can be quite long, typically between 40- and 60-ft long, to allow miscibility to develop dynamically some distance from the injection point.

The tube is first saturated with a known volume of oil. The temperature is then fixed at the reservoir temperature. Gas is injected through the tube to displace the oil and the amounts of gas and oil are recorded with time. The recovery is defined as the ratio of the volume of oil produced to the initial oil volume, otherwise known as the pore volumes recovered, a dimensionless number that quantifies the recovery in terms of initial hydrocarbon volume in place.

The pore volumes recovered are plotted against the pressure at a fixed time, say at 1.2 hydrocarbon pore volumes injected. The procedure is then repeated for higher pressures. The pressure that corresponds to a break or sharp change in the oil recovery at 1.2 pore volumes of injection, when plotted against the injection pressure or the lowest pressure at which the recovery is about 90–95% is often used to define the minimum pressure for miscibility for any oil–gas system (Figure 1.4). These definitions are somewhat arbitrary, however, so there is some uncertainty in the measured MMP. Nevertheless, if the reservoir pressure exceeds this "MMP" good recovery of contacted oil will likely be obtained.

Slim-tube recoveries are recoveries under idealized 1-D conditions and do not represent the recoveries that we see in the field as a significant

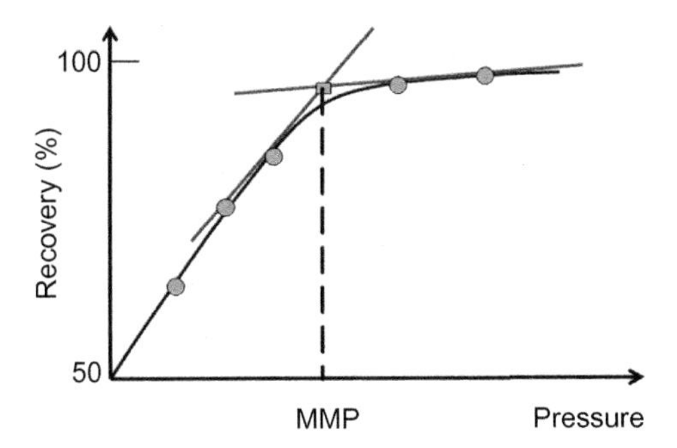

FIGURE 1.4 Example slim-tube recoveries for an oil displaced by pure CO_2. Source: *Data from Yellig and Metcalfe (1980).*

amount of the oil is not contacted by the gas. In addition to channeling caused by reservoir heterogeneity and tonguing by gravitational effects, increased mixing at the field scale by diffusion and dispersion can further degrade the oil recovery by reducing the effective concentration of the gas. A reduction in the gas concentration reduces the displacement efficiency. Slim-tube experiments are also time consuming (and expensive) and require larger sample volumes so that few of them can be made.

1.5.4 Multicontact Test

The interaction of flow and phase behavior is critical to determine the MMP for real fluids. The multicontact test attempts to mimic this dynamic interaction within the reservoir and as the experiments proceed generates physical fluid samples in the laboratory to be analyzed for composition, density, and viscosity.

The procedure begins by mixing gas and oil at constant pressure and temperature so that two equilibrium phases result. The vapor phase, which has greater mobility, flows ahead of the reservoir fluid. The liquid phase is contacted by fresh gas in what is termed a backward contact. Alternatively, the equilibrium vapor phase contacts fresh oil in a forward contact. This process is repeated several times.

Forward or backward contacts are well described in both Lake (1989) and Stalkup (1983). Depending on the nature of the injectant in the multicontact test, only forward or backward contacts are measured, but not both simultaneously. The multicontact test is useful for purely vaporizing or condensing displacements. Miscibility for vaporizing drives is developed by forward contacts whereas miscibility for condensing gas drives is developed by backward contacts. These concepts are discussed in Section 1.6. The conventional multicontact tests are not as useful for combined condensing and vaporizing displacements as is discussed in Section 1.6.

1.5.5 Fluid Characterization Using an Equation-of-State

PVT and gas injection experiments only provide the minimum amount of essential data but do not provide the full range of data required by compositional simulation. Thus, numerical fluid descriptions using EOS representations of the reservoir and injection fluid are used to fill in these data gaps. These models are tuned (or calibrated) to the available experimental data in a complicated and often subjective process. The difficulty arises because oils consist of hundreds or thousands of components and isomers, many of which cannot be precisely identified and quantified experimentally, at least in a routine sense. However, there are standard methods that reduce the number of components based on the boiling point ranges for the C_{6+} range of components. Components in a predefined boiling point range are lumped together as a single carbon number (SCN) component.

Even such simplification or SCN grouping is not enough for most compositional simulators. Compositional simulation with EOS use is computationally intensive and necessitates even fewer components, generally less than 15. Many SCN components and isomers are then lumped into pseudocomponents. Reduced methods that allow for more components are currently being developed to speed up flash calculations in compositional simulation, although they have yet to gain wide acceptance (Okuno et al., 2010).

Each pseudocomponent must be assigned values for EOS properties such as critical temperature and pressure. Because the components are not known precisely, these properties are tuned to match the measured laboratory data to get as good of a fit as possible without adjusting too many parameters or changing their values outside of a reasonable range. Over fitting can reduce the predictability of the EOS model for data outside the range of the experiments. Procedures for fluid characterization and EOS development are outlined in detail by Danesh (1998), Firoozabadi (1999), Pedersen and Christensen (2007), and Whitson and Brule (2000). In addition to tuning of the basic PVT properties, which most of the conventional tuning recipes cover, Egwuenu et al. (2005) showed that tuning to the MMP can significantly improve the fluid characterization for gas or gas injection processes.

1.6 MMP AND DISPLACEMENT MECHANISMS

The primary means of recovery in gas floods is by means of miscibility development between the gas and reservoir oil. There are three types of displacements that can occur depending on the pressure. The first type is first-contact miscibility (FCM) where the oil and gas are miscible when mixed in all proportions. This type is very difficult to achieve with many gases, but when it does nearly all of the oil contacted is extracted. Swelling tests will directly give us the minimum pressure for FCM for a given oil–gas displacement.

Miscibility can also be developed within the reservoir by a multicontact (MCM) process where gas and oil mix in repeated contacts that are either forward, backward, or a combination of both (Johns et al., 1993). Miscibility in MCM floods is developed when the phase compositions that form in each contact move toward a critical point.

The last type of gas flood is an immiscible one, which has more limited, but with varying degrees of mass transfer between phases. In a strict sense, the term "immiscible gas flood" is really a misnomer because gas will always extract some oil components. A true immiscible gas flood is the limit at which solubility of oil in gas phase is very small or negligible (i.e., at low pressures), which in a way is the analogue of water displacing oil. The displacement efficiency improves as a gas flood becomes more miscible (Figure 1.4).

1.6.1 Simplified Ternary Representation of Displacement Mechanisms

Figure 1.5 shows a simplified phase behavior at constant reservoir temperature and pressure where only three pseudocomponents or analog components are present (light = C_1, intermediate = C_2, and heavyweight = C_3 components). As analog components, for example, C_1 could be methane, C_2 butane, and C_3 decane. In plots of this kind, all possible compositions that form as gas and oil are mixed must lie on a line segment drawn between them. This line is called the mixing or dilution line.

Injection of lean gas (mostly C_1) with the oil in Figure 1.5 is "immiscible" because the mixing line between them traverses a large section of the two-phase region. There is some extraction of oil components in this case, but recoveries in a slim-tube experiment would be low at 1.2 PVI. A gas enriched with sufficient C_2, however, is multicontact miscible (MCM) because the gas composition will be outside the region of tie-line extensions. A gas composition that lies on a tie-line extension in the limit of the critical point is at its minimum miscibility enrichment (MME). Contacts of this gas with equilibrium oil would eventually reach the critical point. In addition,

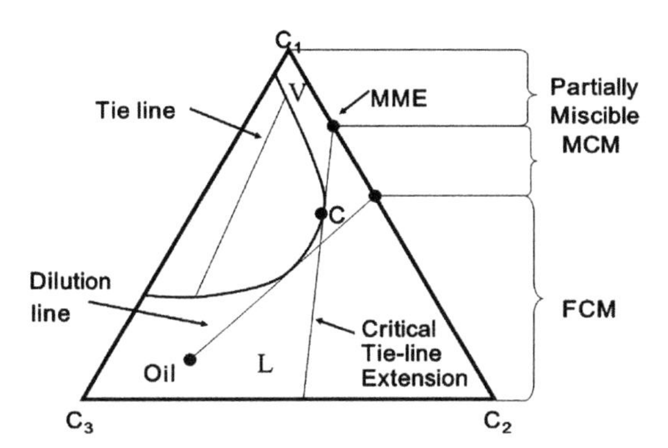

FIGURE 1.5 Ternary representation of condensing gas drive process.

such a displacement causes the intermediate component in the gas to condense into the equilibrium oil, causing the trapped oil to swell, which helps recovery. This process is known as a condensing drive, where miscibility is developed at the trailing edge of the displacements by backward contacts.

A gas enriched even more with C_2 can become first contact miscible (FCM) with the oil because its mixing line (dilution line) does not intersect the two-phase region at all or is just tangent to the two-phase envelope as shown in Figure 1.5. This case is not achievable for most reservoirs and injection gases.

Displacement of a lighter oil (more C_2 in the oil) can be multicontact miscible even for a lean gas in Figure 1.5 as long as the oil lies outside the

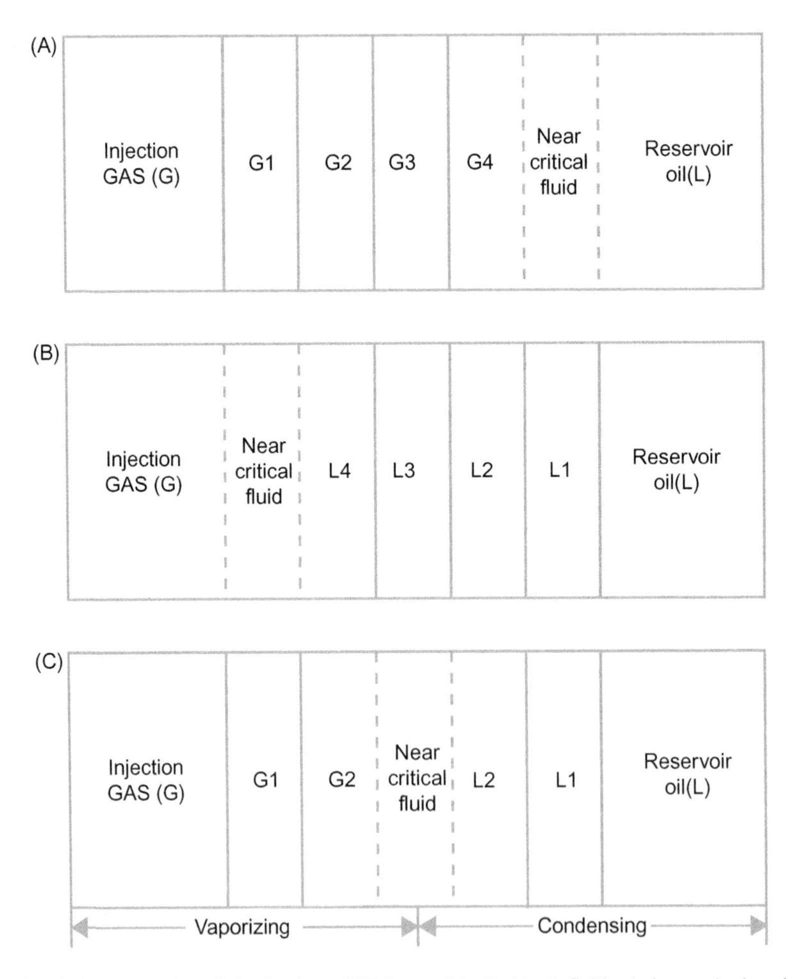

FIGURE 1.6 Location of the low/zero IFT (near-critical/critical) fluid relative to the location of the injection and production wells.

region of tie-line extensions (i.e. oil composition is on or to the right of the critical tie-line extension in Figure 1.5). In this type of displacement, known as a vaporizing drive, the intermediate component in the oil is vaporized into the equilibrium vapor phase. Vaporizing drives develop miscibility at the leading edge of the displacement because two-phase compositions that result from these forward contacts approach the critical point there.

While helpful, the theory described by ternary displacements is overly simplified and can only explain forward and backward miscibility development (Figure 1.6A and B). The mechanism that is a combination of both is discussed next.

1.6.2 Displacement Mechanisms for Field Gas Floods

The simplified model of Figure 1.5 does not adequately represent field displacements where more than three components are present. Most real CO_2 and enriched gas displacements have features of both condensing and vaporizing (CV) drives (Figure 1.6C), in that a complex combination of both forward and backward contacts intersects a critical locus first. In CV displacements miscibility is developed between the condensing and vaporizing portions as shown in Figure 1.6C. Swelling of oil occurs near the front of the displacement followed by a series of vaporizing fronts, where heavier and heavier components in the oil are vaporized in a chromatographic-like separation. Injection of very lean gases, such as pure nitrogen or pure methane, however, is primarily a vaporizing drive.

1.6.3 Determination of MMP

The MMP is one of the most important design considerations for a gas flood. There are several proven methods to determine the MMP:

- Slim-tube and multicontact experiments (see Section 1.2);
- Mixing cell methods;
- Empirical correlations;
- Compositional simulation of slim-tube displacements; and
- Analytical methods using EOS and the method of characteristics (MOC).

As discussed in Section 1.5, experiments are very useful, but are costly and time consuming to perform. Mixing cell methods based on fluid characterizations for MMP determination can yield good MMP estimates (Johns et al. 2010). Empirical correlations for determining the MMP can be good depending on whether the reservoir fluid is similar to those used to develop the correlation (Jarrell et al., 2002; Yuan et al., 2005). Analytical methods using the MOC rely on an accurate EOS that is properly tuned to available PVT data (Dindoruk et al., 1997; Johns et al., 1993; Orr et al., 1993; Yuan and Johns, 2005). The advantage of these methods is that they are quick and cheap, but the calculation methodology must be done carefully. One-dimensional compositional simulation as a way to estimate the MMP is also cheap but is computationally more intensive than mixing cell or MOC methods, and its result is dependent on correcting for the size of the grid blocks (numerical dispersion error). Similar to analytical solution models (MOC) simulation methods require a good PVT description and EOS characterization of the gas and the oil.

Some fluids are more sensitive to dispersion, whether physical or numerical. For gas–oil systems that are sensitive to physical dispersion, the gas flood should be operated well above the MMP to achieve good displacement efficiency (Solano et al., 2001).

The CO_2 MMP generally increases with increasing temperature and with less gas enrichment. Hot reservoirs therefore tend to have larger MMPs, while adding CO_2 to lean gases can significantly decrease the MMP. The MMP for nitrogen injection, however, can either increase or decrease with temperature (Dindoruk et al., 1997; Sebastian and Lawrence, 1992).

1.7 FIELD CASES

We consider three field cases that provide important lessons on the application of gas flooding for EOR. As discussed previously, the recovery in the field depends on both volumetric sweep and displacement efficiency. Summaries of gas floods performed can be found in Manrique et al., 2007 and Christensen et al., 2001.

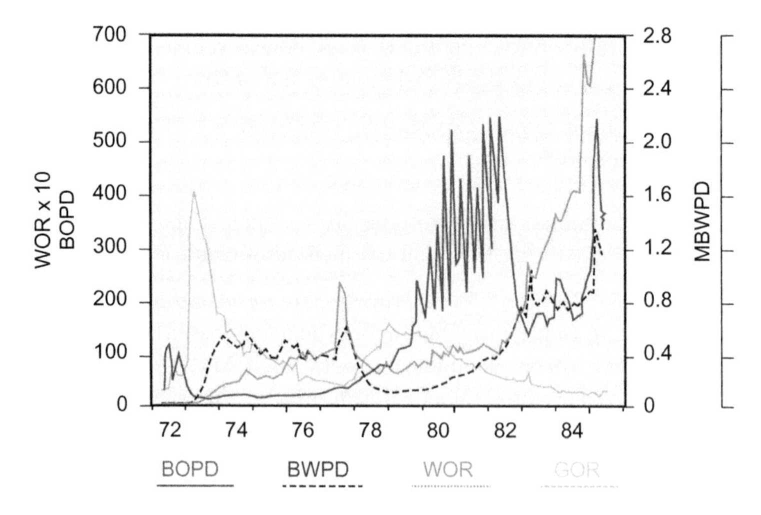

FIGURE 1.7 Oil and water rates and water and gas–oil ratios for Slaughter Estate Unit Pilot. Source: *Data from Stein et al. (1992).*

1.7.1 Slaughter Estate Unit CO_2 Flood

This miscible flood pilot is in the West Texas San Andres dolomite and is an example of a gas flood with very good oil recovery. The permeability is low averaging around 4 mD at a depth of about 5000 ft. Good recovery was obtained in a waterflood in the early 1970s prior to gas flooding (Figure 1.7).

The gas injected contained 72 mol% CO_2 and 28 mol% H_2S. The MMP of approximately 1000 psia with this gas and moderate API oil (32°API) is substantially less than the average reservoir pressure of 2000 psia. Thus, this flood is MCM. The acid gas was eventually replaced with a chase gas consisting of mostly nitrogen and then water. Water was injected alternately with the acid gas with a WAG ratio of about 1.0. A 25% hydrocarbon pore volume (HCPV) slug of acid gas was injected. The chase gas was also alternated with water, and eventually the gas–water ratio was reduced to 0.7 to improve vertical sweep.

The cycles shown in Figure 1.7 for the chase gas correspond to the cycles in WAG injection. During WAG, water injectivity losses of about 50% were experienced owing to trapping of gas as water is injected. Furthermore, there is evidence of gas channeling as the GOR began to climb at the same time or just prior to the oil production.

Incremental tertiary recovery was 19.6% OOIP, which is largely the result of good WAG management and the use of H_2S in the gas (Stein et al., 1992). H_2S, although very dangerous, is a very good miscible agent. When added to the primary and secondary recovery (waterflood) of about 50%, the total recovery in this pilot is expected to be around 70% OOIP, which is well above the average for most fields. Because of the great success of the pilot, the unit was gas flooded field wide. The field-wide flood has also been successful although a higher gas–water ratio was used.

1.7.2 Immiscible Weeks Island Gravity Stable CO_2 Flood

A pilot test of the “S” sand at the Weeks Island, Louisiana field was performed in the early 1980s. This sand, which is up against a salt dome, is highly dipping (30° dip as shown in Figure 1.8) and is very permeable both vertically and horizontally. The initial reservoir pressure for this sand was 5100 psia at a reservoir temperature of 225°F, but at the time of the pilot the pressure was lower. The pilot lasted 6.7 years and consisted of one up-dip injector and two producers about 260 ft down-dip as shown in Figure 1.9. Following a waterflood, gas was injected up-dip so that gravity would stabilize the front in a relatively horizontal interface. The main idea of this gravity stable flood is that the gas–oil contact (oil bank) will move down vertically recovering oil and displacing it to the down-dip production wells. This process can be highly efficient (good volumetric sweep) as long as there is good vertical permeability, and the gas interface is stable and moves vertically downward.

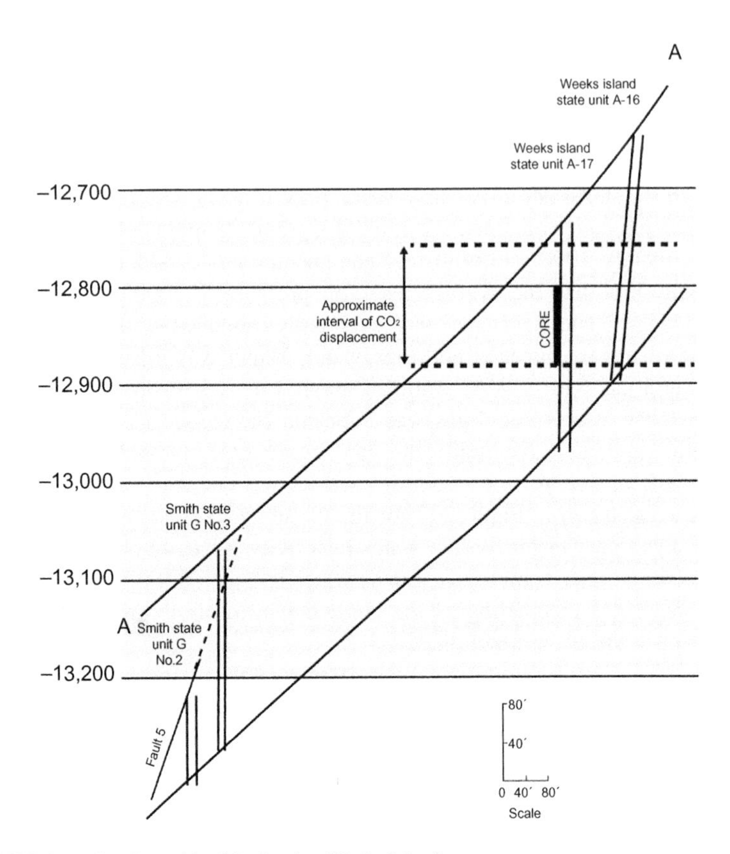

FIGURE 1.8 Gravity stable CO_2 flood at Weeks Island.

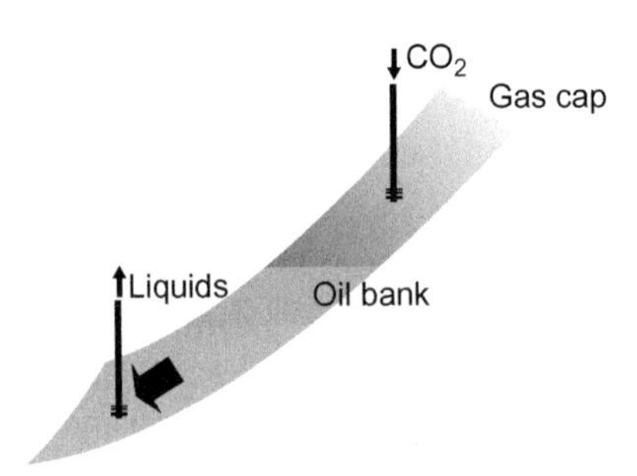

FIGURE 1.9 Illustration of oil bank movement during gravity stable flood at Weeks Island.

The gas injected at Weeks Island was a mixture of CO_2 and about 5% plant gas. The plant gas was used to "lighten" the CO_2 so that the gas–oil interface is more stable. Injection of plant gas with CO_2, however, was found unnecessary to ensure a gravity stable flood at Weeks Island as CO_2 was effectively diluted by dissolved gas (methane) from the reservoir oil.

At the reservoir temperature of 225°F and pressure at gas injection, the flood was immiscible, not miscible. Nevertheless a pressure core taken in zones where the gas traversed were nearly "white" with average oil saturations in the CO_2 swept zone of approximately 1.9% (Figure 1.8). This low oil saturation value is lower than miscible flood residuals, Sorm, that are typically observed due to oil-filled bypassed pores. The unexpectedly good recovery demonstrates that even immiscible floods when properly designed can achieve good extraction of oil components by gravity drainage.

A subsequent commercial test of the gravity stable process was not as successful largely because of significant water influx down-dip of the production wells. Injection of CO_2 largely pressurized the gas cap, but did not cause the gas–oil interface to move vertically downward. A gravity stable process like this would be very effective as long as water influx is relatively small. Perhaps one solution could have been outrunning the aquifer with water production wells or trying to plug off water influx.

One difficult problem also encountered was the production of the thin oil bank owing to both gas and water coning. The second producer was not planned but was drilled to measure saturations in the oil bank and to speed oil bank capture.

Immiscible gas floods in general can achieve better displacement efficiency as a secondary recovery method if gravity override is controlled, than for water floods owing to decreased oil viscosity, oil swelling, interfacial tension lowering, extraction of oil components, and the potential for gravity drainage as occurred at Weeks Island. Immiscible gas floods could also be a good alternative for reservoirs with injectivity issues when water is used. Two main disadvantages of immiscible gas flooding over waterflooding are the potential for poor sweep due to its adverse mobility ratio, and gravity override due to higher contrast between oil and gas gravities.

1.7.3 Jay Little Escambia Creek Nitrogen Flood

The Jay field near the Alabama–Florida border is one of the few nitrogen floods ever conducted. The reservoir is in the Smackover carbonate at a depth of 15,000 ft. Nitrogen is a good miscible gas in this reservoir because of its very light sour crude (50°API), and high reservoir pressure around 7850 psia. The formation permeability averages 20 mD. A significant advantage of nitrogen is that it is readily available via separation from the air, is relatively cheap, and does not cause corrosion unlike CO_2. Nitrogen was injected using a WAG ratio near 4.0, which is greater than typical.

The overall recovery at Jay is expected to be near 60% OOIP. Incremental recovery beyond waterflood recovery from miscible nitrogen injection is forecast to be around 10% OOIP (Figure 1.10). The high primary and secondary recovery of around 50% OOIP is likely the result of low vertical permeability coupled with good horizontal permeability in the dolomite

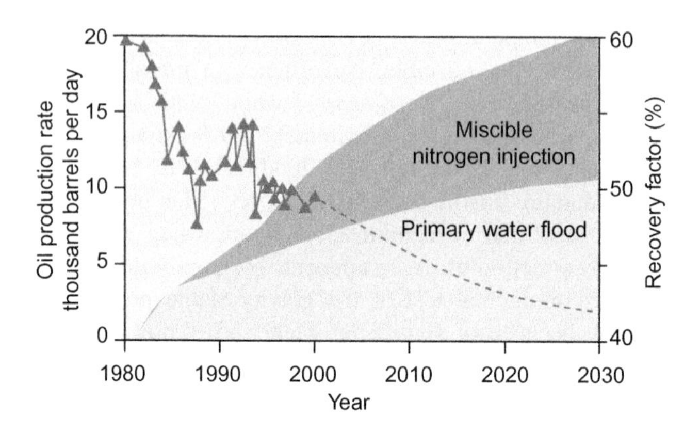

FIGURE 1.10 Oil recovery from Jay's miscible nitrogen flood. Miscible nitrogen injection is expected to give an incremental recovery of 10% OOIP over waterflooding alone. Source: *Data from Lawrence et al. (2002).*

facies (Lawrence et al., 2002). Low vertical permeability caused by shale lenses or in this case cemented zones associated with thin stylolites, is an ideal candidate for both water and gas flooding as fluids are less able to segregate vertically so that gravity override is reduced. This is especially true in this flood since nitrogen has very low density compared to the resident fluids, and would likely have gone to the top of the reservoir otherwise.

1.7.4 Overview of Field Experience

Gas flooding technology is well developed and has demonstrated good recoveries in the field (Manrique et al., 2007). Recoveries from both immiscible and miscible gas flooding vary from around 5–20% OOIP, with an average of around 10% incremental OOIP for miscible gas floods (Christensen et al., 2001). Tertiary immiscible gas flooding recoveries are less on average, around 6% OOIP. Although recovery by gas flooding is very economic at these levels, 55% OOIP still remains on average post miscible gas flooding assuming 65% OOIP prior to gas flooding. The significant amount of oil that remains is largely the result of gas channeling through the formation owing to large gas mobility, reservoir heterogeneity, dispersion (mixing) and gravity effects. Channeling also results in early breakthrough of the gas (gas), typically around the same time that oil is produced. This is in contrast to surfactant/polymer flooding, which almost always exhibits an oil bank prior to surfactant breakthrough. Poor volumetric sweep is not as much of a problem for surfactant-polymer floods or water floods, which have more favorable mobility ratios. Nevertheless, miscible flooding is generally very economic and less complex than chemical flooding, especially for deeper reservoirs that are more technically challenging for surfactant/polymer floods.

Besides poor sweep and early breakthrough of gas, several problems have been observed in implementing gas floods (Jarrell et al., 2002). Corrosion of injection and production facilities can occur when CO_2 is injected. Asphaltenes, hydrate, and scaling formation have also been a problem, but these problems can be overcome. Reduced water injectivity following a gas cycle during WAG has also been reported to be a significant problem in some fields. This is likely the result of relative permeability changes as fluids are injected (hysteresis/gas trapping). Gas injectivity is generally not as much of a problem, and in some cases, gas injectivity has increased with time owing to dissolution of carbonate rocks with CO_2, as well as cleanup of some of the organic materials (i.e., asphaltic substances) around the wells.

Most gas floods use WAG. Of those the WAG ratio is typically around 1.0 (Christensen et al., 2001). Total volumes of gas injected (slug sizes) are mostly in the range from 0.1 to 0.3 PVI. Injection of more gas, however, can yield larger recoveries, although at a cost of diminishing returns (less oil recovered per volume of gas injected). The effectiveness of gas floods is often measured in terms of gas utilization factors, which are the volume of gas injected per volume of oil recovered. Utilization factors less than 10 MCF/STB are very efficient.

One of the most important aspects of operating gas floods is reservoir management. Reservoir management of a field is a life-cycle process that requires good data acquisition and reservoir surveillance. Extensive core data, geologic descriptions, and good reservoir simulation models are necessary to understand how to best implement a gas flood process.

1.8 CONCLUDING REMARKS

Gas flooding is a mature technology that has demonstrated commercial success since the early 1980s. This chapter described the fundamental parameters and processes that are critical to the design of a gas flood, and showed through several case studies how those parameters affected field displacements. Although gas flooding is mature, achieving good recovery from a gas flood requires understanding the impact of these key factors on sweep and displacement efficiency.

Abbreviations

CV combined condensing and vaporizing drive
FCM first-contact miscibility
MCM multicontact miscibility
MME minimum miscibility enrichment
MMP minimum miscibility pressure
MOC method of characteristics
OOIP original oil in place
PVI pore volumes injected

REFERENCES

Amyx, J.W., Bass, D.M., Whiting, R.L., 1960. Petroleum Reservoir Engineering: Physical Properties. McGraw-Hill, New York. 0-07-001600-3.

Christensen, J.R., Stenby, E.H., and Skauge, A. (2001). Review of WAG field experience. SPE Res. Eval. Eng. 4 (2), 97–106.

Dake, L.P., 1994. The Practice of Reservoir Engineering. Elsevier, Amsterdam.

Danesh, A., 1998. PVT and Phase Behavior of Petroleum Reservoir Fluids. Developments in Petroleum Science No. 47, Elsevier, Amsterdam.

Dindoruk, B., Orr Jr., F.M., Johns, R.T., 1997. Theory of multicontact miscible displacement with nitrogen. SPE J. 2 (3), 268–279.

Egwuenu, A.M., Johns, R.T., Li, Y. Improved fluid characterization for miscible gas floods. SPE/IADC No. 94034, Fourteenth Europec Biennial Conference, Madrid, Spain, 13–16 June 2005.

Firoozabadi, A., 1999. Thermodynamics of Hydrocarbon Reservoirs. McGraw-Hill.0-07-022071-9

Green, D.W., Willhite, G.P., 1998. Enhanced oil recovery. SPE.

Jarrell, P.M, Fox, C.E., Stein, M.H., Webb, S.T. Practical Aspects of CO_2 Flooding. SPE Monograph Series, vol. 22, ISBN 1-55563-096-0, Richardson, TX, 2002.

Johns, R.T., Dindoruk, B., Orr Jr, F.M., 1993. Analytical theory of combined condensing/vaporizing gas drives. SPE Adv. Tech. Series 1 (2), 7–16.

Johns, R.T., Ahmadi, K., Zhou, D., Yan, M., 2010. A practical method for minimum miscibility pressure estimation of contaminated CO_2 mixtures. SPE Res. Eval. Eng. 13 (5), 764–772.

Lake, L.W., 1989. Enhanced Oil Recovery. Prentice Hall, Upper Saddle River, New Jersey. 07458.

Lawrence, J.J., Maer, N.K., Stern, D. Jay, 2002. Nitrogen Tertiary Recovery Study: Managing a Mature Field, SPE 78527, Presented at the Tenth Abu Dhabi International Petroleum Exhibition and Conference, 13–16 October 2002. Abu Dhabi, UAE.

Manrique, E.J., Muci, V.E., Gurfinkel, M.E., 2007. EOR field experiences in carbonate reservoirs in the United States. SPE Res. Eval. Eng. 10 (6), 667–686.

McCain Jr., W.D., 1990. The Properties of Petroleum Fluids, second ed. Penwell, Tulsa, OK.

Okuno, R., Johns, R.T., Sepehrnoori, K., 2010. Three-phase flash in compositional simulation using a reduced method. SPE J. 15 (3), 689–703.

Orr Jr., F.M., Johns, R.T., Dindoruk, B., 1993. Development of miscibility in four component CO_2 floods. SPE Res. Eval. Eng. 8 (2), 135–142.

Pedersen, K.S., Christensen, P.L., 2007. Phase Behavior of Petroleum Reservoir Fluids. CRC Press, Boco Raton, FL.0-8247-0694-3.

Sebastian, H.M., Lawrence, D.D. SPE/DOE Eighth Symposium on Enhanced Oil Recovery, SPE 24134, Tulsa, OK, 1992.

Solano, R., Johns, R.T., Lake, L.W., 2001. Impact of reservoir mixing on recovery in enriched-gas drives above the minimum miscibility enrichment. SPE Res. Eval. Eng. 4 (5), 358–365.

Stalkup Jr., F.I., 1983. Miscible Displacement. SPE Monograph Series. pp. 204. ISBN-10:1555630405.

Standing, M.B., 1977. Volumetric and Phase Behavior of Oil Field Hydrocarbon Systems. Society of Petroleum Engineers, ISBN-10: 0895203006.

Stein, M.H., Frey, D.D., Walker, R.D., 1992. Slaughter estate unit CO_2 flood: comparison between pilot and field-scale performance. JPT September.

Whitson, C.H., Brule, M.R., 2000. Phase behavior. SPE Monogr. 20, 240, ISBN: 1555630871.

Yellig, W.F., Metcalfe, R.S., 1980. Determination and prediction of CO_2 minimum miscibility pressures. J. Pet. Tech 32, 160–168.

Yuan, H., Johns, R.T., 2005. Simplified method for calculation of minimum miscibility pressure or enrichment. SPE J. 10 (4), 416–425.

Yuan, H., Johns, R.T., Egwuenu, A.M., Dindoruk, B., 2005. Improved MMP correlation for CO_2 Floods using analytical gas flooding theory. SPE Res. Eval. Eng. 8 (5), 418–425.

Chapter 2

Enhanced Oil Recovery by Using CO_2 Foams: Fundamentals and Field Applications

S. Lee and S.I. Kam
Craft and Hawkins Department of Petroleum Engineering, Louisiana State University, Patrick F. Taylor Hall, Baton Rouge, LA 70803, USA

2.1 FOAM FUNDAMENTALS

This section describes general features associated with CO_2-foam processes. The individual topics include, but not limited to, why CO_2 is the material of interest, what makes CO_2 foams special compared to foams made of other gas phases, what foam does within the context of mobility reduction and overcoming reservoir heterogeneity, what the characteristics of foam rheology in porous media are, and how foam rheology can be modeled and simulated. Not to mention, understanding the fundamentals of foams in porous media is crucial to the optimal and successful operation of foam-assisted enhanced oil recovery (EOR) field processes.

2.1.1 Why CO_2 Is so Popular in Recent Years?

Due to global warming, many chemicals have been added to the list of the National Greenhouse Gas Emission Inventories. They largely consist of two different categories of greenhouse gases: (1) naturally occurring chemical species such as water vapor, carbon dioxide (CO_2), methane (CH_4), nitrous oxide(N_2O), and ozone (O_3) and (2) chemical products resulting from industrial activities such as fluorine, chlorine, bromine, chlorofluorocarbons (CFCs), and hydrochlorofluorocarbons (HCFCs). Among these, CO_2 is in the global spotlight because it is the largest source of US greenhouse gas emissions, followed by CH_4 and N_2O. For example, it is believed that the CO_2 emission caused by human activities represents about 83.0% of total greenhouse gas emissions in the United States in 2009 (US Greenhouse Gas Inventory Report, 2011).

Enhanced Oil Recovery Field Case Studies.

This becomes a major motivation of the recent boom in CO_2-driven EOR which benefits human society with capabilities of boosting up hydrocarbon production as well as capturing and storing CO_2 in petroleum-bearing underground geological structures (Dooley et al., 2010).

2.1.2 Why CO_2 Is of Interest Compared to Other Gases?

What is so special about CO_2-driven EOR processes, compared to other gases such as nitrogen, methane, and ethane? Firstly, CO_2 forms a dense or supercritical phase at typical reservoir pressure and temperature conditions. Note that the critical pressure (P_{crit}) of CO_2 is 73.0 atm (or 1073 psia) and the critical temperature (T_{crit}) of CO_2 is 31.0°C (or 87.8°F) (Chang, 1994). CO_2 is called "supercritical" if reservoir pressure is greater than P_{crit} and reservoir temperature is greater than T_{crit}, while it is called "dense" if reservoir pressure is greater than P_{crit} but reservoir temperature is lesser than T_{crit}. This dense or supercritical CO_2 has a characteristic of high density and viscosity, compared to other gases, which makes the displacement front more stable by naturally mitigating gravity segregation and viscous fingering to some degree during gas injection EOR. Secondly, taking one more step further, if injected CO_2 creates a miscible flooding with the reservoir fluids by satisfying miscibility condition, then the interfacial tension becomes negligible and there is no oil trapped by capillary forces (Holm and Josendal, 1974). This implies that the remaining oil saturation prior to CO_2 injection can be ideally reduced down to almost zero during miscible CO_2 flooding, boosting up the amount of oil recoverable. Lastly, if the injected CO_2 mixes with and dissolved into reservoir oils, the volume of oleic phase increases. This swelling effect, combined with pressure surge, yields more oil production (Yellig and Metcalfe, 1980).

2.1.3 Why CO_2 Is Injected as Foams?

Almost all gas injection processes suffer from phenomena associated with gravity segregation, viscous fingering, and channeling, eventually leading to poor volumetric sweep efficiency. Although these limitations may seem to be important to a lesser degree compared to other gases, there are still very challenging issues for dense or supercritical CO_2 injection.

Needless to say, these limitations resulting from gas injection are equally applicable if field conditions are not met and thus CO_2 forms a gas phase. For example, CO_2 injection into a depleted reservoir may have injection pressure lower than the critical pressure; CO_2 injection into a heavy oil reservoir may not allow CO_2 to be miscible with reservoir fluids, or the process of CO_2 mixing with reservoir oils may not be first-contact miscible and thus may take a significant amount of time; in some remote areas, CO_2 may need to be injected with other flue gases, which makes the injected gas mixture difficult to be dissolved into the reservoir oils due to the volatile components (especially methane); and reservoir temperature may be too high such that the minimum miscibility pressure (MMP) cannot be achieved easily. By injecting foamed CO_2—whether CO_2 is a gas phase, a dense phase, or a supercritical phase—the EOR process can significantly improve both aerial and vertical sweep efficiencies. This benefit comes from the reduction in gas mobility by the presence of thin foam films (so-called lamellae), which in turn is originated from the reduction in gas viscosity and relative gas permeability.

2.1.4 Foam in Porous Media: Creation and Coalescence Mechanisms

In contrast with bulk foams whose height progressively decays with time in a stationary container, foams injected in porous media undergo dynamic mechanisms of *in situ* lamella creation and coalescence. Previous studies identified four major lamella creation mechanisms such as leave behind, snap-off, mobilization and division, and gas evolution, and a major lamella coalescence mechanism based on limiting capillary pressure and disjoining pressure. Details about these mechanisms are available elsewhere (Kovscek and Radke, 1994; Rossen, 1996).

Since CO_2 is highly miscible with and easily diffusive in reservoir fluids, these generation and coalescence mechanisms might be more complicated compared to other gas phases such as nitrogen and hydrocarbon gases. Such an attempt was made by break-and-reform mechanism proposed in earlier foam studies as pointed out by Rossen et al. (1995).

Just like any other chemical flooding treatments, an extreme reservoir temperature may cause surfactant chemicals to thermally degrade and lamella coalescence to prevail so that foam propagation deep into the reservoir could be challenging (Kam et al., 2007a).

2.1.5 Foam in Porous Media: Three Foam States and Foam Generation

A snapshot of foam flow in porous media typically leads to three different situations as illustrated in Figure 2.1. Firstly, there are no foams present initially, or pre-existing foams are destabilized and destroyed, for example, in a high-capillary pressure environment, in a strongly oil-wet formation, and in a medium with high oil saturation (Figure 2.1A). Foam flow in this case then is no other than conventional gas–liquid two-phase flow with no foam films present. This results in high saturation of water, filling smaller pores. Secondly, the presence of numerous foam films forms very fine-textured foams (Figure 2.1C), which is referred to as strong foams. Once formed, strong foams may increase effective foam viscosity (or decrease relative gas mobility, equivalently) by up to several orders of magnitude, exhibiting a dramatic increase in pressure gradient or reduction in water saturation.

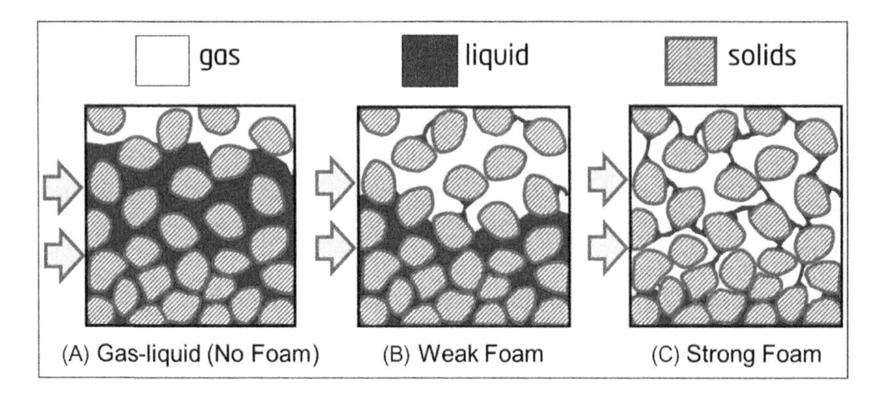

FIGURE 2.1 Schematic figures comparing conventional gas–liquid two-phase flow (no foam), weak foam, and strong foam in porous media.

Lastly, weak foams can be formed by exhibiting a moderate increase in effective foam viscosity, typically by less than a few orders of magnitude, leading to a moderate increase in pressure gradient or reduction in water saturation (Figure 2.1B).

In typical coreflood experiments, the transition from weak-foam to strong-foam state (i.e., commonly referred to as "foam generation") often occurs abruptly, accompanied by a drastic increase in pressure drop. As illustrated by Figure 2.2A with actual experimental data, laboratory foam generation experiments can be carried out at fixed foam quality (or, gas fraction or f_g, equivalently) varying the total injection rate. This particular example is with a Berea sandstone and the injection foam quality of 0.80: initially, at time $t = 0$, the experiment began with coinjection of nitrogen gas and brine solution at the total injection rate (Q_t) of 0.25 cc/min; at time $t =$ A, the brine was simply switched to the surfactant solution keeping other conditions identical; and at subsequent time $t =$ B, C, and D, Q_t was then raised to 0.50, 0.80, and 1.30 cc/min in a step-by-step manner. The results show that the pressure drops measured across the core (Δp_{total}) and in four sections along the core (Δp_1, Δp_2, Δp_3, and Δp_4) increased dramatically when Q_t reached 1.30 cc/min. The similar experimental data are presented in Figure 2.2B in terms of the steady-state Δp_{total} as a function of total injection rates, showing a discontinuity in its path near the transition from weak foams to strong foams (i.e., foam generation).

Recent studies (Gauglitz et al., 2002; Kam and Rossen, 2003; Kam et al., 2007b) show that this discontinuity in pressure (or, effective foam viscosity) at the onset of foam generation takes place only in the experiments with fixed injection rates. Once experiments are conducted with controlled pressure drops across the system (cf. type 1 and type 3 experiments of Gauglitz et al., 2002), the entire S-shaped path from weak-foam state to strong-foam state, interconnected by the intermediate state, can be revealed as shown by

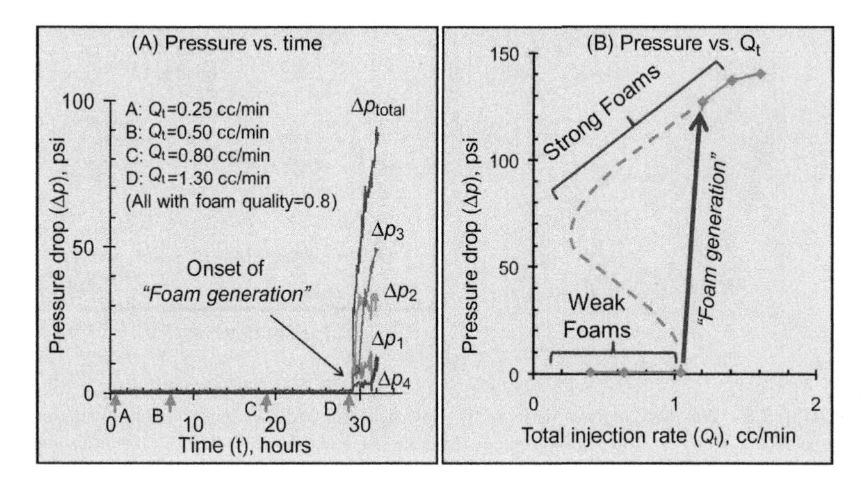

FIGURE 2.2 Typical response in laboratory foam flow experiments showing the onset of foam generation.

the dotted line in Figure 2.2B. The surface constructed by these *S*-shaped curves reminisces the catastrophe theory, which has long been popular in many social, scientific, and engineering studies in 1970s and 1980s, showing the presence of multivalued solutions. Such a catastrophic foam behavior characterized by a sudden and unpredictable shift, often associated with mathematical singularity, is referred to as foam catastrophe theory (Afsharpoor et al., 2010; Kam, 2008).

2.1.6 Foam in Porous Media: Two Strong-Foam Regimes—High-Quality and Low-Quality Regimes

Once the strong-foam state is obtained, foam is shown to follow two distinct flow behaviors, near-Newtonian or slightly shear-thickening behavior in the high-quality regime and highly shear-thinning behavior in the low-quality regime as shown in Figure 2.3. This original two flow regime concept was presented by Osterloh and Jante (1992) in their 2-ft long sandpack flow tests, further confirmed by Alvarez et al. (2001) in a wide range of porous media and experimental conditions, and incorporated into different types of modeling studies (Cheng et al., 2002; Kam and Rossen, 2003). The dramatic difference in pressure contours plotted as a function of gas velocity on the y axis and liquid velocity on the x axis is due to the different governing mechanisms in the two flow regimes: (i) near-vertical pressure contours in the high-quality regime due to the feedback mechanism keeping water saturation near limiting water saturation (or capillary pressure near limiting capillary pressure) (Khatib et al., 1988) and (ii) near-horizontal pressure contours in the

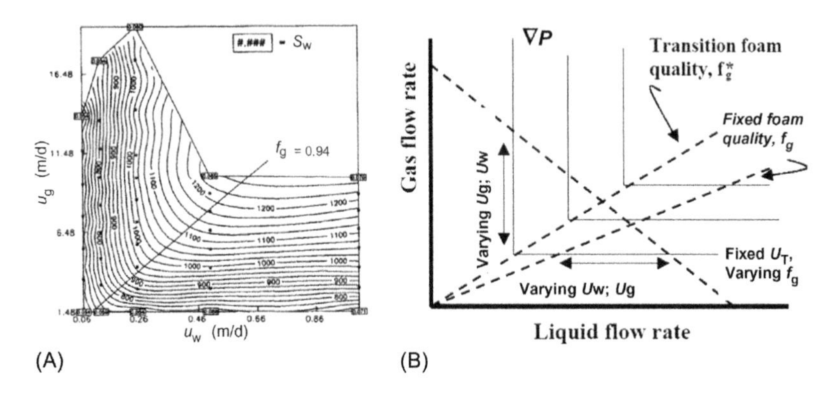

FIGURE 2.3 Two flow regimes defining the flow behavior and rheology of strong foams in porous media. (A) Two flow regime contours from Osterloh and Jante (1992). (B) Identification of general trend (Alvarez et al., 2001).

low-quality regime due to bubble trapping and mobilization, which in turn is related to the relative permeability effect (Rossen and Wang, 1999).

2.1.7 Modeling Foams in Porous Media

There are several different versions of foam modeling and simulation available in the literature. Two most popular approaches are based on "local steady-state modeling" and "mechanistic foam modeling." The major difference between the two lies in the attitude that how the complex dynamic mechanisms of lamella creation and coalescence are handled: the local steady-state modeling, typically combined with fractional flow analysis, has an assumption (among many) that an equilibrium state is attained instantaneously such that the physics of foam propagation is based on a predetermined effective foam viscosity, or reduction in gas mobility often called "mobility reduction factor" (MRF) (Mayberry et al., 2008; Rossen and Zhou, 1995); while the mechanistic foam modeling/simulation chases up with the changes in the rates of lamella creation and coalescence, and let the model/simulator decide foam texture and then resulting foam viscosity and relative permeability at given flow conditions following bubble population balance (Afsharpoor et al., 2010; Bertin et al., 1998; Falls et al., 1988; Kam, 2008; Kovscek et al., 1995; Myers and Radke, 2000). It is generally accepted that the local steady-state modeling is handy and versatile enough to capture the nature of foam displacement processes conveniently if transient behaviors are negligible, while the mechanistic foam modeling/simulation has a unique capability of dealing with true foam physics, especially time-dependent mechanisms and inlet effects.

In fact, these two different types of approaches can be combined together through the use of fractional flow analysis with foam MRF as a function of bubble population balance as shown in Kam and Rossen (2003). Dholkawala et al. (2007) further developed this concept with foam catastrophe theory, presenting velocity-dependent fractional flow curves as shown in Figure 2.4: at low total injection velocity, the fractional flow curve is weak-foam dominant; at high total injection velocity, the fractional flow curve is strong-foam dominant; and, in between, the curve changes its shape showing all three different foam states. The study by Dholkawala et al. (2007), for the first time, presents a fractional flow "surface" (i.e., water saturation (S_w) and water fractional flow (f_w) on the x and y axes with total injection velocity on the z axis), as shown in Figure 2.5, in order to accommodate and visualize complex and dynamic mechanisms of lamella creation and coalescence. The use of three-dimensional fractional flow surface is believed to help identify the dimensionality-dependent foam flow characteristics, showing the implications in foam EOR in

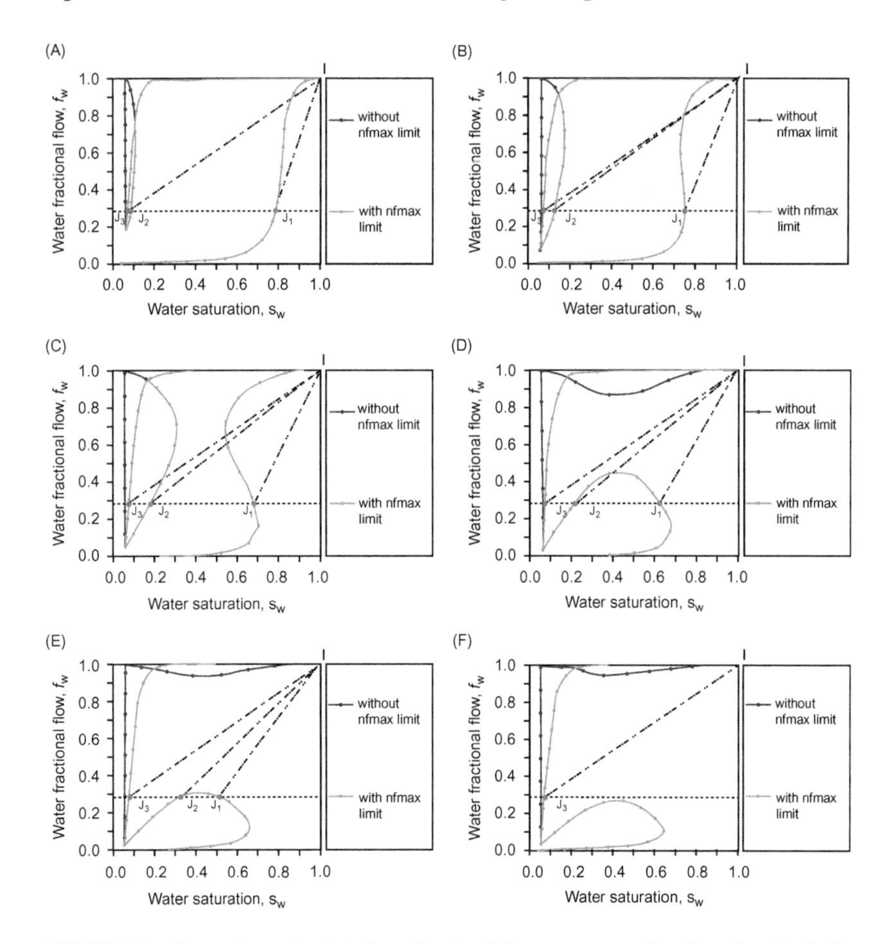

FIGURE 2.4 Change in mechanistic foam fractional flow curves as a function of total velocity (u_t): (A) 1.72 ft/d, (B) 2.91 ft/d, (C) 4.2 ft/d, (D) 4.91 ft/d, (E) 5.89 ft/d, and (F) 6.37 ft/d (Dholkawala et al., 2007).

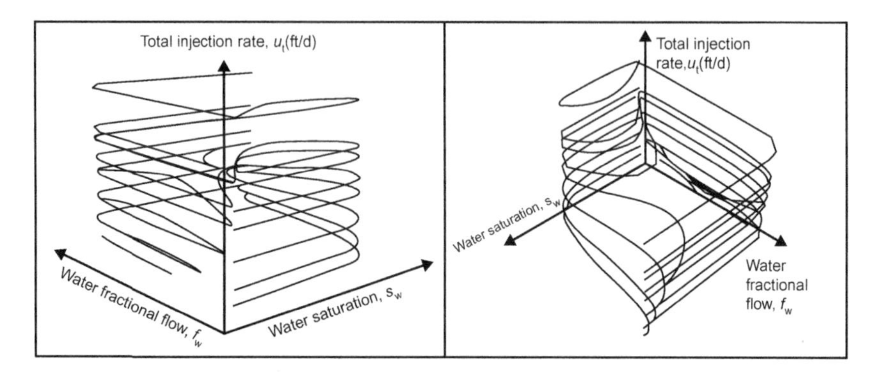

FIGURE 2.5 Three-dimensional fractional flow "surface" showing velocity-dependent foam rheology following foam catastrophe (Dholkawala et al., 2007).

radial geometries in which the total "velocity," not flow rate, decreases rapidly as foam moves away from the injection well.

2.1.8 Foam Injection Methods and Gravity Segregation

Foams can be injected into a reservoir mainly in two different ways for EOR purposes, although some variations are possible: (i) injection of both gas and surfactant solutions together (or, so-called coinjection process), which typically causes pregenerated foams in injection tubing and wellbore to enter the formation and (ii) multiple cycles of alternating gas with surfactant injection (or, commonly called surfactant-alternating-gas (SAG) process), which causes a fluctuation in capillary pressure by repeating drainage and imbibition, and naturally helps creation of fine-textured foams during the process. Because bulk foam rheology is different from foam rheology in porous media due to its low capillary pressure environment (Bogdanovic et al., 2009; Gajbhiye and Kam, 2011), caution should be taken when these two are mingled together as shown in foam treatments in fractured reservoirs, where foams in fractures tend to follow bulk foam rheology while foams in matrix tends to follow foam rheology in porous media.

Any gas and water injection into a reservoir bearing oils ultimately faces gravity segregation (Jenkins, 1984; Stone, 1982), as shown in Figure 2.6, because the low-density gas phase tends to migrate upward and the high-density water phase tends to slump down toward the bottom of the reservoir. The injected mixture of gas and liquid is separated completely after traveling over a certain distance (shown by "L_g" in Figure 2.6), which is related to the characteristics of the formation, fluid properties, and injection conditions. This concept is also shown to be valid with foams (Rossen and van Duijn, 2004)—an increase in pressure gradient during foam injection, however, greatly increases the traveling distances before segregation, and essentially improves sweep efficiency. The level of gravity segregation is obviously very sensitive depending on if

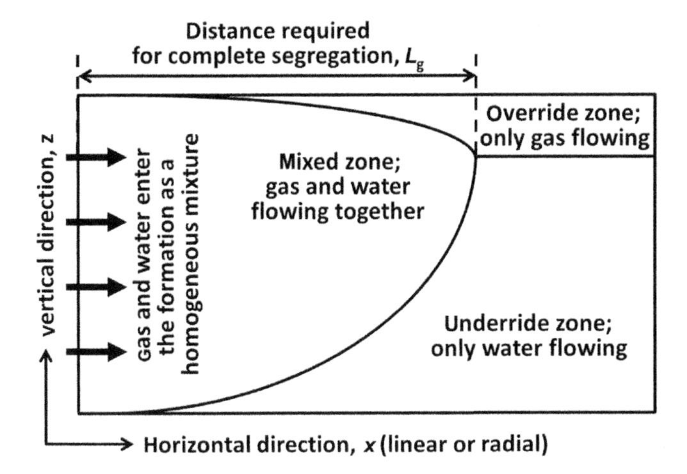

FIGURE 2.6 Schematic representation of three uniform zones during gas–liquid coinjection. Source: *Redrawn from Rossen and van Duijn (2004).*

CO_2 exists as a gas phase or a supercritical (or dense) phase and if formation wettability and reservoir fluids are detrimental to foam stability.

2.1.9 CO_2-Foam Coreflood Experiments

Typical CO_2-foam experiments are involved in the injection of multiple phases, such as reservoir brine, oil, and gas (to mimic the initial reservoir conditions), injected water and gas (to mimic the secondary recovery for pressure maintenance), and injected CO_2 and surfactants (to mimic the tertiary CO_2-foam injection). For supercritical CO_2-foam injection, the coreflood experiments require elevated pressure and temperature above the critical pressure and temperature, which necessitate high pressure and temperature equipment. The use of positive-displacement pumps to pressurize CO_2 prior to injection is an essential step.

Figure 2.7 shows a schematic of such an experimental setup. There are a few injection pumps dedicated to the injection of CO_2, water, and oil phases. CO_2 and water (containing surfactants) are forced into a foam generator to artificially generate fine-textured foams. If the purpose of experiments is to investigate the creation of foams *in situ*, the path to the foam generator can be bypassed. A synthetic porous disk or medium may serve as a good foam generator. The fluid is injected into a coreholder which has multiple pressure taps so that how effective foam viscosity changes as foam propagates from the inlet to the outlet could be monitored. The backpressure regulator downstream of the coreholder regulates the minimum pressure imposed to the system. A separator is installed in the effluent in order to keep track of water

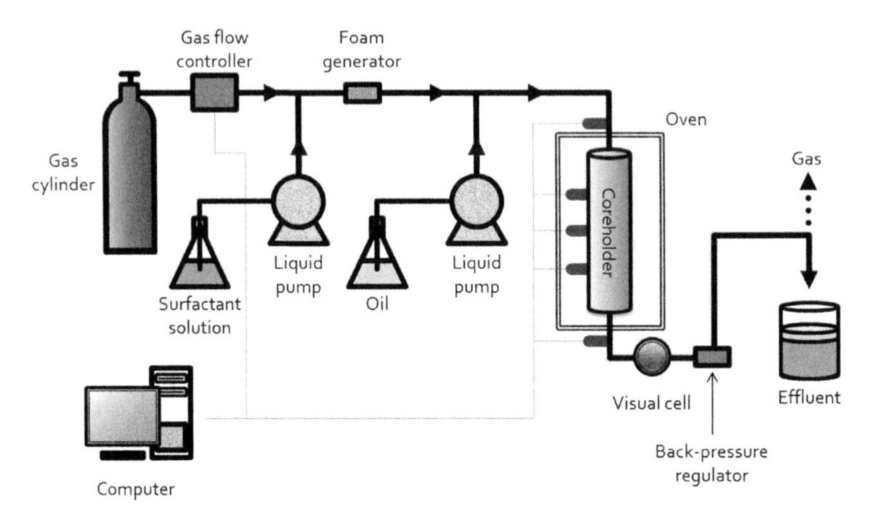

FIGURE 2.7 Typical laboratory setup for CO_2-foam experiments.

and oil production history. The entire coreholder system may need to be positioned in an oven for experiments at elevated temperatures.

2.1.10 Effect of Subsurface Heterogeneity—Limiting Capillary Pressure and Limiting Water Saturation

One of the most intriguing foam properties in porous media is its capability to overcome subsurface heterogeneity. Permeability contrast, often regarded as the most important aspect of subsurface heterogeneity, leads to a poor sweep efficiency resulting from early breakthrough. Conceptually, foam's capability to overcome the heterogeneity is endowed by its sensitivity to capillary pressure (Khatib et al., 1988): foam films are more stable in high-permeability layers due to the low capillary pressure environment, while foam films are less stable in low-permeability layers due to the high capillary pressure environment. Putting these two together, foams injected into the system with permeability variations result in the diversion of subsequent injected flow into the low-permeability layers by blocking the high-permeability layers with stable foams.

The key to understanding this phenomenon is the role of limiting capillary pressure (P_c^*) above which foam films cannot sustain due to too high capillary pressure (or, limiting water saturation (S_w^*) below which foam films cannot survive due to the lack of aqueous phase, equivalently). Suppose there are two layers with higher absolute permeability k_1 and lower absolute permeability k_2. Then the capillary pressure curve for each layer can be delineated as shown in Figure 2.8. If the limiting capillary pressures for both

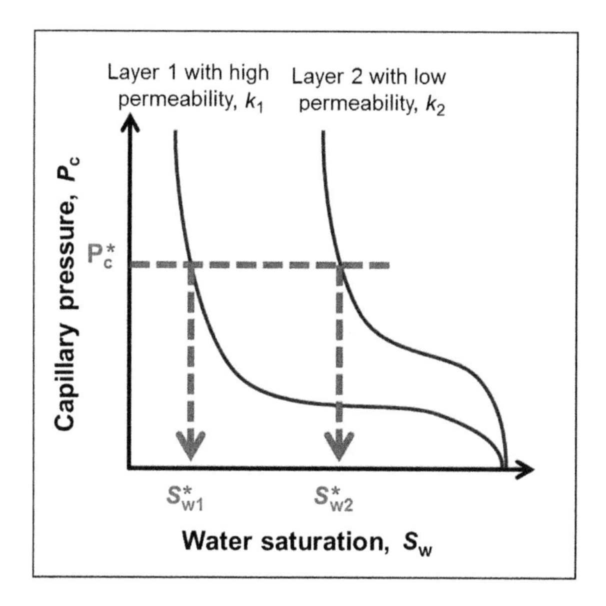

FIGURE 2.8 The effect of absolute permeability on limiting water saturation.

layers are given by P_c^* (i.e., horizontal dotted line in Figure 2.8), the corresponding limiting water saturation is higher in layer 2, or $S_{w1}^* < S_{w2}^*$. This implies that for a two-layer system in capillary equilibrium, the high-permeability layer (layer 1 with k_1) tends to have more stable foams due to lower limiting water saturation (S_{w1}^*). This allows stable foams to reside in the high-permeability layer, diverting subsequent fluids into the low-permeability layer. Although the limiting capillary pressure appears to decrease with absolute permeability, it is not clear what kind of functional relationships they have, especially in a permeability range relevant to petroleum reservoirs (Rossen and Lu, 1997).

It should be noted that this type of foam-diversion process is more relevant to strong foams in the high-quality regime governed by limiting capillary pressure; strong foams in the low-quality regime may not necessarily follow the same trend due to different governing mechanism—bubble trapping and mobilization. More specifically, unlike foams in the high-quality regime exhibiting near-Newtonian or slightly shear-thickening behavior (Alvarez et al., 2001), foams in the low-quality regime are not believed to be as effective in diversion process because of highly shear-thickening rheology.

Another important aspect of foam-diversion process is the role of foam generation. The minimum pressure gradient (∇P^{min}) to mobilize foam films, and thus to result in foam generation, scales with the inverse of absolute permeability (k) or, $\nabla P^{min} \propto k^{-1}$ (Afsharpoor et al., 2010; Gauglitz et al., 2002). This implies that the pressure gradient required to achieve fine-textured strong foams becomes lower with increasing absolute permeability, meaning that obtaining strong-foam state is more readily achievable in the high-permeability layer than the low-permeability layer, essentially helping foam-diversion process.

2.1.11 Foam–Oil Interactions

Earlier studies put significant efforts into identifying the interactions between foams and oils. Two major tools to evaluate foam's sensitivity to reservoir oils are the entering coefficient and the spreading coefficient which are expressed in terms of interfacial tension between different phases as well summarized by Schramm (1994). An oil phase is shown to be detrimental to foam stability if the oil can enter the gas–water interface (i.e., positive entering coefficient) and spread (i.e., positive spreading coefficient).

For foams in porous media, there is one more component to be accounted for in addition to the entering and spreading coefficients. Because fluids are distributed in the media such that the wetting phase occupies small pores and the nonwetting phase occupies large pores, a small quantity of oil may not deteriorate foam stability significantly due to the lack of interactions between foams and oils during the flow. This becomes the origin of critical oil saturation, below which foam behavior is not affected dramatically even in presence of oils, but above which foam collapses abruptly (Dalland et al., 1994; David and Radke, 1990; Koczo et al., 1992). Once foam becomes unstable above the critical oil saturation, the flow is similar to three-phase gas-oil-water flow with no foam. More discussions on the modeling of foam–oil interactions during foam flow in porous media can be found elsewhere (Mayberry et al., 2008; Namdar-Zanganeh et al., 2011).

2.2 FOAM FIELD APPLICATIONS

This section covers examples of field CO_2-foam EOR processes that have widely been cited and referenced. Some of the earlier field examples using steam foams and nitrogen foams in conjunction with CO_2 foams are also included to share the history of overall foam EOR processes.

2.2.1 The First Foam Field Applications, Siggins Field, Illinois

The first foam field application in the literature was conducted in the Siggins field, located near Casey, Illinois, from 1964 to 1967 (Holm, 1970) by using air as the gas phase. The entire field had about 900 producing wells, with a well density of 8–10 wells per 40 acres, from which a total of 13,700,000 gross barrels of oil (or 3425 bbl/acre) was produced by 1920. Many of the wells had been gradually abandoned since then until 1940 when the oil production continuously dropped down to less than 100 bbl/day. When waterflooding was initiated 1 year later, oil production from the field was improved, making an oil production peak of 3000 bbl/day in October 1949. A report in January 1955 shows that the gross oil production from 552 producing wells was 2411 bbl/day and the cumulative oil production by waterflooding was 7,838,023 bbl (i.e., an overall average of 3290 bbl/acre). Since 1955, the producing water–oil ratio (WOR) had sharply increased to about 25:1.

Two phases of foam injection were performed in the field: the first phase was started in October 1964, and completed in March 1966; and the second phase was conducted from August 1966 to June 1967. A surfactant named O.K. Liquid (modified ammonium lauryl sulfate) was identified as the foaming agent from a series of laboratory tests using more than 100 surfactant candidates initially. After several cycles of SAG processes in the field, it was observed that (i) the injection profile was improved such that the surfactant solution was introduced into different layers more uniformly and (ii) the WOR was reduced down to 12:1 at the production wells. The injectivity was shown to be fairly reasonable as expected. It was estimated from further analysis that the mobility of injected air was reduced by 50% which presumably mitigated the channeling of air to production wells through high-permeability layers. The mobility of surfactant solution was also reduced to about 35% of its original value because of a reduction in water saturation during foam propagation.

2.2.2 Steam Foam EOR, Midway Sunset Field, California

Another well-known major field application of foam-assisted EOR in early days was performed in the Midway Sunset field, located in the San Joaquin Valley, 40 miles west of Bakersfield, California, in 1974 (Fitch and Minter, 1976). This field application consists of a series of trials (i.e., the first test followed by two more field tests over the next decades), by using steam foams in order to boost up heavy oil production.

In the first trial, the target was the recovery of heavy oil with 8–16°API gravity. The reservoir was shallow (average depth of 1400 ft) and had very favorable average formation properties such as absolute permeability of 3 Darcy, porosity of 35%, and thickness of 400 ft. The formations of major interest were made up of massive unconsolidated Potter sands in which individual grain sizes varied significantly from fine silts to large cobbles. COR-180, used as the foaming agent and chemical diverter to generate stable foam with high-temperature steam later, was injected as an aqueous solution at the wellhead. More specifically, steam with 75% quality was first injected for approximately 24 hours to stabilize the bottomhole temperature, because thermal stability at 600°F was a key factor for the success of the field application. After the thermal stabilization, 80 gallons of COR-180 surfactant solution was injected as an initial slug, and then steam was injected again for 24–72 hours. Finally, 80–100 gallons of surfactant slug was then injected. On average, 2 gallons per minute (GPM) injection rate of surfactant chemical with steam at an equivalent 33 GPM water rate were introduced to stimulate 21 wells over 2 years beginning in 1973. An increase in injection pressure was observed during foam injection which distributed temperature uniformly across the reservoir. This field trial was believed to be successful leading to a substantial incremental oil production.

The second steam foam field trial was performed in 1983 in two sections of the Midway Sunset field (Ploeg and Duerksen, 1985): Section 15A, located near the northern end of the field, consisting of the Potter sands (about 500 ft thickness and 20° dip angle to the east with average porosity and permeability 0.365 and 3.9 Darcy); and Section 26C, located in the lower central part of the field, consisting of the upper Monarch sands (about 600 ft thick and 20° dip angle to the northwest with average porosity and permeability 0.29 and 1.4 Darcy). Total incremental oil production in two sections was 53,000 bbl and 15,000 bbl, respectively. Additional foam field trials were also followed by using Enordet AOS-1618TM surfactant (C16–18 alpha-olefin sodium sulfonate) to correct gross steam override and improve the drive.

The third steam foam trial in the Midway Sunset Field was performed at Dome-Tumbador in the Midway Sunset Field (Mohammadi et al. 1989; Mohammadi and Tenzer, 1990). The reservoir consisted of four productive zones which were poorly sorted with particle size ranging from silt to gravel. Also, there were intraformational silts of relatively impermeable sand layers which separated these productive zones. This pilot test area was composed of four inverted five-spot patterns with nine producers and four injectors. Two objectives motivated this pilot project at Dome-Tumbador: (i) correcting gross steam override and improve the drive and (ii) understanding mechanisms to explain liquid displacement by steam foam propagation. For the field trial, several foaming agents were tested and Enordet AOS-1618TM, a C16–18 alpha-olefin sodium sulfonate, was selected as the best foamer.

Since January 1985, foam treatment had been performed with simultaneous injection of AOS-1618TM surfactant solution in brine and nitrogen with steam for about 40 months. After 3 years of the start of foam injection, the concentration of surfactant solution was reduced from 0.51 wt% to 0.24 wt% to test foam effectiveness at lower surfactant concentration. Shortly after the beginning of foam treatment, the oil production from the pilot increased and sustained over the foam injection period, which then followed by a gradual reduction in oil production. As a result of foam injection, a total of incremental oil production was estimated to be 6.0% of original oil-in-place (OOIP), and heat redistribution was observed together with a rapid expansion of areal and vertical sweep improvements.

2.2.3 CO_2/N_2 Foam Injection in Wilmington, California (1984)

The Wilmington field, located in southern California and operated by Long Beach Oil Development Co. and Unocal Corp, had a relatively shallow reservoir depth of about 2300 ft with initial reservoir pressure and temperature of 900–1100 psi and 130°F, average porosity of 24–26%, and oil gravity of 13–14°API (Holm and Garrison, 1988). The reservoir with 57 wells drilled for injection and production purposes mainly comprised of three layers in various thickness and permeability ranges. The estimated vertical thickness was about 70, 45, and 45 ft from the top to the bottom layers, respectively. The permeability ranged from 100 to 1000 mD, with the top layer much more permeable than the other two layers.

As a part of secondary recovery process, water injection had been performed into the reservoir for 21 years since 1961. Due to the high water cut of 94%, which is a typical symptom of the last stage of water injection secondary recovery, CO_2 injection was considered as tertiary recovery method. When the mixture of CO_2 and N_2 was introduced, alternating with water, into the tar zone of Fault Block V in 1982, the displacement process was shown to be immiscible because of high MMP of the injected CO_2/N_2 and reservoir oil system, which was believed to be more than 3000 psi. Although a rise in oil production was observed right after the immiscible CO_2 flood, the process was not impressive, primarily because of excessive gas production caused by early breakthrough. This was thought to occur obviously through the high-permeability top layer. The following near-wellbore treatment to improve the distribution of injected gas did not have successful results because of completion problem and zone communication beyond the wellbores. These became the major motivation of CO_2-foam injection in the Wilmington field, more specifically, in order to improve the distribution and sweep efficiency of the injected CO_2 by reducing gas channeling in the top layer.

Laboratory studies selected Alipal CD-128 foaming agent the best candidate. This commercial surfactant product was an ammonium salt of linear alcohol ethoxylate sulfate (about 60% active) containing water and small amount of alcohol. The selection criteria were the stability of resulting foams in various oil-field brines, foam mobility, and diversion ability in layers with differing permeabilities at reservoir temperature and pressure. A tracer test, performed prior to the foam injection, showed that most of the injected gas and water during the water-alternating-gas (WAG) process entered the most permeable top layer and only a very small fraction of the fluids entered the other two bottom layers. The foam injection project performed in 1984 started with a preflush of a solution of 2 bbl of formaldehyde in 5000 bbl of water at a rate of 2500 bbl/day, followed by the injection of 4 cycles of SAG with Alipal CD-128 surfactant solution with CO_2/N_2 gas. After checking gas injection profiles and conducting pressure falloff tests, additional 4 more cycles of Alipal CD-128 solution with CO_2/N_2 gas were injected. Gas injection profiles and pressure falloff tests were carried out once again. Finally, brine, at a rate of 57,720 bbl/day for about 20 days, and CO_2/N_2 gas were subsequently injected as chaser fluids.

This CO_2/N_2 foam project, consisting of eight cycles of SAG injection with 1% Alipal CD-128 in formation water and CO_2/N_2 gas, consumed about 21,000 bbl of Alipal CD-128 surfactant solution to inject 90% quality foams, was shown to be successful, diverting injected foams into the bottom layers with relatively low permeability. More specific observations include the following aspects: (1) a significant reduction of injected gas into the top high-permeability layer (i.e., 99% before foam treatment vs. 57% after foam treatment), (2) an increase in injected gas into the middle low-permeability layer (i.e., less than 1% before vs. up to 43% after), (3) an improvement in skin factor from 10 to 1.5 in the test well, and (4) a reduced in-depth gas mobility in the top layer caused by the propagation of stable foams. Overall, although it was not perfect, the channeling of injected fluids was mitigated effectively by using CO_2 foams and the project was shown to have achieved the primary goals.

2.2.4 CO_2-Foam Injection in Rock Creek, Virginia (1984–1985)

The Rock Creek field was located in south central West Virginia and discovered in 1906 (Heller et al., 1985). Productive area of the field was about 11,000 acres. The field underwent the following histories briefly. First, gas drive utilizing low-pressure gas recycle was applied as a major production method in 1935 and recovered only about 10% of the original oil in place (OOIP), and then six different secondary recovery projects with three different methods were implemented in the field but unsuccessful. Three of the six attempts were separate waterfloods which began in 1950s and 1960s but were not economically viable due to high water cuts. Steam injection was continued after that, but was also unsuccessful because of high heat loss and low injectivity. After these primary and secondary recovery trials, the possibility of CO_2 flooding was considered by Pennzoil. As a first step, waterflood began for initial pressurization in 1977 and then CO_2 was injected into the 30 ft-thick target zone in 1979. As a result of CO_2 injection, the average oil saturation in the upper 10 ft zone decreased to an average of about 4% but the residual oil saturation in the lower 20 ft zone decreased only to about 19%. Based on the results, it was concluded that this poor sweep was due to nonuniform displacement and fingering resulting from the high mobility of CO_2.

In 1981, a new project was initiated to determine the potential of thickened CO_2 for additional oil recovery. This CO_2-foam test was conducted as a joint project of the US Department of Energy (DOE), the New Mexico Petroleum Recovery Research Center (PRRC), and Pennzoil. From laboratory test, it was found that a steady-state relative mobility of 0.2–0.4 cp^{-1} was achieved by 80% quality CO_2 foam with a 0.05% Alipal CD-128 solution in brine, and permanent adsorption was 5.6×10^{-4} lb of surfactant per bulk cubic foot of rock. The procedure of this trial consisted of three steps: (i) injection of 0.1% surfactant slug for preadsorption prior to the CO_2-foam injection, (2) coinjection of liquid CO_2 and 0.05% Alipal CD-128 solution, and (3) injection of formation water to displace injected CO_2 foam.

In August 1984, the CO_2-foam injection began with the injection of tracer (ammonium thiocyanate, NH_4CNS). A total of 33 lb of the tracer in 11 gallons of water was injected for 48 hours, and then a short rinse period to avoid any interactions with Alipal CD-128 was followed. During this preadsorption process, there was an operational difficulty on the injection pump to maintain a steady rate so injected amount of surfactant was much less than the amount desired during first month. Also, for unidentified reasons, the injectivity of the surfactant solution continuously decreased. When injecting CO_2 foam on November 6, there were control problems that caused the foam quality not to be maintained as desired. In addition to this problem, the injectivity continued to decrease. Therefore, injection of CO_2 foam was terminated temporarily on December 22, 1984. However, during January and February, 1985, about 1813 bbl of CO_2 foam was injected successfully, and the injection of CO_2 foam was resumed in June during which the injectivity remained higher so that the amount of CO_2 foam was injected as planned by August 4. On August 14, the injection of chase water started.

Unfortunately, this field trial did not observe any oil bank clearly, probably due to unpredicted and unexplained technical difficulties, especially during the injection of surfactant slug. Hence, it was concluded that this unsatisfactory result might have been caused by (i) the amount of injected fluids was not enough, (ii) oil remaining in the reservoir prior to foam injection was not enough to form an oil bank, and/or (iii) CO_2 foam did not propagate deep into the reservoir because of several reasons such as severe surfactant adsorption into the formation rock, thermal/chemical degradation of chemicals in harsh reservoir conditions, and incompatible rock and fluid properties failing to lower the mobility.

2.2.5 CO_2-Foam Injection in Rangely Weber Sand Unit, Colorado (1988–1990)

The Rangely Weber Sand Unit (RWSU), operated by Chevron Inc., was located in Rio Blanco County in northwestern Colorado and discovered in 1933 (Graue and Zana, 1981). The OOIP in the field was 1.6 billion stb (stock tank barrels), and the oil production from the field peaked at 82,000 bbl/day in mid-1956 (Hervey and Iakoakis, 1991). Waterflood began in 1958, and it was very successful resulting in the estimated total oil recovery of about 789 million stb, or 50% OOIP in addition to about 332 million stb (21.0% OOIP) produced from primary recovery.

As a tertiary recovery, a miscible CO_2 flood on a 1:1 WAG was started in October 1986. Prior to CO_2 flood, the field overall experienced a mature waterflood with a cumulative recovery of 44.0% OOIP and an average WOR of 16. As a result of the miscible CO_2 flood, about 9.2 million stb of incremental oil was produced through June 1989.

However, a major concern of the CO_2 flood was high CO_2 production caused by a thief zone between injectors and producers. The performance of individual patterns varied widely due to reservoir heterogeneity and hydraulic fractures around most of the injectors. The fractures and high-permeability thief zones were accountable for the high CO_2 production rates. Therefore, CO_2-foam trial was considered as a method to block some of the thief zones and thereby improve sweep efficiency and CO_2 utilization (Jonas et al., 1990).

An area with a direct line drive pattern was selected for the CO_2-foam field test. The project began at the end of a water cycle in April 1989. Prior to the CO_2-foam treatment, CO_2 and brine were injected first simultaneously for a day to confirm baseline injectivity data and water was injected for 3 days. During the foam treatment, 12,000 bbl of surfactant slug was injected at 0.46% concentration on average and then followed by 55,000 bbl of 78% quality CO_2 foam. The foam was coinjected through a coarse mixing device in the tubing.

During initial foam injection, the bottomhole pressure was reached as high as 3933 psi, therefore the injection rate was reduced from 3303 bbl/day to 1925 bbl/day in order to maintain the surface pressure around 1700 psi and avoid fracturing the formations. Chevron Chaser CD-1040 was used for this treatment. After the foam injection, a CO_2 chase fluid was injected until the CO_2 injectivity at the injector or CO_2 production rate at the producer reached pre-foam level. A large water slug was followed after that. The ratio between the CO_2 chase and the water slug was about a cumulative average of 1:1 WAG ratio. In January 1990, a second coinjection of CO_2 and water with no surfactant was performed for 2 months to evaluate the production responses more critically.

As a result, CO_2 injectivity during both foam injection and CO_2 chase period was lower than during normal CO_2 injection. Also, CO_2 production decreased to less than 784 MMscf/day in McLaughlin 4, one of the wells in the test pattern which produced more than 2 MMscf/day gas within 10 days during CO_2 miscible flood. The oil production in the meantime was slightly increased.

2.2.6 CO_2-Foam Injection in North Ward-Estes, Texas (1990–1991)

The North Ward-Estes field in Ward and Winkler Counties in Texas, operated by Chevron Inc., is known as one of the most successful foam applications (Chou et al., 1992; Turta and Singhal, 1998; Winzinger et al., 1991). The productive formation of this area was Yates which was composed of fine-grained sandstones to siltstones with a Dykstra-Parsons coefficient of about 0.85 and separated by dense dolomite beds. Most productive parts of the field were developed with wells on 10-acre spacing, and the rest of the field was drilled on 20-acre spacing. Primary production began in 1929, and the field had been waterflooded since 1955.

At the start of the CO_2 flood in 1989, the oil-in-place for the project area was estimated to be around 77 million barrels, or 54% of OOIP. Since the beginning of CO_2 flood, however, poor sweep efficiency was a major concern because early CO_2 breakthrough and low oil production were observed. Among several methods considered to improve the CO_2 conformance, foam injection was chosen as the most promising way because foam was thought to selectively delay gas flow in the high-permeability layers and divert CO_2 from thief zones to other unswept zones. It was estimated that the poor sweep efficiency was especially caused by poor vertical conformance due to large permeability variations between sand layers. Prior to CO_2-foam field test, layer pattern was investigated based on injection profiles through cased-hole injectors. Finally, well 82WC (injector) and well 1020 (producer) were chosen for the foam field test.

Laboratory experiments selected Chaser CD-1040, an alpha-olefin sulfonate, as foaming agent due to its ability to generate higher foam resistance factor (RF) than others. Also, the evaluation of two foam injection methods (coinjection and SAG injection) showed that neither of the two methods manifested superiority in the field. Therefore, the SAG process was applied in the field, repetitions of one-day surfactant injection and one-day CO_2 injection, because it seemed more convenient and viable in field operations technically and economically. To avoid the delay of foam generation and propagation during SAG injection, surfactant preflush was performed into the reservoir and saturated the targeted regions.

From March 1990 to July 1991, a total of four large cycles of SAG was performed. Shortly after the first foam injection, the gas production rate decreased dramatically, and then was back up during the subsequent CO_2 chase injection. As soon as the next foam injection started, however, the gas production rate started to drop. Oil production at well 1020 increased from less than 1 bbl/day to 15 bbl/day during the first foam injection. It recorded a high of 80 bbl/day in May 1991, but decreased to around 15 bbl/day after June 1991.

This foam treatment reduced CO_2 injectivity by 40–85% for 1–6 months and the duration of injectivity reduction depended on several factors such as the amount of surfactant injected and the CO_2 injection pressure and rate after foam injection. Based on the injection and production responses, it was believed that foam successfully diverted CO_2 from the thief zone to unswept regions. The economic analysis of WAG, waterflood, and SAG showed that SAG was the most economical with $118,300 gains and waterflood was also viable with $6600 gains, while WAG was not promising with $415,300 loss, as shown in Chou et al. (1992) with details. These economic calculations, however, may not be directly applicable today due to dramatic changes in oil prices, operating costs, water processing costs, and CO_2 price.

2.2.7 CO_2-Foam Injection in the East Vacuum Grayburg/San Andres Unit, New Mexico (1991–1993)

The East Vacuum Grayburg/San Andres Unit (EVGSAU), operated by Phillips Petroleum Company, is one of the several large units within the Vacuum field area and located about 15 miles northwest of Hobbs in Lea County, New Mexico (Harpole and Hallenbeck, 1996; Turta and Singhal, 1998). Although the field was discovered in 1929, notable development did not start until 1938 due to the lack of transportation facilities and a low demand for crude oil (Martin et al., 1994). Initial development of the field was completed by 1941, and waterfloods began in 1958 using 80-acre inverted nine-spot patterns and was steadily extended across the field.

The full-scale miscible CO_2 injection, in three major operational WAG blocks, began in September 1985 with 45 wells in 80-acre WAG injection patterns. The designated area of this CO_2 project covered about 5000 acres (about 70% of the total EVGSAU area) and contained about 260 MMstb OOIP (about 87% of OOIP for the total unit). Each WAG block had 4 months of CO_2 injection and 8 months of water injection and WAG ratio was 2:1. As a result, around 30 MMstb or 11.5% of the 260 MMstb OOIP of incremental oil was produced from the CO_2 project area. During the CO_2 project, however, several local problems had arisen, thereby project performance was declined—first, the reservoir pressure in one of the three areas dropped below the MMP; second, severe breakthroughs were observed in several patterns; and third, one of the areas showed a drastic permeability contrast between upper and lower zones, leading to poor overall sweep efficiency.

In September 1989, field verification CO_2-foam project was jointly funded by the EVGSAU Working Interest Owners (WIO), the US Department of Energy (DOE), and the State of New Mexico (Martin et al., 1995). The EVGSAU WIO initiated a 4-year cooperative foam field trial project with the New Mexico PRRC in 1990 as a potential solution to the problem of CO_2 channeling in the field. The objective of the project was to achieve in-depth diversion of injected CO_2 through *in situ* generation of low-mobility CO_2 foam within the high-permeability channels. Based on laboratory foam evaluation data, Chevron Chaser CD-1045 was used for the pilot test and 2500 ppm of CD-1045 was injected for both adsorption pad and surfactant solution for SAG cycle. In order to collect a base case response, a rapid WAG process was performed from September to December 1991 after which water was injected for 3 months. In the second and third months of the water injection, 77,000 reservoir barrels of 2500 ppm surfactant solution was injected to satisfy the adsorption requirement. Based on operational constraints, the five cycles of SAG were designed and each cycle consisted of 3 days of surfactant solution injection at 1000 bbl/day followed by 12 days of CO_2 injection (in a total of 12,000 reservoir barrels approximately), which provided about 80% foam quality. In mid-July 1992, the rapid SAG began following injection of the 3-month surfactant adsorption slug. Since injection pressure reached the maximum allowable pressure for CO_2 (1800 psi) at the beginning of the third SAG cycle, the injection rate was reduced to keep pressure within the allowable limit.

The outcome of the foam pilot test exhibited that (i) the apparent *in situ* mobility of CO_2 by foam was estimated to be about only one-third of that during WAG process and (ii) incremental oil production was observed in three of the eight producers in the target pattern. Since the first foam field pilot showed favorable responses, the second round of foam pilot test was initiated in June 1993 by using the same conditions. Because of operation problems, however, the second test was aborted after two cycles of foam generation. Economic evaluation of these two foam injection projects showed that the total incremental oil was about 19,160 bbl, allowing the net yield of the project to be around $48,258 after accounting for other costs including surface facilities, compression costs, surfactant chemicals, loyalty, and so on (Martin et al., 1995).

2.2.8 CO_2-Foam Injection in East Mallet Unit, Texas, and McElmo Creek Unit, Utah (1991–1994)

Four field trials of CO_2-foam injection were performed by Mobil to reduce CO_2 channeling at the East Mallet Unit (EMU) in Hockley County, Texas, and the McElmo Creek Unit (MCU) in San Juan County, Utah, in early 1990s (Hoefner et al., 1995). The main objectives were to verify the economic potential of the foam processes and to develop CO_2-foam technique to be used by operation companies for commercial purposes. As a first step, extensive laboratory studies were conducted to quantify foam performance at reservoir conditions including interactions between injected chemicals and reservoir rock and fluids. In the following second step, two small-scale field pilot tests were tried to check CO_2-foam injectivity. Both CD-128 and CD-1045 surfactants were identified from the screening tests and applied in the field.

The EMU, a part of the San Andres carbonate formation of the Slaughter field in the Permian Basin, was first developed in the 1940s. The unit covered about 2480 acreage of drainage area, and had a measured permeability ranging from 0.01 to 28 mD. It was once treated by miscible CO_2 flooding from 1989, with 41 injection wells and 82 producers, giving the average well spacing around 20 acres with chicken-wire patterns. This treatment resulted in about 2000 bbl/day oil production. From preliminary laboratory flooding experiments with the EMU reservoir core, it was observed that the RF during foam injection, in both SAG and coinjection methods, was low ranging from 3 to 10, but did not show any negative interactions between the surfactants and field samples.

In the EMU, two CO_2-foam field trials were conducted in attempts to delay or reduce CO_2 production, improve oil production, and compare two foam injection methods (i.e., coinjection vs. SAG). When a total of 20,200 lb of active surfactant was injected as a part of SAG processes in the first CO_2-foam field trial in 1991, the CO_2 production rate in the production well was reduced by about 50%, from 500 to 250 MMscf/day. This reduced CO_2 production rate had been continued since the last SAG cycle, implying that the mobility of injected gas was reduced significant. When it comes to oil production rate, the overall oil production rate from the nearby producing wells increased to 22% (16 bbl/day)–31% (22 bbl/day) in terms of incremental oil production rate. In the second CO_2-foam trial in the EMU, a total of 26,900 lb of active surfactant was injected as a part of both coinjection and SAG. More specifically, a total of nine cycles of coinjection were performed, by varying injection foam quality systematically from 80% to 60%, 30%, and finally 20% because, based on laboratory data, reducing foam quality tended to increase injectivity. Coinjection was shown to be more difficult than SAG due to excessively low injectivity, but the injection responses were similar in both cases. The outcome of the second CO_2-foam trial, however, was not obvious compared to the first trial, without showing an identifiable increase in oil production.

The MCU, a part of the Greater Aneth field in the Paradox basin, was discovered in 1957 and had a drainage area of 13,440 acre with three different layers: the Lower Ismay (LI), the Desert Creek I (DCI), and the Desert Creek II (DCII). Among them, LI and DCII were considered as thief zones and, the DCII zone consisted of five sublayers and its permeability ranged from 0.01 to more than 1000 mD. The target zone of CO_2 injection was the DCI zone whose permeability was less than 5 mD. The well pattern was an inverted nine-spot pattern with the average well spacing of about 160 acre. The field was treated by miscible CO_2 flooding in 1985 during which oil production rate was raised to 6000 bbl/day from 120 active production wells and 105 injection wells. CO_2-foam application in MCU was planned following the success in the EMU trials. A foam RF of up to 7 was observed from laboratory flow experiments with the MCU cores at the field conditions. A year before the foam treatment, the injection well was converted from WAG to continuous CO_2 injection. Then a total of 80,500 lb active surfactant was injected using SAG from April 1992 to November 1993. During the CO_2 injection prior to foam injection, oil production rate increased somewhat, but the breakthrough of injected CO_2 was significant. With CO_2-foam injection, the injected gas production rate was reduced by 50% relative to the injected CO_2 rate, and oil production was maintained at the higher level.

The second trial in MCU was performed to reduce excessive gas production in the well R-20 which had a thin (5 ft) thief zone with very high permeability (around 500 mD) in the DCI zone. Although both SAG and coinjection were tried for this second trial (SAG process first in the same manner as previously, followed by continuous coinjection at high foam quality), the operation was terminated because of operational problems associated with coinjection after injecting 33,700 lb active surfactant. The coinjection was successful until that moment, and reduced the amount of surfactant injected and avoided the loss of injectivity, by applying a very low foam injection rate with high foam quality. Further investigation showed that the operational problem was involved in plugging of tubing at the wellhead possibly resulting from freezing or hydrate formation.

As a conclusion, the results from four CO_2-foam field trials in EMU and MCU were positive. This allowed SAG foam EOR processes to be continued in multiple wells for at least 2 more years.

2.3 TYPICAL FIELD RESPONSES DURING CO_2-FOAM APPLICATIONS

This section describes under what circumstances the diversion from high-permeability layers to low-permeability layers takes place and what responses should be expected in terms of injection rates, injection pressure, and effluent history during CO_2-foam field applications.

2.3.1 Diversion from High- to Low-Permeability Layers

The successful use of foams to overcome permeability contrasts and improve sweep efficiency counts on two important features: (1) the onset of foam generation (i.e., transition from weak foams to strong foams) in layers with differing permeabilities and (2) the magnitude of foam strength, typically expressed by mobility reduction factor (MRF) or resistance factor (RF), as a function of absolute permeabilities.

With respect to the first feature, previous studies show that the critical pressure gradient ($\nabla p^{\min}$) required to mobilize foam films and trigger foam generation scales roughly like the inverse of permeability (i.e., $\nabla p^{\min} \sim k^{-1}$) in the case of unconsolidated sand and bead packs as shown in Figure 2.9 (Gauglitz et al., 2002). This is because the pressure gradient to displace foam films is proportional to the inverse of pore throat size over the distance of average pore length. Although this relationship is more complicated in consolidated porous media due to compaction and other foreign materials, it still works reasonably well (for more relevant discussion, see Gauglitz et al., 2002; Ransohoff and Radke, 1988; Rossen and Gauglitz, 1990; Tanzil et al., 2002). This implies that less pressure gradient is needed to cause foam generation in higher permeability layers, which in turn implies that attaining

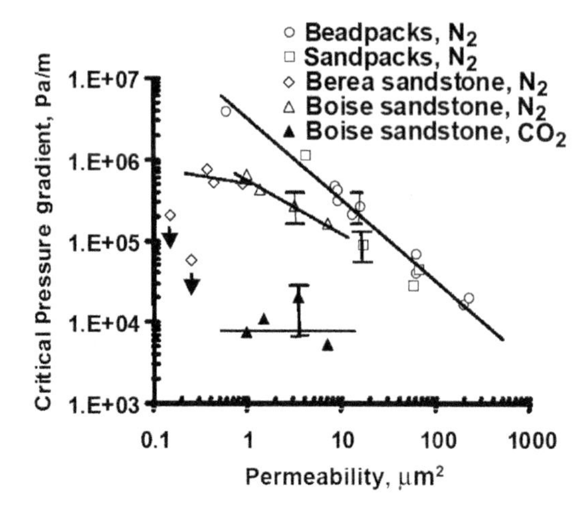

FIGURE 2.9 Minimum pressure gradient for foam generation as a function of permeability for N_2 and CO_2 foams (Gauglitz et al., 2002).

fine-textured foams is easier in higher permeability layers, diverting subsequent flow into lower permeability layers. This topic deserves further investigations, especially for CO_2 foams in various field conditions.

The second feature comes into play when strong foams are placed in both high- and low-permeability layers. Because the higher capillary pressure environment in the lower permeability layer tends to destabilize foam films, foam strength is typically higher in the high-permeability layer. This also helps diverting flowing fluids into the low-permeability layer (Chou et al., 1992; Martin et al., 1995; Lee et al., 1991). Examples can be found from CO_2 laboratory coreflood experiments similar to the one shown in Figure 2.10, where foam RF increases with absolute permeability (See Section 2.1.10 for more details.).

Another important aspect is foam generation at the layer boundaries as pointed out by previous studies (Rossen, 1999; Tanzil et al., 2002). If the permeability of two adjacent layers increases abruptly more than twice at the layer boundary along the flow direction, foams can be created by the fluctuation in capillary pressure. This is a mechanism driven by snap-off, which should be distinguished from the lamella creation mechanisms *within* homogeneous porous media. The mechanism of foam generation at the layer boundary is especially important in foam field applications due to the heterogeneous nature of petroleum reservoirs.

2.3.2 Typical Responses from Successful SAG Processes

The use of SAG processes often involve multiple cycles of CO_2 and surfactant injections. By repeating drainage and imbibition (or, CO_2 and

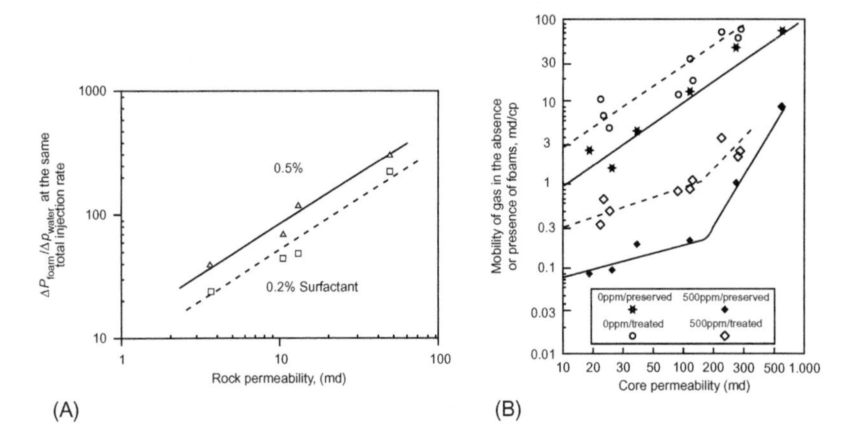

FIGURE 2.10 CO_2-foam mobility as a function of absolute permeability: if the mobility were unaffected by absolute permeability, the slope would be zero (or horizontal) in (A) or a unity (or 45° upward) in a log-log plot of (B). (A) 0.2 and 0.5 % of CD-1040 surfactant solution (B) 0 and 5000 ppm of CD-1045 surfactant solution in preserved and treated cores (101 °F and 2100 psig). Source: *(A) Redrawn from Chou et al. (1992) and (B) redrawn from Martin et al. (1995).*

surfactant injection equivalently) cycles, the capillary pressure near the well moves up and down, respectively. This creates a favorable condition for *in situ* foam generation. If this process works as designed, the injection of CO_2 into the wellbore, which is initially treated by surfactant preflush, first creates stable foams near the wellbore region. Then subsequent CO_2 injection pushes the tailing edge of foam bank and dries out the media. Because the high-mobility CO_2 zone displaces the low-mobility foam bank ahead of it, injection pressure rapidly declines as the foam bank moves away from the wellbore. Note that this reduction in injection pressure is due to the nature of flow geometry (radial, spherical, or a combination of both), and therefore this enhancement in injectivity in field SAG processes does not take place in typical laboratory coreflood experiments in linear geometry. Each surfactant and CO_2 injection cycle can be repeated multiple times.

An example of CO_2-foam SAG process can be found in Hoefner and Evans (1995), which shows 13 cycles of surfactant and CO_2 injection during the period of January through June 1991 in EMU well 31, Hockley County, Texas, as shown in Figure 2.11. The data show that the wellhead pressure sharply increases at the time of CO_2 injection, and then decreases gradually as CO_2 displaces foam bank deeper into the reservoir. The injection of surfactant solution following CO_2 exhibits a sharp reduction in the wellhead pressure due to reduced injection flow rate. Figure 2.12 shows the production history during CO_2-foam SAG in connection with Figure 2.11. The decline curve analysis shows the trends of waterflooding, CO_2-water WAG, and

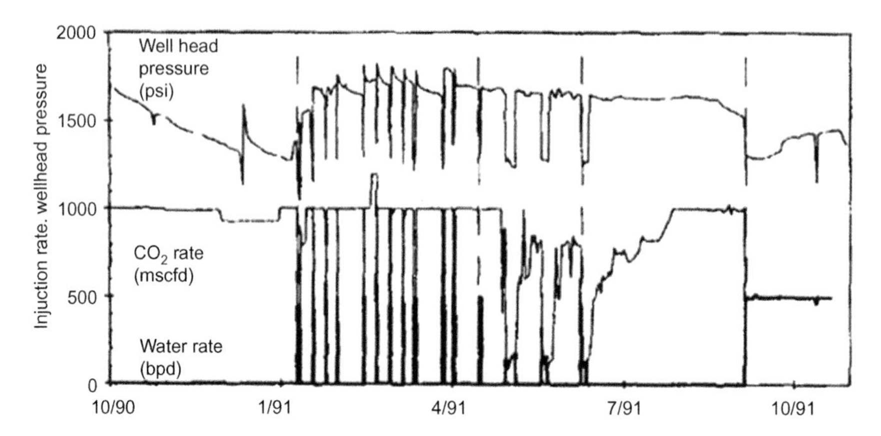

FIGURE 2.11 Response of CO_2-foam SAG process from the EMU unit, Hockley County, Texas (Hoefner and Evans, 1995).

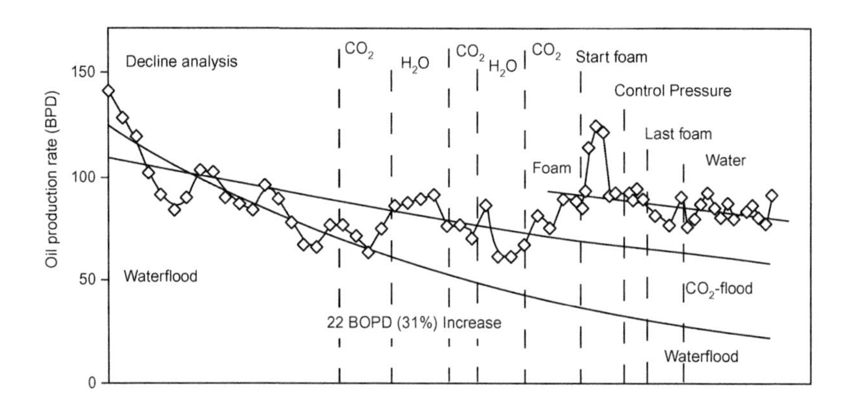

FIGURE 2.12 Production history during CO_2-foam SAG process in comparison with waterflooding and CO_2-water WAG process from the EMU unit, Texas (Hoefner and Evans, 1995).

CO_2-foam SAG processes. As expected, CO_2-foam SAG process is more advantageous than CO_2-water WAG, by improving sweep efficiency.

Another example of CO_2-foam SAG can be found in the East Vacuum Grayburg and San Andres Unit, New Mexico, which used seven rapid cycles during the period of July 1992 through December 1993, as shown in Figure 2.13. The response of injection pressure to water and CO_2 injections is similar to the previous example—a rapid increase (due to foam generation) followed by a slow decline (due to foam bank pushed by continuous CO_2 injection) in injection pressure during CO_2 injections and an immediate reduction in injection pressure during subsequent surfactant injections. The

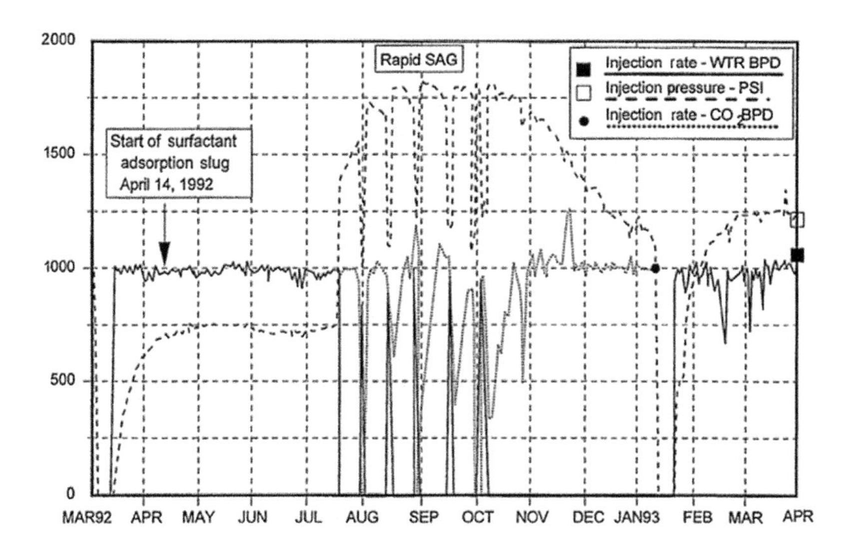

FIGURE 2.13 Response of CO_2-foam SAG process from the EVGSAU, New Mexico. *Redrawn from Martin et al. (1995).*

corresponding gas and oil recovery during CO_2-foam SAG is shown in Figure 2.14, together with the recovery during CO_2-water WAG process. Once again, the results show that the total oil recovery is higher with CO_2-foam SAG than CO_2-water WAG. This improvement is even more pronounced if the oil production is compared at the same amount of CO_2 injected. This again shows the foam's capability to overcome heterogeneity. The use of foams in SAG process, compared to WAG, tends to retard the breakthrough of injected gas, if the process is successful. The GOR during SAG process is shown to be more than 10 times smaller than the GOR during WAG process at the adjacent producing wells. Figure 2.15 shows decline curve analysis which compares implementation of CO_2-foam SAG (solid line), as a revised strategy, with CO_2-water WAG process (dashed line), which demonstrates the advantage of CO_2-foam SAG over CO_2-water WAG process.

These examples reiterate the symptoms that one can observe from successful SAG processes: first, the immediate rise of injection pressure at the beginning of CO_2 injection following surfactant injection can be regarded as pretty good evidence of CO_2 and surfactant solutions indeed flowing into the same layers and creating low-mobility foams *in situ*. The pressure buildup is partly due to the high trapped gas saturation and partly due to the additional resistance to flow in presence of foams (e.g., yield stress, modified viscosity, and relative permeability); second, the exponential decay in injection pressure during continuous CO_2 injection implies that fine-textured foams are being displaced by the injected CO_2, drying out near the wellbore region.

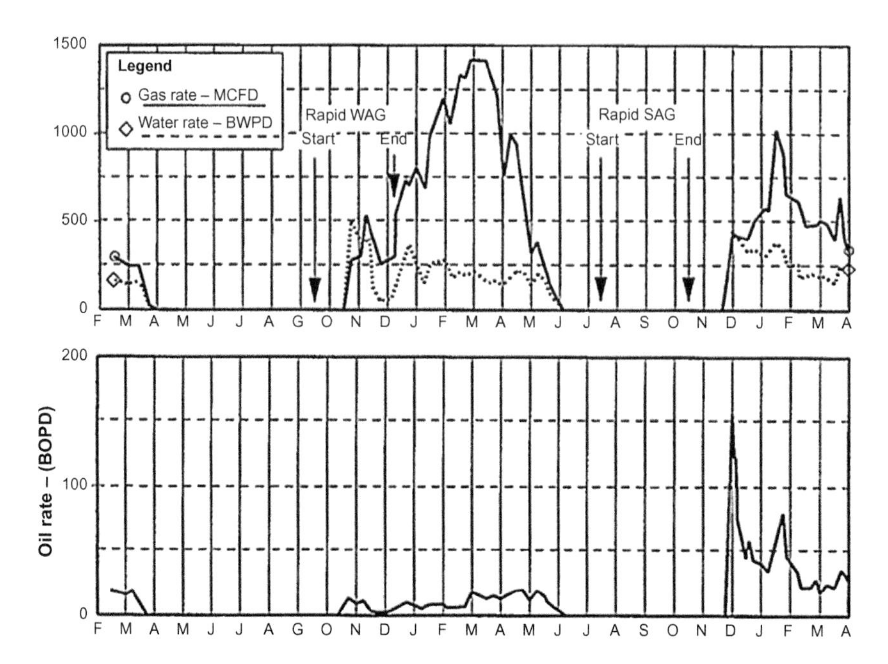

FIGURE 2.14 Production rates of gas and water (top) and oil (bottom) during CO_2-water WAG process versus CO_2-foam SAG process from the EVGSAU, New Mexico (Martin et al., 1995).

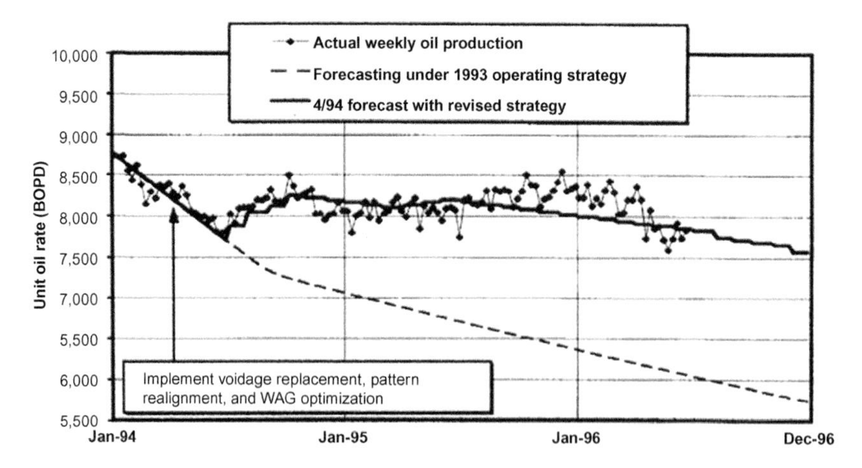

FIGURE 2.15 Decline curve analysis comparing CO_2-water WAG process (dashed line) and CO_2-foam SAG process (solid line) from the EVGSAU, New Mexico (Harpole and Hallenbeck, 1996).

This implies that the region with $S_w > S_w^*$ (away from the wellbore) tends to have stable foams whereas the region with $S_w < S_w^*$ (near the wellbore) tends to have no foams, which is the origin of improved injectivity during SAG process; and lastly, controlling the wellbore pressure not to go over the formation fracturing pressure is not a serious concern in field SAG operations, because the peak in injection pressure is typically achieved during CO_2 injection following surfactant. If monitoring the injection pressure shows a warning sign, then the remedial action can be taken easily simply by reducing CO_2 injection rate or pressure.

2.3.3 Typical Responses from Successful Surfactant–Gas Coinjection Processes

Foams can also be introduced into a reservoir by injecting gas and surfactant solutions simultaneously. This coinjection process is especially useful when the propagation of foams in SAG process is of major concern (e.g., gas and surfactant may flow into different layers in SAG processes, failing to create foams *in situ*). Surfactant preflush is typically performed before pre-generated foams enter the reservoir in order to satisfy surfactant adsorption. Constant and close monitoring of injection well pressure is crucial in order for bottomhole pressure not to go beyond the formation fracturing pressure. Because foams are injected as a single mixture during coinjection, controlling the total injection rate during the entire process is a must. Thorough laboratory experimental studies are required prior to field trials in order to understand foam mobility in a wide range of injection rates. The coinjection is advantageous over the SAG in that there is more room to play with foam properties, if they are engineered from laboratory experiments in advance (especially, optimum foam quality, flow regimes, and injection velocity).

Figure 2.16 shows an example CO_2-foam coinjection process in the RWSU in Colorado (Jonas et al., 1990), showing the injection wellhead pressure, CO_2 flow rate, and water flow rate during the period of 700 hours or about 29 days. After about 120 hours of surfactant preflush (during which the wellhead pressure is maintained slightly over 1000 psi), CO_2 and surfactant solutions are injected together, and the wellhead pressure increases to 1600 psi. As foams propagate deeper into the reservoir, the injection rates were reduced step by step to keep the same wellhead pressure around 1600 psi. This means that the injectivity during CO_2-foam coinjection should be controlled to decrease with time inevitably.

Similar responses can be found with CO_2-foam coinjection field trial at the North Ward-Estes (Chou et al., 1992), as shown in Figure 2.17. This field trial used a combination of CO_2 injection (without surfactant) and CO_2-foam coinjection. During four cycles of CO_2-foam coinjection during the period of July 1990 through January 1992, the bottomhole pressure and the surface pressure fluctuated significantly which required the total injection rates

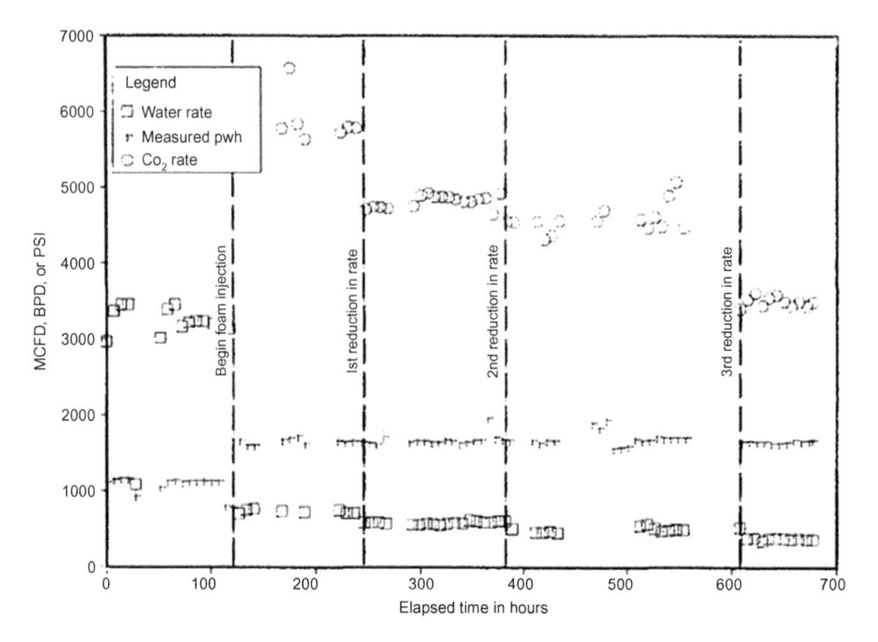

FIGURE 2.16 Injection wellhead pressure, CO_2 flow rate, and water flow rate during CO_2-foam coinjection process following 120 hours of surfactant preflush from the RWSU, Colorado (Jonas et al., 1990).

monitored and adjusted continuously. (Note that each of the foam injection cycle consists of multiple total injection rates in order to control bottomhole pressure.)

2.4 CONCLUSIONS

A number of CO_2-foam field applications demonstrated that foaming injected CO_2 with surfactant solutions could delay the breakthrough of injected fluids and hence enhance oil production dramatically. This improved performance is endowed by the reduction in gas mobility, making the gas phase dispersed in the surfactant-laden liquid phase. Successful foam field applications, however, require careful investigations of foam properties at reservoir conditions and foam propagation rates in the reservoir. These are related to numerous issues including, but not limited to, thermal degradation of injected chemicals, foam–oil interactions, wettability effects on foam stability at the pore wall, severity of reservoir heterogeneity (cf. permeability contrasts, rock type, and mineralogy), foamability of surfactant solutions at reservoir pressure and temperature, surfactant adsorption at rock surface, *in situ* lamella creation and coalescence mechanisms, foam strength (cf. MRF and RF) as a function of injection velocities and injection qualities, foam injection strategies, and so on.

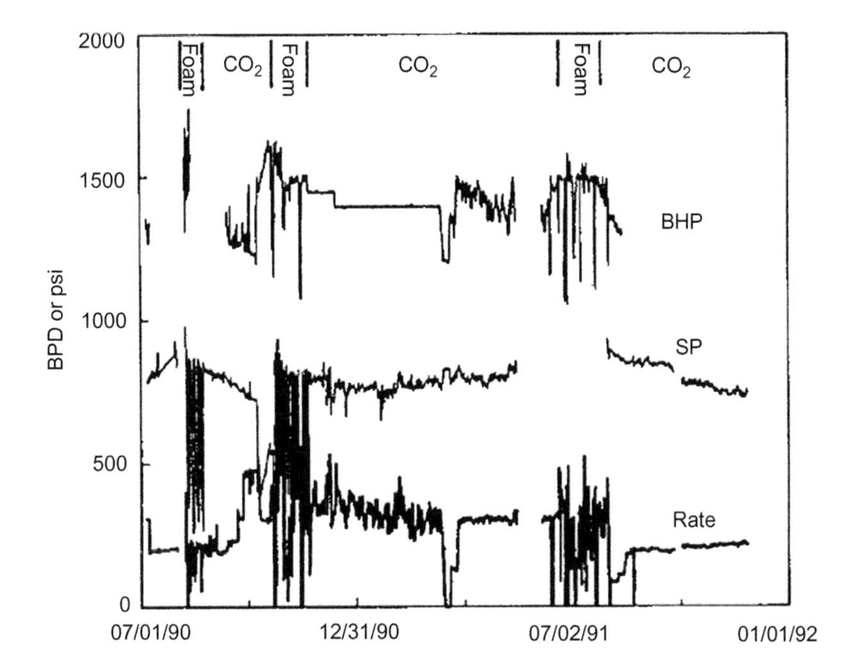

FIGURE 2.17 Surface and bottomhole pressures and total injection rate during a combination of CO_2 and CO_2-foam injections from the North Ward-Estes field, Texas (Chou et al., 1992).

Recent advances in understanding foam physics, especially, two steady-state strong-foam regimes (high-quality regime and low-quality regime) and three foam states (weak-foam, strong-foam, and intermediate states) represented by foam catastrophe theory, are believed to make a significant contribution to the optimum design of foam EOR processes by allowing more reliable foam models. Dimensionality-dependent foam rheological models through the use of foam fractional flow *surface* are expected to greatly improve the accuracy of field-scale reservoir simulation to match injection and production histories and forecast future reservoir performance.

ACKNOWLEDGMENT

We would like to express our appreciation to Professor William R. Rossen, currently with Delft University of Technology, for sharing his thoughts. They are reflected in Figures 2.1, 2.2, and 2.8.

APPENDIX—EXPRESSION OF GAS-MOBILITY REDUCTION IN THE PRESENCE OF FOAMS

There exist several different approaches of reporting the degree of gas-mobility reduction during foam flow in the literature. Some major examples include mobility reduction factor (MRF), gas mobility, relative gas mobility, effective gas viscosity, effective gas relative permeability, and foam resistance factor (RF) (Chou et al., 1992; Kam, 2008; Kam and Rossen, 2003; Kuehne et al., 1992; Mayberry et al., 2008; Namdar-Zanganeh et al., 2011). This section is dedicated to clarifying different terminology used for the same purpose, and presenting the way how to translate from one to another.

A typical gas–liquid two-phase flow can be described by Darcy's equation, for example,

$$\frac{q_q}{A} = \frac{kk_{rg}^{o}(S_w)}{\mu_g^{o}}\frac{\Delta P}{L} = \lambda_g^{o}(S_w)\frac{\Delta P}{L} = k\lambda_{rg}^{o}(S_w)\frac{\Delta P}{L} \tag{A.1}$$

and

$$\frac{q_w}{A} = \frac{kk_{rw}(S_w)}{\mu_w}\frac{\Delta P}{L} = \lambda_w(S_w)\frac{\Delta P}{L} = k\lambda_{rw}(S_w)\frac{\Delta P}{L} \tag{A.2}$$

where, q, k_r, S, μ, λ, and λ_r represent flow rate, relative permeability, saturation, viscosity, mobility, and relative mobility, respectively; ΔP represents the pressure drop across the porous medium with cross-sectional area of A and length of L; subscripts "w" and "g" represent water and gas phases; and superscript "o" represents the state of porous medium with no foams.

Previous studies show that the presence of foam films alters gas-phase mobility significantly, but not liquid-phase flow mobility (Bernard et al., 1965; de Vries and Wit, 1990; Friedmann et al., 1991; Sanchez and Schechter, 1989). This indicates that Darcy's equation for liquid phase is still valid, but the equation for gas phase should be modified in the presence of foams. Taking it into consideration that the mobility consists of relative permeability and viscosity, this modification can be worked out by the following equation:

$$\frac{q_q}{A} = \frac{kk_{rg}^{f}(S_w)}{\mu_g^{f}}\frac{\Delta P}{L} = \lambda_g^{f}(S_w)\frac{\Delta P}{L} = k\lambda_{rg}^{f}(S_w)\frac{\Delta P}{L} \tag{A.3}$$

where, superscript "f" represents the state of porous medium with foams.

Note that the presence of foam films affects relative gas mobility ($\lambda_{rg}^{o} \rightarrow \lambda_{rg}^{f}$) by impacting both relative gas permeability ($k_{rg}^{o} \rightarrow k_{rg}^{f}$) and gas viscosity ($\mu_{rg}^{o} \rightarrow \mu_{rg}^{f}$); those two features are often inseparable, however.

Although physically incorrect, it is sometimes handy for modeling and simulation purposes to include all foam effects into either relative gas permeability term or gas viscosity term, assuming that the other term is not influenced by the presence of foams. This concept can be implemented by using effective gas relative permeability ($k_{rg}^{f'}$) or effective gas viscosity (μ_g^f), i.e.,

$$\frac{q_q}{A} = \frac{kk_{rg}^{f}(S_w)}{\mu_g^f}\frac{\Delta P}{L} = \frac{kk_{rg}^{f'}(S_w)}{\mu_g^o}\frac{\Delta P}{L} = \frac{kk_{rg}^{o}(S_w)}{\mu_g^f}\frac{\Delta P}{L} \tag{A.4}$$

which can further be simplified, if MRF is introduced to take all foam effects into consideration without any modifications to relative gas permeability term or gas viscosity term. Then the equation becomes

$$\frac{q_q}{A} = \frac{kk_{rg}^{f}(S_w)}{\mu_g^f}\frac{\Delta P}{L} = \frac{kk_{rg}^{o}(S_w)}{\mu_g^o \text{ MRF}}\frac{\Delta P}{L} \tag{A.5}$$

Note that is equivalent to effective gas viscosity in presence of foams () in earlier foam studies, for example, Hirasaki and Lawson (1985). Typically the use of MRF is more convenient to incorporate the change in gas mobility into foam modeling and simulations compared with effective gas relative permeability ($k_{rg}^{f'}$) or effective gas viscosity (μ_g^f). It is because MRF is a dimensionless parameter, so the comparison of foam mobility change in different conditions (e.g., different gas phases, surfactant formulations and concentrations, rock permeabilities and wettability, pressures and temperatures, and so on) can be dealt with without impacting other rock and fluid properties. Furthermore, the use of MRF provides perhaps the easiest way to extend any preexisting gas–liquid two-phase flow models and simulators with foam mechanisms. For example, a subroutine to determine the value of MRF as a function of other parameters can be incorporated without making any major changes to the existing multiphase flow simulators. MRF = 1 if foam is not present (or, if pregenerated foams are not sustainable, and hence do not propagate) in the media, and MRF > 1 if foams exist and propagate in the media. Typically, MRF can range from 1 up to as high as more than 100,000.

On the other hand, foam RF, or simply the RF, is typically defined as the ratio of pressure drop with foams (or, pressure drop with gas and surfactant coinjected, equivalently; ΔP^f) to pressure drop without foams (or, pressure drop with gas and water coinjection in the absence of surfactant, equivalently; ΔP^o) at the same gas and liquid flow rates as shown in, for example, Kuehne et al. (1992). In other words,

$$\text{RF} = \frac{\Delta P^f}{\Delta P^o} \tag{A.6}$$

where RF = 1, if no foams are present in the media; and RF > 1, if foams are present and thus increase the resistance to the flow. The use of RF is popular in numerous experimental studies because of its simplicity—calculating RF values do not require any rock or fluid properties other than two measured pressure drops. Put it differently, MRF is a comparison between foam and no-foam cases at the same water saturation (cf. Eq. (A.5)), while RF is a comparison at the same injection condition (cf. Eq. (A.6)). As a result, the RF value cannot be directly used for modeling and simulation purposes without additional adjustments.

Note that Eq. (A.6) can also be written as follows in terms of Darcy's equation for gas by using MRF, i.e.,

$$\mathrm{RF} = \frac{\Delta P^{\mathrm{f}}}{\Delta P^{\mathrm{o}}} = \frac{q_{\mathrm{g}}\mu_{\mathrm{g}}^{\mathrm{o}} L\,\mathrm{MRF}/kk_{\mathrm{rg}}^{\mathrm{o}}(S_{\mathrm{w}}^{\mathrm{f}})}{q_{\mathrm{g}}\mu_{\mathrm{g}}^{\mathrm{o}} L/kk_{\mathrm{rg}}^{\mathrm{o}}(S_{\mathrm{w}}^{\mathrm{o}})} = \mathrm{MRF}\frac{k_{\mathrm{rg}}^{\mathrm{o}}(S_{\mathrm{w}}^{\mathrm{o}})}{k_{\mathrm{rg}}^{\mathrm{o}}(S_{\mathrm{w}}^{\mathrm{f}})} \tag{A.7}$$

or, in terms of Darcy's equation for liquid, i.e.,

$$\mathrm{RF} = \frac{\Delta P^{\mathrm{f}}}{\Delta P^{\mathrm{o}}} = \frac{q_{\mathrm{w}}\mu_{\mathrm{w}} L/kk_{\mathrm{rw}}^{\mathrm{o}}(S_{\mathrm{w}}^{\mathrm{f}})}{q_{\mathrm{w}}\mu_{\mathrm{w}} L/kk_{\mathrm{rw}}^{\mathrm{o}}(S_{\mathrm{w}}^{\mathrm{o}})} = \frac{k_{\mathrm{rw}}(S_{\mathrm{w}}^{\mathrm{o}})}{k_{\mathrm{rw}}(S_{\mathrm{w}}^{\mathrm{f}})} \tag{A.8}$$

for $S_{\mathrm{w}} > S_{\mathrm{w}}^{*}$ at the same gas and liquid injection rates. In general, $S_{\mathrm{w}}^{\mathrm{f}}$ is much smaller than $S_{\mathrm{w}}^{\mathrm{o}}$ in the presence of foams.

It is also worth noting that the reduction in gas mobility is only possible when $P_{\mathrm{c}} < P_{\mathrm{c}}^{*}$ (or $S_{\mathrm{w}} > S_{\mathrm{w}}^{*}$ equivalently); otherwise foams are considered to collapse down rapidly such that there is no reduction in gas mobility in local steady-state modeling (note that $\mathrm{RF} = 1$ and $\mathrm{MRF} = 1$, if $P_{\mathrm{c}} > P_{\mathrm{c}}^{*}$ or $S_{\mathrm{w}} < S_{\mathrm{w}}^{*}$).

The following example is designed to help understand how these different approaches can be related. More detailed calculations are left as an exercise for the readers.

Example: Calculating gas-phase mobility reduction during foam flow

Input Parameter for the Example

Water viscosity, $\mu_{\mathrm{w}} = 1 \times 10^{-3}$ Pa s	Water relative permeability
Gas viscosity, $\mu_{\mathrm{r}}^{\mathrm{o}} = 2 \times 10^{-5}$ Pa s	$k_{\mathrm{rw}} = 0.79\left(\frac{S_{\mathrm{w}} - S_{\mathrm{wr}}}{1 - S_{\mathrm{wr}} - S_{\mathrm{gr}}}\right)^{1.96}$
Limiting water saturation, $S_{\mathrm{w}}^{*} = 0.25$	Gas relative permeability (no foams)
Residual water saturation, $S_{\mathrm{wr}} = 0.15$	$k_{\mathrm{rg}}^{\mathrm{o}} = \left(\frac{S_{\mathrm{g}} - S_{\mathrm{gr}}}{1 - S_{\mathrm{wr}} - S_{\mathrm{gr}}}\right)^{2.29}$
Residual gas saturation, $S_{\mathrm{gr}} = 0.05$	Effective gas relative permeability and effective gas viscosity (with foams)
Absolute permeability, $k = 1 \times 10^{-12}$ m^2	$k_{\mathrm{rg}}^{\mathrm{f}'} = \frac{k_{\mathrm{rg}}^{\mathrm{o}}}{\mathrm{MRF}}$ $\quad \mu_{\mathrm{g}}^{\mathrm{f}'} = \mu_{\mathrm{g}}^{\mathrm{o}}\,\mathrm{MRF}$

Water factional flow (f_{w}) as a function of water saturation (S_{w})

$$f_{w}^{f} = \frac{k_{\mathrm{rw}}/\mu_{\mathrm{w}}}{k_{\mathrm{rw}}/\mu_{\mathrm{w}} + k_{\mathrm{rg}}^{\mathrm{f}}/\mu_{\mathrm{g}}^{\mathrm{f}}} = \frac{k_{\mathrm{rw}}/\mu_{w}}{k_{\mathrm{rw}}/\mu_{\mathrm{w}} + k_{\mathrm{rg}}^{\mathrm{f}'}/\mu_{\mathrm{g}}^{\mathrm{o}}} = \frac{k_{\mathrm{rw}}/\mu_{\mathrm{w}}}{k_{\mathrm{rw}}/\mu_{\mathrm{w}} + k_{\mathrm{rg}}^{\mathrm{o}}/\mu_{\mathrm{g}}^{\mathrm{f}'}}$$

$$= \frac{k_{\mathrm{rw}}/\mu_{\mathrm{w}}}{k_{\mathrm{rw}}/\mu_{\mathrm{w}} + k_{\mathrm{rg}}^{\mathrm{o}}/(\mu_{\mathrm{g}}^{\mathrm{o}}\mathrm{MRF})} = \frac{\lambda_{rw}}{(\lambda_{\mathrm{rw}} + \lambda_{\mathrm{rg}}^{\mathrm{f}})} \text{ with foams}$$

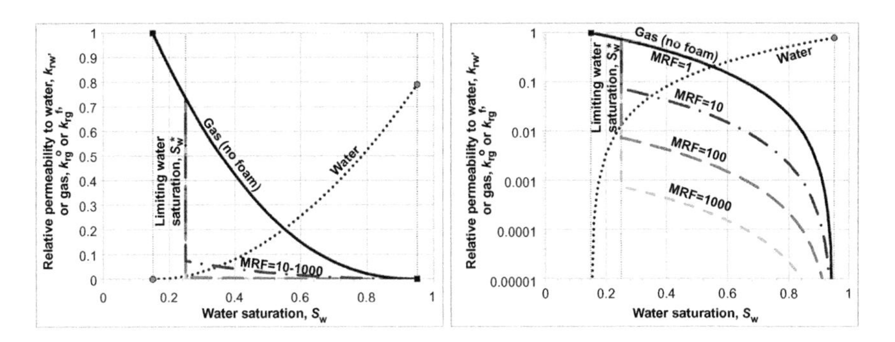

FIGURE EXAMPLE 2.1 Relative permeability curves of water and gas with and without foams at different MRF values (normal and log scales on the y-axis).

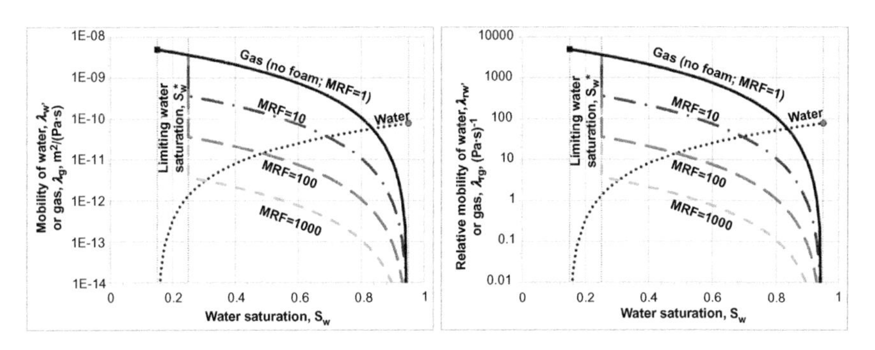

FIGURE EXAMPLE 2.2 Mobility and relative mobility curves of water and gas with and without foams at different MRF values.

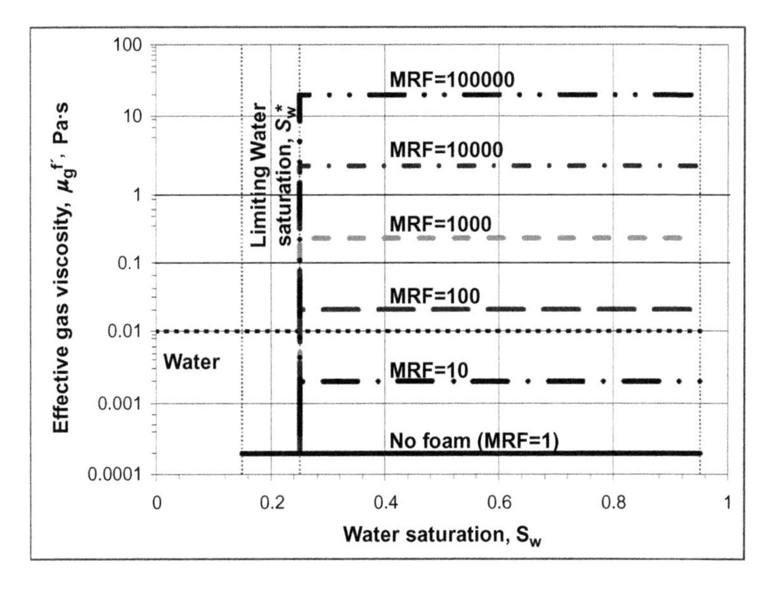

FIGURE EXAMPLE 2.3 Effective gas viscosity with and without foams at different MRF values.

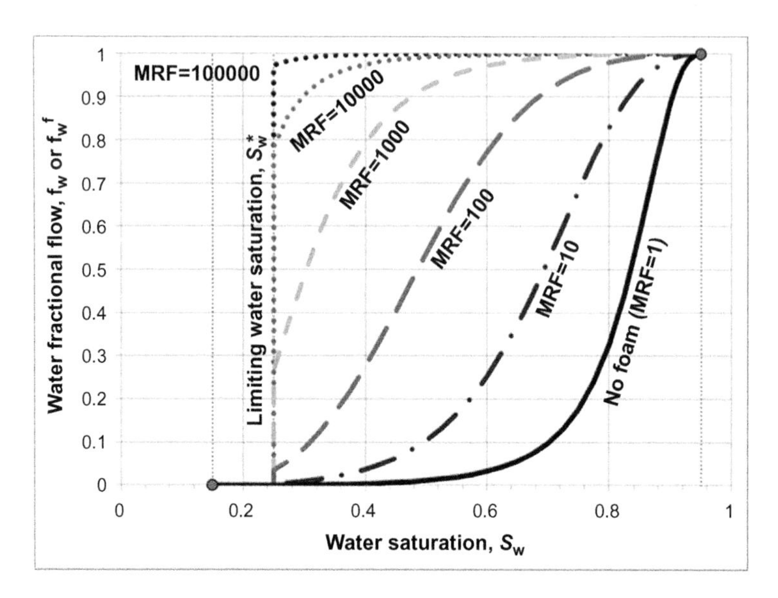

FIGURE EXAMPLE 2.4 Construction of water fractional flow curves as a function of water saturation at different MRF values.

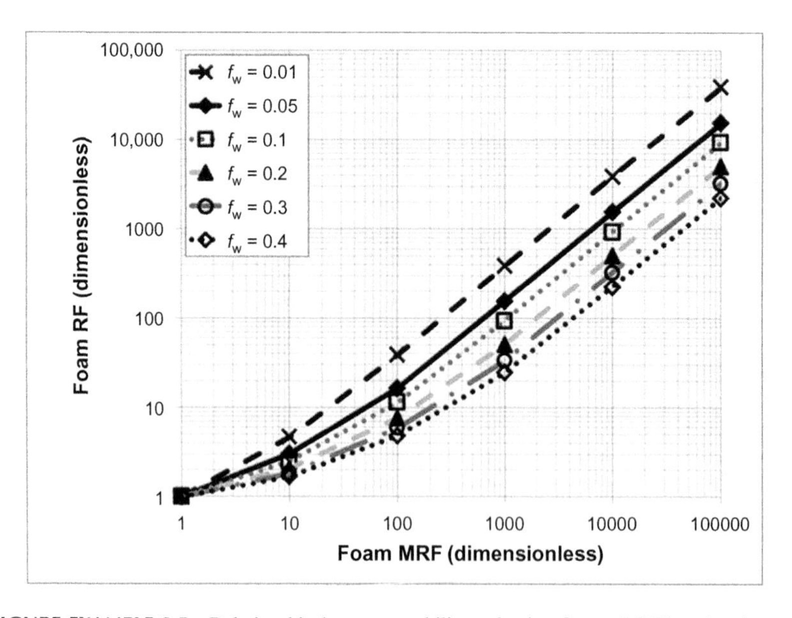

FIGURE EXAMPLE 2.5 Relationship between mobility reduction factor (MRF) and resistance factor (RF) at different injection foam qualities.

REFERENCES

Afsharpoor, A., Lee, G.S., Kam, S.I., 2010. Mechanistic simulation of gas injection during surfactant-alternating-gas (SAG) processes using foam catastrophe theory. Chem. Eng. Sci. 65, 3615–3631.

Alvarez, J.M., Rivas, H., Rossen, W.R., 2001. Unified model for steady-state foam behavior at high and low foam qualities. SPE J. September, 325–333.

Bernard, G.G., Holm, L.W., Jacobs, W.L., 1965. Effect of foam on trapped gas saturation and on permeability of porous media to water. SPE J. December, 195–300.

Bertin, H.J., Quintard, M.Y., Castanier, L.M., 1998. Development of a bubble—population correlation for foam-flow modeling in porous media. SPE J. December, 356–362.

Bogdanovic, M., Gajbhiye, R.N., Kam, S.I., 2009. Experimental study of foam flow in pipes: two distinct flow regimes. Colloids Surf. A Physicochem. Eng. Asp. 344 (4), 56–71.

Chang, R., 1994. fifth ed. Chemistry, 458. McGraw-Hill Inc., New York, NY.

Cheng, L., Kam, S.I., Delshad, M., Rossen, W.R., 2002. Simulation of dynamic foam-acid diversion processes. SPE J. September, 316–324.

Chou, S.I., Vasicek, S.L., Pisio, D.L., Jasek, D.E., Goodgame, J.A., 1992. CO_2 foam field trial at North Ward-Estes. SPE Annual Technical Conference and Exhibition, 4–7 October, Washington, DC.

Dalland, M., Hanssen, J.E., Kristiansen, T.S., 1994. Oil interaction with foams under static and flowing conditions in porous media. Colloids Surf. A Physicochem. Eng. Asp. 82, 129–140.

David, M.J. Radke, C.J., 1990. A pore level investigation of foam oil interactions in porous media. Paper SPE 18089PA.

de Vries, A.S., Wit, K., 1990. Rheology of gas/water foam in the quality range relevant to steam foam. SPERE May, 185–192.

Dholkawala, Z.F., Sarma, H.K., Kam, S.I., 2007. Application of fractional flow theory to foams in porous media. J. Pet. Sci. Eng. 57 (1–2), 152–165.

Dooley, J.J., Dahowski, R.T., Davidson, C.L., 2010. CO_2-driven enhanced oil recovery, PNNL-19557 prepared for the US Department of Energy under Contract DE-AC05-76RL01830.

Falls, A.H., Hirasaki, G.J., Patzek, T.W., Gauglitz, P.A., Miller, D.D., Ratulowski, J., 1988. Development of a mechanistic foam simulator: the population balance and generation by snap-off. SPERE August, 884–892.

Fitch, J.P. Minter, R. B., 1976. Chemical diversion of heat will improve thermal oil recovery. SPE Annual Fall Technical Conference and Exhibition, 3–6 October, New Orleans, LA.

Friedmann, F., Chen, W.H., Gauglitz, P.A., 1991. Experimental and simulation study of high-temperature foam displacement in porous media. SPERE February, 37–45.

Gajbhiye, R.N., Kam, S.I., 2011. Characterization of foam flow in horizontal pipes by using two-flow-regime concept. Chem. Eng. Sci. 66, 1536–1549.

Gauglitz, P.A., Friedmann, F., Kam, S.I., Rossen, W.R., 2002. Foam generation in homogeneous porous media. Chem. Eng. Sci. 57, 4037–4052.

Graue, D.J., Zana, E.T., 1981. Study of a possible CO_2 flood in Rangely field. J. Petrol. Technol. 33 (7).

Harpole, K.J. Hallenbeck, L.D., 1996. East vacuum Grayburg San Andres unit CO_2 flood ten year performance review: evolution of a reservoir management strategy and results of WAG optimization. SPE Annual Technical Conference and Exhibition, 6–9 October, Denver, CO.

Heller, J.P., Boone, D.A., Watts, R.J., 1985. Testing CO_2-foam for mobility control at rock creek. SPE Eastern Regional Meeting, 6–8 November, Morgantown, WV.

Hervey, J.R., Iakovakis, A.C., 1991. Performance review of a miscible CO_2 tertiary project: Rangely Weber Sand unit, Colorado. SPE Reserv. Eng. 6 (2), 163–168.

Hirasaki, G.J., Lawson, J.B., 1985. Mechanisms of foam flow in porous media: apparent viscosity in smooth capillaries. SPE J. 25 (2), 176–190.

Hoefner, M.L., Evans, E.M., Buckles, J.J., Jones, T.A., 1995. CO_2 foam: results from four developmental field trials. SPE Reserv. Eng. 10 (4), 273–281.

Holm, L.W., 1970. Foam injection test in the Siggins field, Illinois. J. Petrol. Technol. 22 (12), 1499–1506.

Holm, L.W., Garrison, W.H., 1988. CO_2 diversion with foam in an immiscible CO_2 field project. SPE Reserv. Eng. 3 (1), 112–118.

Holm, L.W., Josendal, V.A., 1974. Mechanisms of oil displacement by carbon dioxide. J. Petrol. Technol. 26 (12), 1427–1438.

Jenkins, M.K., 1984. An analytical model for water/gas miscible displacements. SPE 12632, SPE/DOE Fourth Symposium on Enhanced Oil Recovery, Tulsa, OK, pp. 15–18.

Jonas, T.M., Chou, S.I., Vasicek, S.L., 1990. Evaluation of a CO_2 foam field trial: Rangely Weber Sand unit. SPE Annual Technical Conference and Exhibition, 23–26 September, New Orleans, LA.

Kam, S.I., 2008. Improved mechanistic foam simulation with foam catastrophe theory. Colloids Surf. A Physicochem. Eng. Asp. 318, 62–77.

Kam, S.I., Rossen, W.R., 2003. A model for foam generation in homogeneous porous media. SPE J. 8 (December), 417–425.

Kam, S.I., Frenier, W.W., Davies, S.N., Rossen, W.R., 2007a. Experimental study of high-temperature foam for acid diversion. J. Pet. Sci. Eng. 58. 138–160.

Kam, S.I., Nguyen, Q.P., Li, Q., Rossen, W.R., 2007b. Dynamic simulation with an improved model for foam generation. SPE J. March, 35–48.

Khatib, Z.I., Hirasaki, G.J., Falls, A.H., 1988. Effect of capillary pressure on coalescence and phase mobilities in foams flowing through porous media. SPERE August, 919–926.

Koczo, K., Lobo, L.A., Wasan, D.T., 1992. Effect of oil on foam stability: Aqueous foams stabilized by emulsions. J. Colloid. Interfer. Sci. 150 (2), 492–506.

Kovscek, A.R. and Radke, C.J., 1994. Fundamentals of foam transport in porous media. In: L.L. Schramm (Ed.), Foams: Fundamentals and Applications in the Petroleum Industry, ACS Advances in Chemistry Series No. 242, American Chemical Society, Washington DC.

Kovscek, A.R., Patzek, T.W., Radke, C.J., 1995. A mechanistic population balance model for transient and steady-state foam flow in Boise sandstone. Chem. Eng. Sci. 50, 3783–3799.

Kuehne, D.L., Frazier, R.H., Cantor, J., Horn Jr., W., 1992. Evaluation of surfactants for CO_2 mobility control in dolomite reservoirs. SPE/DOE Enhanced Oil Recovery Symposium, 22–24 April, Tulsa, OK.

Lee, H.O., Heller, J.P., Hoefer, A.M.W., 1991. Change in apparent viscosity of CO_2 foam with rock permeability. SPERE November, 412–428.

Martin, F.D., Stevens, J.E., Harpole, K.J., 1994. CO_2-foam field test at the east vacuum Grayburg/San Andres unit, SPE/DOE Symposium on Improved Oil Recovery, 17–20 April, Tulsa, OK.

Martin, F.D., Heller, J.P., Weiss, W.W., Stevens, J., Harpole, K.J., Siemers, T., et al., 1995. Field Verification of CO_2 Foam, DOE Report.

Mayberry, D.J., Afsharpoor, A., Kam, S.I., 2008. The use of fractional flow theory for foam displacement in presence of oil. SPE Res. Eval. Eng. 11, 707–718.

Mohammadi, S.S. and Tenzer, J.R., 1990. Steam–foam pilot project at Dome-Tumbador, midway sunset field: part 2, SPE/DOE Enhanced Oil Recovery Symposium, 22–25 April, Tulsa, OK.

Mohammadi, S.S., Van Slyke, D.C., Ganong, B.L., 1989. Steam–foam pilot project in Dome-Tumbador, midway-sunset field. SPE Reserv. Eng. 4 (1), 17–23.

Myers, T.J., Radke, C.J., 2000. Transient foam displacement in the presence of residual oil: experiment and simulation using a population-balance model. Ind. Eng. Chem. Res. 39, 2725–2741.

Namdar-Zanganeh, M., Kam, S.I., La Force, T., Rossen, W.R., 2011. The method of characteristics applied to oil displacement by foam. SPE J. 16 (1), 8–23.

Osterloh, W.T., Jante Jr., M.J.,1992. Effects of gas and liquid velocity on steady-state foam flow at high temperature. SPE/DOE Enhanced Oil Recovery Symposium, 22–24 April, Tulsa, OK.

Ploeg, J.F. Duerksen, J.H., 1985. Two successful steam/foam field tests, sections 15A and 26C, midway-sunset field. SPE California Regional Meeting, 27–29 March, Bakersfield, CA.

Ransohoff, T.C., Radke, C.M., 1988. Laminar flow of a wetting liquid along the corners of a predominantly gas-occupied noncircular pore. J. Colloid. Interfer. Sci. 121, 392–401.

Rossen, W.R., 1996. Foams in enhanced oil recovery. In: Prud'homme, R.K., Khan, S. (Eds.), Foams: Theory, Measurements and Applications. Marcel Dekker, New York, NY.

Rossen, W.R., 1999. Foam generation at layer boundaries in porous media. SPE J. 4 (4), 409–412.

Rossen, W.R., Gauglitz, P.A., 1990. Percolation theory of creation and mobilization of foam in porous media. AIChE J. 36, 1176–1188.

Rossen, W.R., Lu, Q., 1997. Effect of Capillary Crossflow on Foam Improved Oil Recovery. SPE Western Regional Meeting, 25–27 June, Long Beach, CA.

Rossen, W.R., van Duijn, C.J., 2004. Gravity segregation in steady-state horizontal flow in homogeneous reservoirs. J. Pet. Sci. Eng. 43(1–2), 99–111.

Rossen, W.R., Wang, M., 1999. Modeling foams for acid diversion. SPE J. June, 92–100.

Rossen, W.R., Zhou, Z.H., 1995. Modeling foam mobility at the limiting capillary pressure. SPE Adv. Technol. 3, 146.

Rossen, W.R., Zhou, Z.H., Mamun, C.K., 1995. Modeling foam mobility in porous media. J. SPE Adv. Technol. Series 3 (1), 146–153.

Sanchez, J.M., Schechter, R.S., 1989. Surfactant effects on the two-phase flow of steam–water and nitrogen–water through permeable media. J. Petr. Sci. Eng. 3, 185–199.

Schramm, L.L.,1994. Foams: Fundamentals and Applications in the Petroleum Industry. Advances in Chemistry Series No. 242, American Chemical Society, Washington, DC.

Stone, H.L.,1982. Vertical conformance in an alternating water miscible gas flood. SPE 11140, SPE Annual Technology Conference, 26–29 September, New Orleans, LA.

Tanzil, D., Hirasaki, G.J., Miller, C.A., 2002. Mobility of foam in heterogeneous media: flow parallel and perpendicular to stratification. SPE J. 7 (2), 203–212.

Turta, A.T. and Singhal, A.K., 1998. Field foam applications in enhanced oil recovery projects: screening and design aspects. SPE International Oil and Gas Conference and Exhibition in China, 2–6 November, Beijing, China.

US Greenhouse Gas Inventory Report, 2011. Inventory of US Greenhouse Gas Emissions and Sinks 1990–2009, EPA 430-R-11-005.

Winzinger, R., Brink, J.L., Patel, K.S., Davenport, C.B., Patel, Y.R., Thakur, G.C., 1991. Design of a major CO_2 flood, North Ward-Estes Field, Ward County, Texas. SPE Reservoir Engineering 6, 1.

Yellig, W.F., Metcalfe, R.S., 1980. Determination and prediction of CO_2 minimum miscibility pressures. J. Petrol. Technol. 30, 160–168.

Chapter 3

Polymer Flooding—Fundamentals and Field Cases

James J. Sheng
Bob L. Herd Department of Petroleum Engineering, Texas Tech University, Lubbock, TX 79409, USA

3.1 POLYMERS CLASSIFICATION

Basically, two types of polymers are used in enhancing oil recovery: synthetic polymers like partially hydrolyzed polyacrylamide (HPAM) and biopolymers like xanthan. Their derivatives and variations are developed to fit specific needs. HPAM type of polymers are much more widely used than biopolymers (xanthan type), because HPAM has advantages in price and large-scale production. Wang et al. (2006a) believe that HPAM solutions exhibit significantly greater viscoelasticity than xanthan solutions.

Polyacrylamide (PAM) adsorbs strongly on mineral surfaces. Thus, the polymer is partially hydrolyzed to reduce adsorption by reacting PAM with a base, such as sodium or potassium hydroxide or sodium carbonate. Hydrolysis converts some of the amide groups ($CONH_2$) to carboxyl groups (COO^-). Their structures are shown in Figure 3.1. The hydrolysis rate is defined as the mole fraction of amide groups that are converted into carboxyl groups by hydrolysis which is equal to $y/(x+y)$. It ranges from 15% to 35% in commercial products. Hydrolysis of PAM introduces negative charges on the backbones of polymer chains that have large effect on the rheological properties of polymer solution. At low salinities, the negative charges on the polymer backbones repel each other and cause the polymer chains to stretch. When an electrolyte, such as NaCl, is added to a polymer solution, the repulsive forces are shielded by a double layer of electrolytes, thus the stretch is reduced and the viscosity is reduced. The molecular weights (MWs) are in the order of millions of daltons.

A xanthan polymer acts like a semigrid rod and is quite resistant to mechanical degradation. Average reported MWs of xanthan biopolymer used in enhanced oil recovery (EOR) processes range from 1 million to 15 million.

Other PAM-derived polymers used in EOR processes include hydrophobically associating polymer, salinity-tolerant PAM (KYPAM) (Luo et al., 2002), and AMPS (2-acrylamide-2-methyl propane-sulfonate). Some polymers can

Enhanced Oil Recovery Field Case Studies.

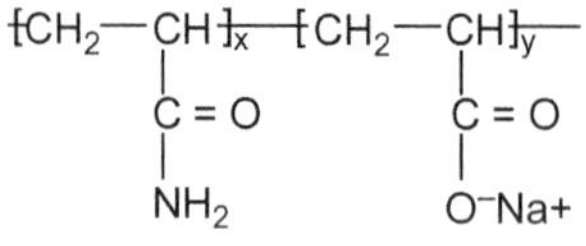

$$-[CH_2-\underset{\underset{NH_2}{|}}{\underset{|}{C=O}}{CH}]_x-[CH_2-\underset{\underset{O^-Na+}{|}}{\underset{|}{C=O}}{CH}]_y-$$

FIGURE 3.1 Structures of amide group and carboxyl group.

change their volumes or viscosities under proper reservoir conditions. For example, BrightWater expands its volume as it contacts high-temperature water. Microball expands as it hydrates. pH-sensitive polymer increases its viscosity above a critical pH. An inverse polymer emulsion is inverted from W/O to O/W type under certain temperature and salinity so that PAM is hydrated and the viscosity is increased in the deep formation. Gels are cross-linked polymers. A detailed introduction is provided by Sheng (2011).

3.2 POLYMER SOLUTION VISCOSITY

The viscosity is the most important parameter for polymer solution. Since HPAM is the most used polymer in EOR, we only discuss HPAM type of polymer properties from now, although most of properties for biopolymers are similar.

3.2.1 Salinity and Concentration Effects

The dependence of polymer solution viscosity at zero shear rate (μ_p^0) on the polymer concentration and on salinity may be described by the Flory–Huggins equation (Flory, 1953):

$$\mu_p^0 = \mu_w(1 + (A_{p1}C_p + A_{p2}C_p^2 + A_{p3}C_p^3)C_{sep}^{S_p}) \tag{3.1}$$

where μ_w is the water viscosity with its unit being the same as μ_p^0, C_p is the polymer concentration in water, A_{p1}, A_{p2}, A_{p3}, and S_p are fitting constants, and C_{sep} is the effective salinity for polymer. The factor $C_{sep}^{S_p}$ allows for dependence of polymer viscosity on salinity and hardness. The effective salinity for polymer, C_{sep}, is given in UTCHEM-9.0 (2000) by

$$C_{sep} = \frac{C_{51} + (\beta_p - 1)C_{61}}{C_{11}} \tag{3.2}$$

where C_{51}, C_{61}, and C_{11} are the anion, divalent, and water concentrations in the aqueous phase, and β_p is measured in the laboratory which is about 10. The unit for C_{51} and C_{61} is meq/mL, and the unit for C_{11} is water volume fraction in the aqueous phase. The commonly used laboratory units for salinity are wt% and ppm (mg/L). These units should be converted to meq/mL in using Eq. (3.2). In principle, any units could be used, as long as they are used consistently in a study. Note that the electrolyte concentrations in the laboratory are commonly expressed in terms of the aqueous phase volume that includes the volume of surfactant and cosolvent in addition to the water. C_{11} in Eq. (3.2) is used to correct the aqueous volume.

3.2.2 Shear Effect

The viscosity of a polymer solution is strongly shear dependent. We may use the power-law model to describe a polymer solution:

$$\mu_p = K\dot{\gamma}^{(n-1)} \tag{3.3}$$

where K is the flow consistency index, n is the flow behavior index, and $\dot{\gamma}$ is the shear rate. In the pseudoplastic region, $n \leq 1$ (typically $n = 0.4-0.7$). At different concentrations, n has little changes, but K changes. For a Newtonian fluid $n = 1$ and K is simply the constant viscosity, μ.

A more general model is the Carreau equation (Bird et al., 1987; Carreau, 1972):

$$\mu_p - \mu_\infty = (\mu_p^0 - \mu_\infty)[1 + (\lambda\dot{\gamma})^\alpha]^{(n-1)/\alpha} \tag{3.4}$$

where μ_∞ is the limiting viscosity at the high (approaching infinite) shear limit and is generally taken as the water viscosity μ_w, λ and n are polymer-specific empirical constants; and α is generally taken to be 2. μ_p^0 and $\dot{\gamma}$ are as defined earlier. Practically, μ and μ_p^0 are much higher than μ_∞ and $(\lambda\dot{\gamma})^\alpha$ is much larger than 1. Thus, Eq. (3.4) becomes the power-law equation of the form $\mu_p = \mu_p^0(\lambda\dot{\gamma})^{n-1}$ which describes the viscosity at the intermediate and high shear rate regions. At the low shear rate region, $\mu_p = \mu_p^0$, as shown in Figure 3.2.

3.2.3 pH Effect

pH affects hydrolysis. pH will increase when an alkali is added. Initially, polymer viscosity may increase because of hydrolysis. However, adding an alkali eventually will result in the decrease of HPAM viscosity because the salt effect of the alkali is dominant compared with the pH effect on hydrolysis. Figure 3.3 shows such an example.

3.3 POLYMER FLOW BEHAVIOR IN POROUS MEDIA

In this section, we discuss polymer viscosity, polymer retention, and rock permeability reduction in porous media.

3.3.1 Polymer Viscosity in Porous Media

We cannot directly measure polymer solution viscosity in porous media. We can only estimate it from the Darcy equation if coreflood tests are conducted. Because polymer viscosity is strongly shear dependent, we have to run many

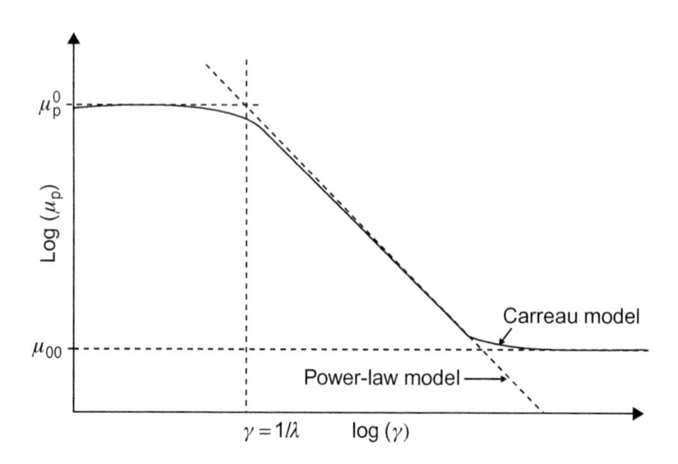

FIGURE 3.2 Comparison of the Carreau and power models.

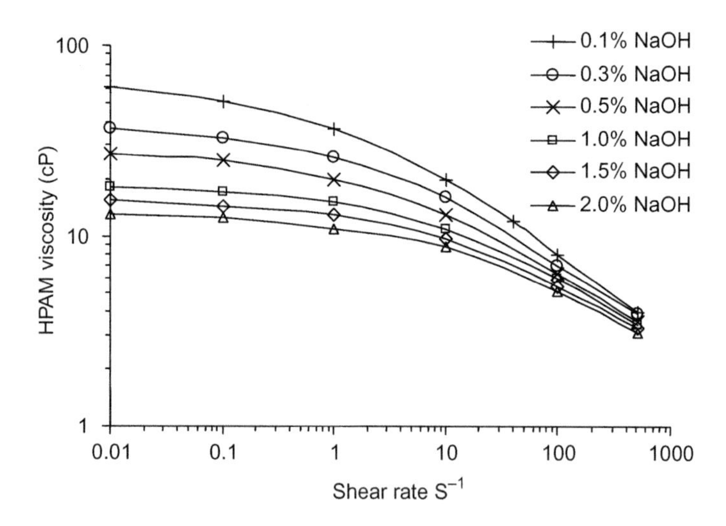

FIGURE 3.3 Alkaline effect on HPAM solution viscosity (Kang, 2001).

coreflood tests to obtain the viscosities at different flow rates (shear rates). Obviously, this is expensive. Ideally, we wish to have a model to estimate polymer viscosities at different flow rates in porous media from the bulk solution viscosities at different shear rates. In practice, we cannot simply do that. Instead, we need to run a few coreflood tests or at least one test to calibrate the model. This approach is presented next.

We want to estimate polymer solution viscosities at different flow rate using the bulk solution viscosities at different shear rates. Our first question is how to convert the flow rates into shear rates. The converted shear rate is called equivalent shear rate which is equivalent to that in the bulk viscometer.

Assume the flow in porous media can be represented by capillary flow and the polymer solution viscosity can follow the power-law equation, the equivalent shear rate ($\dot{\gamma}_{eq}$) may be calculated using the following equation (Sheng, 2011):

$$\dot{\gamma}_{eq} = \gamma\left(\frac{3n+1}{4n}\right)\frac{4u}{\sqrt{8k\phi}} \tag{3.5}$$

where u ($=\overline{v}/\phi$) is the Darcy velocity in porous media, k is the permeability, ϕ is the porosity, and n is the exponent in the power-law equation. In the above equation, γ is an adjustable parameter which is used to fit coreflood experimental data to the bulk viscosities. The unit should be consistent. For example, if the shear rate is in s^{-1}, u is in m/s, k is in m^2, and ϕ is in fraction.

The procedures to calculate the equivalent shear rate in porous media may be summarized as follows.

1. Measure the bulk viscosity of a polymer solution (μ_b) at different shear rates. Thus, we have μ_b versus $\dot{\gamma}$. K and n are obtained by fitting the data to the power-law equation:

$$\mu_b = K(\dot{\gamma})^{(n-1)} \tag{3.6}$$

2. Conduct coreflood tests with the polymer solution at a few injection rates. Measure the pressure drop, Δp, corresponding to each injection rate (velocity u). The core permeability and porosity are measured before the coreflood tests. Calculate the apparent viscosity in the core, μ_{core}, using the Darcy equation at each injection rate (u), so we have μ_{core} versus u.
3. Find the corresponding shear rate, $\dot{\gamma}$, at μ_b equal to μ_{core}. Then find γ from Eq. (3.5) using $\dot{\gamma}_{eq}$equal to this calculated $\dot{\gamma}$ and the injection rate u.
4. Repeat the step 3 if several coreflood tests are conducted. Choose a compromised γ so that the coreflood data, μ_{core} versus $\dot{\gamma}_{eq}$, match the viscometric bulk viscosity data, μ_b versus $\dot{\gamma}$.

3.3.2 Polymer Retention

Polymer retention includes adsorption, mechanical trapping, and hydrodynamic retention. These different mechanisms were discussed by Willhite and Dominguez (1977). Mechanical entrapment and hydrodynamic retention are related and only occur in flow through porous media. Mechanical entrapment occurs when large polymer molecules flow through relatively small pore throats. Hydrodynamic retention is probably not a large contributor in the total retention and can be neglected in field applications because of low flow velocity. Adsorption is a fundamental property of the polymer–rock surface–solvent system and is the most important mechanism. Since it is difficult to differentiate these three mechanisms in dynamic flood tests, we may simply use the term retention to describe the polymer loss, sometimes just using the term adsorption. Adsorption refers to the interaction between the polymer molecules and the solid surface. This interaction causes polymer molecules to be bound to the surface of the solid mainly by physical adsorption, van der Waals and hydrogen bonding.

Generally, we use the Langmuir-type isotherm to describe the polymer adsorption. The Langmuir-type isotherm is given by

$$\hat{C}_p = \frac{a_p C_p}{1 + b_p C_p} \tag{3.7}$$

where C_p is the equilibrium polymer concentration in the rock-polymer solution system, a_p and b_p are empirical constants. The unit of b_p must be the reciprocal of the unit of C_p. a_p is dimensionless. Note that C_p and $\hat{C}_p$ must be in the same unit. a_p is defined as

$$a_p = (a_{p1} + a_{p2} C_{sep}) \left(\frac{k_{ref}}{k} \right)^{0.5} \tag{3.8}$$

where a_{p1} and a_{p2} are input or fitting parameters, C_{sep} is the effective salinity, k is the permeability, and k_{ref} is the reference permeability of the rock used in the laboratory measurement. Eqs. (3.7) and (3.8) take into account the salinity, polymer concentration, and permeability.

Here are some observations on polymer retention. Polymer adsorption in static bulk tests could be quite different from that in dynamic coreflood tests. This is because mechanical trapping and hydrodynamic trapping occur in corefood tests, while disaggregated grains in static bulk tests have higher surface areas resulting in higher adsorption.

The adsorption level of HPAM on the calcium carbonate is much higher than that on the silica surface. This may be owing to the fact that carbonate rock surfaces are positively charged, while the silica surfaces are negatively charged.

As Eqs. (3.7) and (3.8) show, the adsorption increases with the salinity. Adding a low concentration of divalent calcium ion, Ca^{2+}, promotes HPAM adsorption on silica, because the divalent ions compress the size of the flexible HPAM molecules and reduce the static repulsion between the polymer carboxyl group and the silica surface.

Polymer retention decreases with permeability. This is because mechanical trapping in a low permeability rock is higher than that in a high permeability rock, and low permeability rocks generally have high clay content.

3.3.3 Inaccessible Pore Volume

When polymer molecular sizes are larger than some pores in a porous medium, the polymer molecules cannot flow through those pores. The volume of those pores which cannot be accessed by polymer molecules is called inaccessible pore volume (IPV). Because of IPV, a polymer solution will sweep through less pore volume and thus it takes less time for polymer to reach to the effluent end than a nonadsorptive tracer. On the other hand, the polymer breakthrough is delayed because of polymer retention. These two factors will counteract each other. In reality, the first factor is more significant, and the polymer will overtake the tracer.

3.3.4 Permeability Reduction

Permeability reduction or pore blocking is caused by polymer adsorption. The permeability reduction is defined by permeability reduction factor (F_{kr}):

$$F_{kr} = \frac{\text{rock permeability when water flows}}{\text{rock permeability when aqueous polymer solution flows}} = \frac{k_w}{k_p} \tag{3.9}$$

Since the polymer permeability reduction process is considered to be an irreversible process, even when the reservoir is under post-polymer waterflooding, the permeability reduction still remains. This is called residual permeability reduction factor and is defined as

$$F_{krr} = \frac{\text{rock permeability to water before polymer flow}}{\text{rock permeability to water after polymer flow}} \tag{3.10}$$

The permeability to the aqueous phase is reduced by polymer injection, but it is hardly reduced to the other components or other phases (Schneider and Owens, 1982; White et al., 1973). Therefore, the permeability itself cannot be changed in a numerical simulator. Instead, we have to modify the polymer solution viscosity by F_{kr} and the water viscosity after polymer flooding by F_{krr} (Bondor et al., 1972; UTCHEM-9.0, 2000).

Pang et al. (1998) found that the higher the polymer MW, the higher the F_{krr} is. When the MW is the same, F_{krr} is higher when the polymer has a wide MW distribution. Huang et al. (1998a) observed that the permeability reduction factor (F_{kr}) increases with higher injection velocity and lower temperature.

Although higher MW results in higher F_{krr} and even higher oil recovery factor, MW used must be limited by formation permeability. Figure 3.4 shows the highest MW at different permeabilities. The data connected with solid lines are from Zhang and Yang (1998), the data with the empty triangle points are from Wang et al. (2006b), and the data marked with solid points unconnected are from Niu et al. (2006). These data show that lower MW polymer is need for a low permeability formation. For example, Wang et al. show that polymer with MW of 2.4 million can satisfy the need of reservoirs with permeability of 20 mD, while 5.5 million can meet the demand of reservoirs with permeability of 50 mD, and 10 million is suitable to reservoirs of permeability of 200 mD. Apparently, the data from Zhang and Yang

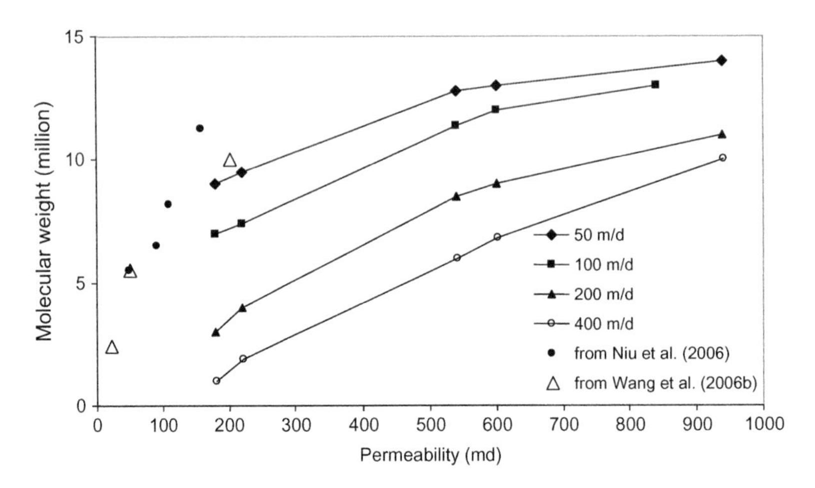

FIGURE 3.4 Polymer MW limits.

are more conservative. Zhang and Yang considered the rates flowing through perforation. The rates of 50, 100, 200, and 400 m/D corresponded to the rates of 2.25, 4.5, 9.0, and 18 m^3/D m, respectively, through 0.008 m diameter holes with 10 holes per meter.

3.3.5 Relative Permeabilities in Polymer Flooding

Experiments (Chen and Chen, 2002; Schneider and Owens, 1982; Taber and Martin, 1983) showed that the water relative permeability is significantly reduced, while the oil relative permeability is relatively unchanged after polymer flooding. It was suggested that this is caused by segregation of oil and water pathways (Liang et al., 1995). This is the mechanism of disproportionate permeability reduction (DPR).

3.4 MECHANISMS OF POLYMER FLOODING

The main mechanism of polymer flooding is the increased viscosity of polymer solution so that the mobility ratio of the displacing polymer solution to the displaced fluids ahead is reduced and the viscous fingering is reduced. When the viscous fingering is reduced, the sweep efficiency is improved.

The mechanism of increased displacing fluid viscosity can be quantified using the Buckley-Leverett (1942) theory. Figure 3.5 shows two fractional flow curves. One is for a waterflooding case with the viscosity ratio of water to oil 0.1, the other one is for a polymer flooding case with the viscosity ratio of polymer to oil 1. From the fractional flow curve, we can estimate the average water saturation at breakthrough by drawing a tangent from the connate water saturation S_{wc} (0.2 in this case) and intersecting the horizontal

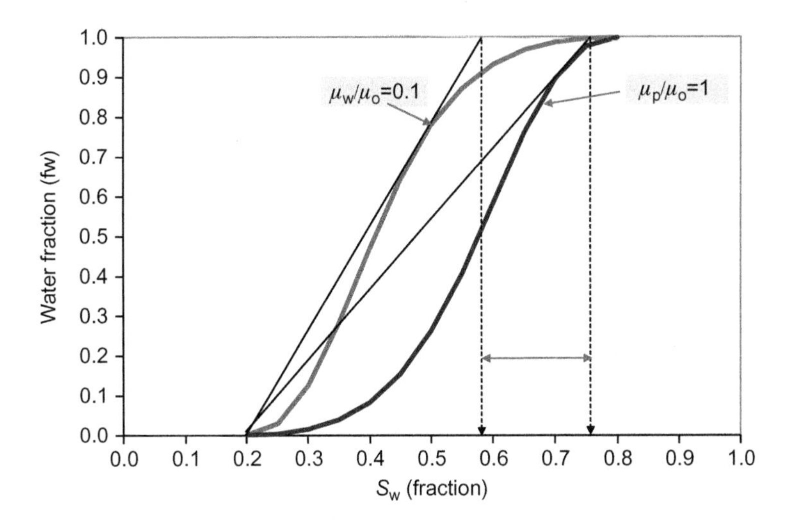

FIGURE 3.5 Effect of viscosity ratio on the fractional flow curve.

line of $f_w = 1$, and the corresponding water saturation is the average water saturation. Here, f_w is the water cut in the producing fluid. From Figure 3.5, the average water saturation in the waterflooding case is 0.58, whereas the average water saturation in the polymer flooding case is 0.76. The difference is 0.18. In other words, by simply increasing the viscosity of displacing fluid, the oil recovery factor can be increased by 18% at breakthrough.

When polymer is injected in vertical heterogeneous layers, crossflow between layers improves polymer allocation in the vertical layers so that the vertical sweep efficiency is improved. This mechanism is detailed in Sorbie (1991).

Another mechanism is related to polymer viscoelastic behavior. The interfacial viscosity between polymer and oil is higher than that between oil and water. The shear stress is proportional to the interfacial viscosity. Due to polymer viscoelastic properties, there is normal stress between oil and polymer solution, in addition to shear stress. Thus, polymer exerts a larger pull force on oil droplets or oil films. Therefore, oil can be "pushed and pulled" out of dead-end pores. Thus, residual oil saturation is decreased. This mechanism was rarely discussed until recently (Sheng, 2011; Wang et al., 2001).

One economic impact of polymer flooding which has been less discussed is the reduced amount of water injected and produced, compared with waterflooding. Because polymer improves the mobility ratio and sweep efficiency, less water is injected and less water is produced. In some situations like an offshore environment and desert area, water and the treatment of water could be costly (Sheng, 2011).

3.5 POLYMER MIXING

Polymer can be delivered in liquid emulsion, water solution, solid powders, or bars. When polymer is in liquid emulsion or water solution, it can be added to injection water using a pump. When it is in solid powder, several processes are needed to prepare the polymer solution: proration—dispersion—maturation—transportation—filtration—storage (Figure 3.6). Proration is metering solid polymer and water to be dispersed. Polymer is delivered to the feeder which can filter impurities. Dispersion is a process to dissolve high MW polymer into water. The dispersed polymer (concentrated solution) is transported to a maturation tank where the mixer is rotating. The maturation takes 0.5–24 h (Huang et al., 1998b; Liu et al., 2006). The concentrated solution is transported to the storage tank through two filters to remove impurities and "fish eye balls." Screw pump is used for transporting polymer solution to reduce mechanic shearing. Plunger displacement pump is used to inject polymer solution. Another unit is called static mixer. It is a special unit and is installed in pipes to change fluid flow direction so that the fluids can be fully mixed. Unlike a dynamic (rotary) mixer, the static mixer does not move as the name implies. When polymer is in bars, it must be grinded into powder first for mixing.

3.6 SCREENING CRITERIA

The screening criteria were proposed by several authors or groups by 1980s, later revisited by Taber et al. (1997a,b). Technology advances will update these criteria. For example, oil viscosity should be low. But people started to inject polymer in high-viscous oil reservoirs (Moe Soe Let et al., 2012; Wassmuth et al., 2009). More laboratory research has been done

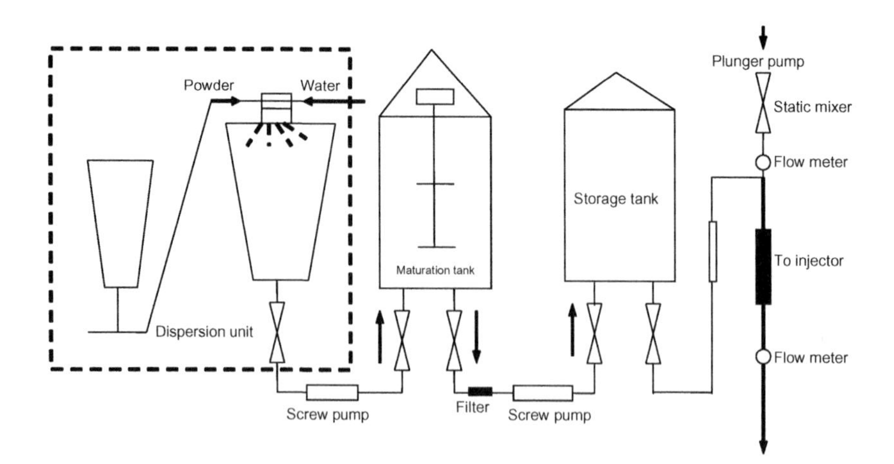

FIGURE 3.6 Schematic of a typical facility to prepare polymer solution.

(Seright, 2010; Wassmuth et al., 2007). The screening criteria can only provide initial evaluation for a possible application. For a specific application, laboratory tests must be conducted.

3.7 FIELD PERFORMANCE AND FIELD CASES

We will first summarize some overall field performance before presenting individual field cases. We present polymer flooding cases in a very heterogeneous reservoir (Xiaermen field in China), using high MW and high concentration polymer in three blocks in the Daqing field, in three heavy oil reservoirs (the East Bodo reservoir in Canada, the Tambaredjo field in Suriname, and the Marmul field in Oman), in a carbonate reservoir (the Vacuum field in New Mexico), and using movable gel for post-polymer conformance control in the Bei-Yi-Qu-Duan-Xi block in Daqing.

3.7.1 Overall Field Performance

The survey of field polymer flooding projects sponsored by the Department of Energy (DOE) (Manning et al., 1983) found that the average incremental oil recovery is 2.91% original oil in place (OOIP), 4.02% remaining oil in place (OIP), 24.83 stock tank barrels (STB) per acre-ft and 2.34 STB oil per lb polymer injected for the field projects based on conservative estimates of actual oil recovery. Based on optimistic estimates, however, the average incremental oil recovery is 3.85% OOIP, 4.22% remaining OIP, 34.4 STB per acre-ft and 3.74 STB oil per lb polymer injected. Their statistical analysis yielded statistically significant regression coefficients for 10 oil recovery functions with the following reservoir and fluid identifiers as independent variables:

1. Porosity
2. Permeability
3. Water-to-oil mobility ratio
4. Polymer-to-oil mobility ratio
5. Polymer solution pore volume
6. Oil viscosity
7. Water-to-oil ratio at start
8. Average polymer concentration
9. Mobile oil saturation
10. Net pay-to-gross pay ratio.

No statistically significant regression coefficients were obtained for permeability variation.

China has carried out many polymer flooding projects in recent years. According to the author's survey (unpublished), the incremental oil recovery from Chinese polymer flooding projects was 8.9%.

In terms of amount of polymer injected, the 1976 US National Petroleum Council (NPC) study reported 125 mg/L·PV (pore volume). In contrast, the amount of polymer reported in the 1984 NPC study was increased to 240 mg/L·PV, but it is still much lower than 400–500 mg/L·PV used in the Daqing projects in 2000s (Sheng, 2011).

Most polymer flooding projects were conducted in sandstone reservoirs. But many polymer projects were carried out in carbonate reservoirs as well according to Manrique et al. (2007) survey. HPAM polymers were more often used than biopolymers. Because HPAM type of polymers are anionic polymers, one may think the polymer adsorption in carbonates would be higher than that in sandstone reservoirs. Limited published adsorption data does not seem to support this perception.

3.7.2 Polymer Flooding in a Very Heterogeneous Reservoir

This case discusses the in-depth profile modification before polymer injection in a very heterogeneous reservoir (Deng et al., 1998). The reservoir was in the H2II zone in the Xiaermen field in China. The zone was a thick complicated fault-block reservoir, controlled by faults in the north and south, and connected with edge water in the west and the east. The formation was unconsolidated. The average porosity was 23.7% and the average permeability was 4.78 D. The permeability variation coefficient was 0.91 and the oil viscosity was 70 mPa·s. The reservoir temperature was 50°C. The formation water salinity was 2127 mg/L. The formation thickness was 19.9 m and its net pay thickness was 14.2 m.

The zone was under edge water drive and irregular inner point water drive. Owing to complex structure, imperfect flooding pattern, severe heterogeneity and high water–oil mobility ratio of 27, the water cut went up quickly. By August 1996, the cumulative oil production was only 24.9%, but the water cut was 89%. In different areas, there were higher remaining oil saturations near faults in the northern and central parts, and lower oil saturations near the oil-water-contact (OWC) in the west and around the injector areas. Vertically, the poorest waterflooding behavior existed in the top of the reservoir. Treatments using suspended particle and TTP-910 modifiers had no improvement to stop channeling.

To solve the problems, in-depth profile modification using high-strength modifier was conducted to adjust the flow profile and the direction of fluid stream before polymer flooding. Profile control treatments were also carried out in some channeling wells and intervals during polymer injection to prevent polymer from channeling and keep the flooding effective. For polymer flooding, the injection sequence optimized through numerical simulation was 0.05 PV 1200 ppm, 0.25 PV 1000 ppm, and 0.06 PV 600 ppm. High MW (23 million daltons) HPAM was mixed with the produced water.

The polymer flooding pattern had 7 injectors and 18 producers. The distance between an injector and a producer was 150–360 m. The polymer was injected into 7 injectors sequentially from September 1996 to April 1997. The prior profile modification treatment was carried out in 6 injectors and the treatment depths were 25.1–39.9 m. The wellhead viscosities were 55 and 40 mPa·s for the frontal slug and main slug, and the *in situ* effective viscosities were estimated to be 16.5 and 12 mPa·s.

Comparing with waterflooding, the wellhead pressure of injectors increased rapidly in the early stage of polymer flooding and went up slowly afterward. The increase of wellhead pressures was commonly 0.9–3.5 MPa, on an average of 2.3 MPa. The injectivity index of injectors decreased by 20–75% compared with that before polymer injection. After a period of polymer injection (about 0.02–0.04 PV), the injectivity index tended to become unchanged. Both the increase in pressure and the decrease in injectivity index implied that the flow resistance around injection well increased and vertical permeability profile was modified after profile modification and polymer injection.

After 0.08 PV of cumulative polymer injection, the liquid rate and production index began to rise slowly. During the injection of 0.02–0.09 PV, oil production rate increased quickly with a rapid decrease in water cut. According to the prediction from numerical simulation, the ultimate incremental oil recovery by polymer flooding would be up to 10% OOIP.

3.7.3 Polymer Flooding Using High MW and High Concentration Polymer

The pilot tests using high MW and high concentration polymer were started in 2001 in three blocks in the Daqing field: WN block, WM block, and E N1 block (Zhang, 2011). The MW was 25 million daltons, and the injected concentration was about 2000 mg/L. The distance from an injector to a producer was 237–250 m. The permeabilities were 625–912 mD, and the formation thickness was 10.1–17.9 m. These pilot areas were under conventional polymer flooding (16 million MW and 1000 mg/L). After injecting polymer of 25 million MW and about 2000 mg/L, the water cut was reduced by 25–30%.

Having obtained positive results from these pilots, a large-scale polymer injection was applied to the E W N1 block in January 2009. The area was 7.75 km^2 with 160,688 million tons of oil in place and 296,838 million m^3. There were 238 producers and 204 injectors in the block. The effective thickness was 12.2 m. The distance from an injector to a producer was about 125 m. The formation permeability was 652 mD.

The polymer MW was 25 million and the injected concentration was 2030 mg/L. The injected polymer solution was 0.61 PV. So the mount of polymer injected was 1238 mg/L·PV. It was reported that water cut was significantly reduced and the incremental oil recovery was more than 10%.

3.7.4 Polymer Flooding in Heavy Oil Reservoirs

In this section, we present two cases of polymer injection in heavy oil reservoirs: the East Bodo reservoir in Canada and the Tambaredjo field in Suriname.

Polymer Flooding in the East Bodo Reservoir, Canada

The East Bodo reservoir in Alberta was produced from the Lloydminster formation which is part of the Lower Cretaceous Mannville Group (Wassmuth et al., 2009). The porosity was 27–30% and the permeability was 1000 mD. The reservoir oil viscosity was 600–2000 mPa·s (14° API). For the formation water, the total dissolved solid (TDS) content ranged from 25,000 to 29,000 ppm with hardness concentrations (Ca^{2+} and Mg^{2+}) of 350–650 ppm. In the pilot area, there were 13 producers and 1 injector. The average thickness was 3.2 m.

After coreflood tests, history matching the coreflood tests and having conducted field simulation study, a pilot test was conducted. The pilot was in a mature waterflood area of the highest injectivity for the field. The polymer injection was initiated in May 2006. It was expected that the injected polymer solution of 1500 ppm would result in 25 mPa·s. Apparently, the reservoir solution viscosity was 10 mPa·s at maximum. So later a fresher water source (TDS = 3700 ppm) was used, and the solution viscosity at surface of 1500 ppm was 60 mPa·s at surface. The polymer concentration at the nearest producing wells were about 100 ppm.

After fill-up, the injection pressure reached 6000 kPa at 200 m^3/D of polymer. Previously, a similar injection pressure was achieved with water at a rate of 250 m^3/D. The pilot performance indicated that for polymer injection in the heavy oil reservoir, horizontal wells helped to alleviate injectivity problem.

Polymer Flooding in the Tambaredjo Field, Suriname

This section presents a case of polymer flooding a heavy oil reservoir in the Tambaredjo field in Suriname (Staatsolie's Sarah Maria pilot) (Moe Soe Let et al., 2012). The pilot had three injectors and nine offset producers. The produced oil viscosity ranged from 1260 to 3057 mPa·s with an average of 1728 mPa·s. The reservoir "foamy oil" (Sheng et al., 1999) viscosity was believed to be 300–600 mPa·s. The average permeability of the sand exceeded 4 D with significant heterogeneity (permeability contrast >10:1). The prepared polymer solutions (1000 ppm SNF Flopaam 3630S in Sarah Maria water of 400–500 ppm TDS) has a viscosity of 50 mPa·s (ambient temperature and 7.3 s^{-1}) at the mixing facility and 45 mPa·s at the closest injection well. Because the injected polymer solution is lower than oil viscosity, obvious fingering was observed in the pilot. It was expected that increasing polymer concentration would improve the performance and they were testing this concept when the paper was written in 2012.

The nine production wells surrounding the injection wells produced 10–60% of the injected polymer concentration. Oil rates in producer were increased while the water cuts were decreased. However, the responses from polymer injection were modest. It was interpreted from calculated injectivity using polymer viscosity at surface that horizontal fractures were formed by polymer injection. However, severe channeling was not witnessed. What could cause these two phenomena was that near wellbore fractures were formed.

The dissolved oxygen levels were ambient (3–8 ppm) throughout the mixing and injection process. Although high dissolved oxygen is not a good general practice, it was argued that the high oxygen levels might be acceptable for the Sarah Maria pilot conditions. The augment is from the experience at Daqing where ambient levels of dissolved oxygen were also present through the mixing and injection process. The Daqing sand contained about 0.25% pyrite and 0.5% siderite. It effectively removed any dissolved oxygen within 1 day and a short distance after polymer enters the reservoir (Seright et al., 2010). A similar result was expected for this pilot, because X-ray diffraction (XRD) analysis showed significant amounts (up to 12%) of siderite and pyrite in some cores.

3.7.5 Polymer Flooding in the Marmul Field, Oman

In the Marmul case, a polymer pilot was carried out in 1986 and the pilot results were promising. However, because of the low oil price at that time, the field expansion was not carried out until 2010. This case shows that an EOR application is sensitive to oil price.

The Marmul field is one of the largest oil fields in the Sultanate of Oman. Its main producing formation is the sandstone Al Khalata formation. Its thickness is 65 ft. The formation porosity is 26–34% and the permeability ranges 8–25 D. The formation temperature is 46°C. The formation salinity is 7404 ppm. The oil viscosity is 40–120 mPa·s and the water viscosity is 0.64 mPa·s (Teeuw et al., 1983).

This field was first discovered in 1956 and was fully brought on stream in 1980. The field was naturally supported by an active edge water aquifer. Waterflooding was still practiced because some of the updip wells could not receive good aquifer support. During the 1980s, pilot studies for steam injection and polymer flooding were conducted. And it was found that polymer flooding was the most suitable EOR method for this field.

The polymer flooding pilot began in May 1986 with a single inverted 5-spot pattern. The distance from the injector to a producer was 140 m. A water preflush of 0.23 PV was first injected, followed with a 0.63 PV of 1000 ppm polymer injection and finally a 0.34 PV of water post-flush slug.

The injection water had 600 ppm salinity. The polymer used was HPAM type and the designed viscosity of polymer solution was 15 mPa·s at the reservoir temperature and a shear rate of 8.1 s^{-1}. This viscosity corresponded to the mobility ratio of 2. After polymer injection, oil cut was increased. During the pilot testing period, 59% OOIP was recovered. The pilot was interpreted as a success (Koning et al., 1988).

The large-scale field application of polymer flooding did not start until February 2010 (24 years after the pilot test). For this large-scale field application, there are 27 injectors including 20 inverted 9-spot patterns, 4 inverted 5-spot patterns, and 3 patterns with horizontal injectors. The total injection rate in the Marmul field is about 82,000 STB/D. The average injection rate per injector is about 3000 STB/D. The step rate tests showed that the target injection rates can be achieved below fracture pressure. The produced formation water is used in the full-scale polymer flood project.

The project is ongoing. By April 2012, the project has been 2 years. The increase in oil rates, decrease in liquid rate, and decrease in water cut had been observed (Al-Saadi et al., 2012).

3.7.6 Polymer Flooding in a Carbonate Reservoir—Vacuum Field, New Mexico

The Vacuum field in the Grayburg-San Andres formations is located in Lea County, NM. The Grayburg-San Andres are dolomite formations. The porosity ranged from 10.6% to 11.6%, and the permeability ranged from 8.5 to 21 mD. The viscosity was 0.88 mPa·s which was close to water viscosity of 0.87 mPa·s. The reservoir temperature was 100–105°F (Hovendick, 1989).

The field began production in 1939. Water injection was initiated in May 1, 1983. Polymer (HPAM) injection was started just 3 months later in August. Oil production peaked in November and remained relatively flat for the next 22 months before declining. Injection-profile surveys showed that fluid injection profile was uniform, and polymer injection promoted a more efficient sweep. Polymer was injected at a low concentration of 50 ppm. Simulation results showed that a low-concentration polymer flood was just as effective as a higher concentration flood as long as the total amount of polymer injected was the same; and the earlier to start polymer injection, the higher the ultimate oil recovery.

3.8 POST-POLYMER CONFORMANCE CONTROL USING MOVABLE GEL

Cross-linked polymer-like bulk gel used in water shutoff has very poor flowability; the viscosity is very high (>10,000 mPa·s). Uncross-linked polymer is used to increase water viscosity. A movable gel is used in between; it has the intermediate viscosity, and more importantly, it can flow under some pressure gradient. Colloidal dispersion gel (CDG) is a typical gel used in these situations.

CDG is made of low concentrations of polymer and cross-linkers. Cross-linkers are the metals, such as aluminum citrate and chromium. Polymer concentrations range from 100 to 1200 mg/L, normally 400 to 800 mg/L. The ratio of polymer to cross-linkers is 30–60. Sometimes, this type of gel is called a low-concentration cross-linked polymer. In such concentration range, there is not enough polymer to form a continuous network, so a conventional bulk-type gel cannot form. Instead, a solution of separate gel bundles forms, in which a mixture of predominantly intramolecular and minimal intermolecular cross-links connect relatively small numbers of molecules. By contrast, in a bulk gel, the cross-links form a continuous network of polymer molecules, through predominantly intermolecular cross-links. A field case is presented next.

The pilot was within a large-scale polymer flooding area (Bei-Yi-Qu-Duan-Xi) in Daqing. It consisted of 6 injectors and 12 producers forming an inverted 5-spot pattern. The area was 0.75 km^2. The thickness was 15.85 m and the effective thickness was 13.29 m. The permeability was 841 mD (Feng, 2007). The oil viscosity was 9–10 mPa·s. The reservoir temperature was 45°C.

Before polymer flooding, 0.66 PV water had been injected with a recovery factor of 28.5%. The water cut was 88%. Polymer injection was started in January 1993 and ended in April 1997. The viscosity of polymer solution at wellheads was 30–40 mPa·s. When polymer injection was ended, the water cut rose and oil rate fell quickly at low-water wells. One year after the post-polymer water drive, the water cut at most of wells was higher than 94.5%. Therefore, gel injection replaced post-polymer drive on August 27, 1999, and ended July 17, 2001.

The gel formula was 600 mg/L HAPM with 12 million daltons MW and 1% mass aluminum citrate. The ratio of polymer to cross-linker was 30. The designed injection volume was 0.30 PV.

Because injection pressure was noticed to be rising slowly, the ratio of polymer to cross-linker was decreased to 20 and the injection rate was increased in April 2000. To avoid past-water fingering into the gel slug, after 0.283 PV injection on June 15, 2001, the polymer concentration was increased from 600 to 2500 mg/L. The wellhead viscosity was 359 mPa·s, and 0.0037 PV of high concentration slug was injected followed by post-gel water drive.

At the end of gel injection (July 17, 2001), total of 0.2868 PV of gel was injected which contained 473.83 tons of dry polymer and 20.54 tons of cross-link. The average polymer concentration was 630 mg/L and the average crosslink concentration was 27.6 mg/L with the ratio 22.8:1. Wellhead sampling showed that the gel formed in 10 days, and the viscosity maintained 10–20 days, then decreased slowly afterward.

Compared with polymer injection, when gel was injected, the wellhead injection pressure increased by 4.14 MPa, the permeability significantly decreased, the residual permeability reduction factor increased from 1.326 to 1.403, and water intake thickness increased to 19%. The gel injection resulted in 137 tons/D oil rate increase, and 8.9% water cut decrease from the 12 producers. It was expected that 3.18% incremental oil recovery could be obtained.

REFERENCES

Al-Saadi, F., Amri, B., Nofli, S., Wunnik, J., Jaspers, H., Harthi, S., et al., 2012. Polymer flooding in a large field in south Oman—initial results and future plans, Paper SPE 154665 Presented at the SPE EOR Conference at Oil and Gas West Asia, 16–18 April, Muscat, Oman.

Bird, R.B., Armstrong, R.C., Hassager, O., 1987. Dynamics of Polymeric Liquids. vol. 1. Fluid Mechanics, second ed. Wiley, Chichester.

Bondor, P.L., Hirasaki, G.J., Tham, M.J., 1972. Mathematical simulation of polymer flooding in complex reservoirs. SPE J. October, 369–382.

Buckley, S.E., Leverett, M.C., 1942. Mechanism of fluid displacements in sands. Trans. AIME 146, 107–116.

Carreau, P.J., 1972. Rheological equations from molecular network theories. Trans. Soc. Rheol. 16 (1), 99–127 (Ph.D. dissertation, University of Wisconsin, Madison, 1968).

Chen, T.-R., Chen, Z., 2002. Multiphase physicochemical flow in porous media and measurement of relative permeability curves. In: Yu, J.-Y., Song, W.-C., Li, Z.-P. (Eds.), Fundamentals and Advances in Combined Chemical Flooding. China Petrochemical Press, Beijing.

Deng, Z., Tang, J., Xie, F., He, J. 1998. A case of the commercial polymer flooding under the complicated reservoir characteristics, Paper SPE 50007 Presented at the SPE Asia Pacific Oil and Gas Conference and Exhibition, 12–14 October, Perth, Australia.

Feng, Q.H., 2007. Theory and Technology of Deep Conformance Control Post-Polymer Flooding. University of Petroleum (China) Press, Beijing.

Flory, P.J., 1953. Principles of Polymer Chemistry. Cornell University Press, Ithaca, NY.

Hovendick, M.D., 1989. Development and results of the Hale/Mable leases cooperative polymer EOR injection project, vacuum (Grayburg-San Andres) field, Lea County, New Mexico. SPERE 4 (3), 363–372.

Huang, P.-C., Li, D.-M., Xing, H.-B., Li, H.-S., Li, C.-J., Liu, S.-F., 1998a. Prediction of HPAM permeability reduction factor. In: Gang, Q.-L. (Ed.), Chemical Flooding Symposium—Research Results During the Eighth Five-Year Period (1991–1995), vol. 1. Petroleum Industry Press, Beijing, pp. 240–245.

Huang, S., Liu, J.-H., Xie, X.-Q., 1998b. Improved dissolution and maturation unit and its effect on polymer solubility. In: Gang, Q.-L. (Ed.), Chemical Flooding Symposium—Research Results During the Eighth Five-Year Period (1991–1995), vol. 1. Petroleum Industry Press, Beijing, pp. 409–413.

Kang, W.-L., 2001. Study of Chemical Interactions and Drive Mechanisms in Daqing ASP Flooding. Petroleum Industry Press, Beijing.

Koning, E., Mentzer, E., Heemskerk, J., 1988. Evaluation of a pilot polymer flood in the Marmul field, Oman, Paper SPE 18092 Presented at the 63rd Annual Technical Conference and Exhibition, 2–5 October, Houston, TX.

Liang, J.-T., Sun, H., Seright, R.S., 1995. Why do gels reduce water permeability more than oil permeability? SPERE 10 (4), 282–286.

Liu, H., Wang, Y., Liu, Y.Z., 2006. Techniques of polymer solution mixing, transport and injection, and oil production. In: Shen, P.-P., Liu, Y.-Z., Liu, H.-R. (Eds.), Enhanced Oil Recovery—Polymer Flooding. Petroleum Industry Press, Beijing, pp. 157–181.

Luo, J.-H., Bu, R.-Y., Wang, P.-M., Bai, F.-L., Zhang, Y., Yang, J.-B., et al., 2002. Properties of KYPAM, a salinity-resistant polymer used in EOR. Oilfield Chem. 19 (1), 64–67.

Manning, R.K., Pope, G.A., Lake, L.W., Paul, G.W., 1983. A Technical Survey of Polymer Flooding Projects, DOE report under DOE Contract No. DE-AC19-80BC10327.

Manrique, E.J., Muci, V.E., Gurfinkel, M.E. 2007. EOR Field Experiences in Carbonate Reservoirs in the United States, SPEREE 10(6), 667–686.

Moe Soe Let, K.P., Manichand, R.N., Seright, R.S., 2012. Polymer flooding a ~500-cp oil, Paper SPE 154567 Presented at the SPE Improved Oil Recovery Symposium, 14–18 April, Tulsa, OK.

Niu, J.-G., Chen, P., Shao, Z.-B., Wang, D.-M., Sun, G., Li, Y., 2006. Research and development of polymer enhanced oil recovery. In: Cao, H.-Q. (Ed.), Research and Development of Enhanced Oil Recovery in Daqing. Petroleum Industry Press, Beijing, pp. 227–325.

Pang, Z.-W., Li, J.-L., Shi, S.-K., Li, Y., Liu, H.-B., Chen, J.-S., et al., 1998. Effect of polymer molecular weight on residual permeability reduction factor. In: Gang, Q.-L. (Ed.), Chemical Flooding Symposium—Research Results During the Eighth Five-Year Period (1991–1995), vol. 1. Petroleum Industry Press, Beijing, pp. 138–149.

Schneider, F.N., Owens, W.W., 1982. Steady state measurements of relative permeability of polymer/oil systems. SPE J. February, 79–86.

Seright, R.S., 2010. Potential for polymer flooding viscous oils. SPEREE 13 (6), 730–740.

Seright, R.S., Campbell, A.R., Mozley, P.S., Han, P., 2010. Stability of partially hydrolyzed polyacrylamides at elevated temperatures in the absence of divalent cations. SPE J. 15 (2), 341–348.

Sheng, J.J., 2011. Modern Chemical Enhanced Oil Recovery: Theory and Practice. Elsevier, Burlington, MA.

Sheng, J.J., Maini, B.B., Hayes, R.E., Tortike, W.S., 1999. Critical review of foamy oil flow. Transport in Porous Media 35 (2), 157–187.

Sorbie, K.S., 1991. Polymer-Improved Oil Recovery. CRC Press, Boca Raton, FL.

Taber, J.J., Martin, F.D., 1983. Technical screening guides for the enhanced recovery of oil, Paper SPE 12069 Presented at the SPE Annual Technical Conference and Exhibition, 5–8 October, San Francisco, CA.

Taber, J.J., Martin, F.D., Seright, R.S., 1997a. EOR screening criteria revisited—part 1: introduction to screening criteria and enhanced recovery field projects. SPEREE August, 189–198.

Taber, J.J., Martin, F.D., Seright, R.S., 1997b. EOR screening criteria revisited–part 2: applications and impact of oil prices. SPEREE August, 199–205.

Teeuw, D., Rond, D., Martin, J., 1983. Design of a pilot polymer flood in the Marmul field, Oman, Paper SPE 11504 Presented at the SPE Middle East Oil Technical Conference, 14–17 March, Manama, Bahrain.

UTCHEM, 2000. Technical Documentation for UTCHEM-9.0, A Three-Dimensional Chemical Flood Simulator, July, Austin, TX.

Wang, D.M., Cheng, J.-C., Xia, F., Li, Q., Shi, J.P., 2001. Viscous-elastic fluids can mobilize oil remaining after water-flood by force parallel to the oil–water interface, Paper SPE 72123 Presented at the SPE Asia Pacific Improved Oil Recovery Conference, 8–9 October, Kuala Lumpur, Malaysia.

Wang, D.-M., Han, P., Shao, Z., Chen, J., Seright, R.S., 2006a. Sweep improvement options for the Daqing oil field, Paper SPE 99441 Presented at the SPE/DOE Symposium on Improved Oil Recovery, 22–26 April, Tulsa, OK.

Wang, J., Wang, D.-M., Sui, X.G., Zeng, H.-M., Bai, W.-G., 2006b. Combining small well spacing with polymer flooding to improve oil recovery of marginal reservoirs, Paper SPE 96946 Presented at the SPE/DOE Symposium on Improved Oil Recovery, 22–26 April, Tulsa, OK.

Wassmuth, F.R., Green, K., Arnold, W., Cameron, N., 2009. Polymer flood application to improve heavy oil recovery at East Bodo. JCPT 48 (2), 55–61.

Wassmuth, F.R., Green, K., Hodgins, L., Turta, A.T., 2007. Polymer flood technology for heavy oil recovery, Paper CIM 2007-182 Presented at the Canadian International Petroleum Conference, June 12–14, Calgary, AB.

White, J.L., Goddard, J.E., Phillips, H.M., 1973. Use of polymers to control water production in oil wells. JPT February, 143–150.

Willhite, G.P., Dominguez, J.G., 1977. Mechanisms of polymer retention in porous media. In: Shah, D.O., Schechter, R.S. (Eds.), Improved Oil Recovery by Surfactant and Polymer Flooding. Academic Press, New York, NY, pp. 511–554.

Zhang, X., 2011. Application of polymer flooding with high molecular weight and concentration in heterogeneous reservoirs, Paper SPE 144251 Presented at the SPE Enhanced Oil Recovery Conference, 19–21 July, Kuala Lumpur, Malaysia.

Zhang, J-.Y., Yang, P.-H., 1998. HPAM molecular weight compatibility with rock permeability. In: Gang, Q.-L. (Ed.), Chemical Flooding Symposium—Research Results During the Eighth Five-Year Period (1991–1995), vol. 1. Petroleum Industry Press, Beijing, pp. 150–154.

Chapter 4

Polymer Flooding Practice in Daqing

Dongmei Wang
Harold Hamm School of Geology and Geological Engineering, University of North Dakota, Grand Forks, ND 58202, USA

4.1 MECHANISM

Early back to 1994, Jiang et al. (1994) summarized the mechanism of polymer floods based on Daqing's laboratory research (which began in 1960s) and pilot trials (which began in 1970s). Jiang et al. (1994) thought polymer flooding not only improved mobility ratio between oil and water, but also improved the water intake profile. Sections 4.1.1 and 4.1.2 describe the mechanism of polymer flooding from Jiang's view.

4.1.1 Mobility Control

Generally, for a water drive within homogeneous reservoir, an unfavorable mobility ratio often exists because the injected water viscosity is lower than the oil viscosity. This result will induce the fraction of water phase (water cut) during liquid production to rise rapidly. As a consequence, the sweep efficiency will be very low, due to viscous fingering. However, by injecting a viscous polymer solution, the mobility ratio can be improved.

For waterflooding, the fraction of water phase (water cut) after water breakthrough can be expressed as Eq. (4.1):

$$f_w = \frac{\lambda_w}{\lambda o + \lambda w} = \frac{(kk_{rw}/\mu_w)}{(kk_{rw}/\mu_w + kk_{ro}/\mu_o)} = \frac{1}{1 + (\mu_o/\mu_w \cdot k_{rw}/k_{ro})} \tag{4.1}$$

where f_w is the water fraction in liquid production or water cut, λ_o is the oil mobility, λ_w is the water mobility, k is the rock absolute permeability, k_{ro} is the relative permeability in oil phase, k_{rw} is the relative permeability in water phase, μ_o is the oil viscosity, and μ_w is the water viscosity.

Equation (4.1) indicates that the viscosity ratio between oil and water (μ_o/μ_w) strongly affects the rate of water cut (fraction of water phase) increase. In other words, a large viscosity ratio leads to fast water cut

Enhanced Oil Recovery Field Case Studies.

increases even though the water saturation is not high. This result will cause the oil field to terminate its production due to the ultimate water cut. The final oil recovery obtained may be far less than the ultimate displacement efficiency which could be achieved otherwise. In contrast, a lower viscosity ratio between oil and water will delay increases in the water cut. Thus, water cut achieves a given value when the water saturation is higher. Consequently, higher actual oil displacement efficiency can be achieved.

Assume a homogeneous reservoir with an initial oil saturation S_o of 0.8, and a connate water saturation S_{wr} of 0.2. If residual oil saturation is 0.3, the ultimate displacement efficiency E_D will be 62.5%. We also assume the water saturation S_w in a reservoir is 0.52 at water breakthrough. Then, a relationship of oil fraction and water saturation can be described in Table 4.1 based on Eqs. (4.2) and (4.3):

$$k_{rw} = 1.6(S_w - 0.2)^2 \tag{4.2}$$

$$k_{ro} = 0.8 - 1.132(0.8 - S_o)^{0.5} \tag{4.3}$$

Based on Table 4.1, a figure can be plotted of water cut versus average water saturation. As Figure 4.1 shows, if $\mu_o/\mu_w = 15$, the water cut reaches 93.9% at water breakthrough. Between this time and the time when the water cut reaches 98%, the average water saturation only increases from 0.52 to 0.6, and the oil displacement efficiency (E_D) is only 50%. However, if $\mu_o/\mu_w = 1$, the water cut reaches only 50.6% at water breakthrough and the average reservoir water saturation achieves 0.69 at 98% water cut. Thus, the oil displacement efficiency is 61.3% which is higher than that in waterflooding.

4.1.2 Profile Modification

Water intake profile modification by polymer flooding results in the increase in swept volume—providing another mechanism for polymer enhanced oil recovery (EOR). Seright et al. (2003) and Sorbie and Seright (1992) demonstrated that when fluids can freely cross flow between oil strata, the rate of movement of a polymer front is independent of permeability, so long as the reciprocal of the mobility ratio is greater than the permeability contrast between the strata.

TABLE 4.1 Water Fractions Versus Water Saturation

S_w	0.52	0.55	0.58	0.6	0.62	0.65	0.68	0.70
k_{rw}	0.164	0.196	0.231	0.256	0.282	0.324	0.369	0.400
k_{ro}	0.160	0.130	0.100	0.084	0.066	0.041	0.016	0.000
$f_w(\mu_o/\mu_w = 15)$	0.939	0.958	0.972	0.979	0.985	0.992	0.997	1.000
$f_w(\mu_o/\mu_w = 1)$	0.506	0.601	0.698	0.753	0.810	0.888	0.958	1.000

Profile modification by polymer flooding only takes effect in heterogeneous reservoirs. For a waterflood, the injected water usually advances along the displacement front in an uneven fashion between different permeability layers. The waterfront advances rapidly in high permeable layer and slowly in lower permeable layers. Because the water viscosity is usually less than the oil viscosity, the fingering problem will become more severe. If the water has already broken through in a high permeable layer, and the distance of front advancement is still very short in the low permeable layer, oil in the low permeable layers cannot be produced efficiently.

However, for polymer flooding, the injected water viscosity is increased, so that the mobility ratio can be improved significantly, thereby the uneven displacement in different layers is reduced. In other words, with a higher viscous polymer solution, the front advanced distance by water drive in low permeable layer can be expanded even if breakthrough occurred in the high permeable layer. Thus, the profile can be modified to expand the sweep volume. This theory can be demonstrated by the following example calculation.

Assume there are two layers in an oil reservoir with permeabilities of k_1 and k_2, and $k_1/k_2 = 5$. Under the same conditions with Section 4.1.1, by using Eq. (4.4), without considering gravity effect or cross flow, the ratio of water intake rates between the two layers can be obtained before the water breakthrough in the high permeable layer:

$$\frac{q_1}{q_2} = \frac{\lambda_1}{\lambda_2} = \frac{(k_1 k_{\mathrm{rw1}}/\mu_{\mathrm{w}} + k_1 k_{\mathrm{ro1}}/\mu_{\mathrm{o}})}{(k_2 k_{\mathrm{rw2}}/\mu_{\mathrm{w}} + k_2 k_{\mathrm{ro2}}/\mu_{\mathrm{o}})} = \frac{k_1}{k_2} \cdot \frac{(\mu_{\mathrm{o}}/\mu_{\mathrm{w}}(k_{\mathrm{rw1}} + k_{\mathrm{ro1}}))}{(\mu_{\mathrm{o}}/\mu_{\mathrm{w}}(k_{\mathrm{rw2}} + k_{\mathrm{ro2}}))} \tag{4.4}$$

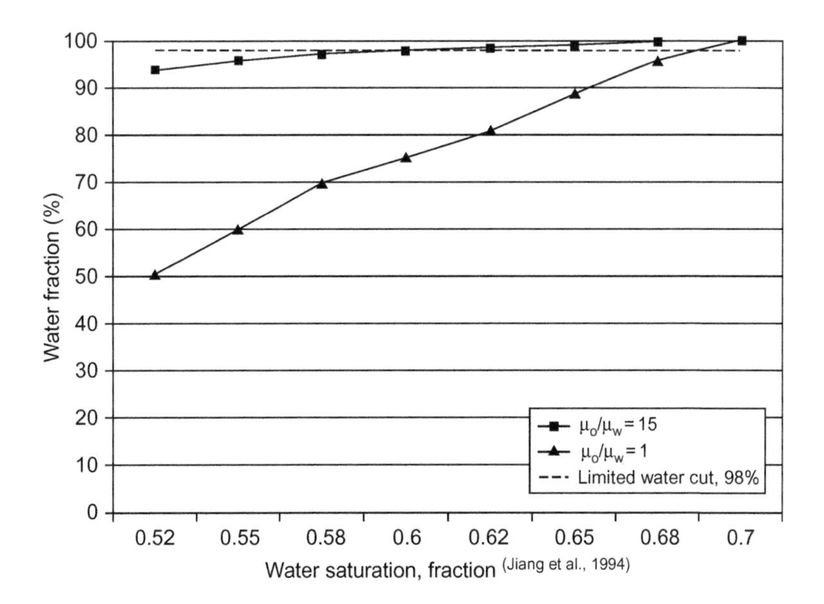

FIGURE 4.1 Water cut changes versus water saturation.

where q_1, q_2 are the instantaneous water intake rates in layers 1 and 2, respectively; λ_1, λ_2 are the fluid mobility in layers 1 and 2, respectively; k_{rw1}, k_{ro1} are the water and oil relative permeability in layer 1, and k_{rw2}, k_{ro2} are the water and oil relative permeability in layer 2, respectively, μ_w, μ_o are the water and oil viscosities, respectively.

Table 4.2 provides water intake ratios between the two permeable layers when the viscosity ratio is 15. As Table 4.2 denoted, in the very beginning of water injection, the water intake in high permeability layer is five times that in low permeability layer. However, when the water saturation in the high permeability layer reaches 0.4, the water intake ratio rises to 10.13. Subsequently, at water breakthrough in layer 1 (high permeability layer), water intake is 21.58 times higher in layer 1 than in layer 2.

In contrast, when the water phase viscosity is increased so that $\mu_o/\mu_w = 1$, the water intake ratio is improved (decreasing from 5.00 to 3.42) as the flood proceeded (Table 4.3).

4.1.3 Microscopic Mechanism

In recent years, a number of literature reports claimed new findings for microscopic oil displacement mechanisms. These theories suggested that polymer flooding enhances oil recovery not only from volumetric sweep improvement but also by increasing the microscopic oil displacement (Wang et al., 2004). According to this mechanism, three types of residual oil remained in the reservoir from microscopic viewpoints: they are named cluster residual oil, insular residual oil, and blind residual oil. Research indicated that polymer solution viscosity can be expressed in three types when it displaces the oil phase: ordinary viscosity, interfacial viscosity, and normal-stress or elongational

TABLE 4.2 Water Intake Ratio with High Viscosity Ratio Presents ($\mu_o/\mu_w = 15$)

S_{w1}	k_{rw1}	k_{ro1}	S_{w2}	k_{rw2}	k_{ro2}	q_1/q_2
0.20	0	0.8	0.2	0	0.8	5.00
0.30	0.016	0.442	0.22	0.001	0.640	5.22
0.35	0.036	0.362	0.23	0.001	0.604	7.29
0.40	0.064	0.294	0.24	0.003	0.574	10.13
0.45	0.100	0.234	0.245	0.003	0.560	14.33
0.52	0.164	0.160	0.250	0.004	0.547	21.58

viscosity. Under the combined effects of these three viscosities, polymer flooding not only enhances the sweep efficiency but also increases the oil displacement efficiency within the swept area. The mechanism can be viewed by the following aspects:

1. Ordinary polymer viscosity raises resistance factor in the reservoir. This viscosity reduces the mobility ratio between oil and water, and it is the major mechanism to displace the remaining unswept oil by waterflooding, as well as cluster residual oil.
2. Polymer interfacial viscosity pulls the insular and film-shape residual oil out from trapped locations.
3. Polymer elongational viscosity results from elasticity of the polymer solution, and through normal-stress action, reduces the residual oil that is trapped in blind locations (dead ends).

4.2 RESERVOIR SCREENING

In the past decades, reservoir screening criteria for polymer flooding were adopted from the 1984 National Petroleum Council report (Bailey, 1984) and revised EOR screening criteria by Taber et al. (1997). However, as the oil price increased but polymer prices remained modest, the reservoir condition screening criteria were modified. Among the reservoir properties, several aspects should be of concern when selecting the reservoir candidate for polymer flooding, such as reservoir type, reservoir temperature, reservoir viscosity, reservoir permeability, and formation water salinity as indicated in Table 4.4. The following sections will introduce these aspects of polymer process screening.

4.2.1 Reservoir Type

So far, most successful polymer projects have been implemented within sandstone reservoirs. The typical example is the large-scale applications of

TABLE 4.3 Water Intake Ratio with low Viscosity Ratio Presents ($\mu_o/\mu_w = 1$)

S_{w1}	k_{rw1}	k_{ro1}	S_{w2}	k_{rw2}	k_{ro2}	q_1/q_2
0.20	0	0.8	0.2	0	0.8	5.00
0.30	0.016	0.0442	0.22	0.001	0.64	3.57
0.35	0.036	0.362	0.23	0.001	0.604	3.29
0.40	0.064	0.294	0.25	0.004	0.574	3.25
0.45	0.100	0.234	0.27	0.008	0.500	3.29
0.52	0.164	0.160	0.29	0.013	0.460	3.42

polymer flooding at Daqing, where 10–12% average incremental oil recoveries were obtained between 1996 and 2010. The typical polymer projects demonstrate that the sandstone reservoir type is still the preferred target for polymer project.

However, current studies for unconventional heavy oil with higher oil viscosity (Seright, 2010; Wassmuth et al., 2009) demonstrated that polymer flooding is a possibility for substantially more viscous oils. Based on the preliminary assessment of Tambaridjo (Suriname) heavy oil (400 cp) polymer flooding pilot test in 2010, an increase in the pattern oil production and water cut reduction was observed after polymer flooding injection (Manichand et al., 2010).

4.2.2 Reservoir Temperature

Key factors affecting polymer stability are oxidative degradation and hydrolysis, followed by precipitation with divalent cations. Polymer degradation becomes more severe as temperature increases, especially above 70°C. Hydrolyzed polyacrylamid (HPAM) can be reasonably stable (viscosity half-life of 8 years at 100°C) if no dissolved oxygen or divalent cations are present (Seright, 2010). For more common conditions (where divalent cations are present) new polymers are becoming available that are more stable at elevated temperatures. However, these polymers tend to be expensive.

4.2.3 Reservoir Permeability

Reservoir permeability is another key factor that affects the propagation of a polymer solution. The effectiveness of a polymer flood is affected

TABLE 4.4 Reservoir Screening for Polymer Flooding

Parameters	Suggested in Literature	Daqing
Dykstra–Parsons coefficient of permeability variation (V_{DP} or V_k)	0.7 ± 0.1	0.5–0.8
Average permeability, mD	>10	100–600
Reservoir temperature, °C	<100	45
Formation water salinity, mg/L	<10,000	3,000–7,000, >10,000 (Shengli)
Makeup water salinity, mg/L	<1,000	60–1,200
Water cut, %	Low water cut	90–96
Oil viscosity, mPa·s (cP)	10–150	6–9
Recovery, %	Low	35–45
Well spacing, m/well pattern	200–300/5-spot	150–300/5-spot

significantly by the polymer MW. A matching relationship should exist between polymer MW and reservoir permeability when the polymer product is selected. That is to say, MW must be small enough so that the polymer can enter and propagate effectively through the reservoir rock. For a given rock permeability and pore throat size, a threshold MW exists, above which polymers exhibit difficulty with propagation. In order to avoid pore blocking by polymer molecules, the ratio of pore throat radius to the root mean square (RMS) radius of gyration of the polymer should be greater than 5 (Chen et al., 2001). Table 4.5 provides core flooding results that match relations between polymer MW and reservoir permeability.

4.2.4 Reservoir Heterogeneity

Reservoir heterogeneity is measured by the dispersion or scatter of permeability values. A homogeneous reservoir has a permeability variation that approaches zero, while an extremely heterogeneous reservoir has a permeability variation that approaches one. The heterogeneity between layers or within layers can be improved by polymer floods. Figure 4.2 (Qi, 1998) denotes the EOR factor versus formation permeability under Daqing reservoir conditions. Also, using Eq. (4.5), for the Daqing reservoir, the EOR factor achieved the peak value when V_k is 0.72. The EOR decreases when V_k is larger or lower than this value. The reason why polymer modification in the high permeability layers becomes less effective is because V_k becomes too high (see Figure 4.2) (Green and Willhite, 1998):

$$V_k = \frac{k_{50} - k_\sigma}{k_{50}} \tag{4.5}$$

where V_k (V_{DP}) is the permeability variation, k_{50} is the permeability value at the 50th percentile, and k_σ is the permeability at the 84.1 percentile.

Also, based on Qi's (1998) studies on permeability variation effects on EOR by polymer flooding at Daqing and Dagang reservoir conditions, the incremental oil recovery depends on the permeability variation with the geological sediment sequence and polymer bank size, as well as the

TABLE 4.5 Core Permeability Versus Polymer MW of Daqing Polymer

MW, Da, 10^6	R_p, μm	k_{water}, mD	R_{50}, μm^2	R_{50}/R_p
8.20	0.162	110	0.83	5.1
11.30	0.220	160	1.39	6.3
17.50	0.245	260	1.24	5.3
28.00	0.312	316	1.59	5.1

Here R_p is the polymer molecule radius of gyration, R_{50} is the median pore radius.

permeability ratio between horizontal and vertical directions of the reservoir. This should be considered by reservoir engineers when they perform reservoir screening.

4.2.5 Oil Viscosity

For an oil field under waterflood, the oil viscosity dominates the mobility ratio between the oil phase and the water phase. However, the mobility ratio can be improved by injecting a viscous polymer solution. Figure 4.3 demonstrates that the ideal oil recovery incremental can be obtained when the reservoir oil viscosity ranges from 10 to 100 mPa·s. The mobility ratio decreased and made favorable when the viscosity of the injected polymer solution was typically 35–40 mPa·s at Daqing's reservoir condition (where the average permeability ranged from 400 to 1000 mD).

4.2.6 Formation Water Salinity

Formation water salinity has a strong effect on polymer viscosity, especially for HPAM. Polymer solution viscosity decreases with salinity. Polymer viscosity is sensitive to the cation content of water solution: Ca^{2+}, Mg^{2+}, Fe^{3+}, etc., far more than K^{+}, Na^{+}. High divalent or trivalent content in the formation water may cause polymer participation. Lower polymer viscosity will lead to poor mobility control by polymer processes. Numerical simulation studies also demonstrated that EOR factors at Daqing decreased from

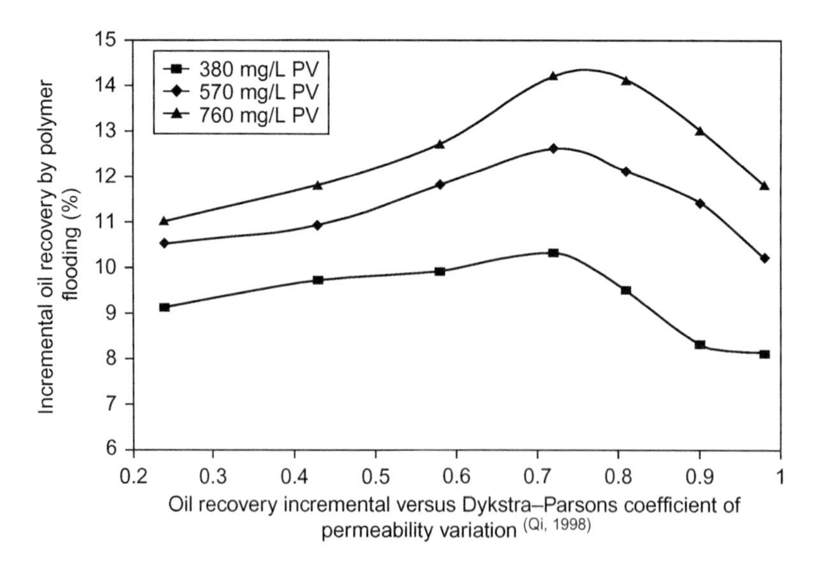

FIGURE 4.2 Permeability variation versus oil recovery.

30% to 50% when salinity increased from 2500 to 10,000 mg/L (Wang et al., 2008a,b).

To avoid the negative effect of high salinity on polymer viscosity, a pre-flushed by low water salinity preceding polymer flooding is suggested. Also, xanthan has high tolerance to salinity compared to HPAM for *small-scale polymer projects*. However, with a lower MW and higher price, this kind of biopolymer might increase the total cost for a large-scale project.

On the other hand, the makeup water for polymer solutions affects the polymer viscosity directly. Higher salinity makeup water leads to a big loss in polymer viscosity. Water quantity control should be taken seriously.

4.3 KEY POINTS OF POLYMER FLOOD DESIGN

If a reservoir meets the screening criteria for polymer flooding, a procedure for designing a polymer project should include steps for detailed lithology characterization and reservoir description, oil strata integration and well pattern design, production analysis and development of the target oil zones, injection sequence optimization, development index prediction, and economical benefit evaluation, as well as project implementation requirements (Wang et al., 2007a,b).

For a polymer project, injection sequence and injection formulation are key design factors. These points include using gel treatments (water shutoff) before polymer injection and zone isolation is of value if severe

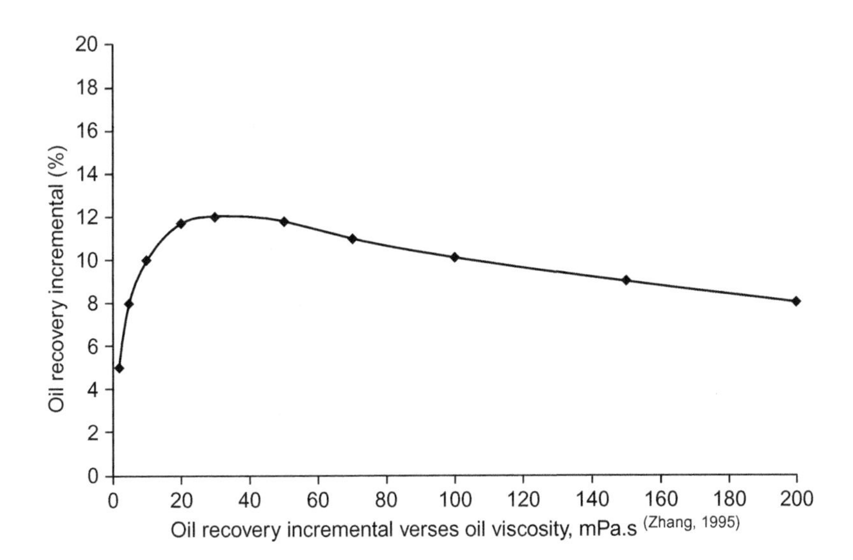

FIGURE 4.3 Oil viscosity versus oil recovery.

heterogeneity (i.e., fractures) exists within the reservoir. On the other hand, the optimal polymer solution viscosity, MW, bank size, concentrations, and injection rate, and well spacing and injection pressure should be considered to obtain the maximum oil recovery.

4.3.1 Well Pattern Design and Combination of Oil Strata

The connectivity factor between injector and producers, permeability differential, and well spacing is very important for the well pattern design. Designs should vary to accommodate the differences in geological properties for diverse parts of the field.

Connectivity Factor

Based on large-scale applications by polymer flooding at Daqing, the effectiveness of polymer flooding is dominated by the connectivity factor: the less interwell continuity, the lower the ultimate oil recovery and incremental oil.

Connectivity factor ($CONF_p$), which also can be taken as controlled degree by polymer flooding, is defined as the pore volume accessed by polymer solution (V_p) divided by the total pore volume of the oil zones (V_t) (Fu et al., 2004) (see Eq. (4.6)):

$$CONF_p = V_p/V_t \tag{4.6}$$

and

$$V_p = \sum_{j=1}^{m}\left[\sum_{i=1}^{n}(S_{pi} \cdot H_{pi} \cdot \phi)\right] \tag{4.7}$$

where $CONF_p$ is the controlled degree by polymer flooding, %; V_p is the pore volume that can be accessed by polymer molecule, m^3; V_t is the total pore volume of reservoir, m^3; S_{pi} is the controlled area by the well pattern at i well group in the j zone, m^2; H_{pi} is the connected net thickness between injectors and producers which can be accessed by the polymer molecule, m; and ϕ is the porosity, fraction.

Numerical simulation indicates that the incremental oil recovery by polymer flooding declines noticeably when the connectivity factor is reduced below 70%. The connectivity factor should be greater than 70% when designing the well pattern and when choosing the polymer MW.

Permeability Differential

When selecting the target oil zones for polymer flooding, oil strata with similar reservoir properties should be combined to promote uniform sweep of the zones and get the maximum oil recovery. To achieve these goals for polymer flooding, permeability differential should be less than 5 for a given set of oil strata while the combined thickness should be at least 5 m (Wang et al., 2009).

Based on the numerical simulation (Wang et al., 2009), the less the permeability differential, the lower water cut obtained. Under the same injection parameters, the lowest water cut of polymer process can be reduced from 69.8% when permeability differential is 5 to 62.8% when permeability differential is 2.

Well Pattern

According to Li and Chen (1995) research, the well pattern has a relatively small effect on the incremental oil recovery by polymer flooding. Table 4.6 provides an EOR comparison of various well patterns based on Li and Chen numerical simulations. The results indicate that the incremental recovery is 10.9% for a line-drive pattern and 10.6% for an inverted 9-spot. For a 5-spot, the incremental oil recovery is 10.3%. However, the injection volume will be three times more for the inverted 9-spot than for the 5-spot—leading to a temptation to inject above the fracture pressure when using the inverted 9-spot pattern. Also, the connectivity factor will be much smaller with a line pattern than with the 5-spot. Therefore, the 5-spot pattern appears to be attractive (Wang et al., 2009).

Well Spacing

Injection pressure and injection rate also need to be considered when choosing well spacing. Eq. (4.8) (based on empirical data at Daqing) indicates that the maximum allowable injection pressure ($p_{\max}$) increases with the square of well spacing (l). Consequently, changing well spacing will result in a change of injection pressure and production time. Based on Eq. (4.8), a smaller well spacing allows a larger injection rate.

$$p_{\max} = \frac{l^2 \cdot \varphi \cdot q}{180 N_{\min}} \tag{4.8}$$

where $p_{\max}$ is the highest allowable wellhead pressure, mPa; l is the distance between injector and producer, m; ϕ is the porosity, %; $N_{\min}$ is the lowest apparent water intake index, $m^3/d\, m\, MPa$; and q is the injection rate, PV/year.

Considered the interwell continuity, the well spacing is suggested to be from 200 to 250 m for oil zones with average permeability above $300-400 \times 10^{-3}\ \mu m^2$ and net pay above 5 m. For oil zones with the average permeability above $100-200 \times 10^{-3}\ \mu m^2$ and the net pay of 1–5, 150–175 m is an ideal well spacing (Wang et al., 2009).

TABLE 4.6 Well Pattern Versus Oil Recovery

Well Pattern	$\Delta\eta$—EOR, %
Line in positive	10.6
Line in diagonal	10.9
5-spot	10.3
4-spot	10.1
9-spot	10.0
Inverted 9-spot	10.6

Parameters: 5 layers, net pay = 12 m, $V_k = 0.70$, $\phi = 0.26$, $k = 101$, 260, 491, 938, and $3207 \times 10^{-3}\ \mu m^2$.

4.3.2 Injection Sequence Options

Profile Modification Before Polymer Injection

Under some circumstances, gel treatments or other types of "profile modification" methods may be of value before implementation of a polymer or chemical flood (Wang et al., 2008a,b). If fractures cause severe channeling, gel treatments can greatly enhance reservoir sweep if applied before injection of large volumes of expensive polymer (Seright et al., 2003; Wang et al., 2002a,b, 2006). Also, if one or more high permeability strata are watered out, there may be considerable value in applying profile modification methods before starting the polymer flooding or other EOR project.

Numerical simulation (Wang et al., 2008a,b) demonstrated that oil recovery can be enhanced 2–4% original oil in place (OOIP) with 0.1 pore volume using profile modification before polymer injection, if layers with no cross flow between layers exists. As expected, the benefits from profile modification decrease if it is implemented toward the middle or late phase of polymer injection (Chen et al., 2004; Trantham et al., 1980).

Based on Daqing field experience with profile modification, candidate wells typically have layers with a high water cut, high water saturation, and a large difference in water intake from layers. Additional criteria used to identify wells that are candidates for profile modification include:

1. The start pressure of polymer injection is lower than the average level for total injectors in the area.
2. The pressure injection index, PI, is lower than the average value in the pilot area (Qiao et al., 2000). PI is defined by Eq. (4.9):

$$\mathrm{PI} = \frac{1}{t}\int_0^t p(t)\mathrm{d}t \tag{4.9}$$

where p(t) is the well pressure after the injector is shut in for time t.

3. Injection pressures are lower than the average level and the water cut at the offset production wells is higher than the average level.

Separate Layer Injection

Based on Daqing's experience, a method is advised to improve this sweep problem when cross flow does not occur. Based on theoretical studies and practical results from Daqing pilot tests (Wu and Chen, 2005), separate layer injection was found to improve flow profiles, reservoir sweep efficiency, and injection rates, and can reduce the water cut in production wells. Numerical simulation studies reveal that the efficiency of polymer flooding depends significantly on the permeability differential between layers and when separate layer injection occurs.

An example based on numerical simulation is provided in Table 4.7, where the permeability differential was 2.5 and flooding occurred until 98% water cut was reached. In this case, the incremental recovery using layer separation was 2.04% more than the case with no layer separation.

Theoretical studies and pilot tests revealed that the conditions which favor separate layer injection at Daqing include (Wu and Chen, 2005): (1) the permeability differential between oil zones ≥2.5; (2) the net pay for the lower permeability oil zones should account for at least 30% of the total net pay; (3) layers should be separated by at least 1 m and should show consistent lateral continuity between wells.

4.3.3 Injection Formulation

Polymer MW

An appropriate polymer product is very important to target oil zones. When a polymer product is selected, it should satisfy the technical requirements for the petroleum industry, including hydrolysis degree, solids content, and MW. Among them, polymer MW is the key parameter that affects polymer flooding effectiveness. Polymers with higher MW provide greater viscosity and leads to high oil recovery. Core flood simulation (Wang et al., 2008a,b) verifies this expectation for the cases of constant polymer slug volume and concentration (Table 4.8).

Based on laboratory tests with a fixed polymer solution, volume injected confirmed that oil recovery increases with increased polymer MW. The reason is simply that for a given polymer concentration, solution viscosity and sweep efficiency increase with increased polymer MW. In other words, to recover a given volume of oil, less polymer is needed using a high MW polymer than a low MW polymer.

Two factors should be considered when choosing the polymer MW. On one hand, choose the polymer with the highest MW practical to minimize the polymer cost. On the other hand, the MW must be small enough so that the polymer can enter and propagate effectively through the reservoir rock. For a given rock permeability and pore throat size, a threshold MW exists, above which polymers exhibit difficulty in propagation. Mechanical entrapment can significantly retard polymer propagation if the pore throat size and permeability are too small. Thus, depending on MW and permeability differential, this effect can reduce sweep efficiency. A trade-off must be made in choosing the highest MW polymer that will not exhibit pore plugging or significant mechanical entrapment in the less permeable zones.

Table 4.9 lists resistance factors (F_r) and residual resistance factors (F_{rr}) for different combinations of polymer MW and core permeability. The reservoir cores used in Table 4.9 were from a large-scale trial (BEX site) at Daqing.

Based on laboratory results and practical experience at Daqing, a medium polymer MW (12–16 million daltons) is applicable for oil zones with the average permeability greater than 0.1 μm^2 and net pay greater than 1 m.

TABLE 4.7 Comparison Between Separate Layer Injection and Without Treatment

Injection Method	Layer	D_{znet}, m	K_{eff}, mD	f_w, %	η_u, %
	1	5	400	98.0	53.36
Separated	2	5	1000	98.0	53.34
	Combined	**10**	**700**	**98.0**	**53.35**
Regular	1	5	400	94.0	45.33
	2	5	1000	99.6	57.29
	Combined	**10**	**700**	**98.0**	**51.31**

η_u: ultimate oil recovery of OOIP, %.

TABLE 4.8 Oil Recovery Versus Polymer MW

MW, Da, 10^6	η_p, %	η_u, %
5.50	10.6	43.3
11.00	17.9	51.8
18.60	22.6	54.8

Total injected polymer mass: 570 mg/L PV; polymer concentration: 1000 mg/L. 3 layers; heterogeneity: $V_k = 0.72$.

A high polymer MW (17–25 million daltons) is appropriate for oil zones with the average permeability greater than 0.4 μm^2 (Wang et al., 2009).

Polymer Solution Viscosity and Concentration

The polymer solution viscosity is a key parameter to improve the mobility ratio between oil and water. As injection viscosity increases, the effectiveness of polymer flooding increases. The viscosity can be affected by a number of factors such as polymer MW, polymer concentration, and degree of HPAM hydrolysis, temperature, salinity, and hardness. When designing the viscosity of polymer flooding project, all of above factors should be considered.

The effectiveness of a polymer flood is directly determined by the magnitude of the polymer viscosity. The viscosity depends on the quality of the water used for dilution. A change in water quality directly affects the polymer solution viscosity (Wu et al., 2007).

For a medium MW HPAM polymer, a relationship between the injection polymer concentration and solution viscosity can be seen in Figure 4.4 for 15 million daltons polymer MW under different formation salinity or TDS (total dissolved solids). These plots can be used during project design for the effective permeability ranges from 0.1 to 0.3 μm^2 if the reservoir temperature is 45°C. The plots were valuable in adjusting polymer concentrations to respond to the change in water quality (salinity). In this application, for a medium MW polymer (12–16 million daltons), 40 mPa·s was recommended. This viscosity level was sufficient to overcome (1) the unfavorable mobility ratio (i.e., 9.4) and (2) permeability differential up to 4:1.

For a high MW polymer (17–25 million daltons) or extra high MW polymer (25–38 million daltons), 50 mPa·s viscosity could be cost-effective. For new polymers that provide special fluid properties, additional laboratory investigations are needed before implementing in a polymer flood.

Polymer concentration determines the polymer solution viscosity and the size of the required polymer slug. The polymer solution concentration dominates every index that changes during the course of polymer flooding.

TABLE 4.9 F_r and F_{rr} for Different k_{air} and MW (Wang et al., 2008a)

MW, Da, 10^6	K_{air}, μm^2	F_r	F_{rr}	Note
	0.498	8.5	3.2	
15.0	0.235	10.1	4.1	
	0.239	7.75	5.0	Plugged
20.0	1.000	27	4.7	
38.0	1.500	53	3.6	

1. Higher polymer concentrations cause greater reductions in water cut and can shorten the time required for polymer flooding. For a certain range, they can also lead to an earlier response in the production wells, a faster decrease in water cut, a greater decrease in water cut, less required pore volumes of polymer, and less required volume of water injected during the overall period of polymer flooding. Table 4.10 provides the effectiveness of polymer flooding as a function of polymer concentration when the injected polymer mass is 640 mg/L PV. As polymer concentration increases, EOR increases and the minimum in water cut during polymer flooding decreases. However, consideration should also be given to the fact that higher concentrations will cause higher injection pressures and lower injectivity. When designing polymer concentrations for a field application, technical feasibility and reservoir conditions should be considered.
2. Using slugs with higher polymer concentration. First, effectiveness can be improved by injecting polymer solutions with higher concentrations during the initial period of polymer flooding. The increase in effectiveness comes from the wells or the units that experienced in-depth vertical sweep improvement during the early stages of polymer flooding. Second, the increase in water cut during the third stage of polymer flooding (i.e., after the minimum in water cut) can be controlled effectively using injection of higher polymer concentrations. Based on the two injection stations where high polymer concentrations were injected in the Daqing

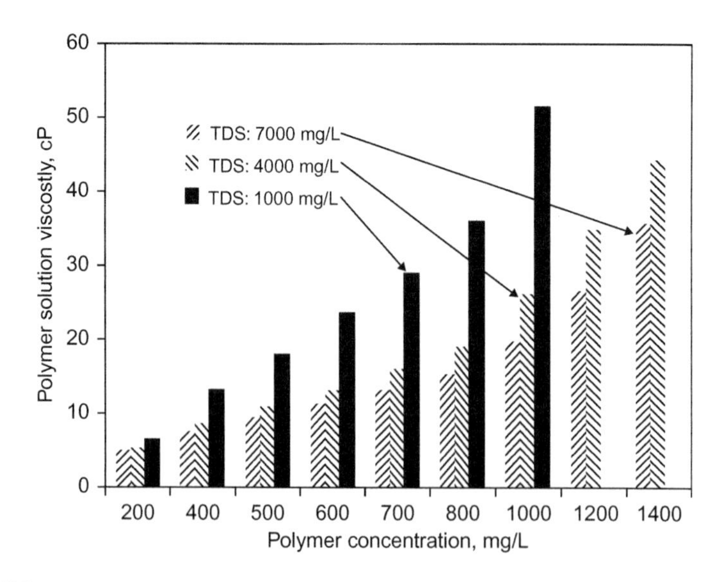

FIGURE 4.4 Relationship between polymer concentration and viscosity.

field, the water intake profile became much more uniform after injecting 2200–2500 mg/L polymer solution in 2004 (Yang et al., 2004).

Polymer Volume

An important mechanism of polymer flooding is to improve the mobility between oil and water and to increase the swept volume. Based on theory (Green and Willhite, 1998; Jiang et al., 1994), oil recovery efficiency decreases with increased mobility of the injected fluid. Consequently to avoid fingering, a continuous polymer flood could be used instead of a waterflood. However, because polymer solutions are more expensive than water, economics limit the volume of polymer that should be injected.

Based on theoretical research and practical experiences (Shao et al., 2001; Wang et al., 2009), the polymer volume should be determined by the gross water cut of the flooding unit. Generally, when the gross water cut achieves 92–94%, the polymer injection should be stopped.

For large polymer banks, polymer was produced from wells after the water cut increased back up to 92%. So, more extended injection of polymer hurts income and economics because the produced polymer is wasted.

Table 4.11 provides the incremental oil (expressed as tons of oil per ton of polymer injected). Based on our economic evaluation, optimum effectiveness can be obtained if a suitable time to end polymer injection is chosen, followed by a water-injection stage. For Daqing, the optimum polymer slug size ranged from 640 to 700 mg/L PV.

To better understand this optimum effectiveness, consider these two points (trade-offs). First, field data revealed that the rate of increase in water cut in Table 4.11 was notably less for the polymer mass of 640 mg/L PV than those for the higher polymer masses (Shao et al., 2001).

Second, numerical simulation and our economic evaluation revealed that when income from the polymer project matched the investment (i.e., the “breakeven point”), the incremental oil was 55 tons of oil per ton of polymer (about 25.5 USD/bbl), and the polymer mass was 750 mg/L PV. Of course,

TABLE 4.10 Ultimate Recovery and EOR Versus Polymer Concentration

Polymer Concentration, mg/L	Lowest Water Cut, %	EOR, %
600	87.1	7.69
800	85.0	9.64
1,000	83.1	9.95
1,200	82.4	10.01
1,500	81.0	10.15

the optimum polymer mass depends on oil price. With the current high oil prices, greater polymer masses could be attractive (Wang et al., 2009).

Injection Rate

The polymer solution injection rate is another key factor in the project design. It determines the oil production rates. Previous researches stated that the magnitude of the injection rate has little effect on the final recovery. It also has a minor effect on the fraction of the injected polymer mass that is ultimately produced (Wang et al., 2009). However, the injection rate has a significant effect on the cumulative production time. Lower injection rates lead to longer production times, higher rates may increase shear degradation of the polymer. So when we design a project, the injection rate should be optimized.

Figure 4.5 shows how reservoir pressure changes with the injection rate after the completion of polymer injection. As expected, the average reservoir pressure near the injectors increases as the injection rate increases while decreasing near production wells. Also, higher injection rates can cause a larger disparity between injection and production. Injection rates must be controlled (i.e., not too high) to minimize polymer flow out of the pattern or out of the target zones.

In summary, the injection rate affects the whole development and effectiveness of polymer flooding. Equation (4.8) can be used to relate the highest pressure at the injection wellhead, the average individual injection rate with the polymer, and the average apparent water intake index for different reservoir conditions. In general, the injection rate should not exceed the reservoir fracture pressure.

To maximize the term of oil production and maximize ultimate production, the injection rate should be maintained from 0.14 to 0.16 PV/year for

TABLE 4.11 Incremental Recovery Versus Polymer Mass

Polymer bank size, mg/L PV	Incremental by Polymer, tons of oil per ton of polymer mass	Ultimate Recovery, %	Rate of Water Cut Increased, %/mg/L PV	Rate of Recovery Increased, %/mg/L PV
524	78	50.74		
640	65	50.93	0.0438	0.0142
681	59	51.20	0.0523	0.0151
760	55	53.26	0.0584	0.0118
855	48	54.28	0.0647	0.0107
950	40	55.10	0.0720	0.0086

250 m well spacing and 0.16–0.20 PV/year for 150 to 175 m well spacing. Injection rates should generally be within these ranges unless special circumstances or reservoir conditions necessitate changes.

4.3.4 Individual Production and Injection Rate Allocation

Injection and production rates in every flooded unit should be properly balanced to achieve optimum sweep. For a polymer flood, this process requires special attention to injection rates and polymer concentrations for individual wells. The following principles should be applied for allocations of production rate and injection rate for individual wells (Wang et al., 2007a,b).

1. For those central wells with high mobile oil saturations, proper balancing of injection and production is needed, often involving an increase in injection rates, to ensure that the timing of oil production coincides with that of other patterns.
2. For wells near a fault, the injection rate and polymer concentration should often be lower than the average design, especially if the injection pressure is higher than the average.
3. For some wells, the reservoir properties may not be favorable. For example, water saturations are too high near line-drive wells associated with the first waterflood pattern at Daqing. Also, oil zones are thin and permeabilities are low in the areas where sediments were deposited by

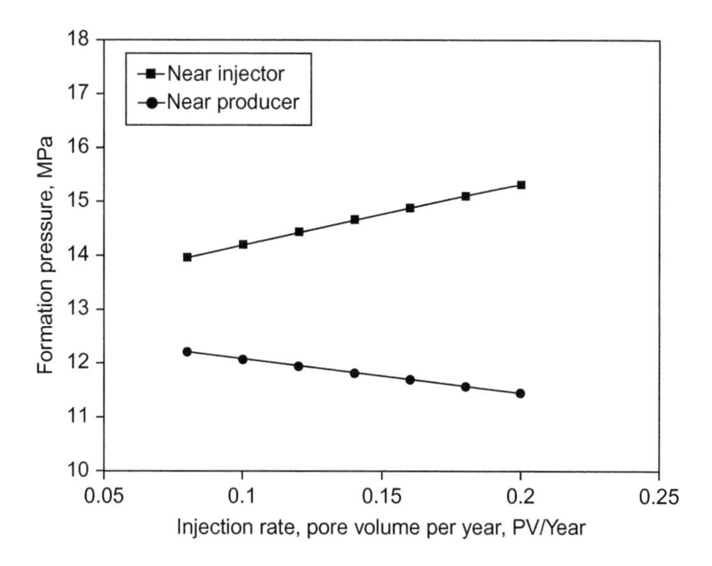

FIGURE 4.5 Injection rate versus formation pressure.

an ancient river. Low permeabilities are also associated with other depositional features.

4. For wells with low injection pressure, high water throughput, a heterogeneous water intake profile, or areas known for channeling, profile modification should be applied before polymer injection.
5. Injection and production rates should be balanced throughout the project area.

Additionally, development prediction for polymer flooding is important for a polymer project design. Basically, three approaches can be used for development index prediction:

1. Prediction by dynamic performance of waterflood before polymer injection or after a period of polymer flooding. This method is relatively accurate for a production index, especially for those after a certain duration of polymer flooding. However, only a few indexes are effective for this method, such as water cut, oil, and water production rates.
2. Fractional flow method. This method is usually applies the reservoir engineering equations to predict 1D or 2D sweep efficiency in an areal well pattern. The disadvantage of this method is that it is not well suited for multilayers with vertical cross flow potential.
3. Numerical simulation. This is most commonly used to perform a polymer project design, using 3D reservoir models. For this method, a clear geological and lithological description is needed, and a highly consistent history match must be available for the waterflood before polymer injection.

4.4 POLYMER FLOODING DYNAMIC PERFORMANCE

Compared with waterflooding, polymer flooding decreases water cut and increases oil production. Dynamic performance primarily refers to changes in water cut, polymer injectivity, and liquid productivity. Water cut and liquid production are the major indexes that are used for evaluation. It is also very important to be aware of these dynamic performance measures for solving problems during polymer injection in the practical oil field application.

4.4.1 Stages and Dynamic Behavior of Polymer Flooding Process

Based on the practical field applications, five stages can be characterized for the entire polymer flooding process (Guo et al., 2002; Liao and Shao, 2004; Shao et al., 2005; Wang et al., 2009) as shown in Figure 4.6.

First stage—Initial stage of polymer flood: In this stage, the water cut has not yet started to decrease. The stage ranges from the very beginning of polymer injection typically to 0.05 PV. During this period, the polymer solution has not begun to work. The injection pressure rises quickly.

Second stage—Response stage: In this stage, a decrease in water cut can be seen. It typically occurs from 0.05 to 0.20 PV of polymer injection. During this period, the polymer solution penetrates deep into the formation of medium and higher permeable layers and pore throats, and forms the oil bank. The polymer front advances further, and the displacement of oil and water is improved. Typically, about 15% of the EOR is produced during this stage.

Third stage—Stable stage: In this stage, the water cut change is relatively stable. The minimum water cut was observed during this period. This stage typically lasts from 0.20 to 0.40 PV of polymer injection. During this time, the polymer solution penetrates deep into the formation of lower and medium permeability layers; the injection profile is improved compared with the second stage. The oil production rate reaches its peak value, and about 40% to the total EOR is produced during this stage. Oil production begins to decrease and the produced polymer concentration begins to increase.

Fourth stage—Water cut increases. This stage typically lasts from 0.40 to 0.70 PV of polymer injection. Areal sweep reaches its maximum, oil production declines, and the produced polymer concentration and the injection pressure follow steady trends. About 30% of the total EOR is produced during this stage.

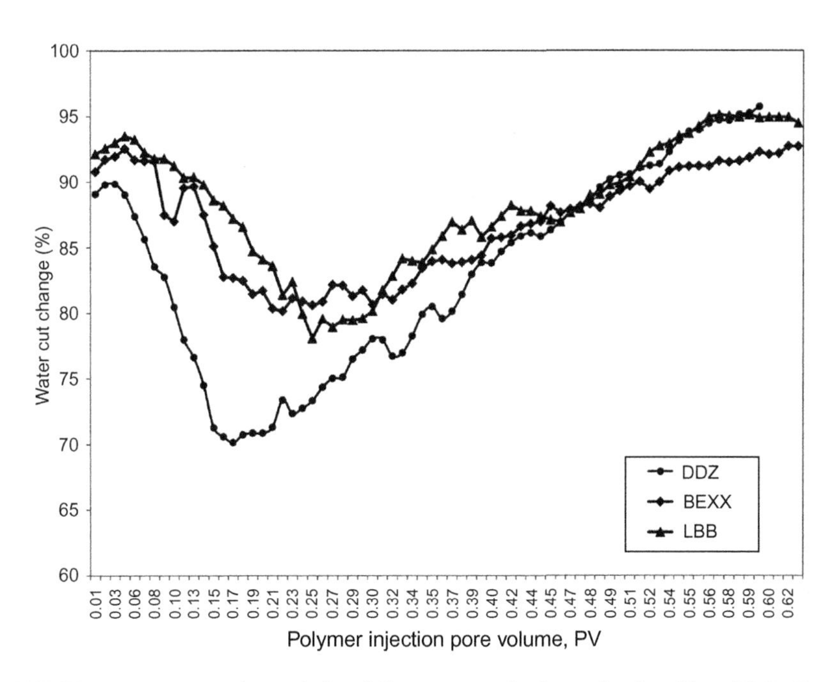

FIGURE 4.6 Water cut change during different stages of polymer flooding. Here, DDZ= The area of Duandong zhong; BEXX= The area of Beisanxixi; LBB= Lamadian Beibu.

Fifth stage—Follow-up water drive. This stage lasts from the end of polymer injection to the point where water cut reaches 98%. Water cut increases continually, and the fluid production capability increases slightly. The EOR produced during this stage is around 10–12% of the total.

4.4.2 Problems and Treatments During Different Phases

Although sweep volume can be improved by polymer flooding compared with waterflooding, inefficient polymer application can occur when implementing a polymer project. For example:

1. A small range of injection pressure may occur, along with less liquid production at the initial stage of polymer injection (when water cut is still increasing).
2. Poor connectivity between oil strata, poor injectivity, large decreases in liquid production, and large pressure differences may occur in production wells when water cut decreases or is stable. High flow pressures may occur within production wells, as well as unresponsive wells at the corners or edges of patterns.
3. Differences exist between producers for the polymer volume injected and the change in water cut. This leads to asymmetry responses in oil production wells, and, to rapid water cut increases for some oil wells.
4. Differences in polymer volume injected may occur if injection is switched back to water at different times.

Three aspects may be responsible for the above issues, besides the heterogeneity and well pattern. First, oil saturation before polymer injection may be low in the low permeability layers due to serious interference between layers. Second, water intake profile reversal may occur during polymer injection. Sweep may be poor within thick high permeability zones (Wang et al., 2007a,b). Third, large well-to-well variations in water cut may occur at the late stage of polymer flooding.

According to Daqing experience on more than 40 large-scale polymer applications in the last decade, analysis suggested a need to focus on the issues occurred in different stages of polymer flooding, as illustrated in Table 4.12. Numerical simulations and practical applications demonstrate that these actions or measurements work effectively for solving certain problems.

4.5 SURFACE FACILITIES

Surface facilities including polymer hydration and mixing, transportation and injection, and treatment of produced fluids are important during polymer flooding (Lu et al., 2007). This section focuses on two topics: polymer solution mixing (including selection of the water source and polymer solution makeup), injection and produced water treatment (including oily water treatment).

TABLE 4.12 Issues and Solutions During Polymer Flooding

Stage	Issues	Solutions
Initial polymer injection	Big permeability differential High permeable layer or high water out layer exist Low injection pressure in injection well High water cut in production well	Separate layer injection Preslug injection with high polymer MW Profile modification in-depth
Middle of polymer injection	Low injection pressure in injection well Poor injection profile improvement in low permeable layer Large decline range on liquid production Poor response in production wells Layers interference	Injection parameters adjustment Injection system adjustment Hydraulic fracture in production well Shut off in wells without separate layer injection Hydraulic fracture to stimulate injectors
Late polymer injection	Poor injection profile improvement High injection pressure Fast rate on water cut increases in production wells High polymer concentration production Big difference of polymer volume among various zones	Injection parameters adjustment Injection system adjustment Hydraulic fracture in production well Hydraulic fracture to stimulate injectors Shut off high permeable layers
Water follow-up	Big difference on water cut among zones Big difference on polymer concentration production Difference exists in the increase rate of water cut	Shut off high water cut layer in oil well Injection rate adjustment Shut in the ultrahigh water cut well Injection water cyclic Switch to injection water in different time

4.5.1 Mixing and Injection

Figure 4.7 provides an illustration of polymer mixing and injection work stations. During these processes, chemical stability, mechanical degradation, and biological degradation affect polymers and polymer flood performance. To maintain *chemical stability* of polymer, good water quality, an effective protective package (chelating agent), stainless steel pipeline, and nonmetal tanks are very necessary for the water used in mixing the polymer solution. The influence of dissolved salts, such as Ca^{2+}, Mg^{2+}, and Fe^{2+}, are particularly important because their presence lessens the effectiveness of the viscosifying agent. To maintain *mechanical stability* of polymer, pipeline flow rates should be sufficiently low, and devices, such as electrical-magnetic flow meters, mixers, valves, pumps, and filters, should not allow high shear or high-pressure gradients that degrade polymer molecules. To prevent *biological degradation* of polymer, use a protective package (bactericide), such as formaldehyde. This is typically solved using a biocide preflush and an ongoing biocide injection. It should be noted, however, that biocides are prone to adsorption by reservoir rock and dissolution in the oil. Both of these effects reduce the effectiveness of biocides.

Based on Daqing's experience, most viscosity loss occurred from the high-pressure injection pumps and mixing system to the near-wellbore—amounting to about 70% of the total loss before 1996 (Zhang, 1995). However, as the facilities function improvement and advanced technologies, the viscosity loss has already decreased to 50–60%. Consistent with the other work (Seright, 1983), the greatest restriction to flow and the greatest mechanical degradation occurred from the entrance to the porous rock at the high velocities on injection sand face. Concerning this problem, necessary measurements should be adopted in the process of injection to reduce the viscosity loss.

4.5.2 Produced Water Treatment

Figure 4.8 shows an illustration of fluid production treatment during polymer flooding. Produced water treatment from polymer flooding is not only related to produced water utilization but also to environment protection.

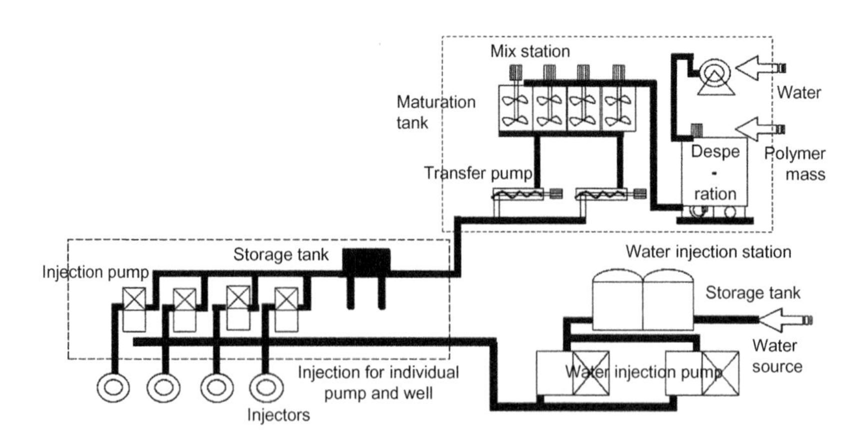

FIGURE 4.7 Flow chart of mixing and injection.

Since some polymer exists in the produced water, the viscosity of the produced water will be increased and make it difficult to separate oil and water. Based on Daqing's experience, three treatment processes can be applied: (1) natural settling by gravity, (2) flocculation settling, and (3) pressure boosting pump.

After being treated, produced water from polymer flooding and dehydration stations can be reused and reinjected in new well patterns. For the locations where the produced water quantity cannot satisfy the injection requirements, underground water and surface water are treated to reach the quality required (Liu et al., 2006; Xia et al., 2001).

According to Daqing criteria, some specifications should be followed during water treatment. First, for gravity settling tanks, the oil content should not be greater than 1000 mg/L in the inlet water, and 30–50% of the oil content and suspended solids should be eliminated in the outlet stream. Second, coagulation settling tanks are used after the gravity settling stage (secondary settling tanks). The objective is to raise the oil–water density difference, to increase oil drop floatation speed and the suspended solid settling speed, and to reduce settling time and promote oil removal efficiency (by adding coagulant into the tank). After this stage, the oil content and suspended solid content should be less than 50 mg/L within the outlet water.

4.6 A FIELD CASE

A large-scale application by polymer flooding—BSXX—is discussed in this section. After nearly 30 years of waterflooding, polymer flooding began in October 1998. A 5-spot well pattern (with 250 m well spacing) was adopted for most of the area, although a small line-drive part was included; 149 total wells were used for the polymer process, including 71 injectors and 78 producers; 20 wells existed before the project.

4.6.1 Well Pattern and Oil Strata Combination

BSXX has a fluvial delta-lacustrine facies with wide distribution of large thickness and high permeability as described in the following paragraph.

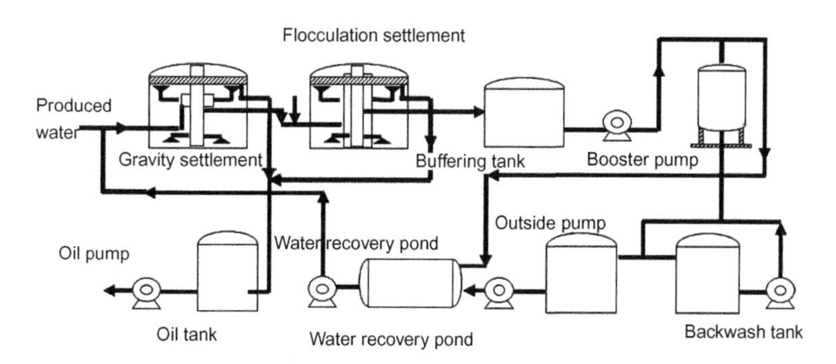

FIGURE 4.8 Flow chart of produced water treatment.

Among the vertical oil zones, PI1-4 was the most developed, with a similar thickness throughout the entire area. In these oil zones, 88.3% of the interval belongs to the rock particles deposited in a "fining upward" sequence. This sedimentary property was favorable for polymer flooding. However, severe heterogeneity exists throughout the layers. Four oil zones were identified for polymer flooding: PI1, PI2, PI3, and PI4.

Additionally, seven faults were developed in the north–south direction. The largest fault distance was 1.4 m, with dip degree of 48°. Among the seven faults, four had a large effect on polymer flooding. The average thickness of the target oil zone was 16.8 m, with 12.8 m of net pay. Permeability ranged from 74 to 1200 mD. Before the polymer process was applied in PI1-4, the average water cut in this area was around 90%, with a number of wells were over 95%. Oil recovery was around 30%. Figure 4.9 provides the well pattern design for polymer flooding with 5-spot and 250 m well spacing. According to the reserve and oil-bearing area of each individual well, the estimated interwell continuity is 72%.

4.6.2 Polymer Injection Case Design

Polymer Injection Formulations Design

Based on core flooding experiments and the matching theory about polymer MW with formation permeability, the optimum polymer MW was determined for PI1-4 oil zones of BSXX. Polymer injection formulation including viscosity, concentration, polymer bank size, and injection rate were

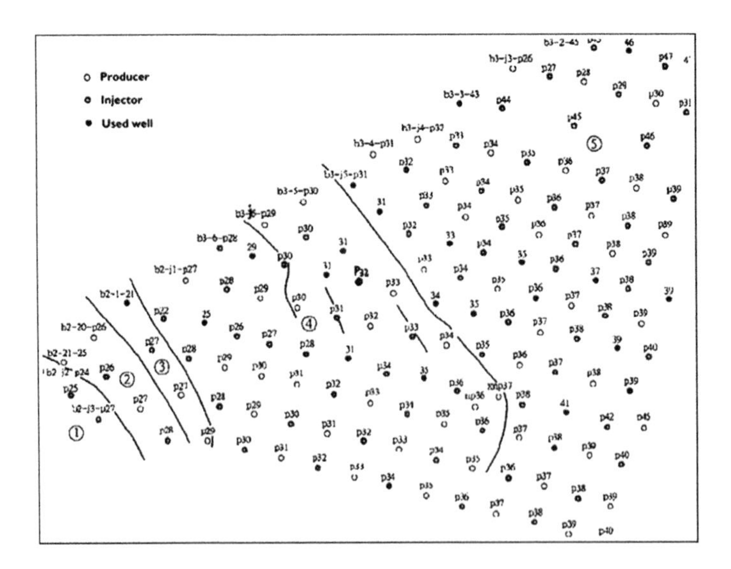

FIGURE 4.9 Well pattern design for polymer flooding at BSXX.

optimized by numerical simulation and analolog with similar pilot tests. Each injection parameter for individual units and wells was adjusted according to the well location and practical geological properties. Table 4.13 provides an average polymer injection formulation for the target oil zones of polymer process employed in numerical simulation.

Individual Well Injection and Production Volume Design

Based on injection and production allocation principles, considering the injection ability of the surface facility, a key point is stated in this section: at a given injection rate, the injection volume is allocated for each unit (the injection volume of four injectors is equal to the production volume of the one producer). Also, considering injectivity for individual wells within a given fault block, 22 wells were treated with gel or other types of "profile modification" due to the severe channeling before polymer injection. The injection volume for entire 71 injection wells was designed at 10,110 m^3 per day, and the average individual injector was 142.39 m^3 per day.

Considering the balance of entire area, each region (five regions divided by faults), as well as the each pattern (four injectors and one producer of 5-spot) between injection and production, the production volume for 78 injection wells were designed to be 10,110 m^3 per day, the average individual producer was 129.62 m^3 (111.47 ton) per day.

4.6.3 Polymer Performance Prediction

Numerical Simulation Model

Based on a 3D reservoir model, a numerical simulation model for polymer flooding prediction used grid blocks with total nodes of $63 \times 54 \times 4 = 13{,}608$. Additionally, based on the fault distributions, five subregions were divided into sealed areas. The purpose was to obtain more accurate data for individual well allocation and development performance prediction. The numbers of injectors and produces and other physical–chemical parameters were also from the field data and laboratory experiments. Here, the numerical simulator employed was VIP-POLYMER™ from Landmark.

TABLE 4.13 Polymer Injection Parameter Design

Polymer Parameter	Result
MW, 10^6 Da	12.0
Bank size, mg/L · PV	570
Concentration, mg/L	800–1000
Injection rate, PV/year	0.14–0.15

History Matching of Waterflooding Before Polymer Injection Using the New Well Patterns

Basically, the history matching involved the date from the very beginning of the target oil field development to the date before polymer injection, using the old well patterns. During this phase, the period of waterflooding simulation is very important using the new well patterns for the polymer flood. Based on the practical oil reserve, oil-bearing area and total well numbers, the numerical simulation model was built with grid blocks of total nodes: $63 \times 54 \times 4 = 13{,}608$, as mentioned earlier. Before polymer injection, the oil production and water cut history match were simulated for the entire area and for each region, and each waterflooding pattern. The duration of simulation for the waterflood was from January 1998 to September 1998 (before polymer injection). By history simulation, the reserve of target oil zones was 1367.13×10^4 tons, with a relative error of 0.01% compared with practical estimates. The pore volume was simulated as 24.76×10^6 m^3, with relative error of 0.1%. The simulated cumulative oil production was 1367.13×10^4 tons, with relative error of 1.21%. Also the simulated water cut of the entire area was 90.79%, compared to the actual value of 90.63%. All of the relative errors of the simulated parameters were not more than 2%.

Performance Prediction for Waterflooding

The waterflooding performance was history-matched using the new well patterns for the polymer flood. According to the numerical simulation results, the water cut in the area BSXX reached 98% after injection 1.612 PV of water. At that time, the predicted recovery (as of October 1998) was 8.50% OOIP and cumulative oil production was 108.07×10^4 tons. The ultimate oil recovery (for continued waterflooding) was 40.22%.

Performance Prediction for Polymer Flooding

Polymer flooding simulation was also performed for the new well pattern. According to the numerical simulation results, the water cut of the area BSXX reached 98% after injecting 1.113 PV of polymer. By that time (98% of water cut), the predicted oil recovery was 19.74% OOIP, with cumulative oil production of 261.58×10^4 tons. The ultimate oil recovery was projected to be 51.46%. The lowest predicted water cut value was noted to be 76.43% at 0.310 PV of polymer solution injected. Polymer injection was stopped at 0.57 PV injection.

Compared with waterflooding, the incremental oil recovery was 11.24% OOIP or 153.51×10^4 tons of cumulative oil. The oil incremental per ton mass of polymer was 108.9 tons. The amount of water saved was 0.653 PV.

Simulation predicted that a polymer flood would provide a financial internal rate of return (FIRR) of 17.81%, a financial net pay alue (FNPV) of 107,705.89 × 10^4 RMB (130,230.69 × 10^4 USD, based on exchange rates in 1998), and the investment period for pay back was 4.16 years after taxes. The results indicated that polymer project is feasible and profitable.

4.6.4 Polymer Performance Evaluation

By the end of October 2002, the polymer project had been active for 4 years with the cumulative oil production of 330.2 × 10^4 tons, and was highly profitable. Compared with the numerical simulation prediction of 76.43% for the lowest water cut predicted to occur at 0.273 PV) closely matched the actual lowest water cut of 76.12% (observed at 0.256 PV). The 87% water cut that was predicted for October 2002 closely matched the actual water cut of 86.31% at that time. Although the response time deviated slightly from the numerical simulation, the basic trends were consistent with the prediction (Figure 4.10) (Wang, 2007).

4.7 CONCLUSIONS

For a sandstone reservoir with low temperature, low viscosity, low formation water salinity and low content of divalent ions, relatively high oil saturation remaining in the reservoir after waterflooding, and suitable reservoir

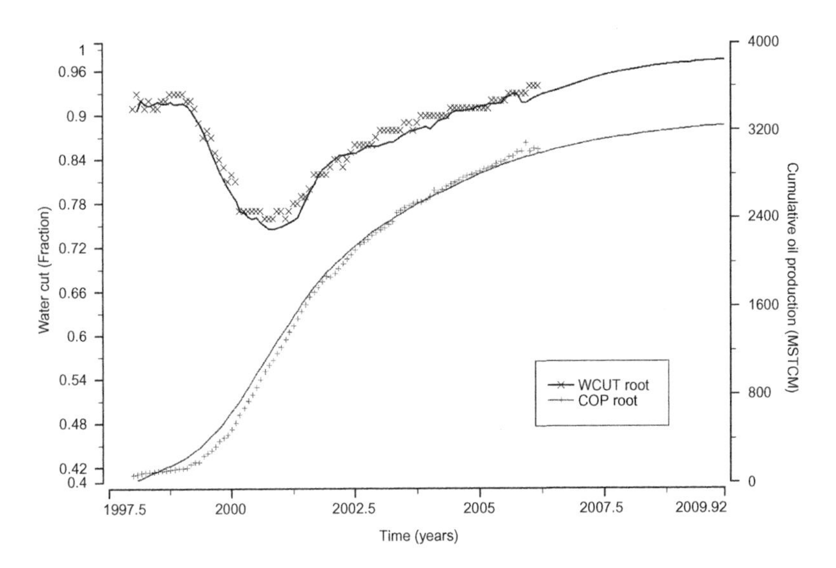

FIGURE 4.10 Water cut and oil production by polymer flooding at BSXX.

heterogeneity (the Dykstra–Parsons coefficient of permeability variation ranged from 0.4 to 0.7), the following results can be highlighted:

1. Favorable reservoir conditions for mobility improvement by polymer flooding using high polymer MWs and large bank sizes lead to a large incremental oil recovery.
2. To achieve an effective polymer flood, the well pattern design and the combination of oil strata must be optimized. To obtain the best benefit, the connectivity factor should be above 70%, and the permeabilty differential should not be greater than 5 in a single unit of flooded zones.
3. Using profile modification in the higher permeability layers and separate layer injection for wells with significant permeability differential between layers and no cross flow, oil recovery can be enhanced 2–4% OOIP over polymer flooding alone.
4. Economics and injectivity behavior can favor changing the polymer MW and polymer concentration during the course of injecting the polymer slug. Polymers with MWs from 12 to 38 million daltons were supplied to meet the requirements for different reservoir geological conditions. The optimum polymer injection volume varied around 0.57–0.7 PV, depending on the water cut (92–94%) in the different flooding units. The average polymer concentration can be designed about 1000 mg/L, but for an individual injection station, it could be 2000 mg/L or more.
5. Understanding the dynamic performance during the entire polymer flooding process can help with problem treatment during different stages of polymer flooding.
6. The selection of water source for making up polymer solution and the produced water treatment should follow certain criteria to assure the effectiveness of polymer flooding.

NOMENCLATURE

$\mathbf{CONF_p}$ controlled degree by polymer flooding, %
$\mathbf{D_{znet}}$ net zone thickness, m
$\mathbf{E_D}$ oil displacement efficiency
$\mathbf{f_w}$ water fraction in liquid production
$\mathbf{F_r}$ resistance factor
$\mathbf{F_{rr}}$ residual resistance factor (permeability before/after polymer placement)
H_{pi} interwell continuity of net pay between injectors and producers that can be accessed by the polymer molecules, m
k_{50} permeability value at the 50th percentile, mD
$\mathbf{k_{air}}$ permeability to air, mD
k_d permeable ratio between zones, fraction
$\mathbf{k_{eff}}$ effective permeability, mD

$\mathbf{k_{water}}$ permeability to water, mD
$k_{ó}$ permeability at the 84.1 percentile, mD
k rock absolute permeability
k_{ro} relative permeability in oil phase
k_{rw} relative permeability in water phase
$\mathbf{k_{rw1}}$ water relative permeability in layer 1
$\mathbf{k_{rw2}}$ water relative permeability in layer 2
$\mathbf{k_{ro1}}$ oil relative permeability in layer 1
$\mathbf{k_{ro2}}$ oil relative permeability in layer 2
l distance between injector and producer, m
$\mathbf{N_{min}}$ minimum apparent water intake index, m^3/d m MPa (injection volume per day per net pay divided by pressure drop)
$\mathbf{N_p}$ reserve of target oil field, m^3
$\mathbf{p_{max}}$ highest wellhead pressure, mPa
p(t) well pressure after the injector is shut in for time t
P pressure, mPa
PI pressure index for an injector, mPa
PV injcted pore volumes, fraction
$\mathbf{\Delta p}$ pressure difference from wellbore to formation, mPa
$\mathbf{q_1}$ instantaneous water intake volume in layer 1
$\mathbf{q_2}$ instantaneous water intake volume in layer 2
q injection or production rate, PV/year
$\mathbf{Q_t}$ oil production at time t, ton
$\mathbf{r_f}$ resistant factor, fraction
$\mathbf{r_{ff}}$ residual resistant factor, fraction
R_p polymer molecule radius of gyration, μm^2
R_{50} median pore radius, μm^2
S_p amount of polymer retention, ppm
S_{pi} interwell continuity of areas of well patterns i in the oil zone j, m^2
$\mathbf{S_w}$ water saturation
$\mathbf{S_o}$ oil saturation
$\mathbf{S_{cw}}$ connate water saturation
$\mathbf{S_{ro}}$ residual oil saturation
t time, min
V_{DP} Dykstra–Parsons coefficient of permeability variation, fraction
V_k Dykstra–Parsons coefficient of permeability variation, fraction
V_p pore volume that can be accessed by polymer molecules, m^3
V_t total pore volume of the reservoir, m^3
$\Delta\eta$ oil recovery incremental between polymer flood and waterflood, %
η_p oil recovery factor by polymer flood, %
η_w oil recovery factor by waterflood, %
η_u oil recovery factor by waterflood, %
ϕ porosity, fraction
λ_1 fluid mobility in layer 1
λ_2 fluid mobility in layer 2
λ_o oil mobility
λ_w water mobility

μ_o oil viscosity
μ_w water viscosity

ABBREVIATIONS

EOR enhanced oil recovery
HPAM hydrolyzed polyacrylamid
MW molecular weight
RMS root mean square
PF polymer flooding
WF waterflooding

SI METRIC CONVERSION FACTORS

cp × 1.0*	E − 03 = Pa·s
bbl × 7.31	E + 00 = ton
ft × 3.048*	E − 01 = m
in. × 2.54*	E + 00 = cm
mD × 9.869 · 233	E − 04 = μm^2
psi × 6.894 · 757	E + 00 = kPa
ppm × 1.0	E + 00 = mg/L

*Conversion is exact.

REFERENCES

Bailey, R.E., 1984. Enhanced Oil Recovery, NPC, Industry Advisory Committee to the US Secretary of Energy Washington, DC.

Chen, J.C., Wang, D.M., Wu, J.Z., 2001. Optimum on molecular weight of polymer for oil displacement. Acta Petrolei Sinica 21 (1), 103–106.

Chen, F.M., Niu, J.G., Chen, P., Wang, J.Y., 2004. Summarization on the technology of modification profile in-depth in Daqing. J. Pet. Geol. Oilfield Dev. Daqing 23 (5), 97–99.

Fu, T.Y., Cao, F., Shao, Z.B., 2004. Calculation method of connectivity factor for polymer flooding. J. Pet. Geol. Oilfield Dev. Daqing, 23 (3), 81–82.

Green, D.W., Willhite, G.P., 1998. Enhanced Oil Recovery, Spe Text Book Series, vol. 6, 92.

Guo, W.K., Cheng, J.C., Liao, G.Z., 2002. The current situation on EOR technique and development trend in Daqing. J. Pet. Geol. Oilfield Dev. Daqing 21 (3), 1–6.

Jiang, Y.L., Ji, P., Han, P.H., Yang, J.C., Zhang, L.X., 1994. Polymer Flooding Optimization, vol. 12. Petroleum Industry Press, Beijing, 3–5.

Li, R.Z., Chen, F.M., 1995. Reasonable well pattern and spacing for polymer flooding in Daqing, Yearly Report 1995.

Liao Niu, J.G., Shao, Z.B., 2004. The effectiveness and evaluation for industrialized sites by polymer flooding in Daqing. J. Pet. Geol. Oilfield Dev. Daqing 23 (1), 48–51.

Liu, H., Tang, S.S., Li, X.S., Zeng, L., 2006. Techniques of Re-Injecting 100% of Produced Water in Daqing Oilfield, Paper SPE 100986 Presented at International Oil and Gas Conference and Exhibition in China, 5–7 December, Beijing, China.

Lu, L., Gao, B.Y., Yue, Q.Y., 2007. Flocculation of waste water produced in polymer flooding. Chin. J. Environ. Sci. 28 (4), 761–765.

Manichand, R., Mogoiión, J.L., Bergwijn, S., 2010. Preliminary assessment of Tambaridjo heavy oilfield polymer flooding pilot test, Paper SPE 138728 Presented at SPE Latin American and Caribbean Petroleum Engineering Conference, 1–3 December, Lima, Peru.

Qi, L.Q., 1998. Numerical simulation study on polymer flooding engineering. Pet. Ind. Press1–5. ISBN 7-5021-2430-6.

Qiao, E.W., Li, Y.K., Liu, P.L., Kuang, Y.J., 2000. Application of PI decision technique in PuCheng oilfield. Drill. Prod. 23 (5), 25–28.

Seright, R.S., 1983. The effects of mechanical degradation and viscoelastic behavior on injectivity of polyacrylamide solutions. SPEJ 23 (3), 475–485.

Seright, R.S., 2010. Potential for polymer flooding reservoirs for viscous oil, Paper SPE 129899 Presented at SPE Improved Oil Recovery Symposium, 24–28 April, Tulsa, OK.

Seright, R.S., Lane, R.H., Sydansk, R.D., 2003. A strategy for attacking excess water production. SPEPF 18 (3), 158–169.

Shao, Z.B., Fu, T.Y., Wang, D.M., 2001. The determinate method of reasonable polymer volume. J. Pet. Geol. Oilfield Dev. Daqing 20 (2), 60–62.

Shao, Z.B., Chen, P., Wang, D.M., 2005. Study of the dynamic rules for polymer flooding in industrial sites in Daqing. Thesis Collect. EOR Technol. 12, 1–8.

Sorbie, K.S., Seright, R.S., 1992. Gel placement in heterogeneous systems with cross flow, Paper SPE 24192 Presented at SPE/DOE Symposium on Enhanced Oil Recovery, 22–24 April, Tulsa, OK.

Taber, J.J., Martin, F.D., Seright, R.S., 1997. EOR screen criteria revisited part 1: Introduction to screen criteria and enhanced recovery field projects. SPEREE 12 (3), 189–198.

Trantham, J.C., Threlkeld, C.B., Patternson, H.L., 1980. Reservoir description for a surfactant/polymer pilot in a fractured, oil-wet reservoir—north Burbank unit tract 97. JPT, 1647–1656.

Wang, D.M., 2007. Reservoir engineering analysis during polymer flooding and effectiveness improvement study, PhD dissertation, Beijing.

Wang, D.M., Chen, J.C., Wu, J.Z., Wang, G., 2002a. Experience learned after production of more than 300 million barrels of oil by polymer flooding in Daqing oil field, Paper SPE 77693 Presented at SPE Annual Technical Conference and Exhibition, 29 September–2 October, San Antonio, TX.

Wang, D.M., Qian, J., Gu, G.S., Xu, X.P., 2002b. The development project design of polymer flooding for eastern in Sazhong in Daqing. Yearly Report 12, 34–35.

Wang, D.M., Jiang, Y.L., Wang, Y., 2004. Viscous-elastic polymer fluids rheology and its effect upon production equipment. SPE 77496-PA. SPE Prod. Facility, 209–216.

Wang, D.M., Han, P.H., Shao, Z.B., Seright, R.S., 2006. Sweep improvement options for the Daqing Oil Field, Paper SPE 99441 Presented at the SPE/DOE Symposium on Improved Oil Recovery, 22–26 April Tulsa, OK.

Wang, D.M., Gao, S.L., Sun, H.L., Zhang, J.X., Ma, M.R., 2007a. Industrial criteria, technical requirement of development project design for polymer flooding, SY/T 6683-2007. Pet. Ind. Press, 155021.6077.

Wang, D.M., Han, D.K., Hou, W.H., Cao, R.B., Wu, L.J., 2007b. The types and changing laws on the profile reversal during the period of polymer flooding. J. Pet. Geol. Oilfield Dev. Daqing 26 (4), 96–99.

Wang, D.M., Seright, R.S., Shao, Z.B., Wang, J.M., 2008a. Key aspects of project design for polymer flooding. SPEREE 11 (6), 1117–1124.

Wang, D.M., Han, P.H., Shao, Z.B., Seright, R.S., 2008b. Sweep improvement options for the Daqing oil field. SPEREE 11 (1), 18–26.

Wang, D.M., Dong, H.Z., Lv, C.S., 2009. Review of practical experience by polymer flooding at Daqing. SPEREE 12 (3), 470–476.

Wassmuth, F.R., Arnold, W., Green, K., 2009. Polymer flooding application to improve heavy oil recovery at East Bodo. JCPT 48 (2), 55–61 (PSC:09-02-55).

Wu, L.J., Chen, P., 2005. Study of injection parameters for separate layers during the period of polymer flooding. J. Pet. Geol. Oilfield Dev. Daqing 24 (4), 75–77.

Wu, W.X., Wang, D.M., Jiang, H.F., 2007. Effect of the visco-elasticity of displacing fluids on the relationship of capillary number and displacement efficiency in weak oil-wet cores, Paper SPE 109228 Presented at SPE Asia Pacific Oil and Gas Conference and Exhibition, 30 October–1 November, Jakarta, Indonesia.

Xia, F.J., Zhang, B.L., Deng, S.B., 2001. Study on disposal process for produced water of polymer flooding. J. Env. Prot. Oil/Gas Field 11 (3).

Yang, F.L., Wang, D.M., Yang, X.Z., Chen, Q.H., Zhang, L., 2004. High concentration polymer flooding is successful, Paper SPE 88454 Presented at SPE Asia Pacific Oil and Gas Conference and Exhibition, 18–20 October, Perth, Australia.

Zhang, J.C., 1995. M. Tertiary Recovery. Petroleum Industry Press, Beijing, 23–24.

Chapter 5

Surfactant–Polymer Flooding

James J. Sheng
Bob L. Herd Department of Petroleum Engineering, Texas Tech University, Lubbock, TX 79409, USA

5.1 INTRODUCTION

Without polymer in the surfactant slug, the surfactant will finger into the oil bank and the reservoir sweep will be very poor, and the surfactant causes the water relative permeability to increase. This increase must be counterbalanced by decreasing the aqueous mobility with polymer. Furthermore, the polymer in both the surfactant slug and the drive slug helps mitigate the effects of permeability variation and improves the overall sweep efficiency in the reservoir. Therefore, including polymer in a surfactant slug is almost compulsory for maintaining a favorable mobility ratio. In reality, we hardly implemented surfactant flooding alone without adding polymer. Therefore, this chapter discusses about surfactant–polymer (SP) flooding as a combined chemical enhanced oil recovery (EOR) method. The fundamentals of polymer flooding have been discussed in the preceding polymer chapters. This chapter only presents the fundamentals of surfactant flooding. Field cases of SP flooding are presented.

5.2 SURFACTANTS

The term "surfactant" is a blend of surface-acting agents. Surfactants are usually organic compounds that are amphiphilic, meaning they are composed of a hydrocarbon chain (hydrophobic group, the "tail") and a polar hydrophilic group (the "head"). As a result, they are soluble in both an organic solvent and water. They adsorb on or concentrate at a surface or fluid–fluid interface to alter the surface properties significantly; in particular, they reduce surface tension or interfacial tension (IFT). Surfactants may be classified according to the ionic nature of the head group as anionic, cationic, nonionic, and zwitterionic. Anionics are most widely used in chemical EOR processes because they exhibit relatively low adsorption on sandstone rocks whose surface charge is negative. Nonionics primarily serve as cosurfactants to improve system phase behavior. Although they are more tolerant of high salinity, their function to reduce IFT is not as good as anionics. Quite often, a mixture of anionics and nonionics is used to increase the tolerance to salinity. Cationics can strongly adsorb in the sandstone rocks, therefore they are generally not used in sandstone reservoirs. Instead, they can be used in carbonate rocks to change wettability. Anionics can also change wettability, and cationics are more expensive than anionics. Therefore, cationics are not as widely used as anionics.

5.2.1 Parameters to Characterize Surfactants

The parameters to characterize surfactants include hydrophilic-lipophilic balance (HLB) (Griffin, 1949, 1954), critical micelle concentration (CMC), Krafft point, solubilization ratio, R-ratio (Bourrel and Schechter, 1988), and packing number. The most used parameters are CMC and solubilization ratio which are introduced next.

CMC is defined as the concentration of a surfactant above which micelles are spontaneously formed. Upon introduction of surfactants (or any surface active materials) into the system, they will initially partition into the interface, reducing the system free energy by (a) lowering the energy of the interface and (b) removing the hydrophobic parts of the surfactants from contacts with water. Subsequently, when the surface coverage by the surfactants increases and the surface free energy (surface tension) has decreased, the surfactants start aggregating into micelles, thus again decreasing the system-free energy by decreasing the contact area of hydrophobic parts of the surfactants with water. Upon reaching CMC, any further addition of surfactants will just increase the number of micelles (in the ideal case), as shown in Figure 5.1, schematic distribution of surfactant molecules in solution at the concentrations below and above CMC. Before reaching the CMC, the surface tension decreases sharply with the concentration of the surfactant. After reaching the CMC, the surface tension stays more or less constant. For a given system,

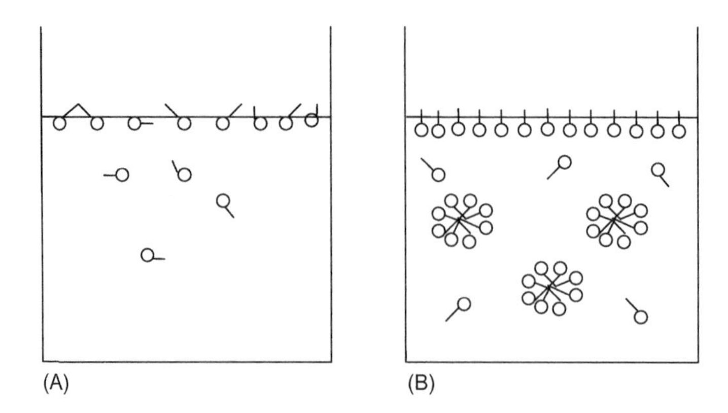

FIGURE 5.1 Distribution of surfactant molecules in solution (A) below and (B) above CMC.

micellization occurs over a narrow concentration range. This concentration is small, about 10^{-5}–10^{-4} mol/L for surfactants used in EOR (Green and Willhite, 1998). In other words, CMC is in the range of a few to tens of parts per million. There is a misconception that a higher surfactant concentration will result in a lower IFT. It is clear by looking at Figure 5.1B that more surfactant molecules will be dissolved inside the bulk liquid and cannot further adsorb at the interface when the concentration is above CMC. Therefore, the IFT cannot be further reduced.

One parameter which relates CMC is Krafft temperature. The Krafft temperature, or critical micelle temperature, is the minimum temperature at which surfactants form micelles. Below the Krafft temperature, micelles cannot form.

Solubilization ratio for oil (water) is defined as the ratio of the solubilized oil (water) volume to the surfactant volume in the microemulsion phase. Solubilization ratio is closely related to IFT, as formulated by Huh (1979). When the solubilization ratio for oil is equal to that for water, the IFT reaches its minimum.

5.3 TYPES OF MICROEMULSIONS

Surfactant solution phase behavior is strongly affected by the salinity of brine. In general, increasing the salinity of brine decreases the solubility of anionic surfactant in the brine. The surfactant is driven out of the brine as the electrolyte concentration is increased. Figure 5.2 shows that as the salinity is increased, the surfactant moves from the aqueous phase to the oleic phase. At a low salinity, the typical surfactant exhibits good aqueous-phase solubility. The oil phase is essentially free of surfactant. Some oil is solubilized in the cores of micelles. The system has two phases: an excess oil phase and a water-external microemulsion phase. Because the microemulsion is aqueous and is denser than the oil phase, it resides below the oil phase and is called a lower-phase microemulsion. At a high salinity, the system separates into an oil-external microemulsion and an excess water phase. In this case, the microemulsion is called an upper-phase microemulsion. At some intermediate range of salinities, the system could have three phases: an excess oil phase, a microemulsion phase, and an excess water phase. In this case, the microemulsion phase resides in the middle and is called a middle-phase microemulsion (Healy et al., 1976). Other names to describe these three types of microemulsions are presented in Figure 5.2.

Surfactant–brine–oil phase behavior is conventionally illustrated on a ternary diagram as shown in Figure 5.2. *If* the top apex of the ternary diagram represents the surfactant pseudocomponent, the lower left represents water, and the lower right represents oil, the tie lines within the lower microemulsion environment have negative slopes. Therefore, the phase environment is called type II(−) because there are two phases in the system and the

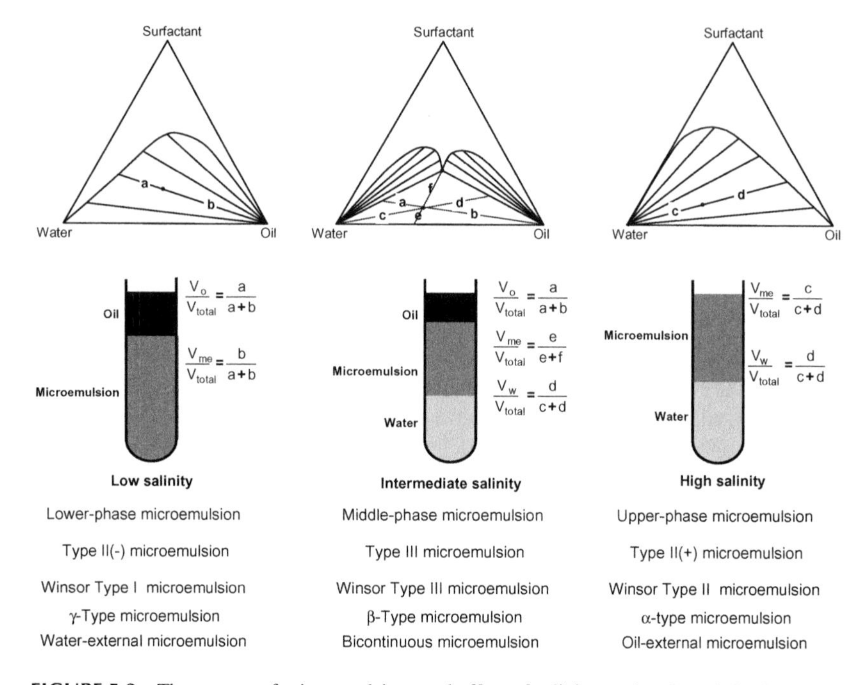

FIGURE 5.2 Three types of microemulsions and effect of salinity on the phase behavior.

slopes of tie lines are negative. Similarly, type II(+) and type III are used to describe the upper- and middle-phase environments, respectively (Nelson and Pope, 1978). However, if the apex representations are changed, for example, if the water and oil positions are exchanged, the original representations of type II(−) and type II(+) will be changed. Winsor (1954) used the names of type I, II, and III microemulsions. Fleming et al. (1978) used γ, β, and α.

5.4 PHASE BEHAVIOR TESTS

Phase behavior tests are conducted in small tubes called pipettes. Therefore, the phase behavior tests, sometimes, are called pipette tests. The phase behavior tests include aqueous stability test, salinity scan, and oil scan. The main objective of phase behavior tests is to find the chemical formula for a specific application.

Any precipitate, liquid crystal, or a second liquid phase can lead to nonuniform distribution of injected material and nonuniform transport owing to phase trapping or different mobilities of coexisting phases. Therefore, we need to first check whether the surfactant solution with cosolvent(s) is transparent without adding oil. The solution should be transparent (clear) up to or higher than the salinity at which you intend to inject the solution. If the solution is clear up to this salinity, problems mentioned above would not appear, because the solution will be more stable after meeting with the oil *in situ*. If the solution is hazy or there is any precipitation, chemicals must be reselected. Such test is called aqueous stability test.

If the solution is clear, change the salinity until a maximum solubility ratio, V_o/V_s, is reached. The corresponding salinity is the optimum salinity for the selected surfactant system. If the solubility ratio is greater than 10, the IFT between the surfactant solution and oil would be in the order of 10^{-3} mN/m, according to the Huh equation, as we will discuss it later. If this surfactant solution is stable, we may use this solution for coreflood tests. If it is less than 10, we may have to reselect the surfactants and/or the cosolvents to repeat the tests.

As discussed earlier, the microemulsion changes from type II(−) to type III to type II(+), as the salinity is increased. Such test is called salinity scan. Generally, the water–oil ratio (WOR) in salinity scan is one or a fixed value.

5.5 INTERFACIAL TENSION

Healy et al. (1976) observed that a large number of anionic surfactant systems exhibited good correlations between IFT and solubilization parameter. Based on this observation, Huh developed a theoretical relationship between the solubilization parameter and IFT for a middle-phase microemulsion (type III). His equations are

$$\sigma_{\mathrm{mw}} = \frac{C_{\mathrm{Hw}}}{\left(V_{\mathrm{wm}}/V_{\mathrm{sm}}\right)^2} \tag{5.1}$$

$$\sigma_{\mathrm{mo}} = \frac{C_{\mathrm{Ho}}}{\left(V_{\mathrm{om}}/V_{\mathrm{sm}}\right)^2} \tag{5.2}$$

where V_{sm}, V_{wm}, and V_{om} are the surfactant volume, water volume, and oil volume in the microemulsion phase, C_{Hw} and C_{Ho} are empirical constants determined experimentally, mN/m. In practice, we use the same value for C_{Hw} and C_{Ho} which ranges from 0.1 to 0.35, and a typical value may be 0.3 if experimental data are available. Using the Huh equations, we can quickly estimate IFT without having to physically measure IFT for screening purpose. In other words, by simply observing the phase volumes in pipettes rather than measuring IFT for each pipette is really an advantage. Many factors affect the IFT. We always have to conduct screening tests to select surfactants and other chemical concentrations for a specific crude oil in practice.

5.6 VISCOSITY OF MICROEMULSION

UTCHEM (2000) is the name of a chemical flooding simulator developed by University of Texas at Austin. In UTCHEM, liquid phase viscosities are modeled as a function of pure component viscosities and the phase concentrations of the organic, water, and surfactant:

$$\mu_j = C_{1j}\mu_w \exp[\alpha_1(C_{2j} + C_{3j})] + C_{2j}\mu_o \exp[\alpha_2(C_{1j} + C_{3j})] + C_{3j}\alpha_3 \exp[\alpha_4 C_{1j} + \alpha_5 C_{2j}] \tag{5.3}$$

where $j = 1$ for aqueous phase, 2 for oleic phase, and 3 for microemulsion phase. The α parameters are determined by matching laboratory microemulsion viscosities at several compositions. In the absence of surfactant and polymer, aqueous and oleic phase viscosities reduce to pure water and oil viscosities (μ_w and μ_o), respectively. When polymer is present, μ_w is replaced by polymer viscosity (μ_p).

5.7 CAPILLARY NUMBER

Analysis of the pore-doublet model yields the following dimensionless grouping of parameters (Moore and Slobod 1955), which is a ratio of the viscous to capillary force:

$$N_C = \frac{F_v}{F_c} = \frac{v\mu}{\sigma \cos \theta} \tag{5.4}$$

where F_v and F_c are viscous and capillary forces, respectively, v is the pore flow velocity of the *displacing* fluid in their derivation, μ is the *displacing* fluid viscosity, and σ is the IFT between the displacing and displaced phases. The dimensionless group is called capillary number, N_C. There are several variations of Eq. (5.4). The flow velocity in the equation may be replaced by the Darcy velocity, u; and $\cos \theta$ may be dropped. Based on the analysis of limited experimental data, Sheng (2011) proposed that the following definition should be used:

$$N_C = \frac{k\left|\vec{\nabla}\Phi_p\right|}{\sigma} \tag{5.5}$$

where Φ_p is the potential of displacing fluid.

Similar to the capillary number, the concept of trapping number has been introduced, which includes the gravity effect. The problem is that different formulas have been developed and presented in the literature. A further investigation is needed to clarify the difference (Sheng, 2011). Therefore, we will not discuss it further in this chapter.

5.8 CAPILLARY DESATURATION CURVE

Many experimental data show that as the capillary number is increased, the residual saturation will be reduced. The general relationship between the residual saturation and a local capillary number is called capillary desaturation curve (CDC). In UTCHEM, the form of Eq. (5.6) is used:

$$S_{\text{pr}} = S_{\text{pr}}^{(N_{\text{C}})_{\max}} + \left(S_{\text{pr}}^{(N_{\text{C}})_{\text{C}}} - S_{\text{pr}}^{(N_{\text{C}})_{\max}}\right)\frac{1}{1 + T_{\text{p}}N_{\text{C}}} \tag{5.6}$$

where S_{pr} is the phase residual saturation, the subscript p denotes the phase which could be water, oil, or microemulsion, the superscript $(N_{\text{C}})_{\text{C}}$ and $(N_{\text{C}})_{\max}$ means at a critical capillary number and a maximum desaturation capillary number, N_{C} is the capillary number, and T_{p} is the parameter used to fitting the laboratory measurements. Note that the definition of capillary number used in the above equation must be the same as that used in the simulation model. One example of CDC using Eq. (5.6) is shown in Figure 5.3. In this figure, the normalized saturations ($S_{\text{pr}}/S_{\text{pr,max}}$) are presented. If several points of residual saturation versus capillary number are measured in laboratory, we can use these data to fit Eq. (5.6). Note that the microemulsion CDC curve lies in the right, the water CDC in the middle, and the oil CDC in the left. In this case, the microemulsion is the most wetting phase, the oil is most nonwetting phase, and the water is in between.

5.9 RELATIVE PERMEABILITY

In surfactant-related processes, the IFT is reduced. As IFT is reduced, the capillary number is increased which leads to reduced residual saturations. Obviously, residual saturation reduction directly changes relative permeabilities. A number of authors reported their research results, as reviewed by

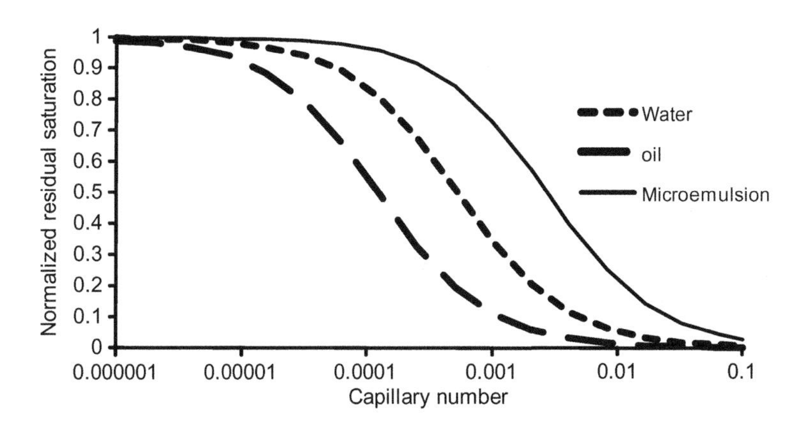

FIGURE 5.3 Example of CDC.

Amaefule and Handy (1982) and by Cinar et al. (2007). The general observations are that the relative permeabilities tend to increase and have less curvature as the IFT decreases or the capillary number increases.

5.10 SURFACTANT RETENTION

Surfactant retention in reservoirs depends on surfactant type, surfactant equivalent weight, surfactant concentration, rock minerals, clay content, temperature, pH, redox condition, flow rate of the solution, etc. Surfactant retention can be broken down into precipitation, adsorption, and phase trapping based on mechanisms. However, it is difficult to separate the surfactant loss from each mechanism. Therefore, we usually report surfactant retention as the total surfactant loss, unless particularly specified.

When we introduced phase behavior tests earlier, we mentioned aqueous stability tests. The main objective of aqueous stability tests is to eliminate surfactant precipitation problem. The surfactant solubility depends on salinity, concentration, etc. During aqueous stability tests, the surfactant solution becomes opaque, up to some salinity, showing the surfactant start to aggregate or even precipitate. When divalent or multivalent ions exist in the solution, the salinity to start precipitation is much lower. If the surfactant concentration is increased, the solution will become opaque. For some surfactants, when the concentration is further increased, the solution becomes clear again. Further increasing surfactant concentration, the precipitation occurs again. In other words, there is a mechanism of precipitation–dissolution–reprecipitation.

Adsorption of surfactant on reservoir rock can be determined by static tests (batch equilibrium tests on crushed core grains) and dynamic tests (coreflood) in laboratory.

Surfactant adsorption is strongly affected by the redox condition of the system. Laboratory cores typically have been exposed to oxygen and are in an aerobic state. It is possible that those data are higher than those in reservoir conditions. One important factor which could reduce surfactant adsorption is pH, which is an important mechanism in alkaline surfactant flooding.

Surfactant adsorption isotherms are very complex in general. The amount adsorbed generally increases with surfactant concentration in the solution, and it reaches a plateau at some sufficiently large surfactant concentration. It has been found that a Langmuir-type isotherm can be used to capture the essential features of the adsorption isotherm:

$$\hat{C}_3 = \frac{a_3 C_3}{1 + b_3 C_3} \tag{5.7}$$

where C_3 is the equilibrium concentration in the solution system. a_3 and b_3 are empirical constants. The unit of b_3 is the reciprocal of the unit of C_3. a_3

is dimensionless. Note that C_3 and $\hat{C}_3$ must be in the same unit. a_3 is defined as

$$a_3 = (a_{31} + a_{32}C_{se})\left(\frac{k_{ref}}{k}\right)^{0.5} \quad (5.8)$$

where a_{31} and a_{32} are fitting parameters, C_{se} is the effective salinity, k is the permeability, and k_{ref} is the reference permeability of the rock used in the laboratory measurement. Adsorption is considered to be irreversible with concentration but reversible with salinity.

Surfactant phase trapping could be due to mechanical trapping, phase partitioning, and hydrodynamic trapping. It is related to multiphase flow. The mechanisms are complex, and the magnitude of surfactant loss due to phase trapping could be quite different depending on multiphase flow conditions. Phase trapping is related to types of microemulsion. In Winsor II microemulsion, the IFT in the rear of the microemulsion slug could be high; and chase water is an aqueous phase whereas the Winsor II microemulsion is oil-external phase whose viscosity could be higher than the chase water. Thus, the chase water can easily bypass the microemulsion phase, resulting in phase trapping. It has been observed that surfactant phase trapping is much lower in Winsor I environment where the microemulsion is water-external phase which can be displaced miscibly by the chase water.

5.11 SP INTERACTIONS

SP interactions and compatibility are summarized as follows.

1. Surfactant can stay in aqueous phase, oleic phase, or middle microemulsion phase. However, essentially all the polymer in an SP solution stays in the most aqueous phase, no matter where the surfactant is (Nelson, 1981; Szabo, 1979).
2. Little difference is observed in the IFT values with and without polymer.
3. Practically, surfactant does not significantly change the viscosity of hydrolyzed polyacrylamide (HPAM).
4. Generally, polymer will flow ahead of surfactant owing to the polymer inaccessible pore volume (PV). The sites available for surfactant adsorption will be reduced. If a surfactant slug is injected ahead of a polymer slug, some of the adsorption sites may be covered by surfactant molecules. Thus, polymer adsorption will be reduced. This is called competitive adsorption.
5. Apparently, experimental data (Chen and Pu, 2006; Gogarty, 1983a,b; Murtada and Marx, 1982) show that the polymer preflush improved the vertical conformance of the surfactant solution so that recovery was increased.

5.12 DISPLACEMENT MECHANISMS

The key mechanism for surfactant flooding is the improved displacement efficiency due to the ultralow IFT effect. Ultralow IFT results in a high capillary number which leads to a low residual oil saturation. In surfactant flooding, emulsions, either oil in water (O/W) or water in oil (W/O), form owing to low IFT. These emulsion drops coalesce to form an oil bank ahead of the surfactant front. This oil bank becomes movable. The key mechanism of polymer flooding is improved sweep efficiency. Thus, the key mechanism of SP flooding is the synergy of those mechanisms.

5.13 SCREENING CRITERIA

The main criteria for SP flooding are salinity and temperature which dictate the conditions surfactants and polymers can be used. The formation of water chlorides should be less than 20,000 ppm and divalent ions should be less than 500 ppm; the reservoir temperature should be less than 93°C; the oil viscosity should be less than 35 cP; the permeability should be greater than 10 mD (Taber et al., 1997). These values are not universally agreed. Different criteria have been proposed. For example, the low-permeability limit should be 50 mD, and the oil viscosity could be higher. Most of the SP projects conducted in United States used 5–20 acres spacing in a confined, inverted five-spot pattern; and positive economics could be obtained for 80–100 acres, if the remaining oil in place is greater than one million barrels (DeBons et al., 2004).

5.14 FIELD PERFORMANCE DATA

Figure 5.4 shows the correlation of the oil recovery as percent of waterflooding remaining oil versus surfactant quantity used in 11 SP field pilots in the United States. The surfactant quantity is the product of injection PV in fraction and surfactant concentration in percent. The surfactants included cosurfactants and cosolvents if they were used. Note that the surfactant quantity used in the field projects was up to 1, whereas it was about 0.1–0.12 in the recent 10 Chinese alkaline surfactant–polymer (ASP) projects (Sheng, 2011). Therefore, the surfactant quantity used in modern chemical EOR projects is about one-tenth of that used before. Another important observation in Figure 5.4 is that the recovery efficiency at 0.5 surfactant quantity in the field pilots is about half that in the laboratory. This is probably that the sweep efficiency in the laboratory is higher than that in the field pilots.

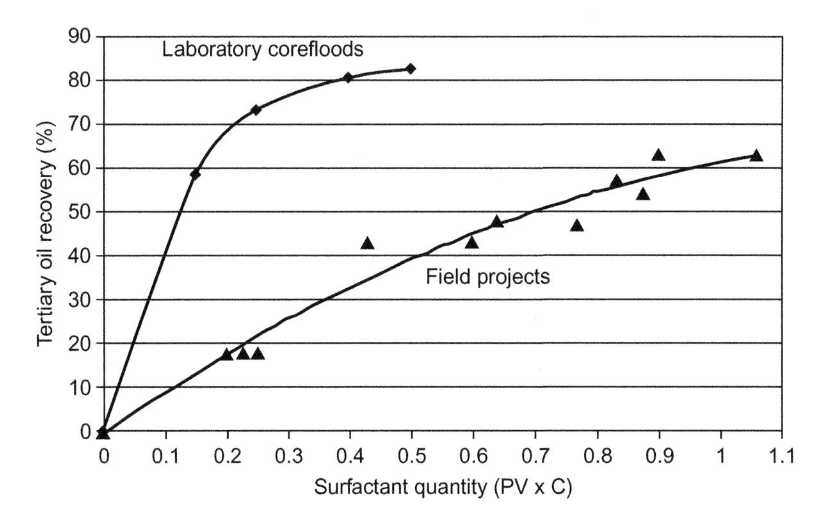

FIGURE. 5.4 Laboratory and successful field project recovery efficiency versus surfactant quantity (DeBons et al., 2004).

5.15 FIELD CASES

In modern chemical EOR, most of the projects are polymer flooding and ASP flooding. Few SP flooding projects have been carried out. More SP projects presented in this section were conducted before 1990. The field projects presented include low-tension waterflooding (Loma Novia, Wichita County Regular field), sequential micellar/polymer (M/P) flooding (El Dorado, Sloss), M/P flooding (Torchlight and Delaware-Childers), Minas SP project preparation and SP flooding (Gudong).

5.15.1 Loma Novia Field Low-Tension Waterflooding

In the mid-1960s, Mobil began a low-tension field test in a watered-out portion of the Loma Novia field, Duval County, Texas (Foster, 1973). On a 5-acre pattern, the well array consisted of four injectors, one producer, and two observation wells placed at 100 and 200 ft from an injector on the injector–producer line. The reservoir contained naphthenic crude, and the high surface area constituents in the sand were kaolinite and sodium montmorillonite in about 4.0% and 5.5% by weight, respectively. Residual oil saturation was estimated at 0.20 from logs, although the oil content from cores averaged considerably less.

A commercial petroleum sulfonate blend with an equivalent weight of about 415 Daltons was used for the test. The lowest tension occurred at 0.5% sodium chloride and 0.15% petroleum sulfonate. Although the petroleum sulfonate was compatible with the reservoir brine, a slug of salt water was injected first, otherwise rather large amounts of sacrificial chemicals (sodium carbonate and sodium tripolyphosphate) were needed because of the high clay content of the sand, and the presence of the chemicals in the surfactant slug would have been detrimental. The surfactant slug had a volume of 0.12 PV and contained 1.1% sodium chloride, 1.9% petroleum sulfonate, and small concentrations of sodium carbonate and sodium tripolyphosphate. The chase water slug had a volume of 0.10 PV and contained 0.33% sodium chloride and small amounts of the sacrificial chemicals. No water thickener was employed in this slug. Oil appeared in samples taken at the closest observation well 36 days after surfactant injection and persisted for the next 84 days. The peak oil cut was 20%. The estimated sweep efficiency between the injector and the first observation well was about 90%.

At the end of test, four core holes were drilled. It was found that the vertical sweep efficiency was poor even though the pay was only 10–12 ft thick, and the areal sweep efficiency was also poor in the regions along the bisectors of injection–producer lines. The post analysis indicated a need for mobility control.

5.15.2 Wichita County Regular Field Low-Tension Waterflooding

A low-concentration surfactant slug was injected from October 1975 to June 1976 in Mobil's West Burkburnett waterflood, Wichita County Regular field, Texas, followed by biopolymer drive until April 1978. Freshwater drives continued (Talash and Strange, 1982). The low-tension waterflood (LTWF) encompassed ten 20-acre five-spot patterns.

In 1971, all the leases in the waterflood project either had become uneconomical to operate or were being projected to reach an economic limit by 1972 or 1973. A plan was developed for application of the LTWF process in 10 adjacent, 20-acre patterns with 10 injectors, 20 producers, and 2 observation wells. The project began in November 1973 with a freshwater preflush. The injected slugs were: 0.2 PV preflush for 451 days, 0.15 PV pretreat with 2000 ppm Na_2CO_3 and 1000 ppm $Na_5P_3O_{10}$ for 257 days, 0.15 PV surfactant of 1.86% sulfonate, 2000 ppm Na_2CO_3 and 2000 ppm $Na_5P_3O_{10}$ for 248 days, 0.1 PV 500 ppm concentrated biopolymer for 188 days, and 0.2 PV 500 to 200 ppm tapered biopolymer for 290 days. The injection rates were 1350–3000 B/D for 10 injection wells.

In early 1974, the first set of falloff and step-rate tests indicated that the formation near the 10 injectors was fractured to various degrees. It was felt that these fractures, with wing lengths ranging from 35 to 110 ft, would not seriously reduce the sweep efficiency in the 20-acre patterns. The second and third sets of falloff tests, conducted midway through polymer slug and near the end of tapered polymer slug, indicated that the wellbore fractures in each of the 10 injectors had been greatly extended, in some cases to more than

400 ft. This drastically changed the flow paths of the injected slugs, which resulted in the polymer slug not displacing the surfactant slug as intended.

Two observation wells, located 87 and 199 ft, respectively, away from an injector, were extremely valuable in monitoring salinity decline caused by freshwater injection, chemical transport and consumption, and oil bank development. Data obtained from the two observation wells are summarized briefly in the following points:

1. The freshwater slug effectively displaced the hostile brine (16% total dissolved solids, TDS) in the permeable zones of interest.
2. The pretreatment, sacrificial chemicals (sodium carbonate and sodium tripolyphosphate) were consumed before reaching the second observation well. Posttest base-exchange studies in the laboratory indicated that high divalent ion concentrations on the reservoir rock caused excessive consumption of the pretreated chemicals.
3. Surfactant concentrations at both observation wells did not reach injected concentrations.
4. Polymer concentrations measured at both observation wells did not reach injected concentrations.
5. A tertiary oil bank developed was observed, with peak oil cuts ranging from 16% to 20%.
6. Pulsed neutron logs taken at both observation wells after completion of the chemical slugs indicated that oil saturations approached zero.

In addition to various laboratory studies, many reservoir engineering studies were undertaken. They are categorized as follows:

1. Injection well evaluations
   a. Falloff
   b. Step-rate
   c. Multiple rate
   d. Reservoir pressure determination
2. Rate response (interference) tests
3. Pressure-wave tests
4. Reservoir simulation studies
5. Observation well evaluations
   a. Tracking of injected chemicals
   b. Oil cuts
   c. Pulsed neutron capture logging
6. Posttest core analysis
   a. Reservoir oil determinations
   b. Rock properties correlations
   c. Surfactant consumptions
7. Individual well performance evaluations and forecast.

In terms of production performance, most of the producers yielded incremental oil. It was expected that a recovery factor of about 24% original oil in place (OOIP) would be obtained from this LTWF area.

5.15.3 El Dorado M/P Pilot

Cities Service Oil Company, in cooperation with the Energy Research and Development Administration, conducted a field demonstration project of M/P flooding in the El Dorado field of Butler County, Kansas (Coffman and Rosenwald, 1975; Miller and Richmond, 1978). This demonstration project was to test the *sequential* M/P flooding in a field that had undergone complete primary and waterflooding depletion operations. The field was abandoned before the M/P project. The estimated oil saturation before M/P was 0.307–0.333. The project was designed to allow a side-by-side comparison of two M/P floods in the same field so that reservoir conditions for the two M/P demonstration floods were as nearly identical as possible. The two flood areas were the northern Chesney and the southern Hegberg pilot areas. In the northern Chesney area, aqueous surfactant system (water-external system) followed by xanthan gum solution was implemented. In the southern Hegberg area, an oil-external micellar system followed by partially hydrolyzed polyacrylamide solution was tested. From laboratory tests, the amount of surfactant plus cosolvent per barrel of oil produced for oil-external micellar system was less than that for aqueous surfactant system. However, such field test results comparing these two micellar systems were not seen.

The original pattern configuration was four contiguous nine-spot patterns in each 25.6-acre pattern. Subsequent detailed modeling and pressure-transient study indicated that the optimum pattern should be an array of four 6.4-acre five-spot patterns. Each pattern consisted of nine injection, four production, 12 monitoring, and two observation wells. The repetition of patterns was used to minimize the possibility of erroneous results owing to heterogeneity in the reservoir.

Reservoir Description and Production History

The multizone El Dorado field was discovered in 1915. The shallow formation was composed of sand interbedded with shale. The reservoir thickness is about 17.5–18.4 ft. The porosity ranged up to 30% with the average 20%, and the permeabilities varied from 159 to 1500 mD. The reservoir temperature at the selected site was 69°F. The oil viscosity was 4.77 cP, and the acid number was 0.4. The formation water salinity was 82,600 ppm with calcium 1900 ppm, magnesium 1600, and barium 500 ppm.

The reservoir went through primary production, air drive and water drive production to depletion, and abandonment. An extensive air drive pilot test was initiated in 1924. The success of the pilot was followed by the full-scale air drive development starting in 1926. A pilot waterflood was begun in 1947. Favorable response was observed in all offset producers within 6 months. The rapid extension of waterflood was started in 1950 and terminated in 1971.

M/P Injection

A designed injection sequence for the two patterns (areas) consisted of five phases: (1) pretreatment, (2) preflush, (3) micellar solution, (4) polymer solution, and (5) drive water. Injection of the pretreatment fluids (aqueous salt solutions) began in all 18 injection wells of both the Chesney and the Hegberg patterns on November 18, 1975. A severe injection-rate decrease was experienced in both patterns after reaching the pressure limitation for fracture extension of 0.74 psi/ft during the first week of injection. Several causes of the initial injectivity problems were identified or considered likely. Barium sulfate formation, fines movement, and poor water quality were among the causes. Because of the initial injectivity problems and continuing problems in this area, a water-quality monitoring program received considerable emphasis. The water-testing program was set up to monitor injected fluids on a daily basis.

Low injectivity continued, even after many causes of injectivity loss had been eliminated. A well-stimulated program was begun in February 1976. The stimulation procedure, which was moderately successful, consisted of the following staged solvent–acid treatment:

1. 200 gal xylene,
2. 250 gal hydrochloric acid (HCl) with demulsifier,
3. 1000 gal hydrofluoric acid (HF) with demulsifier,
4. 250 gal hydrochloric acid (HCl) with demulsifier.

As the weather became warmer, bacterial growth also contributed to the injectivity problem. Accordingly, a program of monitoring and controlling bacterial growth was started in March 1976.

The preflush phase for the Hegberg pattern began on June 20, 1976. The preflush fluid was an aqueous solution of sodium hydroxide and sodium silicate having a pH of 12–13. Injection of the micellar fluid began in the Hegberg pattern on March 22, 1977. The micellar fluid contained sulfonates, crude oil, one cosurfactant, and an aqueous salt solution. The micellar fluid was injected in alternate slugs, called micellar water and micellar oil (soluble oil). Polymer solution followed the injection of micellar fluids. A polyacrylamide was used in the Hegberg pattern and a polysaccharide (biopolymer) was used in the Chesney pattern. The oil cut increase was observed in the two test patterns.

Observation Well Program

The observation wells provided points within the flood area for fluid sampling and periodic logging. Daily fluid samples were taken. Analyses were made for weekly samples. The routine analyses were (1) sodium chloride content, (2) total hardness content, (3) WOR, (4) pH, (5) surfactant concentration, and (6) iron concentration.

Design Criteria of Injection Facility

Gas blankets were necessary on all vessels as a precaution against dissolved oxygen. Oxygen removal was necessary to maintain the integrity of chemicals rather than for the prevention of corrosion. Iron from piping, connections, and vessels also would be detrimental to chemical fluids. Consequently, Fiberglas and polyvinylchloride piping was used extensively. When steel piping was required, it was coated internally with a powder fusion-applied epoxy. All steel tanks were coated internally with a spray-on phenolic epoxy system that was allowed to air dry. The fluid distribution system used high-pressure Fiberglas lines. Fiberglas-lined steel tubing was used in the injection wells. The only deterioration of the Fiberglas was in those sections carrying high-concentrated caustic solutions.

The overall plant facility was designed for maximum flexibility. Because final chemical specifications were received after most of the mechanical design was completed, some over-design was incorporated intentionally. Transfer pumps and storage vessels were arranged to provide steady-state fluid flow through the plant. Blending mechanisms were set up for continuous mixing, in lieu of batch-mixing methods. Continuous mixing may be necessary in any field-wide expansion of the processes. And an on-site laboratory was considered a necessity.

5.15.4 Sloss M/P Pilot

Amoco Production Company conducted the micellar pilot in the Sloss field, Kimball County, Nebraska (Wanosik et al., 1978). The well arrangement was a single nine-acre normal five-spot pattern. The porosity was 17%, the permeability was 80 mD, the formation water was fresh (TDS = 2500 ppm, hardness = 50 ppm) and low salinity water was available (TDS = 260 ppm, hardness = 25 ppm). The reservoir temperature was 200°F.

The design sequence of injection consisted of four fluids: preflush water, micellar, polymer, and chase water. The motilities of the micelle and the polymer were adjusted so that they were approximately equivalent to the mobility of the tertiary oil–water bank.

Laboratory Study

A series of phase-stability tests was made to mixing various quantities of sulfonate, cosurfactant, polymer, and various brines and to observe fluid stability and solubilization characteristics. Based on these extensive laboratory studies (Trushenski, 1977; Trushenski et al., 1974), the principal components finally selected for these tests were Amoco Mahogany AA sulfonate (62% active), isopropyl alcohol (IPA), and Dow pusher 700 polymer. These were chosen on the basis of availability, performance, and stability. The formulation of a type III microemulsion range consisting of 4.5:1 bulk AA:IPA, 92% Sloss freshwater plus 14,600 ppm added sodium chloride was used in mobility tests, adsorption tests, and fresh state core tests. The subsequent phase-stability tests with the field micellar concentrate showed that the added salinity should be 12,000 ppm.

To estimate surfactant adsorption, two methods were used. In the first method, several PV of micellar fluid were passed through a core of 1 in. in diameter and 2.5 in. in length. The loss was 1.7 lb/bbl PV. In the second method, 0.2 PV of micellar fluid was used. The loss was 0.6 lb/bbl PV. The averaged value of 1.2 lb/bbl PV was used for field calculation. The adsorption values measured on cleaned and dried cores were about 0.5 lb/bbl PV higher than the fresh state cores. This was presumably owing to the activation of surface sites during the cleaning process.

A series of tests in Berea cores in 2 and 4 ft long and 2 in. in diameter were made at 200°F. A total of 20–28% PVs of incremental oil recovery were observed in a large range of salinities.

Reservoir Characterization

To be able to adequately interpret pilot performance, a good reservoir description is required. Therefore, prior to initiation of the micellar injection, a comprehensive program was conducted to obtain a reservoir description. The program consisted of coring, logging, production tests, pressure transient tests, pulse tests, and tracer injection. In addition, a geological study was made and thus the detailed reservoir description of the area was obtained.

Field Operation and Pilot Performance

Preflush water was injected to raise the salinity of the in-place fluid to ~12,000 ppm. The micellar injection was initiated on February 26, 1977. The polymer injection commenced on March 30, 1977.

Although the temperature of the Sloss reservoir is 200°F, the flood water is at 50°F which lowered the injection temperature. At temperatures below 40°F, the micellar fluid was completely unstable and would not even pass through a 25 micron filter. Above 40°F, the fluid would pass through a 25 micron filter. However, a phase would settle out upon standing. As the temperature is raised above 130°F, the fluid was cloudy but relatively stable.

Above 160°F the fluid was stable and clear. To avoid instability problem, the micellar fluid was heated before injection.

During the placement of micellar slug and the first 2 months of polymer injection (cumulative injection of 77,000 lb), the operations were smooth and almost trouble-free. The predicted injection rates were in reasonable agreement with the observed field behavior. Analysis of field data indicated that the effective viscosity of micellar solution was 4 cP and the effective viscosity of the polymer solution was about 7 cP. The micellar viscosity was based on simulation results and the polymer viscosity was based on both pressure falloff tests and simulation results.

In early June 1977, a Kobe pump in the producer malfunctioned and the plant was intermittent in operation for about 2 weeks. During the time, a substantial decline in injectivity occurred. Each injected well was swabbed and acidized, and epoxy-coated tubing was installed. A substantial amount of iron sulfide and unhydrated polymer was recovered during swabbing. A faulty biocide pump was repaired and the polymer blender was adjusted. Surface facilities were also acidized to remove iron sulfide particles. Acidizing improved injectivity markedly. However, after a short amount of polymer injection (15,000 bbl), plugging again became evident. The surface lines which were bare steel were suspected of contributing to the injectivity problem. The lines were given "acrolean" soak and a substantial amount of iron sulfide was removed. A complete epoxy-coated surface facility was then installed. During the installation, the wells were place on a low rate of freshwater injection.

Following replacement of the surface facility, the pilot injectors were each acidized. Polymer injection was resumed in the mid-December. The injectivity in the mid-January 1978 is ~60% of the predicted. Some damage at the sand face of the injection wells still appeared to exist.

Prior to micellar injection, the average oil rate of the producer was 6 STB/D at a WOR of 180. Tertiary oil response occurred at the producer after a cumulative injection of 27,000 bbl of micellar solution. At this time, the WOR declined to 60 and the oil rate increased to 11 STB/D. The oil rate continued to increase during the remaining micellar injection and the first two months of polymer injection. After a cumulative injection of 77,000 bbl, the oil rate was about 50 STB/D and the WOR was ~11. By that time, the injectivity problem occurred owing to iron sulfide and unhydrated polymer accumulation on the sand face. Then the production rate was reduced to maintain injection-to-production ratio close to 1.

5.15.5 Torchlight M/P Pilot

The Torchlight M/P pilot was conducted in a normal isolated five-spot pattern in the Tensleep formation of the Torchlight field in Big Horn County,

Wyoming. The pattern area included 6.4 acres. The average pay thickness was 31 ft and the product of porosity thickness was 4.4 ft.

Injection of the high-salinity preflush began in January 1977 and continued through July 1981. The preflush consisted of low-salinity Torchlight Tensleep injection (TTI) water (7.88 mN (about 460 ppm) total salinity and 5.19 mN hardness) supplemented with 0.308 N (18,000 ppm) NaCl. The high-salinity preflush was intended to provide a salinity gradient for the ensuring micellar fluid bank. The remaining oil saturation before the micellar flood was 0.34, and the residual oil saturation was 0.2.

Micellar fluid injection began August 9, 1981. The total PV injected was 0.18 PV. The micellar fluid consisted of 3.4 wt% active Amoco 151 (polybutene sulfonate), 0.8 wt% Shell Neodol 25-3S, 5 wt% *n*-butanol, 1200 ppm Cyanatrol® 950-S polyacrylamide, and 0.3 wt% formalin in TTI. Of the bulk sulfonate (51% active), 19.4 wt% consisted of inorganic salts, primarily Na_2SO_4. The micellar bank was followed by 0.165 PV of phase-control fluid. This fluid consisted of 1200 ppm Cyanatrol 950-S, 2.5 wt% *n*-butanol, and 0.6 wt% active Neodol 25-3S in 5000-ppm-NaCl-supplemented TTI. The phase-control bank was intended to eliminate any adverse interactions between the micellar and mobility-controlled fluids.

About 0.8 PV of 1200 ppm Cyanatrol 950-S in TTI water was initially intended to follow the phase-control fluid. Severe injectivity problems, however, precluded injection of this bank despite several remedial workovers of the pilot injectors. In July 1986, the pilot operation was suspended. Before termination, sulfonate and tracer production in produced fluids peaked and were declining. At suspension, cumulative 0.05 PV oil had been produced. It was expected that 0.18 PV of oil would be produced. Hence, the pilot performance fell well short of expectation.

Two observation wells were monitored throughout the pilot lifetime. These wells were located on a straight line between injector and producer. Analysis of samples withdrawn from an observation well indicated full development of the microemulsion bank with phase transition from the upper- to middle-phase environment. Very little clean micellar fluid was observed at the rear of the chemical bank. That meant the micellar bank had been essentially depleted at this point. Previous laboratory core testing did not predict this result.

Further laboratory studies were conducted to investigate the cause of poor pilot performance (Raterman, 1990). It was found that poor oil displacement efficiency for the Torchlight micellar fluid in laboratory field cores and the field pilot was owing to the tendency for the fluid to change from the lower- to upper-phase microemulsion environment. The upper-phase environment was characterized for the most part by high IFTs and, thus low displacement efficiency. The upper-phase regime was promoted by cation exchange and the mixing of micellar fluids with the high-hardness/salinity preflush. And the generation of viscous polymer coacervates and surfactant macroemulsions, in conjunction with dispersion, promoted the growth of large mixing zones.

When the entire slug changed to the upper-phase environment, chemical loss by phase trapping could be substantial. It was found that slug sizes of about 0.4 PV were required for the complete processing of laboratory field cores. Therefore, the displacement in the Torchlight pilot was not efficient because of large chemical requirement. The laboratory studies suggested that an insufficient slug of micellar fluid was used in the pilot.

5.15.6 Delaware-Childers M/P Project

The producing horizon was the oil-wet Bartlesville sandstone in NE Oklahoma at a depth of 620–700 ft. The thickness was 52 ft. The average porosity was 21% and the average permeability was 100 mD. The estimated oil saturation was 32–36%. The oil viscosity at the reservoir temperature of 86°F was 9.6 cP. The formation was oil-wet. The mineralogical analysis of a Mary Costen core showed that 50% quartz, 12% carbonates (primarily calcite), 10% clay (primarily kaolinite), 5% feldspars, 3% anhydrite and gypsum, and 20% others (mica, limonite, hematite, etc.).

The total dissolved solids (TDSs) of formation water were 100,000 ppm in 1938. At the start of the preflush in 1975, the TDSs were 11,000 ppm. But at the end of the field test in 1980, the TDSs were 7000 ppm. This shows the effect of waterflooding. The TDSs were 8800 ppm in the preflush, 8200 ppm in the micellar slug, and 2900 ppm in the polymer slug (Thomas et al., 1982).

A 2.5-acre inverted five-spot bounded by producers was under chemical flood. The nearby area was under air injection in 1930 and discontinued in October 1954 in favor of waterflooding. Injection of ~0.1 PV of micellar fluid was started on April 28, 1976. And injection of about 0.4 PV of polymer was completed in August 1979 followed by freshwater injection. The micellar formation was Amoco® Floodaid 141, a mixture of petroleum sulfonate blended with Cosurfactant 121, an ethoxylated alcohol. The polymer used was Nal-flo B (originally called Instapol Q-41-F) with 8–10 million molecular weight.

The incremental oil production was small, and the oil saturation near the evaluation wells was not significantly reduced but redistributed. The project was not technically or economically a success. A small amount of tertiary oil was produced by M/P injection.

5.15.7 Minas SP Project Preparation

The Minas surfactant project is still under study. This case shows that a long-time preparation may be needed for a field project. The presentation of this case is based on the papers by Bou-Mikael et al. (2000), Cheng et al. (2012), and Harman and Salem (1994).

Reservoir and Performance Description

Minas is the largest oil field in Southeast Asia with the OOIP of about 9 billion barrels. The field was discovered in 1944 and was placed in production in 1952. The field production peaked at 440,000 bbl/day in 1973. The Minas structure is a broad, gently dipping, and NW–SE trending anticline which is ~28 km long and 7–13 km wide. The formation was designated as the A1, A2, B1, B2, and D sands. The maximum original vertical oil column was 480 ft (150 m). The average porosity of the pay zones was about 26%. The permeability was about 4 Darcies. The oil viscosity was 3 cP. The original reservoir pressure was 930 psig, and the reservoir temperature was 200°F. The formation water salinity was less than 3000 ppm. The aquifer was found not as strong as the early production indicated. Waterflooding was implemented starting in 1972.

Up to 2000, the infill well spacing was about 24 acres, the water cut was 97%, and more than 50% oil had been produced. About 4.5 billion barrels of oil remained and were the EOR target. In 1994, a tertiary EOR screening effort identified light oil steam flood (LOSF) and SP flood processes for evaluation.

Surfactant Field Trials

The first surfactant field trial area was selected in the southern part of the field. The field trial area consists of a 4.3 acre five-spot pattern that includes 4 injectors, 1 central producer, 4 observation wells, 5 sampling wells, and 2 postflood core wells. Four of the five sampling wells and all four observation wells were located two-thirds of the way between the injection wells and the central production well. The fifth sampling well was placed closer to the injector in one quadrant to monitor the early progress of the SP front (Bou-Mikael et al., 2000).

The observation wells were completed with fiberglass casing to allow routine monitoring of saturation changes using open-hole tools such as magnetic resonance and deep induction logs. The producing wells in Minas typically were completed with high rate, electric submersible pumps. For this project, the sampling wells were equipped with low rate, rod pumps to prevent shearing of polymer solutions. The injection sequence was waterflooding to a residual oil saturation, surfactant flooding, polymer flooding, and post-waterflooding.

Numerous laboratory coreflood tests were performed to determine the optimum surfactant and polymer slug size and concentration, in addition to the tests to measure polymer viscosity, permeability reduction factor and residual permeability reduction factor, IFT, and CDC. The following slug sizes and concentrations were recommended for field-testing:

1. Surfactant: 0.25 PV and 2% concentration,
2. Polymer: 0.50 PV, and 1250 ppm tapered to 500 ppm for two injectors, and 900 ppm tapered to 350 ppm for the other two injectors,
3. Chase water: 1 PV.

Two surfactants (lignin II and synthetic petroleum sulfonate) and four polymers were designed for the field trial. Drilling wells started in the first quarter of 1998. Preflush waterflooding started on May 1, 1999 until March 1, 2000. Surfactant injection in the A1 sand started on May 10, 2000 and completed in February 2002. This trial is called Surfactant Field Trial 1 (SFT-1). The SFT-1 project indicated promising results and the need to proceed with further evaluation (Cheng et al., 2012). Another surfactant field trial, SFT-2, was planned in 2012. This SFT-2 project is a seven-spot 4.5-acre pattern with one central producer and six surrounding chemical injection wells. Additionally, there are six outside water injectors for the hydraulic control purpose and four chemical sampling wells.

Surveillance and Monitoring Program

The baseline reservoir surveillance program prior to surfactant injection included pulse testing, tracer testing, profile logging, and standard well testing. A pulse test was conducted in the A2 sand in April 1999 to determine well connectivity and reservoir transmissibility. A pressure pulse was generated by pumping 2000 barrels of water per day in the central producer while monitoring the pressure response in the four surrounding injectors. Sensitive crystal quartz gauges (CQGs) with a pressure resolution of ± 0.01 psi were used to monitor the pulse. The results of the pulse test confirmed well connectivity.

A presurfactant tracer test was conducted in the A1 sand in March 2000. Four different fluorobenzenoic acid (FBA) tracers were injected independently in the four injection wells during the water preflush phase in order to determine directional flow, timing of frontal advance, and pattern drift. FBA tracers were special chemical tracers that had a very low detection level (parts per billion) and conceptually had low adsorption on the reservoir rock. The results of this test provided conclusive evidence on well connectivity, fluid movement, and frontal advance from each injection well.

The SE and SW quadrants in the pattern had a consistent tracer response after 20 and 23 days, while the NE and NW quadrants did not show conclusive evidence of tracer response, suggesting the possibility of fluid drifting outside the pattern. This information would help guide decisions regarding preferred polymer placement in each quadrant and adjusting the surfactant and polymer injection rates.

It was designed that low-concentration chemical tracers would be pumped simultaneously with the surfactant for the full duration of the surfactant injection phase. This long-term introduction of tracer into the reservoir would improve the estimate of the sweep efficiency and would yield valuable data for calibrating the flow simulation model, estimating surfactant adsorption and estimating capture ratio. The capture ratio is defined as the fractional volume of the injected fluid sweeping inside the pattern area. This information would be critically important for designing field expansion.

Two interwell tracer tests (ITTs) were run later. The first one (ITT-1) was run from November 10, 2009 to February 25, 2010. The second one (ITT-2) was started on November 10, 2010 and was still ongoing by February 2011. Before ITT, other tests were conducted which included single and interwell pressure transient tests, pulse test, reservoir drift test, and injection test. The ITT data were analyzed using analytical models and reservoir simulation models. The reservoir flow model was history-matching of ITT data which was very helpful to correctly design the next surfactant field trial.

The interpretation of the ITT-1 results indicated various operational and reservoir properties that would have likely led to failure of the surfactant pilot. Hydraulic control of the SFT pattern was not achieved; in fact, less than 20% of one tracer was recovered. Unexpected communication between the target sand and the underlying sands outside the pattern also contributed to lower tracer recovery and low sweep efficiency.

During the water preflush phase, workovers were conducted to isolate the top A1 sand and install plastic-coated tubing in the injection wells to minimize polymer contact with bare steel. Injection tests were conducted with the rig on site to ensure that all perforations were open and capable of taking fluid at the design rate of 2300 bbl/day. Injectivity surveys were run in the wells to determine the character of the injection profile across the perforated interval of the A1 sand. Although all perforations were open, only one of the four injection wells showed even distribution of fluid across the perforated interval. The remaining injection wells exhibited unbalanced injection profiles where roughly one-third of the perforations were not taking injection. Polymer injection was expected to improve the injection profiles.

The monitoring program was designed to ensure clarity in responsibilities, frequency of data collection, and procedures to be followed. Table 5.1 is a designed monitoring program.

5.15.8 SP Flooding in the Gudong Field, China

Although ASP demonstrated the highest potential to increase oil recovery, there are two main problems: (1) scale and precipitation caused by alkaline reaction and (2) difficult to treat produced emulsions enhanced by alkaline solution. Therefore, more research efforts are directed toward to optimize SP flooding. The presented SP flooding case is one example of such effort.

The pilot SP test was initiated in September 2003. The pilot was located at the southern $NG5^4$-6^1 layer of the west block of the Gudong field. The

TABLE 5.1 Monitoring Program in Minas Surfactant Field Trial

Location	Test Type	Frequency	Special Tools or Procedures
Production well	Well test		
	1-Rate	Once a week for 24 h	
	2-Water cut	Once a week for 24 h	
	3-Surfactant C %	Thrice a week	HPLC*/titration
	4-Oil analysis	Weekly	
	5-O-W IFT	Thrice a week	Spinning-drop tensiometer
Sampling wells	Well test		
	1-Rate	Once a week for 24 h	
	2-Water cut	Once a week for 24 h	
	3-Surfactant C %	Thrice a week	High-performance liquid chromatography (HPLC)
	4-Oil analysis	Weekly	
	5-O-W IFT	Thrice a week	Spinning-drop tensiometer
Pressure well	Sonolog	Monthly	Sonolog gun
Logging wells	Oil saturation	0, 1.5, 2.5, 3.5, 4.5, 6 months	Slim tool and deep induction
Injection wells	1-Rate	Daily	
	2-Pressure	Daily	
	3-PLT	After completion	
	4-Polymer C%	Once a day	
Tracer test	Chemical tracers	Per service program	
Water in mixer	1-Polymer C%	Each batch	Funnel test
	2-Fe and cond test	Monthly	Titration/probe

area was 0.94 km^2. There were 9 injectors and 16 producers. The reservoir temperature was 68°C. The *in situ* oil viscosity was 45 mPa·s. The average permeability was 1320 mD. The salinity of formation water was 8207 mg/L with 231 mg/L Ca^{2+} and Mg^{2+}. Before the pilot was started, the water cut in the central block area was 98.3% in August 2003. The oil recovery factor was 35.2% at that time (Li et al., 2012).

The SP formula designed was 0.3% Mahogany sulfonate + 0.1% polyoxyethylene + 1700 ppm partially hydrolyzed polyacrylamide. The solution viscosity was 22 mPa·s. The IFT was 2.95×10^{-3} mN/m.

The injection scheme was as follows:

1. Preslug injection of 0.075 PV of 2040 mg/L polymer solution was started on September 11, 2003, for profile control.
2. Main slug: 0.495 PV of 1717 mg/L polymer, 0.44% Mahogany sulfonate, and 0.15% polyoxyethylene #1, injection began on June 1, 2004.
3. Postslug: 0.07 PV of 1600 mg/L polymer. The injection began on April 8, 2009 and ended in January 2010.

The oil rate response occurred after 0.04 PV of the main slug injection. The peak oil rate increased from 10.4 ton/day during waterflooding to 127.5 ton/day. The water cut decreased 37.9% from 98.3% to 60.4%. The incremental oil recovery reached 16.7 by December 2011.

REFERENCES

Amaefule, J.O., Handy, L.L., 1982. The effect of interfacial tension on relative oil and water permeabilities of consolidated porous media. SPE J. June, 371–381.

Bou-Mikael, S., Asmadi, F., Marwoto, D., Cease, C., 2000. Minas surfactant field trial tests two newly designed surfactants with high EOR potential. Paper SPE 64288 Presented at the Asia Pacific Oil and Gas Conference and Exhibition, 16–18 October, Brisbane, Australia.

Bourrel, M., Schechter, R.S., 1988. Microemulsions and Related Systems, Formulation, Solvency, and Physical Properties. Marcel Dekker, Inc., New York, NY.

Chen, T.-P., Pu, C.-S., 2006. Study on ultralow interfacial tension chemical flooding in low permeability reservoirs. JXSYU 21 (3), 30–33.

Cheng, H., Shook, G.M., Taimur, M., Dwarakanath, V., Smith, B.R., 2012. Interwell tracer tests to optimize operating conditions for a surfactant field trial: design, evaluation and implications. SPEREE 15 (2), 229–242.

Cinar, Y., Marquez, S., Orr Jr., F.M., 2007. Effect of IFT variation and wettability on three-phase relative permeability. SPEREE June, 211–220.

Coffman, C.L., Rosenwald, G.W., 1975. The El Dorado micellar–polymer project. Paper SPE 5408 Presented at the SPE Oklahoma City Regional Meeting, 24–25 March 1975, Oklahoma City, OK.

DeBons, F.E., Braun, R.W., Ledoux, W.A., 2004. A guide to chemical oil recovery for the independent operator. Paper SPE 89382 Presented at the SPE/DOE Symposium on Improved Oil Recovery, 17–21 April, Tulsa, OK.

Fleming, P.D.III, Sitton, D.M., Hessert, J.E., Vinatieri, J.E., Boneau, D.F., 1978. Phase properties of oil-recovery systems containing petroleum sulfonates. Paper SPE 7576 Presented at the SPE Fifty-Third Annual Technical Conference and Exhibition, 1–3 October, Houston, TX.

Foster, W.R., 1973. A low-tension waterflooding process. J. Pet. Technol. 25 (2), 205–210.

Gogarty, W.B., 1983a. Enhanced oil recovery through the use of chemicals—part 1. J. Pet. Technol. 35 (9), 1581–1590.

Gogarty, W.B., 1983b. Enhanced oil recovery through the use of chemicals—part 2. J. Pet. Technol. 35 (10), 1767–1775.

Green, D.W., Willhite, G.P., 1998. Enhanced oil recovery. Soc. Petrol. Eng.

Griffin, W.C., 1949. Classification of surface-active agents by "HLB". J. Soc. Cosmet. Chem. 1, 311.

Griffin, W.C., 1954. Calculation of HLB values of non-ionic surfactants. J. Soc. Cosmet. Chem. 5, 259.

Harman, Salem, E.A., 1994. Reservoir management in minas field: cooperative efforts between host country and operator to increase ultimate recovery. Paper Presented at the Fourteenth World Petroleum Congress, May 29–June 1, Stavanger, Norway.

Healy, R.N., Reed, R.L., Stenmark, D.G., 1976. Multiphase microemulsion systems. SPE J. June, 147–160 (Trans., AIME, 261).

Huh, C., 1979. Interfacial tension and solubilizing ability of a microemulsion phase that coexists with oil and brine. J. Colloid Interface Sci. 71, 408–428.

Li, Z.-Q., Zhang, A.-M., Cui, X.-L., Zhang, L., Guo, L.-L., Shan, L.-T., 2012. A successful pilot of dilute surfactant–polymer flooding in Shengli oilfield. Paper SPE 154034 Presented at the SPE Improved Oil Recovery Symposium, 14–18 April, Tulsa, OK.

Miller, R.J., Richmond, C.N., 1978. El Dorado micellar–polymer project facility. J. Pet. Technol. 30 (1), 26–32.

Moore, T.F., Slobod, R.C., 1955. Displacement of oil by water—effect of wettability, rate, and viscosity on recovery. Paper SPE 502-G Presented at the SPE Annual Fall Meeting, 2–5 October, New Orleans, LA.

Murtada, H., Marx, C., 1982. Evaluation of the low tension flood process for high-salinity reservoirs—laboratory investigation under reservoir conditions. SPE J. 22 (6), 831–846.

Nelson, R.C., 1981. Further studies on phase relationships in chemical flooding. In: Shah, D.O. (Ed.), Surface Phenomena in Enhanced Oil Recovery. Plenum Press, New York, NY, pp. 73–104.

Nelson, R.C., Pope, G.A., 1978. Phase relationships in chemical flooding. SPE J. October, 325–338 (Trans. AIME, 265).

Raterman, K.T., 1990. A mechanistic interpretation of the torchlight micellar/polymer pilot. SPERE 5 (4), 466–549.

Sheng, J.J., 2011. Modern Chemical Enhanced Oil Recovery—Theory and Practice. Elsevier, Burlington, MA.

Szabo, M.T., 1979. An evaluation of water-soluble polymers for secondary oil recovery, parts I and II. J. Pet. Technol. May, 553–570.

Taber, J.J., Martin, F.D., Seright, R.S., 1997. EOR screening criteria revisited–part 1: introduction to screening criteria and enhanced recovery field projects. SPEREE (August), 189–198.

Talash, A.W., Strange, L.K., 1982. Summary of performance and evaluations in the West Burkburnett Chemical Waterflood Project. J. Pet. Technol. 34 (11), 2495–2502.

Thomas, R.D., Spence, K.L., Burtch, F.W., Lorenz, P.B., 1982. Performance of DOE's micellar–polymer project in Northwest Oklahoma. Paper SPE 10724 Presented at the SPE Enhanced Oil Recovery Symposium, 4–7 April, Tulsa, OK.

Trushenski, S.P., 1977. Micellar flooding: surfactant–polymer interaction. In: Shah, D.O., Schechter, R.S. (Eds.), Improved Oil Recovery by Surfactant and Polymer Flooding. Academic Press, New York, NY, pp. 555–575.

Trushenski, S.P., Dauben, D.L., Parrish, D.R., 1974. Micellar flooding fluid propagation, interaction, and mobility. SPE J. December, 633–642 (Trans., AIME, 257).

UTCHEM, 2000. Technical documentation for UTCHEM-9.0. A Three-Dimensional Chemical Flood Simulator, Austin, July.

Wanosik, J.L., Treiber, L.E., Myal, F.R., Calvin, J.W., 1978. Sloss micellar pilot: project design and performance. Paper SPE 7092 Presented at the SPE Symposium on Improved Methods of Oil Recovery, 16–17 April 1978, Tulsa, OK.

Winsor, P.A., 1954. Solvent Properties of Amphiphilic Compounds. Butterworth Scientific Publication, London.

Chapter 6

Alkaline Flooding

James J. Sheng
Bob L. Herd Department of Petroleum Engineering, Texas Tech University, Lubbock, TX 79409, USA

6.1 INTRODUCTION

Among several chemical methods, alkaline flooding is probably the cheapest method. Overall, however, past field applications did not show significantly high incremental oil recovery over waterflooding. It would be our interest to further study this method for enhanced oil recovery. Currently, there seems a research interest to investigate the alkaline injection in heavy oil reservoirs, because heavy oils have high contents of organic acids (saponifiable components) to react with alkalis so that surfactants called soap are generated *in situ*.

In this chapter, we will briefly present the fundamentals of alkaline flooding which include comparison of alkalis used in alkaline flooding, alkaline reactions with crude oil, water and reservoir rock, and alkaline flooding mechanisms. Instead of presenting complex formulas used in alkaline simulation, we choose to present some typical simulation results about the solution pH, generated surfactant concentration, interfacial tension (IFT), salinity, etc., to help the reader understand the significance of alkaline reactions. Typical field injection data like alkaline injection concentrations and volumes are discussed, and a typical field monitoring program is presented. Interestingly, although there have been many active field projects using chemical EOR methods in China, we did not see a new field application using alkalis alone. Instead, many field projects have been carried out by combining alkalis with other chemicals like polymer and surfactants. Chemical EOR methods have not been active in the Western world in recent years. Therefore, for the field cases, we only present several cases from Russia, Hungary, India and the United States which were conducted before 1990.

6.2 COMPARISON OF ALKALIS USED IN ALKALINE FLOODING

Alkaline flooding is also called caustic flooding. Alkalis used in alkaline-related EOR include sodium hydroxide, sodium carbonate, sodium orthosilicate, sodium tripolyphosphate, sodium metaborate, ammonium hydroxide,

Enhanced Oil Recovery Field Case Studies.

and ammonium carbonate. The most often used alkalis are sodium hydroxide, sodium carbonate, and sodium orthosilicate. Sodium hydroxide generates OH^- by dissociation, the other two through the formation of weakly dissociating acids (silicic and carbonic acid, respectively) that remove free H^+ ions from solution. In terms of the effectiveness to reduce IFT, it has been observed that there is little difference among the commonly used alkalis (Burk, 1987; Campbell, 1982). In the water of high hardness, the IFT is more significantly reduced when sodium orthosilicate was used. This is because of the formation of calcium or magnesium silicates which are much less soluble than calcium or magnesium hydroxides. Thus the water hardness is reduced (Campbell, 1977; Novosad et al., 1981). The domain over which silicate scales form overlays those of other scales. Thus the resulting precipitate or scale may be a complex mixture of compounds. The silicate scales are generally amorphous in nature and highly hydrated.

Owing to the emulsion and scaling problems observed in Chinese field applications, there is a tendency to use weaker alkalis such as sodium carbonate instead of sodium hydroxide.

To minimize the corrosion problem and scale problem associated with inorganic alkalis such as sodium hydroxide and sodium carbonate, an organic alkali was proposed (Berger and Lee, 2006). Metaborate was proposed to sequester divalent cations such as Ca^{++} and to prevent precipitation (Flaaten et al., 2008). No field tests were reported for these alkalis.

6.3 ALKALINE REACTIONS

The alkaline reactions discussed in this section include reactions with crude oil, rock and water.

6.3.1 Alkaline Reaction with Crude Oil

In alkaline flooding, the injected alkali reacts with the saponifiable components in the reservoir crude oil. These saponifiable components are described as petroleum acids (naphthenic acids). The main fractions are carboxylic acids (Shuler et al., 1989). It is assumed that a highly oil-soluble single pseudo-acid component (HA) in oil. The alkali–oil chemistry is described by partitioning of this pseudo-acid component between the oleic and aqueous phases,

$$HA_o \leftrightarrow HA_w \tag{6.1}$$

and a subsequent aqueous hydrolysis in the presence of alkali to produce a soluble anionic surfactant A^- (its component is conventionally denoted by $RCOO^-$) (deZabala et al., 1982):

$$HA_w \leftrightarrow H^- + A^- \tag{6.2}$$

Here HA denotes a single acid species, A denotes a long organic chain, and the subscripts o and w denote oleic and aqueous phases, respectively.

Because of the alkaline reaction with crude oil, the IFT between the oil and water becomes lower. This lower IFT makes the emulsions more stable.

6.3.2 Alkaline Interaction with Rock

When clays originally in equilibrium with formation water are contacted with alkaline solution, the surface will attempt to equilibrate with its new environment, and ions will start exchanging between the solid surfaces and the alkaline solution. Ions present on the clays originally include hydrogen. As the pH of the solution is increased, hydrogen ions on the surface react with hydroxide ions in the flood solution, lowering the pH of the alkaline solution. By this reaction, the base present in the alkaline solution is consumed as the alkaline solution moves through the reservoir. One of such hydrogen exchange is:

$$\text{H-X} + \text{Na}^+ + \text{OH}^- \leftrightarrow \text{Na-X} + \text{H}_2\text{O} \quad (6.3)$$

where X denotes mineral-base exchange sites. Similarly, for Na^+–Ca^{2+} exchange, we have

$$2\text{Na-X} + \text{Ca}^{2+} \leftrightarrow \text{Ca-X} + 2\text{Na}^+ \quad (6.4)$$

Divalents such as calcium and magnesium ions are also present in clays. When calcium-free alkaline salt water contacts the clays, for example, calcium ions on the rock surface will exchange for sodium ions in the alkaline solution. Then calcium precipitation may occur. Thus, reaction with calcium on the clays also consumes the alkaline solution as it moves through the reservoir. Examples of such cation exchanges are

$$\text{Ca-X}_2 + 2\text{Na}^+ + 2\text{OH}^- \leftrightarrow 2(\text{Na-X}) + \text{Ca(OH)}_2 \quad (6.5)$$

$$\text{Ca-X}_2 + 2\text{Na}^+ + \text{CO}_3^{2-} \leftrightarrow 2(\text{Na-X}) + \text{CaCO}_3 \quad (6.6)$$

Contrary to ion exchange which is a fast reversible process, the dissolution of rock minerals by alkalis is a long-term irreversible kinetic process. To determine the minimum alkali requirement, a core is flushed with alkali solution. Then the alkali consumption includes ion exchange and dissolution.

Because of the complex rock mineralogy in most petroleum reservoirs, the number of possible reactions with alkalis is large. Reservoir rock reaction is believed to be by far the largest contributor to alkali consumption. Some early studies on caustic consumption were published Ehrlich and Wygal (1977) for different types of minerals. They found high consumption rates for most clays but relatively low reaction rates for quartz, calcite, and dolomite.

6.3.3 Alkaline Reaction with water

The primary reaction of the alkali with the reservoir water is to reduce the activity of multivalent cations such as calcium and magnesium in the oilfield brines. Upon contact of the alkalis with these ions, precipitates of calcium and magnesium hydroxide, carbonate, or silicate may form depending on pH, ion concentrations, temperature, etc. If properly located, these precipitates can cause diversion of flow within the reservoir, leading to better contact of the injected fluid with the less permeable, less flooded flow channels. This then may contribute to improved recovery.

From the above discussion, total alkali consumption includes

$$C_{\mathrm{i}} - C(t) = \Delta C_{\mathrm{o}} - \Delta C_{\mathrm{w}} - \Delta C_{\mathrm{e}} - \Delta C_{\mathrm{D}} \tag{6.7}$$

where C_{i} and $C(t)$ are the initial and the existing (current) concentrations, respectively, ΔC_{o} is the alkali consumption for the alkali to react with the crude oil to generate soap, ΔC_{w} is the alkali consumption caused by alkali reaction with the multivalent ions in the formation water, ΔC_{e} is the alkali consumption during the ion exchange between the alkali solution and the rock, and ΔC_{D} is the alkali consumption during dissolution reaction between the alkali and the rock.

6.4 RECOVERY MECHANISMS

In alkaline flooding, emulsification is an important mechanism. Emulsification and entrainment in which the crude oil is emulsified *in situ* and entrained by the flowing aqueous alkali due to IFT reduction (Subkow, 1942). The conditions for this mechanism to occur are high pH, low acid number, low salinity, and O/W emulsion size less than pore throat diameter.

In emulsification and entrapment, the sweep efficiency is imposed by the action of emulsified oil droplets blocking the smaller pore throats (Jennings et al., 1974). The conditions for this mechanism to occur are high pH, moderate acid number, low salinity, and O/W emulsion size more than pore throat diameter.

Emulsification and coalescence are related to spontaneously formed unstable W/O emulsion (Castor et al., 1981) or mixed emulsion. The isolated oil droplets are emulsified after contacting with alkaline solution. The emulsified droplets coalescence with each other to become larger droplets while they move in the pores, because the films of W/O emulsion are not rigid and can be easily ruptured and coalescence to become larger. Some of the emulsified droplets are stopped at pore throats. Therefore, the mechanisms of oil recovery are to increase sweep efficiency and to increase coalescence of oil drops into a continuous oil bank.

All these mechanisms are related to emulsification. In an alkaline experiment, if the color of produced fluid is dark brown, and the water color is dark yellow, the oil is emulsified. It has been observed that in such experiments the oil recovery was in the range of 18–22%. If water and oil come out of the core alternatively and the water is clear, then the oil is not emulsified. The oil recovery factors in such experiments were in the range of 14–16% (Cheng et al., 2001). In other words, emulsification increased the oil recovery factor by about 5%. Many wells in Daqing alkaline-surfactant-polymer (ASP) applications show that if the produced fluids are more emulsified, the decrease in water cut will be higher.

Other mechanisms include wettability alteration, mobility control by divalent precipitates (Sarem, 1974), coinjection or alternate injection of alkaline solution and gas to improve sweep efficiency.

6.5 FIELD INJECTION DATA

Alkaline concentration and volume injected appear to vary depending on the recovery mechanism to be used. Concentrations are generally lowest for the emulsification mechanisms, from about 0.001–0.500 wt.%. Higher concentrations ranging from about 0.5–3.0 wt.%, or even as high as 15.0 wt.%, usually have been required for wettability reversal. Most agree that a slug of alkaline solution can be nearly as effective as continuous injection, although the mechanism of emulsification and entrainment will require a sufficient volume to ensure production of the alkaline emulsion. If the volume is small and the alkali is consumed by rock reaction, the emulsion may become trapped again before it reaches the producing wells. Other mechanisms in which mobility-ratio improvement plays an important role appear to require a slug size no more than about 0.1–0.3 PV to be effective (Johnson, 1976).

Mayer et al. (1983) summarized alkaline flooding field projects until 1983. We analyzed the data of injection concentrations and slug sizes from these projects using the statistical method. Figure 6.1 shows the cumulative percentage versus alkali concentration. This figure shows that the average alkali concentration (at 50% cumulative) is about 0.5 wt.%. For most of the field projects, alkali concentration is less than 1.0 wt.%. However, higher concentrations have been used in more recent Chinese field projects. Figure 6.2 shows the cumulative percentage versus injected alkali slug with the data sources same as in Figure 6.1. At the 50% cumulative percentage, the injected alkali solution slug is about 15% PV. Figure 6.3 shows the cumulative percentage versus the total amount of injected alkali which is presented by the product of concentration (%) and slug size (%PV). At the 50% cumulative percentage, the product is 17. For most of those field

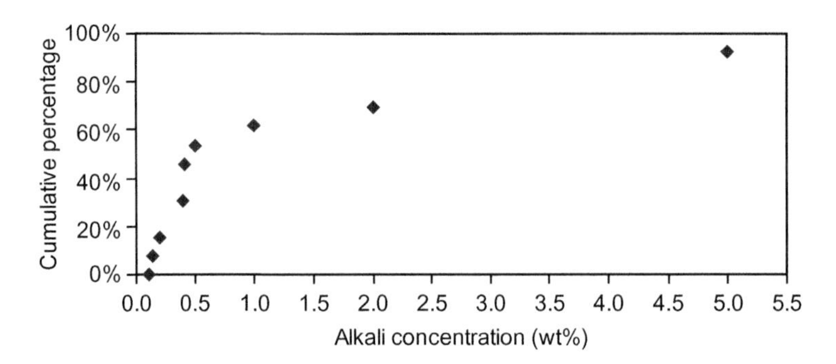

FIGURE 6.1 Cumulative percentage versus alkali concentration for field projects.

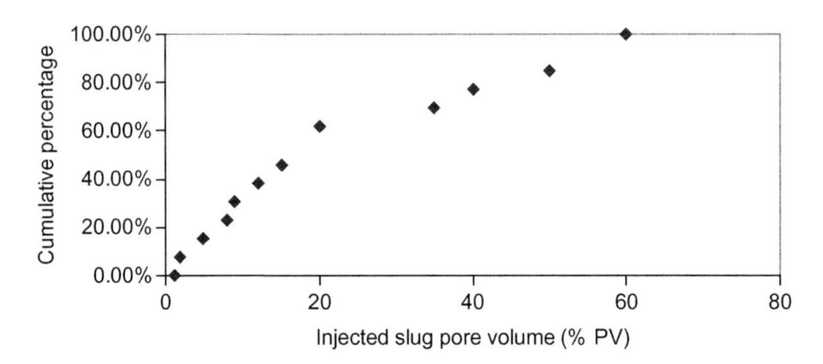

FIGURE 6.2 Cumulative percentage versus injected alkali slug for field projects.

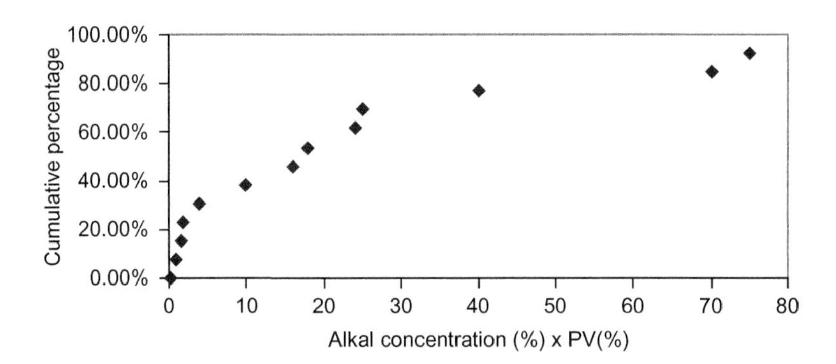

FIGURE 6.3 Cumulative percentage versus total amount of injected alkali.

projects, the incremental oil recovery factors were 1–2% with only two cases having incremental recovery factors of 5–8%.

These low incremental oil recovery factors are consistent with the several facts about alkaline flooding: it is difficult to obtain ultralow IFT (a very low alkali concentration is required, which is not practical); the low IFT range is narrow; mobility control is limited, etc. Considering the synergy among alkaline flooding, surfactant flooding and/or polymer flooding, it appears that alkaline flooding without adding surfactant or polymer has lost its attraction, at least in light oil reservoirs.

For their surveyed field projects, the oil API gravities were low in most cases, and the oil viscosities were 1–220 mPa·s, with most of the case having low viscosities. The reservoir temperatures were 72–205°F. The salinities in the injection water were less than 1000 ppm in the early cases, but higher salinities were used in later cases. Sodium hydroxide, sodium orthosilicate, and sodium carbonate were used.

6.6 APPLICATION CONDITIONS OF ALKALINE FLOODING

Injection wells should be within oil zones not in the peripheral aquifer to avoid alkaline consumption by reacting with divalents. However, if the bottom water has high content of divalents, alkaline solution may be injected to form precipitates so that bottom water coning may be mitigated by precipitates. Such reservoir should not have gas cap. If the expected mechanism is not to use precipitates for the mobility-control purpose, divalent ion exchange capacity should be less than 5 meq/kg, and the *in situ* pH should be greater than 6.5 (French and Burchfield, 1990).

For reservoirs with high acid number oils, alkaline flooding can be executed at any development stage. However, for reservoirs with low acid number oils, alkaline flooding in an earlier stage performs better. In this case, remaining oil saturation should be higher than 0.4. There should be no temperature limitation for alkaline flooding. The alkaline process is not recommended if CO_2 content is high. And low-pH alkalis should be considered in carbonate reservoirs (French and Burchfield, 1990).

Alkaline consumption is mainly due to the existence of clays by chemical reaction and ion exchange. Thus, clay content should not exceed 15–25%. Especially, the reservoir should contain little (<0.1 wt.%) or no gypsum. Formation permeability should be greater than 100 mD. However, it was reported that alkaline flooding has been implemented in a field of the permeability as low as 20 mD (Doe et al., 1987). But an alkaline injection project should not be carried out in a reservoir where a water injection project has faced injectivity problem.

Oil viscosity should be less than 50–100 cP. However, these days alkaline–surfactant injection into very high viscous oils has attracted more and more interests. Formation water salinity should be less than 20%.

Well spacing should be $(2–36) \times 10^4$ m^2/well. Because of alkaline consumption, the well spacing should not be too large. However, the well spacing should be selected based on economic calculation.

When designing an alkaline flooding project, the following facts may be taken into account:

- Alkaline assumption in field is higher than that in laboratory, because the contacting time of alkalis with rocks in the former is much longer than that in the latter.
- The oil recovery factor in field is generally lower than that in laboratory.
- When alkaline flooding is combined with other methods, such polymer flooding, surfactant flooding, hydrocarbon gas injection, thermal recovery, etc, much better effects will be obtained.
- Alkaline injection could cause scale problems in the reservoir, wellbore, and surface facility and equipment.
- When alkaline solution contacts with oil, stable emulsions may be formed. This will increase the cost to treatment of produced fluids on surface.
- A small spacing should be designed for a field pilot so that the pilot results can be obtained in a shorter time. A series of monitoring and evaluation wells should be drilled along the displacement front.
- When designing injected alkaline concentration and slug size, first determine the optimum alkaline concentration corresponding to the lowest IFT. Then determine the alkaline consumption experimentally. The injected alkaline concentration should be the optimum concentration plus the concentration to satisfy alkaline consumption.
- To achieve good results for an alkaline project, the IFT should be less than 0.01 mN/m. If the IFT is higher than 0.1 mN/m, modified alkaline flooding methods should be considered, such as thermal alkaline flooding, gas alkaline flooding, mobility-controlled caustic flood (MCCF) to take advantage of precipitation (Sarem, 1974), or a method combined with surfactant or polymer.

The following are some facility and safety requirements from translated Russian articles.

- The materials of the equipment must not react with alkalis. Stainless steel is recommended. The equipment and pipelines used to transport alkaline solution should be placed on the concrete ground. The concrete layer should be 0.15 m higher than the earth surface.
- The equipment should be flushed periodically to clean precipitates. The flushing water should be accumulated in a disposal container. The volume of flushing water should be three times the pipeline volume.
- Water disposal system in the facility area should be a closed-loop system, although the system outside the facility area could be an open system. The disposal of rain should be designed based on the historically highest rain fall.

- The facility area should be fenced with the fence height not lower than 2 m. The width of the access to the fenced should be at least 4.5 m. The distance between the fence and housing and equipment should be 6 m. In the facility area, it is prohibited to install a water supply source.

6.7 FIELD CASES

In this section, we present two mobility-control cases in Russia, one case using high alkaline concentration in Hungary, one caustic-flooding case in India, three cases in the United States, and one case in a Canadian heavy oil field.

6.7.1 Russian Tpexozephoe Field (Abbreviated as Field T)

This case shows the mobility control by alkaline injection. This case report is based on the author's collected articles. The references are not available. The alkaline pilot in the block III of Field T was one of the first field trials in Russia. Field T had four blocks. The best block was the block III. The producing layers were upper Π and lower Π layers. The lower block Π was the main producing layer. The block III was an isolated block so that alkaline solution would not flow outside the test zone. The reservoir and fluid properties are described in Table 6.1.

TABLE 6.1 Pilot Layer Reservoir and Fluid Data

Effective thickness (m)	15–20
Original oil saturation (fraction)	0.754
Porosity (%)	17
Permeability (mD)	205
Formation temperature (°C)	78(?)
Oil density (g/cm^3)	0.735
Oil formation volume factor (m^3/m^3)	1.26
Oil viscosity in place (mPa·s)	1.03
Bubble point (mpa)	8.91
Solution gas at 20°C (m^3/ton oil)	60.87
Water viscosity (mPa·s)	0.5
Cl^- (mg/L)	9121
SO_4^{2-} (mg/L)	12.4
HCO_3^- (mg/L)	2293.6
Ca^{2+} (mg/L)	153.6
Mg^{2+} (mg/L)	43.2
$Na^+ + K^+$ (mg/L)	6521

Development History

The development of the field was started in 1965 using peripheral water injection. The waterflooding performance was low. To enhance peripheral waterflooding, six injectors were drilled inside the field. Nineteen to twenty producers were in the pilot area. Before the alkaline injection, the pilot area produced 41% OOIP, and the water cut was above 80%.

Laboratory Study

A model oil consisting of 80% crude oil and 20% kerosene was used in the laboratory. Under 78°C, the model oil had a viscosity of 1.25 mPa·s and density of 0.782 g/cm^3 which are close to the oil properties in the ground. The flow velocity in the laboratory was same as that in the reservoir. Core experiments showed that the oil displacement efficiency was increased by 1% and 6% over fresh waterflooding, if 0.05% and 0.1% alkaline concentration were used, respectively. This indicated that the optimum alkaline concentration should be 0.1%. At 0.1% alkaline concentration, the IFT was 0.46 mN/m. Considering alkaline consumption resulting from the reactions with the crude oil, formation water and rock, 0.25% alkaline concentration was designed.

Because the formation water contained a high concentration of HCO_3^-, the IFT between the formation water and oil was 4–6 mN/m which is lower than the IFT between the fresh water and oil of 17–20 mN/m. Core flood experiments in another laboratory showed that the formation water had 5% higher oil displacement efficiency than fresh water. Further alkaline flooding obtained 2.5% more than the formation waterflooding.

Monitoring Program

For almost all producers, oil rate, liquid rate, bottom-hole flow pressure and formation pressure were measured or analyzed. Water samples were collected from the producers at least once a week to analyze the alkaline concentration in the produced fluid. For injection wells, water intake rate, bottom-hole pressure and formation pressure were measured or analyzed.

A tracer test was started in December 1976. Six hundred kilograms of ammonium thiocyanate was injected. Fluid samples were collected 1–2 times a week. By May 1977, the tracer was observed at the first row (closest to injectors) of producers. The initial tracer concentration was 0.065 mg/L, increased to 0.3–0.4 mg/L in the next 2 months. By September 1977, the third row of producers started to show low concentration of tracers. At this time, the tracer concentration started to decline at the first row of producers. By December 1977, the third row of producers stopped showing tracers. Later, the second row of producers stopped showing tracers. By March 1978, no tracer was produced any more. By that time, total 6% of injected tracers were produced.

One month after the tracers were produced, a low alkaline concentration of 20 mg/L was observed. Later this concentration was stable at 40–80 mg/L.

The injected alkaline concentration was 3330 mg/L. The highest alkaline concentrations at two producers were 800 mg/L or higher. The injected alkaline solution broke through at these two wells through high-permeability channels.

Field Performance

Another laboratory study on the interaction of the crude oil with alkaline solution indicated that the interaction was not much significant than that between the crude oil and water. Probably mobility control caused by water hardness was important. It was estimated that 1 g/L of precipitates could occur when alkaline solution was mixed with the formation water. Such precipitates might help improve sweep efficiency. To avoid blockage occurring in the equipment and near wellbore region, a slug of fresh water was added between the alkaline solution and water. The injection program was: 5 days of alkaline solution, 4 days of fresh water followed by 5 days of produced connate water from the separators. The alkaline concentration was 0.65%. This injection program lasted until April 1, 1980. By that time, alkaline injection was stopped in the block III. After the change in the injection program, the alkaline concentrations at the producers maintained at 100–150 mg/L.

It is well known that the equilibrium is established among formation fluids during water injection:

$$CO_2 + H_2O \leftrightarrow H_2CO_3 \leftrightarrow H^+ + HCO_3^- \leftrightarrow 2H^+ + CO_3^{2-} \tag{6.8}$$

The equilibrium moves to the right side when alkaline solution is injected in the formation. There are two reactions:

$$NaOH + CO_2 \leftrightarrow NaHCO_3 \tag{6.9}$$

$$NaHCO_3 + NaOH \leftrightarrow Na_2CO_3 + H_2O \tag{6.10}$$

The fractions of carbonate ions at different pH are shown in Figure 6.4.

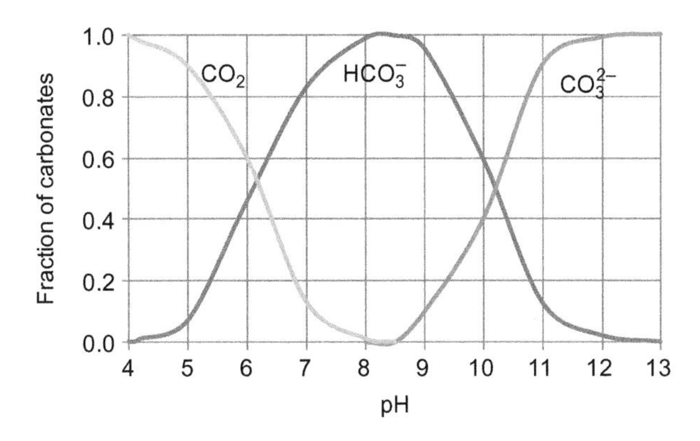

FIGURE 6.4 Fractions of carbonates versus solution pH at 70°C.

From this figure, we can see that CO_2 can only survive in the solution when the pH is less than 8.3–8.4. Increasing pH, CO_3^{2-} increases, which combines with Ca^{2+} so that the precipitates form. For this precipitation process to occur, pH must be at 9–9.5 level. When the pH is greater than 10.5, $Mg(OH)_2$ precipitates. In addition to reaction with formation water, injected alkaline solution must react with rocks. Only after consumption of alkali with formation water and rocks, can the remaining alkali react with the crude oil to reduce the IFT.

By taking samples of the produced fluid and analyzing the pH and carbonate concentrations, it was observed that the alternate injection of alkaline solution and produced formation water improved mobility control. However, such improvement was not significant because the precipitates were not enough to reduce the permeability of the channels. In this pilot area, 1 ton of NaOH produced 14 tons of incremental oil.

Another alkaline pilot was conducted in the other area of the same field. Waterflooding was started in the pilot area in 1966. Before the alkaline injection in nine wells in 1980, the water cut was 73%. The alkaline concentration was increased from 0.6% to 1.6%. The pilot might be divided into five blocks. The alkaline injection was effective in three blocks, but was not successful in the other two blocks because of high clay content (>20%). In the successful three blocks, 1 ton of alkali injected produced 10 tons of incremental oil.

6.7.2 Russian Шагпр-Гоаеан Field (Abbreviated as Field W)

This case also shows the mobility control by alkaline injection. Alkaline flooding was started in Field W in 1978 (Buchenkov et al., 1986). To shorten the distance between an injector and a producer, additional producers were drilled 50–150 m away from a central injector. The test area was 225,000 m^2. The porosity and permeability were 22% and 766 mD, respectively. The oil viscosity was 38 mPa·s. The formation water salinity was 90–120 g/L of $CaCl_2$ type.

The oil/water IFT was 30 mN/m, whereas the IFT between oil and 1% NaOH solution was 1 mN/m. Core flood tests showed that 0.25 PV of 1% NaOH solution produced 14% more oil owing to IFT reduction and wettability alteration.

From August 1978 to June 1979, 13,700 m^3 of alkaline solution was injected in the #1 test zone, followed by fresh water injection. Alkaline injection was resumed in December 1983. At that time the water cut had reached 90%. The vertical sweep efficiency was increased from 0.36 for fresh water injection to 0.91 for alkaline injection. The cumulative ratio of water to oil for alkaline injection was 1.35 times lower than that for fresh water injection.

Alkaline injection was injected in the #2 test zone starting in August 1983. There were 5 injectors and 25 producers in inverted seven-spot patterns. The test area was 3,120,000 m^3. The water cut was 82%. Before alkaline injection, the produced water of $CaCl_2$ type was reinjected. To avoid precipitation, a slug of fresh water was injected. There were two layers of high permeability-contrast (1:23) in this zone. After alkaline injection, the intake profile was changed to 1:3. Injection of fresh water slug was able to avoid precipitation near injection wellbores, but the precipitation occurred away from the wellbores. In this test zone, 1 ton of alkali injected produced 15 tons of incremental oil.

Typically, 3–4 months after alkaline injection, it was observed that Ca^{2+} ion concentration decreased and HCO_3^- ion concentration increased at production wells.

6.7.3 Hungarian H Field

This is a case of high alkaline concentration used. The Hungarian H field started to produce oil in 1951. By 1954, the oil production reached 3000 tons/day. The field was under aquifer drive combined with solution gas drive. There was water coning problem. It was a limestone reservoir with high-density fractures. The matrix permeability was 2–3%. It was oil-wet. The reservoir temperature was 78–148°C. The oil viscosity was 80–30,000 mPa·s. The formation water salinity was 1000–2000 mg/L. The field was depleted and the water cut reached 90–95% before alkaline injection.

Laboratory experiments showed that after NH_4OH injection, the permeability was doubled, and the water saturation was increased by 3–10%. Correspondingly, the oil saturation was decreased. The wettability was changed toward more water-wet. Experimental data showed that when 0.03 PV of 25% ammonium hydroxide was injected, 8% of incremental oil was obtained in a core without connate water, and 11% of incremental oil was obtained with connate water. If 4.5 PV of ammonium hydroxide solution was injected continuously, 12% incremental oil was obtained.

In the field, 24% ammonium hydroxide solution was injected to treat two production wells. Six hundred thirty-one tons and 114 tons of NH_3 were injected in these wells. Two thousand four hundred cubic metres and 350 m^3 of incremental oil were obtained respectively. Forty percent of injected ammonium hydroxide was recovered from on the first well, and 20% was recovered from the second well.

An ammonium hydroxide solution of 25% concentration was injected into three injection wells at a well injection of 35–40 m^3/d in 1975. Total 3233 m^3/d was injected. After the alkaline slug formed, fresh water was injected. The water injection capacity was reduced from 500–600 m^3/d to 300 m^3/d. The liquid production rates from the production wells increased.

Oil rates increased or recovered for some wells. Eighteen percent of injected ammonium hydroxide was restored from producers within 18 months.

6.7.4 North Gujarat Oil Field, India

The pilot data of the caustic flood project in the North Gujarat field are shown in Table 6.2 (Goyal et al., 1991). The field was discovered in 1962. The field had produced 5% OOIP before the initiation of the pilot project. The initial reservoir pressure of 135 kg/cm^2 had decreased to 70 kg/cm^2 by that time.

Laboratory study indicated that a minimum IFT occurred at 0.25% caustic solution, and injection of 0.1–0.2 PV caustic solution could yield 10% incremental oil recovery over waterflooding. Based on the promising laboratory results, a pilot was planned and commissioned in December 1987. The pilot pattern was an inverted five-spot pattern having an area of 10 acres (0.0405 km^2). The distance between the injector to a producer was 142 m.

TABLE 6.2 Reservoir and Fluid Data in the North Gujarat Pilot

Pilot area (km^2)	0.0405
Effective thickness (m)	6.8
Pore volume (m^3)	59,200
Original oil saturation (fraction)	0.7
Porosity (%)	21.5
Permeability (mD)	800
Formation temperature (°C)	84
Oil density at 20°C (g/cm^3)	0.93
Oil formation volume factor (m^3/m^3)	1.06
Oil viscosity in place (mPa·s)	10
Acid number (mg KOH/g oil)	1.5
Solution gas at 20°C (m^3/m^3)	20
Asphaltenes (wt.%)	0.7
Resins (wt.%)	2.1–4.3
Wax (wt.%)	5.7–12.6
Residues (wt.%)	18.5
pH	8.71
Cl^- (mg/L)	1280
CO_3^{2-} (mg/L)	120
SO_4^{2-} (mg/L)	235
HCO_3^- (mg/L)	671
Ca^{2+} (mg/L)	24
Mg^{2+} (mg/L)	98.4
$Na^+ + K^+$ (mg/L)	1220

The total capital investment for this pilot was US$0.3 million. The operation and maintenance costs were estimated to be $5000.

A total 0.2 PV caustic solution was injected at 100 m^3/d without any injectivity problem. A slight increase in injection pressure of the order of 20 kg/cm^2 was observed. But the pressure increase was short-lived, which indicated *in situ* emulsification and demulsification.

Since the pattern was unconfined, it was difficult to measure the oil recovery. It was estimated that 1.5 PV of 0.25% caustic solution would yield 38% incremental oil recovery over waterflooding. This value is probably too high for an alkaline flooding project.

6.7.5 Whittier Field in California

An experimental secondary recovery project by alkaline flooding in the Whittier Field was begun in October 1966. The objective was to improve oil recovery from an existing waterflood through the use of caustic injection (Graue and Johnson, 1974).

The net oil sand of two zones averaged about 37 and 100 ft, respectively. The project began with 45 producers and 4 injectors. Later one producer was converted into an injector. The reservoir and fluid data are shown in Table 6.3.

A slug of 0.2 PV of 0.2 wt.% sodium hydroxide was injected in fresh, nonsaline water. The incremental oil recovery was 5–7%. The chemical cost of the incremental oil recovered ranged from $0.36/bbl to $0.48/bbl. The relationship between the producing water–oil ratio (WOR) and cumulative oil production in the fraction of pore volume in the laboratory flood was found to be similar to that in the field pilot area. Thus the field performance

TABLE 6.3 Reservoir and Fluid Data in the Whittier Field

Lithology	Sandstone
Pattern type	Line drive
Project area (acres)	63
Well spacing (acres)	1–2
Formation depth (ft)	1500–2100
Net sand thickness (m)	137
Porosity (%)	30
Permeability (mD)	320–495
Formation temperature (°C)	48.9
Oil density (g/cm^3)	0.93
Oil viscosity in place (mPa·s)	40
Oil saturation at the start of flood (fraction)	0.51
Dykstra–Parsons coefficient	0.66–0.74

might be estimated from the laboratory performance. No excess sand or emulsion production occurred during the caustic field trial.

A tracer study was conducted. The objective of the tracer study was to establish whether caustic flooding changed the sweep efficiency of the flood and to determine the locations of thief zones. Chloride, bromide, nitrate, tritium, and cobalt-60 tracers were injected before and after caustic injection.

Water from nearly all of the wells in the pilot area was analyzed for these tracers. Only two wells (Well 102 and Well 121) showed tracer in the two tests (before and after caustic injection). They were both near the injector Well 48. In each tracer test, Well 102 showed tracer about 40 days after it was injected into Well 48. Both tests showed that about 1% of the fluid injected into Well 48 was produced from Well 102. Tracer was detected in Well 121 in less than 10 days in the test before the caustic injection and in 8 days after. About 4% of the fluid injected into Well 48 was produced from Well 121. These scanty data indicate that the water flow through these zones was not significantly changed by the caustic injection. The data gave no indication regarding the area sweep improvement by the caustic injection. However, spinner surveys showed that injection profiles in injection wells were significantly improved.

6.7.6 Torrance Field in California

An alkaline flooding system was designed for application in the main reservoir zone of the Joughin Unit in the Torrance field which is located between the cities of Redondo Beach and Wilmington in Los Angeles County, California. It seems that the designed field project was not completed. We present this case because extensive laboratory and simulation studies were conducted to design the largest alkaline flood ever attempted in the world by 1984 (Konopnicki and Zambrano, 1984). The designed flood encompassed the major portion of the unit and consisted of 12 inverted nine-spot patterns with a caustic injection rate of 38,000 BPD (6042 m^3/d). A 0.30 PV preflush injection which consisted of softened fresh water began on June 30, 1981 to reduce the divalent ion content and salinity of the reservoir. A 0.16 PV caustic slug of 1.2% sodium orthosilicate was scheduled to begin in early 1985. The caustic injection was planned for 2 years. The pilot data are shown in Table 6.4.

Oil production started in the Joughin Unit in 1923 and peaked in 1924 at 8100 BOPD (1288 m^3/d). The initial recovery mechanism was pressure depletion and limited water encroachment. The recovery from the depletion in the pilot area (Section B) was 29%.

Water injection started in December 1971 with fresh water being injected in a peripheral pattern arrangement at approximately 43,000 BWPD (6837 m^3/d) into 16 wells with an average surface injection pressure of

TABLE 6.4 Reservoir and Fluid Data in Section B

Pilot area (km^2)	2.44
Formation depth (m)	566
Effective thickness (m)	37
Pore volume (m^3)	27,984
Original oil saturation (fraction)	0.676
Porosity (%)	30.93
Permeability (mD)	220
Formation temperature (°C)	74
Oil density (g/cm^3)	0.279
Oil viscosity in place (mPa·s)	17
Acid number (mg KOH/g oil)	0.68–0.96
Waterflood mobility ratio	2.76

1600 psig. The incremental oil recovery from the peripheral waterflood was estimated in Section B of the Joughin Unit to be 6.3%.

In 1981, the peripheral waterflood was converted to inverted nine-spot patterns in preparation for the caustic flood. This conversion was expected to yield 4.7% OOIP. And the caustic flood was estimated to result in 1.75% OOIP. The ultimate recovery including caustic injection in the pilot area (Section B) was 41.8%.

IFT measurements demonstrated that extremely low IFTs could be achieved with a dilute solution of sodium orthosilicate in softened fresh water and reservoir brine with the total dissolved solids (TDS) less than 500 ppm. The lowest IFT were measured at 100% fresh water. Core floods utilizing sodium orthosilicate solutions of 0.8% and 1.2% resulted in reduction in the residual oil saturation of up to 9.4% PV over waterflooding alone. Laboratory data were included in a chemical simulation model to upscale to the field for prediction. The studies indicated that the injection of 0.16 PV of 1.2% sodium orthosilicate would yield an incremental oil recovery of 1875 MBO (298,125 m^3) or 1.75% OOIP.

Three continuous flow caustic consumption tests were conducted in 76-cm tubes packed with the reservoir core material. The caustic consumption rates were 15, 18 and 24 meq alkalinity per 100 g of core corresponding to 0.4, 0.8, and 1.2% sodium orthosilicate solutions. All these three tubes were flooded with 10 PV of caustic solution.

6.7.7 Wilmington Field in California

An alkaline flood EOR project was conducted in the Ranger Zone of the Wilmington Field. The Wilmington Field was one of the major fields in the

United States and was typical of many of the Pliocene–Miocene reservoirs in the Los Angeles basin. The city of Long Beach (Unit Operator of the Long Beach Unit), THUMS Long Beach Co., and the US DOE initiated alkaline injection in the Ranger Zone of Fault Block VII in March 1980 (Mayer and Breit, 1986). The Ranger Zone was a layered section of alternating sands and shales with an average gross thickness of 259 m and a net sand thickness of approximately 91 m. The pilot covered three areas. The 93-acre central area consisted of 11 producers bracketed by two four-well injection rows and two major faults. This area was covered by the cost-shared contract between DOE and the city of Long Beach. Another was the northern area of 59 acres which included six producers. The third was the 41-acre southern area which contained six producers. Both the northern and southern areas were bounded on the east and west by the same faults as the central area. The detailed reservoir and fluid data are shown in Table 6.5.

Mini-Injection Test

A one-well injectivity test (mini-injection test) was conducted during the period of June 6–August 17, 1977, in preparation for the full scale pilot.

TABLE 6.5 Reservoir and Fluid Data in the Ranger Zone of Wilmington Field

Lithology	Sandstone
Pattern type	Staggered line
Well spacing (acres)	~12
Formation depth (ft)	2225–2800
Gross sand thickness (m)	259
Net sand thickness (m)	97
Porosity (%)	25
Permeability (mD)	240
Formation temperature (°C)	51.7
Oil density (g/cm^3)	0.953
Oil viscosity in place (mPa·s)	23
Acid number (mg KOH/g oil)	0.86–2.05
Oil saturation at the start of flood	0.51
Dykstra–Parsons coefficient	0.7
Na^+ (mg/L)	6938
NH_4^+ (mg/L)	149
Ca^{2+} (mg/L)	355
Mg^{2+} (mg/L)	315
Cl^- (mg/L)	11,911
HCO_3^- (mg/L)	1306
$B_4O_7^{2-}$ (mg/L)	123

The test was conducted in Well B-836I. The mini-injection test consisted of: (1) preflush of 62,000 barrels of softened water with 1% sodium chloride beginning June 6, 1977, (2) alkaline injection of 237,000 barrels of 0.1% sodium hydroxide in softened fresh water beginning June 20, 1970. The following conclusions were derived from the test:

1. Injectivity of the preflush and alkaline fluids were satisfactory.
2. Profile surveys indicated no significant change in fluid flow distributions.
3. Chemical analyses of backflow samples indicated a substantial increase of fluid alkalinity. However, it was not determined if the increased alkalinity was due to NaOH or the conversion of HCO_3^- to CO_3^{2-} in the reservoir.
4. No determination could be made on the amount of consumption of the alkaline fluids.
5. The fill in the wellbore at the end of the test appeared to come from backflow as opposed to the formation of precipitates from chemical reactions.
6. The logistics for the handling of chemicals and the injection of fluids was satisfactory.

Chemical Injection Program

The pilot area of 193 acres was preflushed with approximately 0.1018 PV of softened fresh water with 0.9756 wt.% of sodium chloride added from January 1979 through March 27, 1980. The purpose of the preflush was to remove excessive hardness from the reservoir and to establish a salinity level which allowed the injected alkaline fluids to efficiently displace oil. The alkaline injection with 0.3917 wt.% of sodium orthosilicate in softened fresh water and 0.6703 wt.% of sodium chloride extended from March 27, 1980 to December 6, 1983. The slug volume was 0.3624 PV. The postflush of soft water with 0.7048 wt.% salt began December 6, 1983, and continued until June 26, 1984. The postflush volume was 0.0555 PV (Krumrine et al., 1985). The specifications of the types, concentrations and volumes of chemicals injected were based on laboratory testing, mini-injection test and simulation studies (Mayer and Breit, 1986).

Laboratory tests showed that alkaline consumption ranged from 8.6 to 16.9 meq/100 g rock at 51.7°C over a test period of 36 days. The high side consumption of 16.9 meq/100 g rock is equivalent to 28.6 pounds/barrel of pore space when converted. When scaled to the project pattern, the injected quantity of orthosilicate (65,000,000 pounds) was computed to be adsorbed with 0.2% of the pore volume. This indicates that the alkaline slug would penetrate very little of the reservoir and the oil recovery would be poor. Simulation studies using a kinetically controlled adsorption loss showed that the recovery from alkaline injection would be about the same as that for waterflooding. The selected 0.3917% concentration of the injected orthosilicate was based on the laboratory data that this concentration corresponded to the minimum IFT.

Sodium orthosilicate was preferred to sodium hydroxide because of better oil recovery, greater resistance to hardness, and lower IFT. The choice of sodium orthosilicate appeared justified upon the basis of laboratory tests. It was recognized that this choice could lead to problems with magnesium silicate scale. It was not realized, however, the degree to which the scaling problems would occur. It was assumed that the fresh water preflush would remove much of the hardness and thereby minimize scale problems.

Overall, the softened water preflush was not effective. The following were some of the indications:

1. The hardness of produced water remained high even after large volume of softened water injection. Typically, hardness (Ca^{2+} and Mg^{2+}) concentrations in the range of 200–300 ppm existed during the later stages of the project. It was recognized that the concentrations as measured in the producing well reflected the flow from various sublayers within the Range Zone. Thus, it is likely that hardness concentrations were lower in the more permeable zones where greater softened water injection had occurred. Regardless, the high hardness levels, coupled with a lack of oil response, suggest that the preflush was not effective in removing hardness.
2. Scale problem was severe in a number of the producing wells. Scale formation would have been less severe if the preflush had been effective in reducing hardness of the connate water.

In general, preflushing is generally ineffective in removing hardness from the reservoir, even if large volumes of softened and/or less saline water had been used in prior waterflooding.

Scaling Problems

In the formation, many reactions may occur that will alter the injected slug significantly. These include dissolution, mixing, neutralization, and ion exchange. Such reactions may lead to beneficial fluid diversion as precipitates form and block high-permeability channels. At the production wells, however, precipitation and deposition phenomena are undesirable because scales may form that may restrict production and foul well equipment.

The oil recovery from this project was poor. However, water production was reduced in some wells. The key factors for the lower than expected oil recovery were attributed to high consumption of the alkaline fluids and to wellbore plugging problems. High consumption was due in part to the mixing of alkaline fluids with the water of high hardness and to reactions of the alkaline fluids with the reservoir rock. The plugging problem was due to the formation of calcium carbonate scale, magnesium silicate scale, and to silicate-containing precipitates.

Workovers to remove the damage were partially successful. The performance might have been improved by several different procedures. These improvements include the use of a different preflush design and polymers for mobility control (Dauben et al., 1987). However, these were considered to be of secondary importance since the alkaline consumption appeared to be the dominant factor affecting the performance.

Beginning in early 1981, the water rate in a number of producing wells which were closest to the alkaline injection wells began to decline. The decrease affected 16 wells. The decrease amounted to 5000–7000 B/D of water out of an initial total of 15,000–16,000 B/D of water. The decrease generally began before chemical changes were observed in the produced water. Small-volume acidizing of the wells did not restore productivity. As a result, the decrease in water rate was attributed to partial blockage of flow channels in the formation away from the wellbores. The WOR of these wells did not improve but remain essentially unchanged. As a result, a ±300-B/D oil-rate loss was incurred in the pilot.

The chemistry of each well's effluent was measured monthly. These tests served as the diagnostic of the producing-well problems that were encountered. Well B-105 showed increase in silica in the produced water and a loss of water productivity with no increase in pH. The increase in silica content was generally accompanied by carbonate-scaling problems. The calcium carbonate deposition usually occurred in and around the pump but not in the liner. Hydrochloric acid was used for tubing and pump cleanup, while a continuous annual sidestream treatment was used for carbonate-scale inhibition. This treatment allowed wells of this type to be produced in a satisfactory manner.

Well B-712 is an example of a well in which the silica increased for a time without a pH increase. During this period, the calcium carbonate scale problems described for Well B-105 were found and successfully alleviated. However, when the silica increased further, along with a large jump in pH and corresponding drops in hardness and salinity, a siliceous scale was detected. This material was analyzed to be mixed magnesium silicate and calcium carbonate. The well was subsequently treated specifically to overcome this magnesium silicate problem, but the pump ran only a short time before failing. Pump scaling was the cause of this failure.

Note that in any of the pilot-area wells where pump scaling or plugging of any type was encountered, the problem manifested itself basically as (1) a seizing of the stages in the centrifugal bottom-hole pump, (2) plugging of the pump intake, (3) some external pump scaling, and in a few cases (4) internal scaling of a few bottom joints of the tubing string. Except for well-pulling work where a mechanical failure of the casing/liner was detected, there was only one well-servicing job where more than minimal fill had to be removed from the liner. For more details, see Krumrine et al. (1985).

Interestingly, plans were prepared for the installation of additional production handling facilities if emulsion problems were encountered. It was found that these facilities were not needed.

6.7.8 Court Bakken Heavy Oil Reservoir in Saskatchewan, Canada

The Court Bakken heavy oil reservoir was discovered in 1982 and produced in the same year. The waterflooding was started in 1988 with 40-acre well spacing in irregular patterns. The water cut in 2007 was 95% with 20 injectors and 28 producers. It was predicted that waterflooding recovery factor was 30%. Some of the reservoir and fluid data are presented in Table 6.6 (Xie et al., 2008).

A single-well test was conducted in two phases. In the first phase, water was injected into the test well to reduce oil saturation in the near wellbore region to waterflood residual oil saturation. In the second phase, a slug of 5000 ppm caustic solution was injected followed by a slug of push water. The residual oil saturations, measured by chemical tracer tests and RST (reservoir saturation tool) logs, after waterflooding and caustic flooding were 0.28 and 0.17, respectively.

After laboratory tests, single-well tracer tests, and reservoir simulation study, a pilot was designed. The pilot consisted of an injector, two existing producers and one new producer. Injection of 0.4 PV of 0.5 wt.% caustic solution (NaOH) was planned. The following monitoring programs were planned.

- Radioactive tracer survey to determine the injected fluid flow path and distribution among pilot producers. One tracer test would be conducted during the preflush period. Another tracer test would be conducted after the caustic flood front passed through the three pilot producers. The tracer

TABLE 6.6 Reservoir and Fluid Data in the Court Bakken Reservoir

Original reservoir pressure (kPa)	6620
Original reservoir temperature (°C)	31
Well spacing (acres)	40
Formation depth (m)	870
Gross sand thickness (m)	2–8
Net sand thickness (m)	7.6
Porosity (%)	29
Permeability (mD)	2100
Oil density (°API)	17
Oil viscosity in place (mPa·s)	155
Acid number (mg KOH/g oil)	0.77

tests would determine the caustic consumption and to estimate sweep efficiencies before and after caustic flooding.
- RST logs to determine residual oil saturation.
- Injector spinner surveys planned before and after the caustic flood.
- Pressure transient tests would be conducted on the injector and producers.
- Oil and water cut monitoring on a frequent and regular basis to determine the arrival of the oil bank, if any, as well as to determine the effect of caustic on the long-term performance of the producers.
- pH and alkalinity monitoring on a frequent and regular basis at the pilot producers and the offset observation well to determine arrival of caustic front, the sweep efficiency and caustic consumption.
- Scaling and corrosion monitoring to evaluate and mitigate the effect of caustic solutions on wells.

Caustic solution injection started on June 30, 2006. A sector model simulation predicted an incremental oil recovery of 9% OOIP.

6.8 CONCLUSIONS

Based on what are presented in this chapter, we may summarize the following observations regarding alkaline flooding EOR.

1. Sodium hydroxide was most often used alkali in the early alkaline flooding projects. Sodium orthosilicate was used if the main objective of alkaline injection was to achieve mobility control by using precipitates. Weak alkalis such as sodium carbonate are preferred to avoid scaling and produced emulsion problems.
2. Apparently, the mobility-control mechanism was emphasized in Russian projects. And the difficulty to treat produced emulsions was not an issue.
3. Combined chemical EOR methods are very active in China, but no single field project using alkaline flooding alone was reported in the literature.
4. More research is needed for alkaline flooding, as the process is completely unproved in field applications (Doe et al., 1987).

REFERENCES

Berger, P.D., Lee, C.H., 2006. Improved ASP process using organic alkali. Paper SPE 99581 Presented at the 2006 SPE/DOE Symposium on Improved Oil Recovery, Tulsa, OK, 22–26 April.

Burk, J.H., 1987. Comparison of sodium carbonate, sodium hydroxide, and sodium orthosilicate for EOR. SPERE February, 9–16.

Buchenkov, L.N., Gorbunov, A.T., Zhdanov, S.A., et al., 1986. Field studies of the alkaline flooding method. VNIIOENG.

Campbell, T.C., 1977. A comparison of sodium orthosilicate and sodium hydroxide for alkaline waterflooding. Paper SPE 6514 Presented at the 1977 SPE California Regional Meeting, Bakersfield, CA, 13–15 April.

Campbell, T.C., 1982. The role of alkaline chemicals in the recovery of low-gravity crude oils. JPT November, 2510–2516.

Castor, T.P., Somerton, W.H., Kelly, J.F., 1981. Recovery mechanisms of alkaline flooding. In: Shah, D.O. (Ed.), Surface Phenomena in Enhanced Oil Recovery. Plenum Press, New York, NY, pp. 249–291.

Cheng, J.-C., Liao, G.-Z., Yang, Z.-Y., Li, Q., Yao, Y.-M., Xu, D.-P., 2001. Overview of Daqing ASP pilots. Pet. Geol. Oilfield Dev. Daqing 20 (2), 46–49.

Dauben, D.L., Easterly, R.A., Western, M.M., 1987. An evaluation of the alkaline waterflooding demonstration project, Ranger Zone Wilmington Field, California, Report DOE/BC/10830-5, US DOE, Bartlesville, OK (May).

deZabala, E.F., Vislocky, J.M., Rubin, E., Radke, C.J., 1982. A chemical theory for linear alkaline flooding. SPEJ April, 245–258.

Doe, P.H., Carey, B.S., Helmuth, E.S., 1987. The 1984 Natl. Petroleum Council study on EOR: chemical processes. JPT 39 (8), 976–980.

Ehrlich, R., Wygal, R.J., 1977. Interaction of crude oil and rock properties with the recovery of oil by caustic waterflooding. SPEJ August, 263–279.

Flaaten, A.K., Nguyen, Q.P., Zhang, J.-Y., Mohammadi, H., Pope, G.A., 2008. Chemical flooding without the need for soft water. Paper SPE 116754 Presented at the SPE Annual Technical Conference and Exhibition, Denver, CO, 21–24 September.

French, T.R., Burchfield, T.E., 1990. Design and optimization of alkaline flooding formulations. Paper SPE 20238 Presented the SPE/DOE Enhanced Oil Recovery Symposium, Tulsa, OK, 22–25 April.

Goyal, K.L., Gupta, R.K., Nanda, S.K., Paul, A.K., Sharma, P., 1991. Performance evaluation of caustic pilot project of north Gujarat oil field, India. Paper SPE 21410 Presented at the Middle East Oil Show, Bahrain, 16–19 November.

Graue, D.J., Johnson Jr., C.E., 1974. Field trial of caustic flooding process. JPT 26 (12), 1353–1358.

Jennings Jr., H.Y., Johnson Jr., C.E., McAuliffe, C.D., 1974. A caustic waterflooding process for heavy oils. JPT December, 1344–1352.

Johnson Jr., C.E., 1976. Status of caustic and emulsion methods. JPT January, 85–92.

Konopnicki, D.T., Zambrano, L.G., 1984. Application of the alkaline flooding process in the Torrance field. Paper SPE 12701 Presented at the SPE Enhanced Oil Recovery Symposium, Tulsa, OK, 15–18 April.

Krumrine, P.H., Mayer, E.H., Brock, G.F., 1985. Scale formation during alkaline flooding. JPT 37 (8), 1466–1474.

Mayer, E.H., Berg, R.L., Carmichael, J.D., Weinbrandt, R.M., 1983. Alkaline injection for enhanced oil recovery—a status report. JPT January, 209–221.

Mayer, E.H., Breit, V.S., 1986. Alkaline flood prediction studies, Ranger VII pilot, Wilmington Field, CA. SPERE 1 (1), 9–22.

Novosad, Z., et al. 1981. Comparison of oil recovery potential of sodium orthosilicate and sodium hydroxide for the Wainwright Reservoir, Alberta. Petroleum Recovery Institute Report No. 81–10, Alberta, Canada (July).

Sarem, A.M., 1974. Secondary and tertiary recovery of oil by MCCF (mobility-controlled caustic flooding) process. Paper SPE 4901 Presented at the SPE–AIME 44th Annual California Regional Meeting, San Francisco, CA, 4–5 April.

Shuler, P.J., Kuehne, D.L., Lerner, R.M., 1989. Improving chemical flood efficiency with micellar/alkaline/polymer process. JPT January, 80–88.

Subkow, P., 1942. Process for the removal of bitumen from bituminous deposits, US Patent No. 2,288,857, 7 July.

Xie, J., Chung, B., Leung, L., 2008. Design and implementation of a caustic flooding EOR pilot at Court Bakken heavy oil reservoir. Paper 117221 Presented at the International Thermal Operations and Heavy Oil Symposium, Calgary, AB, Canada, 20–23 October.

Chapter 7

Alkaline-Polymer Flooding

James J. Sheng

Bob L. Herd Department of Petroleum Engineering, Texas Tech University, Lubbock, TX 79409, USA

7.1 INTRODUCTION

As reported in the chapter of alkaline flooding, past field applications did not show significantly high incremental oil recovery over waterflooding for several reasons. One possible reason is lack of mobility control in alkaline flooding alone. The combination of alkaline and polymer floods seems to be a better option. In this chapter, we first present the interaction and synergy between alkali and polymer, then several field applications.

7.2 INTERACTIONS BETWEEN ALKALI AND POLYMER

There are some interactions between alkali and polymer when they are combined. Sheng (2011) presented detailed discussion of these interactions. Some of the main conclusions are summarized here.

Since an alkali has salt effect, adding an alkali will reduce polymer solution viscosity. However, an alkali accelerates or increases the hydrolysis of polymer solution. Therefore, adding an alkali may increase polymer solution viscosity. In general, the salt effect is greater. Therefore, the polymer viscosity decreases with alkaline concentration, as shown in Figure 7.1 as an example.

There is no consensus regarding the polymer effect on alkaline solution/oil interfacial tension (IFT). Generally, it is believed that polymer has little effect on the IFT.

Laboratory test results show that alkaline consumption in an alkaline-polymer (AP) system is lower than that in the alkaline solution itself. This is probably because polymer covers some of rock surfaces to reduce alkali–rock contact. In an AP system, alkali competes with polymer for positive sites. Thus polymer adsorption is reduced because the rock surfaces become more negative-charged (Krumrine and Falcone, 1987).

7.3 SYNERGY BETWEEN ALKALI AND POLYMER

Alkaline reaction with crude oil results in soap generation, wettability alteration, and emulsification. Polymer provides required mobility control. AP

Enhanced Oil Recovery Field Case Studies.

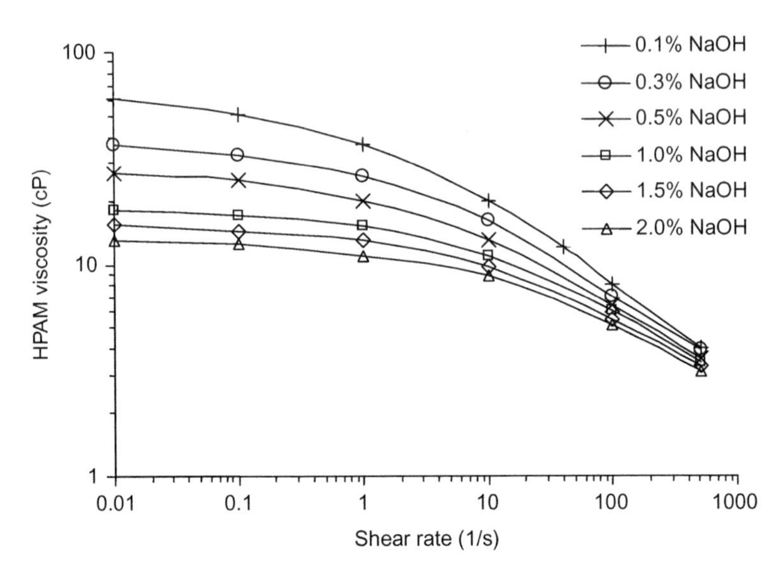

FIGURE 7.1 Alkaline effect on polymer (1000 mg/L 1275 A) viscosity (Kang, 2001).

flooding can displace more residual oil than individual alkaline flooding or polymer flooding.

Some Russian and Chinese researchers (Chen et al., 1999) observed in laboratory that as pH was increased, polymer adsorption was reduced. But this phenomenon has not been documented as well as for surfactant reduction (Labrid, 1991). Meanwhile, polymer adsorption reduces alkali reaction with rocks.

As we know, adding alkali in a polymer solution will reduce the polymer solution viscosity. We may take advantage of this fact in low injectivity wells. Initially, the polymer solution with alkali has a low viscosity. As the alkali is consumed by reacting with formation water and rocks, the polymer solution viscosity will become higher than the initial value. Thus, initially the injectivity and later the sweep efficiency are improved. Note that polymer concentration will also be reduced by adsorption. The final effect is determined by the balance between the two effects of reduced alkaline and polymer concentrations.

The synergy between alkaline and polymer flooding may be summarized as follows.

1. Alkaline polymer can reduce polymer adsorption and alkaline consumption as well.
2. Polymer makes the AP solution more viscous to improve sweep efficiency. Thus polymer "brings" alkaline solution to the oil zone where the alkaline cannot go without polymer. More oil can be displaced by lowered IFT due to alkaline-generated soap. In other words, alkaline and

polymer work together to improve both sweep efficiency and displacement efficiency.

3. Alkaline may reduce polymer viscosity in the near-wellbore region, so that the injectivity is improved.

7.4 FIELD AP APPLICATIONS

In this section, we will present the AP field cases in Almy Sands (Isenhour Unit), Moorcroft West and Thompson Creek in Wyoming, David Lloydminster "A" Pool and Etzikom in Canada, Xing-28 Block (Liaohe Field) and Yangsanmu in China.

7.4.1 Almy Sands (Isenhour Unit) in Wyoming, USA

In this case, cationic and anionic polymer was injected before AP flooding to improve sweep efficiency.

In an effort to reduce residual oil saturation in the swept area in the Almy Sands in Sublette County, Wyoming, a fieldwide AP flood was initiated in September 1980 at the Isenhour Unit.

The M-42 reservoir of the Almy Sand series, a tertiary sand deposition in a river bed was subjected to eight stages of diagenesis prior to hydrocarbon employment. Feldspar overgrowths, clay and calcite cement, emplacement of abundant chlorite and kaolinite clays as well as calcite reduced primary intergranular porosity. The detailed reservoir and fluid data are shown in Table 7.1.

On the basis of core testing, crude oil alkaline-agent screening determined that soda ash (Na_2CO_3) and anionic polymer would be most beneficial in reducing IFT and mobilizing residual oil. A cationic and anionic polyacrylamide spearhead preceded the addition of the alkaline agent.

TABLE 7.1 Reservoir and Fluid Data in the Ranger Zone of Almy Sands Isenhour Unit

Lithology	Sandstone
Pilot area (acres)	173
Formation depth (ft)	3700
Net sand thickness (m)	15
Porosity (%)	15.5
Permeability (mD)	21
Formation temperature (°C)	36.1
Oil density (g/cm^3)	0.81
Oil viscosity in place (mPa·s)	2.8
Oil saturation at the start of flood	0.35

Individual injection wells were presoaked for 2 weeks with 10,000 lbm (4536 kg) KCl in 500 bbl (79.5 m^3) water. Cationic polyacrylamide followed to coat the near-wellbore area, thus preventing fresh water effects on the formation clays. This soak-and-coat technique had been the standard clay stabilization practice in the Almy Sands since September 1969. A 0.043 pore volume (PV) of cationic spearhead was injected. Anionic polyacrylamide followed to provide sweep improvement. This process was called the "CAT-AN" process in which cationic polymer was injected before anionic polymer (Mack, 1978).

A 20:1 dry weight ratio of soda ash to a dry blend of anionic polymer with inorganic wetting and stabilizing agents were injected over an 8-month period. Anionic polyacrylamide was injected to provide continuing sweep improvement and as a mobility buffer behind the alkaline slug. The total anionic polyacrylamide and soda ash with a polymer-blend slug was about 0.372 PV. Use of tripolyphosphate/anionic-polymer blend continued to provide long-term wettability control and alkaline slug stability with approximately 0.069 PV injected up to January 1, 1985.

On the basis of actual chemical usage by January 1, 1985, and the engineered oil recovery, the program chemical cost was $1.12/STB oil. For individual well performance, see Doll (1988).

7.4.2 Moorcroft West in Wyoming, USA

In this case, KOH was used. The Moorcroft West Minnelusa Sand Unit was located in the northeastern corner of the Powder River Basin approximately 15 miles northwest of Moorcroft, Wyoming. The unit produced from a confined reservoir providing an ideal setting for evaluation of EOR technologies because there was little possibility for off pattern operations to influence interpretation of results. There were two wells: Texas Trials No. 1 and Evans No. 1-12 (Bala et al., 1992).

The Moorcroft West Minnelusa Sand Unit was estimated to contain 686,000 bbls. The primary producing mechanism was rock and fluid expansion (there is very little associated gas and no apparent active water drive in this reservoir). No water/oil contact was evident at a well. The primary oil recovery averaged 11% of original oil in place (OOIP).

The porosity and water saturation of the Moorcroft West reservoir as determined from log calculations were 15.2% and 28.1%, respectively. The estimated permeability for the field averaged 114 mD (6–150 mD), and the Dykstra–Parsons permeability variation coefficient for Moorcroft West was 0.65. The average sand thickness was 8.4 ft. The reservoir temperature was 48.9°C. The oil density was 0.92 containing 27.23% of aromatic, 7.69% pentane soluble, 7.6% pentane insoluble components, and 55.44% aliphatic. The oil viscosity was about 42 mPa·s at the reservoir temperature. The injected and produced water analysis is shown in Table 7.2.

TABLE 7.2 Produced and Injected Water Analysis in Moorcroft West

Water	Produced	Injected
Formation	Minnelusa A	Unknown
Depth (ft)	5900	200
Sample point	Texas Trail #1 Treater	Wellhead
pH	7.34 ± 0.47	7.3
Total dissolved solids (TDS)	42,370 ± 15,313	366
Hardness (as $CaCO_3$)	4522 ± 1234	21
Na^+ (ppm)	16,837 ± 3044	93
Ca^{2+} (ppm)	999 ± 95	31.3
Mg^{2+} (ppm)	536 ± 76	14.6
Fe^{total} (ppm)	5.8 ± 4.2	0
Cl^- (ppm)	26,895 ± 3256	1.94
SO_4^{2-} (ppm)	1763 ± 519	101

In April 1989, the Evans well was converted for injection using a temporary water injection system. Water for injection was obtained using unburied piping (approximately 600 ft) connected from an existing livestock supply well to a 400 barrel holding tank (filled on meter with filtration). Injection was initiated on vacuum from the tank into the injection well. In August, 1989 after 12,000 bbls (1.2% PV) of water were injected, the production at the Texas Trails No. 1 increased from 3 to 6 BOPD with no increase in produced water volume.

In October 1989, a volumetric sweep improvement program based on cationic and anionic polymers to treat reservoir heterogeneities was initiated. Cationic polymer (dry granular cationic polyacrylamide) was added under vacuum at 25 lbs/day in 120 bbls (about 600 mg/L) injection water. The injection by vacuum ended in November 1989. Cationic polymer was followed by anionic polymer (dry granular anionic polyacrylamide) in April 1990 at 10 lbs/day in 100 bbls (286 mg/L) injection water. The targeted polymer injection volume was 0.06 PV. The anionic polymer was intended to improve the adverse mobility ratio of 7 experienced at Moorcroft West.

As of June 1991, the oil production was 38 BOPD from the Minnelusa A sand (the main producing sands were the upper portion of the Minnelusa formation which contained A and B sands). The injection rate averaged 45 BPD of polymer solution at 1100 psig surface pressure. The primary production from this reservoir was low (approximately 5% OOIP) and the field production had dropped to 4 BOPD prior to the initiation of water injection. Small volumes of water had been produced throughout the life of the reservoir, but there was no evidence of breakthrough of the injected fluid by June 1991.

The cumulative injection was 56,947 BW or 5.8% PV by June 1991. A total of 4500 lbs of cationic polymer at an average concentration of 606 mg/L and 1700 lbs of anionic polymer at an average concentration of 284 mg/L had been injected. Chemical cost was $1.46/bbl of oil produced since the initiation of water injection. The recovery was 49,950 bbl oil, or 7.3% OOIP. The oil recovery since the initiation of water injection was 15,978 bbls of oil, or 2.3% OOIP. An original investment of $75,000 included initial purchase, unitization, completion of the injection well, the purchase and installation of the injection plant, and all associated costs through November 1990. These costs were returned by July 1991, approximately 14 months after initiating water injection.

In September 1991, an AP process was designed to begin. The objective was to reduce residual oil saturation by reducing the IFT of the oil and the brine was initiated. The AP process was based on the injection of 0.2 PV of 0.2 wt.% potassium hydroxide and 225 mg/L anionic polymer. This design was developed through the laboratory analysis which indicated the IFT reduction up to 100-fold or greater by the addition of KOH. Biocide (1,2-dibromo -2,4-dicyanobutane) was injected as a slug once a month at 2 lbs in 45 bbls water (127 mg/mL). Biocide application was designed to prevent microbial contamination at injection facilities and tubulars.

The IFT between the Moorcroft West crude and these solutions were 100 times lower than that of oil/water: (1) 0.2–3.0 wt.% KOH, (2) 1.5–3.0 wt.% of Na_2CO_3, or (3) 0.2–3.0 wt.% of cationic quaternary amine surfactant and sodium carbonate with the surfactant–soda ash ratio of 1:1. The lowest IFT was reached at 0.2 wt.% KOH. KOH was as effective as the combined 1:1 solution of surfactant and soda ash, but the associated cost of KOH was six times less. If Na_2CO_3 (3.0 wt.%) was used, the cost would be three times higher than that of KOH. Unfortunately, the performance data of AP flooding are not found in the literature.

7.4.3 Thompson Creek Field in Wyoming, USA

An AP project was carried out in the Thompson Creek Field. The flood area was 26,000 MBBL PV. The reservoir temperature was 31°C. The oil density was 0.94. The oil viscosity was 300–515 cP, and the mobility ratio was 13.6. The permeability was 1400 mD. Waterflooding began in July 2002 and AP flooding began in May 2004. The long-term prediction based on the short-period performance after the AP flooding showed more oil production but less water production.

7.4.4 David Lloydminster "A" Pool in Canada

An AP flood was designed for the David Lloydminster "A" pool in the east central Alberta. The David Lloydminster "A" pool was originally under waterflood from November 1978 to June 1987, and 31.2% OOIP had been recovered.

The David AP project was located approximately 100 km south of Lloydminster, Alberta. The pool was in the Lloydminster formation which was an unconsolidated (partially cemented) sand of the Upper Cretaceous period and located at a depth of about 758 m KB. Eighteen producers and seven injectors were arranged in irregular patterns. The detailed reservoir and fluid data are shown in Table 7.3.

In December 1983, the operator, Dome Petroleum, submitted an application to the Energy Resources Conservation Board (ERCB) of Alberta for an EOR scheme which involved injection of 0.5 PV of 2 wt.% caustic (NaOH) and infill drilling of 22 wells. A 0.1 PV of soft water (preflush) was to be injected to condition the reservoir prior to caustic injection. A postflush of 0.2 PV of soft water was to be injected and a total project life of 6 years was anticipated. As a result of the falling oil price, 14 infill wells instead of 22 were drilled by January 1986.

Further laboratory tests showed incremental oil by using alkali in conjunction with polymer and surfactant. The laboratory tests and the progression of the process from caustic injection only, to sodium carbonate-polymer to sodium-surfactant-polymer have been described in the literature (Manji and Stasiuk, 1988).

Two radial corefloods were conducted to determine the difference in recoveries between an AP system and an alkaline-surfactant-polymer (ASP) system. The cumulative oil recovery of 83% OOIP by ASP was higher than that of 67% by AP. However, the initial oil saturation in the ASP core was

TABLE 7.3 Reservoir and Fluid Data in the David Lloydminster "A" Pool

Lithology	Sandstone
Project area (acres)	474
Formation depth (m KB)	758
Net sand thickness (m)	3.2
Project rock volume (10^6 m^3)	6.07
Pattern rock volume (10^6 m^3)	3.98
OOIP (10^6 m^3)	1.486
Well spacing (acres)	20
Porosity (%)	29
Permeability (mD)	1400
Initial reservoir pressure (kPa)	5841
Reservoir temperature (°C)	30.6
Oil density (g/cm^3)	0.9182
Oil viscosity in place (mPa·s)	34.1

0.87 compared to 0.67 for the AP core. Other studies by Dome Petroleum showed that the oil recovery was higher when the initial oil saturation was higher. Therefore, superior performance by adding surfactants was not conclusive. The IFT was in the order of 0.1 mN/m for the AP solution, and 0.01–0.001 mN/m for the solution with surfactant added.

AP flooding was initiated in the David Pool during June 1987. At that time the oil cut was 40%. The injected chemical solution was 1.0 wt.% sodium carbonate and 800 ppm Alcoflood 1175 (polyacrylamide polymer). The injection sequence was 0.213 PV of AP solution followed by 0.041 PV of a tapered polymer concentration. Water followed the tapered polymer slug in December 1990. AP injection produced extra 21.1% OOIP. The project cost was C$7,284,000 for a chemical mixing plant with water softening, drilling of 14 new wells, purchase of chemicals and engineering design expenses. The incremental chemical cost per produced barrel of oil was C $1.41/bbl or US$0.99/bbl. The total cost per incremental barrel of oil was C $3.53/bbl or US$2.48/bbl (Pitts et al., 2004; Wyatt et al., 2004).

7.4.5 Etzikom Field in Alberta, Canada

An AP project was carried out by Husky Oil in the Etzikom B and C Pools. The reservoir depth was 3150 ft. The reservoir temperature was 30°C. The oil density was 0.94. The oil viscosity was 100 cP, and the water viscosity was 0.8 cP with the mobility ratio of 17.2. The reservoir thicknesses of the B Pool and C Pool were 26 and 22 ft, respectively. The porosities of the B Pool and C Pool were 23.4 and 19.7%, respectively. The permeability was 1000 mD. Waterflooding began in 1971 and AP flooding began in 2000. After chemical injection, the oil rate reached the historical maximum rate in the C Pool. The details of the project were not found.

7.4.6 Xing-28 Block, Liaohe Field, China

This is an AP case in a high water-cut reservoir having a gas cap and edge water. An AP pilot test was conducted in the Xing-28 Block in Liaohe Field. The pilot was in the western part of Xing-28 Block. Xing-28 block had 2.05 km^2, reservoir thickness of 3 m, and OOIP of 0.96 million tons. It had an anticline structure. It had a gas cap of 1.08 km^2 with the thickness of 2.6 m. It also had edge water. The reservoir depth was 1650–1730 m. The formation porosity was 0.276 and permeability of 2063 mD. The formation was water-wet sandstone.

The oil in place had a density of 0.8174 g/cm^3 and viscosity of 6.3 mPa·s at the reservoir temperature of 56.6°C. The paraffin content in the oil was 4.14%. The initial reservoir pressure was 17.29 MPa. The formation water TDS was 3112 mg/L with Ca^{2+} and Mg^{2+} of 14 mg/L.

The oil production was started in September 1971, and the water injection was started in November 1974. In December 1994 (just before the AP pilot test), the water cut was 96.2% and the oil recovery was 46.75%.

Several studies were conducted including reservoir modeling to investigate residual oil saturation distribution, to select well pattern, to optimize chemical injection formula, and to predict AP performance.

Based on the study work, the selected central well pattern included three injectors and one observation (production) well. The well spacing (interwell distance) was 160–190 m. Four peripheral observation (production) wells were around the three injectors. These wells were well connected. The central pilot area covered 0.037 km^2 with thickness of 7.4 m. The rock had 6.3% carbonate content and 2.5% clay content.

The optimized formula was 2% Na_2CO_3 and 1000 mg/L 1175 A. The study also showed that the economics of AP flood was better than that of ASP flood. Therefore, the AP option was selected for this pilot.

The AP pilot test was conducted from January 1995 to August 1998. The AP flood increased the oil recovery by 1.98 OOIP for the whole pilot area and 18.5% for the central well area. From January 1995 to the time the water cut reached 98.0%, the oil recovery was 3.34% (OOIP) for the whole pilot area. However, it was found that the AP flood conducted in this pilot area was not economically attractive owing to larger amount of capital investment and the low oil price at that time. For a more detailed description of this project, see Zhang et al. (1999) and Sheng (2011).

7.4.7 Yangsanmu in China

This case presents an AP pilot in a viscous oil reservoir with high water cut. Treated produced water was used in mixing AP solution. The pilot was carried out in the NgII well group, Block 3, Yangsanmu field in China (Yang et al., 2010).

There were 4 injectors and 18 producers. The reservoir porosity was 31%, permeability 830 mD. The reservoir temperature was 62°C and the oil and water viscosities at this temperature were 120 and 0.5 mPa·s, respectively. The acid number was 1.1–2.6 mg KOH/g oil.

The selected formula was 1% Na_2CO_3 and 1000 ppm polymer. The designed injection volumes were 0.16 PV AP slug and two tapered polymer slugs of 0.06 PV and 0.03 PV. The adsorption for 0.5% solution was 1.2645 mg/g sand, and the adsorption of 0.5% sodium carbonate and polymer solution was 0.8962 mg/g sand. The IFT reached 10^{-2} mN/m within 0.3–0.5% sodium carbonate.

The actual AP flood was started in March 1999 and ended in May 2001. The actual injection concentrations were 0.84% alkali and 1094 ppm polymer. The changes in concentrations were made because of scale problem. The subsequent polymer injection was ended in April 2004.

Oil rate increased from 74.6 tons/d to 124 tons/d, and water cut decreased from 97% to 94.3% 7 months after AP injection in October 1999.

7.5 CONCLUDING REMARKS

Apparently, AP trials were applied to relatively high viscous oil reservoirs. An effort was made to collect more AP field cases. However, there have not been many field trials or applications so far.

REFERENCES

Bala, G.A., Duvall, M.L., Jackson, J.D., Larsen, D.C., 1992. A flexible low-cost approach to improving oil recovery from a (very) small Minnelusa Sand reservoir in Crook County, Wyoming. Paper SPE 24122 Presented at the SPE/DOE Enhanced Oil Recovery Symposium, Tulsa, OK, 22–24 April.

Chen, Z.-Y., Sun, W., Yang, H.-J., 1999. Mechanistic study of alkaline–polymer flooding in Yangsanmu field. J. Northwest University (Natural Science Edition) 29 (3), 237–240.

Doll, T.E., 1988. An update of the polymer-augmented alkaline flood at the Isenhour unit, Sublette County, Wyoming. SPERE 3 (2), 604–608.

Kang, W.-L., 2001. Study of Chemical Interactions and Drive Mechanisms in Daqing ASP Flooding. Petroleum Industry Press, Beijing, China.

Krumrine, P.H., Falcone, J.S., 1987. Beyond alkaline flooding: Design of complete chemical systems. Paper SPE 16280 Presented at the SPE International Symposium on Oilfield Chemistry, San Antonio, TX, 4–6 February.

Labrid, J., 1991. The use of alkaline agents in enhanced oil recovery processes. In: Bavière, M. (Ed.), Basic Concepts in Enhanced Oil Recovery Processes, SCI, pp. 123–155.

Mack, J.C., 1978. Improved oil recovery—product to process. Paper SPE 7179 Presented at the SPE Rocky Mountain Regional Meeting, Cody, WY, 17–19 May.

Manji, K.H., Stasiuk, B.W., 1988. Design considerations for Dome's David alkali/polymer flood. JCPT 27 (3), 49–54.

Pitts, M.J., Wyatt, K., Surkalo, H., 2004. Alkaline-polymer flooding of the David Pool, Lloydminster Alberta. Paper SPE 89386 Presented at the SPE/DOE Symposium on Improved Oil Recovery, Tulsa, OK, 17–21 April.

Sheng, J.J., 2011. Modern Enhanced Oil Recovery—Theory and Practice. Elsevier, Burlington, MA.

Wyatt, K., Pitts, M.J., Surkalo, H. 2004. Field chemical flood performance comparison with laboratory displacement in reservoir core. Paper SPE 89385 Presented at the SPE/DOE Symposium on Improved Oil Recovery, Tulsa, OK, 17–21 April.

Yang, D.-H., Wang, J.-Q., Jing, L.-X., Feng, Q.-X., Ma, X.-P., 2010. Case study of alkali—polymer flooding with treated produced water. Paper SPE 129554 Presented at the SPE EOR Conference at Oil & Gas West Asia, Muscat, Oman, 11–13 April.

Zhang, J.-F., Wang, K.-L., He, F.-Y., Zhang, F.-L., 1999. Ultimate evaluation of the alkali/polymer combination flooding pilot test in XingLongTai oil field. Paper SPE 57291 Presented at the SPE Asia Pacific Improved Oil Recovery Conference, Kuala Lumpur, 25–26 October.

Chapter 8

Alkaline-Surfactant Flooding

James J. Sheng

Bob L. Herd Department of Petroleum Engineering, Texas Tech University, Lubbock, TX 79409, USA

8.1 INTRODUCTION

When an alkali is added in a surfactant solution, the salinity in the solution is increased because alkali provides electrolyte. The increased salinity combined with the lower optimum salinity of the generated soap results in different optimum salinity than that of the surfactant solution alone. Addition of an alkali reduces surfactant adsorption. This chapter will first discuss these interactions. After that, two field trials of alkaline-surfactant flooding will be presented.

8.2 INTERACTIONS AND SYNERGIES BETWEEN ALKALI AND SURFACTANT

In this section, we will discuss several interactions and synergies between alkali and surfactant.

8.2.1 Alkaline Salt Effect

Alkalis can also provide electrolyte to a surfactant solution. However, the relationship between alkaline concentration and the salinity provided is complex. In other words, adding 1 meq/mL of alkali may not equivalently add 1 meq/mL of salinity.

8.2.2 Effect on Optimum Salinity and Solubilization Ratio

Generally, an injected surfactant is more hydrophilic than a soap. Thus the optimum salinity of the soap is lower than that of the synthetic surfactant. Then what is the optimum salinity of a system of a soap and a surfactant?

Based on experimental data, Salager et al. (1979) proposed the following logarithmic mixing rule for optimum salinities:

$$\ln(C_{\text{se,m}}^{\text{opt}}) = X_1 \ln(C_{\text{se,1}}^{\text{opt}}) + X_2 \ln(C_{\text{se,2}}^{\text{opt}}) \tag{8.1}$$

where $C_{\text{se,m}}^{\text{opt}}$, $C_{\text{se,1}}^{\text{opt}}$, and $C_{\text{se,2}}^{\text{opt}}$ are the optimum salinities of the mixture and surfactant components 1 and 2, respectively, X_1 and X_2 are the mole fractions of surfactants 1 and 2, respectively. Puerto and Gale (1977) used a linear mixing rule on optimum salinity to fit their data.

For other parameters, we also need a mixing rule. Mohammadi et al. (2008) found that for the optimum solubilization ratios both logarithmic and linear mixing rules are satisfied:

$$R_{23,m}^{\text{opt}} = X_1 R_{23,1}^{\text{opt}} + X_2 R_{23,2}^{\text{opt}} \tag{8.2}$$

or

$$\ln(R_{23,m}^{\text{opt}}) = X_1 \ln(R_{23,1}^{\text{opt}}) + X_2 \ln(R_{23,2}^{\text{opt}}) \tag{8.3}$$

where $R_{23,m}^{\text{opt}}$, $R_{23,1}^{\text{opt}}$, and $R_{23,2}^{\text{opt}}$ are the optimum solubilization ratio of the mixture and surfactant components 1 and 2, respectively. There is no agreement whether a linear mixing rule or logarithmic mixing rule should be used.

Generally, the amount of injected alkali is high enough to react with all the naphthenic acid in the crude oil. More soap would likely be generated in a system with higher oil content. Thus the phase behavior of the two-surfactant system depends on the water–oil ratio (WOR). Zhang et al. (2006) found that if the optimum salinity of a system is plotted versus the mole ratio of the soap to the surfactant, the effect of WOR can be removed. In other words, the relationship between the optimum salinity versus the mole ratio is independent of WOR.

8.2.3 Synergy Between Soap and Surfactant to Improve Phase Behavior

Many investigators have observed that the lowest interfacial tensions (IFTs) between a crude oil and an alkali solution frequently occur at very low concentrations of alkali (Nelson et al., 1984), for example, in the region of 0.1 wt.% (Green and Willhite, 1998). To reach the lowest IFT, we have to use a low alkaline concentration. However, if we use a low alkaline concentration, the alkaline bank may not be able to propagate through the reservoir because of higher alkaline consumption.

Nutting (1925) was the first to notice this problem, and suggested use of weaker bases like sodium carbonates and silicates for improving waterflood performance. This problem can be resolved by adding synthetic surfactants in the alkaline solution.

If the alkaline concentration is above what is required for the minimum IFT of an alkali-generated soap solution, the system becomes over-optimum. By adding a synthetic surfactant, the optimum salinity of the mixture system of the soap and the surfactant becomes higher than the initial optimum salinity of the soap solution, according to a mixing rule such as Eq. (8.1), because the optimum salinity of a synthetic surfactant is generally higher than that of a soap. Sometimes, the new optimum salinity of the mixture system may not be lower than the salinity of the mixture system. Then the mixture system will become under-optimum. In this case, an extra amount of salt needs to be added.

Figure 8.1 shows an example. In the figure, the shaded areas represent type III region. Above and below the shaded areas are type II and type I regions, respectively. The numbers in the shaded areas are cosurfactant concentrations. Nelson et al. (1984) used the term cosurfactant for the injected synthetic surfactant, NEODOL 25-3S. In the horizontal axes, both the petroleum soap concentration and volume percent oil in the test tubes are marked. These two variables are proportional to each other because for a higher percentage of oil higher moles of soap will be generated. Here it is assumed that the amount of alkali is enough to react with the crude oil. In this figure,

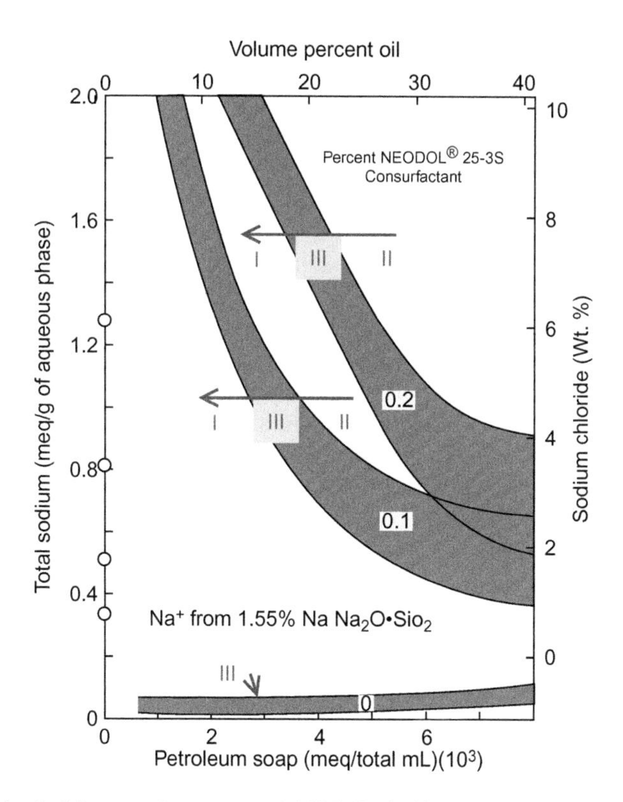

FIGURE 8.1 Activity maps for a system of 1.55% $Na_2O \cdot SiO_2$ and 0, 0.1 and 0.2% NEODOL 25-3S and a Gulf Coast crude oil at 30.2°C (Nelson et al., 1984).

the alkali used was 1.55% $Na_2O \cdot SiO_2$. In the vertical axes, sodium chloride concentrations are marked to represent the salinity.

In the case of 1.55% $Na_2O \cdot SiO_2$ solution only contacting with the crude oil, the salinity from this alkaline solution is above the optimum salinity (above the shaded type III region). The system is above-optimum. This is not desirable because the IFT will not be at its lowest.

As shown in Figure 8.1, when 0.1% surfactant (NEODOL 25-3S, 60% active) is added in the alkaline-oil system, the optimum salinity of the new system is above the salinity 1.55% $Na_2O \cdot SiO_2$ solution provides, because, again, the optimum salinity of the surfactant is higher. So the new system is at under-optimum. To make the system at an optimum condition (type III), more salt (NaCl) needs to be added in the system. Therefore, adding synthetic surfactant can improve the phase behavior of the original alkaline-oil system. The similar phenomenon occurs when 0.2% surfactant is added.

Furthermore, Figure 8.1 shows that when the oil saturation is high (in the right side of the figure), the system is type II (above-optimum); and when the oil saturation is low (in the left side of the figure), the system is type I (under-optimum). The high oil saturation (or high soap concentration) corresponds to the zone at the displacement front, and the low oil saturation (or low soap concentration) corresponds to the zone behind the displacement front. This figure shows that the system transitions from a type II to type III to type I environment, as it moves from the displacement front (downstream) to the upstream. This is desirable because a type II environment has an oil external microemulsion which has higher viscosity so that the surfactant will not break through fast. In the upstream (injection end), the system is a type I environment so that the subsequent aqueous solution can miscibly displace this microemulsion system. In the middle zone, the system is a type III environment so that the IFTs are lowest.

The shape of the active region in the presence of a synthetic surfactant, as shown in Figure 8.1, is typical of this type of activity map. An activity map is to show at what range of concentrations in the system and how the chemical flood will work. Nelson et al. (1984) used an activity map in Figure 8.1 to show the active region as a function of petroleum soap concentration and salinity when the alkali type, alkali concentration, surfactant type, surfactant concentration, type of oil, and temperature were fixed. If the concentration of synthetic surfactant is constant, moving from the right to the left on the map increases the synthetic surfactant-to-petroleum soap ratio. The salinity requirement (optimum salinity) for the system is increased. As the concentration of petroleum soap goes to zero, the active region rises to the salinity requirement of the synthetic surfactant.

When constructing an activity map, glass sample tubes (pipettes) containing oil, aqueous alkaline, and surfactant solution are equilibrated at the test temperature with periodic shaking. For the oil volume, we generally start with WOR = 1, then reduce it so that WORs are up to 10 or 20. The chemical concentrations are varied around our target concentrations. This type of pipette tests are called oil scan.

Sometimes we conduct this type of pipette tests by changing alkali concentration while the salinity is fixed. Then the activity map can be presented by alkali concentration versus oil volume in vol.% or the ratio of oil volume (in vol.%) to the volume of surfactant in vol.% or wt.%, schematically shown in Figure 8.2.

8.2.4 Effect on IFT

Figure 8.3 shows another example of soap-surfactant synergy. It shows IFT between Yates oil and the microemulsion which was formed by 0.2% 4:1 mixture by weight of NEODOL 67-7PO sulfate and internal olefin sulfonate 15–18, with the WOR = 3 (Liu, 2007). Not only are the IFT values lower, but also the width of low IFT region ($<10^{-2}$ mN/m) is much wider with sodium carbonate added than with the case without alkali. Martin and Oxley (1985) attributed such behavior to ionization of the carboxylic acid by alkali.

8.2.5 Effect on Surfactant Adsorption

The primary mechanism for the adsorption of anionic surfactants on sandstone and carbonate formation material is the ionic attraction between mineral sites and surfactant anion (Zhang and Somasundaran, 2006). The mineral will become increasingly negatively charged with an increase in pH and thereby possibly retard the adsorption of an anionic surfactant. Another mechanism for alkaline additives to reduce surfactant retention may be due to removal of multivalent ions.

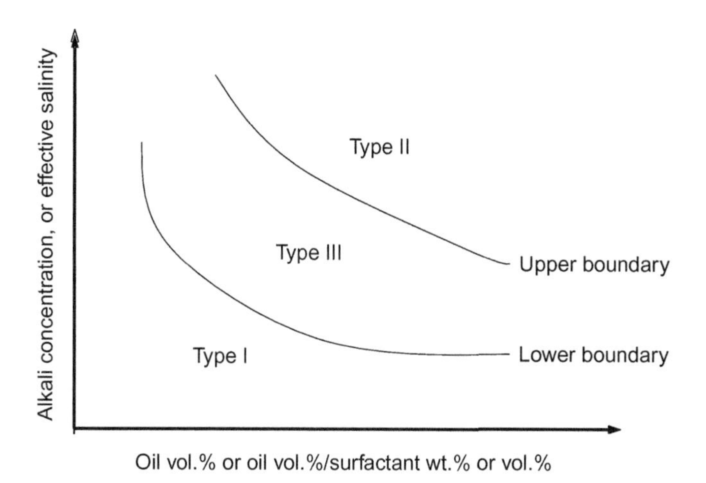

FIGURE 8.2 Variations of activity maps.

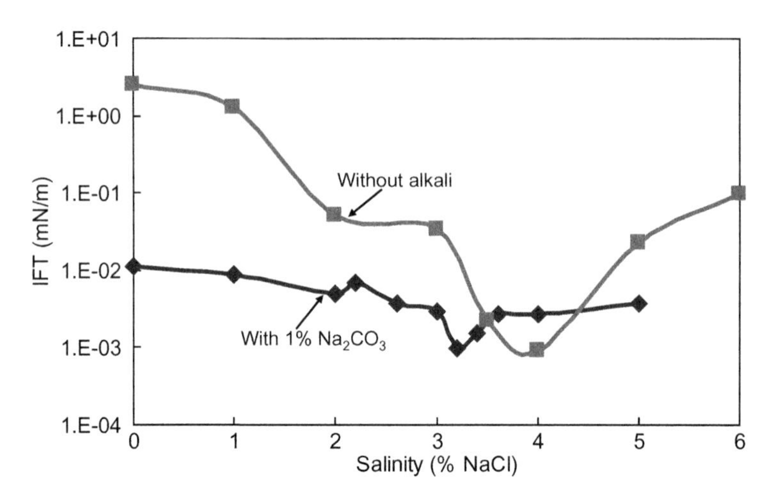

FIGURE 8.3 IFT between Yates oil and the microemulsion (Liu, 2007).

Hirasaki and Zhang (2004) found that the potential determining ions (CO_3^{2-}) can change the surface charge and reduce the anionic surfactant adsorption on calcite. Carbonate formations and sandstone cementing material can be calcite or dolomite. These two minerals have an isoelectric point (or point of zero charge (PZC)) of about pH = 9. In these cases, carbonate ion as well as calcium and magnesium ions are more significant potential determining ions (Hirasaki et al., 2008). In the case of seawater injection into fractured chalk formation, sulfate ions adsorb on the chalk surfaces to alter the surface potential (Austad et al., 2005). In this case, sulfate ion is the potential determining ion.

8.3 SIMULATED RESULTS OF AN ALKALINE-SURFACTANT SYSTEM

Sheng (2011) provided detailed procedures to simulate phase behavior of an alkaline-surfactant system and presented sensitivity study results. Here we only present the simulation results of the fraction of acid component converted to soap and the fraction of soap in the total surfactants for a typical case.

First we would be interested in seeing how much acid content in the crude oil is converted into soap. Figure 8.4 shows the converted fraction of acid into soap at different alkali concentrations. It shows that at 15 wt.% sodium carbonate, less than half of the acid component is converted into soap. In practice, alkaline concentration is less than 2%. Then only about a quarter of the acid can be converted into soap according to Figure 8.4.

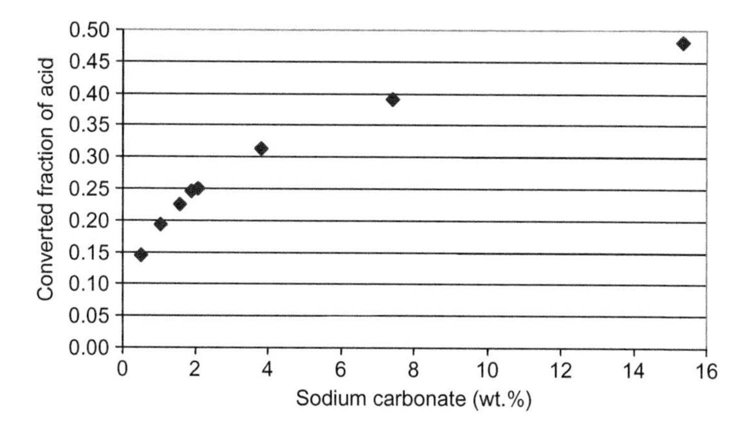

FIGURE 8.4 Converted fraction of acid into soap at different alkali concentrations.

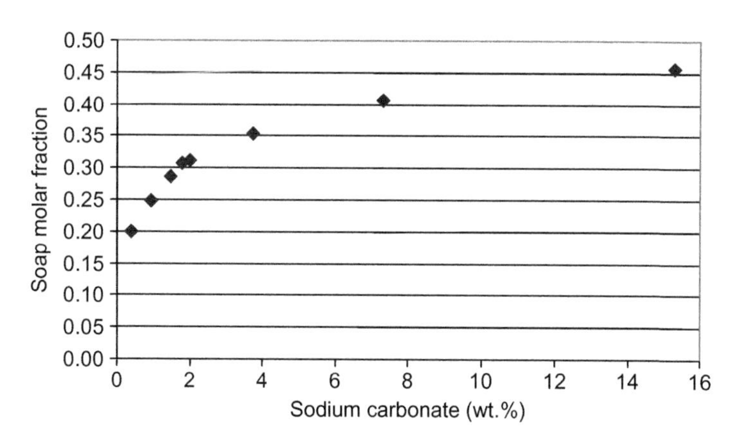

FIGURE 8.5 Molar fraction of soap in the total amount of surfactants at different alkali concentrations.

Next question is: what is the molar fraction of soap in the total surfactant? Figure 8.5 shows the molar fraction of generated soap in the total moles of surfactants at different alkali concentrations. The surfactant concentration is 1%. It shows that at 15 wt.% sodium carbonate, the soap molar fraction is less than 0.5. In practice, the alkaline concentration is less than 2%, the generated soap molar fraction is less than 0.3.

8.4 FIELD CASES

Several groups (Bryan and Kantzas, 2009; Kumar et al., 2010; Liu et al., 2006) conducted the studies on alkaline-surfactant flooding in heavy oil reservoirs in laboratory. Some studies directly targeted at field applications, for example, study for the Hepler field in Kansas (French, 1994). However, not many pilots were conducted in fields, partly owing to the low oil price. Probably more is owing to the fact that mobility control using polymer is essential in alkaline-surfactant flooding. We present two field cases next. Some alkaline-surfactant stimulation cases are presented in Chapter 12, Surfactant-enhanced oil recovery in carbonate reservoirs.

8.4.1 Big Sinking Field in East Kentucky

The Big Sinking field was a mature waterflood in the Appalachian Basin of Eastern Kentucky. The field was a tight oil reservoir. There were two problems: (1) water cut was too high; (2) water injection rate was restricted which limited the rates of fluid production including oil rate. Therefore, one objective of injecting alkaline-surfactant solution was to reduce oil residual saturation so that the effective permeability to water is increased. Once the water effective permeability was increased, the injection rate can be lifted and thus the fluid production rates including oil rate might be increased.

The selected pilot area was in the EL Rogers tract. The average thickness was 25 ft. The average permeability was 45 mD. Laboratory study selected 1.5% sodium hydroxide and 0.1% (active concentration) ORS-62HF surfactant as an alkaline-surfactant (AS) solution. The single well injectivity test program was as follows.

- Preflush: 1500 bbls of fresh water (started on September 5, 2003)
- Main slug: 1500 bbls of AS solution (penetration radius about 25 ft)
- Postflush: fresh water injection

The injectivity test showed that the injectivity after AS injection was about 2.19 times that before injection.

8.4.2 White Castle Field in Louisiana

The pilot test was part of chemical EOR pilot test which was planned to use polymer as a mobility control agent. The test was carried out in Q sand whose reservoir and fluid properties are shown in Table 8.1 (Falls et al., 1994).

Before the AS pilot, Well 85 was the only well completed in the Q sand. A new producer (Well 267) was drilled. This producer was watered out. To determine the PV affected by the flood, characterize flow patterns, and separate the hard formation brine from the alkaline slug, the Q sand was preflooded with NaCl brine. The first (initially intended to be the only one) preflood consisted of injecting 8% NaCl solution followed by 3% brine.

Because of its potential to guard against gravel-pack dissolution and protect surfactants from magnesium, the silicate slug was initially chosen. However, laboratory floods showed a potential injectivity problem with the

TABLE 8.1 Reservoir and Fluid Properties of the White Q-Sand Reservoir

Lithology	Deltaic Miocene Unconsolidated Sandstone
Permeability	1 Darcy
Porosity	31%
Cation exchange capacity	0.95 meq/100 g rock
Thickness	5–35 ft
Area	~1 acre
Dip angle	~45°
Reservoir temperature	147°F
Reservoir viscosity	2.8 cP
Oil acid number	~1.5 mg KOH/g oil
Residual oil saturation	0.2

TABLE 8.2 Compositions of AS Solutions (Active wt.%)

	Carbonate Slug	Silicate Slug
Sodium carbonate	2.53	2.03
Sodium silicate		0.21
Internal olefin sulfonates (IOS)	0.44	0.45
Sodium chloride	0.4	0.4
NEODOL 25-12	0.06	0.06
Sodium thiocyanate	0.05	0.05
Water	96.52	96.8

silicate solution. The carbonate slug was also developed. The compositions of these solutions are presented in Table 8.2.

Because of the concern with silicate injectivity, 1065 bbl of carbonate slug (solution) was first injected to establish an injectivity baseline. The silicate slug was then injected. It was found that the injectivity decreased obviously when the silicate slug was injected. Therefore, only carbonate slug was injected after 2100 bbl of silicate slug was injected. To ensure that chemical consumption could be determined from production data, the designed amount of AS injected was about twice the consumption in the reservoir.

During the implementation, acidizing had to be performed because of formation damage. Oil recovery was estimated to be at least 38% of the remaining oil. Log data indicated virtually 100% displacement efficiency and 50% vertical sweep efficiency without polymer for mobility control.

REFERENCES

Austad, T., Strand, S., Høgnesen, E.J., Zhang, P., 2005. Seawater as IOR fluid in fractured chalk. Paper SPE 93000 Presented at SPE International Symposium on Oilfield Chemistry, Woodlands, TX, 2–4 February.

Bryan, J., Kantzas, A., 2009. Potential for alkali-surfactant flooding in heavy oil reservoirs through oil-in-water emulsification. JCPT 48 (2), 37–46.

Falls, A.H., Thigpen, D.R., Nelson, R.C., Ciaston, J.W., Lawson, J.B., Good, P.A., et al., 1994. Field test of cosurfactant-enhanced alkaline flooding. SPERE 9 (3), 217–223.

French, T., 1994. Evaluation of a site on Hepler oil field for a surfactant-enhanced alkaline chemical flooding project and supporting research, report NIPER/BDM-0073.

Green, D.W., Willhite, G.P., 1998. Enhanced Oil Recovery. Society of Petroleum Engineers, Richardson, TX.

Hirasaki, G.J., Zhang, D.L., 2004. Surface chemistry of oil recovery from fractured, oil-wet, carbonate formations. SPEJ June, 151–162.

Hirasaki, G.J., Miller, C.A., Puerto, M., 2008. Recent advances in surfactant EOR. Paper SPE 115386 Presented at the SPE Annual Technical Conference and Exhibition, Denver, CO, 21–24 September. The Revised Version Presented at the International Petroleum Technology Conference, Kuala Lumpur, Malaysia, 3–5 December.

Kumar, R., Dao, E., Mohanty, K.K., 2010. Emulsion flooding of heavy oil. Paper SPE 129914 Presented at SPE Improved Oil Recovery Symposium, Tulsa, OK, 24–28 April.

Liu, S., 2007. Alkaline Surfactant Polymer Enhanced Oil Recovery Processes. Rice University, Ph.D. dissertation.

Liu, Q., Dong, M., Ma, S., 2006. Alkaline/surfactant flood potential in western Canadian heavy oil reservoirs. Paper SPE 99791 Presented at the SPE/DOE Symposium on Improved Oil Recovery, Tulsa, OK, 22–26 April.

Martin, F.D., Oxley, J.C., 1985. Effect of various chemicals on phase behavior of surfactant/brine/oil mixtures. Paper SPE 13575 Presented at the International Symposium on Oilfield and Geothermal Chemistry, Phoenix, AZ, 9–11 April.

Mohammadi, H., Delshad, M., Pope, G., 2008. Mechanistic modeling of alkaline/surfactant/polymer floods. Paper SPE 110212 Presented at SPE/DOE Improved Oil Recovery Symposium Held in Tulsa, OK, 19–23 April.

Nelson, R.C., Lawson, J.B., Thigpen, D.R., Stegemeier, G.L., 1984. Cosurfactant-enhanced alkaline flooding. Paper SPE 12672 Presented at the SPE/DOE Fourth Symposium on Enhanced Oil Recovery Held in Tulsa, OK, 15–18 April.

Nutting, P.G., 1925. Chemical problems in the water driving of petroleum from oil sands. Ind. Eng. Chem. 17, 1035–1036.

Puerto, M.C., Gale, W.W., 1977. Estimation of optimal salinity and solubilization parameters for alkylorthoxylene sulfonate mixtures. SPEJ 17 (3), 193–200.

Salager, J.L., Bourrel, M., Schechter, R.S., Wade, W.H., 1979. Mixing rules for optimum phase-behavior formulations of surfactant/oil/water systems. SPEJ October, 271–278.

Sheng, J.J., 2011. Modern Chemical Enhanced Oil Recovery—Theory and Practice. Elsevier, Burlington, MA.

Zhang, R., Somasundaran, P., 2006. Advances in adsorption of surfactants and their mixtures at solid/solution interfaces. Adv. Colloid Interface Sci. 123–126, 213–229.

Zhang, D.L., Liu, S., Puerto, M., Miller, C.A., Hirasaki, G.J., 2006. Wettability alteration and spontaneous imbibition in oil-wet carbonate formations. J. Pet. Sci. Eng. 52, 213–226.

Chapter 9

ASP Fundamentals and Field Cases Outside China

James J. Sheng

Bob L. Herd Department of Petroleum Engineering, Texas Tech University, Lubbock, TX 79409, USA

9.1 INTRODUCTION

Individual chemical flooding processes, alkaline flooding, surfactant flooding and polymer flooding, can be combined differently. Two-component combinations, surfactant-polymer (SP), alkaline-polymer (AP), and alkaline-surfactant (AS), have been presented in the preceding chapters. In this chapter, we will present the three-component combination, alkaline-surfactant-polymer (ASP). We will focus on the practical issues of ASP process. These issues include produced emulsion, scaling, and chromatographic separation.

Eleven pilot tests and field applications in China have been presented earlier (Sheng, 2011). In this chapter, we will present several ASP cases of pilot tests and applications in North America.

9.2 SYNERGIES AND INTERACTIONS OF ASP

An incomplete list of the synergies and interactions of ASP may be summarized as follows.

1. Alkali reacts with crude oil to generate soap.
2. Alkaline injection reduces surfactant adsorption.
3. The combination of soap and synthetic surfactant results in low interfacial tension (IFT) in a wider range of salinity and a favorable salinity gradient.
4. Generally it is assumed that polymer has little effect on IFT.
5. Soap and surfactant make emulsions stable through reduced IFT. Emulsions improve the sweep efficiency.
6. There is a competition of adsorption sites between polymer and surfactant. Therefore, addition of polymer reduces surfactant adsorption, or vice versa.
7. Addition of polymer improves the sweep efficiency of ASP solution.

8. Polymer may help to stabilize emulsions owing to its high viscosity to retard coalescence.

9.3 PRACTICAL ISSUES OF ASP FLOODING

This section discusses some issues resulting from ASP applications, including produced emulsion, chromatographic separation, precipitation, and scaling.

9.3.1 Produced Emulsions

There exist mixed types of emulsions in an ASP system. In phase behavior pipette tests, the emulsions discussed are microemulsions. However, even in pipettes, other types of emulsions can be present in the early time. Because of their unstable nature, macroemulsions (or most of them) are phase-separated after a long time of equilibrium. What we focus in pipette tests is the microemulsion at equilibrium. Now we address emulsions in general.

According to their structures, there are four types of emulsions: W/O, O/W, W/O/W, and O/W/O. Sometimes, W/O/W and/or O/W/O are called multiple types. In waterflooding, most of the produced emulsions are W/O type, with some O/W. In chemical flooding, the type of emulsion depends on the types of chemicals used and their concentrations, water and oil properties. It also depends on the water/oil ratio (WOR). Generally, W/O type of emulsions are formed at low WOR. As the water cut in emulsions increases, W/O type will be transferred to O/W type. In other words, when one phase volume is much larger, then this phase will be a continuous phase.

The water cut at which W/O type emulsions are transferred to O/W type emulsions is called type transferring point or critical water cut. Table 9.1 lists the critical water cuts for several emulsions at which the emulsions are transferred from W/O type to O/W type.

From Table 9.1, we can see that adding surfactant and polymer reduced their critical water cuts below 50%, while 1.2% alkali did not reduce the water/oil critical water cut. Table 9.1 indicates that under ASP flood saturations (high water saturation), most likely, O/W emulsion will be formed.

TABLE 9.1 Critical Water Cuts from W/O type to O/W Type (Kang, 2001)

Emulsion System	Critical Water Cut
Water, low-concentration ASP	50
1.2% NaOH– or Na_2CO_3–crude oil	50
0.3% surfactant (ORS-41)–crude oil	10
0.12% hydrolyzed polyacrylamide (HPAM)–crude oil	20

According to their droplet sizes, emulsions have macroemulsion, miniemulsion, microemulsion, and micelle. Table 9.2 lists the typical sizes and shapes of the aggregates in a common alkaline-surfactant-oil system (Kang, 2001).

Emulsion stability may be described by the half-life of the emulsion similar to the concept used for foam stability (Sheng et al., 1997). The half-life corresponds to the time at which the emulsion volume has decayed to the half of its initial volume. Generally, W/O emulsion was much more stable than O/W emulsion. More factors that affect emulsion stability were discussed by Sheng (2011).

9.3.2 Chromatographic Separation of Alkali, Surfactant and Polymer

Figure 9.1 is the effluent concentration histories of an ASP slug injection. The vertical axis is the relative concentration of polymer, alkali, and surfactant, the effluent concentrations relative to its respective injection concentrations. The horizontal axis is the injection pore volume.

First we can see that polymer broke through first, then alkali followed by surfactant. Second, each maximum relative concentration depended on its retention or consumption in the pore medium. The maximum polymer concentration was 1, the maximum alkali concentration was 0.9, and the maximum surfactant concentration was 0.09 in this case. Third, their concentration ratios in the system were constantly changing. In other words, the chemical injection concentrations will not be proportionally decreased.

Because of the inaccessible pore volume (IPV), polymer may even transport faster than the aqueous phase. Polymer will break through earlier than surfactant or alkali. In general, actual effluent concentrations and breakthrough times depend on their individual balance between the injection concentration and the retention.

To have the ASP synergy, the three components should transport at the same velocity. We cannot get rid of IPV or change retention too much to solve the separation problem. What we can do is to change the injection concentrations. Here is a simulated example. Initially, the injection scheme is

TABLE 9.2 Typical Sizes and Shapes of the Aggregates in an AS System

	Size (μm)	Shape	Stability
Micelle	<0.01	Spherical, cylindrical, disk-like, lamellar	Stable
Microemulsion	0.01–0.1	Spherical	Stable
Miniemulsion	0.1–0.5	Spherical	Unstable
Macroemulsion	0.5–50	Spherical	Unstable

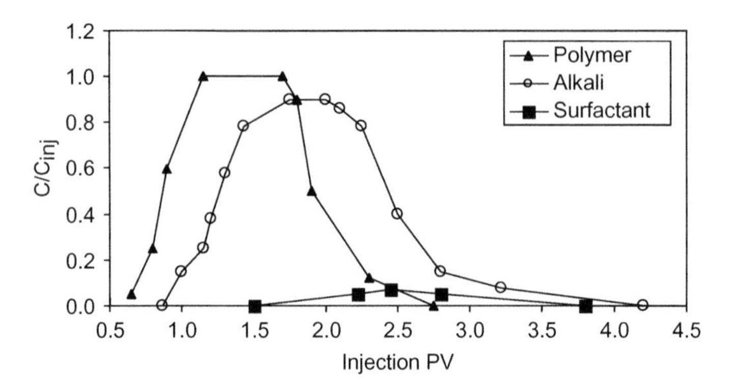

FIGURE 9.1 Effluent concentration histories of polymer, alkali, and surfactant (Huang and Yu, 2002).

0.5 PV slug of 0.7% Na_2CO_3, 1% surfactant and 0.15% polymer. Figure 9.2 shows the pH which represents alkaline concentration, and surfactant and polymer concentrations of the produced fluid. Polymer breaks through first, followed by surfactant and alkali.

If we only change the alkaline concentration from 0.7% to 1.8%, keeping the other injection concentrations and the slug volume same as the initial case, the produced concentrations are shown in Figure 9.3. Now the alkali and surfactant break through the production end at the same time, although polymer still·transports ahead of them. This example demonstrates that we can change injection concentrations to solve separation problem.

9.3.3 Precipitation and Scale Problems

When an alkaline solution is injected into the formation, the concentrations of OH^-, CO_3^{2-}, and SiO_3^{2-} are increased. The increase in OH^- is from the injected alkaline solution. The increase in CO_3^{2-} is from HCO_3^- because high OH^- makes the reservoir an alkaline environment and converts HCO_3^- into CO_3^{2-}. SiO_3^{2-} is due to the reaction between the injected alkali and formation minerals. If the seawater is injected, SO_4^{2-} is increased. Divalents like Ca^{2+} and Mg^{2+} are from the formation water, cation exchange and reactions between the injected solution and rock minerals. Sometimes, there is Al^{3+}. Several inorganic scales and precipitates can be formed from these ions. Frequent operation failures of production wells due to these problems were observed in Daqing. The scaling and precipitates may cause formation damage.

9.4 AMOUNTS OF CHEMICALS INJECTED IN CHINESE FIELD ASP PROJECTS

From 1990s to early 2000s, most of the field pilots and field applications were conducted in China. The average injection concentrations of alkali,

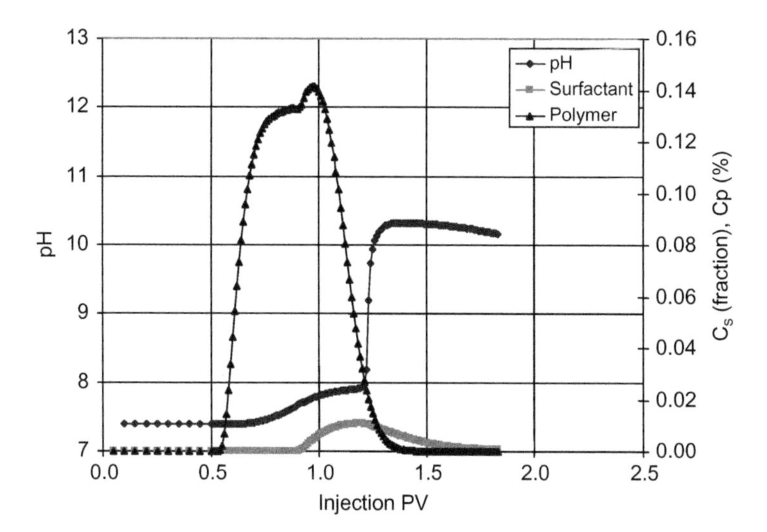

FIGURE 9.2 Concentration histories of the produced fluid at the initial injection concentration.

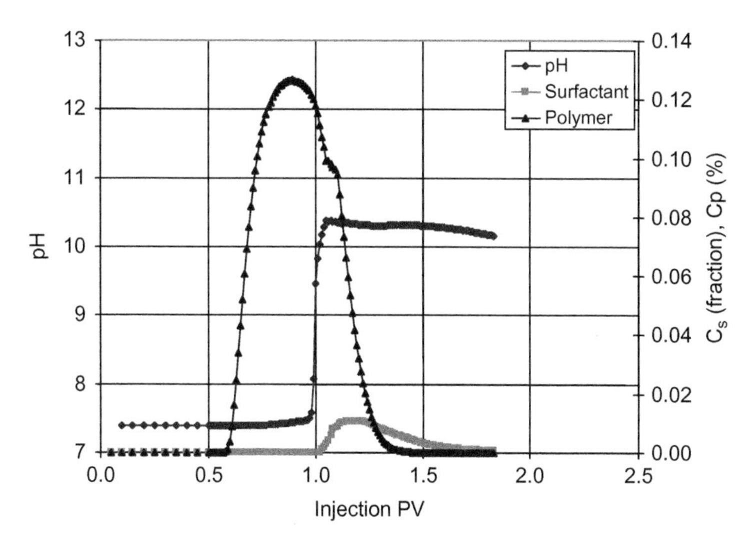

FIGURE 9.3 Concentration histories of the produced fluid when alkaline injection concentration is increased.

surfactant, and polymer were 1.28%, 0.28%, and 0.15%, respectively. For the micellar flood or micellar-polymer flood projects in 1980s, the injected surfactant concentrations were more than a few percent. In these ASP projects, the surfactant concentrations used were one order of magnitude lower. Corresponding to those chemical concentrations, the average injected pore volume was 41.8% PV.

The total amounts of alkali, surfactant, and polymer injected in Chinese ASP projects were 53.44, 11.53, and 8.08. Their unit is the product of injection volume in PV% and the concentration in %. These field data may serve as a reference or a guide for other field projects.

9.5 OVERALL ASP FIELD PERFORMANCE

Field performance as well as laboratory coreflood data shows that mobility control is essential. In a chemical flooding, it is always good to have polymer included. It was reported that the viscosity ratio of chemical fluid to crude oil should be more than 4 (Zhu et al., 2012).

Our analysis of the ASP projects in China shows that the incremental oil recovery over waterflooding is 18.9% on the average (Sheng, 2012).

Although ASP outperformed any other combinations of alkaline, surfactant, and polymer flooding, the problems with produced emulsions (difficult to demulsify or increased cost), scaling and corrosion have led the industry to seek alkaline-free options like SP process. Or a weak alkali is preferred to a strong alkali. It was reported that the decrease of liquid production was not only related to the increase of the viscosity of displacing fluid but also related to emulsification and scaling after injection of ASP slug (Zhu et al., 2012).

9.6 ASP EXAMPLES OF FIELD PILOTS AND APPLICATIONS

The field results of Chinese ASP projects have been summarized in the preceding section. A more detailed description of 11 Chinese cases was presented in Sheng (2011). In this section, we only present field cases outside China. These projects are the Lawrence field in Illinois, the Cambridge Minnelusa field, the West Kiehl field and Tanner field in Wyoming, and Lagomar LVA-6/9/21 area in Venezuela.

9.6.1 Lawrence Field in Illinois

This ASP project, run by Rex Energy, is a very aggressive pilot. The project was initiated in July 2009. The ASP slug was injected in August 2010 (Sharma et al., 2012). This 15-acre pilot is in the Middagh site within the Lawrence field. The location was selected so that the facilities designed for previous chemical flood pilots in the field could be used to minimize the cost for new infrastructure. The reservoir is in the Bridgeport sand. The reservoir pressure in the pilot area prior to waterflooding was 71 psig. The water cut was approaching 99%. Some of the reservoir and fluid data are shown in Table 9.3.

The pilot consists of six regular five-spot patterns including 12 injectors and 6 producers with additional injectors in the periphery of the pattern.

Injection Formula

The oil saturation before the pilot was close to the residual oil saturation. Therefore, polymer flooding was not justified. Although the oil has a low acid number, alkali was added to reduce surfactant adsorption. The selected ASP formula was 1% surfactant, 0.75% cosolvent, 2200 ppm HPAM and 1 wt.% Na_2CO_3. The formation water of 1.6% total dissolved solids (TDS) was softened.

Field Injection

A preflush of 3.5% NaCl softened water was carried out to reduce Ba^{2+} concentration. Injection of 0.25 PV ASP slug was initiated in August 2010. The slug also contained an oxygen scavenger (bisulfitc) and Ethylenediaminetetraacetic acid (EDTA) to sequester ferrous ions. Polymer drive was started in December 2010. By February 2012, about 1 PV of polymer solution had been injected. The viscosity of polymer solution at a salinity lower than that in the ASP slug was about 20 cP.

The fresh water was filtered through a 1-μ filter and then softened. The softening process included a primary softener and a secondary polisher with resin. The fresh water was mixed with soda ash in mixing tanks. Surfactants were then added at the outlet of the soda ash tanks. Brine and ammonia bisulfite was then added. EDTA was injected into the soda ash mixing tanks and the polymer mixing tanks. This solution was then mixed with the mother polymer solution and went through a static mixer and 25-μ filters before injection.

TABLE 9.3 Reservoir and Fluid Data in the Bridgeport Sand

Pilot area (acres)	17
Injector-producer distance (ft)	230
Formation depth (ft)	900
Porosity (%)	20
Permeability (mD)	200
Formation temperature (°F)	80
Formation brine salinity (wt.%)	1.6
Oil viscosity in place (mPa·s)	11
Oil saturation at the start of flood	0.31

Production Response

The oil cut for the pilot wells gradually increased from 1% to 12%. When oil banks reached producers, the total production rate decreased. For most wells, a tertiary oil bank arrived producers 0.5–0.7 PV after ASP injection. No severe produced emulsion or scale problems were encountered. The project is ongoing.

9.6.2 Cambridge Minnelusa Field in Wyoming

This is an ASP secondary recovery case. The Cambridge field is located in Crook County, Wyoming. The field was discovered in 1989 with 4875 MSTB of original oil in place. The field produced 31 cP, 20°API gravity crude oil from the Permian Minnelusa upper "B" sand at 2139 m. The Minnelusa upper B reservoir was a friable, eolian sandstone with modest amounts of dolomite and anhydrite cement and was a preserved remnant of a highly dissected coastal dune complex. Dolomite and anhydrite cement were the main chemical adsorbing sites of the Cambridge sand.

The reservoir temperature was 55.6°C and the average thickness was 8.75 m. The crude oil formation volume factor was 1.03 with a bubble point of 85 psi. The average porosity and permeability were 18% and 834 mD, respectively. The connate water saturation was 31.6% with an initial reservoir pressure of 1792 psi.

Water injection began in January 1993 in the block Federal 32-28. ASP injection started 1 month later. So the ASP flood was in the secondary recovery. It ended in September 1996. The total ASP injection was approximately 0.252 total reservoir pore volume. Injection of 0.244 PV polymer drive solution began in October 1996. The total polymer injection was 0.496 PV. The final water drive began in May 2000.

The chemical concentrations injected into the field were 1.25 wt.% Na_2CO_3, 0.1 wt.% active Petrostep B-100, and 1475 mg/L Alcoflood 1175 A. The water for chemical dissolution in the laboratory study and the field application was the Fox Hills water. The water had 1400-mg/L TDS water that was essentially a 50:50 mixture of sodium sulfate and sodium carbonate with less than 10-mg/L hardness.

The ASP injection sequence increased the oil cut by as much as 36%. The ultimate incremental oil was 1.143 million bbls at a cost of $2.42 per barrel. The average initial oil saturation of the radial corefloods was 0.731 with the average waterflood residual oil saturation being 0.402. The final oil saturation after chemical flooding was 0.186.

Polymer was not produced from all producers. Two wells produced surfactant, but the results were questionable because low concentration (50 mg/L) was within the detection limit. No alkali was detected at any producer.

Facilities

The injection plant, mixing facilities, artificial lift, and produced fluid handling were based on simplicity and dependability. The plant design parameters were 3000 psi capacity, 1500 bbl/d of ASP solution injection rate, 1000 bbl/d of produced water disposal rate, and 40–45 days of chemical storage. The plant was designed to minimize operator attention. Once the plant was running, the required operator time was 1 h or less per day. For more detailed facility designs for polymer, alkali and surfactant, see Vargo et al., 2000.

Rod pumping was the preferred method of artificial lift for the Cambridge field. Each pump was governed with stand-alone computerized pump-off controllers. Flow line pressure switches protected against overpressure and shut the well down if a leak was detected. Vibration and high-tank kill switches were used to minimize problems. Flow lines and components of the fluid-handling system were composites, polyethylene, or internally coated steel. Downhole tubing and rod strings were carbon steel.

Corrosion, Demulsification, and Bactericide Program

From January 1992 to April 1994, bimonthly downhole treatments of a water-soluble corrosion inhibitor were performed at the Federal 23-28. Treatments were expanded to each well as water was produced. The corrosion inhibitor was changed to an oil-soluble and highly water dispersible compound in February 1997 to provide better protection and the program was switched to continuous downhole injection behind the annulus.

Injection and produced water qualities were monitored for oxygen, sulfide, sulfate-reducing bacteria, and oil carryover. Owing to good mechanical facility design and proper maintenance, oxygen intrusion had not been a problem. Oil carryover varied over the life of the flood, appearing to increase with polymer production. Variation of residence times in the flow splitter, treaters, and produced water tanks helped reduce oil carryover.

Beginning January 1993, a genera- use demulsifier at 120 mg/L was continuously injected into the production well flow lines. During the early months of 1994, treater upsets and reject oil were experienced from emulsion pad buildups due to increasing oil and water production. In March 1994 demulsifier concentration was increased to 250 mg/L and treater temperatures increased from 48.9°C to 60.0°C. Ultimately, a demulsifier specific for Cambridge was developed with continuous injection into the flow lines of the wells beginning in October 1995. When polymer production began, a tight emulsion was produced which necessitated increasing the demulsifier concentration to 475 mg/L. Demulsifier was injected into the annulus of the production wells beginning November 1996 providing longer treatment time which allowed a concentration reduction to 240 mg/L. In October 1998, further testing indicated that demulsifier concentration could be reduced to

90 mg/L and still had good oil and water separation at a lower heater treater temperature of 48.9°C.

Sulfate-reducing bacteria, acid-producing bacteria, and aerobic bacteria were evaluated for their role in corrosion, fouling of oil–water separators, and plugging of the disposal well, the Federal 22-28. The initial bacterial remediation consisted of bimonthly gluteraldehyde treatments of the treaters and production wells beginning March 1993. When the flow splitter system was installed in November 1995, gluteraldehyde was replaced with a quaternary amine and treatment was shifted to a flow splitter.

Project Economics

The facility costs beyond a typical Minnelusa waterflood plant were $160,000 which included chemical storage equipment, chemical mixing equipment, filtration, and labor for assembly. The total chemical cost was $2,518,000. The incremental maintenance, electrical, and pumping costs were estimated to be $1000 per month or $89,000 during the entire period of chemical injection. The total incremental cost was $2,767,000. The cost per incremental barrel of oil was $2.42.

9.6.3 West Kiehl Field in Wyoming

In the Minnelusa Lower "B" sand at the West Kiehl unit, the oil density was 24°API and the viscosity was 17 cP at the reservoir temperature of 120°F at a depth of 6630 ft. The water viscosity was 0.5 cP. The average porosity and average permeability were 23% and 345 mD, respectively. The average thickness was 11 ft. The produced water had 46,480 mg/L TDS.

Four EOR methods were evaluated for this reservoir: Conventional waterflood, polymer flood, AP flood, and ASP. Although polymer flooding was expected to take less production time and inject less water than waterflooding, the first three methods basically had the same oil recovery of 40%, whereas the estimated ultimate oil recovery for ASP was 56% (Clark et al., 1993).

The ASP solution consisted of 0.8% sodium carbonate, 0.1% Petrostep B-100 and 0.105% Pusher 700. It was planned to inject 0.25 PV ASP, 0.25 PV polymer followed by chase water until an economic limit. The injection water from Fox Hills was very fresh (806 mg/kg).

Water injection began in the mid-September 1987. On December 3, 1987, sodium carbonate injection began. Surfactant was added to the injection water in addition to the sodium carbonate on December 17, 1987. On January 28, 1988, polymer was added to the injected AS solution until June 22, 1990 when surfactant was discontinued. Soda ash was discontinued on July 5, 1990. Polymer injection continued at the designed concentration until April 25, 1991 when a taper polymer concentration began.

No injected chemical production had been observed in the unit wells or the Kottraba wells (extended wells in the north of the unit) as of November 1991. No changes in the Hall plot slopes were observed once a stable slope was achieved until water breakthrough occurred. ASP was injected without damage to the Minnelusa formation. The projected cost was $2.13 per incremental barrel of oil. For individual well performance, see Meyers et al. (1992).

9.6.4 Tanner Field in Wyoming

Tanner field in Campbell County, Wyoming, is a Minnelusa B sand with one injection well and two production wells, a small field. The well spacing was roughly 40 acres. The oil density was 21°API and the viscosity was 11 cP at the reservoir temperature of 175°F at a depth of 8915 ft. The average porosity and average permeability were 20% and 200 mD, respectively. The average thickness was 25 ft. The produced water had 66,800 mg/L TDS with 2000 mg/L calcium and magnesium ions.

The primary production began in April 1991. The waterflooding began in October 1997 and continued through April 2000 when the water cut was 57%. During this period, 0.245 PV of water was injected. ASP was started in May 2000. A solution of 1.0 wt.% sodium hydroxide plus 0.1 wt.% active ORS-41HF plus 1000 mg/L Alcoflood 1275 A dissolved in Fox Hills water (1000 mg/L TDS) was injected through January 2005. During this period, 0.251 PV of the ASP solution was injected. The IFT was on the order of 10^{-3} mN/m. A tapered concentration polymer drive began in February 2005. Radial coreflood tests showed that the chemical retention were 0.02–0.03 mg/g rock for ORS-41, 0.1–0.3 mg/g rock for NaOH, and 0.02–0.04 mg/g for Alcoflood 1275 A.

The incremental oil recovery was 10% OOIP by December 2005. The final incremental oil recovery was estimated to be 17%. Chemical, facilities, and design cost per incremental barrel of oil was estimated to be $5.85. The oil price used was $40/bbl. The Tanner project is unique in that the ASP injection began after a short waterflood when the water cut was 57% (Pitts et al., 2006).

9.6.5 Lagomar LVA-6/9/21 Area in Venezuela

The ASP pilot was carried out at the Lagomar VLA-6/9/21 area, C4 member, Lake Maracaibo, Venezuela after a detailed laboratory study (Manrique et al., 2000) which is similar to a study that evaluated the ASP potential in the La Salina field (Hernandez et al., 2003). A single well chemical tracer test was conducted to determine ASP efficiency (Hernandez et al., 2002). The test well, VLA-1325, was perforated in 1998 for the test. The test area was in an offshore environment. The water cut in this area before ASP

TABLE 9.4 Reservoir and Fluid Data in the VLA-6/9/21 Area

Depth of test zone (ft)	6048–6070
Porosity (%)	24
Permeability (mD)	58-1815
Formation temperature (°F)	194
Formation brine salinity (ppm)	4200
Oil viscosity in place (mPa·s)	2.5
Acid number (mg KOH/g oil)	0.04

injection was 95%. Some of C4 reservoir and fluid properties in the VLA-6/9/21 are described in Table 9.4.

The designed ASP solution was ensured to have the IFT below 9×10^{-3} mN/m and the viscosity at least 2 mPa·s above the oil viscosity. The ASP formula used in the pilot test was 0.35 PV (1850 bbls) of 5000 ppm sodium carbonate, 2000 ppm active petroleum sulfonate, 1000 ppm HPAM and 200 ppm thiourea (to prevent polymer oxidation), followed by 0.15 PV (740 bbls) of 1000 ppm HPAM. Finally, 1.2 PV (6000 bbls) of fresh water displaced the ASP slug.

The tracer test results showed that the oil saturations before and after the ASP injection were 0.31 and 0.16, respectively. Thus the ASP efficiency to recovery remaining oil was 48%.

REFERENCES

Clark, S.R., Pitts, M.J., Smith, S.M., 1993. Design and application of an alkaline-surfactant-polymer recovery system to the West Kiehl field. APE Adv. Technol. Ser. 1 (1), 172–179.

Hernandez, C., Chacon, L., Anselmi, L., Angulo, R., Manrique, E., Romero, E., et al., 2002. Single well chemical tracer test to determine ASP injection efficiency at Lagomar VLA-6/9/21 area, C4 member, Lake Maracaibo, Venezuela. Paper 75122 Presented at the SPE/DOE Improved Oil Recovery Symposium, Tulsa, OK, 13–17 April.

Hernandez, C., Chacon, L., Anselmi, L., Baldonedo, A., Qi, J., Dowling, P.C., et al., 2003. ASP system design for an offshore application in La Salina field, Lake Maracaibo. SPEREE 6 (3), 147–156.

Huang, Y.-Z., Yu, D.-S., 2002. Mechanisms and principles of physicochemical flow. In: Yu, J.-Y., Song, W.-C., Li, Z.-P., et al., Fundamentals and Advances in Combined Chemical Flooding, Chapter 15. China Petrochemical Press.

Kang, W.-L., 2001. Study of Chemical Interactions and Drive Mechanisms in Daqing ASP Flooding. Petroleum Industry Press, Beijing, China.

Manrique, E., De Carvajal, G., Anselmi, L., Romero, C., Chacon, L., 2000. Alkali/surfactant/polymer at VLA 6/9/21 field in Maracaibo Lake: experimental results and pilot project design. Paper SPE 59363 Presented at the SPE/DOE Improved Oil Recovery Symposium, Tulsa, OK, 3–5 April.

Meyers, J.J., Pitts, M.J., Wyatt, K., 1992. Alkaline-surfactant-polymer flood of the West Kiehl, Minnelusa unit. Paper SPE 24144 Presented at the SPE/DOE Enhanced Oil Recovery Symposium, Tulsa, OK, 22–24 April.

Pitts, M.J., Dowling, P., Wyatt, K., Surtek, A., Adams, C., 2006. Alkaline-surfactant-polymer flood of the Tanner field. Paper SPE 100004 Presented at the 2006 SPE/DOE Symposium on Improved Oil Recovery, Tulsa, OK., 22–26 April

Sharma, A., Azizi-Yarand, A., Clayton, B., Baker, G., McKinney, P., Britton, C., et al., 2012. The design and execution of an alkaline-surfactant-polymer pilot test. Paper SPE 154318 Presented at the SPE Improved Oil Recovery Symposium, Tulsa, OK, 14–18 April.

Sheng, J.J., 2011. Modern Chemical Enhanced Oil Recovery—Theory and Practice. Elsevier, Burlington, MA.

Sheng, J.J., 2012. Alkaline-surfactant-polymer flooding (ASP)—principles and applications. Paper Presented at the Fifty-Ninth Annual Southwestern Petroleum Short Course, Lubbock, TX, 18–19 March; Proceedings, 221–234

Sheng, J.J., Maini, B.B., Hayes, R.E., Tortike, W.S., 1997. Experimental study of foamy oil stability. J. Can. Pet. Tech. 36 (4), 31–37.

Vargo, J., Turner, J., Vergnani, B., Pitts, M.J., Wyatt, K., Surkalo, H., et al., 2000. Alkaline-surfactant-polymer flooding of the Cambridge Minnelusa field. SPEREE 3 (6), 552–558.

Zhu, Y.-Y., Hou, Q.-F., Liu, W.-D., Ma, D.-S., Liao, G.-Z., 2012. Recent progress and effects analysis of ASP flooding field tests. Paper SPE 151285 Presented at the SPE Improved Oil Recovery Symposium, Tulsa, OK, 14–18 April.

Chapter 10

ASP Process and Field Results

Harry L. Chang
Chemor Tech International, LLC, 4105 W. Spring Creek Parkway, #606, Plano, TX 75024, USA

Laboratory experimental and simulation studies including basic approaches, types, and methods used will be discussed with an emphasis on the relationships between laboratory data and simulation studies using mechanistic models. Major field tests, applications, results, and interpretations will be reviewed and discussed. These reviews and discussions will provide a basis for recommendations on extending laboratory results to field implementation and for addressing some key design issues.

In addition to reviewing a few field examples, including pilot test designs and results, some assessments and lessons learned in the past 40 years will be presented. Based on these lessons learned, a suggested focus on future laboratory research to improve the process design and offer guidance on successful field applications including commercialization will be outlined. Conclusions and recommendations for the design and implementation of ASP field projects are provided.

10.1 INTRODUCTION

It was recognized in the United States as early as in the 1920s that certain alkaline agents would react with acidic components in the crude oil to generate surfactants *in situ* to improve oil recovery. It was also recognized later that the minimum acid content required in the crude oil for the process to be effective is about 0.3 mg KOH/g of oil in order to generate sufficient surfactants or soaps *in situ* to lower the interfacial tensions (IFTs) between the oil and the aqueous slug to below 10^{-3} dyne/cm so that the residual oil saturation in the porous media may be reduced down to nearly zero.

Similar to the surfactant–polymer (SP) process, polymer is also required in the alkaline slug to improve the mobility control. In some cases, where the *in situ* generated soaps are not sufficient in the reservoir to account for retention, additional surfactants with an amount less than 0.5 wt% are then added into the alkaline slug to enhance the displacement process. Accordingly, the process is called alkaline-surfactant–polymer (ASP) flooding and is the most complex chemical injection method.

Enhanced Oil Recovery Field Case Studies.

The commonly used alkaline chemicals are sodium hydroxide (NaOH), sodium carbonate (Na_2CO_3), and sodium orthosilicate ($Na_2O \cdot SiO_2$). Most field tests conducted prior to 2000 used NaOH because it is the most effective reagent when it comes into contact with acidic crudes. This is the reason why the process was also called caustic flooding. With the addition of NaOH in the injection slug, problems in scale formation in the injection fluid and in the produced fluid have been experienced since 1925. In this chapter, we will briefly discuss the background of the ASP process and review the mechanisms involved, key laboratory studies, important field examples, and some design issues for successful field implementations.

10.2 BACKGROUND

The first alkaline flooding was conducted in the Bradford Field, Pennsylvania (Nutting, 1925, 1928). But extensive laboratory studies were not conducted until the late 1960s and early 1970s (Castor et al., 1979; Farmanian et al., 1978; Seifert and Howells, 1969; Seifert and Teeter (1970); Song et al., 1995). Since then, many field tests had been conducted as either alkaline or caustic flooding both with polymer and without polymer over the next 50 years, approximately. A status report by Mayer et al. (1983) gave updates on completed field tests, listed field tests in progress at that time, and noted future planned studies or field tests thereby providing a good summary of field activities prior to 1980.

The US National Petroleum Council (NPC) conducted two enhanced oil-recovery (EOR) potentials in the United States, one in the late 1970s and then another in the early 1980s. Two reports were issued, the first one in 1976 (NPC Study, 1976) and the second in 1984 (NPC Study, 1984). In these two reports, alkaline flooding mechanisms and field tests conducted, along with simple economic analysis and future potentials in the United States, were documented. Up to 1984, there were 48 field tests conducted with most of them being in California and Texas. Table 10.1 shows these field tests with reservoir and crude oil properties. Some of these tests were conducted with alkaline chemicals alone and some with polymers. In some cases, several tests were conducted in the same field.

Field tests conducted prior to 1984 showed some successes but also demonstrated that (1) alkaline consumption can be greater in the field than anticipated from laboratory studies, (2) use of polymer with alkaline can provide improved mobility control, and (3) well plugging can be a potential problem due to the formation of scales in the injection slug. No field-wide expansion or commercialization was implemented in this period due to the low oil price environment, low incremental recovery, and operational problems associated with caustic, the most used alkaline chemical at that time.

After the first international oil crisis, the US Department of Energy (DOE) provided a significant amount of funding for field trials of EOR with

TABLE 10.1 Alkaline Project Conducted Before 1980 in the United States

No.	Field	State	Permeability (mD)	Oil Gravity (°API)	Oil Viscosity (cP)	Res. Temperature (°F)	Reservoir Salinity (ppm TDS)
1	Alba-SE	TX	–	16	–	–	–
2	Alba, West	TX	500	13	750		83,300
3	Bell Creek	MT	900	42	2.5	100	5,000
4	Big Sinking	KY	–	–	–	–	–
5	Sison Basin	WY	144	16	220	85	
6	Brea-Olinda	CA		16	90	135	–
7	Burnt Hollow	WY	300	15	1,000,000	70	–
8	Charmousca	TX	500	20	6	119	10,400
9	Cresent Heights	CA	550	28	1.6	190	–
10	Cyclone Canyon	WY	530	22	134	70	–
11	Dominguez	CA	175	30	1.5	155	–
12	Golden Trend	OK	100	43	0.5	138	1,90,000
13	Goose Creek	TX	–	–	75	112	–
14	Harrisburg	NE	119	–	1.5	200	8,500
15	Hospah Sand	NM	634	30	15	75	–
16	Huntington Beach	CA	200	22	15	165	–
17	Interstate	KS	–	–	–	–	–
18	Isenhour Unit	WY	108	43	1.04	97	5,000
19	Isenhour Unit	WY	21	43	2.8	97	–
20	Kern River	CA	2,000	13.5	1,000	90	–
21	Midway Sunset	CA	450	22.5	180	87	15,000
22	N. Ward-Estes	TX	20	34	2.3	94	–

(*Continued*)

TABLE 10.1 (Continued)

No.	Field	State	Permeability (mD)	Oil Gravity (°API)	Oil Viscosity (cP)	Res. Temperature (°F)	Reservoir Salinity (ppm TDS)
23	N. Ward-Estes	TX	25	32	1.4	86	–
24	N. Ward-Estes-7798	TX	39	32	1.4	86	–
25	Nebo Hemphill	LA	2,470	21	126	91	67,600
26	Orcutt Hill	CA	70	22	6	160	–
27	Orcutt Hill	CA	71	22	8	168	–
28	Orcutt Hill	Ca	70	22	8	168	14,300
29	Puerto Chiquito	NM		34	3.43	109	29,000
30	Quarantine Bay	LA	220	33.2	1.45	185	1,38,000
31	Saddle Ridge	WY	–	–	–	–	–
32	San Miguelito	CA	36	30	0.7	205	31,165
33	Sharp Minnulusa	WY	160	26	10.9	229	1,30,000
34	Singleton	NE	280	40	1.5	160	–
35	Smackover	AR	2,000	20	75	110	–
36	Taylor-Ina	TX	–	–	–	–	–
37	Taylor-Ina	TX	300	24	200	80	36,000
38	Toborg	TX	216	22	82	76	2,580
39	Torrance	CA	1,000	15	17	–	–
40	Tyro-Overlook	KS	56	31	25	70	1,23,000
41	Van-Carroll Unit	TX	100	34.2	2.3	135	–
42	Van-S. Lewisville	TX	100	34.2	2.3	135	–
43	Van-SW Lewisville	TX	100	34.2	2.3	135	–
44	West Perryton	TX	19	38	0.8	168	–
45	Whitter	CA	388	20	40	120	3,500
46	Whitter	CA	388	20	40	120	–
47	Wilmington	CA	238	28	23	125	–
48	Wilmington	CA	1,000	18	70	125	–

alkaline flooding being one of the chemical EOR (CEOR) processes funded by DOE. Several companies conducted field pilots which included the Wilmington field (Mayer et al., 1983) and the Huntington Beach field (Weinbrandt, 1979). Unfortunately, none of these tests showed promising results.

More field activity was implemented between 1990 and 2000 with a total of approximately 30 worldwide. A large number of tests were conducted in China with six in Daqing (Wang et al., 1999a,b), two in Shengli (Qu et al., 1998; Wang et al., 1997), and one in Karamay (Qiao et al., 2000). Then Daqing conducted more small-scale pilot tests in the early 2000s (Chang et al., 2006) and started several large-scale field tests in the mid-2000s (Cheng et al., 2008; Li et al., 2008; Wang et al., 2008). Other ASP field activities are mostly in Canada. Most of these field tests showed significant improvements in recovery efficiencies over the past several decades with maximum up to about 25% of the original oil in place (OOIP).

Even with such high incremental recovery efficiencies, large-scale expansions were only attempted after 2005 in China and Canada. To date, real commercial practices have not been successfully implemented. The lack of commercialization to date is related to the difficulties in treating the injection and produced fluids and to the understanding of the complex interactions between alkaline chemicals and reservoir solids. More detailed discussions on some of these projects will be given in Section 10.5.

10.3 LABORATORY STUDIES AND MECHANISTIC MODELING

10.3.1 Laboratory Studies

The laboratory study in ASP flooding is a multistage process and can be divided into static and dynamic types. The multistage process includes the acid analysis of the crude oil, the alkaline-oil interfacial visual tests, the alkaline-surfactant (AS) formulation tests with phase behavior studies and IFT measurements, the polymer stability tests, the ASP slug optimization-stability-clarity tests, and finally, the coreflood experiments.

Since the properties of the dead oil used in laboratory studies are not the same as the live oil in the reservoir and also because live fluid experiments are difficult to perform in the laboratory, diluted oils with C-7 to C-10 light hydrocarbons, including hexane, octane, iso-octane, decane, and decalin, were used to mimic live oil viscosities in the reservoir (Southwick et al., 2010; Wade et al., 1976). It has been reported (Nelson, 1983; Roshanfekr, et al., 2009) that the optimal salinity of the ASP-dead oil microemulsion has a tendency to shift to a lower value in the ASP-live oil microemulation system. Modeling of the equivalent alkane carbon number (EACN) concept also has been conducted (Han et al., 2006a; Roshanfekr et al., 2011). One should also be aware of the possible shift in optimal salinity that may result from

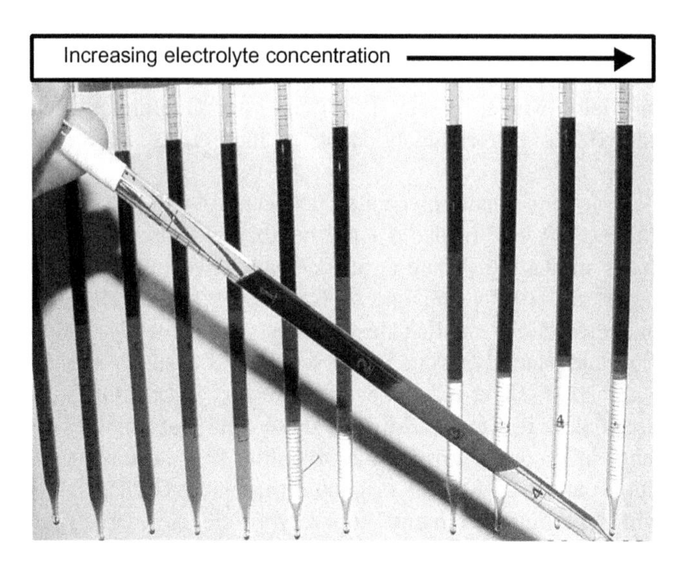

FIGURE 10.1 Winsor type III middle-phase microemulsion and the interfacial fluidity.

the dilution process that lowers the total acid content in the crude oil. But such a change is insignificant because large volumes, 20–50%, of oil phase are used in the phase behavior studies.

Two key objectives have to be accomplished in the static experiments: the aqueous formulation clarity and stability, and large middle-phase microemulsions (Winsor, 1954, 1968) with high fluidity as shown in Figure 10.1. A clear and stable ASP slug will ensure a good injectivity in the field and no phase separation on the surface or in the reservoir. A large middle-phase microemulsion will ensure the high solubilization ratios and an ultra-low IFT between the ASP slug and the oil phase. The high fluidity will ensure that there is no presence of any immobile gels or liquid crystals when the ASP slug makes contact with the oil in the reservoir.

From the analyses of the surfactant solubilization in oil and water, as shown in the equations below, a diagram of the solubilization ratios can be created and the solubilization parameter, that is, the intersection of the solubilization ratios of the water and the oil, may be obtained.

$$\sigma_o = V_o / V_s \tag{10.1}$$

$$\sigma_w = V_o / V_s \tag{10.2}$$

where, σ_o = oil solubilization ratio, σ_w = water solubilization ratio, V_o = volume of oil, V_w = volume of water, and V_s = volume of surfactants.

An example of the solubilization parameter is shown in Figure 10.2.

From the solubilization parameter, the IFT may be calculated using the Chun Huh correlation (1979) as shown below.

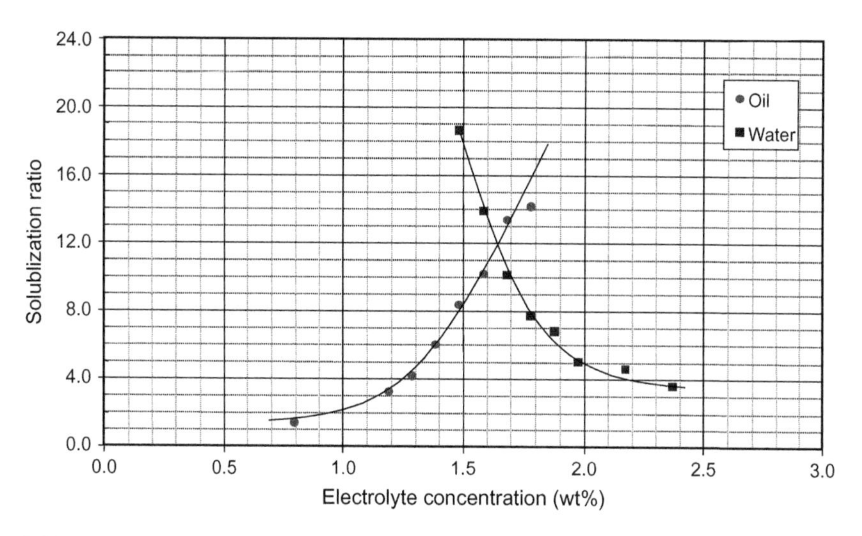

FIGURE 10.2 Solubilization ratios.

$$\gamma = C/(\sigma)^2 \tag{10.3}$$

where C = a correlation constant of 0.3, γ = IFT, and σ = solubilization parameter.

As shown in Figure 10.3 (Glinsmann, 1979), a solubilization parameter of >9 (i.e., IFT $< 10^{-3}$ dynes/cm) would be required for an efficient ASP formulation. Detailed experimental procedures with interpretation of the phase behavior and the correlation of the phase behavior with IFT and their relationships to corefloods may be found in the literature (Chun, 1979; Flaaten et al., 2008a,b; Healy and Reed, 1974; Pandey, 2010; Scriven, et al., 1976). Detailed discussion of these subjects is beyond the scope of this chapter. Due to the uncertainty of the optimal salinity of an ASP system determined with a diluted dead oil, it would be desirable to conduct phase behavior tests in a high pressure PVT cell using a live oil or a simulation study (Han et al., 2006a; Roshanfekr et al., 2011) to confirm the optimal salinity before the implementation of a field project (Southwick et al., 2010).

The evaluation of suitable polymers should be conducted when the AS formulation is completed. Details of the polymer evaluation tasks will not be discussed here, but basic experiments such as rheological properties, filtration tests, and long-term stability should be used in the evaluation process. After the addition of polymer to the AS system, a clear and stable ASP formulation needs to be obtained.

Because of the highly efficient displacement process, adequate mobility control is extremely important in the design process (Hou et al., 2006). If we examine the relative mobility of the highly efficient displacement chemical flooding process as shown in Figure 10.4 (Gogarty et al, 1970) or the

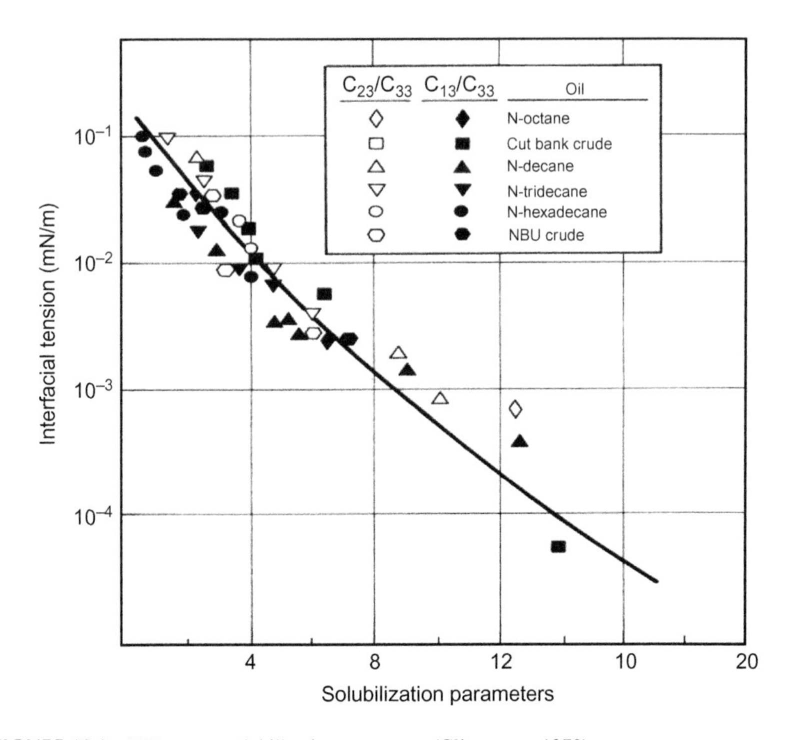

FIGURE 10.3 IFT versus solubilization parameter (Glinsmann, 1979).

pressure increase in the water/oil bank (Flaaten et al., 2008a,b) as shown in Figure 10.5, it is clear that high viscous chemical slug and polymer drive are required to achieve the high sweep efficiency. Total mobilities of various banks may be measured in laboratory linear corefloods.

The final stage of laboratory studies is the linear coreflood experiment. At least two coreflood experiments need to be conducted, one using an outcrop cores such as the Berea sandstone and one with reservoir cores, if available. It is preferable to conduct a live oil coreflood in reservoir cores to confirm the recovery efficiency before the implementation of a field project.

The linear coreflood experiments are typically conducted in a 2 in. diameter by 12 in. long cores prepared in an epoxy resin or in a core holder with multiple pressure taps along the holder. A minimal length of 12 in. is required in the coreflood experiment to minimize the dispersion effect. The flow of the flood will be in the upward vertical direction to minimize the gravity effect. Multiple pressure taps on the core need to be installed to monitor the flow behavior of the oil bank, ASP slug, and the polymer drive.

Recovery efficiency will be obtained as well as the analysis of produced chemical concentrations, pH change, and chloride concentrations. Pressure data, oil recovery and final residual oil saturation, chemical concentrations,

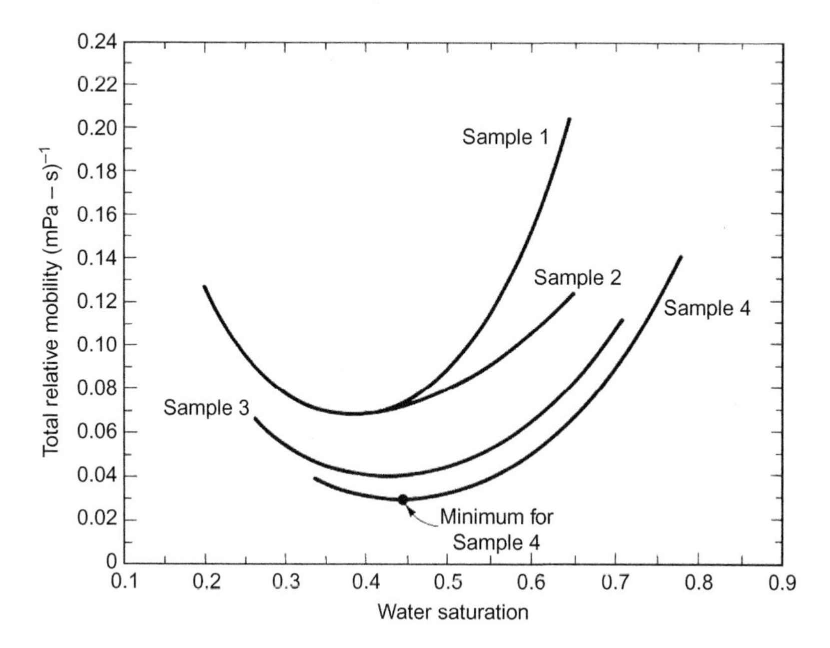

FIGURE 10.4 Total relative mobilities (Gogarty et al., 1970).

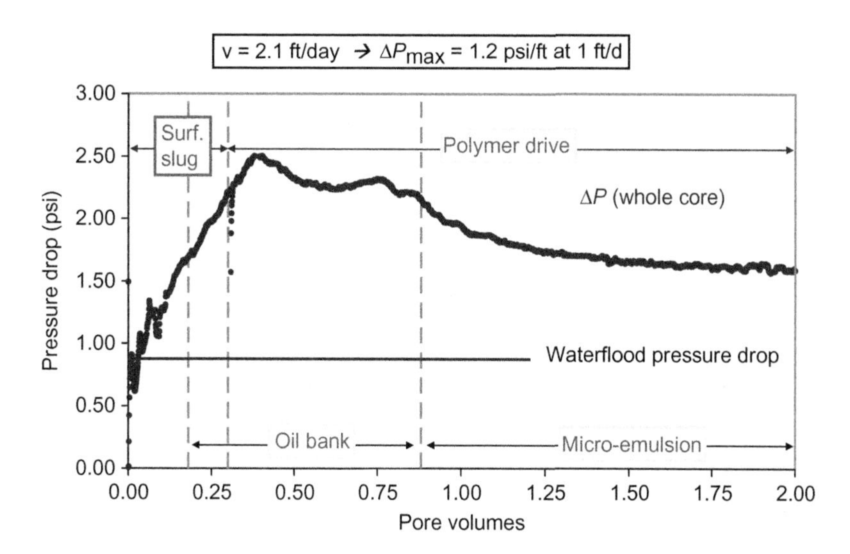

FIGURE 10.5 Typical coreflood pressure gradient (Flaaten et al., 2008).

and pH behavior will be used in a coreflood simulation with a mechanistic simulator. Successful coreflood would require a good oil bank, that is, high oil cut with most of the oil produced before reaching 1.2 pore volume from

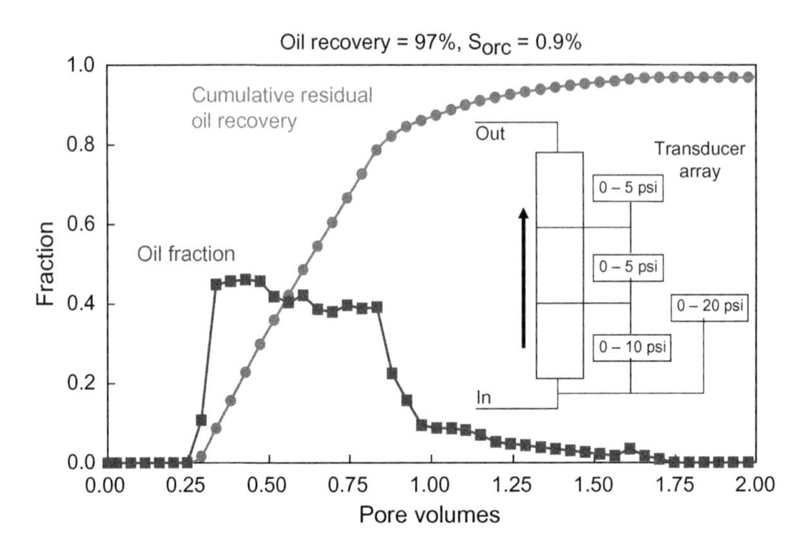

FIGURE 10.6 Oil recovery from a sandstone core.

the beginning of the chemical slug injection. Chemical concentration analyses would provide information on the chemical retention and transport in the porous media (Li et al., 2009a). Typical coreflood results and a schematic diagram of a coreflood experimental unit are shown in Figure 10.6.

10.3.2 Mechanistic Modeling

The components of mechanistic modeling include model set-up and coreflood simulation (Farajzadeh et al., 2011; Kalra et al., 2011; Karpan et al., 2011), small-scale or 2-D field simulation (Moreno et al., 2003; Zerpa et al., 2004), and large-scale field simulation. To set up a mechanistic model for core-scale simulation, key mechanisms including phase behavior, IFT and capillary desaturation, rheological properties of the ASP system, reactions and interactions of chemical species between fluids and between reservoir rocks and fluids, and so on have to be included. In addition, physical characterization of the simulation system, such as layers and grid-blocks for either the linear core or a small sector of a reservoir also has to be determined.

At the present time, many commercial reservoir simulators do not have full mechanistic descriptions of the ASP system, so it will be difficult to model all behavior and phenomena in the displacement process. The most encompassing mechanistic model for the ASP process is UTCHEM, a chemical flood simulator developed by the University of Texas (Delshad et al., 1998, 2011; John et al., 2005; Mohammadi et al., 2009). This model can be used to calibrate the commercial simulator for large-scale reservoir simulation studies.

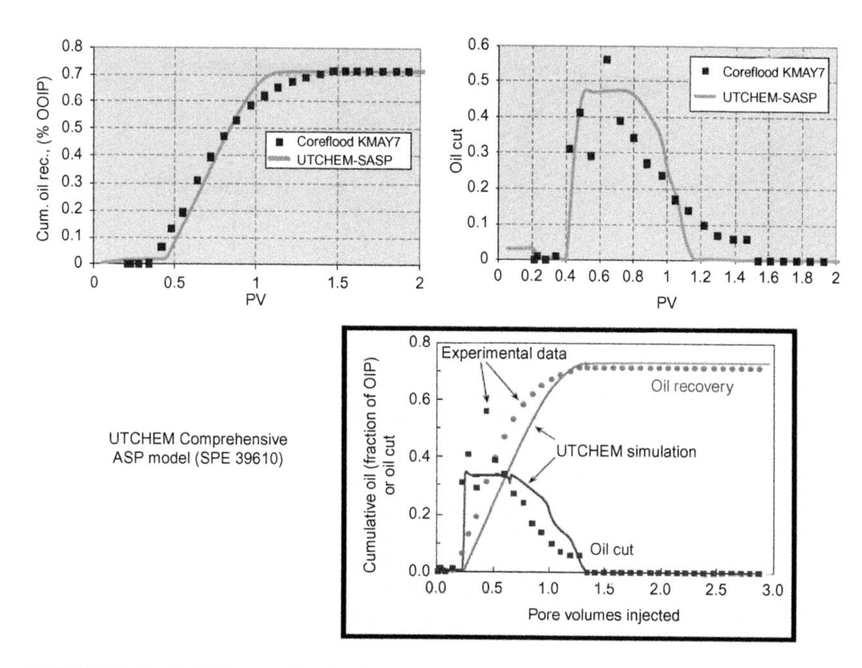

FIGURE 10.7 Modeling coreflood—Karamay ASP Project.

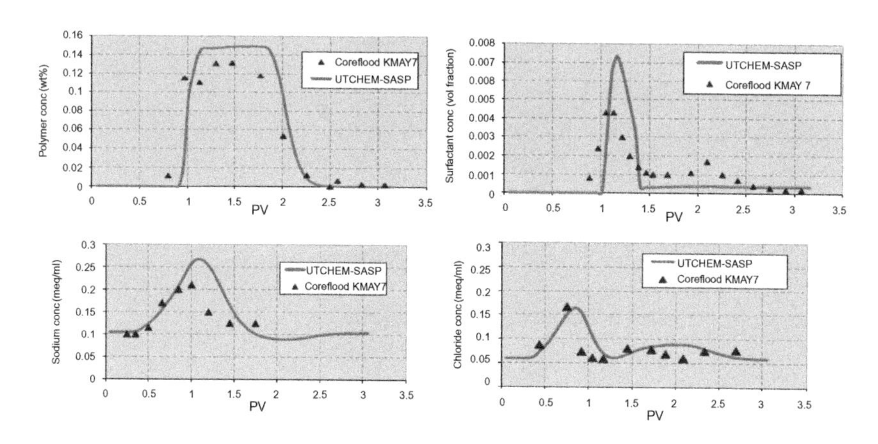

FIGURE 10.8 Modeling coreflood—Karamay ASP Project.

Examples of coreflood simulation results for the Karamay ASP laboratory studies are shown in Figures 10.7 and 10.8, and pilot-scale simulation can be found in previous publications (Pandey et al., 2008a; Qiao et al., 2000; Yuan et al., 1995). Figures 10.9 and 10.10 show simulation results of the pilot test in Daqing and in Karamay, respectively. In the mechanistic simulation, the

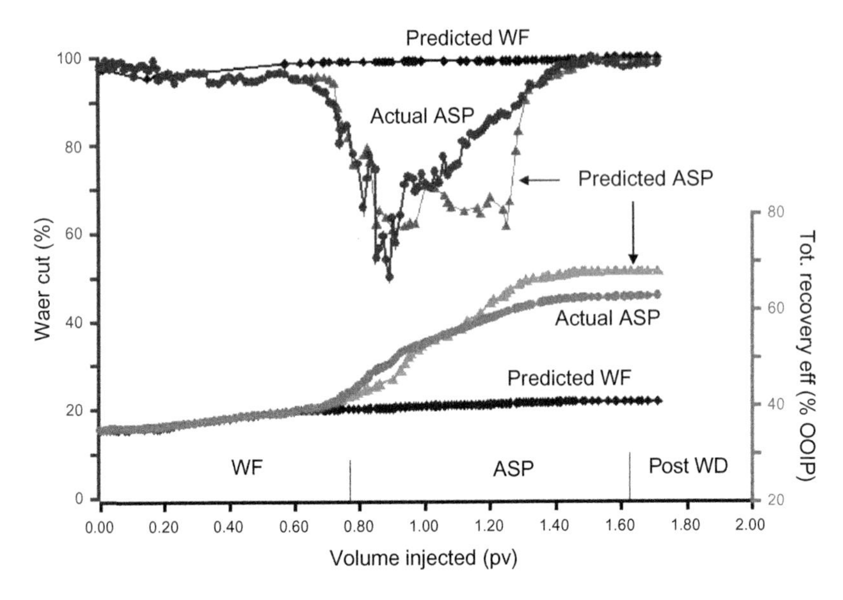

FIGURE 10.9 Daqing B1-FBX ASP pilot test results central production wells.

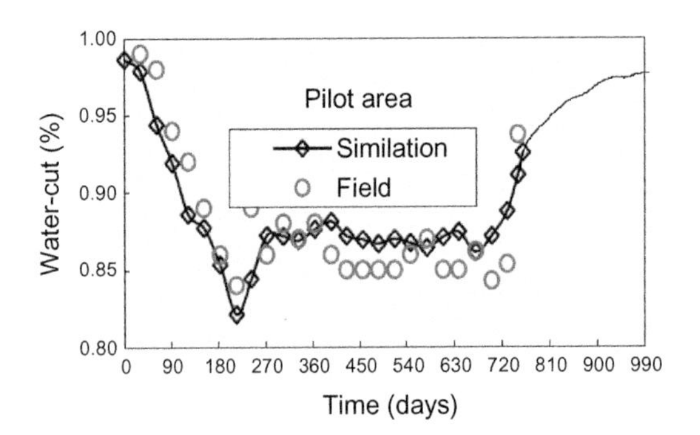

FIGURE 10.10 Modeling of Karamay pilot area performance.

coreflood match should focus on the chemical transport and residual oil saturation. Specific types of matches include (1) cumulative oil recovery, (2) residual oil saturation, (3) chemical profiles of surfactants and polymer in the effluent, (4) pH and salinity in the effluent, and (5) the pressure behavior. Matching the coreflood data to the simulated results would ensure the correct input parameters for the field-scale simulation. Chemical consumptions under dynamic conditions would be obtained for chemical slug designs and economic evaluations (Somerton and Radke, 1979; Wei et al., 2011).

Once the history matching of coreflood experiments is accomplished, the model can be used for small-scale or 2-D field pilot simulation studies. The pilot-scale simulation should focus on chemical slug designs, particularly the slug size and the mobility control. Then the model can be used for sensitivity studies on field performance and economic forecasts.

The final stage is to use the fully mechanistic model to calibrate a commercial simulator that is capable of performing large-scale or full-field simulation studies (Pandey et al., 2008b).

10.3.3 Other Laboratory Studies and Field Experiments

In recent years, many publications have indicated that radial coreflood in the laboratory would be useful for the design of field pilot tests. Since this type of experiment is conducted in a relatively thick core with only a few inches in diameter, the dispersion effect (Stoll et al., 2010) and the uncertainty in vertical sweep make the interpretation of the results difficult.

Single well chemical tracer tests (SWCTT) have been used in recent years in the field to evaluate the ASP process performance. The single well tracer test (Bragg et al., 1976) is a powerful tool for residual oil saturation determination in the reservoir and can cover a much larger distance than the laboratory cores. It also has been applied in the field in recent years to determine the injectivity and the displacement efficiency of the ASP process (Abdul Manap et al., 2011; Oyemade et al., 2010; Zubari and Babu Sivakumar, 2003), but other information such as oil recovery, microscopic displacement efficiency, and chemical concentrations in produced fluids cannot be obtained. Although the SWCTT will cost much less than a pilot test with its multiple wells and larger distances between injection and production wells, several issues still need to be considered before applying such tests to evaluate the effectiveness of an ASP system in the field.

- Distance: Most field data from observation wells indicated that (1) the ASP process is very sensitive to well-spacing (Zhu et al., 2012) and (2) most failures due to faulty design, i.e., slug breakdown or high chemical losses, occur beyond the distance reachable by tracers used in SWCTT. Displacement efficiencies are usually very high around the injection wells and may extend beyond 30–50 m. Therefore, a well-designed ASP formulation should be highly efficient in the first part of the reservoir that comes into contact with the chemical slug, as indicated by the laboratory linear coreflood data. The main concern will be the heterogeneity of the reservoir and the slug integrity beyond 50+ meters away from the injection well. This is the reason that most observation wells are designed to be located beyond one-third of the distance between injection and production wells in the pilot test.
- Confinement and heterogeneity: In many pilot tests today, multiwell tracer tests usually are conducted to detect the confinement and heterogeneity in the reservoir. SWCTTs are conducted to determine the residual

oil saturation after waterflood and also after chemical injection. In most cases, these tests were without the use of polymer. Because the ASP slug would have a much higher viscosity and higher displacement efficiency than the water used in the tracer test, the flow path will be different with fluids of different viscosities.

With the combination of reservoir heterogeneity and the contrast of viscous fluids, the areas covered by the SWCTT may not be able to provide an accurate representation of residual oil saturations (S_{orw}) prior to chemical flood or the residual oil saturations after the injection of a small ASP slug (S_{orc}) in the reservoir.

- Interpretation: With respect to the concerns noted above, field data taken from SWCTTs should be carefully interpreted prior to field implementation. Definitely, one should not conclude that a successful SWCTT with an ASP slug would be sufficient to warrant field-wide expansion without conducting a pilot test to gather information for the actual production of oil and chemicals in the production wells (Pandey et al., 2008; Reppert et al., 1990). One should also be careful not to confuse the "single well tracer test" with the "single well pilot test" which uses one injection well and multiple observation wells in longer distances without production wells (Abdul Manap et al., 2011; Finol et al., 2012).

10.4 THE SCREENING PROCESS

The screening criteria for the ASP process have noticeably changed over the last decade due to improvements in application and chemical technology that broadened the range of potential ASP applications. These criteria that have changed include the oil viscosity, the temperature, and the lithology.

1. Oil viscosity: One of the most significant changes in the ASP screening criteria is in the acceptable range of oil viscosity (Jennings et al., 1974). For many years, the application of all chemical processes was limited to reservoirs having oil viscosities up to about 200 cP. Today, the application of the ASP process can be extended to oils with viscosities over 1000 cP (DeFerrer, 1977; Kumar and Mohanty, 2010; Walker, 2011; Zhang et al., 2012). The upper limit has not yet been fully explored, but it could be as high as 5000 cP at reservoir conditions. Reasons for the broadening of application limits are (1) the low recovery by conventional methods in heavy oil reservoirs—high oil saturation for EOR, (2) the advances in drilling technology enabling the industry to use horizontal wells for injection and production, (3) the highly acidic nature of high viscous heavy oils, and (4) the higher oil prices in tandem with relatively low chemical costs.
2. Reservoir temperature: The temperature limit of the ASP flooding is mainly constrained by the temperature limits of the polymer because

most surfactants are stable at high temperatures, except for some sulfates. In recent years, improved polymers that tolerate higher temperatures have become available, and, therefore, the temperature limit for ASP flooding is higher depending upon the type of polymer or other mobility control agents used in the ASP slug. Recent laboratory studies have shown that polymer stabilities are dependent on other chemicals in the system and high-quality polyacrylamide polymers are stable at temperatures of at least 100°C with calcium ions below 200 ppm. Copolymers of acrylamide and sodium 2-acrylamido-2-methyl-propane sulfonate (AMPS) may also be used at higher temperatures (Levitt and Pope, 2008).

3. Lithology: All CEOR processes were limited to sandstone reservoirs until recent years where SP and ASP processes have been considered in carbonate reservoirs in Saudi Arabia (Al-Hashim et al., 1996) and West Texas (Levitt et al., 2011). Laboratory studies have been initiated in Mexico to use the ASP process in naturally fractured carbonate reservoirs in the Gulf Coast area. Although there are no major successes of ASP flooding in carbonate reservoirs to date, the industry is continuing to explore new territory for ASP applications. With more new chemicals that are available, applications of ASP in carbonate reservoirs may be viable in the future. Concerns in applying the ASP process in carbonate reservoirs are the reservoir heterogeneity, the low permeability, and the high chemical consumption.
4. Acid number: Most screening criteria for ASP applications require a minimum acid number of 0.3 mg KOH/g of oil. With the successful ASP project in the Daqing oil field in China, it appears that this process may also apply to reservoirs with low acidic crude oils without the benefits from the petroleum soaps generated *in situ*.
5. Other criteria: All other selection criteria such as acid content, permeability, and the requirement of freshwater with low divalent calcium (Ca^{++}) and magnesium (Mg^{++}) content are about the same as they have been in the past. One should note that it is not necessary to have all divalent cations removed from the ASP slug make up water using an expensive water softening unit (WSU). Previously proposed criteria limited the Ca^{++} and Mg^{++} to less than 10 ppm, but field data has shown that the alkaline solution can tolerate divalent cation contents to about 80 ppm, as shown in the Karamay oil field and in all ASP tests conducted in the Daqing oil field. Other options are to select alkaline chemicals such as sodium metaborate that have a higher tolerance to these divalent cations (Flaaten et al., 2008a,b). The addition of sequestering or chelating chemicals such as sodium tripolyphosphate (STPP) (Qiao et al., 2000) or ethylene diamine tetra acetic acid (EDTA) would also prevent the formation of scales or precipitation of $CaCO_3$ or $MgCO_3$ in injection fluids.

10.5 FIELD APPLICATIONS AND RESULTS

Field implementation of alkaline flooding without the addition of surfactants and polymer was initiated as early as 1925 (Nutting, 1925, 1928). Approximately 50 field trials, as shown in Table 10.1, were implemented prior to 1980 after it was discovered that surfactants could be generated *in situ* by interactions between alkaline chemicals and petroleum acids in the crude oil. Most of these tests were conducted in California because of the high acid content in most of the intermediate and heavy oil reservoirs in the region. In some cases, several tests were conducted in one field, such as in Orcutt Hill, Whittier, and Wilmington. Most of these tests showed some improvement in oil recovery but not enough to justify commercialization.

The need for polymer to improve mobility control and the addition of surfactants to enhance performance was realized in the laboratory from the 1980s forward but was not widely practiced in the field until the late 1990s. Testing ASP in the field was delayed due to the relatively high costs of polymer and surfactants and the low oil prices at that time. Since then, the application of alkaline chemicals in the field has mostly changed to ASP. Results from the ASP field trials showed significant improvement in the recovery efficiency with the highest being up to about 26% of OOIP (Qiao et al., 2000; Qu et al., 1998; Wang et al., 1999a,b). During the last 30 years, most ASP tests were conducted in the Daqing oil field, Shengli oil field, and Karamay oil field in China. A summary of the most significant field projects from 1980 to 2000 is shown in Table 10.1.

Beginning in late 1990, several ASP field projects were also conducted in the United States in smaller fields in the Minnelusa formation as shown in Table 10.2. These tests used very low concentrations of surfactants in the order of 0.1% by weight (Jay et al., 2000; Pitts, 2006) and were mostly conducted as the secondary recovery process. Although significant amounts of oil have been recovered in these applications, the amount of true incremental recovery by ASP over that of waterflood remains questionable since many of these applications were implemented in the early stages of waterflood and because polymers were also used for the improvement of mobility control. In order to logically evaluate the incremental recovery from the ASP process, a baseline of the ultimate waterflood and polymer flood recoveries must first be established. Subsequently then, the incremental recovery may be assessed. One other uncertainty here is the ambiguity of whether the incremental oil recovered was due to the sweep improvement by the use of polymer or a result of the actual ASP process. Because of the very small amount of surfactants ($\sim$0.1% active weight) used in these applications and the lack of analyses of surfactants in the produced fluid, the true contribution and significance of the additional surfactants in these tests is not clear.

Of all ASP field tests, three areas, Daqing, Shengli, and Karamay in China, will be discussed in detail. The ASP projects conducted in Daqing are

TABLE 10.2 ASP Project Conducted from 1980 to Early 2000

No.	Field/Block	State/ Country	Chemicals	Oil Gravity (°API)	Oil Viscosity (cP)	Stage of Flood	Inc. Rec. (%OOIP)	Note
1	Adena	CO, USA	Soda Ash	43	0.42	Tertiary	–	
2	Cambridge	WY, USA	Soda Ash	20	25	Secondary	28	Single injection well
3	Cressford	AB, Canada	AP	–	–	Secondary	–	Information not available
4	Daqing, S-ZX	Daqing, China	ASP	36	9–11	High water cut	21	See Table 10.4 for slug designs
5	Daqing, X5-Z	Daqing, China	ASP	36	9–11	High water cut	25	See Table 10.4 for slug designs
6	Daqing, X2-X	Daqing, China	ASP	36	9–11	High water cut	19	See Table 10.4 for slug designs
7	Daqing, S-B	Daqing, China	ASP	36	9–11	High water cut	23	See Table 10.4 for slug designs
8	Daqing, B1-FBX	Daqing, China	ASP	36	9–11	High water cut	21	See Table 10.4 for slug designs
9	Daqing, X2-Z	Daqing, China	ASP	36	9–11	High water cut	18	See Table 10.4 for slug designs
10	Daqing, ZB-B2-Z	Daqing, China	ASP	36	9–11	High water cut	NA	See Table 10.4 for slug designs
11	Daqing, S-B3X	Daqing, China	ASP	36	9–11	High water cut	NA	See Table 10.4 for slug designs
12	David	AB, Canada	–	23	–	Tertiary	–	
13	Enigma	WY, USA	–	–	–	–	–	
14	Isenhaur	WY, USA	AP	43	2.8	Secondary	12	High temperature, 200°C
15	Etzikorn	AB, Canada	AP	19	39–100	Secondary	12	NaOH

(*Continued*)

TABLE 10.2 (Continued)

No.	Field/Block	State/ Country	Chemicals	Oil Gravity (°API)	Oil Viscosity (cP)	Stage of Flood	Inc. Rec. (%OOIP)	Note
16	Gudong	China	ASP	17	41	Tertiary	26	No expansion
17	Gudao	China	ASP	–	46	Tertiary	16	No expansion
18	Karamay	China	ASP	30	53	Tertiary	25	No expansion
19	Mellot Ranch	WY, USA	ASP	21	28	Tertiary	NA	Two injection wells
20	Tanner	WY, USA	ASP	21	11	Secondary	18	Single injection well test
21	Viraj	India	ASP	18.9	50	~85% Water cut	NA	Four 5-spot pattern
22	West Kiehl	WY, USA	ASP	24	17	Secondary	21	
23	West Moorcroft	WY, USA	AP	22	20	Secondary	15	
24	White Castle	LA, USA	AS	29	3	Tertiary	10	

unique due to the fact that they were applied to a high wax and low acid crude oil. Additionally, there have been more than 10 field tests conducted in the last 15 years. The ASP process conducted in the Shengli oil field was for high viscosity and high acidic crudes in reservoirs with high salinity brines. The project was successful, but with the high divalent cation contents (>2000 ppm) in the reservoir and also in the injection brine, the application of the ASP process was not continued. The Karamay ASP project was conducted in a 50+ year old conglomerate water-flooded reservoir containing an oil with an acid content over 0.8 mg KOH/g. The pilot wells were all newly drilled, and the pre-ASP water injection showed nearly 100% water cut.

10.5.1 ASP Flooding in the Daqing Oil Field

Daqing had conducted more than 10 ASP pilot tests between 1994 and 2010 (Chang et al., 2006; Wang et al., 1999a,b; Zhu et al., 2012). Table 10.3 lists the most recently updated ASP field projects conducted in the Daqing oil field. The size of these tests varied with well-spacing, defined as the distance between the injection wells to production wells here, ranging from 164 to 820 ft. Update and analysis of all ASP field tests (Gao et al., 2010; Zhu et al., 2012) showed that there is definitely a correlation between recovery efficiency and well-spacing with higher recovery efficiencies (20–25% OOIP) for tests conducted in well-spacing of less than 410 ft and lower recovery efficiencies (18–20% OOIP) for well-spacing between 574 and 820 ft.

In the ASP design process, Daqing used two unique concepts including (1) a small high concentration polymer ahead of the ASP slug and (2) a lower concentration of alkaline and surfactants in the first part of the polymer drive. The polymer preflush is to provide a better sweep in displacing the reservoir brine and some profile adjustments. The lower concentrations of alkaline and surfactant in the polymer drive is to provide extra assurance of sufficient chemicals in the entire displacement process.

The slug designs for most pilot tests conducted prior to 2000 are shown in Table 10.4. Several types of surfactants including alkyl benzene sulfonates (ABS), petroleum sulfonates, lignosulfonates, petroleum carboxylates, and biologically produced surfactants (Wang et al., 1999a,b, 2009) were tested. HPAM polymers with different molecular weights were used in the preflush, ASP slug, and driving slug. In most cases, alkaline concentrations were in the range from 1 to 1.5 wt% and surfactant concentration was about 0.3 wt%. Polymer concentrations ranged from 1500 to 2000 ppm, much higher as compared to the polymer flooding (~1100 ppm) practiced in the same field.

Among all ASP field tests conducted in the Daqing oil field prior to 2000, one of the most significant tests was conducted in 1997 in the B1-FBX formation (Yang et al., 2003) that led to several large-scale field projects

TABLE 10.3 Ongoing or Planned ASP Pilot, Including Updated Daqing Large-Scale Tests

No.	Field	Country	Res. Temperature (°F)	Viscosity (cP)	Stage of Flood	Inc. Rec. (%OOIP)	Inj./Prod. Wells	Note
1	Daqing, S5-Z	China	113	9–11	Water cut: >85%	19	29/39	Projected recovery
2	Daqing, N1-E-II	China	113	9–11	Water cut: >85%	23	49/63	Projected recovery
3	Daqing, N2-W-II	China	113	9–11	Water cut: >85%	19	35/44	Projected recovery
4	Daqing, S6	China	113	9–11	Water cut: >85%	NA	144/160	–
5	Daqing, X1-2	China	113	9–11	Water cut: >85%	NA	112/143	–
6	Daqing, X6-E1	China	113	9–11	Water cut: >85%	NA	102/129	–
7	Daqing, X6-E2	China	113	9–11	Water cut: >85%	NA	105/109	–
8	Mangala	India	145.4	15	After PF	NA	¼	PF in progress, ASP next
9	Rex Energy	USA	80	11	Tertiary	NA	6/12	Pilot in progress, plan to expand
10	Elk Hill	USA	180–240	0.4–1.2	High water cut	NA	NA	Pilot complete, plan to expand
11	St. Joseph	Malaysia	126	1.5	High water cut	NA	NA	Offshore, planning
12	Angsi	Malaysia	246	0.3	High water cut	NA	NA	Offshore, single well tracer
13	Jhalora (K-IV)	India	179.6	30–50	Water cut: 80–85%	NA	1/6	Ongoing, plan to expand
14	Taber S. Mannville B	Canada	NA	NA	High water cut	NA	24/52	Large-scale field projects
15	Crowsnest	Canada	NA	NA	High water cut	NA	NA	Design stage
16	Marmul	Oman	Medium	90	High water cut	NA	¼	Planning stage

Note: NA = not available or not applicable.

TABLE 10.4 ASP Slug Formulation—First Eight Pilot Test Conducted in the Daqing Oil Field

Test Block	Chemicals	Polymer Preflush		ASP Slug I		ASP Slug II		Polymer Drive	
		Size (V_p)	Concentration (wt%)	Size (V_p)	Concentration (wt%)	Size (V_p)	Concentration (wt%)	Size (V_p)	Concentration (wt%)
Daqing, S-ZX	Soda Ash	0.00	0.00	0.00	0.00	0.30	1.25	0.00	0.00
	Surfactant 1	0.00	0.00	0.00	0.00	0.30	0.30	0.00	0.00
	HPAM	0.00	0.00	0.00	0.00	0.30	0.13	0.29	0.08
Daqing, X5-Z	Caustic soda	0.00	0.00	0.30	1.20	0.00	0.00	0.00	0.00
	Surfactant 2	0.00	0.00	0.30	0.30	0.00	0.00	0.00	0.00
	HPAM	0.00	0.00	0.30	0.12	0.18	0.12	0.09/0.03	0.08/0.04
Daqing, X2-X	Caustic soda	0.00	0.00	0.35	1.20	0.10	1.20	0.00	0.00
	Surfactant 2	0.00	0.00	0.35	0.30	0.10	0.10	0.00	0.00
	HPAM	0.04	0.15	0.35	0.23	0.10	0.18	0.25	0.08
Daqing, S-B	Caustic soda	0.00	0.00	0.33	1.20	0.15	1.20	0.00	0.00
	Surfactant 3	0.00	0.00	0.33	0.35	0.15	0.10	0.00	0.00
	HPAM	0.00	0.00	0.33	0.18	0.15	0.18	0.25	0.08

(*Continued*)

TABLE 10.4 (Continued)

Test Block	Chemicals	Polymer Preflush		ASP Slug I		ASP Slug II		Polymer Drive	
		Size (V_p)	Concentration (wt%)	Size (V_p)	Concentration (wt%)	Size (V_p)	Concentration (wt%)	Size (V_p)	Concentration (wt%)
Daqing, B1-FBX	Caustic soda	0.00	0.00	0.30	1.20	0.15	1.20	0.00	0.00
	Surfactant 2	0.00	0.00	0.30	0.30	0.15	0.10	0.00	0.00
	HPAM	0.00	0.00	0.30	0.20	0.15	0.14	0.05/0.05/01	0.09/0.07/0.06
Daqing, X2-Z	Caustic soda	0.00	0.00	0.35	1.00	0.10	1.00	0.00	0.00
	Surfactant 4	0.00	0.00	0.35	0.20	0.10	0.10	0.00	0.00
	HPAM	0.04	0.15	0.35	0.18	0.10	0.16	0.01/0.01	0.01/0.063
Daqing, ZB-B2-Z	Caustic soda	0.00	0.00	0.35	1.60	0.15	1.40	0.00	0.00
	Surfactant 4	0.00	0.00	0.35	0.30	0.15	0.10	0.00	0.00
	HPAM	0.04	0.14	0.35	0.18	0.15	0.18	0.20	0.14
Daqing, S-B3X	Caustic soda	0.00	0.00	0.35	1.20	0.10	1.20	0.00	0.00
	Surfactant 4	0.00	0.00	0.35	0.25	0.10	0.10	0.00	0.00
	HPAM	0.04	0.14	0.35	0.16	0.10	0.16	0.20	0.10

initiated in 2005. As it was discussed above, ASP projects are sensitive to well-spacing. This project was conducted in a larger well-spacing, 820 ft, with an objective to evaluate the performance of the ASP process in a well-spacing comparable to current waterflood operations. There were 6 injection wells and 12 producing wells. The recovery efficiency of this test was ~20% OOIP with a maximum water-cut reduction from 90% to 50%, as shown in Figure 10.9. This test was a foundation for Daqing to continue experimentation of the ASP process on a larger scale, starting in 2005.

10.5.2 ASP Flooding in the Shengli Oil Field

Shengli started experimental research in ASP flooding in the early 1980s and the first small well-spacing field test began in 1992 in the Gudong reservoir. Incremental recovery was reported to be 26% OOIP (Qu et al., 1998)

The second ASP pilot test was conducted in 1997 in the western part of the Gudao reservoir in an area of 150 AC. The reservoir temperature is about 160°F. The oil viscosity is about 46 cP and has an acid content above 0.5 mg KOH/g of oil. The well-spacing and net pay were 695 ft and 53 ft, respectively. The reservoir is a channel-sand deposit with average porosity and permeability of 32% and 1520 mD, respectively. The pilot area has 6 injection wells and 10 producing wells. The waterflood recovery efficiency was estimated to be about 22% of the OOIP in the pilot area prior to ASP flooding.

The ASP process was conducted in a 3-slug sequence.

1. Preflush: A 0.1-V_p 2000 ppm polymer solution was injected for 306 days.
2. ASP slug: A total of 0.3-V_p ASP slug containing 1.2% Na_2CO_3, 0.2% surfactant A, 0.1% surfactant B, and 1700 ppm polymer was injected for 948 days.
3. Polymer drive: A 0.05-V_p 1500 ppm polymer solution was injected for 158 days.

The injection of chemical slugs was completed in 2002. The oil rate increased from 630 B/D to 1490 B/D at the peak and corresponding water cut was 83%, decreased from 96% in waterflooding. The total incremental recovery was 15.5% OOIP. Although the ASP field trials conducted in Shengli showed significant incremental oil recovery, the project was not expanded due to the difficulties and costs associated in handling scales in the injection water and emulsions in production fluids.

10.5.3 ASP Flooding in the Karamay Oil Field

An ASP pilot test was conducted in Karamay in 1995 in a heterogeneous conglomerate reservoir with a well-spacing of 164 ft and four 5-spot patterns (Qiao et al., 2000). A three-slug process was designed with (a) a 0.40 V_p of 1.5% NaCl brine for preflush, (b) a 0.34 V_p of ASP slug containing 1.4%

Na_2CO_3, 0.3% crude oil sulfonates, and 0.13% HPAM polymer, and (c) a 0.15 V_p of 0.1% polymer and 0.4% NaCl drive fluid.

The waterflood (WF) recovery efficiency in the pilot area prior to ASP slug injection was about 50% OOIP at 99% water-cut. The ASP slug was injected from July 1996 to June 1997 with continued water drive to early 1999. The increased recovery started after about 0.04 V_p of the ASP slug had been injected and peaked when approximately 0.2 V_p of the ASP slug had been injected, resulting in a six-fold increase in oil rate and a 99% to 79% water-cut reduction. Incremental recovery in the central well was 25% OOIP. Severe emulsions in production fluids were observed, and difficulties were encountered in breaking the emulsions.

10.5.4 Other Field Test Results

Other than China, many other field tests were also conducted after 1995 in the United States, Canada, South America, and India. These include the David Pool (Pitts et al., 2004), Etzikom, Taber South, La Salina (Guerra et al., 2007), Cambridge (Jay et al., 2000), Tenner (Pitts, 2006), White Castle (Falls et al., 1994), Mellot Ranch, Viraj (Pratap and Gauma, 2004), and Jhalora (Jain et al., 2012) as shown in Table 10.2. Both Etzikom and Taber South are large field tests conducted in Canada, but some of them, the Etzikom and David Pool projects, only used alkaline polymer (AP) without surfactant. The White Castle test was an AS test without the use of the polymer. As discussed before, most of the field projects conducted in the Minnelusa formation in the United States were relatively small with one or two injection wells in the secondary stage; very low-surfactant concentrations were used, and the interpretation of results is difficult. More discussions on this topic will be given in Sections 10.5 and 10.10.

The Etzihom alkaline polymer project in Canada is a large-scale (18 injection wells) application which began in 2000 and used a low concentration of NaOH, 0.5%, without the addition of any surfactants. Significant incremental recovery (13–15% OOIP) was observed. Questions remaining with respect to this field project are (1) how to clearly differentiate the contribution to the incremental recovery due to the use of polymer from that of using the alkaline injection and (2) whether using a higher concentration of alkaline with the addition of some surfactants in the injection fluid would yield a higher oil recovery.

The Taber South Mannville B ASP flood is also a large-scale project in Canada. Sodium hydroxide was used in the ASP formulation. Although detailed publication of information on this project is not available, scales in the produced fluids would be expected.

Two most recently completed ASP pilot tests in the United States were the Elk Hills in California and the Bridgeport formation in the Illinois basin. Official reports on results of the Elk Hills project are not available, but results from the Bridgeport test were presented recently (Sharma et al., 2012). Although the final results are not available, interim production data showed significant incremental recovery. Two major issues are the decrease in fluid productivity and some severe chemical channeling to production wells outside of the pattern. The results are encouraging enough for the operator to consider a phase 1 expansion in the near future.

10.6 INTERPRETATION OF FIELD TEST RESULTS

The interpretation of field test results includes three components: the assessment of oil recovery efficiency, the interpretation of recovery mechanisms, and the process application.

10.6.1 Assessment of Oil Recovery Efficiency

As previously established, it is important to be able to assess the individual contributions of the sweep improvement and displacement efficiency. Much less ambiguity exists when the EOR process is applied in a tertiary stage, but results from secondary floods with CEOR can be confusing.

As shown in Table 10.2, fewer than half of the ASP tests that have been conducted in the last 30 years were in true tertiary stage. Furthermore, there were no detailed analyses of the benefits resulting from improving the sweep efficiency or displacement efficiency. Discussions on the value of assessing various factors are given below.

Daqing ASP Flooding

With the new interpretation of the reduction of waterflood residual oil saturation by the viscoelastic behavior of high concentration polymer solutions (Guo et al., 2011; Ma et al., 2010; Wang et al., 2011; Yang et al., 2006), the data should be evaluated to determine the actual benefit realized by the surfactant element of the ASP process. As shown in Table 10.4, the total polymer used in most of the ASP pilot tests should be approximately equal to or more than the amount of polymer used in the high concentration polymer floods recently implemented. A question that should be addressed is, "What is the incremental recovery of the ASP pilot tests in the past when compared to that of the high concentration polymer floods?"

Daqing technical staff has addressed answers to the question above in recent years (Hou et al., 2001a,b; Ma et al., 2010), but the focus of future CEOR in the Daqing oil field is still not clear. From the technology standpoint, it appears that three approaches may be considered in the giant field including the continuation of the low concentration polymer flood in low permeability areas, the application of high concentration polymer flood in high permeability areas, and the use of weak alkaline ASP or SP in all areas if significant amounts of incremental oil can be recovered over the high concentration polymer flood. But one needs to clarify the cost-benefits and applicability of each process when making decisions on the investment. For example, if the ASP flood did not recover a significantly greater amount of oil as compared to the high concentration polymer flood, then there is no need to invest the extra money in chemical costs for alkaline chemicals and, particularly, surfactants. Furthermore, the costs for treating produced fluids from polymer floods would be much lower than those from the ASP floods.

Etzihom AP Project

This project conducted in Canada is one of the most successful CEOR projects with recovery efficiencies between 10% and 15% OOIP. The sodium hydroxide concentration used is about ~0.5 wt%. The range of incremental recovery efficiency is similar to some of the most successful polymer floods. So the question is "Can a straight polymer flood obtain the same recovery?" If so, the cost of using sodium hydroxide may be saved.

ASP in Minnelusa Formation, USA—Secondary Application of the ASP Process with Low-Surfactant Concentration

As it was discussed before, some of the ASP successes in the Minnelusa formation in the United States used very low-surfactant concentrations (~0.1 wt%) but showed significant increased oil recovery. Investments in these projects are relatively low, project economics are good, and no significant production problems were reported. However, successes of these small-scale ASP projects are difficult to interpret due to the use of polymer in these floods, since it is known that polymer flooding can also significantly improve the amount of oil recovery, over 10% of OOIP, as shown in polymer flooding projects conducted in China. Unfortunately, detailed analyses to assess the contributions of polymer versus ASP or AP in these floods are not available. Without such assessments, optimization and planning for future projects such as these are difficult.

Commercial Implication

A comprehensive understanding of the recovery process can have a significant impact on commercial implementation. For example, if an ASP or AP process applied to a reservoir is successful with a significant amount of incremental recovery but such incremental recovery may also be recovered by a simple polymer injection, then the chemical costs and associated operating expenses for handling the alkaline and surfactant chemicals are unnecessary. In most successful cases, the chemical costs for polymer flood is about 30–50% of what it would be for the ASP or AP process. This is even more significant when such investments would be made for field-wide expansions.

The issue of incremental investment versus incremental benefit has been emphasized in recent years, and some operators are trying to conduct either side-by-side comparison of ASP versus polymer, such as the Marmul polymer flood and the ASP flood planned by PDO (Finol et al., 2012). Another approach is to conduct a pilot within a pilot with two stages of injection, polymer flood first to assess the incremental recovery, followed by an ASP flood to justify the incremental investments (Pandey et al., 2012). More considerations will be addressed in Section 10.10.

10.6.2 Interpretation of Recovery Mechanisms

Many investigators have studied the recovery mechanisms in the past (Arihara et al., 1999; Castor et al., 1979; Cooke et al., 1974; Edinga and McCaffery, 1979; Ehrlich and Wygal, 1977; Johnson, 1975; Leach et al., 1962, Raimodi et al., 1976; Shen et al., 2009; Stegemeier 1977; Tong et al., 1998; Wang et al., 2007). Since there could be several complex mechanisms in the ASP process including the IFT reduction, emulsification, and wettability alteration, each chemical-crude oil system may have different controlling mechanisms that require different combinations of the ASP chemicals. In some cases, only alkaline was used for wettability alteration (Emery et al., 1970, Graue and Johnson, 1974; Han et al., 2006b). In other cases, emulsification may have been the controlling mechanism (Kang and Wang, 2001; Lei et al., 2008; McAuliffe, 1972; Radke and Somerton, 1978; Wasan et al., 1979). Some designs only used alkaline and polymer such as in the Etzikom AP process and the MCCF process (Sarem, 1974). Today, it is well understood that the use of all three chemicals, an alkaline, at least one surfactant, and a polymer, results in the most efficient design. In fact, in some cases, it is beneficial to add a cosolvent and a cosurfactant (Sahni et al., 2010) into the ASP system to improve its robustness and performance.

All processes described above will only be effective with highly acidic crude oils because they depend on surfactants generated *in situ* from interactions between the injected alkaline chemicals and the petroleum acids in the crude oil. The mechanism of the ASP process applied in Daqing is somewhat different from other ASP flooding systems with the primary source of the surfactant being injected. *In situ* generated surfactants in this oil field are negligible due to the low acid content of the oil, <0.1 mg KOH/g of oil.

Historically, high pH SP flooding has been practiced (Holm and Robertson, 1978; Sui et al., 2000). Therefore, it appears that the ASP process conducted in the Daqing oil field is dependent primarily on the decreased surfactant retention in the reservoir and the improvements in lowering the IFT, as opposed to relying on the *in situ* generation of petroleum soaps.

10.6.3 Process Application

Based on the results of ASP tests conducted in the Daqing oil field in China, one may conclude that the ASP process can also be applied effectively in reservoirs with crude oils that have low acid content. However, the mechanisms are different when compared to conventional ASP applications because additional synergies and benefits may be realized by taking the advantage of the *in situ* generated petroleum soaps. Therefore, regardless of the detailed mechanistic analysis, it is known that alkaline chemicals play a very important role in CEOR, but they can also create problems in the production side if the amount used is excessive and the type is less desirable.

10.7 LESSONS LEARNED

According to publications, over 100 pilot tests have been conducted in the last 50 + years using the AP of ASP. Additionally, several hundred papers have also been written documenting laboratory studies on chemicals, formulation, mechanisms, reactions, etc. If we carefully examine the cumulative information published in the literature, we should be able to learn from past mistakes made in the industry and from the accomplishments achieved in R & D laboratories in order to improve the design and implementation of the ASP process. Below are summaries of few key lessons learned in field implementation.

1. Polymer preflush: In most cases, a preflush may be required to minimize the interaction between chemicals in the injection slug and the reservoir brine (Campbell, 1979; Dabbous and Elkins, 1976; Qiao et al., 2000). By using a higher salinity brine with low divalent cations in a preflush slug, a salinity gradient that is required for an efficient displacement process can be created (Levitt et al., 2006). If a polymer is used in preflush slug, better sweep will be achieved and some profile modification can also be accomplished (Dabbous and Elkins, 1976).
2. Injection water treatment: Good injection water quality is one of the most important requirements to ensure good injectivity and success in CEOR (Chen et al., 2007). Many operators use the water softening process to reduce the divalent cation content to less than 10 ppm. However, contrary to this practice, the ASP projects conducted in Daqing did not use water softening for the injection water, and the divalent cations, such as Mg^{++} and Ca^{++}, were above 10 ppm. The ASP process conducted in Karamay was similar to this approach, although STPP was added for sequestration. Therefore, the ASP system should always be tested for possible scale formation when the divalent cations are less than 100 ppm, and the option of using sequestration to overcome the scale issue on the injection side should be considered before investing a large amount of CAPEX on water softening equipment. Note that the scale formation is a slow process at room temperature and, sometimes, it may take 24 h to form. The other option is to design an ASP slug that will not form any scales in the injection system (Flaaten et al., 2008a,b).

3. Produced fluid treatment: Field experiences in Karamay, Daqing, and S. Taber have shown that emulsions and/or silicate/carbonate scales were formed in the produced fluid during the ASP process (Krumrine et al., 1985; Wang et al., 2004). Additional chemicals and equipment are required to treat the produced emulsions, but the issue of scales is more difficult to solve. Both mechanical and chemical inhibitors have been tried in the field with limited success (Arensdorf et al., 2010; Bataweel and Nasr-El-Din, 2011; Cao et al., 2007; Cheng et al., 2011a; Karazincir et al., 2011; Li et al., 2009b; Sonne et al., 2012; Qing et al., 2002; Yang et al., 2011). Studies have shown that silicate scales are more difficult to treat than the carbonate scales. Given that these two concerns exist, it would be prudent to ascertain how to effectively treat the produced emulsions (Di et al., 2001; Nguyen et al., 2011) and scales by conducting extensive laboratory research using real fluids/rock systems to simulate the produced fluid in order to develop treatment methods before field implementation. Moreover, the silicate scales are produced mainly because of the use of sodium hydroxide. Therefore, future field implementations should consider avoiding the use of sodium hydroxide in the ASP slug.
4. Decrease in productivity: Many field tests conducted in the past have shown significant loss of productivity. Daqing reported the productivity decrease in the range from about 38% to 71% as compared with the pre-ASP waterflood (Wang et al., 1999a,b). The production rate in the Bridgeport ASP flood also lost about 65% from the peak production level (Sharma et al., 2012). Several reasons include the oil–water bank relative permeability effect, the emulsification, the polymer, and the scale. The ASP slug design needs to take these effects into consideration in order to minimize the loss of productivity in the future.
5. Design and optimization of ASP slugs: In all the mechanistic aspects we discussed and have worked within the last 50 + years, it appears that the phase behavior with a large middle-phase microemulsion and the mobility controls are the two most important design requirements. In the design process, mobility control can be properly designed by adding polymer into the ASP slug. But the design of an efficient slug with ultra-low IFT and/or a large middle-phase microemulsion is more complicated. The conventional method is to measure the IFT of the AS system with the crude oil, but the IFT is a transient behavior and the volume ratio of the AS solution to the crude oil is so large that it makes the IFT data questionable because the acids in the tiny oil droplet are so small.

   In many alkaline–acidic crude oil systems, a macroemulsion may be formed spontaneously (Kang and Wang, 2001; McAuliffe, 1972; Shah et al., 1978; Wasan et al., 1979). These macroemulsions will coalesce over a time period to a thermodynamically stable equilibrated microemulsion at the optimal salinity for a given ratio of oil to aqueous AS formulation. Microemulsions with fast equilibration and low viscosity are preferred over both macroemulsions and high viscous–slow coalescence microemulsions to avoid phase trapping and high retention in the reservoir (Nelson and Pope, 1978; Levitt et al., 2006). The final ASP slug has to be clear and stable without precipitation, phase separation, or colloidal appearance and effective in displacement efficiency with an S_{orc} below 5%.

Although a laboratory coreflood with properly designed slug size, injection sequence, mobility control, and salinity gradient can be very successful, the design and optimization for any field project should be conducted with mechanistic simulation. In most cases, a pilot test design should consider the risk associated with allocating only marginal resources, such as cutting down on chemical costs, against overdesigning the process to ensure success in the pilot test. It is always easy to cut back the overdesigned parameters when a pilot test is successful, but one may never understand the reason for a failure and may never have a second chance if a pilot test fails due to the under design.

6. New chemicals: In addition to improvements in the chemical slug formulation, new chemicals, surfactants in particular, have been developed in recent years. As discussed before in Section 10.4, high temperature and high salinity polymers, low- or nonscale forming alkaline chemicals, cosurfactants, and cosolvents (Sahni et al., 2010) were used in recent ASP formulations. But the most advanced progress in chemicals is in the area of high-performance surfactants (Adkins et al., 2010, 2012; Barnes et al., 2010, 2012; Levitt et al., 2009; Yang et al., 2010; Zhao et al., 2008). These surfactants are now available for designing more robust ASP slugs and for harsher environments. Detailed discussion on ASP chemicals are beyond the scope of this chapter.

10.8 FUTURE OUTLOOK AND FOCUS

The ASP process is on the verge of being commercialized in the near future. Several companies are currently planning to expand their pilot projects in the United States and in other countries as well. The focus for future pilot tests and commercialization will be to improve the injection/production facilities and to enhance the chemical slug design in order to minimize the production of emulsions and scales.

1. Facilities: On the injection side, improvements will have to be made in the polymer handling facilities including (1) being capable of making the higher concentration polymer mother solution, (2) eliminating expensive and bulky storage/maturation units, (3) minimizing the mechanical shear degradation, and (4) improving the wellhead distribution system by adopting the single pump multiwell injection system.

On the production side, the emulsion treatment requires very large facilities including floatation, mechanical separation, chemical de-emulsification, and filtration. Unfortunately, there is no completely efficient method to date to handle the scales from production wells. The searching for low-cost remedies is continuing (Lo et al., 2011). Future focus on R & D should emphasize the design of better methods and facilities for de-emulsification and efficient ways for handling the scales in the produced fluid.

2. New slug designs: With the reality of the emulsion and scale issues in the produced fluids, improvements in facility designs and chemical methods for produced fluid treatment are urgent needs. However, equally critical is the research into ASP chemical slug formulations that can be used in ASP systems that will not cause severe scale and emulsion problems in the produced fluid. Laboratory results have shown that the practice of adding cosurfactant (Nelson et al., 1984) and/or cosolvents (Sahni et al., 2010) into the ASP slug may (1) reduce severe emulsion production, (2) eliminate the phase trapping in the reservoir, (3) enhance the phase behavior, and (4) lower the IFT. New laboratory tests have also shown that cosolvents might be used to replace the primary surfactant in the ASP slug in reservoirs containing high acidic crude oils.

   Other approaches such as replacing the conventional alkaline chemicals (Berger and Lee, 2006; Campbell, 1977), such as sodium hydroxide and sodium carbonate, with softer alkaline chemicals may also be considered. Adding sequestering or chelating chemicals into the conventional ASP injection slug may also alleviate some of the production problems (Cheng et al., 2011b).

3. Offshore applications: One encouraging fact is that ASP flooding has also been considered for offshore reservoirs in recent years such as those in Malaysia (Chon et al., 2011; Lo et al., 2011; Othman et al., 2007) and South America (Hernandez et al., 2001, 2003). Offshore ASP tests have not been conducted in the past due to the high costs associated with the requirement of massive injection and production facilities and the complex logistics in the handling and delivering of chemicals. The focus on the application of the ASP flooding in offshore reservoirs should be (1) developing manageable (weight and size) injection and production facilities, (2) improving the storage and delivering logistics and chemical handling for all chemicals, and (3) simplifying the operational procedures in the injection and production systems. For example, developing a continuous polymer mixing unit without maturation would reduce the space and weight requirement on the platform.

   Due to the limited injection and production wells and the relatively shorter field life in offshore operations, an accurate reservoir description is more difficult to obtain than it would be for old onshore fields. Yet, an adequate reservoir description is a key factor in any EOR project.

Therefore, the difficulty in obtaining this fundamental information presents an extra challenge for the design and implementation of an ASP project offshore.

Cooperation between operators, offshore logistics and equipment companies, and ASP injection/production facilities manufacturers is imperative in order to develop new systems that will make this task possible. Other issues that need extra consideration are the health, safety, and environment (HSE), surveillance and monitoring, and large capacities due to the fact that injection volumes are usually much larger for offshore than for onshore.

4. Reservoir issues: All reservoir issues need to be carefully evaluated for all EOR projects, particularly for CEOR due to higher costs of the chemical slug and relatively smaller slug size. Other than the regular screening parameters such as lithology, temperature, permeability and so on, reservoirs with strong edge and/or bottom water drive, large gas cap, severe dipping, and high thickness require additional efforts to evaluate the impacts of these factors on flood performances prior to field implementation. Extensive simulation studies to strategize the well location and type and pattern orientation are critical tasks in the evaluation process.

5. Chemical cost: Costs in EOR projects generally can be categorized in two major areas, CAPEX and OPEX. CAPEX is usually related to investments in equipment such as injection and production facilities. OPEX is usually related to expenses in daily operations such as injected chemicals and produced fluid treatments. There are other complex expenses such as drilling and completion, taxes, and royalties. The discussion in this chapter is limited to chemical costs only because they can be easily related to the incremental oil recovery as the cost per incremental barrel. In most cases, the range of the chemical cost is in the range from \$10 to \$20 per barrel incremental recovery with assumptions shown below.

Chemical slugs:

0.05 V_p preflush, 1500 ppm HPAM

- 0.30 V_p ASP slug, 1.5% Na_2CO_3, 0.3% surfactants, 2000 ppm HPAM
- 0.30 V_p polymer drive, 1500 ppm HPAM

Chemical cost:

- Na_2CO_3 = \$0.25/lb
- Surfactants = \$4/lb (100% active)
- HPAM = \$3/lb (90% active)

Incremental recovery: 20% OOIP

Other chemicals such as stabilizing agents for polymer, oxygen scavenger, and biocides are not included. As shown in Table 10.5, the chemical cost is \$16.54 per barrel of incremental oil. Using a spreadsheet, the cost can be calculated easily by entering the unit chemical cost, slug size, chemical concentration, and the incremental recovery efficiency.

TABLE 10.5 ASP Process Chemical Costs in US$/bbl Oil

Chemicals/Designs	Chemical Slugs		
	Preflush	ASP Slug	Polymer Drive
Slug size (P_v)	0.05	0.3	0.3
HPAM concentration (ppm)	1500	2000	1500
HPAM costs[a] (US$/lb)	1.5	1.5	1.5
Surfactants (ppm)	–	3000	–
Surfactants[a] (US$/lb)	–	2	–
Alkaline agents (ppm)	–	15,000	–
Alkaline agents[a] (US$/lb)	–	0.25	–
Recovery efficiency (%OOIP)		20	
Chemical slug costs (US$/bbl)	0.79	6.56	0.79
Chemical costs/bbl inc. oil (US$)		16.54	

[a]*All chemical costs are based on 100% active and current prices.*

10.9 CONCLUSIONS

- It has been proved that >20% OOIP incremental recoveries can be obtained with the ASP process, but optimal slug design with sufficient mobility control are needed for a field success.
- ASP slugs with alkaline concentrations above 1.0%, surfactant concentrations about 0.3%, and polymer concentrations higher than >1000 ppm are effective in most field operations.
- Sodium carbonate is preferred over sodium hydroxide for conventional alkaline chemicals.
- Small-scale tests appear to be more effective than large-scale tests due to the reservoir heterogeneity and chemical stability in the displacement process.
- Better ASP systems need to be developed to avoid the production of scales and heavy emulsions.
- ASP processes can be applied to reservoirs with higher temperature and higher salinity, oils with low acid contents, and oils with high viscosities.
- Commercialization of ASP flooding is on the verge of implementation in onshore reservoirs and implementation of pilot tests of the ASP process in offshore reservoirs will be initiated soon.

10.10 RECOMMENDATION ON FIELD PROJECT DESIGNS

1. Initial screening: The first step in the consideration of an ASP flood in the field is to sort out all key information on (1) reservoir properties, (2) oil properties, (3) water composition, and (4) the source and availability of a

freshwater supply. Table 10.6 shows key required data for the initial screening.

2. Reservoir studies: The reservoir study is one of the key tasks in any EOR project. Historically, most field projects have failed due to an insufficient understanding of the reservoir. In order to fully understand the reservoir and to forecast the EOR performance, one should start with a geological model, followed by a review of production history, and the execution of some field inter-well tracer tests, and pulse tests. Without conducting detailed reservoir studies through a joint effort of geoscientists and reservoir engineers, the project is not likely to succeed.

TABLE 10.6 Key Reservoir and Fluid Data for Initial Process Screening

Reservoir
Lithology, availability of cores
Location—onshore or offshore
Depth and dipping
Net pay
Temperature
Permeability, average
Porosity, average
Heterogeneity
Pressure
Gas cap
Bottom/Edge water
Oil—Sample Availability
Gravity
Viscosity at reservoir condition
Gas oil ratio (GOR)
Acid number
Brine—Formation, Injection, Produced
Total dissolved solids (TDS)
Ca^{++}
Mg^{++}
Detailed water analysis
Bacterial type and count
Freshwater availability
Production Data
Flood stage
Water–Oil ratio
Gas–Oil ratio
Original oil in place (OOIP)
Remaining oil in place (ROIP)

3. **Laboratory studies:** The third step is the laboratory evaluation of chemical systems and slug designs including polymer evaluation, acid number determination, alkali–oil interactions, ASP slug stability and optimal salinity determination, chemical transport and displacement efficiencies in linear cores, and mobility requirement. Most of the tasks were outlined and discussed in Section 10.3. In the chemical formulation and coreflood experiments, all studies should be conducted at reservoir temperatures.
4. **Process design and simulation studies:** With favorable laboratory studies in chemical formulations, polymer evaluation, mobility requirements, and high displacement efficiencies ($S_{orc} < 5\%$) in cores, a simulation of coreflood is necessary to establish a mechanistic model using parameters obtained. Subsequently, a field mechanistic model needs to be set up to conduct small-scale simulation studies on ASP performance, slug size designs, mobility designs, sensitivity studies, and economic forecasts. The next step would be to use the mechanistic model to calibrate a commercial simulator for larger scale simulation studies. One should not simply use any type of laboratory floods to design a field-scale project. Simulation with either a mechanistic model or calibrated commercial model should always be used for sensitivity studies, slug design, economic analysis, and process optimization.
5. **Field implementation:** Once the chemical slug is complete, the field implementation would begin. Depending on the size of the project, many logistics issues need to be considered. For example, a large international project may require a tender process that will further delay the project. Major tasks involved in a field implementation are (1) developing an overall plan, (2) identifying chemical and equipment suppliers, (3) carrying out field work such as well designs, drilling and completion, surface preparation, and (4) the tendering process, if required. A field project team including geologists, petrophysicists, engineers, accountants, legal advisors, and administrative staff needs to be set up to handle various tasks. An on-site field laboratory should also be available for quality control and monitoring. In order to optimize the planning, injectivity tests and small-scale tests with different CEOR processes such as polymer and ASP should be considered for some reservoirs.
6. **Facilities and services:** Acquiring field injection and production facilities is one of the major tasks in field implementation because the equipment must handle all injection fluids properly (Weatherill, 2009). Identification of qualified vendors and on-site services is critical.
7. **Monitoring and surveillance:** In all EOR processes, comprehensive monitoring and surveillance programs have to be developed in advance, especially for the complex CEOR processes like ASP. A simple laboratory in the field for water analysis and quality assurance and quality control (QA/QC) of chemicals received and produced fluid analysis is absolutely necessary. Periodical sampling and on-site analyses of injection and

produced fluids in the field laboratory have to be scheduled. Among the other tasks necessary may be scheduling and executing the logistics for sample delivery and also scheduling contracts for more sophisticated analysis by modern R & D laboratories that have high power instruments and equipment, such as HPLC and HRMS.

In addition to the chemical analysis and QA/QC, a data acquisition system for gathering dynamic injection and production data such as pressure, volume, chemical concentrations also needs to be developed. These data are essential for the performance evaluation.

Finally, the data submission procedures, transmission methods, storage and maintenance programs, and integration also have to be considered. Data collection and analysis is critical for pilot tests because such information will be used as the basis or justification for field-wide expansion.

8. Project expansion: After a successful pilot test, a field-wide or staged expansion program needs to be developed. Redesign and optimization with simulation studies based on pilot test data should be considered. A redesign of the field facilities is necessary to simplify the operational complexity, consider more cost-effective and space-saving programs, and develop a large-scale fluid distribution system. Since a much larger amount of data will be gathered for the surveillance and monitoring program, reduction of routine types of data gathering and analysis should be considered.

Lastly, it is critical to conduct a very comprehensive reservoir description and analysis for areas under consideration for the expansion because most failures are due to an inadequate understanding of the reservoir.

A flow diagram to show the steps and estimated timing for the ASP field project design is shown in Figure 10.11. Please note that this simple diagram

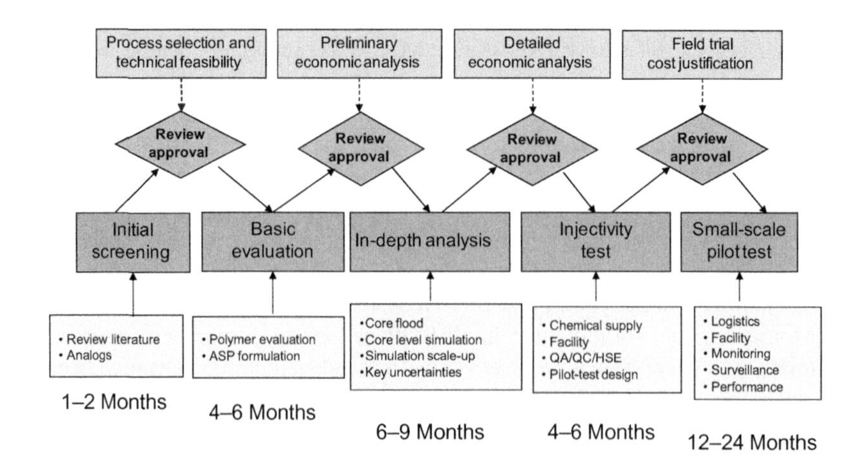

FIGURE 10.11 Flow diagram of the ASP process selection and design.

can be only used as a guideline for the implementation and execution of onshore ASP field projects. Additional preparation and more detailed plans need to be made for actual field project designs and execution. Offshore ASP flooding will be different and will involve several more logistical issues with regard to chemical supplies and the handling of produced fluids.

NOMENCLATURE AND ABBREVIATIONS

2-D two dimensional
ABS alkyl benzene sulfonate
AC acre
AMPS 2-acrylamido-2-methyl-propane sulfonate
AP alkaline polymer
AS alkaline surfactant
ASP alkaline surfactant–polymer
B1-FBX North 1—West Fault Block
CAPEX capital expenses
CEOR chemical enhanced oil recovery
DOE Department of Energy (USA)
EACN equivalent alkane carbon number
EDTA ethylene diamine tetra acetic acid
EOR enhanced oil recovery
HPAM partial hydrolyzed polyacrylamide
HPLC high-pressure liquid chromatography
HRMS high-resolution mass spectrometry
HSE health, safety, and environment
IFT interfacial tension
NPC National Petroleum Council (USA)
OOIP original oil in place
OPEX operating expenses
QA/QC quality assurance and quality control
PF polymer flooding
PVT pressure–volume–temperature
SP surfactant–polymer
SWCTT single well chemical tracer test
TDS total dissolved solids
UTCHEM University of Texas Chemical Flood Simulator
V_p pore volume
WF waterflood
WSU water softening unit

SYMBOLS

σ_o oil solubilization ratio
σ_w water solubilization ratio
V_o volume of oil
V_w volume of water

V_s volume of surfactants
C a correlation constant of 0.3 dyne/cm
γ interfacial tension
σ solubilization parameter

CHEMICALS

Ca^{++} calcium ion
EDTA ethylene diamine tetra acetic acid
KOH potassium hydroxide
Mg^{++} magnesium ion
NaOH sodium hydroxide
Na_2CO_3 sodium carbonate
$Na_2O \cdot SiO_2$ sodium orthosilicate
STPP sodium tripolyphosphate

REFERENCES

Abdul Manap, A.A., Chong, M.O., Sai, R.M., Zainal, S., Yahaya, A.W., Othman, M., 2011. Evaluation of alkali–surfactant effectiveness through single well test pilot in a Malaysian offshore field environment, SPE 144150. SPE Enhanced Oil Recovery Conference, 19–21 July, Kuala Lumpur, Malaysia.

Adkins, S. Liyanage, P.J., Gayani, W.P., Arachchilage, P., Mudiyanselage, T., Weerasooriya, U., et al., 2010. A new process for manufacturing and stabilizing high-performance EOR surfactants at low cost for high-temperature, high-salinity oil reservoirs, 129923-MS. SPE Improved Oil Recovery Symposium, 24–28 April, Tulsa, OK.

Adkins, S. Arachchilage, G.P., Solairaj, S., Lu, J., Weerasooriya, U., Pope, G.A., 2012. Development of thermally stable large-hydrophobe alkoxy carboxylate surfactants, 154256-MS. SPE Improved Oil Recovery Symposium, 14–18 April, Tulsa, OK.

Al-Hashim, H.S., Obiora, V., Al-Yousef, H.Y., Fernandez, F., Nofal, W., 1996. Alkaline surfactant polymer formulation for Saudi Arabian carbonate reservoirs, SPE 35353-MS. SPE/DOE Improved Oil Recovery Symposium, 21–24 April, Tulsa, OK.

Arensdorf, J., Hoster, D., McDougall, D., Yuan, M.D., 2010. Static and dynamic testing of silicate scale inhibitors, SPE 132212. CPS/SPE International Oil & Gas Conference and Exhibition in Beijing, 8–10 June, China.

Arihara, N., Yoneyama, T., Akita, Y., Lu, X.G., 1999. Oil recovery mechanisms of alkali–surfactant–polymer flooding, SPE 54330-MS, SPE Asia Pacific Oil and Gas Conference and Exhibition, 20–22 April, Jakarta, Indonesia.

Barnes, J.R., Dirkzwager, H., Smit, J.R., Smit, J.P., On, A., Reinald, C.N., et al., 2010. Application of internal olefin sulfonates and other surfactants to EOR, part 1: structure—performance relationships for selection at different reservoir conditions, 129766-MS. SPE Improved Oil Recovery Symposium, 24–28 April, Tulsa, OK.

Barnes, J.R., Groen, K., On, A., Dubey, S., Rezhik, C., Buijse, M.A., et al., 2012. Controlled hydrophobe branching to match surfactant to crude composition for chemical EOR, 154084-MS, SPE Improved Oil Recovery Symposium, 14–18 April, Tulsa, OK.

Bataweel, M.A., Nasr-El-Din, H.A., 2011. Alternatives to minimize scale precipitation in carbonate cores caused by alkalis in ASP flooding in high salinity/high temperature applications,

SPE 143155-MS. SPE European Formation Damage Conference, 7–10 June, Noordwijk, The Netherlands.

Berger, P.D., Lee, C.H., 2006. Improved ASP process using organic alkali, SPE 99581-MS. SPE/DOE Symposium on Improved Oil Recovery, 22–26 April, Tulsa, OK.

Bragg, J.R., Carlson, L.O., Atterbury, J.H., 1976. Recent application of the single-well tracer method for measuring residual oil saturation, SPE 5805. SPE Symposium on Improved Oil Recovery, 22–24 March, Tulsa, OK.

Campbell, T.C., 1977. A comparison of sodium orthosilicate and sodium hydroxide for alkaline waterflooding, SPE 6514. SPE Forty-Seventh Annual California Regional Meeting, 13–15 April, Bakersfield, CA.

Campbell, T.C., 1979. Chemical flooding, a comparison between alkaline and soft saline pre-flush systems for removal of hardness ions from reservoir brines, SPE 7873. SPE Oilfield and Geothermal Chemistry Symposium, 22–24 January, Houston, TX.

Cao, G., Liu, H., Shi, G.C., Wang, G.Q., Xiu, Z.W., Ren, H.F., et al., 2007. Technical breakthrough in PCPs' scaling issue of ASP flooding in Daqing oil field, SPE 109165-MS. SPE Annual Technical Conference and Exhibition, 11–14 November, Anaheim, CA.

Castor, T.P., Somerton, W.H., Kelly, J.F., 1979. Recovery mechanisms of alkaline flooding. Paper Presented at Third International Conference on Surface & Colloid Science, Surface Phenomenon in Enhanced Oil Recovery Section, 20–25 August, Stockholm, Sweden.

Chang, H.L., Zhang, Z.Q., Wang, Q.M., Xu, Z.S., Guo, Z.D., Sun, H.Q., et al., 2006. Advances in polymer flooding and alkaline-surfactant–polymer processes developed and applied in the People's Republic of China. J. Pet. Technol. February, 74.

Chen, G.Y., Wu, X.L., Yang, Z.Y., Yu, F.L., Hou, J.B., Wang, X.J., 2007. Study of the effect of injection water quality on the interfacial tension of ASP/crude oil. PETSOC (07-02-06).

Cheng, J.C., Xu, D.P., Bai, W.G., 2008. Commercial ASP flood test in Daqing oil field, SPE 117824-MS. Abu Dhabi International Petroleum Exhibition and Conference, 3–6 November, Abu Dhabi, UAE.

Cheng, J.C., Wu, D., Liu, W.J., Meng, X.C., Sun, F.X., Zhao, F.L., et al., 2011a. Field application of chelants in the handling of ASP-flooding produced fluid, SPE 143704-PA, SPE Projects. Facil. Const. 6 (3), 115 (September).

Cheng, J.C., Zhou, W.F., Zhang, Y.S., Xu, G.T., Ren, C.F., Peng, Z.G., et al., 2011b. Scaling principle and scaling prediction in asp flooding producers in Daqing oil field, SPE 144826-MS. SPE Enhanced Oil Recovery Conference, 19–21 July, Kuala Lumpur, Malaysia.

Chon, F.C., Adamson, G., Sho, W.L., Agarwal, B., Ritom, S., Du K.F., et al., 2011. St. Joseph chemical EOR pilot: a key de-risking step prior to offshore ASP full field implementation, SPE 144594-MS, SPE Enhanced Oil Recovery Conference, 19–21 July, Kuala Lumpur, Malaysia.

Chun, H., 1979. Interfacial tensions and solubilizing ability of a micro-emulsion phase that coexists with oil and brine. J. Colloid Interface Sci. 71 (2), 408.

Cooke Jr., C.E., Williams, R.E., Kolodzie, P.A., 1974. Oil recovery by alkaline waterflooding. J. Pet. Technol. December, 1365.

Dabbous, M.K., Elkins, L.E., 1976. Preinjection of polymer to increase reservoir flooding efficiency, SPE 5836. SPE Symposium on Improved Oil Recovery, 22–24 March, Tulsa, OK.

DeFerrer, M., 1977. Injection of caustic solutions in some Venezuelan crude oils, The Oil Sands of Canada and Venezuela, CIM Special, 17, 696.

Delshad, M., Han, W., Pope, G.A., Sepehrnoori, K., Wu, W., Yang, R., et al., 1998. Alkaline/surfactant/polymer flood predictions for the Karamay oil field, SPE 39610-MS. SPE/DOE Improved Oil Recovery Symposium, 19–22 April, Tulsa, OK.

Delshad, M., Han, C.Y., Veedu, F.K., Pope, G.A., 2011. A simplified model for simulation of alkaline-surfactant–polymer floods, SPE 142105-MS. SPE Reservoir Simulation Symposium, 21–23 February, The Woodlands, TX.

Di, W., Meng, X.C., Zhao, F.L., Zhang, R.Q., Chen, Y., Wang, Q.S., et al., 2001. Emulsification and stabilization of ASP flooding produced liquid, SPE 65390-MS, SPE International Symposium on Oilfield Chemistry, 13–16 February, Houston, TX.

Edinga, K.J., McCaffery, F.G., 1979. Cressford basal Colorado "A" reservoir-caustic flood evaluation, SPE 8199. SPE Fifty-Fourth Annual Fall Meeting, 23–26 September, Las Vegas, NV.

Ehrlich, R., Wygal Jr., R.J., 1977. Interrelation of crude oil and rock properties with the recovery of oil by caustic waterflooding. SPE J. August, 263.

Emery, L.W., Mungan, N., Nocholson, R.W., 1970. Caustic slug injection in the singleton field. J. Pet. Technol. December, 1969.

Falls, A.H., Thigpen, D.R., Nelson, R.C., Ciaston, J.W., Lawson, J.B., Good, P.A., et al., 1994. Field test of co-surfactant enhanced alkaline flooding, SPE 24117-PA. SPE Reserv. Eng. 9 (3), 217.

Farajzadeh, R., Matsuura, T., Van Batenburg, D., Dijk, H., 2011. Detailed modeling of the alkali surfactant polymer (ASP) process by coupling a multi-purpose reservoir simulator to the chemistry package PHREEQC, SPE 143671-MS, SPE 143992-MS. SPE Enhanced Oil Recovery Conference, 19–21 July, Kuala Lumpur, Malaysia.

Farmanian, P.A., Davis, N., Kwan, J.T., Yen, T.F., Weinbrandt, R.M. 1978. Isolation of native petroleum fractions for lowering interfacial tensions in aqueous alkaline system. Paper Presented at Symposium on Chemistry of Oil Recovery, Division of Petroleum Chemistry, ACS, 12–17 March, Anaheim, CA.

Finol, J., Al-Harthy, S., Jaspers, H., Batrani, A., Al-Hadhrami, H., van Wunnik, J., et al., 2012. Alkaline-surfactant–polymer pilot test in southern Oman, SPE 155403-MS. SPE EOR Conference at Oil and Gas West Asia, 16–18 April, Muscat, Oman.

Flaaten, A.K., Nguyen, Q.P., Pope, G.A., Zhang, J.Y., 2008a. A systematic laboratory approach to low-cost, high-performance chemical flooding, SPE 113469-MS. SPE/DOE Symposium on Improved Oil Recovery, 20–23 April, Tulsa, OK.

Flaaten, A.K., Nguyen, Q.P., Zhang, J.Y., Mohammadi, H., Pope, G.A., 2008b. ASP chemical flooding without the need for soft water, SPE 116754-MS. SPE Annual Technical Conference and Exhibition, 21–24 September, Denver, CO.

Gao, S.T., Gao, Q., Ji, L., 2010. Recent progress and evaluation of ASP flooding for EOR in Daqing oil field, SPE 127714-MS. SPE EOR Conference at Oil & Gas West Asia, 11–13 April, Muscat, Oman.

Glinsmann, G.R., 1979. Surfactant flooding with microemulsions formed *in situ*—effect of oil characteristics, SPE 8326. Annual Technical Conference and Exhibition, 23–26 September, Las Vegas, NV.

Graue, D.J., Johnson Jr., C.E., 1974. A field trial of the caustic flooding process. J. Pet. Technol. December, 1353.

Gogarty, W.B., Meabon, H.P., Milton, H.W., 1970. Mobility control design for miscible-type waterfloods using micellar solutions. J. Pet. Technol. January, 141.

Guerra, E., Valero, E., RodrAguez, D., Gutierrez, L., Castillo, M., Espinoza, J., et al., 2007. Improved ASP design using organic compound-surfactant-polymer (OCSP) for La Salina

Field, Maracaibo Lake, SPE 107776-MS. Latin American and Caribbean Petroleum Engineering Conference, 15–18 April, Buenos Aires, Argentina.

Guo, C., Han, P.H., Shao, Z.B., Zhang, X.L., Ma, M.R., Lu, K.W., et al., 2011. History matching method for high concentration viscoelasticity polymer flood pilot in Daqing oil field, SPE 144538. SPE Enhanced Oil Recovery Conference, 19–21 July, Kuala Lumpur, Malaysia.

Han, C., Delshad, M., Pope, G.A., Sepehrnoori, K., 2006a. Coupling EOR compositional and surfactant models in a fully implicit parallel reservoir simulation using EACN concept, SPE103194-MS. SPE Annual Technical Conference and Exhibition, 24–27 September, San Antonio, TX.

Han, D., Yuan, H., Weng, R., Dong, F., 2006b. The effect of wettability on oil recovery by alkaline/surfactant/polymer flooding, SPE 102564-MS. SPE Annual Technical Conference and Exhibition, 24–27 September, San Antonio, TX.

Healy, R.N., Reed, R.L., 1974. Physical–chemical aspects of micro-emulsion flooding, SPE 4583-PA. SPE J. 14 (5), (October).

Hemandez, C., Chacon, L. J., Anselmi, L., Baldonedo, A., Qi, J., Dowling, P. C., et al., 2003. ASP System Design for an Offshore Application in La Salina Field, Lake Maracaibo. SPEREE 6 (3), 147–156.

Hernandez, C., Chacon, L.J., Anselmi, L., Baldonedo, A., Qi, J., Dowling, P.C., et al., 2001. ASP system design for an offshore application in the La Salina Field, Lake Maracaibo, SPE 69544-MS. SPE Latin American and Caribbean Petroleum Engineering Conference, 25–28 March, Buenos Aires, Argentina.

Holm, L.W., Robertson, S.D., 1978. Improved micellar-polymer flooding with high pH chemicals, SPE 7583. SPE Fifty Third Annual Fall Meeting, 1–8 October, Houston, TX.

Hou, J.R., Liu, Z.C., Xia, H.F., 2001a. Viscoelasticity of ASP solution is a more important factor of enhancing displacement efficiency than ultra-low interfacial tension in ASP flooding, SPE 71061-MS. SPE Rocky Mountain Petroleum Technology Conference, 21–23 May, Keystone, CO.

Hou, J.R., Ziu, Z.C., Yue, X.A., Xia, H.F., 2001b. Study of the effect of ASP solution viscoelasticity on displacement efficiency, SPE 71492-MS. SPE Annual Technical Conference and Exhibition, 30 September–3 October, New Orleans, LA.

Hou, J.R., Liu, Z.C., Dong, M.Z., Yue, X.A., Yang, J.Z., 2006. Effect of viscosity of alkaline/surfactant/polymer (ASP) solution on enhanced oil recovery in heterogeneous reservoirs. PETSOC (06-11-03).

Jain, A.K., Dhawan, A.K., Misra, T.R., 2012. ASP flood pilot in Jhalora (K-IV), India—a case study, SPE 153667-MS. SPE Oil and Gas India Conference and Exhibition, 28–30 March, Mumbai, India.

Jay, V., Turner, J., Vergnani, B., Pitts, M.J., Wyatt, K., Surkalo, H., et al., 2000. Alkaline-surfactant–polymer flooding of the Cambridge Minnelusa field, SPE 68285-PA. Reserv. Eng. Eval. 3 (6), 552.

Jennings Jr., H.Y., Johnson Jr., C.E., McAuliffe, C.D., 1974. A caustic waterflooding process for heavy oils. J. Pet. Technol. December, 1344.

John, A., Han, C., Delshad, M., Pope, G.A., Sepehrnoori, K., 2005. A new generation of chemical-flooding simulator, SPE 89436-MS. SPE/DOE Symposium on Improved Oil Recovery, 17–21 April, Tulsa, OK.

Johnson Jr., C.E., 1975. Status of caustic and emulsion methods, SPE 5561. SPE Fiftieth Annual Fall Meeting, 28 September–1 October, Dallas, TX.

Kalra, A., Venkatraman, A., Raney, K., Dindoruk, B., 2011. Prediction and experimental measurements of water-in-oil emulsion viscosities during alkaline surfactant injections, SPE 143992-MS. SPE Enhanced Oil Recovery Conference, 19–21 July, Kuala Lumpur, Malaysia.

Kang, W.L., Wang, D.M., 2001. Emulsification characteristic and de-emulsifiers action for alkaline/surfactant/polymer flooding, SPE 72138-MS. SPE Asia Pacific Improved Oil Recovery Conference, 6–9 October, Kuala Lumpur, Malaysia.

Karazincir, O., Thach, S., Wei, W., Prukop, G., Malik, T., Dwarakanath, V., 2011. Scale formation prevention during ASP flooding, SPE 141410-MS. SPE International Symposium on Oilfield Chemistry, 11–13 April, The Woodlands, TX.

Karpan, V., Farajzadeh, R., Zarubinska, M., Dijk, H., Matsuura, T., Stoll, M., 2011. Selecting the "right" ASP model by history matching coreflood experiments, SPE 144088. SPE Enhanced Oil Recovery Conference, 19–21 July, Kuala Lumpur, Malaysia.

Krumrine, P.H., Mayer, E.H., Brock, G.F., 1985. Scale formation during alkaline flooding. J. Pet. Technol. 37 (8), 1466–1474.

Kumar, R., Mohanty, K.K., 2010. ASP flooding of viscous oils, SPE 135265-MS. SPE Annual Technical Conference and Exhibition, 19–22 September, Florence, Italy.

Leach, R.O., Wagner, O.R., Wood, H.W., Harpke, C.F., 1962. A laboratory and field study of wettability adjustment in water flooding. J. Pet. Technol. February, 206.

Lei, Z.D., Yuan, S.Y., Song, J., Yuan, J.R., Wu, Y.S., 2008. A Mathematical Model for Emulsion Mobilization and Its Effect on EOR during ASP Flooding, paper SPE 113145-MS presented at the SPE/DOE Symposium on Improved Oil Recovery, 20–23 April, Tulsa, OK.

Levitt, D.B., Pope, G.A., 2008. Selection and screening of polymers for enhanced-oil recovery, SPE 113845. SPE IOR Symposium, 19–23 April, Tulsa, OK.

Levitt, D.B., Jackson, A.C., Heinson, C., Britton, L.N., Malik, T., Dwarakanath, V. et al., 2006. Identification and evaluation of high performance EOR surfactants, SPE 100089. SPE Symposium on Improved Oil Recovery, 22–26 April, Tulsa, OK.

Levitt, D.B., Jackson, A.C., Heinson, C., Britton, L.N., Malik, T., Dwarakanath, V., et al., 2009. Identification and evaluation of high-performance EOR surfactants, 100089-PA. SPE Reserv. Eval. Eng. 12 (2), 243.

Levitt, D.B., Dufour, S., Pope, G., Morel, D., Gauer, P., 2011. Design of an ASP flood in a high-temperature, high-salinity, low-permeability carbonate, IPTC 14915-MS. International Petroleum Technology Conference, 7–9 February, Bangkok, Thailand.

Li, H.F., Liao, G.Z., Han, P.H., Yang, Z.N., Wu, X.L., Chen, G.Y., et al. 2008. Alkaline/Surfactant/Polymer (ASP) commercial flooding test in Central Xing2 area of Daqing oil field, SPE 84896-MS. SPE/DOE Symposium on Improved Oil Recovery, 20–23 April, Tulsa, OK.

Li, D.S., Shi, M.Y., Wang, D.M., Li, Z., 2009a. Chromatographic separation of chemicals in alkaline surfactant polymer flooding in reservoir rocks in the Daqing oil field, SPE 121598-MS. SPE International Symposium on Oilfield Chemistry, 20–22 April, The Woodlands, TX.

Li, J.L., Li, T.D., Yan, J.D., Zuo, X.W., Zheng, Y., Yang, F., 2009b. Silicon containing scales forming characteristics and how scaling impacts sucker rod pump in ASP flooding, SPE 122966. SPE Asia Pacific Oil & Gas Conference and Exhibition, 4–6 August, Jakarta, Indonesia.

Lo, S.W., Shahin, G.T., Graham, G., Simpson, C., Kidd, S., 2011. Scale control and inhibitor evaluation of an alkaline surfactant polymer flood, SPE 141551-MS. SPE International Symposium on Oilfield Chemistry, 11–13 April, The Woodlands, TX.

Ma, W.G., Xia, H.F., Yu, H.Y. 2010. The effect of rheological properties of alkali-surfactant–polymer system on residual oil recovery rate after water flooding, SPE 134064-MS. SPE Asia Pacific Oil and Gas Conference and Exhibition, 18–20 October, Brisbane, Queensland, Australia.

Mayer, E.H., Berg, R.L., Carmichael, J.D., Weinbrandt, R.M., 1983. Alkaline injection for enhanced oil recovery—a status report. J. Pet. Technol. 35 (1).

McAuliffe, C.D., 1972. Oil-in-water emulsions improve fluid flow in porous media, SPE 3784. SPE Improved Oil Recovery Symposium, 16–19 April, Tulsa, OK.

Mohammadi, H., Delshad, M., Pope, G.A., 2009. Mechanistic modeling of alkaline/surfactant/polymer floods, SPE 110212-PA. SPE/DOE Symposium on Improved Oil Recovery, 20–23 April, Tulsa, OK.

Moreno, R., Anselmi, L., Coombe, D., Card, C., Cols, I., 2003. Comparative mechanistic simulations to design and ASP field pilot in La Salina, Venezuela, PETSOC 2003-199.

National Petroleum Council Report, 1976. Enhanced Oil Recovery—An Analysis of the Potential From Known Fields in The United States—1976 TO 2000, National Petroleum Council, December, Library of Congress Catalog Card No. 76-62538.

National Petroleum Council Report, 1984. Enhanced Oil Recovery, National Petroleum Council, June, Library of Congress Catalog Card No. 84-061296.

Nelson, R.C., 1983. The effect of live crude on phase behavior and oil-recovery efficiency of surfactant flooding systems, SPE 10677-PA. SPE J. 23 (3).

Nelson, R.C., Pope, G.A., 1978. Phase behavior in chemical flooding, SPE-7079-PA. SPE J. 18 (5), 325.

Nelson, R.C., Lawson, J.B., Thigpen, D.R., Stegemeier, G.L., 1984. Co-surfactant—enhanced alkaline flooding, SPE 12672. SPE Symposium on Improved Oil Recovery, 15–18 April, Tulsa, OK.

Nguyen, D., Sadeghi, N., Houston, C., 2011. Emulsion characteristics and novel demulsifiers for treating chemical EOR induced emulsions, SPE 143987. SPE Enhanced Oil Recovery Conference, 19–21 July, Kuala Lumpur, Malaysia.

Nutting, P.G., 1925. Chemical problems in the water driving of petroleum from oil sands. Ind. Eng. Chem. 17, 1035.

Nutting, P.G., 1927. Soda process for petroleum recovery. Oil Gas J. 25 (45), 76 and 150; Principles underlying soda process. Ibid 25 (50), 32 and 106; Petroleum recovery by soda process. Ibid (1928) (27), 146 and 238.

Othman, M., Chong, M.O., Sai, R.M., Zainal, S., Zakaria, M.S., Yaacob, A.A., 2007. Meeting the challenges in alkaline surfactant pilot project implementation at Angsi field, offshore Malaysia, SPE 109033-MS. SPE Offshore Europe, 4–7 September, Aberdeen, Scotland.

Oyemade, S.N., Al-Harthy, S.A., Jaspers, H.F., Van Wunnik, J., De Fruijf, A., 2010. Alkaline-surfactant–polymer (ASP): single well chemical tracer tests—design, implementation and performance, SPE 130042-MS. SPE EOR Conference at Oil and Gas West Asia, 11–13 April, Muscat, Oman.

Pandey, A., 2010. Refinement of chemical selection for the planned ASP pilot in Mangala field—additional phase behaviour and coreflood studies, SPE 129046-MS. SPE Oil and Gas India Conference and Exhibition, 20–22 January, Mumbai, India.

Pandey, A., Beliveau, D., Corbishley, D.W., Kumar, M.S., 2008a. Design of an ASP pilot for the Mangala field, laboratory evaluations and simulation studies, SPE 113131-MS. SPE Indian Oil and Gas Technical Conference and Exhibition, 4–6 March, Mumbai, India.

Pandey, A., Kumar, M.S., Beliveau, D., Corbishley, D.W., 2008b. Chemical flood simulation of laboratory corefloods for the Mangala field, generating parameters for field-scale simulation, SPE 113347-MS. SPE/DOE Symposium on Improved Oil Recovery, 20–23 April, Tulsa, OK.

Pandey, A., Kumar, M.S., Jha, M.K., Tandon, R., Punnapully, B.S., Kalugin, M., et al., 2012. Chemical EOR pilot in Mangala field, results of initial polymer flood phase, SPE 154159-MS. SPE EOR/IOR Symposium, 14–18 April, Tulsa, OK.

Pitts, M.J., 2006. Alkaline-surfactant–polymer flood of the tanner field, SPE 100004-MS. SPE/DOE Symposium on Improved Oil Recovery, 22–26 April, Tulsa, OK.

Pitts, M.J., Wyatt, K., Surkalo, H., 2004. Alkaline-polymer flooding of the David pool, Lloydminster AB, SPE 89386. SPE/DOE Symposium on Improved Oil Recovery, 17–21 April, Tulsa, OK.

Pratap, M., Gauma, M.S., 2004. Field implementation of alkaline-surfactant–polymer (ASP) flooding, a maiden effort in India, SPE 88455-MS. SPE Asia Pacific Oil and Gas Conference and Exhibition, 18–20 October, Perth, Australia.

Qiao, Q., Gu, H.J., Li, D.W., Dong, L., 2000. The pilot test of ASP combination flooding in Karamay oil field, SPE 64726. International Oil and Gas Conference and Exhibition, 7–10 November, Beijing, China.

Qing, J., Bin, Z., Zhang, R., Zhongxi, C., Yuchun, Z., 2002. Development and application of a silicate scale inhibitor for ASP flooding production scale. Paper SPE 74675-MS. SPE International Symposium on Oilfield Scale, 30–31 January, Aberdeen, UK.

Qu, Z.J., Zhang, Y.G., Zhang, X.S., Dai, J.L., 1998. A successful ASP flooding pilot in Gudong oil field, SPE 39613-MS. SPE/DOE Improved Oil Recovery Symposium, 19–22 April, Tulsa, OK.

Radke, C.J., Somerton, W.H., 1978. Enhanced recovery with mobility and reactive tension agents. Paper B-2. Fourth Annual DOE Symposium, 29–31 August, Tulsa, OK.

Raimodi, P., Gallagher, B.J., Bennett, G.S., Ehrlich, R., Messmer, J.H., 1976. Alkaline water-flooding design and implementation of a field pilot, SPE 5831. SPE Symposium on Improved Oil Recovery, 22–24 March, Tulsa, OK.

Reppert, T.R., Bragg, J.R., Wilkinson, J.R., Snow, T.M., Maer Jr., N.K., and Gale, W.W. 1990. Second Ripley surfactant flood pilot test, SPE/DOE 20219. SPE/DOE Seventh Symposium on Enhanced Oil Recovery, April, Tulsa, OK.

Roshanfekr, M., Johns, R.T., Pope, G.A., Britton, L., Linnemeyer, H., Britton, C., et al., 2009. Effects of pressure, temperature, and solution gas on oil recovery from surfactant polymer floods, SPE 125095. SPE Annual Technical Conference, 4–7 October, New Orleans, LA.

Roshanfekr, M., Johns, R.T., Delshad, M., Pope, G.A., 2011. Modeling of pressure and solution gas for chemical floods, SPE 147473. SPE Annual Technical Conference, 30 October–2 November, Denver, CO.

Sahni, V., Dean, R., Britton, C., Kim, D.H., Seerasooriya, U., Pope, G.A., 2010. The role of co-solvents and co-surfactants in making chemical floods robust, SPE 130007-MS. SPE Improved Oil Recovery Symposium, 24–28 April, Tulsa, OK.

Sarem, A.M., 1974. Secondary and tertiary recovery of oil by MCCF process, SPE 4901. SPE Forty-Fourth California Regional Meeting, 4–5 April, San Francisco, CA.

Scriven, L.E., Anderson, D.R., Bidner, M.S., Davis, H.T., Manning, C.D., 1976. Interfacial tension and phase behavior in surfactant–brine–oil system, SPE 5811-MS. SPE Symposium on Improved Oil Recovery, 22–24 March, Tulsa, OK.

Seifert, W.K., Howells, W.G., 1969. Interfacially active acids in a California crude oil, isolation of carboxylic acids and phenols. Anal. Chem. 41 (4), 554.

Seifert, W.K., Teeter, R.M., 1970. Identification of polycyclic aromatic and heterocyclic crude oil carboxylic acids. Anal. Chem. 42 (7), 750.

Shah, D.O., Bansal, V.K., Chan, K.S., McCallough, R., 1978. The effect of caustic concentration on interfacial charge, interfacial tension and droplet size, a simple test for optimum caustic concentration for crude oils. J. Can. Petrol. Technol. 17 (1).

Sharma, A., Azizi-Yarand, A., Clayton, B., Baker, G., McKinney, P., Britton, C., et al., 2012. The design and execution of an alkaline-surfactant–polymer pilot test. SPE 154318 Eighteenth SPE Improved Oil Recovery Symposium, 14–18 April, Tulsa, OK.

Shen, P.P., Wang, J.L., Yuan, S.Y., Zhong, T.X., Jia, X., 2009. Study of enhanced-oil-recovery mechanism of alkali/surfactant/polymer flooding in porous media from experiments, SPE 126128-PA. SPE J. 14 (2), 237.

Somerton, W.H., Radke, C.J., 1979. Role of clays in enhanced recovery of petroleum. Paper D-7, Fifth Annual DOE Symposium, 22–24 August, Tulsa, OK.

Song, W.C., Yang, C.Z., Han, D.K., Qu, Z.J., Wang, B.Y., Jia, W.L., 1995. Alkaline-surfactant–polymer combination flooding for improving of the oil with high acid value, SPE 29905-MS. International Meeting on Petroleum Engineering, 14–17 November, Beijing, China.

Sonne, J., Miner, K., Kerr, S., 2012. Potential for inhibitor squeeze application for silicate scale control in ASP flood, SPE 154331-MS. SPE International Conference on Oilfield Scale, 30–31 May, Aberdeen, UK).

Southwick, J.G., Svec, Y., Chilek, G., Shahin, G.T., 2010. The effect of live crude on alkaline-surfactant–polymer formulations, implications for final formulation design, SPE 135357-MS. SPE Annual Technical Conference and Exhibition, 19–22 September, Florence, Italy.

Stegemeier, G.L., 1977. Mechanisms of oil entrapment and mobilization in porous media. Paper Presented at the AICHE Symposium on Improved Oil Recovery by Surfactant and Polymer, 12–14 April, Kansas City, MO.

Stoll, W.M., Al-Shureqi, H., Finol, J., Al-Harthy, S.A.A., Oyemade, S., De Krujif, A., et al., 2010. Alkaline-surfactant–polymer flood, from the laboratory to the field, SPE 129164-MS. SPE EOR Conference at Oil & Gas West Asia, 11–13 April, Muscat, Oman.

Sui, J., Yang, C.Z., Yang, Z.Y., Lio, G.Z., Yuan, H., Dai, Z.J., et al., 2000. Surfactant-alkaline-polymer flooding pilot project in non-acidic paraffin oil field in Daqing, SPE 64509-MS. SPE Asia Pacific Oil and Gas Conference and Exhibition, 16–18 October, Brisbane, Australia.

Tong, Z.S., Yang, C.Z., Wu, G.Q., Yuan, H., Yu, L., Tian, G.L., 1998. A study of microscopic flooding mechanism of surfactant/alkali/polymer, SPE 39662-MS. SPE/DOE Improved Oil Recovery Symposium, 19–22 April, Tulsa, OK.

Wade, W.H., Cayias, J.L., Schechter, R.S., 1976. Modeling crude oils for low interfacial tension, SPE 5813-PA. SPE Symposium on Improved Oil Recovery, 22–24 March, Tulsa, OK.

Walker, D.L., 2011. Experimental Investigation of the Effect of Increasing Temperature of ASP Flooding. MS thesis, University of Texas, Austin, TX.

Wang, C.L., Wang, B.Y., Cao, X.L., Li, H.C. 1997. Application and design of alkaline-surfactant–polymer system to close well spacing pilot Gudong oil field, SPE 38321. SPE Western Regional Meeting, 25–27 June, Long Beach, CA.

Wang, D.M., Cheng, J.C., Wu, J.Z., Yang, Z.Y., Yao, Y.M., Li, H.F., 1999a. Summary of ASP pilots in Daqing oil field, SPE 57288. SPE Asia Pacific Improved Oil Recovery Conference, 25–26 October, Kuala Lumpur, Malaysia.

Wang, D.M., Cheng, J.C., Li, Q., Li L.Z., Zhao, C.J., Hong, J.C., 1999b. An alkaline biosurfactant polymer flooding pilots in Daqing oil field, SPE 57304-MS. SPE Asia Pacific Improved Oil Recovery Conference, 25–26 October, Kuala Lumpur, Malaysia.

Wang, Y.P., Liu, J.D., Liu, B., Liu, Y.P., Wang, H.X., Chen, G., 2004. Why does scale form in ASP flood? How to prevent from it? A case study of the technology and application of scaling mechanism and inhibition in ASP flood pilot area of N-1DX block in Daqing, SPE 87469-MS. SPE International Symposium on Oilfield Scale, 26–27 May, Aberdeen, UK.

Wang, J.L., Yuan, S.Y., Shen, P.P., Zhong, T.X., Jia, X., 2007. Understanding of the fluid flow mechanism in porous media of EOR by ASP flooding from physical modeling, 11257-MS. International Petroleum Technology Conference, 4–6 December, Dubai, U.A.E.

Wang, F.L., Yang, Z.Y., Wu, J.Z., Li, Y., Chen, G.Y., Peng, S.K., et al., 2008. Current status and prospects of ASP flooding in Daqing oil fields, SPE 114343-MS. SPE/DOE Symposium on Improved Oil Recovery, 20–23 April, Tulsa, OK.

Wang, D.M., Zhang, Y., Yongjian, L., Hao, C., Guo, M., 2009. The application of surfactin biosurfactant as surfactant coupler in ASP flooding in Daqing oil field, SPE 119666-MS. SPE Middle East Oil and Gas Show and Conference, 15–18 March, Bahrain.

Wang, D.M., Wang, G., Xia, H.F., 2011. Large scale high visco-elastic fluid flooding in the field achieved high recoveries, SPE 144294-MS. SPE Enhanced Oil Recovery Conference, 19–21 July, Kuala Lumpur, Malaysia.

Wasan, D.T., Shah, S.M., Chan, M., Shah, R., 1979. Spontaneous emulsification and the effect of interfacial fluid properties on coalescence and emulsion stability in caustic flooding. ACS Symposium Series No. 91, The Chemistry of Oil Recovery, American Chemical Society.

Weatherill, A., 2009. Surface development aspects of alkali-surfactant–polymer (ASP) flooding, IPTC 13397-MS. International Petroleum Technology Conference, 7–9 December, Doha, Qatar.

Wei, J.G., Tao, J.W., Xin, S.Z., Zhang, Q.J., Zhang, X., Wang, Z.M., 2011. A study on alkali consumption regularity in minerals of reservoirs during alkali (NaOH)/surfactant/polymer flooding in Daqing oil field, SPE 142770. SPE Enhanced Oil Recovery Conference, 19–21 July, Kuala Lumpur, Malaysia.

Weinbrandt, R.M., 1979. Improved oil recovery by alkaline flooding in the Huntington Beach field, Paper C-4. Fifth Annual DOE Symposium, 23–24 August, Tulsa, OK.

Winsor, P.A., 1954. Solvent Properties of Amphiphilic Components. Butterworth, London.

Winsor, P.A., 1968. Binary and multicomponent solutions of amphiphilic components. Chem. Revs. 68 (1).

Yang, X.M., Liao, G.Z., Han, P.H., Yang, Z.Y., Yao, Y.M., 2003. An extended field test study on alkaline-surfactant–polymer flooding in Beiyiduanxi of Daqing oil field, SPE 80532-MS. SPE Asia Pacific Oil and Gas Conference and Exhibition, 9–11 September, Jakarta, Indonesia.

Yang, F.L., Wang, D.M., Wu, W.X., Wu, J.Z., Liu, W.J., Kan, C.L., et al., 2006. A pilot test of high-concentration polymer flooding to further enhance oil recovery, SPE 99354-MS. SPE/DOE Symposium on Improved Oil Recovery, 22–26 April, Tulsa, OK.

Yang, H.T., Britton, C., Pathma, J., Sriram, L., Kim, D.H., Nguyen, Q., et al., 2010. Low-cost, high-performance chemicals for enhanced oil recovery, 129978-MS. SPE Improved Oil Recovery Symposium, 24–28 April, Tulsa, OK).

Yang, Y.H., Zhou, W.F., Shi, G.C., Cao, G., Wang, G.Q., Sun, C.L., et al.. 2011. 17 years development of artificial lift technology in ASP flooding in Daqing oil field, SPE 144893-MS. SPE Enhanced Oil Recovery Conference 19–21 July, Kuala Lumpur, Malaysia.

Yuan, S.Y., Yang, P.H., Dai, Z.Q., Shen, K.Y., 1995. Numerical simulation of alkali/surfactant/polymer flooding, paper SPE 29904-MS presented at the International Meeting on Petroleum Engineering, 14–17 November, Beijing, China.

Zerpa, L.E., Queipo, N.V., Pintos, S., Salager, J.L., 2004. An optimization methodology of alkaline-surfactant–polymer flooding processes using field scale numerical simulation and multiple surrogates, SPE 89387-MS. SPE/DOE Symposium on Improved Oil Recovery, 17–21 April, Tulsa, OK.

Zhang, J.Y., Ravikiran, R., Freiberg, D., Thomas, C., 2012. ASP formulation design for heavy oil, SPE 153570-MS. SPE Improved Oil Recovery Symposium, 14–18 April, Tulsa, OK.

Zhao, P., Jackson, A.C., Britton, C., Kim, D.H., Britton, L.N., Levitt, D.B., et al. 2008. Development of high-performance surfactants for difficult oils, 113432-MS. SPE/DOE Symposium on Improved Oil Recovery, 20–23 April, Tulsa, OK.

Zhu, Y.Y., Hou, Q.F., Liu, W.D., Ma, D.S., Liao, G.Z., 2012. Recent progress and effects analysis of ASP flooding field tests, SPE151285. SPE EOR/IOR Symposium, 14–18 April, Tulsa, OK.

Zubari, H.K., Babu Sivakumar, V.C., 2003. Single well tests do determine the efficiency of alkaline-surfactant injection in a highly oil-wet limestone reservoir. SPE81464-MS, SPE Middle East Oil Show, 9–12 June, Bahrain.

Chapter 11

Foams and Their Applications in Enhancing Oil Recovery

James J. Sheng

Bob L. Herd Department of Petroleum Engineering, Texas Tech University, Lubbock, TX 79409, USA

11.1 INTRODUCTION

One of the objectives of this chapter is to present the fundamentals about foams used in enhancing oil recovery so that the readers can refer to this chapter when they carry out a field foam project. We also briefly discuss some basic scientific concepts that will help the readers to understand foams. After introducing these fundamentals, we focus our discussions on field application related topics, like the modes of foam applications, the factors that need to be considered in design foam projects. Finally, we present several examples of foam applications.

11.2 CHARACTERISTICS OF FOAM

Foam is a dispersion of a relatively large volume of gas in a small volume of liquid. Figure 11.1 shows a schematic of a foam system illustrating different terms to describe a foam. In the figure, a gas phase is separated from a thin liquid film by an interface. Each thin liquid film is sided by two interfaces on this two-dimensional slice. The region encompassed by dotted square and composed of thin liquid film, interfaces and bordered with a junction is called lamella. The junction encompassed by a dotted circle and connected by three lamellae is referred to as the Plateau border. In the presence of oil, an asymmetrical, oil–water–gas film between the oil drops and the gas phase (bubbles) is called pseudoemulsion film (Nikolov et al., 1986). This is the thin liquid film bounded by gas on one side and oil on the other. A foam comprising spherical bubbles separated by relatively thick layers of liquid is referred to as a wet foam or kugelschaum, whereas a foam of polyhedral bubbles separated by thin, plane films is referred to as a dry foam or polyederschaum. Generally, the liquid is water. For some foams, the liquid could be hydrocarbon-based fluids or acids. And these forms are collectively called nonaqueous foams. For some foams, the liquid could be replaced by solids.

Enhanced Oil Recovery Field Case Studies.

Foam can be generated by disturbing a liquid that has a small amount of forming agent (surfactant) and contacts with a gas. Without a foaming agent, foams are unstable and will quickly break. Therefore, foam essentially is a mixture of gas, water, and foaming agent (surfactant). Foam is generally characterized by foam quality and bubble size (Lake, 1989). Foam quality is expressed as the fraction or the percentage of gas volume in the foam. Bubble size is specified in terms of average size (diameter) and bubble size distribution. When discussing forms, foam texture means bubble size.

Foam quality typically ranges from 75% to 90%. Average bubble size and distribution of sizes may vary significantly among foams, from the colloidal size (0.01−0.1 μm) (Lake, 1989) to tenths of millimeters (David and Marsden, 1969). Foam quality is closely related to bubble size. As bubble sizes become larger, foams are more likely to be unstable, and the foam quality would become lower.

A foaming agent (surfactant) is required to generate and stabilize a foam. Like selecting surfactants in surfactant flooding, practically, we need laboratory tests to select foaming agents for a specific application. A foaming agent is selected based on its foaming ability, foam stability, and foam mobility.

11.3 FOAM STABILITY

No foams are thermodynamically stable. Eventually they will collapse. The term stability is used to mean relatively stable in a kinetic sense. The stability of a foam is determined by a number of factors involving both bulk solution and interfacial properties. A method to quantify foam stability is to measure its half-life or the average lifetime (Sheng et al., 1997). Before discussing the foam stability, several concepts specific to foams are introduced

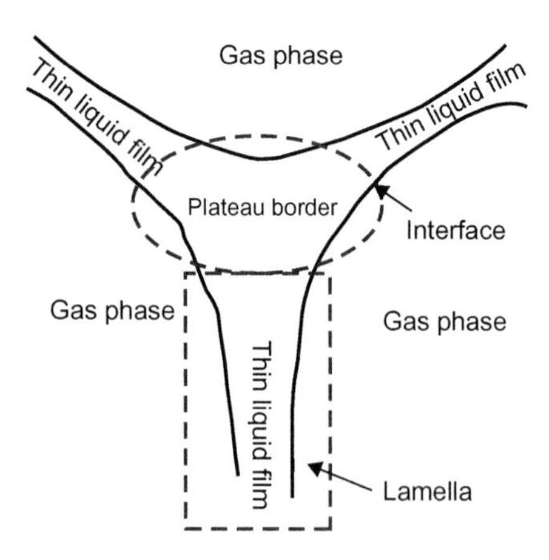

FIGURE 11.1 Schematic of a foam system.

below. Some of the concepts may not be referred in this chapter, but they will be useful when the author reads other references about foams.

Gravity drainage: Foams are separated by thin layers. Liquid gravity will cause the liquid in the liquid layers to drain.

Laplace capillary suction: As shown in Figure 11.1, the gas/liquid interfaces at the Plateau borders are quite curved (smaller radii), while along the thin-film regions, the interfaces are flat (larger radii). Because the capillary pressure is inversely related to the radii of an interface according to the Young–Laplace equation, there is a pressure difference between these two regions. This pressure difference forces liquid to flow toward the Plateau borders and causes thinning of the films.

Marangoni effect: The Marangoni effect (also called the Gibbs–Marangoni effect) is the fluid mass transfer along an interface between two regions due to surface tension gradient. In a foam system, when a surfactant-stabilized liquid film undergoes an expansion, the local surfactant concentration is lowered owing to the increased surface area, and the film becomes thinner. The lower surfactant concentration leads to higher surface tension. And higher surface tension causes contraction of the surface in order to maintain low energy. This surface contraction induces liquid flow in the film from the low-tension region to the higher-tension region. This liquid flow provides the resistance against the thinning of liquid film. In other words, the Marangoni effect due to surface tension gradient helps stabilize a foam system. This effect is also referred to as surface elasticity.

Disjoining pressure Π_d: It arises when two surface layers mutually overlap and is due to the total effect of forces that are different by nature. It is the difference between the normal pressure on the film exerted by the bodies and the pressure in the bulk phase, as shown in Figure 11.2. It is the total of electrical, dispersion, and steric forces. The pressure depends on the thickness of the film, the composition and properties of the

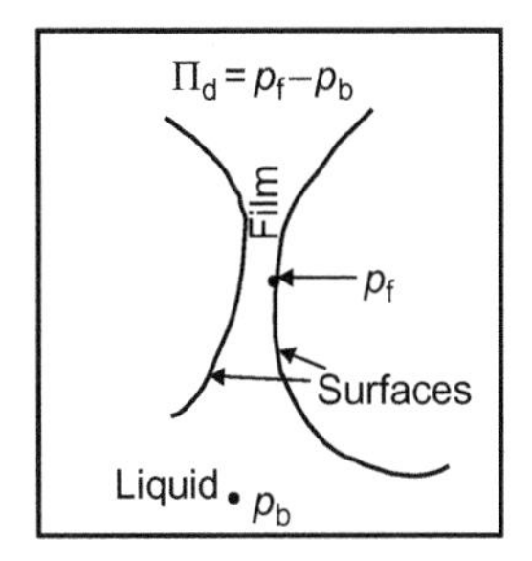

FIGURE 11.2 Schematic description of disjoining pressure.

interacting phases (bodies), and the temperature. Positive values of Π_d reflect net repulsive film forces, and negative values of Π_d indicate net attractive forces. Adsorption of ionic surfactant at each gas–liquid interface of the film causes the excess repulsive forces. Attractive van der Waals forces tend to destabilize the film. The identically charged surfaces repel each other through overlap of their double layer ionic clouds.

Electric double layer and zeta (ζ) potential: Mathematical description of this topic is beyond the scope of this chapter. However, a conceptual description of this topic will be helpful in understanding foams and other chemical enhanced oil recovery (EOR) methods. A schematic description of an (electric) double layer and the corresponding electric potential are shown in Figure 11.3. A double layer is a structure that appears on the surface of an object when it is placed into a liquid. The object might be a solid particle, a gas bubble, a liquid droplet, or a porous body. The double layer refers to two parallel layers of charge surrounding the object. The first layer called Stern layer, the surface charge (either positive or negative), comprises ions adsorbed directly onto the object due to a host of chemical interactions. The second layer is composed of ions attracted to the surface charge via the coulomb force, electrically screening the first layer. This second layer is loosely associated with the object,

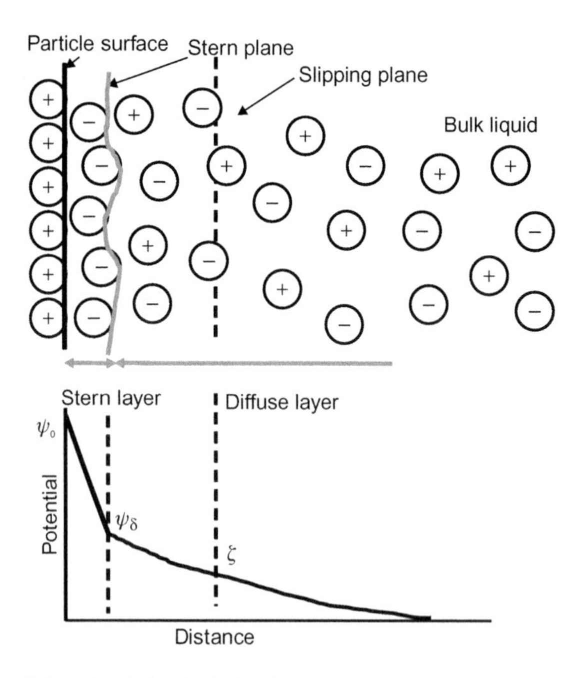

FIGURE 11.3 Schematic of electric double layer.

because it is made of free ions which move in the fluid under the influence of electric attraction and thermal motion rather than being firmly anchored. It is thus called the diffuse layer.

The diffuse layer, or at least part of it, can move. A slipping plane is introduced to separate mobile fluid from fluid that remains attached to the surface. Electric potential at this plane is called electrokinetic potential or zeta potential. It is also denoted as ζ-potential that is a measure of fluid or particle flow. In the figure, ψ_0 is called surface potential, whereas ψ_δ is called Stern potential.

Entering and spreading: To rupture a foam film, an oil/hydrophobic particle droplet must in a first step emerge from the aqueous phase into the gas–water interface during a process called entering. After this entering, some oil from the droplet can spread on the solution–gas interface in a second step. Two coefficients measure the changes in the free energy of the system associated with these two steps. When an oil drop enters the gas–water interface, a new gas–oil interface is created, and the change is measured by the entering coefficient, E (Lecomte et al., 2007):

$$E = \sigma_{wg} + \sigma_{wo} - \sigma_{og} \tag{11.1}$$

A positive E means the surface tension of the antifoam liquid (σ_{og}) is lower than the sum of the surface tension of the foaming liquid (σ_{wg}) and the interfacial tension (IFT) between the antifoam and the foaming liquid (σ_{wo}). The free energy associated with this change is the difference between the energy of the end result (the oil surface tension) and the starting point (the sum of the water surface tension and the oil–water IFT). This value is the opposite of the free energy associated with the entering step.

When an oil drop spreads over the gas–water surface, a new gas–oil surface and water–oil interface are created, and the change is measured by a spreading coefficient, S:

$$S = \sigma_{wg} - \sigma_{wo} - \sigma_{og} \tag{11.2}$$

In this step, the water surface is replaced by an oil surface. A positive S means the surface tension of the foaming liquid (σ_{wg}) is greater than the sum of the surface tension of the antifoam liquid (σ_{og}) and the IFT between the antifoam and the foaming liquid (σ_{wo}). The free energy associated with this change is the difference between the energy of the end result (the sum of the oil–water IFT and the oil surface tension) and the starting point (the water surface tension). The spreading coefficient is the opposite of the free energy change associated with the spreading step.

Foam stability is mainly governed by the liquid drainage rate and the film strength. Some of factors that affect foam stability are discussed in the following.

Interfacial tension (IFT): From the energy required to form bubbles, lower IFT makes easier to form bubbles. However, low IFT (LIFT) cannot

guarantee a stable foam. For a foam to be stable, the film must have a certain level of strength (Wang, 2007). The foam stability also depends on surfactants used.
Diffusion: Gas bubbles have a nonuniform size distribution. Small bubbles have higher pressure than larger bubbles. This pressure difference causes a chemical difference, which in turn causes gas to diffuse through liquid from small bubbles to larger bubbles (Lake, 1989). This causes bubbles to coalesce.
Liquid drainage by gravity: Because of the liquid drainage caused by gravity, the liquid films become thinner. This causes gas bubbles coalesce. From this point of view, it can be understood that low liquid phase saturation would make foams less stable.
Liquid viscosity: It can be understood that higher liquid viscosity will retard liquid drainage. Experimental data also show that higher viscosity leads to more stable foams (Sheng et al., 1997).
Influence of additional phases: Foam stability can be affected by the presence of other dissolved species, an additional liquid phase such as oil in an aqueous foam or fine solids. In these cases, whether the effect is stabilizing or destabilizing depends on several factors. The addition to a foaming system of any soluble substance that can become incorporated into the interface may decrease dynamic foam stability if the substance does any combination of these: increase surface tension, decrease surface elasticity, decrease surface viscosity, or decrease surface potential (Schramm and Wassmuth, 1994).

The presence of dispersed particles can increase or decrease aqueous foam stability. One mechanism for the stability enhancement is the bulk viscosity enhancement that results from having a stable dispersion of particles present in the solution. A second stabilizing mechanism is operative if the particles are not completely water-wet. In this case, particles would tend to collect at the interfaces in the foam where they may add to the mechanical stability of the lamellae (Schramm and Wassmuth, 1994). For example, asphaltene was observed to stabilize foamy oils (Sheng et al., 1997).

Effect of oil: Foam–oil interactions are particularly important in the application of foams for mobility control in enhanced oil recovery. When oil drops enter an aqueous/gas surface and spread on the aqueous/gas surface, the foam breaks. Ross and McBain (1944) postulated that the oil breaks the foam film by spreading on both sides of the foam film, thereby driving out the original film liquid and leaving an oil film that is unstable and breaks easily. Hanssen and Dalland (1990), Manlowe and Radke (1990), and Schramm and Novosad (1990) concluded that the pseudoemulsion film stability is a controlling factor in the stability of three-phase foams within porous media. Researchers (e.g. Friedmann and Jensen, 1986; Hudgins and Chung, 1990; Yang and Reed, 1989) found that from their coreflood experiments, foams could not be generated when the oil saturation is above a "critical foaming oil saturation." Some possibilities for the mechanisms of foam destabilization by a given oil phase include (1) partition of foam-forming surfactants in the oil phase, (2) oil spreading spontaneously on foam lamellae and displacing the foam stabilizing interface, (3) oil emulsifying spontaneously and allowing drops to breach and rapture the stabilizing interface, and (4) oil droplets can be located at some sites in the porous media where bubble snap-off is to occur, thereby hindering generation or regeneration mechanisms. And foams seem to be destabilized most by lighter oils (Schramm, 1994). In real foam applications, oil saturations can cover a wide range. For the applications where foam is injected into low-oil-saturation swept zones for mobility control purpose, the foam with intermediate or low tolerance to oil may be adequate. If foam is to be used as an oil displacing fluid or for gas/oil ratio (GOR) control in producing wells, foam stability to oil is essential.

Bubble sizes: Foam bubbles usually have diameters $>10\ \mu m$ and may be larger than $1000\ \mu m$. Foam stability is not necessarily a function of drop size, although there may be an optimum size for an individual foam type. Some foams that have a bubble size distribution that is heavily weighted toward the smaller sizes will represent the most stable foam (Schramm and Wassmuth, 1994). The effect of bubble sizes may be more generally classified as the effect of foam texture which includes bubble size, shape, and distribution within the foam matrix. Foams will be more stable if bubble sizes are more uniformly distributed.

Effect of wettability: Foams were observed to be less stable in the presence of crude oil and oleophilic solid surfaces compared with the same crude oils and hydrophilic surfaces (Suffridge et al., 1989).

Pressure and temperature: When the pressure is increased, bubbles become smaller. And liquid films become larger and thinner leading to slower liquid drainage. Therefore, higher pressure helps stabilize bubbles. But too high pressure may break bubbles. As the temperature is increased, the surfactant solubility in the liquid phase increases, leading less surfactant in the gas/liquid interface. Also higher temperature increases liquid drainage. In other words, higher temperature destabilizes foams.

Others: The foam stability can be improved by increasing the micellar concentration, decreasing the individual film area (e.g., by decreasing the bubble size), decreasing the electrolyte concentration, decreasing the solubilized oil, or lowering the temperature.

11.4 MECHANISMS OF FOAM FLOODING TO ENHANCE OIL RECOVERY

The mechanisms of foam flooding to enhance oil recovery lie in reducing gas phase permeability. The mechanisms can be explained in terms of the mechanisms of foam formation and decay. Therefore, we first discuss foam formation and decay.

11.4.1 Foam Formation and Decay

There are three fundamental pore-level generation mechanisms: snap-off, lamella division, and leave-behind. Snap-off occurs when a bubble penetrates a pore constriction and a new bubble is formed, as shown in Figure 11.4. This mechanism puts some of the gas into discontinuous form. It can occur repeatedly at the same site, so snap-off at a single site can affect a relatively large portion of the flow field. It is widely believed to be a predominant foam-generation mechanism (Ransohoff and Radke, 1988). The snap-off mechanism influences the flow properties of the gas phase by increasing the discontinuity of the gas phase and by creating lamellae. The generated gas bubbles may lodge at some point in the porous medium, thereby blocking gas pathways to reduce gas permeability.

Lamella division occurs when a lamella (preformed foam) approaches a branch point so that the lamella is divided into two or more lamellae, as shown in Figure 11.5. This mechanism is similar to the snap-off mechanism in that a separate bubble is formed, which can either flow or block gas pathways. And it also occurs numerous times at one site. Snap-off and lamella division mechanisms are in effect at high flow velocities.

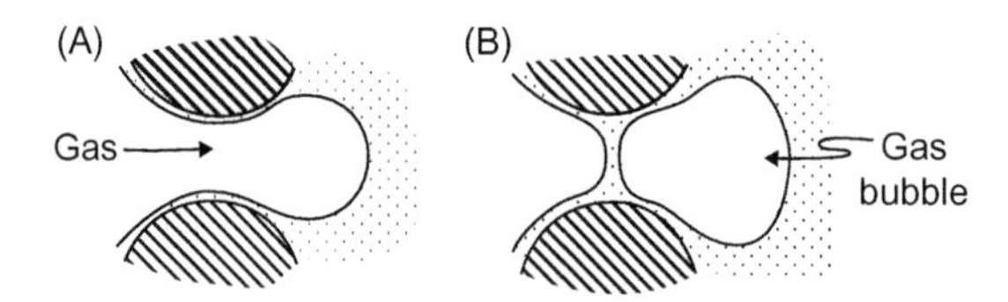

FIGURE 11.4 Schematic of snap-off mechanism showing (A) gas penetrates to a constriction and a new bubble is formed (B) (Ransohoff and Radke, 1988).

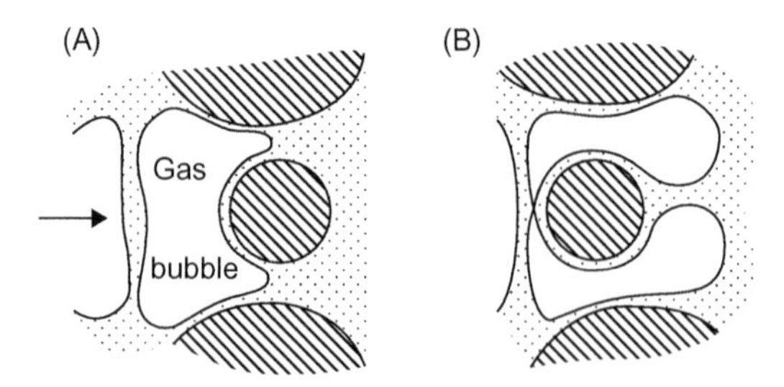

FIGURE 11.5 Schematic of lamella division mechanism showing a lamella is approaching the branch point from (A) and divided gas bubbles formed (B) (Ransohoff and Radke, 1988).

Leave-behind begins as two gas menisci invade adjacent liquid-filled pore bodies from different directions, as shown in Figure 11.6. A lens is left behind as the two menisci converge downstream. By this mechanism, a large number of lamellae are formed to block gas pathways. In effect, the lamellae reduce the relative permeability to gas by creating dead-end pathways and blocking flow channels. This mechanism is important at low velocities and generates relatively weak forms. Leave-behind generates stationary aqueous lenses when aqueous-phase saturation is high. It is a nonrepetitive process that alone cannot account for the large reduction in gas mobility seen with foam. Ransohoff and Randke (1988) found that foam generated solely by leave-behind gave approximately a fivefold reduction in steady-state gas permeability, whereas discontinuous-gas foams created by snap-off resulted in a several hundred-fold reduction in gas permeability (Persoff et al., 1991). According to (Kovscek and Radke, 1994) discussion and some experimental observations (Owete and Brigham, 1987), lamella division mechanism does not appear significant; snap-off is the dominant foam-generation mechanism.

Chambers and Radke (1991) enunciated two basic mechanisms of foam coalescence: capillary suction and gas diffusion. Capillary-suction coalescence is the primary mechanism for lamellae breakage. Because the pressure on the concave side of a curved foam film is higher than that on the convex side, gas diffuses from the concave to the convex side of the film, and dissolves in the liquid. Foam decay or foam coalescence is opposite to foam stability which is discussed earlier. Thus more factors which affect foam coalescence can be derived from the discussion of foam stability.

Note that although foam formation and decay are governed by the mechanisms above, the final distance foams can travel must be dictated by the surfactant retention. The availability of surfactant is a necessary condition. Porous medium work like a filter when gas and liquid flow through it. As discussed above, porous medium plays important roles in foam generation and decay. Foam generation and decay is a dynamic process in which foams are constantly be generated and decayed.

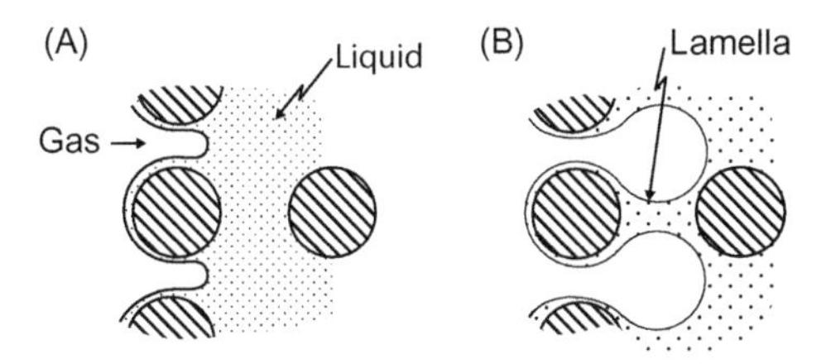

FIGURE 11.6 Schematic of leave-behind mechanism showing gas invasion (A) and forming lens (B) (Ransohoff and Radke, 1988).

11.4.2 Foam Flooding Mechanisms

Foam directly addresses the problem of high gas mobility in porous media. Injected foam appears to be a much more viscous fluid, which helps increase sweep efficiency. The decrease in gas mobility also indirectly reduces gravity segregation, if injection-well pressure can be increased (Shan and Rossen, 2002; Shi and Rossen, 1998). When low-velocity gas flows through surfactant liquid phase, liquid films form. These liquid films partially block pore throats. Thus gas phase relative permeability is reduced. When high-velocity gas flows through the liquid, liquid films break and gas clusters are broken. The broken separate gas bubbles block pore throats. Thus gas phase relative permeability is reduced and the sweep efficiency is improved. When the poor sweep efficiency is caused by reservoir heterogeneity, it can also be improved by using foams. Foams are stronger (with higher apparent viscosity) in high permeability layers compared to lower permeability layers (Hirasaki, 1989). One reason for this is that nonwetting gas preferably stays in high permeability channels. This feature helps to divert injected flow into lower permeability layers.

Surfactants added not only help form foams thus to improve sweep efficiency, but they can also improve displacement efficiency. Foam flow requires a high pressure gradient which helps drive residual oil out by overcoming capillary pressure.

11.5 FOAM FLOW BEHAVIOR

Foam flow behavior is described by one of the following ways.

11.5.1 Foam Viscosity

Increased apparent foam viscosity is calculated using Darcy's law and treating the foam as a single gas phase (Minssieux, 1974). In this calculation, gas permeability is used. Based on the theoretical studies of Bretherton (1961) and Hirasaki and Lawson (1985), the effective viscosity of confined foam is a linear function of bubble density (n_f) and inversely proportional to the foam real velocity (v_f). Mathematically, it is

$$\mu_f = \mu_g + \frac{\alpha n_f}{v_f^c} \tag{11.3}$$

where α, a proportional constant, is a strong function of surfactant properties and c is the empirical exponent which has a theoretical value of ⅓. Foam viscosity is much higher than that of water or gas. Blauer et al. (1974) discussed the foam quality effect on the viscosity. Marfoe et al. (1987) proposed a semiempirical formula to calculate foam viscosity which is a function of gas velocity, gas fraction, surfactant concentration, etc. Later Islam (1990) modified the formula. Owing to the complexity of foam rheology, there are different opinions regarding whether foams are Newtonian or non-Newtonian, shear thinning or shear-thickening. Wang (2007) listed more viscosity models published in the literature.

11.5.2 Relative Permeabilities

Reduced relative permeabilities calculated using their individual single-phase viscosities, flow rates, and pressure drop (Bernard and Holm, 1964). This approach treats the flow of foam as multiphase flow. A number of researchers (Bernard et al., 1965; de Vries and Wit, 1990; Friedmann and Jensen, 1986; Sanchez et al., 1986) experimentally observed that wetting aqueous phase relative permeability is unchanged because the wetting aqueous phase is primarily concentrated in small pores. However, gas relative permeability is significantly reduced partly because trapped foam severely reduces the effective permeability of gas moving through a porous medium by blocking all but the least resistive flow paths. Kovscek and Radke (1994) presented the following equation to describe foam relative permeability (k_{rf}):

$$k_{rf} = k_{rg}^{0}\overline{S}_{f}^{n_g} \tag{11.4}$$

where

$$S_f = X_f(1 - \overline{S}_w) \tag{11.5}$$

$$\overline{S}_w = \frac{S_w - S_{wc}}{1 - S_{wc}} \tag{11.6}$$

$X_f = S_f/S_g$ is the fraction of the foam phase that is flowing, S_{wc} is the connate aqueous-phase saturation, and k_{rg}^{0} is the gas relative permeability at S_{wc}.

11.5.3 Mobility Reduction

Reduced foam mobility calculated using Darcy's law, total volumetric flow rate, and the pressure drop (Heller et al., 1985). Because foam flooding mechanism is mainly due to reduced gas permeability and increased foam viscosity, commercial simulators like STARS and ECLIPSE model foam behavior by using mobility reduction factor. The parameters that are considered to affect significantly to foam behavior in field scale application are surfactant concentration, oil saturation, water saturation, and capillary number (gas viscosity). Based on Kovscek (1988) and Cheng et al. (2000) work, all these parameters are coupled together in mobility reduction factor, M_{rf} (ECLIPSE, 2011) using this equation:

$$M_{rf} = \frac{1}{(1 + M_r F_s F_w F_o F_c)} \tag{11.7}$$

where M_r is the reference mobility reduction factor at a reference surfactant concentration, a reference water saturation, a reference oil saturation and a reference capillary number, F_s, F_w, F_o, and F_c are the mobility reduction factor components due to surfactant concentration, water saturation, oil saturation, and capillary number (gas velocity), respectively. Purwoko (2007) did detailed sensitivity study of these parameters. Note that the reference mobility reduction factor M_r is typically in the range of 5–100 (i.e., >1) in ECLIPSE, whereas the mobility reduction factor, M_{rf}, is actually the foam mobility multiplier and less than or equal to 1. For the detailed description of other mobility reduction factor components, see the ECLIPSE technical manual. In ECLIPSE, foam is modeled as a tracer either in water or gas phase.

Another popular commercial simulator, STARS (steam and additive reservoir simulator), also uses similar approach that ECLIPSE uses. Such model belongs to semiempirical models. Other types of foam models include population balance method (Kovscek and Radke, 1994), percolation model (Chou, 1990; Kharabaf and Yortsos, 1998), fractional flow theory (Zhou and Rossen, 1995). The population model and percolation model better represent foam generation and decay mechanisms in pore levels. However, they are so complex that they have limitations in simulating field performance. Some critic for the fractional flow model is that the fractional flow theory may not applicable to compressible flow.

11.5.4 Flow Resistance Factor

Flow resistance factor defined as the pressure drop measured across a foam generator divided by the pressure drop for the same system without the presence of surfactant (Duerksen, 1986). It was reported that the resistance factor increases with the formation permeability (Khan, 1965; Wang, 2007). This characteristic is beneficial to improve sweep efficiency.

11.6 FOAM APPLICATION MODES

Foam application modes could be CO_2 foam, steam-foam, and foam injection in gas miscible flooding.

11.6.1 CO_2 Foam

The uniqueness of CO_2 foam compared with other foams is its high CO_2 density. Its density is 0.468 g/cm^3 at the critical conditions: $p_c = 1070$ psia and $T_c = 31.04°C$. When CO_2 foam is to be injected with surfactant-containing water for mobility control, there are two injection modes: co-injection and alternative injection (commonly called water alternating gas (WAG)). For the co-injection mode, because of the difference in density between water and CO_2, the two fluids may not enter both the upper and lower parts of the formation at the same relative rates. It seems that CO_2 and surfactant solution should be injected alternatively. This alternative mode is called SAG for surfactant alternated with gas. The important parameters for such mode are the SAG ratio and the cycle time to be used (Heller et al., 1985). Another problem associated with CO_2 foam flooding which must be considered is corrosion.

11.6.2 Steam-Foam

One issue with steam-foam is high temperature. Therefore, it is important to select thermal stable surfactants. Surfactants with sulfate moieties decompose rapidly at temperatures above 100°C, and surfactants stable above 200°C have, almost exclusively, sulfonate groups (Isaacs et al., 1994). Thermal stability of sulfonates increases in the following order (Ziegler, 1988):

Petroleum sulfonates < alpha olefin sulfonates < alkylarylsulfonates

Theoretical considerations, core studies, and field data have demonstrated that foams are more effect when noncondensible gas is present even in small concentrations. The major attraction of injecting surfactants without a noncondensible gas is economic (Isaacs et al., 1994). The surfactants injected can also reduce the water–oil IFT to reduce residual oil saturation.

For field steam-foam applications, surfactants were injected as high-concentration (10 wt% active) or continuously at a lower concentration (0.1–1.0 wt% active). Noncondensible gas (at a concentration of 0.5–1.0 mol.% in the gas phase) or NaCl (1–4 wt% in the aqueous phase) may be co-injected with the surfactant. Extra water may be added to the steam to maintain the liquid fraction at the desired value (typically above 0.01). Assume 0.3 m^3 of incremental oil is produced per kilogram of active surfactant, and the cost of a surfactant is $10/kg the cost of incremental oil would be $33/$m^3$ ($5/bbl). However, in many tests, the incremental oil production is not high enough to be economical (Isaacs et al., 1994).

Steam-foam was also applied to the steam cyclic process as well as in the steam drive process. During a cyclic process, steam initially flows through high permeability zones and depleted the oil in those zones. As the oil is depleted, the flow resistance is even lower. Thus the subsequent steam will continue flowing into those depleted zones if steam-foam is not injected. By injecting steam-foam, because steam-foam has higher resistance, it diverts the subsequently injected steam into the zones of higher oil saturation (less depleted zones). Turta and Singhal (2002) compared the continuous and semicontinuous additives (surfactant and nitrogen) injection in the Venezuela projects and concluded that the on–off injection mode was superior.

11.6.3 Foam Injection in Gas Miscible Flooding

Few foam injection projects in gas miscible flooding have been tried. In the process, gas and surfactant water solution are injected either simultaneously (Chad et al., 1988) or alternatively (Liu et al., 1988), so that foam is formed. Also, the injection pressure is high so that the miscibility between gas and oil is reached. The generated foam is to mitigate gas fingering or channeling and gravity override.

11.6.4 Gas Coning Blocking Foam

Similar to the practice of mitigating water coning by injecting high viscous fluids like gels, gas-blocking foam is used to block gas coning. The idea is this. When there is a gas cap above the oil zone, gas will cone to the producer. Place a foaming solution whose density is between those of gas and oil so that this solution will stay in the zone between the gas cap and the oil. When the producer is open, some gas fingers into the producer. Meanwhile, a cone shape of foam is formed. Because of the low mobility of foam, gas coning is reduced. An US patent was filed (Heuer and Jacocks, 1968). Hanssen (1993a,b) did some laboratory investigation. A few field trials were conducted to assess the effectiveness of foamable EOR surfactants in preferentially reducing the gas production through direct treatment of high GOR producing wells. By reviewing carried projects, Turta and Singhal (2002) concluded that the idea of placing a surfactant solution pad and generation of foam while backflowing well was found totally ineffective in all cases they analyzed. However, injection of foam, preformed at the surface using nitrogen, caused a significant GOR reduction which translated into an increase in oil production in the target well for a period of several weeks (Krause et al., 1992).

11.6.5 Enhanced Foam Flooding

An enhanced foam system is to add polymer in a foam system. Polymer increases the liquid viscosity thus increase the foam stability. Polymer may also reduce the adsorption of foaming agents. Thus the synergy is achieved. Such polymer-enhanced foam flooding was tried in a Shengli field (see Section 11.9.1). It was reported that a similar flooding system called alkaline-surfactant-polymer-foam flooding (ASPF) was tested in China (Yang and Me, 2006). The terms LIFTF (LIFT foam) and alkaline-surfactant-gas (ASG) for similar systems were also used (Sheng, 2011).

11.6.6 Foams for Well Stimulation

N_2 or CO_2 is added in fracturing and/or acidizing fluids to form foam stimulation fluids. Foam stimulation fluids will reduce liquid requirement, reduce formation damaging fluids, improve fluid cleanup, and in certain instances reduce fluid leaking-off and improve proppant-carrying capability. N_2 is injected in gas form while CO_2 is injected as liquid form at surface. CO_2 will become gas in subsurface where the temperature is high. Air or hydrocarbon gas is not used in such stimulation fluids because they pose a potential danger to people and equipment during the treatment.

Another application of foams in well stimulation is the use of foam as a diversion fluid. Because of high viscosity of foams, the foamed stimulation fluids can be diverted from less resistance zones to higher resistance zones. In terms of diverting function, it was found that brine foams sometimes are better than acid foams (Chou, 1991).

11.7 FACTORS THAT NEED TO BE CONSIDERED IN DESIGNING FOAM FLOODING APPLICATIONS

The factors discussed in this section are screening criteria, surfactants and injection modes.

11.7.1 Screening Criteria

The screening criteria specifically for foam flooding in field applications were hardly discussed in the literature. Because surfactants are used in foam flooding, the screening criteria should be similar to those for surfactant flooding, polymer flooding, or gel injection. Those criteria were discussed in Sheng (2011). Some of the important criteria include high permeability heterogeneous reservoirs, low salinity especially low divalent formation water, and relatively short well spacing. Very high temperature (e.g., $>200°C$) poses challenges to foaming agents. Another important condition is that the remaining oil saturation should be low so that stable foams can be formed. Many field foam flooding were applied at residual oil saturations.

Turta and Singhal (2002) stated that the most important factors in foam assisted EOR projects are: (a) manner of foam placement in the reservoir (injection of preformed foam, co-injection foam, and SAG or surfactant alternating gas foam), (b) reservoir pressure, and (c) permeability.

11.7.2 Surfactants

Many factors affect foam stability. More importantly, we need to select the right surfactants. As mentioned earlier, surfactants are needed to foam bubbles and make the generated foam stable. Surfactants that are good at foaming may not be good at reducing IFT. When evaluating and selecting surfactants, we need to consider the following aspects:

- Foaming ability
- Stability

- Thermal stability
- Salinity and multivalent ion resistance
- Compatibility with the formation fluids
- Performance in the presence of oil
- IFT reduction
- Adsorption.

Some surfactants successfully used in field applications include ORS-41 and AOS (Chinese product) at low temperature (~45°C) and low salinity (<10,000 ppm) (Yang and Me, 2006), DP-4 (Chinese product, 60°C, 17,000 ppm salinity, 1000 ppm Ca^{2+} and Mg^{2+}) (Wang, 2007), AGES (Chinese product, 250°C, 50,000 ppm salinity, 5000 ppm Ca^{2+} and Mg^{2+}) (Zhang et al., 2005). Some surfactants that can be used in high-temperature steam-foam flooding include SuntechIV (Sun), DowFax2A (Dow), Neoden14−16 and Neoden 16−18 (Shell), and Stepanflo30 (Stepan) (Zhang et al., 2005). Other good foaming agents are AEO-9, ABS, AES, SDS, AOS Q-9, S-6, and R-5 (Wang, 2007). For the foaming agents used in CO_2 foam, see Heller et al. (1985). Some fluorinated surfactants can form foams that are very stable in the presence of oil (Hanssen and Dalland, 1990; Novosad, 1989). However, they are very expensive.

11.7.3 Injection Mode

It was reported that the resistance factor in a co-injection of gas and liquid was higher than that in an alternate injection, especially in a high gas/liquid ratio. In a single-well trial in the Gudao field and a field trial, because too high injection pressure occurred in the co-injection mode, the alternate injection mode was applied (Wang, 2007).

The laboratory results reported by Yang and Me (2006) indicate that the recovery in co-injection is higher than that in alternate injection, if the same amount of chemicals is used. In alternate injection, smaller slug sizes and higher alternation frequency are better, because too large slug sizes make foam-forming difficult. Some simulation results showed that if the alternate cycle is 10 or 20 days, the recovery is close to that from co-injection mode (Wang, 2007). However, the surfactant-alternating-gas (SAG) was preferable to the co-injection in the Snorre project because the injection pressure was above the fracture pressure in the co-injection mode (Blaker et al., 2002). For designing a steam-foam project (which is a low-pressure foam application), foam quality in the range 45−80% should be considered. In this application, a co-injection of gas and water is employed but the additives (surfactant and noncondensable gas) are injected intermittently (on and off), superimposed on a continuous steam injection. At high pressure, such as in gas miscible flooding, foam application can result in excessive mobility reduction factors, and injectivity reduction. Due to this reason, in these projects, alternate injection of surfactant solution and gas (SAG foam) is favored over a co-injection mode of placement (Turta and Singhal, 2002).

11.8 RESULTS OF FIELD APPLICATION SURVEY

A field application survey was conducted. More than 60 field projects were reviewed. For the reviewed field applications, more than half are about steam-foam, the rest are about CO_2 foam and other foam applications. In this section, we summarize the general results (observations) from this survey.

11.8.1 Locations of Conducted Foam Projects

Steam-foam projects were dominantly carried out in California and Venezuela, with a few in Wyoming, in Canada, China, Norway, all onshore except several ones in North Sea. For the surveyed projects, almost all projects were carried out in sandstone reservoirs with only a few in carbonate reservoirs.

11.8.2 Applicable Reservoir and Process Parameters

Table 11.1 summarizes some reservoir and process parameters for the surveyed field projects. For the reservoir parameters, Table 11.1 provides that these ranges of parameters almost cover all the practical ranges. In other words, foam applications are not limited by conventional reservoir parameters. Surfactant concentration is low. Most of the projects used surfactant concentration <1%. The average surfactant concentration was about 0.5%. In some cases, 0.1% was found sufficient (Ploeg and Duerksen, 1985). The injection of a noncondensable gas at reservoir conditions is desirable to maintain a foam when the steam condenses as it moves away from the injection well into the reservoir. The applicable reservoir temperature and salinity are limited by the conditions that surfactants can be stable.

TABLE 11.1 Range of Reservoir and Process Parameters in the Surveyed Field Projects

Parameters	Range	Notes
Reservoir thickness, m	3–350	Higher for steam-foam
Permeability, mD	1–5000	A wide range
Reservoir temperature, °C	<101	Only a few above 101°C, highest 232°C
Reservoir pressure, MPa	<500	
Oil viscosity, cP	<10,000	
Well spacing, m	30–1500	
Formation water total dissolved solid (TDS), ppm	<180,000	Limited by surfactant
Noncondensate gas in steam	<2%	Desirable for steam-foams
Surfactant concentration	A few percent, generally <1%	

11.8.3 Injection Mode

For gas foams, about two-third projects were in the mode of SAG injection, and the one-third were in co-injection of gas and water. In steam-foam projects, co-injection projects were a little bit more than alternate injection. In the alternate injection, surfactant solution was added intermittently while steam with some liquid water was continuously injected. This mode called slug application method (Eson and Cooke, 1989). It appeared from the survey results that the slug injection mode (alternate injection mode) needed lower amount of surfactants (higher efficiency) than the co-injection mode. This observation is in line with that from Eson and Cooke (1989). Turta and Singhal (2002) stated that the most important factor for the success was the foam injection mode.

11.8.4 Gas Used in Foam

In steam-foam, nitrogen was added as noncondensable gas in 80% of the projects. Air and methane were also used. For the other gas foams, CO_2 was used in more than 50% of the projects. The rest were natural gas, nitrogen, and air.

11.9 INDIVIDUAL FIELD APPLICATIONS

Foams are used to modify flow profile in injection wells in waterflooding. Foams are also used in steam flooding as steam-diverting agents. Gravity override and steam channeling are serious problems in steam flooding. Due to their high flow resistance, foams can be used as diverting agents to make steam move more uniformly through a reservoir.

Typical field applications include aqueous foams for improving steam drive (Djabbarah et al., 1990; Mohammadi et al., 1989; Patzek and Koinis, 1990) and CO_2-flood performance (Hoefner et al., 1995), gelled-foams for plugging high permeability channels (Friedman et al., 1999), foams for prevention or delay of gas or water coning (Aarra and Skauge, 1994), and SAG processes for cleanup of groundwater aquifers (Hirasaki et al., 1997). All of these methods have been tested in both the laboratory and the field. During these processes, the gas phase is generally discontinuous and the liquid phase is continuous. Among the field trials, some used co-injection of surfactant solution and gas, while others used SAG. A few field applications are presented next. Some CO_2 related and steam-foam field cases are presented in Chapter 2.

11.9.1 Single Well Polymer-Enhanced Foam Flooding Test

This section presents an example of combined polymer-foam flooding (Wang, 2007; Zhang et al., 2005). A combined ASPF was presented elsewhere (Sheng, 2011).

Reservoir and Well Description

The tested well, 28-8 as shown in Figure 11.7, was situated in the Guodao field, Shengli Oil Company. There were 12 affected producers in this injection pattern. The pattern covered 0.16 km^2, the effective thickness 20 m, original oil in place (OOIP) 0.34 million tons, permeability 1300–1800 mD, salinity 8379 ppm, and oil viscosity 74 cP. Polymer (3530S from SNFI) injection through the tested well was started in January 2002, and polymer broke through the closet well, 29-506, in September 2002. The produced polymer concentrations were 890 and 1100 ppm in September and October, respectively. Polymer also broke through at well 28–509 in March 2003 with the concentration of 370 ppm. After polymer breakthrough, oil rate was deceased and water cut was increased to about 94%.

Experimental Study

A Ross-Milcs instrument and a sand pack were used to evaluate foam performance. The foaming agent DP-4 was found to be the best. When the foaming solution with 0.5% DP-4 and nitrogen flowed through 1500 mD silicon sand pack (the gas/liquid ratio of 1:2), the resistance factor was above 1500. Adding 1800 ppm polymer (3530S), the combined polymer-foaming solution increased the oil recovery from 54% for waterflooding to 79%.

After the experimental study, the designed scheme for this single well polymer-foam test was as follows.

1. Two slugs injection lasted 180 days, with the first preslug 10 days and the main slug 170 days. Afterward, polymer is injected.

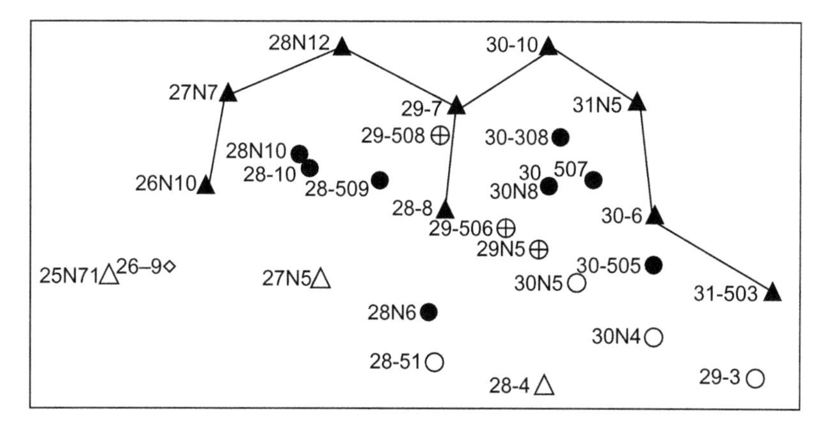

FIGURE 11.7 Well locations in well 28-8 pattern (Wang, 2007).

2. In the first preslug, the foaming concentration was 1.5% and the polymer was 1800 ppm. The polymer and foaming solution was injected at 150 m^3/day.
3. In the main slug, the foaming concentration was 0.75%, and the polymer was 1800 ppm and the gas/liquid ratio of 1. The nitrogen injection rate was 7500 m^3 (standard conditions), and the water rate was 61 m^3/day (produced water rejected).
4. The total nitrogen injection was 1,309,000 m^3 (standard conditions), polymer 28.5 tons, and foaming agent 118 tons.

Test Performance

The preslug began on April 24, 2003. On May 4, nitrogen was injected through the tubing simultaneously with the foaming solution. The initial tubing pressure was 11.5 MPa. It rose to 12.5 MPa one hour after. Owing to the high injection pressure, nitrogen injection was changed through the casing. However, the injection was still high (12.5 MPa). Alternate injection mode was tried. Different gas/liquid ratios (1, 0.6, 0.4, and 0.2) were tried. The wellhead injection pressure changed from 13.5 to 17 MPa. Simultaneous mode was also tried later and the wellhead injection pressure was above 18 MPa. The test was stopped by March 25, 2004. The total injected nitrogen was 1142 m^3, polymer (dry) 57.4 tons, and foaming agent 274.3 tons.

Before foam injection, the water intake profile from the three layers (33, 42, and 44) in April 2003 was 1.37%, 92.56%, and 6.07%, respectively. After foam injection, the profile was 0.73%, 58.14%, and 41.13%, respectively, for the three layers in June 2003. The water intake in layer 44 was significantly improved.

The oil rate for the well 28-8 pattern increased from 75.1 tons/day before foam injection to 140.5 tons/day, and the water cut decreased by 5.7% from 93.8% to 88.1%.

Gas broke through well 29-3 on July 11. The gas rate increased from 376 m^3/day to 1870 m^3/day with a maximum 2900 m^3/day. The casing pressure increased from 0.13 to 4.7 MPa. The produced gas had 30% nitrogen. When gas injection was stopped on August 27, the gas rate decreased and the casing pressure was also reduced. When gas injection was restored on October 21 at lower gas/liquid ratios, gas breakthrough was not observed. In other words, gas breakthrough was controlled by temporarily stopping gas injection and/or injecting gas at lower gas/liquid ratios. This well oil rate increased by 60 tons/day from 10 to 70 tons/day, and the water cut reduced by 50% from 94% to 44%.

Laboratory tests using a single tube (30 cm by 2.5 cm) showed that at low gas/liquid ratios (<2), the resistance factors in the two injection modes were similar. And the optimum gas/liquid ratios to achieve the highest resistance factors were 1–2 in this case. At high gas/liquid ratios (>2), the resistance factors decreased with the gas/liquid ratios. The resistance factors versus gas/liquid ratios showed a Ω shape, as shown in Figure 11.8. Note that the resistance was very high! In high gas/liquid ratios, it may be easier for gas fingering to happen in the alternate mode.

In this single-well test, initially the gas/liquid ratio was one. After one-month injection, gas broke through well 30N5. The gas/liquid ratio was reduced to 0.6, but the gas breakthrough was not stopped. Gas injection was then temporarily stopped and later the ratio was further reduced to 0.2. Gas breakthrough was stopped. When the ratio was increased to 0.44, gas breakthrough happened again. This single-well test showed that the ratio should not be higher than 0.5. Similarly, although Yang and Me (2006) reported an optimum gas/liquid ratio of 4 from a laboratory study, the corresponding field designed gas/liquid was 1, and the actual ratio in the field test was 0.34. It seems that the optimum gas/liquid ratios in field applications could be different from those from laboratory tests.

11.9.2 Nitrogen Foam Flooding in a Heavy Oil Reservoir After Steam and Waterflooding

This section presents a case of foam flooding in heavy oil reservoir after the reservoir has been flooded by water and cycles of steam-soak (Yang et al., 2005).

Reservoir and Well Description

The nitrogen foam flooding test was conducted in the block Jin 90 of the Huan-Xi-Ling field. The reservoir targeted was the Xing I group layer, as described in Table 11.2.

Yang et al. (2004) studied the application conditions of waterflooding in heavy oil reservoirs. The conditions include reservoir depth <2000 m,

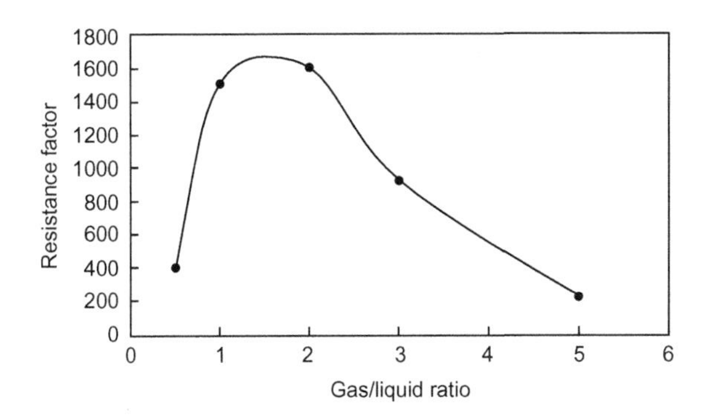

FIGURE 11.8 Resistance factor versus gas/liquid ratio (Wang, 2007).

reservoir thickness >15 m, permeability >250 mD, viscosity <500 cP, and permeability variation coefficient 0.4–0.8. Since foaming flooding accompanied with water injection, these waterflooding conditions must be met. From Table 11.1, we can see that these conditions were met in this reservoir.

The reservoir was produced early in 1985. Its production history is:

1985–1986, production using rod-pumps and steam-soak, recovery factor (RF) 5.91%;

1987–August 1991, steam-soak in the entire reservoir, cumulative RF up to 24.35%;

September 1991–December 2004, steam flooding, waterflooding, and nitrogen foam flooding were carried in Jin 90, but the other blocks in the reservoir were under steam-soak. The cumulative RF was up to 50.8%.

The nitrogen pilot injection was initiated at the well 19-141 pattern in Jin 90 in September 1996. By that time, the cumulative RF was 33%. The pilot test was terminated in May 1999 to change well patterns. During this period (33 months), 5.5% OOIP was produced. From September 1999 to August 2000, the pilot area was expanded to 9 well patterns. The nitrogen foam flooding lasted 54 months by December 2004; 9.75% OOIP was produced during this period.

Expanded Pilot Test Performance

Compared with water injection wells, foam injection wells showed higher injection pressures. The water injection profile was improved. And the oil production rate increased. Because the pressure in the foam flooding area was increased by 2 MPa, the neighboring steam-soak producers benefitted from the increased reservoir pressure and the oil rates from these wells increased. For steam-soak wells, generally, oil rate will be lower at later

TABLE 11.2 Xing I Reservoir and Fluid Data

Area, km^2	1.54
Formation depth (subsea), m	1080
Permeability, mD	1065
Average porosity, %	29.7
Permeability variation coefficient	0.45
OOIP, tons	4,340,000
Effective thickness, m	20.9
Effective/total thickness ratio	0.58
Formation temperature, °C	49.7
Formation water total dissolved solid (TDS), mg/L	2042
Dead oil viscosity at 50°C, mPa · s; oil viscosity in place, mPa · s	462.7; 110–129

cycles. However, some neighboring wells oil rates in the 7th soak cycle was higher than those in the 6th cycle owing to the increased reservoir pressure.

This example demonstrated that foaming flooding could further increase oil production after cycles of steam-soak and waterflooding. The study conducted in this application also showed that foam flooding should not be carried out too late when serious water channeling happened, because sweep efficiency provided by foam injection is limited.

11.9.3 Snorre Foam-Assisted-Water-Alternating-Gas Project

The Snorre foam-assisted-water-alternating-gas (FAWAG) was the world's largest application of foam in the oil industry, injecting about 2000 tons of commercial grade alphaolefin sulfonate (AOS) surfactant and consisting of three injectivity tests, one full-scale SAG test and one full-scale co-injection test. The field application of foam used in WAG as a mobility reduction agent is described in this section, based on the reports from Blaker et al. (2002) and Skauge et al. (2002).

Field Description

The Snorre field is located in the Norwegian sector of the North Sea, 200 km northwest of Bergen. The sea depth in the area is 300–350 m. The reservoir is a massive fluvial deposit within rotated fault blocks. This field was put on stream in August 1992 with waterflooding as the primary recovery mechanism. In February 1994, a down-dip WAG pilot project with two injection wells and three production wells in the central fault block (CFB) was initiated in the Statfjord formation of the field. In 1995 it was decided to extend the WAG injection to cover the three main fault blocks in the field. The injection gas used was identical to the export gas, and was rich in intermediate components. Laboratory studies concluded that the gas was miscible with reservoir oil at pressures above 282 bars. The Snorre oil was originally undersaturated by 260 bars. The permeability was 400–3500 mD. The formation dips 5–9° toward the southwest. The oil viscosity was 0.789 cP in the WFB and 0.687 cP in the CFB. The reservoir temperature is 90°C. The well locations in the test areas are shown in Figure 11.9.

Foam Injection Tests

Gas Shutoff Treatment by Foam at Well P-18 in the CFB

Initially the Snorre field was under WAG flooding. Because of premature gas breakthrough at well P-18 in the CFB, a foam treatment was performed at the well in July 1996. The surfactant used was C14–16 alphaolefin sulfonate which was used in the subsequent foam injections. In the treatment, two cycles of SAG injection and one co-injection were carried out; 8 tons of surfactant was injected in each SAG injection and 16 tons of surfactant was injected in co-injection mode. The surfactant concentrations were 1 or 2 wt%. The foam was placed in the target high permeability layer which was mechanically isolated by a packer. The plug was removed after the treatment. Analysis of the pressure buildup test and tracer test showed that the foam generation during the SAG mode was limited, but a strong foam was generated in the co-injection mode. This treatment resulted in a 50% reduction in GOR for 2 months in this well.

Foam Injection at P-25A in the CFB

The foam injection was started in February 1997 at well P-25A in the CFB of the Snorre field. The pilot area had a down-dip WAG injection from the injection well P-25A (a sidetrack of well P-25) to the producer P-18. Three surfactant slugs with the surfactant concentration of about 0.5wt% were injected in the SAG mode, which was ended with a small slug of co-injection from January to May 1999. Gas injectivity was significantly reduced during the first and partially the second gas cycle. Foam was generated, and the foam zone expanded during the first and partially the second cycle. The first test indicated that injectivity could be a limiting factor, especially for the co-injection mode. However, during the second test some months later, the injectivity was better and the initially planned FAWAG rates could be attained. The pressure transient tests after the second slug confirmed that the injection pressure had been above the fracture pressure during the last period of surfactant injection.

The geological complexity in the CFB and resulting complex gas flow pattern near well P-18 made it difficult to evaluate the foam performance. Gas tracers were applied to monitor the gas breakthrough at P-18. However, logs indicated a secondary gas cap was established in the P-18 area. The

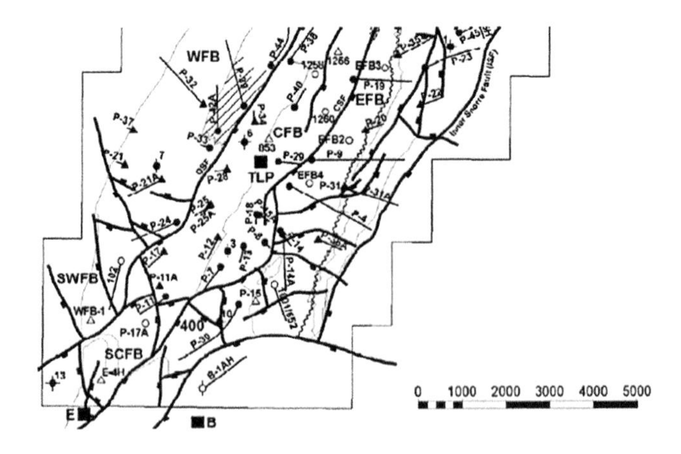

FIGURE 11.9 Well locations in the test areas (Skauge et al., 2002).

secondary gas cap reduced the information that could have been obtained otherwise from the gas tracers.

During the third gas cycle, the gas injectivity values returned to those during normal gas injection. The fact that the injection well was fractured and fractures were present during the second and third surfactant slugs reduced the foam flooding efficiency. It seemed that no mobility control effect was observed after the third surfactant slug injection.

Fractures were dominating during the co-injection at reasonable field rates. The fractures affected the foam transport. The evaluation of the SAG period showed that the radius of the foam zone was limited, or the sweep efficiency was not improved. A reduced gas injectivity was observed after the first and second gas injection period.

The complexity of the CFB pilot and the difficulty in interpretation had limited the possibility to use this pilot as a tool to estimate the FAWAG effect in improved oil recovery and gas-storage in the other areas of the Snorre field. The FAWAG injection on the CFB was stopped in early 1999 when well P-25A suffered gas leak. The FAWAG pilot was changed from the CFB to western fault block (WFB) of the Snorre field.

WAG Injection at Well P-32 in the WFB

The selected well pair was the injector P-32 and the production well P-39. The WFB was put on production during the autumn of 1992. The main drive mechanism was down-dip water alternating gas injection. In the WFB, as shown in Figure 11.8, there were three injectors (P-32, P-37, and P-21) and four producers (P-39, P-33, P-24, and P-42A).

Water was injected into the injector P-32 from March 1996 to November 1998. On November 25, gas injection was started in P-32, and a total of 83.9 MSm^3 gas was injected. Gas broke through the producer P-39 between January 7 and January 25, indicating that the maximum gas travel time P-32 to P-39 was 60 days. The distance between P-32 and P-39 was 1450−1550 m. A second gas slug injection was started on June 7, 1999 (total 105.8 MSm^3 injected). The gas breakthrough time during the second gas injection was 29 days. The reduction in the gas breakthrough time between the first and second gas injection can be caused by the establishment of a trapped gas in the area after the first gas injection. Other arguments for the faster gas breakthrough in the second gas injection period were reduced mass exchange between oil and gas; and the second gas slug entered the path already established during the first gas injection. Gas breakthrough is expected to be earlier in later gas injection of a near miscible WAG injection (Skauge et al., 2002).

FAWAG Injection at Well P-32 in the WFB

The SAG injection period started November 1999 in well P-32. In the first cycle, 15,262 Sm^3 surfactant solution at 0.49 wt% surfactant concentration

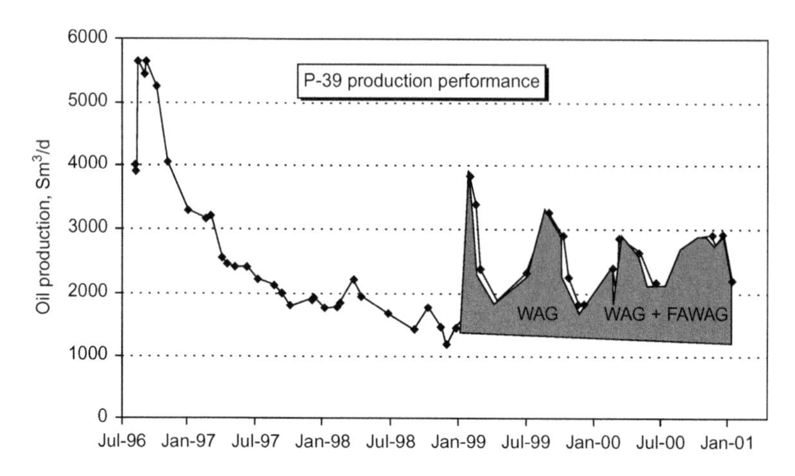

FIGURE 11.10 Well P-39 oil rates during WAG and FAWAG periods (Blaker et al., 2002).

was injected for 9.5 days, followed by about 100 days of gas injection until a predefined injectivity level was reached. In the second cycle, surfactant solution at 0.2 wt% surfactant concentration was injected from February 26 to March 17, 2000 (about 20 days). During the injection, the injection rate was adjusted to ensure that the fracturing pressure was not exceeded. No major operational problems during the injection of foam chemicals and gas were reported. The oil rate increase from well P-39 was observed, as shown in Figure 11.10. A tracer test showed that the breakthrough time from well P-32 to well P-39 was at least 5 months compared with the earlier 29 days in the second cycle and maximum 60 days in the first cycle of the WAG injection period. The production from the WFB also showed that a large volume of gas could be stored, either temporarily or permanently, in the reservoir. It was estimated that the FAWAG treatment could contribute approximately 250,000 Sm^3 of oil, and the cost of the treatment in the WFB was approximately US$1 million.

REFERENCES

Aarra, M.G., Skauge, A., 1994. A Foam pilot in a North Sea oil reservoir: preparation for a production well treatment, paper SPE 28599 presented at the SPE Annual Technical Conference and Exhibition, 25–28 September, New Orleans, LA.

Bernard, G.G., Holm, L.W., 1964. Effect of foam on permeability of porous media to gas. SPEJ September, 267–274 (Trans., AIME, 231).

Bernard, G.G., Holm, L.W., Jacobs, W.L., 1965. Effect of foam on trapped gas saturation and on permeability of porous media to gas. SPEJ 5 (4), 295–300.

Blaker, T., Aarra, M.G., Skauge, A., Rasmussen, L., Celius, H.K., Martinsen, H.A., et al., 2002. Foam for gas mobility control in the Snorre field: the FAWAG project. SPEREE 5 (4), 317–323.

Blauer, R.E., Mitchell, B.J., Kolhlhaas, C.A., 1974. Determination of laminar, turbulent, and transitional foam flow losses in pipes, Paper SPE 4885 Presented at the SPE California Regional Meeting, 4–5 April, San Francisco, CA.

Bretherton, F.P., 1961. The motion of long bubbles in tubes. J. Fluid Mech. 10, 166–188.

Chad, J., Malsalla, P., Novosad, J.J., 1988. Foam forming surfactants in Pembina/Ostracod "G" pool, Paper CIM 88-3940 Presented at the Annual Technical Meeting of the Petroleum Society of CIM, 12–16 June, Calgary, Alberta, Canada.

Chambers, K.T., Radke, C.J., 1991. Capillary phenomena in foam flow through porous media. In: Morrow, N.R. (Ed.), Interfacial Phenomena in Petroleum Recovery. Marcel Dekker, New York, NY, pp. 191–255. (Chapter 6).

Cheng, L., Reme, A.B., Shan, D., Coombe, D.A., Rossen, W.R., 2000. Simulating foam processes and high and low foam qualities, Paper SPE 59287 Presented at the SPE/DOE Symposium on Improved Oil Recovery, 3–5 April, Tulsa, OK.

Chou, S.I., 1990. Percolation theory of foam in porous media, Paper SPE 20239 Presented at the SPE/DOE Enhanced Oil Recovery Symposium,22–25 April, Tulsa, Oklahoma.

Chou, S.I., 1991. Conditions for generating foam in porous media, Paper SPE 22628 Presented at the SPE Annual Technical Conference and Exhibition, 6–9 October, Dallas, TX.

David, A., Marsden Jr., S.S., 1969. The rheology of foam, Paper SPE 2544 Presented at the SPE Annual Meeting 28 September–1 October, Denver, CO.

de Vries, A.S., Wit, K., 1990. Rheology of gas/water foam in the quality range relevant to steam foam. SPERE 5 (2), 185–192.

Djabbarah, N.F., Weber, S.L., Freeman, D.C., Muscatello, J.A., Ashbaugh, J.P., Covington, T.E., 1990. Laboratory design and field demonstration of steam diversion with foam. Paper 20067 Presented at the SPE California Regional Meeting, 4-6 April, Ventura, CA.

Duerksen, J.H., 1986. Laboratory study of foaming surfactants as steam diverting agents. SPERE January, 44–52.

ECLIPSE, 2011. Technical manual, Schlumberger.

Eson, R.L., Cooke, R.W., 1989. A comprehensive analysis of steam foam diverters and application methods, Paper SPE 18785 Presented at the SPE California Regional Meeting, 5–7 April, Bakersfield, CA.

Friedmann, F., Hughes, T.L., Smith, M.E., Hild, G.P., Wilson, A., Davies, S.N., 1999. Development and testing of a foam-gel technology to improve conformance of the Rangely CO_2 flood. SPEREE 2 (1), 4–13.

Friedmann, F., Jensen, J. A., 1986. Some parameters influencing the formation and propagation of foams in porous media, Paper SPE 15087 Presented at the Fifty Sixth SPE California Regional Meeting, 2–4 April, Oakland, CA.

Hanssen, J.E., 1993a. Foam as a gas-blocking agent in petroleum reservoirs I: empirical observations and parametric study. J. Pet. Sci. Eng. 10 (2), 117–133.

Hanssen, J.E., 1993b. Foam as a gas-blocking agent in petroleum reservoirs II: mechanisms of gas blockage by foam. J. Pet. Sci. Eng. 10 (2), 135–156.

Hanssen, J.E., Dalland, M., 1990. Foams for effective gas blockage in the presence of crude oil, Paper SPE 20193 Presented at the SPE/DOE Enhanced Oil Recovery Symposium, 22–25 April, Tulsa, OK.

Heller, J.P., Cheng, L.L., Murty, S.K., 1985. Foam-like dispersions for mobility control in CO_2 floods. SPEJ August, 603–613.

Heuer, G.J., Jacocks, C.L., 1968. Control of gas–oil ratio in producing wells, US Patent 3,368,624.

Hirasaki, G.J., 1989. Paper 19518, Supplement to SPE 19505, The steam-foam process – review of steam-foam process mechanisms.

Hirasaki, G.J., Lawson, J.B., 1985. Mechanisms of foam flow in porous media: apparent viscosity in smooth capillaries. SPEJ 25 (2), 176–190.

Hirasaki, G.J., Miller, C.A., Szafranski, R., Tanzil, D., Lawson, J.B., Meinardus, H., et al., 1997. Field demonstration of the surfactant/foam process for aquifer remediation, Paper SPE 39292 Presented at the SPE Annual Technical Conference and Exhibition, 5–8 October 1997, San Antonio, TX.

Hoefner, M.L., Evans, E.M., Buckles, J.J., Jones, T.A., 1995. CO_2 foam: results from four developmental field trials. SPERE 10 (4), 273–281.

Hudgins, D.A., Chung, T.-H., 1990. Long-distance propagation of foams, Paper SPE 20196 Presented at the SPE/DOE Enhanced Oil Recovery Symposium, 22–25 April, Tulsa, OK.

Isaacs, E.E., Iyory, J., Green, M.K., 1994. Steam-foams for heavy oil and bitumen recovery. In: Schramm, L.L. (Ed.), Foams: Fundamentals and Applications in the Petroleum Industry. American Chemical Society, Washington, DC, pp. 235–258.

Islam, M.R., Farouq Ali, S.M., 1990. Numerical simulation of foam flow in porous media. J. Can. Pet. Tech. 29 (4), 47–51.

Khan, S.A., 1965. The flow of foam through porous media, MS Thesis, Stanford University.

Kharabaf, H., Yortsos, Y.C., 1998. A pore-network model for foam formation and propagation in porous media. SPEJ 3 (1), 42–53.

Kovscek, A.R., 1988. Reservoir simulation of foam displacement processes, Paper Presented at the Seventh UNITAR International Conference on Heavy Crudes and Tar Sands, 27–30 October, Beijing, China.

Kovscek, A.R., Radke, C.J., 1994. Fundamentals of foam transport in porous media. In: Schramm, L.L. (Ed.), Foams in the Petroleum Industry. American Chemical Society, Washington, DC, pp. 115–163.

Krause, R.E., Lane, R.H., Kuehne, D.L., Bain, G.F., 1992. Foam treatment of producing wells to increase oil production at Prudhoe Bay, Paper SPE 24191 Presented at the SPE/DOE Enhanced Oil Recovery Symposium, 22–24 April, Tulsa, OK.

Lake, L.W., 1989. Enhanced Oil Recovery. Prentice-Hall, Englewood Cliffs, NJ.

Lecomte, J.P., et al., 2007. Silicone in the food industries. In: De Jaeger, R., Gleria, M. (Eds.), Inorganic Polymers. Nova Science Publishers, pp. 61–161. (Chapter 2).

Liu, P.C., Besserer, G.J., 1988. Application of foam injection in Triassic pool, Canada: laboratory and field test results, Paper SPE 18080 Presented at the SPE Annual Technical Conference and Exhibition, 2–5 October, Houston, TX.

Manlowe, D.J., Radke, C.J., 1990. A pore-level investigation of foam/oil interactions in porous media. SPERE 5 (4), 495–502.

Marfoe, C.H., Kazemi, H., Ramirez, W.F., 1987. Numerical simulation of foam flow in porous media, Paper SPE 16709 Presented at the SPE Annual Meeting, Dallas, TX.

Minssieux, L., 1974. Oil displacement by foams in relation to their physical properties in porous media. JPT January, 100–108 (Trans., AIME, 257).

Mohammadi, S.S., Van Slyke, D.C., Ganong, B.L., 1989. Steam-foam pilot project in Dome-Tumbador, Midway-Sunset Field. SPERE 4 (1), 7–16.

Nikolov, A.D., Wasan, D.T., Huang, D.W., Edwards, D.A., 1986. The effect of oil on foam stability: mechanisms and implications for oil displacement by foam in porous media, Paper SPE 15443 Presented at the SPE Annual Technical Conference and Exhibition, 5–8 October, New Orleans, LA.

Novosad, J.J., Mannhardt, K., Rendall, A., 1989. The interaction between foam and crude oils, Paper 89-40-29 Presented at the Annual Technical Meeting of the Petroleum Society of CIM, 28–31 May, Banff, Alberta, Canada.

Owete, O.S., Brigham, W.E., 1987. Flow behavior of foam: a porous micromodel study. SPEREE 2 (3), 315–323.

Patzek, T.W., Koinis, M.T., 1990. Kern river steam-foam pilots. JPT 42 (4), 496–503.

Persoff, P., Radke, C.J., Pruess, K., Benson, S.M., Witherspoon, P.A., 1991. A laboratory investigation of foam flow in sandstone at elevated pressure. SPEREE 6 (3), 365–371.

Ploeg, J.F., Duerksen, J.H., 1985. Two successful steam/foam field tests, sections 15A and 26C, midway-sunset field, Paper SPE 13609 Presented at the SPE California Regional Meeting, 27–29 March, Bakersfield, CA.

Purwoko, K., 2007. Evaluation of FAWAG injection on the Snorre field by using STARS simulator, MS Thesis, University of Stavanger.

Ransohoff, T.C., Radke, C.J., 1988. Mechanisms of foam generation in glass-bead packs. SPEREE 3 (2), 573–585.

Ross, S., McBain, J.W., 1944. Ind. Eng. Chem. 36, 570.

Sanchez, J.M., Schechter, R.S., Monsalve, A., 1986. The effect of trace quantities of surfactant on nitrogen/water relative permeabilities, Paper SPE 15446 Presented at the SPE Annual Technical Conference and Exhibition, 5–8 October, New Orleans, LA.

Schramm, L.L., 1994. Foam sensitivity to crude oil in porous media. In: Schramm, L.L. (Ed.), Advances in Chemistry Series 242. American Chemical Society, Washington, DC, pp. 165–197.

Schramm, L.L., Novosad, J.J., 1990. Micro-visualization of foam interactions with a crude oil. Colloids Surf. 46, 21–43.

Schramm, L.L., Wassmuth, F., 1994. Foams: basic principles. In: Schramm, L.L. (Ed.), Foams: Fundamentals and Applications in the Petroleum Industry. American Chemical Society, Washington, DC, pp. 2–45. (Advances in Chemistry Series 242).

Shan, D., Rossen, W.R., 2002. Optimal injection strategies for foam IOR. Paper SPE 75180 Presented at the SPE/DOE Improved Oil Recovery Symposium, 13–17 April, Tulsa, OK.

Sheng, J.J., 2011. Modern Chemical Enhanced Oil Recovery: Theory and Practice. Elsevier, Burlington, MA.

Sheng, J.J., Maini, B.B., Hayes, R.E., Tortike, W.S., 1997. Experimental study of foamy oil stability. J. Can. Petrol. Technol. 36 (4), 31–37.

Shi, J.-X., Rossen, W.R., 1998. Simulation and dimensional analysis of foam processes in porous media. SPEREE 1 (2), 148–154.

Skauge, A., Aarra, M.G., Surguchev, L., Martinsen, H.A., Rasmussen, L., 2002. Foam-assisted WAG: experience from the Snorre field, Paper SPE 75157 Presented at SPE/DOE Improved Oil Recovery Symposium, 13–17 April, Tulsa, Oklahoma.

Suffridge, F.E., Raterman, K.T., Russell, G.C., 1989. Foam performance under reservoir conditions, Paper SPE 19691 Presented at the SPE Annual Technical Conference and Exhibition, 8–11 October, San Antonio, TX.

Turta, A.T., Singhal, A.K., 2002. Field foam applications in enhanced oil recovery projects: screening and design aspects. J. Can. Petrol. Technol. 41, 10.

Wang, Z.-L., 2007. Enhanced Foam Flooding to Improve Oil Recovery. China Science and Technology Press, Beijing, China.

Yang, G.-L., Lin, Y.-Q., Liu, J.-L., Bao, J.-G., 2004. The feasibility study of nitrogen foam flooding in heavy oil reservoirs in Liaohe field, Xinjiang. Petrol. Geol. 25 (2), 188–190.

Yang, G.-L., Lin, Y.-Q., Xu, W.-H., 2005. Lessons from the pilot test of nitrogen foam flooding in the Jin 90 block of Liaohe field. In: Yan, C.-Z., Li, Y. (Eds.), Symposium on Tertiary Recovery. Petroleum Industry Press, Beijing, China, pp. 150–155.

Yang, L., Me, S.-C., 2006. Research and testing of combined oil recovery. In: Cao, F.C. (Ed.), Research and Practice of Improved Oil Recovery in Daqing. Petroleum Industry Press, Beijing, China, pp. 326–351.

Yang, S.H., Reed, R.L., 1989. Mobility control using CO_2 forms, Paper SPE 19689 Presented at the SPE Annual Technical Conference and Exhibition, 8–11 October, San Antonio, TX.

Zhang, L.-M., Zhu, Y.-Y., Zhou, L.-F, 2005. State of the art of foaming agents in high-temperature steam foam flooding. In: Yan, C.-Z., Li, Y. (Eds.), Symposium on Tertiary Recovery. Petroleum Industry Press, Beijing, China, pp. 363–367.

Zhang, X.-S., Wang, Q.-W., Song, X.-W., Zhou, G.-H., Gu, P., Li, X.-L., 2005. Polymer enhanced nitrogen foam injection for EOR in well pattern GD2-28-8, Gudao, Shengli. Oilfield Chem. 22 (4), 366–369, 384.

Zhou, Z.-H., Rossen, W.R., 1995. Applying fractional-flow theory to foam processes at the "limiting capillary pressure". SPE Advanc. Technol. Ser. 3 (1), 154–162.

Ziegler, V.M., 1988. Laboratory investigation of high temperature surfactant flooding. SPERE 3 (2), 586–596.

Chapter 12

Surfactant Enhanced Oil Recovery in Carbonate Reservoirs

James J. Sheng
Bob L. Herd Department of Petroleum Engineering, Texas Tech University, Lubbock, TX 79409, USA

12.1 INTRODUCTION

Currently, more than 85% of world energy consumption comes from fossil fuels and the World Energy Outlook shows that energy demand could rise by 53% between now and 2030 (IEA, 2006). Although most energy experts agree that the world's energy resources are adequate to meet this projected growth, more reserves will be needed. This means the petroleum industry will have to increase recovery factors significantly from all types of reservoirs. Schlumberger Market Analysis 2007 shows that more than 60% of the world's oil and 40% of the world's gas reserves are held in carbonates (Schlumberger, 2007). BP Statistical Review 2007 shows that the Middle East has 61% of the world's proved conventional oil reserves (BP, 2008); approximately 70% of these reserves are in carbonate reservoirs (Schlumberger, 2007). The Middle East also has 41.3% of the world's proved gas reserves (BP, 2008); 90% of these gas reserves lie in carbonate reservoirs (Schlumberger, 2007). It is clear that the relative importance of carbonate reservoirs compared with other types of reserves will increase dramatically during the first half of the twenty-first century. The world has 3000 billion barrels of remaining oil and 3000 trillion SCF gas in place in carbonates. However, due to complex structures, formation heterogeneities and oil-wet/mixed-wet conditions, etc., the oil recovery factor in carbonate reservoirs is very low, probably below 35% on average, and it is lower than that in sandstone reservoirs. Therefore, there is increasing interest to improve hydrocarbon recovery from carbonate reservoirs, as we are facing challenges to make up depleted reserves.

Although there is a great potential to improve oil recovery in carbonate reservoirs, the research in this area is very limited due to technical and economic challenges. Most of the field development schemes in carbonate

Enhanced Oil Recovery Field Case Studies.

reservoirs are limited to waterflooding and gas flooding with low ultimate recovery factors. A few surfactant-related enhanced oil recovery (EOR) methods have been tried in carbonate fields, although more polymer flooding projects were carried out before 1990s.

Chemical EOR research in carbonate reservoirs has been focused on using surfactants to change oil-wet to water-wet to enhance water imbibition into matrix blocks. Wettability alteration results in spontaneous imbibition of water into oil containing matrix, thus driving oil out of matrix. These surfactants include cationics, nonionics, and anionics. It has been found that anionic function to reduce interfacial tension (IFT) and associated buoyancy are very important mechanisms (Sheng, 2012). The problem is that such process is slow. Upscaling from laboratory results to field application needs more research work to be done. If the process is deemed to be slow, forced imbibition has to be applied. The future research should be on the area to optimize different development schemes and EOR methods in carbonates.

In this chapter, we first present the problems with carbonate reservoirs, followed by models of wettability alteration using surfactants. We also discuss the upscaling models related to oil recovery in fractured carbonate reservoirs. Chemicals used in carbonate reservoirs are reviewed. Finally we present several field cases using surfactants to stimulate oil recovery.

12.2 PROBLEMS IN CARBONATE RESERVOIRS

The average recovery factor in both sandstone and carbonate reservoirs is about 35%. The average recovery factor in sandstone reservoirs is higher than in carbonates. Therefore, the average recovery factor in carbonate reservoirs is below 35%. Carbonate reservoirs present a number of specific characteristics posing complex challenges in reservoir characterization, production and management. Carbonate rocks typically have a complex texture and pore network resulting from their depositional history and later diagenesis.

Heterogeneity may exist at all scales—in pores, grains, and textures. The porosities of carbonate rocks can be grouped into three types: connected porosity which is existing porosity between the carbonate grains, vugs which are unconnected pores resulting from the dissolution of calcite by water during diagenesis, and fracture porosity which is caused by stresses following deposition. Diagenesis can create stylolite structures which form horizontal flow barriers, sometimes extending over kilometers within the reservoir, having a dramatic effect on field performance. Fractures can be responsible for water breakthrough, gas coning, and drilling problems such as heavy mud losses and stuck pipe. Together, these three forms of porosities create a very complex path for fluids and directly affect well productivity.

In addition to the variations in porosity, wettability is a further heterogeneous characteristic in carbonates. The great majority of sandstone reservoirs are probably water-wet. However, the aging of carbonate rocks containing water and oil turns initially water-wet rocks into mixed-wet or even oil-wet. This means that oil can adhere to the surface of carbonate rock and it is therefore harder to produce. Most carbonate reservoirs are believed to be mixed-wet or oil-wet.

12.3 MODELS OF WETTABILITY ALTERATION USING SURFACTANTS

One important mechanism using surfactants in carbonate reservoirs is to change wettability from oil-wet to more water-wet. Wettability alteration has been formulated with surfactant adsorption, and relative permeabilities and capillary curves are modified based on the degree of wettability alteration. Delshad et al. (2009) used this parameter to modify capillary curve and relative permeability curves:

$$\omega = \frac{\hat{C}_{\text{surf}}}{C_{\text{surf}} + \hat{C}_{\text{surf}}} \tag{12.1}$$

where ω is the interpolation scaling factor, $\hat{C}_{\text{surf}}$ and C_{surf} represent the adsorbed and equilibrium concentrations of surfactant, respectively. The capillary curve and relative permeability curve are then modified:

$$k_{\text{r}} = \omega k_{\text{r}}^{\text{ww}} + (1 - \omega) k_{\text{r}}^{\text{ow}} \tag{12.2}$$

$$p_{\text{C}} = \omega p_{\text{C}}^{\text{ww}} + (1 - \omega) p_{\text{c}}^{\text{ow}} \tag{12.3}$$

where the superscripts ww and ow denote water-wet and oil-wet conditions, k_{r} is the relative permeability, and p_{c} is the capillary pressure. These equations are proposed based on the assumption that surfactant adsorption on calcite rock surfaces increases water-wetness, although this assumption may not be generally valid. The capillary pressure p_{c} is scaled with the IFT as follows:

$$p_{cjj'}^{\text{ww}} = C_{\text{pc}} \sqrt{\frac{\phi}{k} \frac{\sigma_{jj'}^{\text{ww}}}{\sigma_{jj'}^{\text{ow}}}} \left(1 - \frac{S_j - S_{jr}}{1 - \sum_{j=1}^{3} S_{jr}}\right)^{E_{\text{pc}}}, \; j = 1, 2, 3 \tag{12.4}$$

where $C_{\text{pc}}\sqrt{\phi/k}$ takes also into account the effect of permeability and porosity using the Leverett-J function (Leverett, 1941), ϕ is the porosity and k is permeability, σ is the IFT, S is the saturation at the water-wet condition, the subscript j and j' denote the phase j and the conjugate phase j', respectively, and E_{pc} is the exponent for capillary pressure. The above model is implemented in UTCHEM version 9.95 (UT Austin, 2009). In ECLIPSE 2009 version (Schlumberger, 2009), we input a table of ω versus surfactant adsorption.

Another model explicitly including wetting angle effect was proposed by Adibhatla et al. (2005). In their model, a simple interpolation technique is used to consider the wettability effect on residual saturations and trapping numbers:

$$\frac{S_{rj}^{low} - S_{r,b1}^{low}}{\cos\theta - \cos\theta_0} = \frac{S_{r,b2}^{low} - S_{r,b1}^{low}}{\cos(\pi - \theta_0) - \cos\theta_0} \tag{12.5}$$

$$\frac{\ln T_j - \ln T_{b1}}{\cos\theta - \cos\theta_0} = \frac{\ln T_{b2} - \ln T_{b1}}{\cos(\pi - \theta_0) - \cos\theta_0} \tag{12.6}$$

In the above equations, the superscript "low" refers to the parameter value at a low trapping number. To use the above equations, the residual saturation values of S_{rj}^{low} and T_j for a pair of base phases are needed. These base phases are represented with subscripts "b1" and "b2." Without losing the generality, it is assumed the contact angle of the base phase b1 before wettability alteration is θ_0, and the contact angle of the base phase b2 is $(\pi - \theta_0)$. Note that oil and aqueous phases are not distinguished (a dummy phase j is used). The residual saturation at a low trapping number, S_{rj}^{low}, and the trapping parameter, T_j, for phase j are calculated from the above equations, respectively.

Once S_{rj}^{low} and T_j at the altered contact angle θ are obtained from the above equations, the residual saturation at a different trapping number N_T is calculated by

$$S_{rj} = S_{rj}^{high} + \frac{S_{rj}^{low} - S_{rj}^{high}}{1 + T_j N_T} \tag{12.7}$$

where S_{rj} is the residual saturation of phase j at the trapping number N_T. The superscript "high" refers to the parameter value at a high trapping number. The trapping number is the capillary number including gravity effect which is discussed in detail in Sheng (2011). In this equation, S_{rj}^{high} is typically 0. Given the values of S_{rj}^{low} and T_j (the latter can be obtained by fitting experimental data), Eq. (12.7) yields the desaturation curve (S_{rj} versus N_T) that is similar to the capillary desaturation curve (CDC).

Before we present an end-point k_r of a phase at a trapping number, we need to discuss the relationship between the end-point k_r and the conjugate residual saturation first. According to Delshad et al. (1986),

$$\frac{k_{rj}^{e} - k_{rj}^{e,high}}{k_{rj}^{e,low} - k_{rj}^{e,high}} = \frac{S_{j'r} - S_{j'r}^{high}}{S_{j'r}^{low} - S_{j'r}^{high}} \tag{12.8}$$

where k_{rj}^{e} denotes the end-point relative permeability of phase j, the superscripts low and high correspond to low and high capillary (trapping) numbers, respectively, and the subscript j' denotes the conjugate phase of phase j. In Eq. (12.8), it is assumed that the end-point relative permeability enhancements (and the later exponent decreases) are caused by the residual saturation reduction of the conjugate phase as a function of the trapping number. However, the residual saturation of the conjugate phase may not be a good predictor for the end-point relative permeabilities and exponents, especially when wettability alteration is involved (Anderson, 1987; Fulcher et al., 1985; Masalmesh, 2002; Tang and Firoozabadi, 2002).

Combining Eqs. (12.7) and (12.8), we have

$$\frac{k_{rj}^{e} - k_{rj}^{e,high}}{k_{rj}^{e,low} - k_{rj}^{e,high}} = \frac{1}{1 + T_{j'} N_{Tj'}} \tag{12.9}$$

To derive an end-point relative permeability, k_{rj}^{e}, at a trapping number N_T, we have to consider two factors. One is the effect of trapping number; the other is the effect of wettability alteration. According to Eq. (12.9), the effect of trapping number on k_{rj}^{e} at N_T can be considered using the following equation:

$$\frac{k_{rj}^{e,N_T} - k_{rj}^{e,high}}{k_{rj}^{e,N_{T0}} - k_{rj}^{e,high}} = \frac{1 + T_{j'} N_{T0j'}}{1 + T_{j'} N_{Tj'}} \tag{12.10}$$

where k_{rj}^{e,N_T}, $k_{rj}^{e,N_{T0}}$, $k_{rj}^{e,high}$ correspond to the end-point relative permeabilities at N_T, N_{T0} and a very high trapping number.

To include the effect of wettability, we have

$$k_{rj}^{e,N_{T0}} - k_{r,b1}^{e,N_{T0}} = \frac{\cos\theta - \cos\theta_0}{\cos(\pi - \theta_0) - \cos\theta_0} (k_{r,b2}^{e,N_{T0}} - k_{r,b1}^{e,N_{T0}}) \tag{12.11}$$

Here it is assumed that we have the relative permeability curves measured at a certain trapping number N_{T0} for a pair of base phases with the contact angle θ_0 for the phase b1 and $\pi - \theta_0$ for the phase b2. Putting Eq. (12.11) into Eq. (12.10), we have the relative permeability curves with the trapping number N_T and the contact angle θ:

$$k_{rj}^{e,N_T} = k_{rj}^{e,high} + \left[k_{r,b1}^{e,N_{T0}} + \frac{\cos\theta - \cos\theta_0}{\cos(\pi - \theta_0) - \cos\theta_0} (k_{r,b2}^{e,N_{T0}} - k_{r,b1}^{e,N_{T0}}) - k_{rj}^{e,high} \right] \frac{1 + T_{j'} N_{T0}}{1 + T_{j'} N_T} \tag{12.12}$$

Similarly, the exponents of relative permeabilities are

$$n_j^{N_T} = n_j^{high} + \left[n_{b1}^{N_{T0}} + \frac{\cos\theta - \cos\theta_0}{\cos(\pi - \theta_0) - \cos\theta_0} (n_{b2}^{N_{T0}} - n_{b1}^{N_{T0}}) - n_{rj}^{high} \right] \frac{1 + T_{j'} N_{T0}}{1 + T_{j'} N_T} \tag{12.13}$$

Equations (12.12) and (12.13) are just conceptual models that qualitatively capture the typical trends observed about the effects of trapping number and wettability on relative permeabilities. Note $T_{j'}$ is the trapping parameter of the conjugate phase of phase j and its value is evaluated with Eq. (12.6) using the contact angle $\pi - \theta$, where θ is the contact angle of phase j.

Again, we assume that the end-point value, k_{rj}^{e}, and the exponent n_j, for the phase j are correlated to the residual saturation of the conjugate phase j' through linear interpolation. And the Brooks–Corey model is used to describe the relative permeability

$$k_{rj} = k_{rj}^{e}(\overline{S}_j)^{n_j} \tag{12.14}$$

$$\overline{S}_j = \frac{S_j - S_{jr}}{1 - S_{jr} - S_{j'r}} \tag{12.15}$$

The effects of IFT and contact angle on capillary pressure are described with the following equation:

$$p_c = p_{c0}\frac{\sigma \cos \theta}{\sigma_0 \cos \theta_0} \tag{12.16}$$

where p_c and p_{c0} are the capillary pressures, and σ and σ_0 are the IFT between oil and aqueous phases at the contact angle θ and θ_0, respectively.

12.4 UPSCALING

Either surfactant diffusion process or surfactant induced gravity drainage process through wettability alteration and IFT reduction is slow. Therefore, upscaling the laboratory scale to the field scale becomes very important. Since the pioneering work by Mattax and Kyte (1962) who scaled capillary forced imbibition under specific conditions, several modified formulas have been proposed. Basically, the scaling group for capillary imbibitions is defined in terms of the dimensionless time defined as

$$t_D = \frac{\sigma\sqrt{k/\phi}}{\mu L_c^2}t \tag{12.17}$$

where k is the rock permeability, ϕ is the porosity, σ is the IFT between the wetting and the nonwetting phase, μ is the viscosity, t is the actual time, and L_c is the characteristic length. In the above scaling group, different authors defined μ and L_c differently (Kazemi et al., 1992; Ma et al., 1995; Mattax and Kyte, 1962). Although they used different equations to define these parameters, they all used the squared characteristic length. In other words, the imibibition rate, thus recovery rate and total recovery, is inversely proportional to the squared characteristic length. Zhang et al. (1996) verified Eq. (12.17) in different core dimensions experimentally.

Cuiec et al. (1994) performed experiments in low permeability chalk samples at high IFT and proposed a reference time including the gravity force as the ratio of viscous to gravity forces as

$$t_g = \frac{L_c \mu_o}{k \Delta \rho g} \tag{12.18}$$

where t_g is the gravity reference time, μ_o is the oil viscosity, and $\Delta\rho$ is the density difference between water and oil. Sheng (2012) upscaled a base simulation model into several models by increasing the each dimension size by 2, 5, and 10 times using UTCHEM (version 9.95). The model volumes are increased by 2, 5, and 10 times along each side. According to Eq. (12.18), if we only change the model sizes, the only variable is L_c. Thus we calculate the normalized time by the real time divided by 2 in the case of "Enlarged by $2 \times 2 \times 2$," and similarly in the other cases. The results are shown in Figure 12.1. It shows that the curves of oil recovery factor versus the normalized time for the models of different sizes almost overlap each other. This indicates that the gravity is the dominant mechanism.

Note that corresponding to Eq. (12.18), the dimensionless time can be defined as

$$t_D = \frac{k \Delta \rho g}{L_c \mu_o} t \tag{12.19}$$

From this equation, we can see that the oil recovery rate is inversely proportional to the characteristic length.

Li and Horne (2006) derived a scaling model which incorporates both capillary and gravity forces. The model also contained parameters such as mobility and capillary pressure. They tested their scaling model by using the experimental data published by Schechter et al. (1994), and the fit was

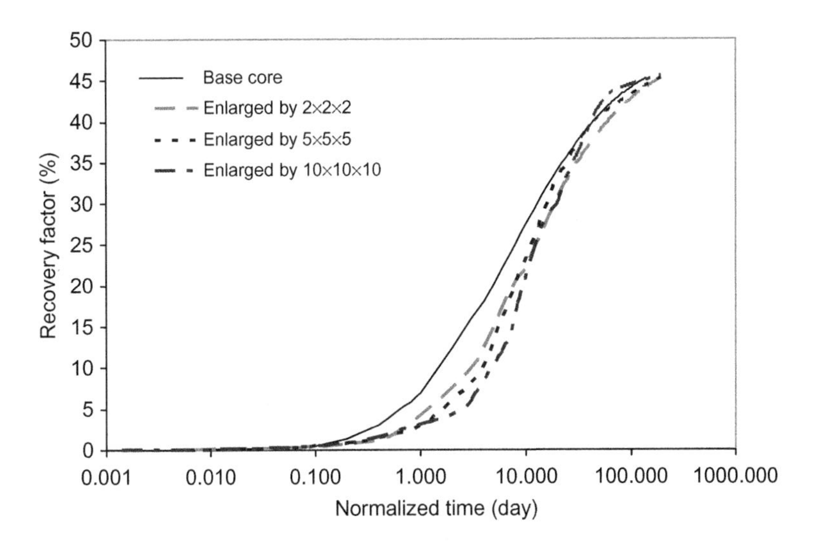

FIGURE 12.1 Oil recovery factor versus normalized time.

surprisingly good when the normalized recovery was plotted versus their defined dimensionless time. Their dimensionless time is

$$t_D = c^2 \left(\frac{M_e^* p_c^* (S_{wf} - S_{wi})}{\phi L_c^2} \right) t \tag{12.20}$$

where S_{wf} is the water saturation at the water front, S_{wi} is the initial water saturation, M_e^* is the effective mobility at the water displacement front, p_c^* is the capillary pressure at the water front, ϕ is the porosity, L_c is the characteristic length, and t is the actual time. The normalized recovery (factor) is

$$R^* = cR \tag{12.21}$$

where c is the ratio of the gravity force to the capillary force (the Bond number) which is defined as

$$c = \frac{\Delta \rho g L_c}{p_c^* (S_{wf} - S_{wi})} \tag{12.22}$$

where $\Delta\rho = \rho_w - \rho_o$ (the density difference between water and oil phases) and g is the acceleration constant. In Li and Horne's derivation, L_c is the length of the core. Here it is generalized to be the characteristic length. Using these notations, the normalized recovery versus the dimensionless time is

$$R^* \frac{dR^*}{dt_D} = 1 - R^* \tag{12.23}$$

Høgnesen et al. (2004) tested the Li and Horne's scaling model using published experimental data for spontaneous imbibitions of aqueous surfactant solution into preferentially oil-wet carbonate reservoirs, which involved wettability alteration. The scaling was performed by plotting the normalized oil recovery versus the dimensionless time. Generally, the experimental data fitted the model surprisingly well. Interestingly, all the tested experimental data scaled well if the heights of the cores were used as the characteristic length in the dimensionless time. That indicates the gravitational force had a very important role in the fluid-flow mechanism.

In our simulated cases (Sheng, 2012), when the IFT was as low as 0.049 mN/m, the gravitational force alone (without wettability alteration) could not produce oil. In the experiments tested by Høgnesen et al., the IFT ranged 0.3–1.0 mN/m. In other words, the IFTs were not at ultralow values. Probably some degree of wettability alteration occurred in those experiments. If the gravitational effect is the dominated mechanism, the oil recovery rate should be scaled with L_c, instead of L_c^2 as in the capillary dominated flow. Further research into the dominating mechanisms and development of correct upscaling models are extremely important to predict field scale EOR potential.

12.5 OIL RECOVERY MECHANISMS IN CARBONATES USING CHEMICALS

One mechanism of surfactant stimulation is wettability alteration from oil-wet to mixed-wet or water-wet. Wettability alteration results in spontaneous imbibition of water into oil containing matrix, thus driving oil out of matrix. Cationics and nonionics work based on this mechanism. Cationic surfactants form ion-pairs with adsorbed organic carboxylates of crude oil and stabilize them into the oil thereby changing the rock surface to water-wet (Adibhatla and Mohanty, 2008; Austad and Standnes, 2003). Austad and his workers used cationics to change wettability from oil-wet to water-wet in carbonate rocks. They found that cationic surfactants of the type $R{-}N^+(CH_3)_3$ were able to desorb organic carboxylates from the chalk surface in an irreversible way. The mechanism of wettability alteration is supposed to take place by an ion-pair formation by the cationic surfactant and negatively charged carboxylates in oil. The mechanism of ion-pair formation is schematically described in Figure 12.2. Due to electrostatic forces, the cationic monomers will interact with adsorbed anionic materials from the crude oil. Some of the adsorbed material at the interface between oil, water, and rock will be desorbed by forming an ion-pair between the cationic surfactant and the negatively charged adsorbed material, mostly carboxylic groups. This ion-pair complex is termed "cat-anionic surfactant," and it is regarded as a stable unit. In addition to electrostatic interactions, the ion-pairs are stabilized by hydrophobic interactions. The ion-pairs are not soluble in the water phase but can be dissolved in the oil phase or in the micelles. As a result, water will penetrate into the pore system, and oil will be expelled from the core through connected pores with high oil saturation in a so-called countercurrent flow mode. Thus, once the adsorbed organic material has been released from the surface, the chalk becomes more water-wet, and the imbibition of water is in fact mostly governed by capillary forces in the case of short cores.

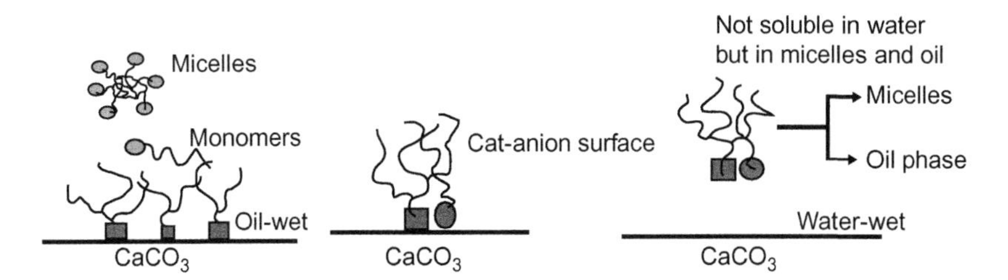

FIGURE 12.2 Mechanism of wettability alteration from oil-wet to water-wet. Large squires represent carboxylate groups, $-COO^-$, small squares represent other polar components, and circles represent cationic ammonium group, $-N^+(CH_3)_3$ (Austad and Standnes, 2003).

Anionic surfactants were not able to desorb anionic organic carboxylates from the crude oil in an irreversible way. Ethoxylated sulfonates with high EO-numbers did, however, displace oil spontaneously in a slow process. The brine imbibed nonuniformly, and the mechanism is suggested to involve the formation of a water-wet bilayer between the oil and the hydrophobic chalk surface. The mechanism of formation of surfactant double layers is shown schematically in Figure 12.3. The EO-surfactant is supposed to adsorb with the hydrophobic part onto the hydrophobic surface of the chalk. The water soluble head-group of the surfactant, the EO-group, and the anionic sulfonate group may decrease the contact angle below 90° by forming a small water zone between the organic coated surface and the oil. In this way, a weak capillary force is then created during the imbibition process. The fact that the imbibition of surfactant solution increases with increasing number of EO-groups supports such a model. The formation of the surfactant double layer must not be regarded as a permanent wettability alteration of the chalk. In fact, it will probably be fully reversible due to the weak hydrophobic bond between the surfactant and the hydrophobic surface. The other anionic surfactants tested did not imbibe any significant amount of water into the oil-wet chalk, confirming that the EO-groups play a very important role regarding the imbibition mechanism.

To reduce anionic surfactant adsorption on carbonate rock surfaces, Hirasaki and Zhang (2004) injected Na_2CO_2 with surfactants. The mechanism is that CO_3^{2-} and HCO_3^- change the rock surface to negative surface. The role of anionic surfactants was to reduce the IFT between oil and brine. Once the IFT is reduced, the gravity drive can be enhanced. Gravity plays the role in oil mobilization. For the gravity effect to function, IFT must be reduced to lower capillary pressure so that oil drops can move upward from the matrix.

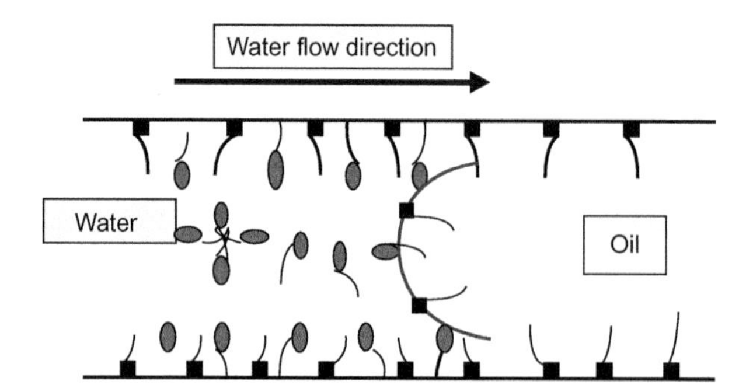

FIGURE 12.3 Schematic illustration of the mechanism of bilayer formation in a pore by EO-sulfonates. The eclipses represent EO-sulfonates and the squares represent the carboxylates in the oil (Standnes and Austad, 2000).

Hirasaki and Zhang (2004) also explained how adding Na_2CO_3 change wettability of carbonate rock surfaces. The zeta potential of the crude oil they used was negative for pH greater than 3. This is because of the dissociation of the naphthenic acids in the crude oil with increasing pH. The surface of calcite was positive for pH less than 9 when the only electrolytes were 0.02 M NaCl and using NaOH or HCl to adjust pH. The opposite charge between the oil/brine and mineral/brine interfaces results in an electrostatic attraction between the two interfaces, which tends to collapse the brine film and bring the oil in direct contact with the mineral surface. Thus, this system can be expected to be nonwater-wet around neutral pH. However, the zeta potential of calcite was negative even to a neutral pH when the brine was 0.1 N Na_2CO_3/$NaHCO_3$ and using HCl to adjust pH. This is because the potential determining ions for the calcite surface are Ca^{2+}, CO_3^{2-}, and HCO_3^-. An excess of the carbonate/bicarbonate anions makes the surface negatively charged. If both the crude oil/brine and brine/calcite interfaces are negatively charged, there will be an electrical repulsion between the two surfaces, which tends to stabilize the brine film between the two surfaces. Therefore, a system with carbonate/bicarbonate ions may be expected to have a preference to be water-wet, compared to that in the absence of carbonate ions.

Xie et al. (2005) compared the spontaneous imbibition rates using nonionics poly-oxyethylene alcohol (POA) and cationics (cocoalkyltrimethyl ammonium chloride, CAC). Their results show that the additional recovery from POA was higher and faster with respect to the scaled time than that from CAC. The IFT of POA solution was 19 times higher than that of CAC solution (5.7 versus 0.3 mN/m). This observation indicates that, ideally, the wettability should be changed to some optimal water-wet condition with respect to rate and extent of recovery while keeping the IFT relatively high for imbibition.

Acidizing is a common practice in carbonate reservoirs which is used to remove oxidized products of iron (iron sulfide). However, no response to surfactant treatment after acidizing was observed. This is probably because at least an outside layer of rock became strongly water-wet and any remaining oil was trapped (Xie et al., 2005).

12.6 CHEMICALS USED IN CARBONATE EOR

For alkalis, sodium tripolyphosphate (STPP) was used in laboratory tests for the Cretaceous Upper Edwards reservoir (Central Texas). STPP was proposed to minimize divalent precipitation, alter wettability, and generate emulsions (Olsen et al., 1990). Sodium carbonate was used in chemical EOR research in carbonate cores (Hirasaki and Zhang, 2004). The main function was to reduce surfactant adsorption. Sodium metaborate was also proposed to minimize divalent precipitation (Flaaten et al., 2010).

Cationics, anionics, and nonionics were all used in research for chemical EOR in carbonate reservoirs. Many cationic surfactants were investigated by the Austad research group (Standnes and Austad, 2000). Some of the surfactants they used include dodecyl trimethyl ammonium bromide (DTAB), n-C_8—$N(CH_3)_3Br$ (C8TAB), n-C_{10}—$N(CH_3)_3Br$ (C10TAB), n-C_{12}—$N(CH_3)_3Br$ (C12TAB), n-C_{16}—$N(CH_3)_3Br$ (C16TAB), n-C_8—Ph—$(EO)_2$—N $(CH_3)_2(CH_2$—Ph)Cl (Hyamine), and n-(C_8—C_{18})—$N(CH_3)_2(CH_2$—Ph)Cl (ADMBACl). Xie et al. (2005) used CAC. Tabatabal et al. (1993) used cetylpyridinium chloride (CPC) and dodecyl pyridinium chloride (DPC).

Some of anionic surfactants used by Seethepalli et al. (2004) were alkyl aryl ethoxylated sulphonates and propoxylated sulfate. Hirasaki and Zhang (2004) used ethoxylated and propoxylated sulfates which were tolerant to divalent ions. These surfactants included CS-330 (sodium dodecyl 3EO sulfate), C12-3PO (sodium dodecyl (Guerbet) 3PO sulfate), TDA-4PO (ammonium iso-tridecyl 4PO sulfate), and ISOFOL14T-4.1PO (sodium tetradecyl (Guerbet) 4PO sulfate).

Standnes and Austad (2000) used n-(C_{12}—C_{15})—$(EO)_{15}$—SO_3Na (S-150), n-C_{13}—$(EO)_8$—SO_3Na (B 1317), n-C_8—$(EO)_3$—SO_3Na (S-74), n-(C_{12}—C_{15})—$(PO)_4$—$(EO)_2$—OSO_3Na (APES), (n-C_8O_2CCH2)(n-C_8O_2C) CH—SO_3Na (Cropol), n-C_8—$(EO)_8$—OCH_2—COONa (Akypo), n-C_9—Ph—$(EO)_x$—PO_3Na (Gafac), n-C_{12}—OSO_3Na (SDS). Nonionic surfactants were used by Chen et al. (2001) and Xie et al. (2005).

12.7 CHEMICAL EOR PROJECTS IN CARBONATE RESERVOIRS

According to the Oil & Gas Journal survey in 2004 (Moritis, 2004), for the total 57 gas injection projects in the United States (water alternate gas (WAG) or continuous injection), 48 projects were CO_2 injection. Among those CO_2 projects, 67% were in Texas carbonates. There were abundant availability of CO_2 in the Texas Permian Basin, and CO_2 price in Texas was low. Other EOR projects which were active in 2004 include 7 air injection projects, 2 N_2 injection, 1 steam flood, and 1 surfactant stimulation (Manrique et al., 2007; Moritis, 2004). These data indicate that not many EOR projects were active in carbonate reservoirs.

For chemical EOR, many polymer projects were conducted in 1960s–1990s. During this period, there were only a few surfactant—polymer (SP) projects, but no alkaline—Surfactant—polymer (ASP) project was reported. From 1990s to 2000s, no chemical flood projects were reported in carbonate reservoirs, only three surfactant stimulations were reported: Mauddud, Cottonwood Creek, and Yates projects. These cases are presented next.

12.7.1 The Mauddud Carbonate in Bahrain

The Mauddud carbonate reservoir was the main oil producing reservoir in the Bahrain oilfield. The Mauddud zone was a 100 ft thick, low-dip, and heterogenous limestone reservoir. Its rock was described as moderately soft to hard, fine to medium grained, fossiliferous, detrital, clean, oil-wet limestone with limited fractures and vugs. The acid number of oil was 0.23–0.64.

The Mauddud reservoir had been producing since 1932 and was in a very mature stage. The dominant recovery mechanism was gravity drainage with crestal gas injection that started in 1938. The reservoir energy was supplemented by aquifers. Due to the reservoir rocks wettability characteristics (preferentially oil-wet nature), the residual oil saturation left behind gas and water fronts ranged from 20% to 70%. Thermal decay time (TDT) saturation logs showed residual oil saturations (S_{or}) of about 43%. The water cut was about 98–99%.

Early attempts to reduce water cut in the waterflooded area was made with a pilot trial using conformance chemicals such as cross-linked gels which failed to improve oil production. Later 6 wells were treated with surfactant washes using carrier fluids such as diesel and xylene. The wells were soaked a few days then produced. S_{or} was reduced by 10–15%. Reservoir saturation tool/carbon-oxygen (RST/CO) logs were used to evaluate S_{or}. Although these jobs were successful, the wells returned to the original water cuts very soon. This gave, however, a positive indication that with chemical treatments it would be possible to strip more oil from the Mauddud rock. It was thereafter decided to study the effectiveness of treating the area with a combination of alkaline and surfactants (AS) (Zubari and Sivakumar, 2003). No further report has been published since.

12.7.2 The Yates Field in Texas

The Yates field in Texas was discovered in 1962. The Yates San Andres reservoir is a naturally fractured dolomite formation. Several improved oil recovery methods had been evaluated in this prolific field with cumulative production greater than 1.3 billion bbl of oil with 30°API. San Andres was a 400 ft thick formation with an average matrix porosity and permeability of 15% and 100 mD, respectively. The oil viscosity was 6 cP and the reservoir temperature was 28°C.

Marathon Oil Company initiated a dilute-surfactant-well stimulation pilot test in the field in the early 1990s. Surfactant slugs were injected into the oil/water transition zone. Once the surfactant slug was injected, the well was shut in (soak time) for a brief period of time. When production resumed, the well showed an increase in the recovery of oil owing to reduction in IFT, gravity segregation of oil and water between the fractures and matrix, and wettability alteration (the latter contributed to a lesser extent) (Chen et al., 2001; Yang and Wadleigh, 2000).

The surfactant used in the Yates pilot was 0.3–0.4% nonionic ethoxy alcohol (Shell 91-8) and 0.35% Stepan CS-460 anionic ethoxy sulfate that were well above the critical micelle concentration (CMC) levels. The surfactant solutions injected were prepared with produced water in the concentrations of 3100–3880 ppm. The reported field results were positive, as evidenced by some pilot wells showing an increase in oil production over 30 bbl/day (Yang and Wadleigh, 2000).

12.7.3 The Cottonwood Creek Field in Wyoming

The Cottonwood Creek field is located in the Bighom Basin of Wyoming. Cottonwood Creek is a dolomitic class II reservoir. The class II reservoirs have low matrix porosities and permeabilities. The matrix provides some storage capacity, and the fractures provide the fluid-flow pathways. Typically, these types of reservoirs produce less than 10% of the original oil in place (OOIP) by primary recovery and exhibit low additional recovery factors during waterflooding. Cottonwood Creek was produced from the dolomitic Phosphoria formation. The reservoir thickness varied from 20 to 100 ft, and the average porosity and permeability were 10% and 16 mD, respectively. The reservoir produced a sour crude oil of 27°API.

Continental Resources initiated single-well surfactant stimulation treatments in Cottonwood Creek in August 1999. Well treatments included the injection of 500–1500 bbl of a surfactant solution slug, depending on the perforated interval. Typically, the injection period lasted 3 days with a 1-week shut-in period (soak time). Surfactant solutions were prepared using the nonionic POA at a concentration of 750 ppm, almost twice the CMC. Initial well treatments used an acid cleanup with HCl (15%) to remove iron sulfide (FeS) from the wellbore to avoid and/or reduce surfactant adsorption. Despite the effort, production results were discouraging. The initial results led to the elimination of the acid pretreatment and an increase of surfactant concentration of up to 1500 ppm (to allow for potential losses by adsorption to FeS) in subsequent surfactant stimulations (Weiss et al., 2006; Xie et al., 2005).

Single-well surfactant soak treatments were made at 23 wells. The general trend was that the oil recovery was increased. However, this increase is not significant. The problem was that 70% of the treated wells failed.

Oil recovery increase in Cottonwood Creek was attributed to wettability alteration (less oil-wet) and not to a reduction in IFT. Experimental IFT measurements of POA solutions (prepared with synthetic brine) with the Cottonwood Creek oil indicated 5.7 dynes/cm at ambient temperature. The minimum amount of surfactant used for a successful treatment was 60 lbm/ft of perforated internal on the basis of the analysis of 23 well treatments reported in the literature (Weiss et al., 2006; Xie et al., 2005).

12.7.4 The Baturaja Formation in the Semoga Field in Indonesia

The Semoga field was discovered in 1996 and is located in the Rimau block in the South Sumatera. The field consists of three prospect formations: Telisa formation (tight sandstone), Baturaja formation (carbonate), and

Talang Akar formation (sandstone). Thc Baturaja formation (BRF) is a carbonate reservoir with a proven volume of about 317,856 acre-ft (77 ft net pay). There were 127 wells: 82 producers, 28 injectors, and 17 shut-in wells.

The production began in 1997 and oil production peaked at 36,200 barrels oil per day (BOPD) in November 2001. Since then the production has declined owing to rising water cut. The average water cut before the surfactant stimulation was 86%, and some wells above 95% or even 100%. A laboratory study showed that the Baturaja formation was oil-wet. Huff and Puff surfactant stimulation was studied for this formation.

In this project, the surfactant was soaked for 7 days to allow a reaction with the hydrocarbon. The radial penetration designed for Well X-1 and Well X-2 was about 21 ft. The injection consisted of three steps:

1. *Preflush*: The purpose of the preflush was to displace the reservoir brine which contained potassium, sodium, calcium, and magnesium ions in the near-wellbore area, therefore avoiding adverse interactions with the chemical solution. The other purpose was to adjust reservoir salinity to favorable conditions for the surfactant; 100 bbl of produced water was injected into each well.
2. *Main-flush*: 9 bbl of surfactant and 451 bbl of water were injected into Well X-1; and 9 bbl of surfactant and 536 bbl of water were injected into Well X-2.
3. *Over-flush*: In this phase, the formation water was injected to displace the rest of the surfactant away from the wellbores at the end of stimulation. In the over-flush, 3 bbl of surfactant and 127 bbl of water were injected into Well X-1; and 0.65 bbl of surfactant and 43.35 bbl of water were injected into Well X-2.

This surfactant stimulation decreased water cut by about 8%, with an increased cumulative oil production of about 5800 bbl over a period of 3 months. An extended study was proposed to further investigate the mechanisms (Rilian et al., 2010).

12.7.5 Cretaceous Upper Edwards Reservoir (Central Texas)

A laboratory study was conducted to study the feasibility of ASP in the Cretaceous Upper Edwards reservoir, located in Central Texas (Olsen et al., 1990). The field was discovered in 1922. Over 950 development wells had been drilled. The water cut was 99%. The reservoir was preferentially oil-wet. The average permeability was 75 mD. The formation water salinity was low (produced TDS = 12,000 ppm). There was no anhydrate or gypsium. The reservoir temperature was 42°C, the acid number was 0.34, and the crude oil viscosity was 3 cP. The ASP formula selected was:

- 0.4–0.5% Sodium tripolyphosphate;
- 2% sodium carbonate;

- 0.2–0.5% Petrostep B100;
- 0.12% Pusher 700E.

The injection scheme was 0.1 PV freshwater, 0.1 PV ASP, 0.2 PV polymer. The ASP flood recovered approximately 45% of the residual oil after waterflooding. No field trial was reported.

12.8 CONCLUDING REMARKS

Field application of injecting surfactants in carbonate reservoirs to stimulate oil recovery has been limited to only a few field cases. Surfactant injection is believed to change wettability from oil-wet or more water-wet and to reduce IFT. It is assumed that the wettability alteration is caused by surfactant adsorption on carbonate rock surfaces. The EOR mechanisms are related to capillary imbibitions and gravity drive enhanced by surfactant injection. Capillary imbibitions and gravity drive could be slow processes. Therefore, upscaling from laboratory results to field scale application is important. Several upscaling models have been proposed. These models and even the drive mechanisms need more research work and field data for validation.

NOMENCLATURE

c ratio of the gravity force to the capillary force, dimensionless
C_{pc} capillary pressure end-point in Eq. (12.4), m/t^2
C_{surf} equilibrium surfactant concentration, m/L^3, vol.%, or mol/L pore volume
$\hat{C}_{surf}$ adsorbed surfactant concentration, m/L^3, vol.%, or mol/L pore volume
g acceleration of gravity, L/t^2
k permeability, L^2, mD, or m^2
k_r relative permeability, fraction or %
L_c characteristic length, L, m, or ft
M_e^* effective mobility at the displacement front (S_{wf}), L^{3}t/m, mD/cP
n exponent to define a relative permeability
N_T trapping number, dimensionless
p_c capillary pressure, m/Lt^2, Pa, or psi
p_c^* capillary pressure at the displacement front, m/Lt^2, Pa, or psi
R recovery factor (total oil produced/the OOIP), fraction or %
R^* normalized recovery factor
S saturation, fraction or %
$\overline{S}$ normalized saturation, fraction or %
S_{wf} water saturation at the displacement front, fraction or %
S_{wi} initial water saturation, fraction or %
t time, t, s or days
t_D dimensionless time
t_g gravity reference time, t, s or days
T trapping parameter, dimensionless

GREEK SYMBOLS

Δ operator that refers to a discrete change
Φ porosity, fraction or %
ρ density, m/L^3, g/cm^3
μ viscosity, m/Lt, mPa s (cP)
σ interfacial tension, m/t^2, mN/m
θ contact angle, degree
ω interpolation scaling factor for p_c and k_r, dimensionless

SUPERSCRIPT

n end-point
high at a high trapping number
low at a low trapping number
ow oil-wet
ww water-wet

SUBSCRIPT

j phase *j*
j' conjugate phase of phase *j*
r residual

REFERENCES

Adibhatla, B., Mohanty, K.K., 2008. Oil recovery from fractured carbonates by surfactant-aided gravity drainage: laboratory experiments and mechanistic simulations. SPEREE February, 119–130.

Adibhatla, B., Sun, X., Mohanty, K.K., 2005. Numerical studies of oil production from initially oil-wet fracture blocks by surfactant brine imbibitions, Paper SPE 97687 Presented at the SPE International Improved Oil Recovery Conference in Asia Pacific, 5–6 December, Kuala Lumpur, Malaysia.

Anderson, W.G., 1987. Wettability literature survey part 5: the effects of wettability on relative permeability. JPT 39 (11), 1453–1468.

Austad, T., Standnes, D.C., 2003. Spontaneous imbibition of water into oil-wet carbonates. J. Petroleum Sci. Technol. (JPSE) 39, 363–376.

BP, 2008 BP Statistical Review of World Energy, June. <http://www.bp.com/liveassets/bp_internet/china/bpchina_english/STAGING/local_assets/downloads_pdfs/statistical_review_of_world_energy_full_review_2008.pdf>.

Chen, H.L., Lucas, L.R., Nogaret, L.A.D., Yang, H.D., Kenyon, D.E., 2001. Laboratory monitoring of surfactant imbibition with computerized tomography. SPEREE February, 16–25.

Cuiec, L.E., Bourbiaux, B., Kalaydjian, F., 1994. Oil recovery by imbibition in low-permeability chalk. SPEFE 9 (3), 200–208.

Delshad, M., Delshad, M., Bhuyan, D., Pope, G.A., Lake, L.W., 1986. Effect of capillary number on the residual saturation of a three-phase micellar solution, Paper SPE 14911 Presented at the SPE Enhanced Oil Recovery Symposium, 20–23 April, Tulsa, OK.

Delshad, M., Fathi Najafabadi, N., Anderson, G.A., Pope, G.A., Sepehrnoori, K., 2009. Modeling wettability alteration by surfactants in naturally fractured reservoirs. SPEREE June, 361–370.

Flaaten, A.K., Nguyen, Q.P., Zhang, J., Mohammadi, H., Pope, G.A., 2010. Alkaline/surfactant/polymer chemical flooding without the need for soft water. SPE J. 15 (1), 184–196.

Fulcher Jr., R.A., Ertekin, T., Stahl, C.D., 1985. Effect of capillary number and its constituents on two-phase relative permeability curves. JPT 37 (2), 249–260.

Hirasaki, G., Zhang., D.L., 2004. Surface chemistry of oil recovery from fractured, oil-wet, carbonate formations. SPE J. June, 151–162.

Høgnesen, E.J., Standnes, D.C., Austad, T., 2004. Scaling spontaneous imbibitions of aqueous surfactant solution into preferential oil-wet carbonates. Energy Fuels 18, 1665–1675.

International Energy Agency (IEA), 2006. World Energy Outlook 2006, 65. <http://www.worldenergyoutlook.org/media/weowebsite/2008-1994/WEO2006.pdf>

Kazemi, H., Gilman, J.R., Elsharkawy, A.M., 1992. Analytical and numerical solution of oil recovery from fractured reservoirs with empirical transfer functions. SPERE 7 (2), 219–227.

Leverett, M.C., 1941. Capillary behavior in porous solids. Trans. AIME 142, 152–169.

Li, K., Horne, R.N., 2006. Generalized scaling approach for spontaneous imbibitions: an analytical model, *SPEREE* (June), 251–258; Paper SPE 77544 Presented at the 2002 SPE Annual Technical Conference and Exhibition, 29 September–2 October, San Antonio, TX.

Ma, S., Zhang, X., Morrow, N.R., 1995. Influence of fluid viscosity on mass transfer between rock matrix and fractures, Paper Presented at the Annual Technical Meeting of the Canadian Society of Petroleum Engineers, 7–9 June, Calgary, Alberta, Canada.

Manrique, E.J., Muci, V.E., Gurfinkel, M.E., 2007. EOR field experiences in carbonate reservoirs in the United States. SPEREE December, 667–686.

Masalmesh, S.K., 2002. The effect of wettability on saturation functions and impact on carbonate reservoirs in the Middle East, Paper 78515 Presented at the Abu Dhabi International Petroleum Exhibition and Conference, 13–16 October, Abu Dhabi, United Arab Emirates.

Mattax, C.C., Kyte, J.R., 1962. Imbibition oil recovery from fractured, water-drive reservoir. SPE J. 2 (2), 177–184.

Moritis, G., 2004. EOR survey. Oil Gas J., 53–65.

Olsen, D.K., Hicks, M.D., Hurd, B.G., Sinnokrot, A.A., Sweigart, C.N., 1990. Design of a novel flooding system for an oil-wet Central Texas carbonate reservoir, Paper SPE/DOE 20224 Presented at the SPE/DOE Seventh Symposium on Enhanced Oil Recovery, 22–25 April, Tulsa, OK.

Rilian, N.A., Sumestry, M., Wahyuningsih, 2010. Surfactant stimulation to increase reserves in carbonate reservoir "A Case Study in Semoga Field", Paper SPE 130060 Presented at the SPE EUROPEC/EAGE Annual Conference and Exhibition, 14–17 June, Barcelona, Spain.

Schechter, D.S., Zhou, D., Orr Jr., F.M., 1994. Low IFT drainage and imbibitions. J. Petroleum Sci. Eng. 11, 283–300.

Schlumberger, 2007. Schlumberger Market Analysis. <http://www.slb.com/~/media/Files/industry_challenges/carbonates/brochures/cb_carbonate_reservoirs_07os003.ashx>.

Schlumberger, 2009. ECLIPSE Technical Manual.

Seethepalli, A., Adibhatla, B., Mohanty, K.K., 2004. Physicochemical interactions during surfactant flooding of fractured carbonate reservoirs. SPE J. December, 411–418.

Sheng, J.J., 2011. Modern Chemical Enhanced Oil Recovery—Theory and Practice. Elsevier, Burlington, MA.

Sheng, J.J., 2012. Comparison of the effects of wettability alteration and IFT reduction on oil recovery in carbonate reservoirs. Asia-Pacific J. Chem. Eng. in press. doi: 10.1002/apj.1640.

Standnes, D.C., Austad, T., 2000. Wettability alteration in chalk 2. Mechanism for wettability alteration from oil-wet to water-wet using surfactants. J. Petroleum Sci. Eng. 28, 123–143.

Tabatabal, A., Gonzalez, M.V., Harwell, J.H., Scamehorn, J.F., 1993. Reducing surfactant adsorption in carbonate reservoirs. SPERE 8 (2), 117–122.

Tang, G.Q., Firoozabadi, A., 2002. Relative permeability modification in gas/liquid systems through wettability alteration to intermediate gas wetting. SPEREE 5 (6), 427–436.

UT Austin, 2009. A three-dimensional chemical flood simulator (UTCHEM, version 9.95), University of Texas at Austin, Austin, TX.

Weiss, W.W., Xie, X., Weiss, J., Subramanium, V., Taylor, A., Edens, F., 2006. Artificial intelligence used to evaluate 23 single-well surfactant-soak treatments. SPEREE June, 209–216.

Xie, X., Weiss, W.W., Tong, Z., Morrow, N.R., 2005. Improved oil recovery from carbonate reservoirs by chemical stimulation. SPE J. September, 276–285.

Yang, H.D., Wadleigh, E.E., 2000. Dilute surfactant IOR—design improvement for massive, fractured carbonate applications, Paper SPE 59009 Presented at the SPE International Petroleum Conference and Exhibition in Mexico, 1–3 February, Villahermosa, Mexico.

Zhang, X., Morrow, N.R., Ma, S., 1996. Experimental verification of a modified scaling group for spontaneous imbibition. SPERE 11 (4), 280–286.

Zubari, H.K., Sivakumar, V.C.B., 2003. Single well tests to determine the efficiency of alkaline–surfactant injection in a highly oil-wet limestone reservoir, Paper SPE 81464 Presented at the Middle East Oil Show, 9–12 June, Bahrain.

Chapter 13

Water-Based EOR in Carbonates and Sandstones: New Chemical Understanding of the EOR Potential Using "Smart Water"

Tor Austad
University of Stavanger, 4036 Stavanger, Norway

13.1 INTRODUCTION

During million of years, a chemical equilibrium between the crude oil, the brine, and the rock, (crude oil–brine–rock, CBR) has been established, which includes all the components of the oil reservoir. The distribution of oil and formation water in the pores of the rock is then fixed at given saturations of oil and water. Due to gradients in the composition of oil and water in an oil reservoir and inhomogeneous rock properties, the relative distribution of oil and water may change a little. The distribution of oil and water in the porous system is linked to the wetting properties of the CBR system, that is, the contact between the rock surface and the two fluids, oil and brine. As a rough characterization of a reservoir rock, the terms preferential water-wet, preferential oil-wet, and neutral wetting condition have been used. The wetting properties of a given CBR system have strongly influenced on the two-phase fluid flow in the porous medium, because they dictate the capillary pressure, P_c, and the relative permeabilities of oil, k_{ro}, and water, k_{rw}. A systematic laboratory study by Jadhunandan and Morrow (1995) showed an optimum in oil recovery by water flooding when the CBR system was slightly water-wet.

Most of the oil reservoirs are today water flooded in order to improve oil recovery. Initially, the main reasons for performing a water flood were to:

- Give pressure support to the reservoir to keep the reservoir pressure above the bubble point pressure of the oil, $P_{res} > P_b$ and
- Displace the oil by water by taking the benefit of viscous forces.

Water flooding has been characterized as a secondary oil recovery process since no specially enhanced oil-recovery (EOR) chemicals were

Enhanced Oil Recovery Field Case Studies.

injected. During systematic laboratory studies by many research groups during the last 20 years on wetting properties on different CBR systems, it has been verified that injected water, which is different in composition compared to the initial formation water, can disturb the established chemical equilibrium of the CBR system. During the process to establish a new chemical equilibrium, the wetting properties will also be changed, which may result in improved oil recovery. If, however, the composition of the injected water is similar to the initial formation water, the chemical equilibrium will be very little affected, only the relative saturations are affected. Thus, we may then say that injection of formation water is a secondary recovery process, but injection of water with a different composition, than the initial formation brine, may change wetting properties and thus acts as a tertiary recovery process. In order to use water injection as a tertiary oil recovery process, the kinetics linked to the wetting properties for the CBR system to establish a new chemical equilibrium must be high enough to take place during the time frame of the oil production. Many geochemical reactions are very slow, which may be negligible during the production time of an oil reservoir. The kinetics of chemical reactions usually increases as the temperature is increased, and therefore, it is not surprising that wettability alteration by water injection may in some cases be sensitive to the reservoir temperature.

In order to be able to evaluate "Smart Water" as an EOR fluid, it is important to improve the chemical understanding of the most important parameters dictating the wetting properties of oil reservoirs.

13.1.1 Wetting in Carbonates

More that 50% of the known petroleum reserves are trapped in carbonate reservoirs, which can be divided into limestone, chalk, and dolomite. The formation water may be of high salinity, and it is usually rich in Ca^{2+}. On average, the oil recovery from carbonates is well below 30% due to low water wetness, natural fractures, low permeability, and inhomogeneous rock properties.

At relevant reservoir conditions, the carbonate surface is positively charged. The carboxylic material in crude oil, as determined by the acid number, AN (mg KOH/g), is the most important wetting parameter for carbonate CBR systems. Crude oil components containing the carboxyl group, $-COOH$, are mostly found in the heavy end fraction of crude oils, i.e., in the resin and asphaltene fraction (Speight, 1999). The bond between the negatively charged carboxylic group, $-COO^-$, and the positively charged sites on carbonate surface is very strong, and the large molecules will cover the carbonate surface. The impact of the AN of the crude oil on the wetting properties of chalk is illustrated by Figure 13.1 showing spontaneous imbibitions of water into chalk cores saturated with oils of different AN. The imbibitions rate and oil recovery decreased dramatically as the AN of the oil increased (Standnes and Austad,

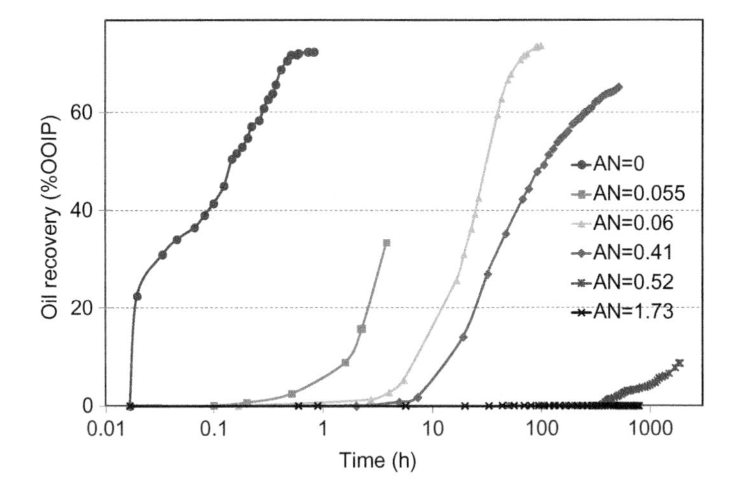

FIGURE 13.1 Spontaneous imbibition into chalk cores saturated with different oils (Standnes and Austad, 2000a).

2000a). The chemical properties of the carboxylic material in the crude oil also affect the wetting properties (Fathi et al., 2011).

The basic components in the crude oil, quantified by the base number, BN (mg KOH/g), play a minor role as wetting parameter. It was, however, observed that the increase in a basic material improved the water wetness of chalk containing oil with a given AN (Puntervold et al., 2007). It was suggested that an acid–base complex could be formed in the crude oil, which to some extent, made the acidic material less active toward the carbonate surface.

Rao (1996) has pointed out that high-temperature carbonate reservoirs appeared to be more water-wet compared to low temperature reservoirs. It is further known that the AN and the reservoir temperature are not independent variables. The AN of the crude oil appears to decrease as the reservoir temperature increases due to increased decarboxylation of the acidic material at high temperatures. The decarboxylation process is even catalyzed by solid $CaCO_3$ (Shimoyama and Johns, 1972).

Pressure may have an effect on wetting properties mostly due to the impact on the solubility of asphaltenic material in the crude. As the pressure decreases toward the bubble point of the oil, the solubility of asphaltenes in the crude oil decreases and they may precipitate and adsorb onto the rock (Buckley, 1995).

The composition of the formation water can affect the wetting properties as well. As discussed later, sulfate is the most active ion regarding wetting properties in carbonates. Due to the high concentration of Ca^{2+} in the formation brine, and especially in combination with high temperatures, the amount of SO_4^{2-} present in the formation water is usually very low due to precipitation of anhydrite, $CaSO_4$(s). The presence of sulfate in the formation water will increase the water wetness of the system (Shariatpanahi et al., 2011).

A new wettability test based on the chromatographic separation between a nonadsorbing tracer, SCN^-, and sulfate, SO_4^{2-}, was developed and found to be very useful to verify changes in the water-wet fraction after exposing the carbonate rock to different fluid systems (Strand et al., 2006b). When plotting the relative concentration of SCN^- and SO_4^{2-} versus the injected pore volume, the area between the effluent curves is directly proportional to the water-wet area of the core, because the chromatographic separation only takes place at the water-wet area.

13.1.2 Wetting in Sandstones

In contrast to carbonates, a sandstone is composed of many different minerals. Minerals of the silica type are negatively charged at the relevant pH range of the formation water. It is, however, the clay minerals that are most strongly adsorbed by polar components from the crude oil. Clays are chemically unique due to the presence of permanent negative charges, and the clays therefore act as cation exchangers. The relative affinity of cations toward the clay surface is regarded to be:

$$Li^+ < Na^+ < K^+ < Mg^{2+} < Ca^{2+} < H^+$$

It should be noticed, that the proton, H^+, is the most reactive cation toward the clay. Even though the concentration of H^+ is very low in the pH range between 6 and 8, it will still play an important role in cation exchange reactions at low salinities. In competition with cations, both basic and acidic material can adsorb onto the clay surface and make the clay preferential oil-wet. The adsorption of basic and acidic material onto the clay is very sensitive to the pH, and it can change dramatically within the pH range: $5 < pH < 8$ (Burgos et al., 2002; Madsen and Lind, 1998; RezaeiDoust et al., 2011). The adsorption of both basic and acidic material from the crude oil appeared to increase as the pH decreased to about 5. Clays are normally not uniformly distributed in an oil reservoir, and therefore, certain areas can be less water-wet than others. These areas may be bypassed in a water flood process, and both microscopic and macroscopic sweep efficiencies are decreased.

13.1.3 Smart Water Flooding

The established initial wetting properties of an oil reservoir are related to the present CBR system. The filling history of an oil reservoir may also play an important role regarding the wetting properties, especially in carbonates. There are examples of limestone reservoirs that are completely oil-wet even though the crude oil present in the reservoir has a very low AN. Experimental work showed that it was impossible to obtain wettability modification in this type of rock material using different techniques (Fathi et al., 2010b). Obviously, the reservoir must have been exposed to organic material before the present oil invaded the reservoir, because the oil and formation brine in contact with a new water-wet carbonate did not change the wetting properties significantly.

For both carbonates and sandstone reservoirs, the oil recovery by injecting original formation water is usually different from the recovery obtained when injecting a water with a different composition from formation water. The oil recovery can both increase and decrease compared to formation water, which is in equilibrium with the CBR system. By using a "Smart Water," the oil recovery can be increased significantly from both carbonates and sandstones. This is illustrated in Figures 13.2 and 13.3.

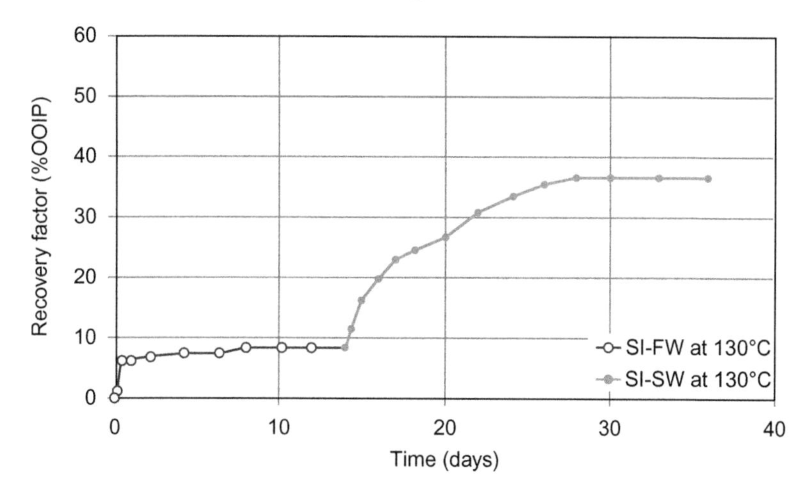

FIGURE 13.2 Spontaneous imbibition of formation water, FW, and seawater, SW, into a reservoir limestone core at 130°C (Ravari et al., 2010).

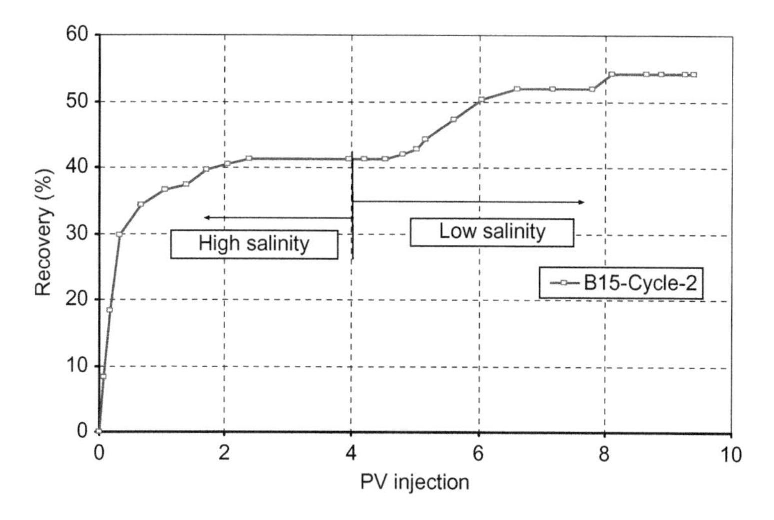

FIGURE 13.3 Low salinity effects in sandstone (Austad et al., 2010).

What is a "Smart Water," and what is the main mechanism for the increase in oil recovery? "Smart Water" can be made by adjusting/optimizing the ion composition of the injected fluid in such a way that the change in the equilibrium of the initial CBR system will modify the initial wetting conditions. Therefore, the oil is more easily displaced from the porous network. As illustrated by Figures 13.2 and 13.3, injection of "Smart Water" can be characterized as a tertiary oil recovery method since extra oil was recovered after performing a secondary water flood with formation water. The physical principle for EOR by "Smart Water" is a change in wetting properties of the CBR system, which has a positive effect on the capillary pressure and relative permeability of oil and water regarding oil recovery. The technique is cheap, environmentally friendly, no expensive chemicals are added, and no injection problems. From an economical point of view, the smartest water should be injected from the start of the water flooding process.

The validation of "Smart Water" as an EOR fluid has been verified both in the laboratory and in the field by several research groups and oil companies during the last 20 years. Extensive research has been performed in order to understand the chemical/physical mechanism for the wettability alteration process taking place at the rock surface, and the mechanism is still under debate in the published literature. Our EOR group at the University of Stavanger has worked with wettability modification in carbonates for nearly 20 years and in sandstones for about 5 years. In the coming sections, I will present a summary of our chemical understanding of the mechanism for wettability modification in carbonates and sandstones and relate it to field experience. In order to be able to evaluate the potential of using "Smart Water," the chemical mechanism must be understood in detail because "Smart Water" does not function in all types of oil reservoirs.

The objectives of the present work are shortly summarized as follows.

- Present documentation of our new chemical understanding of the mechanism for wettability alteration by "Smart Water" resulting in EOR from carbonate and sandstone oil reservoirs.
- The results from field experience will be discussed in relation to the new understanding of the surface chemistry.

13.2 "SMART WATER" IN CARBONATES

13.2.1 Introduction

Cationic surfactants of the type alkyl trimethyl ammonium, R– $N(CH_3)_3^+$, dissolved in seawater (SW) are able to improve water wetness in chalk, limestone, and dolomite (Austad et al., 2008b; Standnes and Austad, 2000b; Standnes et al., 2002). The mechanism for the wettability alteration was suggested to be an interaction between the cationic surfactant monomers and adsorbed negatively charged carboxylic material, forming a cationic–anionic complex, which is released from the surface (Standnes and Austad, 2000b). In a spontaneous imbibition process, the capillary forces are increased due to improved water wetness, but the interfacial tension (IFT) between the oil and surfactant solution is decreased to 0.3−1.0 mN/m. The imbibition rate is therefore low, especially in low permeable rock, as pointed out by Stoll et al. (2007). Laboratory experiments have shown that there is a change in the flow regime from capillary forced imbibition at the start to an imbibition process dominated by gravity forces at a later stage, i.e., a surfactant enhanced gravity drainage (SEGD) process (Masalmeh and Oedai, 2009). The simulation work conducted by Sheng (2012) also confirmed their experimental observation. In a very recent paper, it was verified that the SEGD process could be an interesting EOR process in high permeable, >300 mD, fractured limestone reservoirs due to the increased imbibition rate (Ravari et al., 2011).

It was, however, observed that the imbibition rate of cationic surfactant dissolved in SW was catalyzed by sulfate present in SW (Zhang et al., 2006). We asked the following question: *"Why is injection of SW into the Ekofisk chalk such a tremendous success regarding oil recovery, which is now estimated to approach 55%?"* Ekofisk is mixed wet, highly fractured, and has low matrix permeability, about 2 mD, and the reservoir temperature is high, 130°C. Can SW act as a wettability modifier and increase the water wetness of the matrix blocks? A lot of parametric studies on oil recovery by spontaneous imbibition were performed using outcrop chalk and relevant composition of formation water and SW (Table 13.1).

13.2.2 Reactive Potential Determining Ions

SW contains reactive ions, Ca^{2+}, Mg^{2+}, and SO_4^{2-}, toward the chalk surface, which can act as potential determining ions since they can change the surface charge of $CaCO_3$ (Pierre et al., 1990; Zhang and Austad, 2006). The impact on oil recovery for each of these ions in SW was varied and tested separately.

Effect of SO_4^{2-}: As the concentration of SO_4^{2-} in the imbibing SW varied from 0 to 4 times the concentration in ordinary SW, the oil recovery

TABLE 13.1 Composition of Ekofisk FW and SW as mole/L

Composition	Ekofisk	SW
Na^+	0.685	0.450
K^+	0	0.010
Mg^{2+}	**0.025**	**0.045**
Ca^{2+}	**0.231**	**0.013**
Cl^-	1.197	0.528
HCO_3^-	0	0.002
SO_4^{2-}	**0**	**0.024**

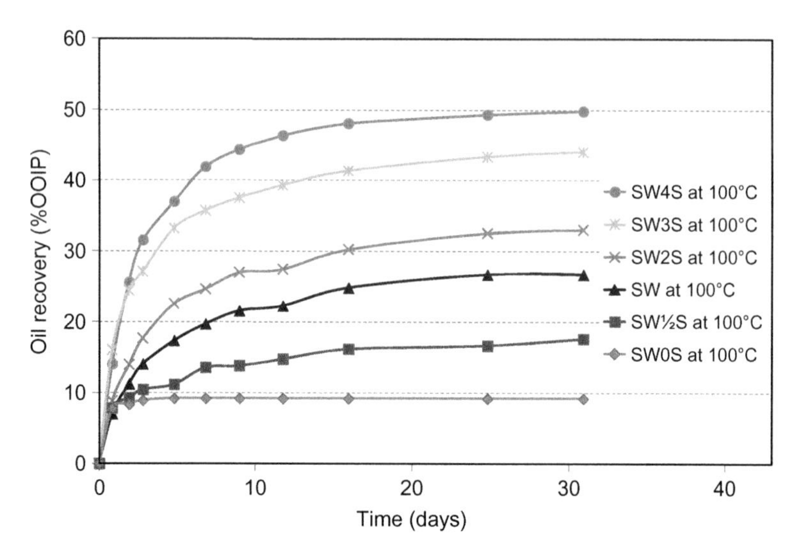

FIGURE 13.4 Oil recovery by spontaneous imbibition of sulfate-modified SW into chalk cores at 100°C (Zhang and Austad, 2006).

increased from less than 10% to about 50% of original oil in place (OOIP) (Figure 13.4). Thus, there was a tremendous effect on oil recovery by just changing the concentration of one single ion.

Effect of Ca^{2+}: Similarly, the concentration of Ca^{2+} in SW was varied from 0 to 4 times the concentration in SW, the oil recovery increased from 28% to 60% after 30 days of imbibition (Figure 13.5). In this case, the concentration of sulfate remained constant and similar to the SW concentration.

The affinity of SO_4^{2-} and Ca^{2+} toward the chalk surface was tested chromatographically by flooding SW spiked with a tracer, SCN^-, through a chalk core at different temperatures. The effluent concentrations of each of the components were recorded and plotted versus the pore volume (PV) injected. As noticed, the adsorption of SO_4^{2-} onto the rock increased as the temperature increased (Figure 13.6A) and coadsorption of Ca^{2+} increased as the concentration of Ca^{2+} in the effluent decreased (Figure 13.6B). Thus, increased adsorption of SO_4^{2-} onto the chalk surface will reduce the positive charge, which causes increase in the affinity of Ca^{2+} due to less electrostatic interaction (Strand et al., 2006a).

The symbiotic effect of Ca^{2+} and Mg^{2+} at the chalk surface was studied by flooding equal concentrations of Ca^{2+}, Mg^{2+}, and SCN^- dissolved in a NaCl solution through chalk cores at 20°C and 130°C. At low temperature, the affinity of Ca^{2+} toward the chalk surface was higher than Mg^{2+} (Figure 13.7A). At 130°C, the concentration of Ca^{2+} in the effluent was much higher than the concentration injected, and Mg^{2+} in the effluent was very strongly retarded, and it did not reach the initial concentration (Figure 13.7B) (Zhang et al., 2007a). The only explanation for this is that Mg^{2+} in SW is able

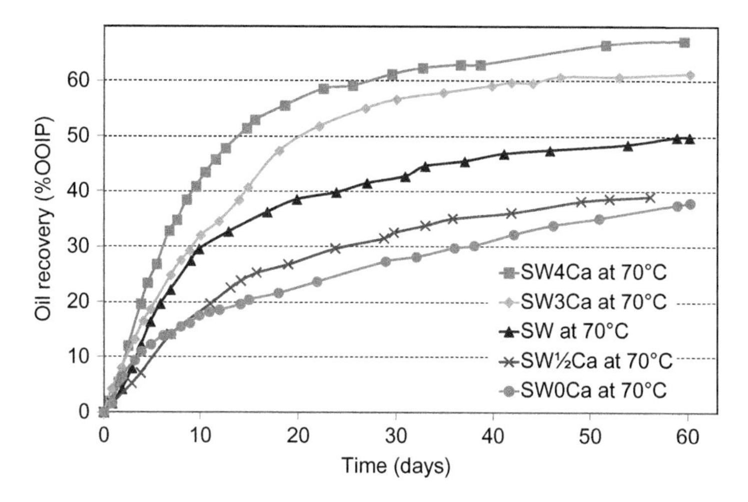

FIGURE 13.5 Oil recovery by spontaneous imbibition of Ca^{2+} modified SW into chalk cores at 70°C (Zhang et al., 2006).

to displace Ca^{2+} from the rock, in a 1:1 reaction, which has been verified experimentally. Whether it is a pure surface substitution or it is formation of $MgCO_3$(s) is not clear. The process is illustrated by the following equilibrium:

$$CaCO_3(s) + Mg^{2+} \leftrightarrow MgCO_3(s) + Ca^{2+}$$

Thus, the reactivity of Mg^{2+} toward the carbonate surface increases dramatically as the temperature is increased beyond 70°C, which appeared as the temperature threshold for observing substitution. The small Mg^{2+} ion has a strong hydration energy, which make it less reactive at low temperature.

In order to illustrate the effect of the potential determining ions and the temperature on oil recovery from chalk cores of low water wetness, four cores were prepared using a NaCl solution as the formation brine, and the following tests were performed as shown in Figure 13.8:

- First, the cores were imbibed with NaCl brine containing different amounts of SO_4^{2-} (0, 1, 2, and 4 times the concentration of SW) at 70°C. The oil recovery was a little less than 10% including thermal expansion.
- The temperature was increased to 100°C, and negligible extra oil was produced.
- When adding either Mg^{2+} or Ca^{2+} similar to the concentration found in SW to the different imbibing fluids, oil recovery increased immediately.
  - The imbibing fluid containing 4 times SO_4^{2-} compared to SW was spiked with Mg^{2+} at 100°C, and the oil recovery increased to about 42% of OOIP (curve with solid circle points). By increasing the temperature to 130°C, the oil recovery exceeded 60% as is usually observed for water-wet cores.

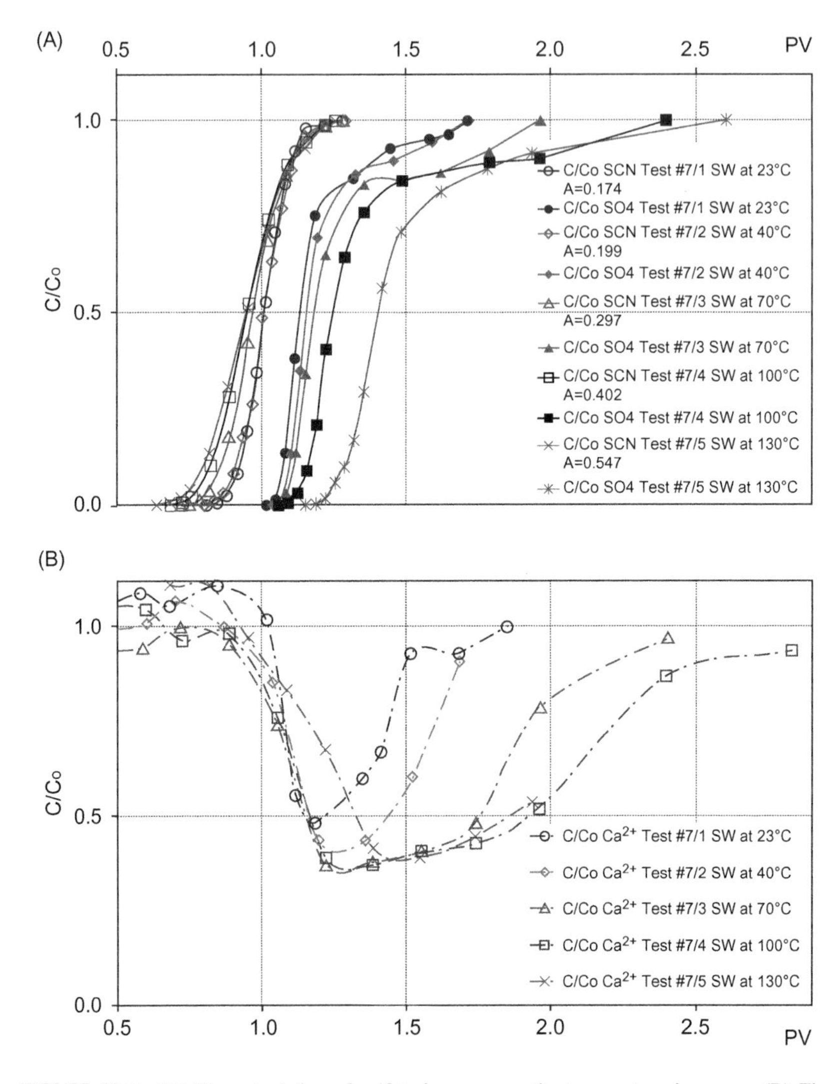

FIGURE 13.6 (A) The retardation of sulfate increases as the temperature increases. (B) The decrease in Ca^{2+} concentration increases as the temperature increases (Strand et al., 2006a).

- A similar test for the core imbibed with the fluid containing SW concentration of SO_4^{2-} resulted in somewhat lower recovery, 30% and 50% of OOIP at 100°C and 130°C, respectively (curve with diamonds).
- The similar test for the core imbibed with the fluid free from SO_4^{2-} only resulted in a relative small increase in oil recovery both at 100°C and 130°C (curve with solid squares).
- The core imbibed with the fluid containing 2 times SO_4^{2-} concentration found in SW was spiked with Ca^{2+}. The oil recovery increased to

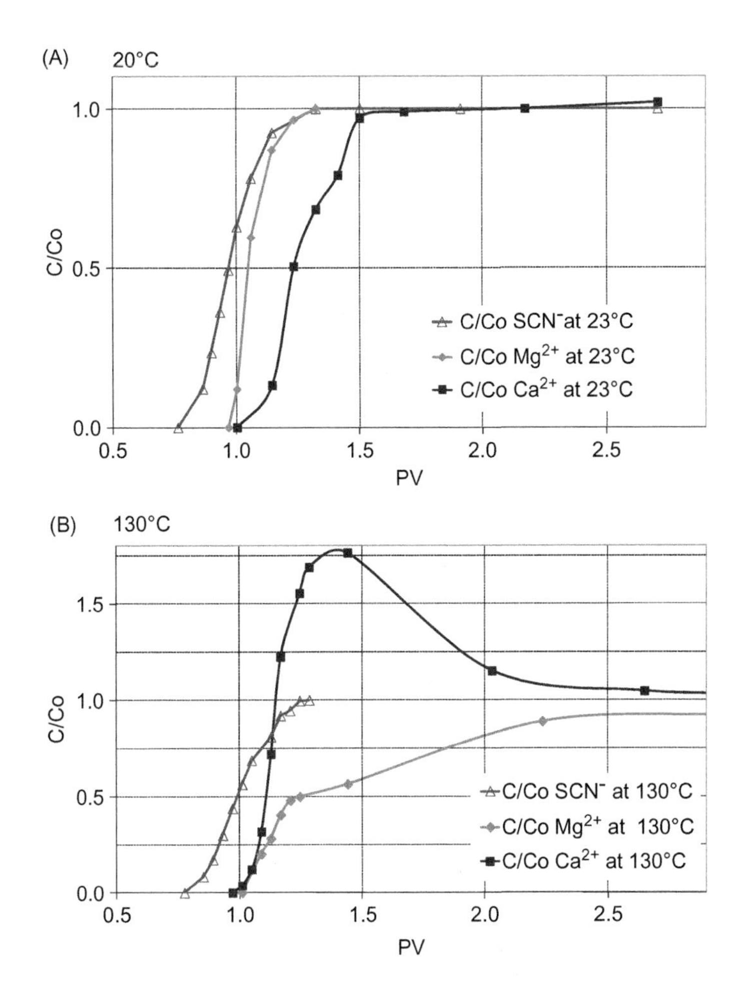

FIGURE 13.7 (A) Ca^{2+} is more strongly adsorbed onto chalk at low temperature. (B) Mg^{2+} is strongly retarded at high temperature, and Ca^{2+} is substituted by Mg^{2+} at the chalk surface (Zhang et al., 2007a).

25% of OOIP at 100°C, but only a very small increase in oil recovery was observed at 130°C. Without Mg^{2+} present, the solubility of anhydrite, $CaSO_4$(s), is drastically decreased, and at 100°C $CaSO_4$(s) is precipitated, and the imbibing solution is further stripped for Ca^{2+} as the temperature is increased to 130°C. Thus, precipitation of $CaSO_4$(s) will decrease the concentration of active ions and block the porous system. Note that the solubility of anhydrite decreases as the temperature increases.

Thus, a "Smart Water" for carbonates must contain potential determining ions, Ca^{2+} and/or Mg^{2+}, SO_4^{2-}, and the symbiotic interaction between the

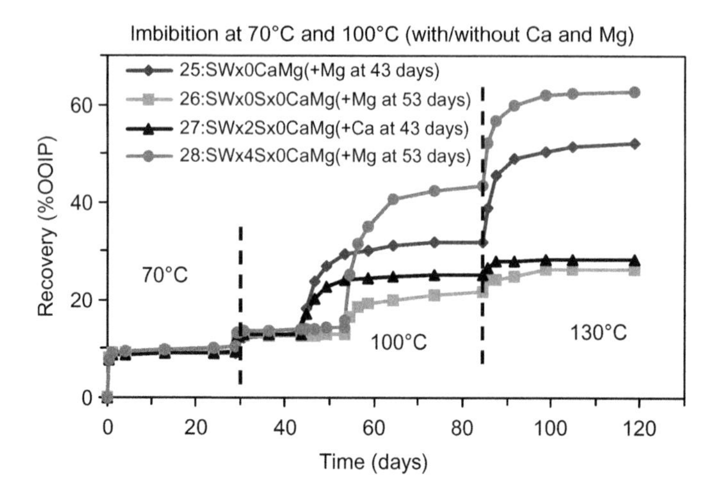

FIGURE 13.8 The effects of potential determining ions and temperature on spontaneous imbibition into preferential oil-wet chalk (Zhang et al., 2007a).

ions and the carbonate surface is very sensitive to the temperature, which should be well above 70°C.

13.2.3 Suggested Mechanism for Wettability Modification

Putting together all the parametric studies performed on chalk samples a chemical mechanism for the wettability alteration process using SW is suggested as schematically illustrated in Figure 13.9 (Zhang et al., 2007a). Initially, the rock was positively charged due to a pH < 9 and high concentration of Ca^{2+} and possibly some Mg^{2+} in the formation water. The concentration of negatively charged potential determining ions like CO_3^{2-} and SO_4^{2-} in the formation water is negligible. As SW is injected into the fractured carbonate reservoir, SO_4^{2-} will adsorb onto the positively charged surface and lower the positive charge. Due to less electrostatic repulsion, the concentration of Ca^{2+} close to the surface is increased, and Ca^{2+} can bind to the negatively charged carboxylic group and release it from the surface. Both the concentration of SO_4^{2-} and Ca^{2+} at the carbonate surface increases as the temperature is increased. As Mg^{2+} is even able to displace Ca^{2+} from the carbonate rock at high temperatures, it should also be able to displace the Ca^{2+}–carboxylate complex from the surface.

13.2.4 Optimization of Injected Water

A relevant question that may be asked is: *Is SW an optimum Smart Water?* As illustrated by Figure 13.9, the reservoir temperature and sulfate are very important parameters for a successful wettability alteration. *Is it possible to*

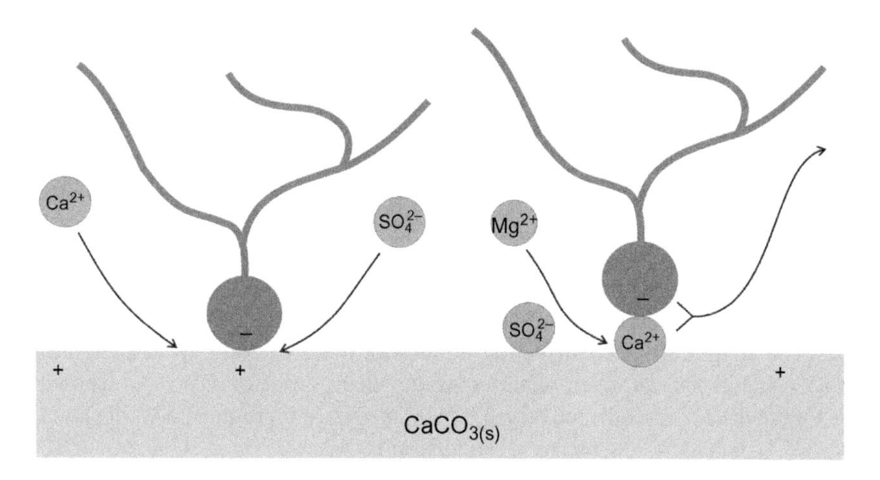

FIGURE 13.9 Illustration of the suggested chemical mechanism for wettability alteration of carbonate by SW (Zhang et al., 2007a).

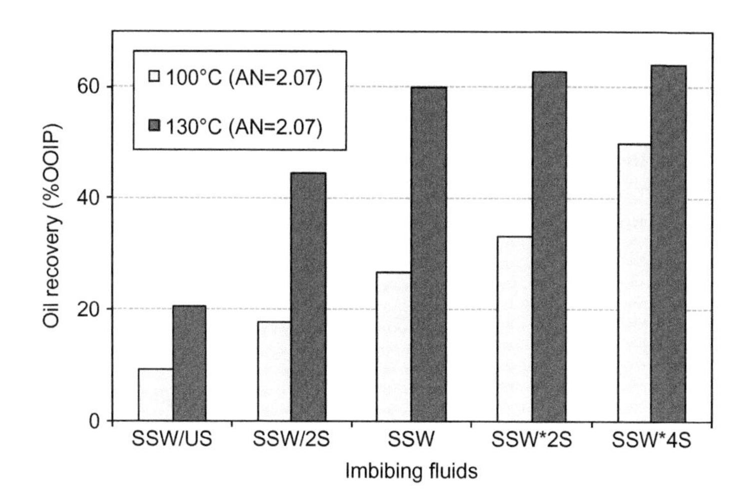

FIGURE 13.10 Maximum oil recovery from chalk cores at 100°C and 130°C when different imbibing fluids were used (SW with varying SO_4^{2-} concentration; Oil: AN = 2.07 mg KOH/g) (Zhang and Austad, 2006).

compensate for a decrease in reservoir temperature by increasing the concentration of sulfate in the imbibing fluid? As illustrated by Figure 13.10, the oil recovery at 130°C is not increased very much by increasing the sulfate concentration beyond the concentration present in SW (Zhang and Austad, 2006). At 100°C, the recovery is nearly doubled when SW was spiked with sulfate corresponding to 4 times SW concentration. Thus, there is a potential for increasing the oil recovery by spiking SW with sulfate in the temperature range to enhance the wettability alteration process. It is not, however, advisable to spike the SW with sulfate at high temperatures because of precipitation of anhydrite, $CaSO_4$(s).

All charged surfaces in contact with a brine will have an excess of ions close to the surface, which is usually called the double layer. If the double layer consists of a lot of ions not active in the wettability alteration process like NaCl, the access of the active ions, Ca^{2+}, Mg^{2+}, and SO_4^{2-} to the surface is partly prevented. The concentration of NaCl in SW is much larger than the concentration of Ca^{2+}, Mg^{2+}, and SO_4^{2-} (Table 13.1). Thus, in line with the suggested mechanism for wettability modification, SW depleted in NaCl should be an even smarter water than ordinary SW. As shown in Figure 13.11, the oil recovery by spontaneous imbibition of SW depleted in NaCl increased the recovery significantly from 37% to 47% of OOIP compared to ordinary SW. An increase in NaCl concentration of SW decreased the oil recovery (Fathi et al., 2010a).

At temperatures below 100°C, SW depleted in NaCl, but spiked with sulfate appeared to be the smartest water regarding oil recovery, as shown in Figure 13.11. Compared to ordinary SW, the oil recovery increased dramatically from 37% to 62% of OOIP by spiking the NaCl depleted SW with 4 times the sulfate concentration in ordinary SW (Fathi et al., 2010a).

Thus, by knowing the chemical mechanism for wettability modification, it is possible to optimize the ion composition in the injected water to maximize the oil recovery. From an economical point of view, it is important to inject the optimized fluid from the start of the water flood.

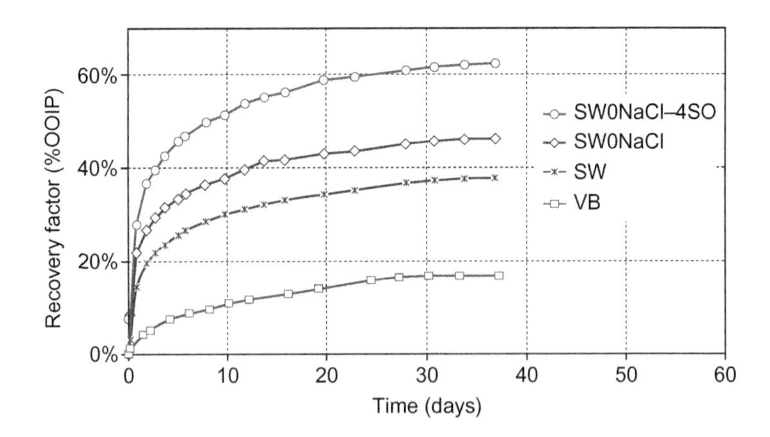

FIGURE 13.11 Spontaneous imbibition into chalk cores at 90°C using different imbibing fluids: formation water, VB, seawater, SW, seawater depleted in NaCl, SW0NaCl, and seawater depleted in NaCl and spiked with 4 times SO_4^{2-} compared to SW, SW0NaCl–4SO_4. Oil with AN = 0.50 mg KOH/g, S_{wi} = 0.10 (Fathi et al., 2010a).

13.2.5 Viscous Flood Versus Spontaneous Imbibitions

In a preferentially oil-wet naturally fractured carbonate reservoir with a great difference in matrix and fracture permeability, the spontaneous imbibition of the injected water into the matrix blocks will be the main mechanism for oil recovery. In that case, positive capillary forces must be created by wettability modification. In a viscous flood, the displacement of oil is believed to be at an optimum at slightly water-wet conditions (Jadhunandan and Morrow, 1995). Due to low capillary forces, the microscopic sweep efficiency can be improved significantly if a wettability alteration is taking place during the flood. The effect of oil recovery by wettability alteration using formation water and SW is illustrated in Figure 13.12. Two cores were first imbibed with formation water (FW). One of the cores was flooded with FW and thereafter with seawater (SW). The other core was imbibed with SW, and then flooded with SW. The oil recovery was plotted in the same diagram versus time. Improved oil recovery was obtained both in a spontaneous imbibition process and also in the viscous flood (Strand et al., 2008).

13.2.6 Environmental Effects

If produced water is not reinjected into the reservoir, it will be an environmental problem because of the content of aromatic carcinogenic material. The produced water from an offshore carbonate reservoir flooded with SW will contain low concentration of sulfate, even after breakthrough of SW,

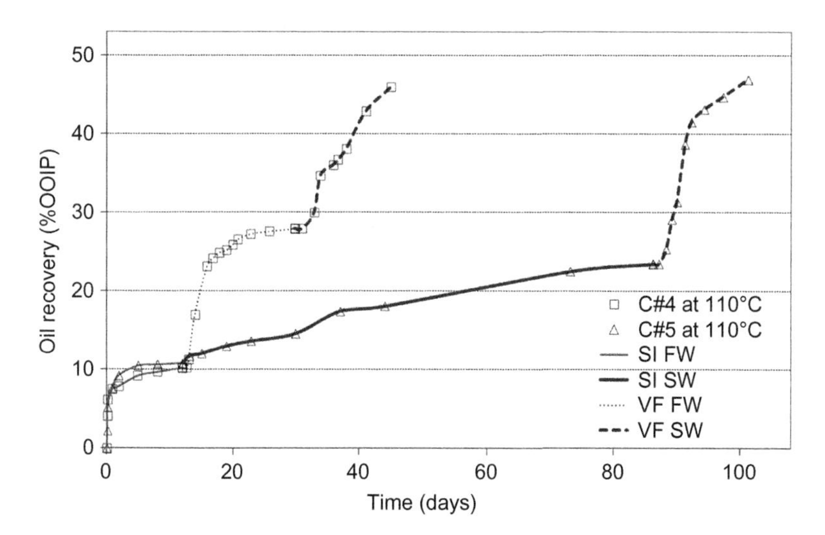

FIGURE 13.12 Improved oil recovery by SW both by spontaneous imbibitions and viscous flood. For comparison, the oil recovery was plotted in the same diagram as a function of time (Strand et al., 2008).

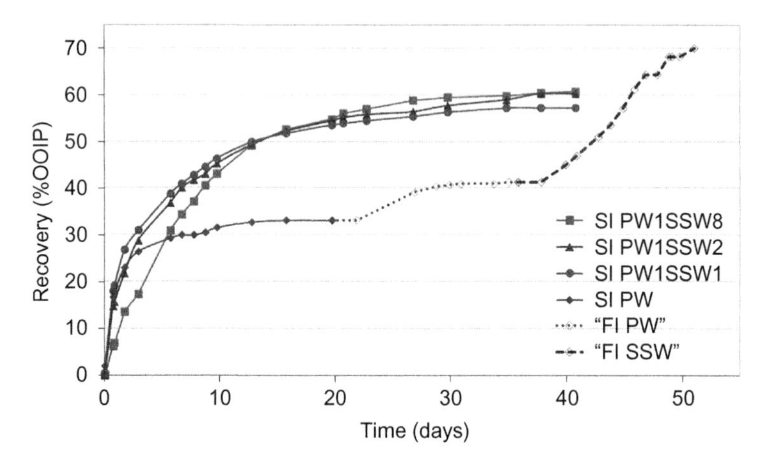

FIGURE 13.13 Spontaneous imbibition of mixtures of produced water, PW, and SW in the volume ratios 1:1, 1:2, and 1:8, respectively. The results are compared using PW: (1) in an imbibition process and (2) in a displacement process. Finally, the core was flooded with SW to verify the EOR potential (Puntervold et al., 2009).

due to adsorption/precipitation of sulfate in the reservoir (Puntervold and Austad, 2008; Puntervold et al., 2009). Thus, reinjection of produced water will not be a smart water causing wettability alteration and improved oil recovery. If, however, the produced water is mixed with SW, the good quality of the injected fluid can be maintained as shown in Figure 13.13.

13.2.7 Smart Water in Limestone

The surface area and reactivity toward potential determining ions varies for different carbonates, i.e., chalk, limestone (reservoir and outcrop), and dolomite. In the previous sections, it was verified that chalk was reactive toward the active ions in SW, and wettability alteration can take place. Recently, it was observed that reservoir limestone cores acted in the same way, although the reactivity toward Ca^{2+}, Mg^{2+}, and SO_4^{2-} varied. The water-wet surface area of a reservoir limestone core which was precleaned with toluene and methanol was increased by about 30% after being flooded with SW at 130°C (Austad et al., 2008b). A low permeable reservoir limestone core was spontaneously imbibed with formation water and then with SW in a secondary and tertiary process. The oil recovery increased drastically as the imbibing fluid was switched from FW to SW, from 8% to 37% of OOIP (Figure 13.2) (Austad et al., 2008b). As for chalk, SW depleted in NaCl appeared to be an even better wettability modifier in limestone than ordinary SW (Shariatpanahi et al., 2012). Outcrop limestone acted completely different from reservoir limestone in wettability alteration studies. Two outcrop limestones were tested, one from France submitted by TOTAL and the Edwards limestone from United States of America. Both limestones acted water-wet, and the wettability was changed in the presence of crude oil. The rock surface appeared nonreactive toward the potential determining ions, and improvements in oil recovery by wettability modification using SW could not be obtained, even at high temperatures (Ravari et al., 2010). As a learning lesson: *Care must be taken in using outcrop limestone as a model for reservoir rock in parametric studies on oil recovery by wettability modification.*

13.2.8 Condition for Low Salinity EOR Effects in Limestone

Until very recently, EOR by low saline water flooding was a phenomenon, which was allocated to sandstones, and not observed for carbonates. Yousef and coworkers (2010) increased the oil recovery from composite limestone cores by successively flooding the cores with Gulf SW and diluted Gulf SW: 2, 10, and 20 times. A significant increase in oil recovery was observed as the injected SW was diluted. On the other hand, no low salinity EOR effect was observed when outcrop chalk cores were imbibed or flooded with diluted SW. In fact, the oil recovery was decreased drastically as SW was diluted due to the decrease in active ions (Fathi et al., 2010a).

A very recent study on reservoir limestone cores containing small amounts of anhydrite, $CaSO_4(s)$, as part of the matrix, showed a low salinity EOR effect when the flooding fluid was switched from FW to 100 times diluted FW (Figure 13.14A) (Austad et al., 2011). The oil recovery increased by 4% of OOIP. The high saline formation water did not contain any sulfate. The concentration of dissolved sulfate in the effluent was quantified and plotted versus the PV injected (Figure 13.14B). In this case, $SO_4^{2-}(aq)$, which acts as a catalytic chemical for the wettability alteration was created *in situ* by dissolution of anhydrite.

In order to discuss the chemical mechanism for the observed low salinity effect in carbonates, the impact of temperature and brine composition on the following equilibrium must be understood:

$$CaSO_4(s) \leftrightarrow Ca^{2+}(aq) + SO_4{}^{2-}(aq) \leftrightarrow Ca^{2+}(ad) + SO_4{}^{2-}(ad)$$

$Ca^{2+}(aq)$ and $SO_4^{2-}(aq)$ are ions dissolved in the pore water, and $Ca^{2+}(ad)$ and $SO_4^{2-}(ad)$ are ions adsorbed onto the carbonate surface. According to a previously reported study, where the impact of sulfate on the initial wetting conditions was studied, the concentration of $SO_4^{2-}(aq)$ appeared to be the key factor determining the wetting properties (Shariatpanahi et al., 2011). Dissolution of anhydrite, $CaSO_4(s)$, which is the source for $SO_4^{2-}(aq)$, depends on the salinity/composition of the brine and the temperature in the following ways:

- The solubility increases as the concentration of Ca^{2+} in FW decreases (common ion effect).
- The solubility decreases as the concentration of NaCl decreases.

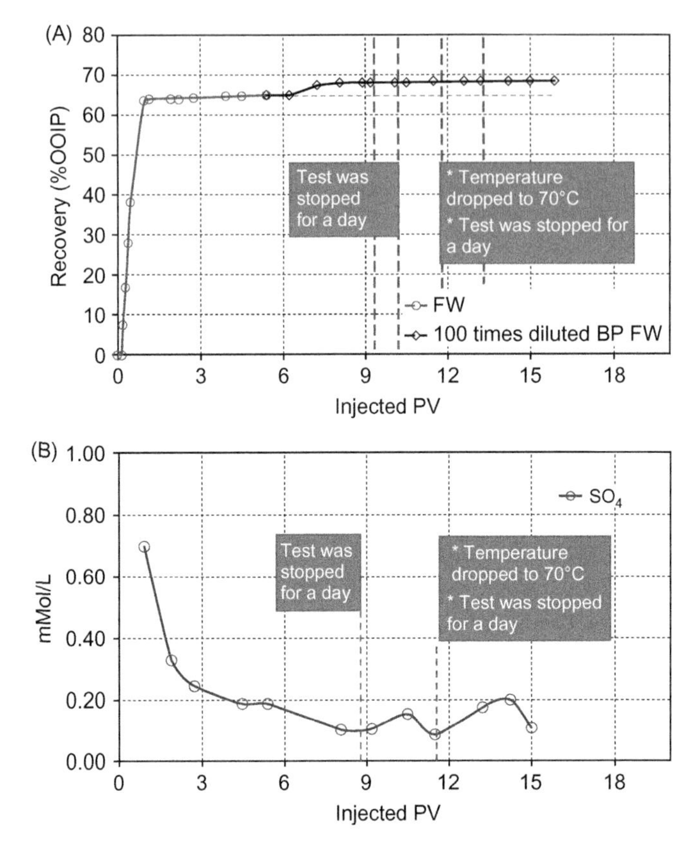

FIGURE 13.14 (A) The core 5B was flooded by FW and 100× diluted FW at 110°C and 70°C. Crude oil AN≈0.70 mg KOH/g. (B) Concentration of SO_4^{2-} in the effluent versus injected PV (Austad et al., 2011).

- The solubility normally decreases as the temperature increases.
- The concentration of SO_4^{2-}(aq) may also decrease as the temperature is increased due to increased adsorption onto the carbonate surface, i.e., the concentration of SO_4^{2-}(ad) increases.

The efficiency of the wettability alteration process also depends on the temperature and the concentration of nonactive salt, NaCl, in the following ways:

- Imbibition rate and ultimate oil recovery increase as the temperature increases.
- Imbibition rate and ultimate oil recovery increase as the concentration of nonactive salt, NaCl, in the imbibing brine decreases.

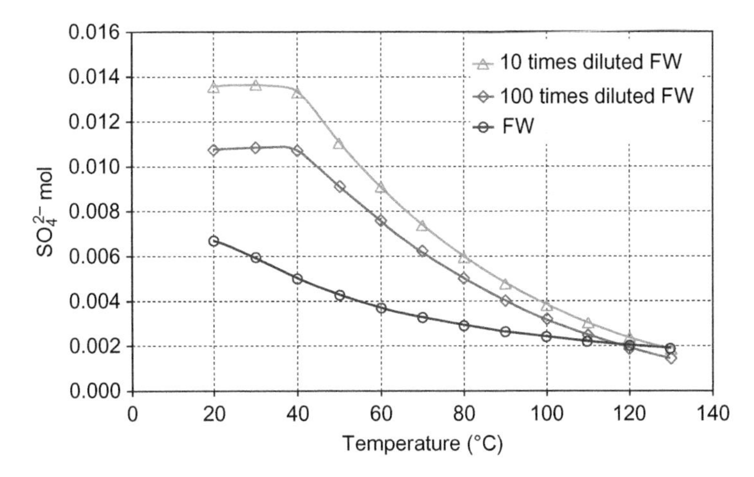

FIGURE 13.15 Dissolution of $CaSO_4$ when exposed to FW, 10 and 100 times diluted FW at different temperatures as modeled by the OLI software (pressure: 10 bar) (Austad et al., 2011).

- Imbibition rate and ultimate oil recovery increase as the concentration of SO_4^{2-}(aq) increases.

Thus, both the effect of temperature and concentration of NaCl is in conflict, that is, the concentration of SO_4^{2-}(aq) decreases as the temperature increases, but the surface reactivity leading to the wettability alteration increases as the temperature increases. Similarly, the concentration of SO_4^{2-}(aq) decreases as the amount of NaCl decreases, but the surface reactivity promoting wettability alteration increases. Therefore, for a given system, there appears to be an optimum temperature window for observing maximum low salinity effect, probably between 90°C and 110°C. It is a combined effect of the presence of SO_4^{2-} and decrease in NaCl concentration, which is important for the low salinity EOR effect in carbonates. This is illustrated by the calculated solubility of anhydrite, $CaSO_4$(s) in FW and diluted FW, 10 and 100 times as shown in Figure 13.15.

We have experienced that much greater low salinity EOR effects from limestone cores can be observed if the formation contained a higher amount of anhydrite, up to 20% of OOIP. Provided that dolomite will respond in a similar way as calcite toward SW as wettability modifier, it is reasonable to believe that the low salinity effect observed by Pu et al. (2008) in a sandstone with low content of clay, but significant amount of dolomite and anhydrite, was linked to a wettability alteration of dolomite. An increase in sulfate concentration and a decrease in the salinity will activate the wettability alteration process. It was also concluded by the authors that the interstitial dolomite crystals probably played a role in the low salinity recovery mechanism.

13.3 "SMART WATER" IN SANDSTONES

13.3.1 Introduction

A great number of laboratory tests by Morrow and coworkers (Tang and Morrow, 1999a,b; Zhang and Morrow, 2006; Zhang et al., 2007b) and also by researchers at British Petroleum (BP) (Lager et al., 2007; Webb et al., 2005a,b) have confirmed that EOR can be obtained when performing a tertiary low salinity water flood, with salinity in the range of 1000–2000 ppm. Thus, a low salinity water may act as a smart EOR fluid in a sandstone oil reservoir.

13.3.2 Conditions for Low Salinity Effects

The listed conditions for observing low salinity effects are mostly related to the systematic experimental work by Tang and Morrow (1999a), but some points are also included from researchers at BP (Lager et al., 2007, 2008a).

- *Porous medium*: Sandstones containing clay minerals.
- *Oil*: Must contain polar components (i.e., acids and/or bases).
- *Formation water*: Must be present, and must contain divalent cations, i.e., Ca^{2+} and Mg^{2+}.
- *Low salinity injection fluid:* The salinity is usually between 1000 and 2000 ppm, but effects have been observed up to 5000 ppm. It is appeared to be sensitive to ionic composition (Ca^{2+} vs. Na^{+}).
- *Produced water:* For a nonbuffered system, the pH of the effluent water usually increases about pH 1–3 units, when switching from high salinity to low salinity fluid. It has not been verified that an increase in pH is needed to observe low salinity effects. In some cases, production of fines has been detected, but low salinity effects have also been observed without visible production of fines.
- *Permeability:* Both an increase and a decrease in differential pressure over the core has been observed by switching from high-to-low salinity fluid, which may indicate a change in permeability.
- *Temperature*: There appears to be no temperature limitations to where low salinity effects can be observed. Most of the reported studies have, however, been performed at temperatures below 100°C.

13.3.3 Suggested Low Salinity Mechanisms

Due to mineralogical properties, it is obvious that a chemical mechanistic study of the low salinity effect in sandstones is much more complicated compared to wettability alteration studies in carbonates by "Smart Water." The low salinity effect is probably a result of different mechanisms/steps acting together. Even though it is generally accepted that the low salinity effects are caused by wettability alteration, some physical mechanisms have also been proposed; such as pore blockage by migration of fines and fluid flow due to osmotic pressure caused by salinity gradients.

Some of the most relevant previously proposed mechanisms are:

- Migration of fines (Tang and Morrow, 1999a),
- Multicomponent ion exchange (MIE) (Lager et al., 2008a),
- Extension of the electrical double layer (Ligthelm et al., 2009).

None of the suggested mechanisms have got a general acceptance in the scientific literature so far. There were always some observed contradictions among the experimental facts.

13.3.4 Improved Chemical Understanding of the Mechanism

Recently, the desorption of organic material from the clay surface by a local increase in pH at the clay–water interface caused by desorption of surface active inorganic cations as the low saline fluid invaded the porous medium was suggested by Austad et al. (2010) to play an important role in the low salinity EOR process. Both acidic and basic crude oil components are released from the surface as the pH is increased from 5–6 to about 8–9. In laboratory experiments, the increase in pH is usually verified, but due to buffering effects in field situations caused by the presence of CO_2 and/or H_2S, an increase in pH is seldom observed in the field. The suggested chemical mechanism for EOR by low salinity water flood was based on three experimental observations:

1. Clay must be present in the sandstone (Tang and Morrow, 1999a).
2. Polar components (acidic and/or basic material) must be present in the crude oil (Tang and Morrow, 1999a).
3. The formation water must contain active ions like Ca^{2+} (Lager et al., 2007).

The suggested mechanism is schematically illustrated in Figure 13.16.

13.3.5 Chemical Verification of the Low Salinity Mechanism

In a recent published paper, the new chemical understanding of the low salinity EOR mechanism was verified by different experiments using both model and real systems, and the most relevant results are listed below (RezaeiDoust et al., 2011):

- *Increase in pH*: Independent of the composition of the low salinity brine, the low salinity EOR effects were comparable, and pH of the effluent increased as the flooding fluid was switched from high to low salinity (Figure 13.17A and B). The differential pressure over the core decreased, i.e., no pressure buildup.

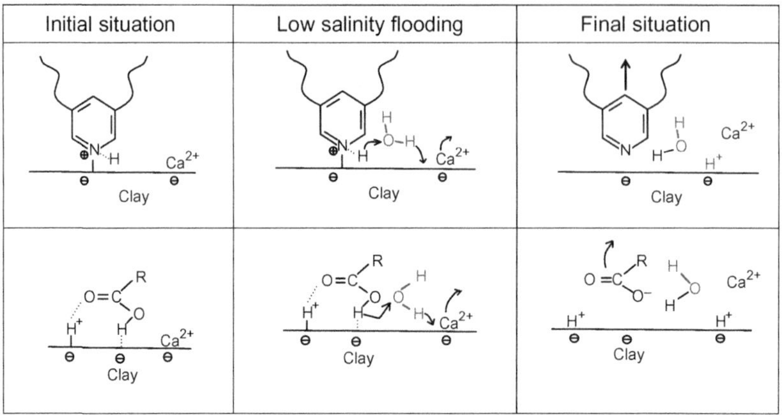

Initial situation	Low salinity flooding	Final situation
N^+···H, Ca^{2+}, Clay	H–O–H, Ca^{2+}, Clay	N, H–O–H, Ca^{2+}, H^+, Clay
O=C(R)O, H^+, H, Ca^{2+}, Clay	O=C(R)O, H^+, H, O, H, Ca^{2+}, Clay	O=C(R)O$^-$, H–O–H, Ca^{2+}, H^+, H^+, Clay

FIGURE 13.16 Proposed mechanism for low salinity EOR effects. Upper: desorption of basic material. Lower: desorption of acidic material. The initial pH at reservoir conditions may be in the range of 5 (Austad et al., 2010).

- *Initial pH of formation water*: In order to observe a low salinity EOR effect by wettability alteration, organic material must be adsorbed onto clay minerals, and the adsorption increases as the pH decreases. The low salinity EOR effect was compared with and without CO_2 initially present in the oil. In the presence of CO_2, the initial pH of the formation water was decreased. As shown in Figure 13.18, the low salinity EOR effect was doubled when CO_2 was present, confirming that the initial pH of the formation water plays a very important role.

- *Crude oil properties*: The low salinity EOR effects from cores saturated with crude oils of completely different/opposite AN and BN (AN = 0.12, BN = 1.78 and AN = 1.82, BN = 0.54 mg KOH/g) were compared (Figure 13.19). The oil recovery in the secondary process using high saline brine was different, but the low salinity EOR effect was quite similar. Thus, both basic and acidic material in the crude oil contributed to the adsorption onto actual clay minerals.

- *Effect of salinity and pH on adsorption*: The adsorption of the basic component Quinoline, which is present in crude oils, onto kaolinite, decreased as the salinity increased at pH = 5 and 8 (Figure 13.20). The change in pH has a much larger effect on the adsorption of basic components onto kaolinite than the change in salinity from 0 to 25,000 ppm, supporting the fact that an increase in pH as the injected water is switched from high to low salinity will desorb organic material from the clay. Due to the competition between the active cations, especially Ca^{2+}, and basic material toward the negatively charged sites on the clay surface at a given pH, the adsorption of quinoline decreased as

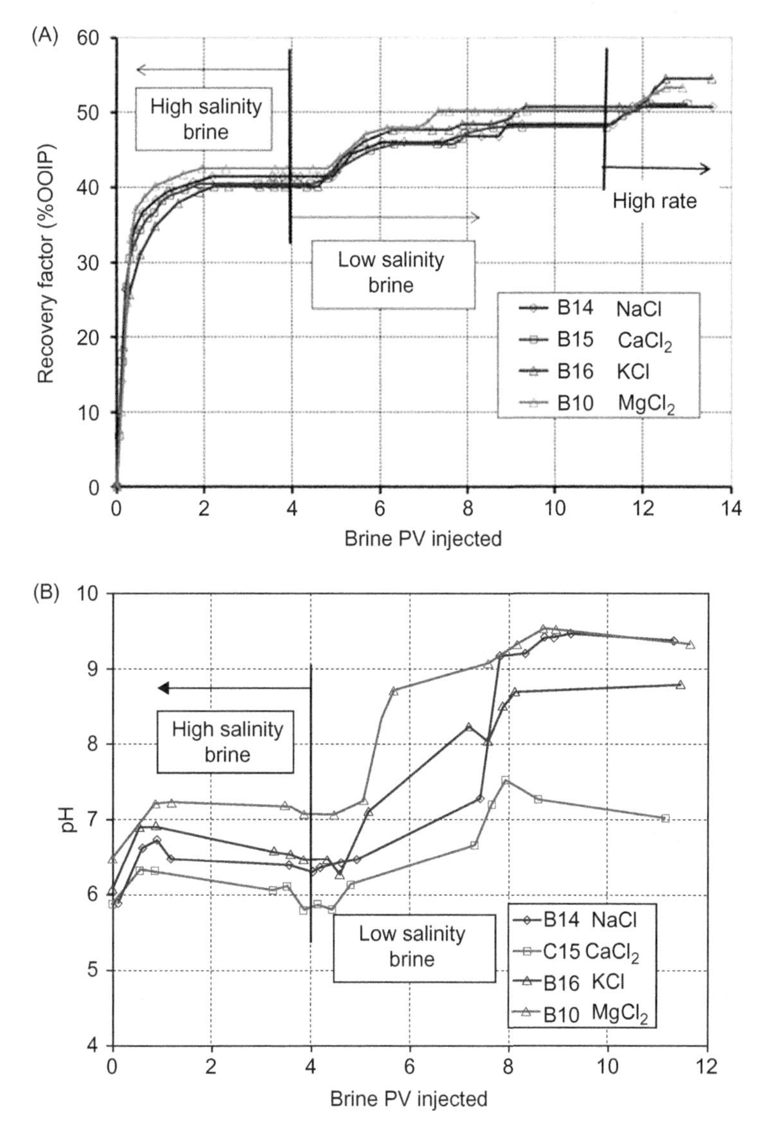

FIGURE 13.17 (A) Crude oil recovery curve for brine composition tests. (B) pH change caused by different low salinity brines (RezaeiDoust et al., 2011).

the salinity increased. This is not in line with a low salinity mechanism based on dilution of the electrical double layer as proposed by Ligthelm et al. (2009). The reversibility in the adsorption of quinoline onto kaolinite at high and low salinities is illustrated in Figure 13.21. Note that at pH = 2.5, the adsorption of the protonated base decreases because protons, H^+, become active in the adsorption onto the clay.

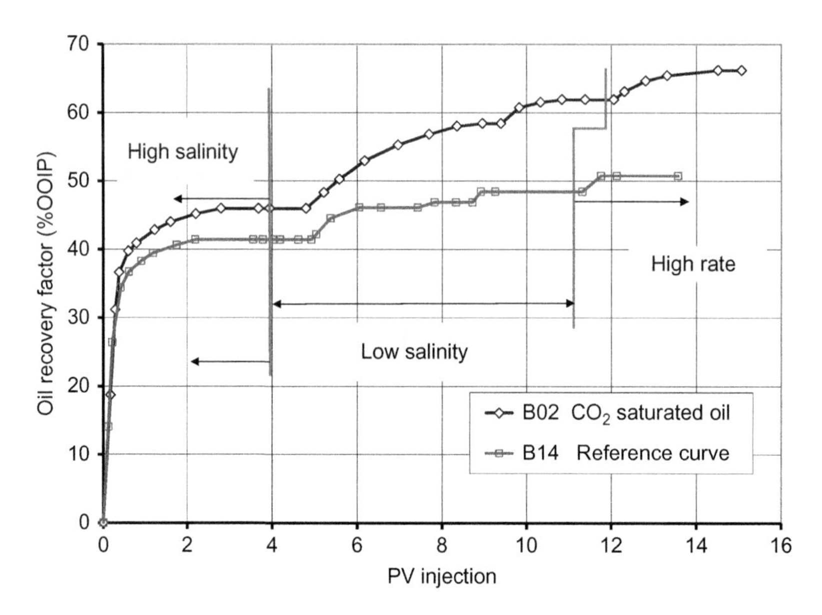

FIGURE 13.18 Effect of CO_2 (initial pH of FW) on oil recovery by a tertiary low salinity flood (RezaeiDoust et al., 2011).

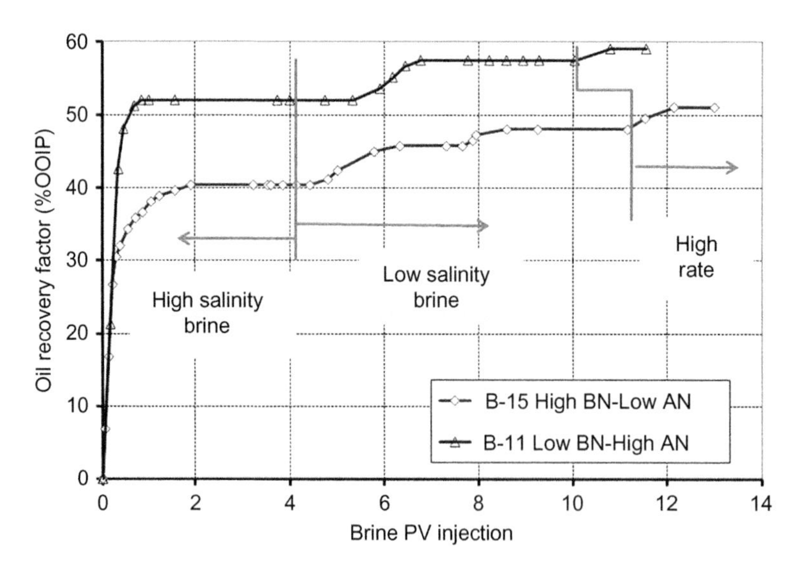

FIGURE 13.19 Tertiary low salinity effect in cores containing a basic (B-15) and an acidic (B-11) crude oil (RezaeiDoust et al., 2011).

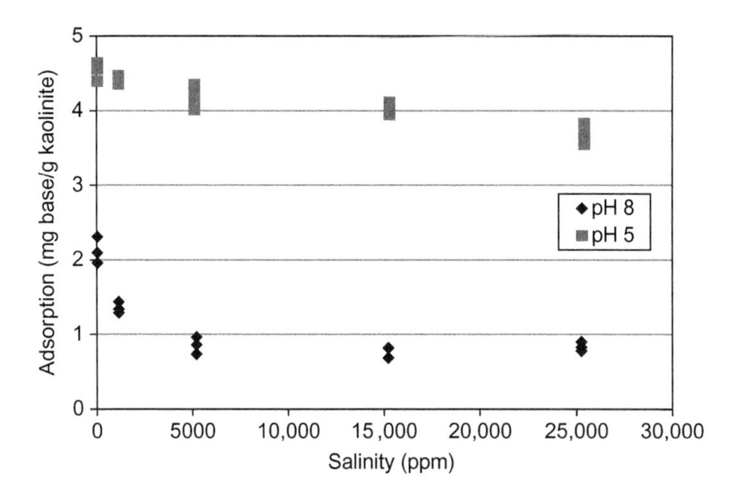

FIGURE 13.20 Adsorption of quinoline onto kaolinite at room temperature as a function of brine salinity at pH 5 and 8 (RezaeiDoust et al., 2011).

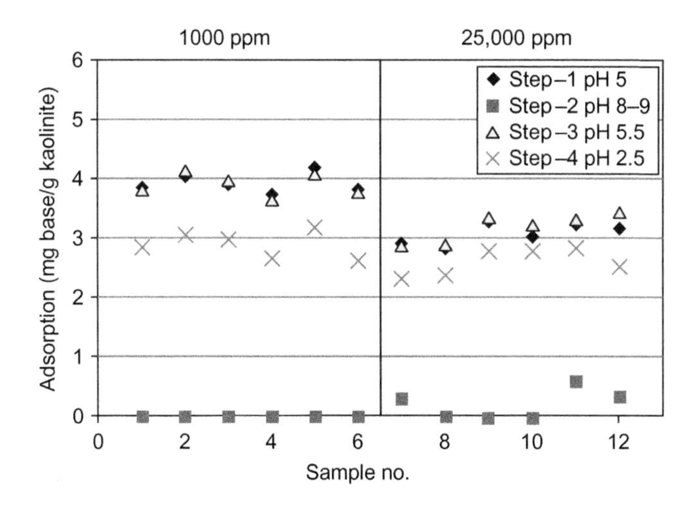

FIGURE 13.21 Adsorption reversibility by pH variations in systems containing quinoline, kaolinite, and brine at room temperature (1000 ppm in samples 1–6 and 25,000 ppm in samples 7–12) (RezaeiDoust et al., 2011).

Proton is the cation with the highest affinity toward the clay surface. Similar observations have also been observed for illite.

Also the adsorption of acidic material present in the crude oil is strongly affected by the pH of the formation brine, as illustrated by the adsorption of benzoic acid onto kaolinite, which showed a decrease in maximum adsorption from 3.7 to 0.1 μmole/m^2 as the pH increased from 5.3 to 8.1

TABLE 13.2 Adsorption of Benzoic Acid onto Kaolinite Using a 0.1 M NaCl Solution at 32°C (Madsen and Lind, 1998)

$pH_{initial}$	Γ_{max} μmole/m^2
5.3	3.7
6.0	1.2
8.1	0.1

(Table 13.2) (Madsen and Lind, 1998). Thus, both basic and acidic materials in the crude oil will adsorb onto clay minerals, and the adsorption decreased as pH increased in the interval 5–9.

13.4 FIELD EXAMPLES AND EOR POSSIBILITIES

The question is now: *Will improved chemical understanding of the EOR mechanism of smart water obtained by systematic laboratory studies in carbonates and sandstones be reflected in the observations experienced from the field?* In the next sections, results from field observations in carbonates and sandstones will be discussed in relation to the suggested mechanisms for EOR by smart water.

13.4.1 Carbonates

The fractured chalk reservoir Ekofisk in the Norwegian sector of the North Sea has been flooded with SW for about 25 years with a tremendous success. The prognoses for the oil recovery have increased as the knowledge of fluid–rock interaction has improved. The prognoses today are slightly above 50%, but it is not surprising if the recovery approaches 55% of OOIP. Since the first work on wetting properties of core material from the different formations: Tor, Lower, and Upper Ekofisk by Torsaeter (1984), which indicated that the Upper Ekofisk was too oil-wet to be water flooded, improved chemical understanding from laboratory tests, and field experience have been obtained, and there is no longer any doubt that SW has an impact on the wetting properties and also on the rock mechanics (Austad et al., 2008a). Due to water weakening of the chalk by SW, the compaction of the reservoir still takes place, about 12 cm/year, even though the reservoir pressure is increased up to the initial pressure. After breakthrough of SW in the production wells, the concentration of Mg^{2+} and SO_4^{2-} in the produced water was reduced and Ca^{2+} was enriched compared to a pure mixture of formation and SW without any chemical reactions taking place in the reservoir (Figure 13.22) (Puntervold et al., 2009). Thus, we see the same fluid–rock

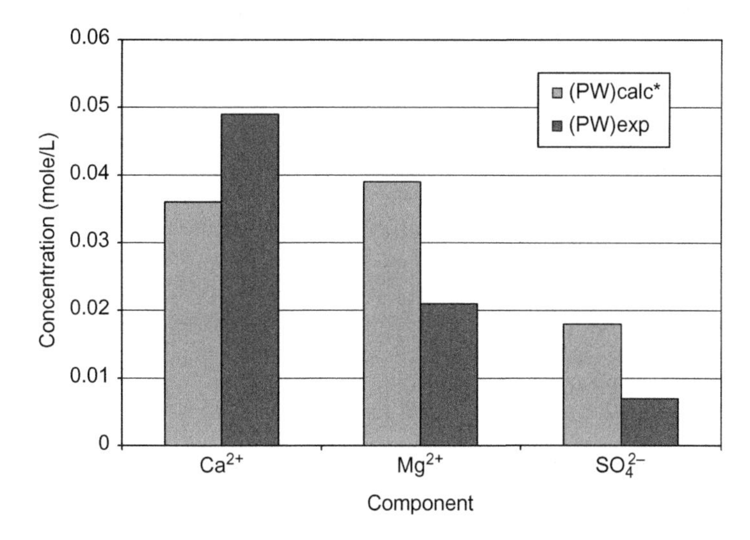

FIGURE 13.22 Calculated and measured component concentration in produced water, PW, linked to substitution of Ca^{2+} by Mg^{2+} at the rock surface, adsorption of SO_4^{2-} onto the rock and precipitation of $CaSO_4$. The PW contained 73.6 vol.% SW and 26.4 vol.% FW (Puntervold et al., 2009).

interaction in the field as is observed in small core samples in the laboratory. The high reservoir temperature in the Ekofisk, 130°C, is excellent for SW to act as a wettability modifier, and the injected SW can imbibe from the fractures into the matrix blocks and both oil and initial formation water will be displaced into the fractures and transported through the fracture system to the producers.

Provided that the most oil-wet Upper Ekofisk has not been flooded with SW yet, laboratory studies indicate that about 10% of OOIP extra oil can be recovered if the SW is modified by removing the nonactive salt, NaCl (Fathi et al., 2010a). Due to the high reservoir temperature the injected SW should not be spiked with sulfate because of precipitation of anhydrite, $CaSO_4$(s).

The other fractured chalk field in the North Sea, the Valhall field, has a reservoir temperature of 90°C, significantly lower than the Ekofisk. The performance of the wettability alteration process with SW is therefore not as good as for Ekofisk, and in general, the Valhall field is much less water-wet than the Ekofisk field. This is also reflected by the acid numbers of the crude oils, about 0.1 and 0.35 mg KOH/g for Ekofisk and Valhall, respectively. Injection of SW into the Valhall field has started, and it was confirmed by systematic studies at reservoir conditions by BP, that SW improved the oil recovery significantly compared to formation water (Webb et al., 2005b). In the spontaneous imbibition process, sulfate-free FW produced oil corresponding to 0.224 PV, while the SW increased the oil recovery to 0.31 PV, which is an increase in oil recovery of 40%. In a forced imbibition process using a differential pressure over the core of 1 psi, the oil recovery increased from 45% PV with FW to 60% PV with SW. Thus, also at 90°C SW acted as a smart EOR fluid in fractured chalk. Due to the lower reservoir temperature compared to that of Ekofisk, significant improvements in oil recovery compared to SW could be obtained by injecting modified SW, i.e., SW depleted in NaCl and spiked with sulfate, 3–4 times the concentration present in ordinary SW. Laboratory experiments indicated an increase in oil recovery compared to SW of 20–25% of OOIP (Fathi et al., 2010a). It is just a question of being able to produce the modified SW in a simple and cheap way.

Great success has also been reported by injecting SW into nonfractured limestone reservoirs. Pipelines are transporting SW from the Gulf and into the desert in Saudi Arabia to be injected into limestone reservoirs with great success in oil recovery. Depending on the nature of the reservoir, improvements in oil recovery could be obtained by modifying/diluting the SW as recently reported by Yousef et al. (2010). As pointed out previously, in order to observe a low salinity EOR effect in carbonates, the matrix must contain anhydrite.

13.4.2 Sandstones

BP has the greatest experience in applying the low salinity technique in the field. Through a number of single well tests (McGuire et al., 2005; Seccombe et al., 2008), and also an interwell test, incremental oil recovery corresponding to 10% of the total pore volume in the swept area has been demonstrated (Seccombe et al., 2010). After being aware of the low salinity EOR effects on oil recovery by water flooding, a historical field evidence study was undertaken by Robertson (Robertson, 2007) looking at different fields in the United States. It was found that the recovery from fields using lower salinity injection water was greater than that using higher salinity injection water. Furthermore, people from Shell have suggested that the Omar field in Syria showed a change in wettability from oil-wet to a water-wet system in a secondary low salinity water flood, leading to an associated incremental oil recovery of 10–15% of the stock tank oil (STO) initially in place (Vledder et al., 2010).

BP has collected data from the field tests like changes in the chemical composition of the produced water, and interpreted these observations to give field evidence for the suggested MIE mechanism (Lager et al., 2008b, 2011; Seccombe et al., 2010). Normally, the concentration of similar and nonreactive ions in the formation and injected water should decrease as breakthrough of the low saline water occurred. However, some ions showed an opposite trend, which pointed to specific injection–formation water interaction or fluid–rock interactions:

- The concentration of iron, Fe^{2+}, increased at the same time as the breakthrough of the low saline water took place, and then it decreased again as the high saline water was produced. It was suggested that the iron coated the clay minerals and formed a bridge between the clay and naphthenic acids, and both were released from the surface in the MIE mechanism.

- A great decrease in the Mg^{2+} concentration occurred, which was too large to be related to dilution. Afterward, as the high salinity water arrived at the producer, the concentration increased again. It was explained by the strong participation of Mg^{2+} in the MIE process, adsorbing onto the clay minerals.
- The concentration of alkalinity in the form of bicarbonate, HCO_3^-, increased during the low saline flooding. It was, however, hard to explain this effect, but it was suggested that it could be associated with increased organic components dissolved in the produced water. Increased dissolution of CO_2 in the brine due to lower brine salinity was also suggested.
- The concentration of H_2S was slightly decreased, and that was suggested to be due to a lower amount of sulfate in the low salinity water, and thereby lower activity of sulfate reducing bacteria (SRB), which is believed to be responsible for the Endicott souring.

The Endicott reservoir contained significant amount of sour gases, CO_2 and H_2S, when the low saline brine was injected, which will act as buffers for the brine. This means that the pH of the formation brine is well below 7, and strong adsorption of acidic and basic material from the crude oil onto the clay minerals will take place, i.e. the clays become partly oil-wet. If it is accepted that a local increase in pH occurred at the clay–brine interface as active cations, Ca^{2+}, etc., were desorbed from the clay surface as the low saline water invaded the pore system, it is very easy to explain all of the above mentioned observations (Austad et al., 2010; RezaeiDoust et al., 2011):

- If some of the iron is present in the formation as FeS, which is not unreasonable, the solubility of FeS(s) is increased by 4–5 orders of magnitude in an alkaline environment due to complex formation, $Fe(OH)^+$.
- The decrease in Mg^{2+} can be related to precipitation of $Mg(OH)_2$ as the pH increased to 8–9.
- The concentration of alkalinity in the brine is a direct increase in dissolution of CO_2 into an alkaline solution. The chemical equilibrium: $CO_2 + H_2O \leftrightarrow [H_2CO_3] \leftrightarrow H^+ + HCO_3^-$ will move to the right as the pH is increased.
- Similarly, the concentration of H_2S will decrease due to an ordinary acid–base reaction: $H_2S + OH^- = HS^- + H_2O$.

Thus, no sophisticated geochemical modeling is needed to explain the observations. It is correct that no increase in pH was observed in the produced water, instead a small decrease was noticed. Due to the great buffering effects of chemical components present in the reservoir, an increase in pH is not expected to be recorded in the produced water.

13.4.3 Statoil Snorre Pilot

Statoil performed a low salinity single well pilot test in the Snorre field recently, even though the laboratory core flooding experiments only indicated a very low EOR effect of 2% of OOIP (Skrettingland et al., 2010). Laboratory tests were performed on the core material from the following formations: Upper Statfjord, Lower Statfjord, and Lunde. The field test, which was performed in the Upper Statfjord formation, confirmed, however, the laboratory investigation showing no significant change in the oil saturation after switching from SW (34,020 ppm) to the low salinity water (440 ppm). In general, the condition for observing low salinity EOR effects should be present in the Snorre formation. The STO oil contained polar components, the acid number was low, AN = 0.02 mg KOH/g, but the base number was quite high, BN = 1.1 mg KOH/g. Although the salinity of the formation water was close to the salinity of SW, 34,300 ppm, the concentration of Ca^{2+} was 3–4 times higher than in SW. Kaolinite was the main clay material, and it varied between 10 and 20 wt%. The reservoir temperature was 90°C. *What was the reason for such a small low salinity EOR effect?*

It was mentioned by Skrettingland et al. (2010) that parameters such as the type of crude oil and the initial wetting conditions could be crucial. The low content of organic acid in the Snorre oil could be one factor that contributed to the small low salinity EOR effect. The authors also noticed an unusually high pH of the produced water, about pH = 10.

A new study (Reinholdtsen et al., 2011) using the core material from the Lunde formation confirmed that the pH of the effluent was quite high when flooding the core with formation water during core preparation with pH about 10. The Snorre STO was saturated by CO_2 at a pressure of 6 bar, to hopefully decrease the pH of the initial formation water during aging. The oil recovery from one of the cores is shown in Figure 13.23. The flooding sequence was: FW, SW, and 500 ppm NaCl as the low salinity fluid. The oil recovery by FW was slightly above 51% of OOIP, and it increased to 55% by SW, but no increase in oil recovery was noticed by the low saline water or by increasing the flooding rate 2 and 4 times. Thus, a small tertiary low salinity EOR effect was noticed by flooding the core with SW, which has a similar salinity as formation water, but it contained 3 times lower concentration of Ca^{2+} compared to that in the FW. In line with the suggested mechanism, it is not the change in salinity, which is most important, it is the change in the concentration of the most active cation, Ca^{2+}. The concentration of Mg^{2+} in SW is 6.4 times higher than that in the FW, but that appeared to be of negligible importance.

Further inspection of the composition of the rock minerals confirmed that the core material from the Snorre field contained significant amounts of plagioclase, 6–35 wt%. Plagioclase is a polysilicate, and albite with the chemical structure: $NaAlSi_3O_8$ is often used as an example. In water, the

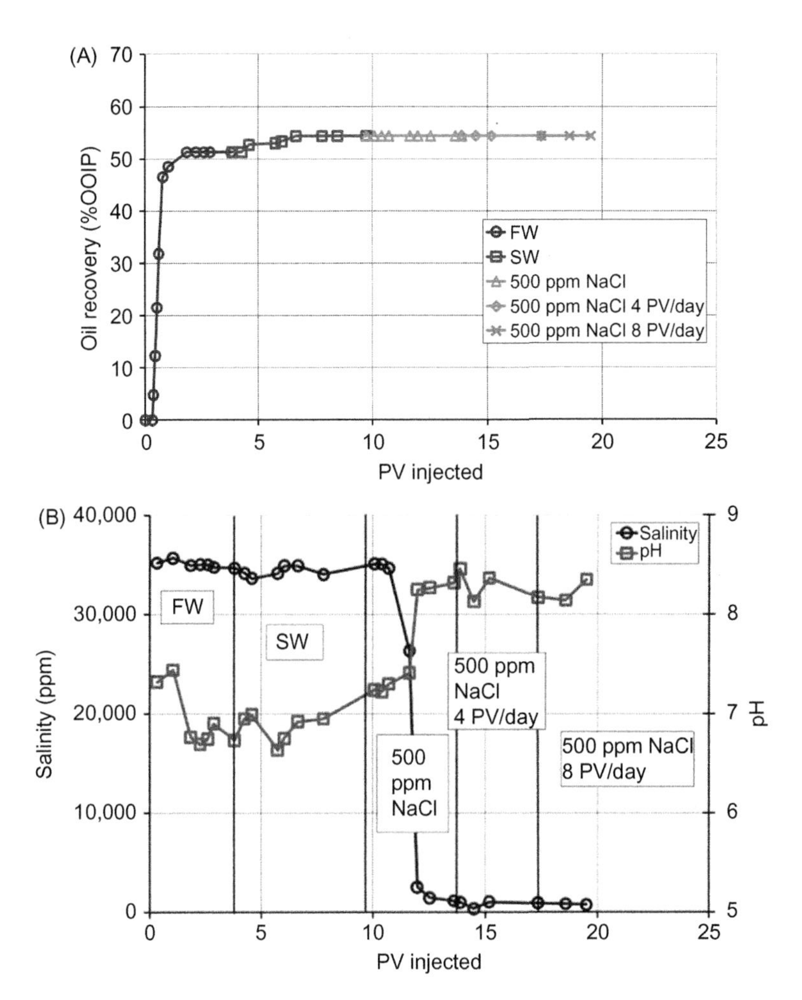

FIGURE 13.23 (A) Oil recovery versus injected PVs. Flooding rate of 2 PV/D. Sequence of injected fluid: (1) FW, (2) SW, (3) low saline water 500 ppm NaCl, and (4) and (5) increased flooding rate. $T_{res} = 90°C$. (B) Salinity and pH versus PVs injected (Reinholdtsen et al., 2011).

plagioclase will give an alkaline solution according to the following reaction:

$$NaAlSi_3O_8 + H_2O \leftrightarrow HAlSi_3O_8 + OH^- + Na^+$$

At moderate salinities, like the Snorre formation water, Na^+ can be substituted by H^+, and the pH of the solution increases. Even though the crude oil was saturated with CO_2, the first brine eluted from the core had a pH well above 7 (Figure 13.23B). Thus, the presence of reactive Plagioclase minerals has a buffering effect on the formation brine, and the equilibrium pH will be above 7, which will decrease adsorption of basic and acidic components from the crude oil onto the clay minerals. The very low AN and the reasonable high BN of the Snorre crude oil indicate that the wetting properties of the clay minerals are mostly dictated by basic material from the crude oil. The fraction of protonated pyridine-like base at $pH > 7$ is low, which will decrease the adsorption of basic material onto the clay (Austad et al., 2010; Burgos et al., 2002). Thus, due to the relatively high pH of the formation water, $pH \approx 7.5$, the clay minerals will act quite water-wet, and the potential for observing low salinity effects becomes small.

Thus, in light of improved chemical understanding of the low salinity EOR mechanism, it should be possible in the future to evaluate the low salinity EOR potential for oil fields.

CONCLUSION

Injection of Smart Water can improve oil recovery significantly from both carbonates and sandstones. In order to design an optimized smart water, detailed chemical knowledge about the CBR interaction is needed and must be obtained from systematic laboratory studies under controlled conditions. The improved chemical understanding of wettability alteration can be used to explain field observations and evaluate possible water-based EOR potential. The work presented in this paper is a small but important step in that direction.

ACKNOWLEDGMENTS

The author acknowledges Dr. Skule Strand and Dr. Tina Puntervold for a fruitful research cooperation for many years. They have together with a large number of Ph.D. students kept the continuity in the research work and also given good comments to the present paper.

REFERENCES

Austad, T., Strand, S., Madland, M.V., Puntervold, T., Korsnes, R.I., 2008a. Seawater in chalk: an EOR and compaction fluid. SPE Reservoir Eval. Eng. 11 (4), 648–654.

Austad, T., Strand, S., Puntervold, T., Ravari, R.R., 2008b. New method to clean carbonate reservoir cores by seawater. Paper SCA2008-15 Presented at the International Symposium of the Society of Core Analysts, 29 October–2 November.

Austad, T., RezaeiDoust, A., Puntervold, T., 2010. Chemical mechanism of low salinity water flooding in sandstone reservoirs. Paper SPE 129767 Prepared for Presentation at the 2010 SPE Improved Oil Recovery Symposium, 24–28 April.

Austad, T., Shariatpanahi, S.F., Strand, S., Black, C.J.J., Webb, K.J., 2011. Condition for low salinity EOR effect in carbonate oil reservoirs. Thirty-Two Annual IEA EOR Symposium and Workshop, 17–19 October.

Buckley, J.S., 1995. Asphaltene precipitation and crude oil wetting—crude oils can alter wettability with or without precipitation of asphaltenes. SPE Adv. Technol. Ser. 3 (1), 53–59.

Burgos, W.D., Pisutpaisal, N., Mazzarese, M.C., Chorover, J., 2002. Adsorption of quinoline to kaolinite and montmorillonite. Environ. Eng. Sci. 19 (2), 59–68.

Fathi, S.J., Austad, T., Strand, S., 2010a. "Smart Water" as wettability modifier in chalk: the effect of salinity and ionic composition. Energy Fuels 24, 2514–2519.

Fathi, S.J., Austad, T., Strand, S., Frank, S., Mogensen, K., 2010b. Evaluation of EOR potentials in an offshore limestone reservoir: a case study. Eleventh International Symposium on Reservoir Wettability, 7–9 September.

Fathi, S.J., Austad, T., Strand, S., 2011. Effect of water-extractable carboxylic acids in crude oil on wettability in carbonates. Energy Fuels 25, 2587–2592.

Jadhunandan, P.P., Morrow, N.R., 1995. Effect of wettability on water flood recovery for crude oil/brine/rock systems. SPE Reservoir Eng. February, 40–46.

Lager, A., Webb, K.J., Black, C.J.J., 2007. Impact of brine chemistry on oil recovery. Paper A24 Presented at the Fourteenth European Symposium on Improved Oil Recovery, 22–24 April.

Lager, A., Webb, K.J., Black, C.J.J., Singleton, M., Sorbie, K.S., 2008a. Low salinity oil recovery—an experimental investigation. Petrophysics 49 (1), 28–35.

Lager, A., Webb, K.J., Collins, I.R., Richmond, D.M., 2008b. LoSal™ enhanced oil recovery: evidence of enhanced oil recovery at the reservoir scale. Paper SPE 113976 Presented at the 2008 SPE/DOE Improved Oil Recovery Symposium, 19–23 April.

Lager, A., Webb, K.J., Seccombe, J.C., 2011. Low salinity water flood, Endicott, Alaska: Geochemical study and field evidence of multicomponent ion exchange. Sixteenth European Symposium on Improved Oil Recovery, 12–14 April.

Ligthelm, D.J., Gronsveld, J., Hofman, J.P., Brussee, N.J., Marcelis, F., van der Linde, H.A., 2009. Novel water flooding strategy by manipulation of injection brine composition. Paper SPE 119835 Presented at the 2009 SPE EUROPEC/EAGE Annual Conference and Exhibition, 8–11 June.

Madsen, L., Lind, I., 1998. Adsorption of carboxylic acids on reservoir minerals from organic and aqueous phase. SPE Reservoir Eval. Eng. 47–51 (February).

Masalmeh, S.K., Oedai, S., 2009. Surfactant enhanced gravity drainage: laboratory experiments and numerical simulation model. Paper SCA2009-06 Presented at the International Symposium of the Society of Core Analysts, 27–30 September.

McGuire, P.L., Chatham, J.R., Paskvan, F.K., Sommer, D.M., Carini, F.H., 2005. Low salinity oil recovery: an exciting new EOR opportunity for Alaska's North slope. Paper SPE 93903 Presented at the 2005 SPE Western Regional Meeting, 30 March–1 April.

Pierre, A., Lamarche, J.M., Mercier, R., Foissy, A., 1990. Calcium as a potential determining ion in aqueous calcite suspensions. J. Dispersion Sci. Technol. 11 (6), 611–635.

Pu, H., Xie, X., Yin, P., Morrow, N.R., 2008. Application of coalbed methane water to oil recovery by low salinity water flooding. Paper SPE 113410 Presented at the 2008 SPE Improved Oil Recovery Symposium, 19–23 April.

Puntervold, T., Austad, T., 2008. Injection of seawater and mixtures with produced water into North Sea chalk formation: impact of fluid–rock interactions on wettability and scale formation. J. Pet. Sci. Eng. 63, 23–33.

Puntervold, T., Strand, S., Austad, T., 2007. Water flooding of carbonate reservoirs: effects of a model base and natural crude oil bases on chalk wettability. Energy Fuels 21 (3), 1606–1616.

Puntervold, T., Strand, S., Austad, T., 2009. Co-injection of seawater and produced water to improve oil recovery from fractured North Sea chalk oil reservoirs. Energy Fuels 23 (5), 2527–2536.

Rao, D.N., 1996. Wettability effects in thermal recovery operations. SPE/DOE Improved Oil Recovery Symposium, 21–24 April.

Ravari, R.R., Strand, S., Austad, T., 2010. Care must be taken to use outcrop limestone cores to mimic reservoir core material in SCAL linked to wettability alteration. Eleventh International Symposium on Reservoir Wettability, 7–9 September.

Ravari, R.R., Strand, S., Austad, T., 2011. Combined surfactant-enhanced gravity drainage (SEGD) of oil and the wettability alteration in carbonates: the effect of rock permeability and interfacial tension (IFT). Energy Fuels 25, 2083–2088.

Reinholdtsen, A.J., RezaeiDoust, A., Strand, S., Austad, T., 2011. Why such a small low salinity EOR—potential from the Snorre formation? Sixteenth European Symposium on Improved Oil Recovery, 12–14 April.

RezaeiDoust, A., Puntervold, T., Austad, T., 2011. Chemical verification of the EOR mechanism by using low saline/smart water in sandstone. Energy Fuels 25, 2151–2162.

Robertson, E.P., 2007. Low-salinity water flooding to improve oil recovery—Historical field evidence. Paper SPE 109965 Presented at the 2007 SPE Annual Technical Conference and Exhibition, 11–14 Nov.

Seccombe, J.C., Lager, A., Webb, K.J., Jerauld, G., Fueg, E., 2008. Improving water flood recovery: LoSal^{TM} EOR field evaluation. Paper SPE 113480 Presented at the 2008 SPE/DOE Improved Oil Recovery Symposium, 19–23 April.

Seccombe, J., Lager, A., Jerauld, G., Jhaveri, B., Buikema, T., Bassler, S., et al., 2010. Demonstration of low-Salinity EOR at interwell scale, Endicott field, Alaska. Paper SPE 129692 Presented at the 2010 SPE Improved Oil Recovery Symposium, 24–28 April.

Shariatpanahi, S.F., Strand, S., Austad, T., 2011. Initial wetting properties of carbonate oil reservoirs: effect of the temperature and presence of sulfate in formation water. Energy Fuels 25 (7), 3021–3028.

Shariatpanahi, S.F., Strand, S., Austad, T., Aksulu, H., 2012. Wettability restoration of a limestone oil reservoir using completely water-wet cores from aqueous zone. Pet. Sci. Technol. 30, 1–9.

Sheng, J.J., 2012. Comparison of the effects of wettability alteration and IFT reduction on oil recovery in carbonate reservoirs. Asia-Pacific J. Chem. Eng. (in press).

Shimoyama, A., Johns, W.D., 1972. Formation of alkanes from fatty acids in the presence of $CaCO_3$. Geochim. Cosmochim. Acta 36, 87–91.

Skrettingland, K., Holt, T., Tweheyo, M.T., Skjevrak, I., 2010. Snorre low salinity water injection—core flooding experiments and single well field pilot. Paper SPE129877 Presented at the 2010 SPE Improved Oil Recovery Symposium, 22–26 April.

Speight, J.G., 1999. The Chemistry and Technology of Petroleum. Chemical Industries. Marcel Dekker, New York, NY.

Standnes, D.C., Austad, T., 2000a. Wettability alteration in chalk. 1. Preparation of core material and oil properties. J. Pet. Sci. Eng. 28 (3), 111–121.

Standnes, D.C., Austad, T., 2000b. Wettability alteration in chalk. 2. Mechanism for wettability alteration from oil-wet to water-wet using surfactants. J. Pet. Sci. Eng. 28 (3), 123–143.

Standnes, D.C., Nogaret, L.A.D., Chen, H.-L., Austad, T., 2002. An evaluation of spontaneous imbibition of water into oil-wet carbonate reservoir cores using a nonionic and a cationic surfactant. Energy Fuels 16 (6), 1557–1564.

Stoll, W.M., Hofman, J.P., Ligthelm, D.J., Faber, M.J., van den Hoek, P.J., 2007. Field-scale wettability modification—The limitations of diffusive surfactant transport. Paper SPE 107095 Presented at the SPE Europec/EAGE Annual Conference and Exhibition, 11–15 June.

Strand, S., Høgnesen, E.J., Austad, T., 2006a. Wettability alteration of carbonates—effects of potential determining ions (Ca^{2+} and SO_4^{2-}) and temperature. Colloids Surf. A Physicochem. Eng. Aspects 275, 1–10.

Strand, S., Standnes, D.C., Austad, T., 2006b. New wettability test for chalk based on chromatographic separation of SCN^- and SO_4^{2-}. J. Pet. Sci. Eng. 52, 187–197.

Strand, S., Puntervold, T., Austad, T., 2008. Effect of temperature on enhanced oil recovery from mixed-wet chalk cores by spontaneous imbibition and forced displacement using seawater. Energy Fuels 22, 3222–3225.

Tang, G., Morrow, N.R., 1999a. Influence of brine composition and fines migration on crude oil/brine/rock interactions and oil recovery. J. Pet. Sci. Eng. 24, 99–111.

Tang, G., Morrow, N.R., 1999b. Oil recovery by water flooding and imbibition—invading brine cation valency and salinity. Paper SCA9911 Presented at the International Symposium of the Society of Core Analysts, 1–4 August.

Torsaeter, O., 1984. An experimental study of water imbibition in chalk from the Ekofisk field. Paper SPE12688 Presented at the SPE/DOE Fourth Symposium on Enhanced Oil Recovery, 15–18 April.

Vledder, P., Fonseca, J.C., Wells, T., Gonzalez, I., Ligthelm, D., 2010. Low salinity water flooding: proof of wettability alteration on a field wide scale. Paper SPE 129564 Presented at the 2010 SPE Improved Oil Recovery Symposium, 24–28 April.

Webb, K.J., Black, C.J.J., Edmonds, I.J., 2005a. Low salinity oil recovery—the role of reservoir condition corefloods. Paper C18 Presented at the Thirteenth European Symposium on Improved Oil Recovery, 25–27 April.

Webb, K.J., Black, C.J.J., Tjetland, G., 2005b. A laboratory study investigating methods for improving oil recovery in carbonates. International Petroleum Technology Conference (IPTC), 21–23 November.

Yousef, A.A., Al-Saleh, S., Al-Kaabi, A., Al-Jawfi, M., 2010. Laboratory investigation of novel oil recovery method for carbonate reservoirs. Paper CSUG/SPE 137634 Presented at the Canadian Unconventional Resources & International Petroleum Conference, 19–21 October.

Zhang, P., Austad, T., 2006. Wettability and oil recovery from carbonates: effects of temperature and potential determining ions. Colloids Surf. A Physicochem. Eng. Aspects 279, 179–187.

Zhang, Y., Morrow, N.R., 2006. Comparison of secondary and tertiary recovery with change in injection brine composition for crude oil/sandstone combinations. Paper SPE 99757 Presented at the 2006 SPE/DOE Symposium on Improved Oil Recovery, 22–26 April.

Zhang, P., Tweheyo, M.T., Austad, T., 2006. Wettability alteration and improved oil recovery in chalk: the effect of calcium in the presence of sulfate. Energy Fuels 20, 2056–2062.

Zhang, P., Tweheyo, M.T., Austad, T., 2007a. Wettability alteration and improved oil recovery by spontaneous imbibition of seawater into chalk: impact of the potential determining ions: Ca^{2+}, Mg^{2+} and SO_4^{2-}. Colloids Surf. A Physicochem. Eng. Aspects 301, 199–208.

Zhang, Y., Xie, X., Morrow, N.R., 2007b. Water flood performance by injection of brine with different salinity for reservoir cores. Paper SPE 109849 Presented at the 2007 SPE Annual Technical Conference and Exhibition, 11–14 November.

Chapter 14

Facility Requirements for Implementing a Chemical EOR Project

John M. Putnam
SNF Holding Company, FLOQUIP Engineering, P.O. Box 250, Riceboro, GA 31323, USA

14.1 INTRODUCTION

This chapter defines and describes the specific surface equipment and facilities required by the various chemical EOR processes—whether polymer-augmented waterflooding, alkaline-surfactant-polymer (ASP) flooding, or surfactant polymer flooding—to be designed and installed on the injection side of the flood. It is helpful to think of the injection facilities as a modified water injection scheme, whereas certain chemical handling, processing, and metering subsystems are integrated with the primary waterflood water processing, storage and injection facilities to form a single, overall facility that is built for purpose. The examples shown represent landside field applications.

Whether considered a pilot or a full-field project, the style of fabrication and construction may differ, but the process flow designs remain relatively the same. For instance, a single injection well (IW) pilot or even a low-rate multiple-well pilot may be designed and fabricated as factory-prefabricated skid-type assemblies. These types of facilities are transported to the field where interconnecting piping and electrical cabling is quickly installed between water supply tanks, liquid waste disposal pits, and the IW flow lines.

This photo, Figure 14.1, shows a factory-prefabricated ASP injection skid that was designed for a 1500 barrels per day (BPD), single IW pilot located in northeastern Colorado. It included all the chemical processing systems within a single 10 ft wide × 40 ft long skid building. The soda ash was delivered by bulk pneumatic truck to the silo in the foreground. The surfactant storage tank, located to the left of the silo and at the end of the plant received bulk deliveries from a Texas manufacturer. The process water tanks are just behind the main plant.

Enhanced Oil Recovery Field Case Studies.

FIGURE 14.1 This factory pre-fabricated ASP injection skid includes all the chemical processing systems within a single 10′ wide X 40′ long skid building. The soda ash was delivered by bulk pneumatic truck to the silo in the foreground. The surfactant storage tank, located to the left of the silo and at the end of the plant received bulk deliveries from a the manufacturer. *(Photo by John M. Putnam)*

Scaling facilities for high injection rates with many IWs may be more economically installed within permanent style buildings, or even multiple injection stations, depending upon the aerial expansion of the field.

Figure 14.2 shows factory-prefabricated 3000 BPD polymer injection facility designed and set up for the harsh arctic environments common in western Canada, including special considerations for module insulation, heating, and ventilation. One module contains the polymer handling and process system and the second contains two plunger style injection pumps and the motor control center and control panel.

Figure 14.3 is an example of a large polymer processing and injection facility constructed on-site in Oman. The equipment skids are anchored to permanent concrete foundations with sunshades constructed overhead. Each of 27 IWs is served by a dedicated high-pressure, positive-displacement injection pump. This plant is set up to receive dry bulk polymer into silos from containers shipped by sea and land to the site. This facility processes and injects over 110,000 BPD at a design pressure up to 1015 psi.

And Figure 14.4 shows another example of a permanent installation. This polymer flood pilot facility is installed in the giant Daqing oilfield in northeastern China. Notice the multiple (3) polymer processing units and the bulk bag handling systems.

This chapter does not address the production side equipment facilities that are sometimes enhanced and modified for chemical EOR projects.

FIGURE 14.2 This pre-fabricated 3000 BPD polymer injection facility is designed and setup for the harsh arctic environments common in western Canada, including special considerations for module insulation, heating, and ventilation. Source: *Photo courtesy of SNF Floerger.*

FIGURE 14.3 Here is an example of a large polymer processing and injection facility constructed on site in Oman. This plant is setup to receive dry bulk polymer into silos from containers shipped by sea and land to the site. This facility processes and injects over 110,000 BPD at a design pressure up to 1015 PSI. Source: *Photo courtesy of SNF Floerger.*

14.2 OVERALL PROJECT REQUIREMENTS

Regardless of the aforementioned fabrication and construction style, the chemical EOR facilities designer needs a clear definition from the operator of the following performance and operating parameters to enable him to properly size the plant and to take into consideration the various logistical and operational factors that are inherent in the operations of a waterflood facility and the chemical processing systems.

Figure 14.5 depicts a useful spreadsheet setup to calculate daily chemical consumption and process flow rates for an ASP flood. This material balance and consumption worksheet shows input for a 10,000 BPD ASP pilot

FIGURE 14.4 Another example of a permanent installation is this polymer flood pilot facility installed in the giant Daqing oilfield in northeastern China. Notice the multiple (3) polymer processing units and the bulk bag handling systems. *(Photo by John M. Putnam)*

implemented for the national oil company of Columbia, S.A. Notice that the ASP formula in the lower left and the reagent details in the lower right. Once the design injection rates are inserted, the material balance and daily chemical consumption values are calculated. These are the basic flow rate data needed to configure metering pump systems and to design on-site storage of the stock chemicals. At the design rate, this project requires the operator to manage and process over 61,000 lb of soda ash everyday.

Figure 14.6 shows calculations for polymer hydration time based on the working volume of the mother solution tank and horsepower requirements for the injection pumps based on rate and design injection pressure. This worksheet provides two essential design and engineering details. (1) The minimum hydration tank volume is calculated by inserting the mother solution concentration, the design hydration time in minutes, the downhole concentration, and the design injection rate. In this example, the polymer system is calibrated to produce a 10,000 ppm mother solution; the injection rate is 10,000 BPD at a downhole concentration of 1500 ppm. The worksheet calculates that a 62.5 bbl hydration tank volume is required when the hydration time interval is 60 min. However, this polymer system is already designed with a 100 bbl hydration tank. Thus, the actual hydration time interval under the same operating parameters is 97 min. This same worksheet also shows the calculated horsepower requirement for each of three injection pumps. The design injection rate and pressure is inserted and the required horsepower of each pump is calculated to be 85, thus 100-HP inverter duty motors meeting other relevant project specifications are installed.

Proportional rate table for ASP chemical injection

Injection pump rate (LPM)		m^3/h	Resulting chem pump rates		
IW No. 1	368.06	22.08	Polymer	165.63	L/min
IW No. 2	368.06	22.08	Alkaline	1159.38	kg/h dry
IW No. 3	368.06	22.08	Surfactant	646.97	L/h
IW No. 4	0.00	0.00	H_20 Treating	8.10	L/h
IW No. 5	0.00	0.00			
IW No. 6	0.00	0.00	Alkaline solution	24.61	m^3/h
IW No. 7	0.00	0.00			
Cum	1104.17	1590 m^3/day			
		103.4 Bar			

Daily consumption of chemical (as supplied)			
2385.00	kg	3.18	750-kg bags
27,825.00	kg	28.95	m^3
15,900.00	kg	15,163.42	L

Annual consumption of chemical (as supplied)		
Year 1 / ton	Year 2 / ton	Project total
871	435	1306
10,156		10,156
5804		5804

Instructions:

1. Enter injection rates in LPM for each injection pump, e.g., 1.45 LPM enter 1.45.
2. Enter a value for each IW even if the well is shut-in and the injection pump is off.
3. The PLC will adjust the polymer, alkaline and surfactant metering pumps to the rate values shown in the "Resulting Chem Pump Rates" table based on the below ASP Formula table.
4. Enter new values in the blue table cells below to change the calculations and material balance data.

Basis for calculations:

ASP Formula

Polymer	1500	mg/L
Alkaline	17,500	mg/L
Surfactant	2000	mg/L
H_20 Treating	55	mg/L

Constants:

					SpG
Polymer stock Concentration	10,000	mg/L	1.000		1.000
Soda ash	4.5%	% active	1.047	kg/L	1.047
Surfactant	20%	% active	1.024	kg/L	1.024
O_2 scavenger	45%	% active			1.000

FIGURE 14.5 This material balance worksheet shows input for a 10,000 BPD alkaline-surfactant-polymer pilot project. Notice the ASP formula in the lower left and the reagent details in the lower right. Once the design injection rates are inserted, the material balance and daily chemical consumption values are calculated. These are the basic flow rate data needed to configure metering pump systems and to design on-site storage of the stock chemicals.

SNF FLOQUIP

Polymer system calculator worksheet

3-injection well polymer flood

Instructions:

1. Enter desired values in the yellow shaded fields.
2. See calculated results in the blue shaded fields.
3. Maximum working volume of skid maturation tank = 16 m^3, thus any value < 16 (F11) equals a longer hydration time than maturation time (C13) input

Stock polymer conc. (% wt.)	1.0000%
Downhole concentration (% wt.)	0.1500%
Downhole concentration (m^3/day)	1590
Minimum maturation time (min)	60
Actual maturatin time (min)	97

Injection pump required power calculator	
Horsepower required	84.99
Kw required	63.38
Rate (m^3/day)	530.00
Pressure (Bar)	103.40

	Metric	bbls	Gallons	ft^3	lbs.
Polymer usage per day (Dry kg)	2385.00				5258.03
Maturation tank working volume m^3	9.94	62.50	2625.21	350.94	
Metering pump nominal rate (LPM @ 200 RPM)	165.63		43.75		
Dilution water rate (LPM)	607.29	3.82	160.43	21.45	
Recommended total tank volume m^3	12.42	78.13	3281.51	438.67	
Recommended metering pump maximum Rate per minute	220.83		58.34		

Rate conversion	BPD	m^3/day
	1	0.16
	250	39.75
	500	79.49
	750	119.24
	1000	158.99
Pressure conversion	PSI	Bar
	1	0.0689476
	1000	68.95
	1750	120.66

FIGURE 14.6 This worksheet calculates the minimum hydration tank volume by inserting the mother solution concentration, the design hydration time in minutes, the down-hole concentration, and the design injection rate. In this example, the polymer system is calibrated to produce a 10,000 PPM mother solution; the injection rate is 10,000 BPD at a down-hole concentration of 1500 PPM. The worksheet calculates that a 62.5 BBL hydration tank volume is required when the hydration time interval is 60 minutes. However, this polymer system is configured with a 100 BBL hydration tank. Thus, the actual hydration time interval under the same operating parameters is 97 minutes. This same worksheet also shows the calculated horsepower requirement for each of 3 injection pumps.

The following items are the minimum design criteria and operating parameters needed by the facilities engineer to properly design and specify the chemical process systems and injection equipment.

- Injected fluid formula or recipe
- Chemical packaging and logistics (i.e., shipment weights, net package weights, available warehousing, availability and order lead times)
- Cumulative daily injection for nominal design
- Individual IW daily injection rate range
- Wellhead injection pressure, maximum design
- Quantity of IWs
- Injection system scheme (i.e., dedicated injection pump per well or multiple-well trunk line distribution)
- Water source and quality
- Geographical location and weather conditions, including seismic zone classification and local building code, air quality, and permitting requirements
- Electrical and control field interfaces with existing utility service and field automation

14.3 MODES OF CHEMICAL EOR INJECTION

Since chemical handling, processing and metering systems are specific to each chemical that makes up the injected fluid's recipe, the facility's process flow schematic must be tailored for that project mode. Furthermore, some chemical injection schemes require different injected fluid recipes at progressive stages during the program. Thus, the injection facility must be designed to accommodate these various ranges of process and injection rates, with and without some of the individual chemical components. For instance, a surfactant-polymer (SP) project may require an initial injection stage of surfactant to be followed by a polymer-only stage. In this case, the facility must be designed to produce and inject the rate as required under the two different modes.

Injection modes for polymer flooding (P), SP, alkaline-polymer (AP), and ASP are discussed and illustrated herein. As a basic principle, each facility design is anchored around a conventional waterflood injection process flow, and the various chemical subsystems are integrated with the main injection facility to form the overall chemical EOR injection plant. For this discussion, a basic waterflood injection facility will include water treatment and storage, water charge or booster pump, final solution filtration, high-pressure positive-displacement injection pumps, and control.

On-site handling and processing requirements for the various chemicals will be discussed following this description of the basic process flow designs for each chemical EOR injection mode.

14.3.1 Polymer Flooding

This block diagram in Figure 14.7 depicts a typical polymer flood process flow diagram (PFD). The polymer handling, processing, and metering system is simply integrated with a conventional waterflood injection layout. Control and automation of the polymer system is interlocked with the injection system for automatic start, stop, and rate control of the polymer mother solution metering pump. The polymer handling, dispersion, batching, and hydration processes are automatically controlled based on demand from the mother solution metering pump.

In this mode, a specialized polymer handling, processing, and metering system is added to the basic waterflood facility design. The polymer system will include a low-shear, positive-displacement-type pump system to meter a calculated flow of hydrated concentrate "mother" solution to the injection pump suction manifold where it is mixed with dilution water to the desired downhole concentration. An inline or static mixer is normally incorporated downstream of the mother solution injection point to enhance mixing and produce a final homogeneous solution prior to entering the final solution filtering skid. The optimum mother solution to dilution water ratio is in the 1:3 range. Commercial polymer hydration systems in use today produce mother solutions in ranges up to 10,000 ppm or 1 wt.%. Depending upon water quality and the desired downhole viscosity, most polymer floods are designed for a range of 500–3000 ppm final solution polymer concentration.

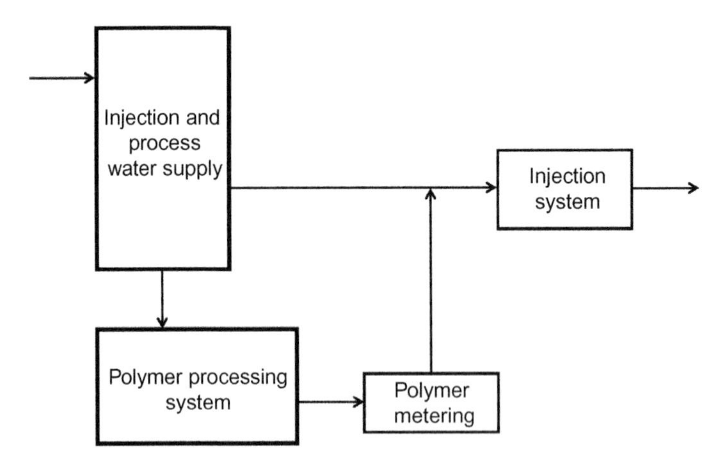

FIGURE 14.7 This block diagram depicts a typical polymer flood process flow diagram (PFD). The polymer handling, processing and metering system is simply integrated with a conventional waterflood injection layout.

14.3.2 Surfactant-Polymer Flooding

Building on the polymer flood PFD, an SP flood facility shown in Figure 14.8 adds the surfactant component(s) and the associated metering pump system integrated with the injection system for automation and proportional rate control that ensures that the desired SP recipe is continuously maintained even if the injection rate changes during daily operations.

Generally, SP programs require simultaneous injection of the surfactant and polymer components within a specified recipe with the surfactant portion being relatively lower concentration compared to the polymer. The overall surfactant slug can also be made up of multiple components that are delivered to the location separately and are then combined on-the-fly in a specified order of mixing to create the final surfactant stream. Most surfactants are available to the operator as liquid concentrates from the chemical manufacturer. Thus, these chemicals require liquid storage and positive displacement type metering pumps at the injection facility to properly ratio them into the dilution water stream, usually upstream of the polymer mother solution addition point.

14.3.3 Alkaline-Polymer Flooding

An AP flood PFD, Figure 14.9, shows the alkaline solution used as the dilution stream for the polymer mother solution. Metering of the soda ash to the dissolving tank is provided through use of a loss-in-weight or gravimetric screwfeeder that is paced by the proportional control system. Thus, soda ash

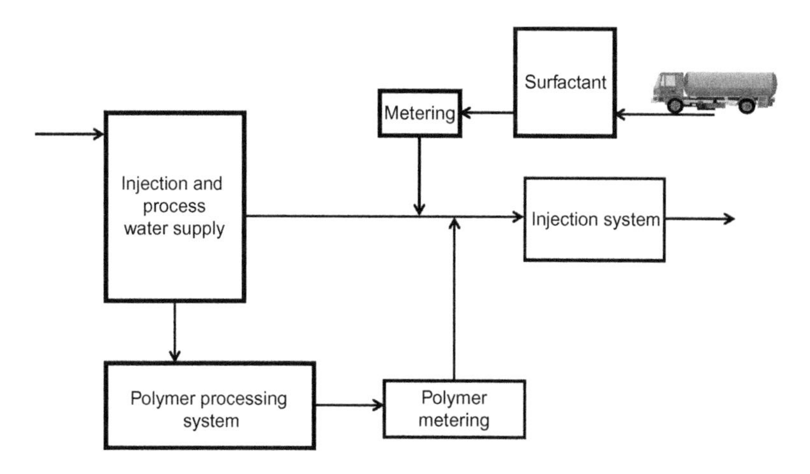

FIGURE 14.8 Building on the polymer flood PFD, a Surfactant-Polymer flood facility adds the surfactant component(s) and the associated metering pump system integrated with the injection system for automation and proportional rate control that ensures that the desired SP recipe is continuously maintained even if the injection rate changes during daily operations.

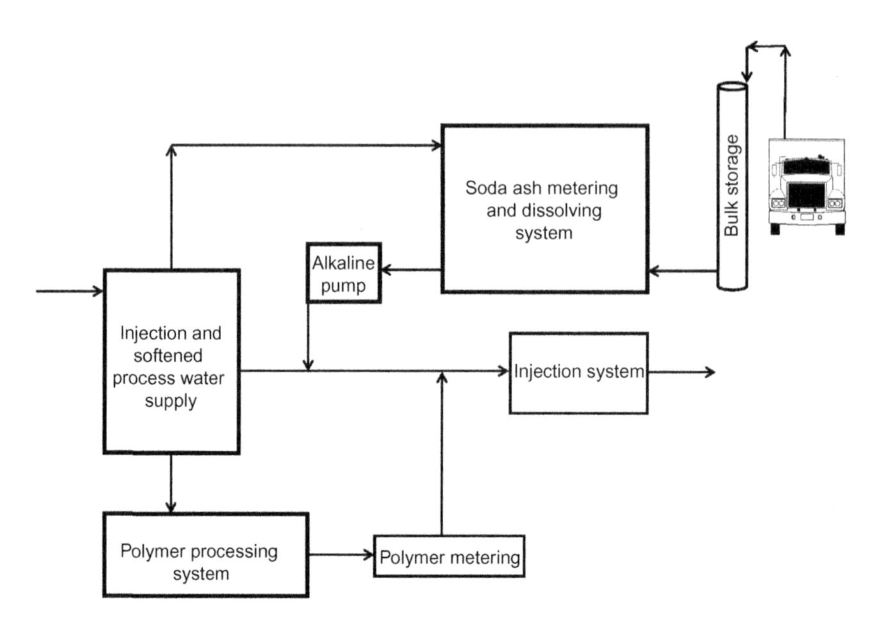

FIGURE 14.9 An Alkaline-Polymer flood process flow diagram shows the alkaline solution used as the dilution stream for the polymer mother solution.

is continuously fed to a dissolving tank and the solution is pumped to the injection system suction manifold as required to maintain suction pressure and meet the cumulative injection rate demand.

In this mode, an alkaline solution is either created from sodium hydroxide stock liquid (NaOH) or by dissolving sodium carbonate (Na_2CO_3) into solution, with the later being the most popular in the field today due to health, safety and environmental considerations, reservoir lithology reasons, and cost. Soda ash may be prepared either by batch method into a concentrate solution that is metered by positive displacement pump into the dilution water and polymer mother solution stream, or it may be flash mixed on-the-fly through a continuous dissolving system, with the resultant solution used as the primary dilution stream for the polymer mother solution. Alkaline solutions derived from soda ash should be filtered to a nominal 10 μm prior to mixing with the polymer mother solution. Compared to a batch process scheme, the on-the-fly continuous method offers the facility designer and project operator a more compact process facility with reduced equipment components, complexity, and a smaller footprint.

Sodium hydroxide or caustic soda is typically delivered from the manufacturer in a 50% concentrate solution. The NaOH is metered into the dilution water stream by a positive displacement pump upstream of the polymer mother solution injection point.

Typical downhole alkaline concentrations range from 0.5 to 3 wt.%.

14.3.4 Alkaline-Surfactant-Polymer

Combining the primary process flows of alkaline, surfactant, polymer, and a conventional waterflood injection scheme results in an ASP PFD similar to the one shown in Figure 14.10 that was developed for a 10,000 BPD in Columbia. This design used a single surfactant component and a batch style soda ash dissolving and metering system. Therefore, all three chemical streams required positive displacement metering pumps in addition to a final water dilution stream to achieve the downhole rate. The chemical stream rates are automatically calculated and controlled based on input of the desired downhole recipe and cumulative injection rate to the facility computer control system. Notice, also, the water softening system to pretreat all the process and injection water.

As the name implies, ASP flooding, includes all three of the above modes combined into a single ASP injection fluid recipe or cocktail. ASP also implies the order of chemical introduction to create the final downhole fluid. If the alkaline has been created using the batch method, it is recommended that the metered concentrate alkaline be introduced to the dilution water stream, followed by surfactant, and finally by the polymer mother solution. In some cases, one may also elect to predilute the surfactant in a ratio of 1:4 prior to mixing with the other components. This step needs to be verified in the lab to determine its necessity.

Many ASP injection programs are staged to inject a defined pore volume ASP slug followed by a defined pore volume of polymer-only (P) injection. Also, in the case of a large field, the ASP project may be designed to inject progressively into patterns throughout the field's acreage. This method allows the operator to manage the ASP project with a lower capital investment and a lower daily operational cost. Furthermore, it also provides an ongoing evaluation of the project's effectiveness without an upfront all-in investment. In this type of ASP/P injection scheme, it is necessary to design the chemical EOR injection facilities so that the operator can simultaneously process and inject the two phases. Therefore, the polymer and injection systems must be designed to process the required fluids when the polymer-only stage starts and the ASP phase moves to the next pattern.

14.4 WATER TREATMENT AND CONDITIONING

Pretreating process and injection water to minimum standards is often required. AP and ASP projects require removal of calcium and magnesium hardness to very low levels (<17 ppm or 1 grain of hardness) to prevent precipitation of these ions due to the high pH solution. Softening low TDS, fresh water can be accomplished with conventional ion exchange softeners, as shown in Figure 14.11. In this example, over 36,000 BPD of fresh well water is softened through this single train. A duplicate softener train provides

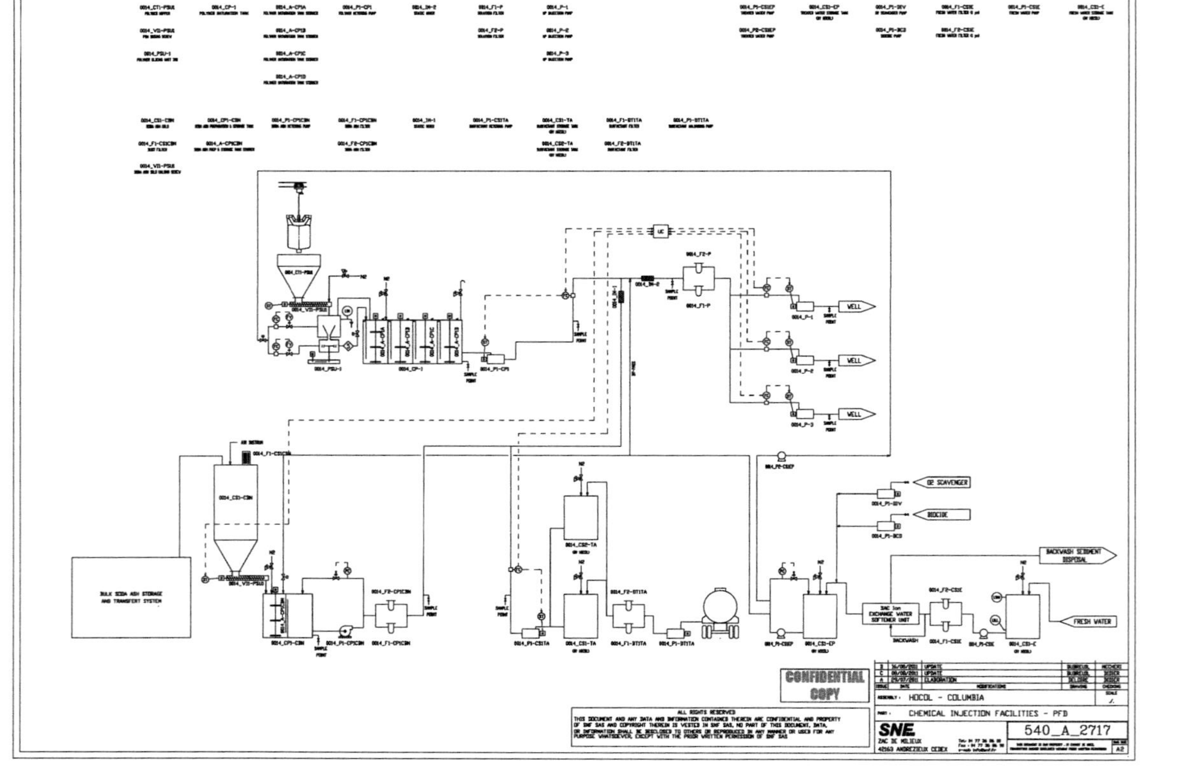

FIGURE 14.10 Combining the primary process flows of alkaline, surfactant, and polymer with the conventional waterflood injection scheme results in an ASP PFD similar to this one developed during the facility design phase for a 10,000 BPD. This design used a single surfactant component and a batch style soda ash dissolving and metering system. Therefore, all three chemical streams required positive displacement metering pumps in addition to a final water dilution stream to achieve the down-hole rate.

FIGURE 14.11 AP and ASP projects require removal of calcium and magnesium hardness to very low levels ($<$ 17 PPM or 1 grain of hardness) to prevent precipitation of these ions due to the high pH solution. Softening low TDS, fresh water can be accomplished with conventional ion exchange softeners.

operational redundancy and permits one train to regenerate and back wash while the other continues softening.

It is outside the scope of this chapter to address the myriad of injection water scenarios and the complexity of handling, treating, and storing certain types of water. Both fresh and produced water sources are successfully used for chemical EOR projects. The level of water treatment and conditioning is determined by the mode of chemical EOR to be implemented. Of course, operational economics, availability of fresh water, disposal of excess produced water, and the EOR reservoir studies all play an important role in the final decision of which water source or sources are to be utilized.

An excellent source for more information about produced water and its use in a chemical EOR project is the Produced Water Society web site, http://producedwatersociety.com.

Some general guidelines, developed through years of field experience, for chemical EOR projects, should be followed to prevent field upsets and project failure. Figure 14.12 lists the primary water characteristics that affect the efficacy of polymer, surfactants, and alkaline chemicals. In all modes of chemical EOR injection, one should also follow the best practices for primary waterflooding. (Bennion et al., 1998)

Note: If Fe^{2+} (ferrous) is present in the mixing or dilution water, it is necessary to treat the water to remove or convert the ferrous. In lieu of pretreatment, one may scavenge the O_2 and then blanket water tanks and chemical mixing equipment with nitrogen to prevent polymer degradation by Fe^{2+}or H_2S in the presence of O_2.

	O_2	Fe^{2+}	H_2S	Total Hardness	TDS	TSS	Skim oil Carryover
Polymer	< 40 PPB if H_2S or Fe^{2+} is present	<1	O_2 scavenger and N_2 blanket required if present	4000	100,000	<20	<5
S-P	Same as above	Same as above	Same as above	4000	100,000	<20	<5
A-P	Same as above	Same as above	Same as above	<17	(a)	<20	<1
ASP	Same as above	Same as above	Same as above	<17	(a)	<20	<1

FIGURE 14.12 Units shown are mg/L and considered the maximum allowable for optimum chemical performance. The unit of O_2 value is PPB.

In the case of AP and ASP projects high TDS is not a limiting factor because the total hardness of water must be reduced to a value <17 ppm (1 grain of hardness) and the high hardness values of high TDS brines and the treatment methods required to treat these brines are often disqualified due to high treatment costs per barrel.

14.5 HANDLING AND PROCESSING EOR CHEMICALS ON-SITE

With the exception of stock liquid surfactant products and sodium hydroxide (caustic soda) liquid alkaline agent, the polymer and soda ash components require installation of specialized handling and processing subsystems at the injection facility. Each of these systems can be segregated into the handling and storage of chemical as supplied by the manufacturer and the process section where the stock dry chemical is processed into solution.

14.5.1 Polymer Handling, Processing, and Metering

The logistics of handling and storing the EOR chemical stock reagents are the primary operational factors requiring an increase of manpower. Accepting dry chemicals in bulk shipments minimizes these daily operational expenses for large projects. Figure 14.13 shows 40,000 lbs. of containerized bulk polymer being transferred to the on-site silo. The polymer processing system automatically conveys the dry product from the silo during the

FIGURE 14.13 Accepting dry chemicals in bulk shipments minimizes daily operational expenses for large projects. This photo shows 40,000 pounds of containerized bulk polymer being transferred to the on-site silo. Source: *Photo courtesy of SNF Floerger.*

process steps to produce the hydrated mother solution. On-site silos for both polymer and soda ash should be purged with dehumidified air or nitrogen to prevent moisture contamination and resulting handling problems.

Polymer manufacturers supply EOR grade products in both dry and liquid emulsion forms. Dry product is available in bulk, semi-bulk bags ($\sim$750 kg net), and small bags (25 kg net). For the purposes of facility design, one may consider dry polymer as 100% active. Liquid emulsion products range from 25% to 30% active within a stable, mineral oil and water emulsion. They are available in bulk, totes, and drums.

Most landside polymer EOR projects utilized the dry form for economical reasons. Regardless of which stock polymer form is delivered, each requires a specialized system that properly hydrates the chosen polymer into a homogeneous, fish-eye free solution.

Dry polymer processing requires a purpose-designed wetting device followed by a properly sized hydration tank arrangement. Key to trouble-free, continuous operations is the polymer wetting device used to initially wet the dry polymer particles and start the hydration process. Three primary types of dry polymer wetting devices, with proven oilfield track records, are commercially available—the polymer particle size reduction type, the solids/liquid eductor, and the solids/air eductor with water curtain. Not all of the commercial systems process dry polymer at the same rate. In fact, depending upon the project throughputs, some of these wetting methods may not meet the flow requirement. The caveat when selecting which dry polymer process method to use is to verify its operating parameters under field conditions and its reliability under oilfield operating conditions.

The SNF FLOQUIP PSU™, or polymer slicing unit, type of dry polymer dispersion system, Figure 14.14, operates at variable flow rates, depending upon model, and achieves accelerated hydration and high mother solution concentrations by reducing the size of the dry polymer particle and increasing available particle surface area during initial wetting. Currently, this type of technology produces hydrated mother solutions at the highest rates when compared to older methods. Continuous production rates exceeding 250 gpm are possible.

The advantage of mother solutions up to 15,000 ppm and accelerated hydration is a reduced footprint for the overall polymer processing and hydration system when compared to the conventional batch systems that process the mother solution to a maximum concentration of 5000 ppm. Figure 14.15 shows the continuous cascading type of hydration tank installed within a compact, factory-prefabricated module. If conventional polymer dispersion technology was used, this tank would have to be six times its volume to produce the same process throughput. Notice the materials of construction are stainless steel and the enclosed tank is outfitted with nitrogen blanketing to prevent degradation of the polymer from exposure to ferrous or H_2S in the presence of oxygen.

FIGURE 14.14 The SNF FLOQUIP PSU, or Polymer Slicing Unit, type of dry polymer dispersion system operates at variable flow rates and achieves accelerated hydration and high mother solution concentrations by reducing the size of the dry polymer particle and increasing available particle surface area during initial wetting. Source: *Photo courtesy of SNF Floerger.*

Figure 14.16 shows the water curtain type of polymer dispersion and wetting system where a solids/air eductor pneumatically pressure coveys the dry polymer to the wetting device that discharges directly to the hydration tank. While these systems are long proven in oilfield environments, they are limited to mother solution concentrations of 5000 ppm and throughput rates up to 100 gpm.

After initial dispersion and wetting, the concentrate mother solution is discharged to a hydration tank or tanks where the polymer undergoes low shear agitation for a specified hydration time interval before metering to the injection system. Under most conditions, system engineers use a 60-min hydration interval when designing the hydration tank volume required for the project. One should note that a specific polymer's optimum hydration time depends upon water quality, temperature, and product molecular weight. Even though a simple lab study with field water provides a baseline curve for viscosity versus hydration time, one may expect the field process equipment to be more efficient in the initial water polymer dispersion and wetting stages when compared to laboratory polymer dissolving procedures.

Once hydrated, the polymer mother solution may be transferred to a day tank where low shear positive displacement pump systems meter the solution

FIGURE 14.15 The advantage of mother solutions up to 15,000 PPM and accelerated hydration is a reduced footprint for the overall polymer processing and hydration system when compared to the conventional batch systems that process the mother solution to a maximum concentration of 5000 PPM. This photo shows the continuous cascading type of hydration tank installed within a compact, factory pre-fabricated module. Source: *Photo courtesy of SNF Floerger.*

FIGURE 14.16 This photo shows the water curtain type of polymer dispersion and wetting system where a solids/air eductor pneumatically pressure coveys the dry polymer to the wetting device that discharges directly to the hydration tank. While these systems are long proven in oil field environments, they are limited to mother solution concentrations of 5000 PPM and throughput rates up to 100 GPM. *(Photo by John M. Putnam)*

to the injection system. Progressing cavity, rotary lobe, plunger, and diaphragm style pumps are best suited for metering the viscous mother solution in a low shear environment. Polymer solutions and fluids are sensitive to mechanical shear degradation caused by excessive pressure differential when flowing through pumps, across valves, through inline mixers, and orifices. Piping and flow lines are sized to prevent turbulent flow, usually 2–4 fps (0.6–1.2 mps). Progressing cavity, rotary lobe, and diaphragm pumps are normally operated at nominal speeds conservatively less than the rated maximum speeds due to the high viscosity and shear sensitivity.

14.5.2 Surfactant Handling and Metering

Liquid stock surfactant products are typically delivered in bulk and stored in a tank farm similar to the one shown in Figure 14.17. Many stock surfactants require insulated and heated tanks to keep product viscosity within specifications for pumping and efficient dilution. In this case, surfactant chemicals may require a minimum loading and delivery temperature. In case of weather or operational delays, the bulk delivery service must be able to circulate engine coolant or have a local steam service to reheat the insulated tanker before delivery to the site.

As stated earlier, the surfactant may be a single stock product delivered from the manufacturer, or multiple surfactant components delivered from separate manufacturers to be combined at the injection facility. Surfactants are typically supplied to the field as concentrate liquids that can be directly metered to the injection system with positive displacement metering pumps. Progressing cavity or diaphragm style pumps are typically specified for surfactant metering.

FIGURE 14.17 Liquid stock surfactant products are typically delivered in bulk and stored in a tank farm similar to the one shown in this photo. Many stock surfactants require insulated and heated tanks to keep product viscosity within specifications for pumping and efficient dilution.

During the facility design phase, the physical and chemical characteristics of the surfactant or multiple surfactant components must be understood and the chemical manufacturer's recommendations for shipping and storage considered. Some surfactant components may be alcohols and require special consideration on location due to fire hazard. And, many surfactant components display high viscosity at low temperatures and require heated storage tanks and insulated flow lines. (Some surfactant components even display a reverse temperature viscosity curve, becoming more viscous with higher temperatures.) Surfactant manufacturers regularly recommend agitation or circulation of their products when stored in bulk tanks. However, some of these products produce foam if the circulation or agitation system is not properly designed and operated.

14.5.3 Alkaline Agent Handling, Processing and Metering

In almost all cases, the optimum delivery and site-handling method for soda ash is bulk. Figure 14.18 shows contractors erecting a 4000-ft^3 factory-prefabricated steel silo for soda ash storage. Pneumatic bulk trucks deliver the dry product without disruption of the soda ash dissolving process. Soda ash should be pneumatically conveyed because mechanical conveying causes dusting and pack setting problems with augers and inclined screws. In geographical locations where bulk deliveries are not possible, bulk bag off-loading systems can still be used to transfer soda ash from 1-ton bags bags to bulk storage silos.

FIGURE 14.18 In almost all cases, the optimum delivery and site handling method for soda ash is bulk. Here, contractors erect a 4000 ft^3 factory pre-fabricated steel silo. Pneumatic bulk trucks deliver the dry product without disruption of the soda ash dissolving process.

Because the alkaline-injected concentration is typically much higher than other chemical components, logistics and handling of dry soda ash requires careful preplanning and facility design to accommodate delivery vehicles and on-site storage all while continuously processing the soda ash into solution for injection. While a batch method of soda ash preparation is valid, it is this author's experience that the continuous flash mixing method of alkaline preparation is best suited for landside oilfield operations. Additionally, bulk delivery of soda ash to site is preferred over any other package. Pneumatic conveyance, whether pressure or vacuum dilute phase, should be utilized to convey the dry soda ash from delivery trucks to bulk storage silos and intermediate hoppers. Mechanically conveying with screws and augers reduces the friable soda ash particle and produces dust and operational problems.

Soft water with a total hardness of <1 grain (17 ppm) must be used in the preparation of the alkaline solution. Otherwise, the high pH solution causes the calcium and magnesium to precipitate and creates subsequent problems with the mixing equipment, filters, pumps and flow lines—not to mention possible well plugging. (Softening water by the pH method is viable for very large, high-rate projects or when having to use high TDS produced water becomes more of a water-treating operation rather than the simple production of an alkaline solution suitable for downhole injection.)

With water temperatures greater than 50°F (10°C), soda ash dissolves with vigorous agitation within 15–30 min. The continuous mixing type dissolving system uses a loss-in-weight or gravimetric screw feeder to meter the required amount of soda ash continuously into the flash mix section of a cascading tank with the solution moving through multiple compartments separated by overflow weirs. The screw feeder output is calculated based on the cumulative rate from the downhole injection system and solution quality is monitored by pH and conductivity data from inline monitors or periodic sampling and testing.

Alkaline solution is pumped from the last tank compartment through a bag filter system (10 μm recommended) with a centrifugal pump to the injection system suction header. In this scenario, the alkaline solution serves as the dilution for the polymer mother solution.

If a batch method of alkaline preparation from soda ash is employed, one must provide additional alkaline solution storage tanks, alkaline transfer pumps, positive displacement metering pumps, and a separate water dilution pump for final dilution of the stock chemical streams. Be certain to consult soda ash dissolving tables for minimum allowable water temperature and maximum soda ash solution concentrations during the design of a batch style system. An excellent handbook detailing soda ash handling and storage is FMC Soda Ash Storage Options and Technical Data. (FMC Corporation, November 2000) This publication can also be found at http://www.fmcchemicals.com/Portals/chem/Content/Docs/Soda%20Ash%20Documents/SodaAshStorageHandling.pdf.

Liquid sodium hydroxide (NaOH) is a highly caustic concentrate that can be stored on-site and metered to the dilution water stream with positive displacement pumps. Diaphragm or sealless pump types are preferred to prevent leakage of the hazardous liquid. Caustic soda requires some very specific storage conditions with regard to temperature and product concentrations. Prior to designing facilities for use of sodium hydroxide, it is recommended to consult a chemical manual and the manufacturers' product bulletins for handling, storage, and pumping. Operating personnel require product specific safety training and personal protective equipment. The handbook, Dow Caustic Soda Solution Handbook, (Dow Chemical Company, August 2010) is an excellent source for handling and storage of NaOH or caustic soda. More information may also be found at http://www.dow.com/causticsoda.

14.6 INJECTION SCHEMES AND STRATEGIES

The final downhole injection system design depends upon many operational and economical factors. If the project's cumulative injection rate is low (under 5000 BPD) and the number of IWs not too numerous (10 or fewer), one may consider a dedicated injection pump for each IW. This type of distributed injection permits the operator to control individual well rates continuously for volumetric balancing across the injection pattern. In this setup, each injection pump is usually equipped with an electronic variable speed drive that provides a wide operating range without the need to make mechanical changes to the pump drive in order to achieve the desired rate. This is especially useful for pilot projects where the injection pressure is subject to change as injection of the viscous fluid is established and incremental rate changes are necessary to maintain a desired injection pressure.

For many chemical EOR projects, a dedicated injection pump is installed for each IW. This type of injection scheme permits the operator to control injection rates independent of each well's pressure response, thus allowing balancing of injection patterns—particularly during a pilot project. Primary rate control is provided through motor speed control using electronic variable speed drives. In Figure 14.19, notice the stainless steel piping and final solution filtration canisters.

However, when integrating a chemical EOR project with an existing waterflood facility or implementing a very large project, in terms of cumulative injection rates, aerial expanse, or number of IWs, it is likely that

FIGURE 14.19 For many chemical EOR projects, a dedicated injection pump is installed for each injection well. This type of injection scheme permits the operator to control injection rates independent of each well's pressure response, thus allowing balancing of injection patterns – particularly during a pilot project. Primary rate control is provided through motor speed control using electronic variable speed drives. *(Photo by John M. Putnam)*

multiple IWs will be served by a common primary pressure injection header through a trunk line distribution system and then branched to the individual IW flow lines. In this scenario, chokes or flow control valves are used to regulate flow during waterflooding. However, polymer fluids cannot be regulated with these conventional methods due to high sensitivity to mechanical shear degradation.

To use this type of injection scheme and still provide rate control to the individual wells, one may consider installing a dedicated high-pressure, polymer mother solution metering pump with discharge downstream of the flow control valve. Thus, the polymer mother solution is metered to the high-pressure dilution stream. With flow rate instrumentation installed on both the high-pressure dilution flow stream and the polymer mother solution metered stream, one can manually or automatically set the desired cumulative rate and proper polymer mother solution rate to achieve the desired downhole recipe.

Regardless of the injection scheme, the viscosity of the polymer fluid usually requires that the maximum operating speed of multiplex plunger style pumps be derated to reduce mechanical shear degradation and improve valve efficiency. Consult the API Standard 674, Positive Displacement Pumps—Reciprocating, (American Petroleum Institute, December 2010, 3rd Edition) for information about fluid viscosity and maximum operating speed. A maximum 250 to 300 rpm is a typical guideline when calculating the $1.2 \times$ nominal rate. Pumps driven with a variable speed drive or variable frequency drive may also require auxiliary electric motor cooling fans and crankcase lubrication systems to operate at very low speeds. Do not attempt to pump polymer fluids through horizontal centrifugal, multistage centrifugal, or conventional centrifugal pumps.

14.7 MATERIALS OF CONSTRUCTION

Industry codes and standards dictate the materials of construction for piping, fittings and valves depending upon the project location and classification of service. Additionally, some operators have their own codes and standards that may go above and beyond the general API, ASME, ASTM, CSA, etc. standards. Facilities that are designed to utilize produced water, high TDS waters, and those with H_2S content may be required to use higher grade piping than those designed for use with fresh water. As a matter of design principle, one should consult the current codes and standards currently in effect for a waterflood facility to be constructed under the same conditions expected for the chemical EOR facility. Doing so saves many projects from cost overruns by preventing replacement of complex piping systems because they were not originally fabricated with the acceptable classes of piping and fittings.

Chemical suppliers provide recommendations for storage vessels and wetted pump components. Generally, lined carbon steel or fiber-reinforced plastic (FRP) tanks suffice for storage of surfactant components and diluted alkaline solutions. Polymer systems are normally fabricated with all stainless steel wetted surfaces for tanks, piping, valves, filter canisters, and pumps. Most factory-prefabricated process facilities utilize stainless steel wetted surfaces for all tanks, piping, fittings, valves, filter canisters, and pumps. However, the selection of other corrosion-resistant materials for the fabrication of the chemical process systems should not be excluded from the designer's allowable materials list; especially for capital cost sensitive projects.

14.8 CONCLUSION

The goal of implementing a chemical EOR project is to transpose the details of laboratory studies and expectations of reservoir engineering to a field scale. Doing so successfully requires careful planning, fit-for-purpose system designs, oilfield-proven mechanical components, and implementation by an experienced chemical EOR field team. Regarding facility design, fabrication, installation, and commissioning, the operator's task remains to select a vendor and service provider from a narrow field of qualified and experienced suppliers.

It is important to remember that while the chemical EOR process facilities will be installed on the injection side of the flood, this does not mean that even the best waterflood operator can immediately adapt to handling and processing chemicals that are completely foreign to the typical waterflood operation. In fact, a chemical EOR project brings systems and processes completely outside the scope of the normal, day-to-day, waterflood operator's routine and knowledge base. Therefore, do not underestimate the daily manpower requirements to operate a chemical flood and invest wisely in the thorough training of field supervisors and operators by an experienced service provider.

REFERENCES

Dow Caustic Soda Solution Handbook, August 2010.

FMC Soda Ash Storage Options and Technical Data, November 2000.

Injection water quality—a key factor to successful waterflooding. J. Can. Petrol. Tech. 1998 (revised).

Positive Displacement Pumps—Reciprocating, API Standard 674. (Edition 3).

Chapter 15

Steam Flooding

James J. Sheng
Bob L. Herd Department of Petroleum Engineering, Texas Tech University, Lubbock, TX 79409, USA

15.1 THERMAL PROPERTIES AND ENERGY CONCEPTS

We first review thermal properties of rock and fluids followed by energy concepts. Then the general practice in steam flooding projects is discussed, and field cases are presented.

15.1.1 Heat Capacity (C)

Heat capacity (usually denoted by the capital C, often with subscripts), or thermal capacity, is the measurable physical quantity that characterizes the amount of heat required to change a substance's temperature by a given amount. In the International System of Units (SI), heat capacity is expressed in the unit of joule(s) (J) per kelvin (K). The unit of kJ/°C is often used in thermal recovery. C_V and C_p denote the heat capacity at constant volume and constant pressure, respectively.

Derived quantities that specify heat capacity as an intensive property, independent of the size of a sample, are the molar heat capacity, which is the heat capacity per mole of a pure substance, and the specific heat capacity, often simply called specific heat and still denoted by the capital C, which is the heat capacity per unit mass of a material (e.g., kJ/kg °C). Sometimes, it is more convenient to express the heat capacity of a substance on the basis of unit volume instead of unit mass. It then is called the volumetric heat capacity M which is equal to the ρC, where ρ is the bulk density of the material.

15.1.2 Latent Heat (L_v)

Latent heat is the amount of energy in the form of heat released or absorbed by a substance during a change of phase (i.e., solid, liquid, or gas). In thermal recovery, latent heat refers the heat released by water during the change

Enhanced Oil Recovery Field Case Studies.

from steam to hot water at the same temperature. The unit is kJ/kg, for example.

15.1.3 Sensible Heat

In contrast to a latent heat that is hidden during heat exchange, sensible heat can be observed as a change of temperature. Sensible heat may be expressed by the product of the material mass (m) with its specific heat capacity (C) and the change in temperature ($T - T_r$): $H_{sensible} = mC\ (T - T_r)$, T_r is a reference temperature. For a unit mass of liquid water, any heat change will cause temperature change. Therefore, the sensible heat change can be described by $h_w = C_w(T - T_r)$. H and h_w will be defined later in Section 15.1.6.

15.1.4 Total Volumetric Heat Capacity

To estimate how much heat is needed to heat a reservoir, we need to know the total volumetric heat capacity of the reservoir (M_R). For a reservoir of porosity ϕ filled with a nonvolatile oil, water, and a gas phase containing steam and noncondensable gas, the amount of heat required to increase the temperature of a bulk volume of formation V_b by a small amount of ΔT, and at a constant pressure, is

$$Q = V_b M_R \Delta T \tag{15.1}$$

where M_R is the isobaric volumetric heat capacity of the bulk, fluid-filled reservoir.

However, the increase of the heat content of this reservoir by an increase in temperature ΔT is

$$\begin{aligned} Q = {} & (1-\phi)\rho_r C_r \Delta T + \phi S_o \rho_o C_o \Delta T + \phi S_w \rho_w C_w \Delta T \\ & + \phi S_g(\rho_g C_g f_g \Delta T + (1-f_g)(\rho_s C_w \Delta T + L_v \rho_s)) \end{aligned} \tag{15.2}$$

where f_g is the volume fraction of noncondensable gas in the vapor phase; C is the heat capacity per unit of mass for a unit temperature change; ρ is the density; S is the saturation, the subscripts r, o, w, g, and s denote rock, oil, water, gas, and steam, respectively; ϕ is porosity in fraction.

By equating the above two Qs, we can find the total volumetric heat capacity:

$$\begin{aligned} M_R = {} & (1-\phi)\rho_r C_r + \phi S_o \rho_o C_o + \phi S_w \rho_w C_w \\ & + \phi S_g\left(\rho_g C_g f_g + (1-f_g)\left(\rho_s C_w + \frac{L_v \rho_s}{\Delta T}\right)\right) \end{aligned} \tag{15.3}$$

Note that the steam contribution has two terms. One is the latent heat of vaporization $\rho_s L_v/\Delta T$, and the other one is the sensible heat $\rho_s C_w$. L_v is the average within ΔT.

15.1.5 Thermal Diffusivity (α)

Thermal or heat diffusivity is defined as the ratio of the thermal conductivity to the volumetric heat capacity:

$$\alpha = \frac{\lambda}{\rho C} \tag{15.4}$$

where λ is the thermal conductivity (coefficient) of a material, and ρ is the density. Notice the similarity in form of hydraulic diffusivity $\eta = k/(\mu \phi c_t)$, where k is the permeability, μ is the viscosity, ϕ is the porosity, and c_t is the total compressibility.

15.1.6 Enthalpy (H, h)

Enthalpy (H) is a measure of the total energy of a thermodynamic system. Its unit is in Joule (J) or kJ. Mathematically, it can be expressed by $H = U + pV$, U is the internal energy, p is the pressure at the boundary of the system and its environment, V is the volume of the system. The total enthalpy, H, of a system cannot be measured directly. Thus, change in enthalpy, ΔH, is a more useful quantity than its absolute value. h is the enthalpy per unit mass of material (e.g., kJ/kg). In thermal recovery, $h_s = L_w + h_w = L_w + C_w(T - T_r)$. Here the subscript s stands for steam, and w for water.

15.1.7 Vapor Pressure, Saturation Pressure, and Saturation Temperature

At any system pressure and temperature, a liquid has some vapor pressure. When the vapor pressure of the liquid equals the system pressure, this vapor pressure is the saturation pressure, and the corresponding temperature is the saturation temperature which means the boiling point.

15.1.8 Steam Quality

Steam quality is the amount of the steam (vapor), by weight, expressed as a fraction (or percent) of the total mass of liquid and vapor. Note that foam quality in foam flooding is expressed as the fraction or the percentage of gas volume in the foam.

15.1.9 Temperature-Dependent Oil Viscosity

The dependency of oil viscosity on temperature is described by the Andrade (1930) equation:

$$\mu_o = A \exp(B/T) \tag{15.5}$$

where T is in absolute degrees, and A and B are empirical constants determined by measurements.

15.1.10 Gravitational Potential Energy

When accounting only for mass, gravity, and altitude, the potential energy is:

$$E_g = mgh \tag{15.6}$$

where E_g is the potential energy of the object, m is the mass of the object, g is the acceleration due to gravity, and h is the altitude of the object relative to a reference datum. If m is expressed in kilograms, g in meters per second squared and h in meters and U will be calculated in joules.

The contribution of the potential energy of the total energy in a typical thermal recovery project is small except where the change in altitude is large. This is not to say that the contribution due to gravity on the potential gradient in the Darcy equation is negligible (Prats, 1982).

15.1.11 Kinetic Energy

In classical mechanics, the kinetic of an object is given by

$$E_k = \tfrac{1}{2}mv^2 \tag{15.7}$$

where v is the velocity of an object. If the Darcy velocity, u, is used, v should be replaced by (u/ϕ). Kinetic energy contributions are usually greatest near the wellbore where the fluid velocities are largest. But the contribution of kinetic energy to the energy balance of a reservoir is, for practical purposes, negligible (Prats, 1982).

15.1.12 Total Energy

The total energy of an object is then

$$E_t = mh + E_g + E_k \tag{15.8}$$

15.2 MODES OF HEAT TRANSFER

The fundamental modes of heat transfer are conduction or diffusion, convection and radiation.

15.2.1 Heat Conduction

In heat conduction or diffusion, the transfer of energy occurs between the objects that are in physical contact by molecular collision. It is governed by Fourier's law:

$$u_{\lambda x} = -\lambda \frac{\partial T}{\partial x} \tag{15.9}$$

where $u_{\lambda x}$ is the rate of heat transfer by conduction in the positive x-direction per unit cross-sectional area normal to the x-direction, λ is the thermal conductivity (coefficient) of the material, and T is the temperature. This equation is similar to the Darcy equation in form:

$$u_x = -\left(\frac{k}{\mu}\right)\frac{\partial p}{\partial x} \tag{15.10}$$

15.2.2 Heat Convection

In heat convection, the transfer of energy occurs between an object and its environment due to fluid motion. In other words, the heat is carried by a flowing fluid from one place to another. Therefore, it is related to the fluid velocity and heat capacity. Mathematically, it is:

$$u_{Cx} = u_x M_R(T - T_r) = u_x \rho C(T - T_r) \tag{15.11}$$

where u_{Cx} is the rate of heat by convection in the x-direction, u_x is the Darcy flow velocity in the x-direction, and other symbols are defined earlier. Where fluids flow in porous media, the heat transfer by convection is the dominant.

The above equation to calculate heat loss is expressed in terms of fluid velocity. Alternatively, it can be expressed by

$$u_{Cx} = h_c(T - T_r) \tag{15.12}$$

where h_c is the heat-transfer coefficient for convection.

15.2.3 Thermal Radiation

In thermal radiation, heat is transferred to or from a body by means of the emission or absorption of electromagnetic radiation. The rate of radiation heat transfer from a heated surface per unit surface area is given the Stefan–Boltzmann law as

$$u_r = \sigma\varepsilon(T^4 - T_r^4) \tag{15.13}$$

where σ is the Stefan–Boltzmann constant (1.713×10^{-9} Btu/ft^2 h °R^4), the temperature T is in degree Fahrenheit, and ε is the emissivity of the surface.

The emissivity is dimensionless, equal to one for a black body and 0 for a perfectly reflecting body. Thermal radiation is not considered to be an important heat-transfer mechanism in porous media.

15.3 HEAT LOSSES

Heat losses in a thermal project include those from surface pipes and wellbore, those from the heated reservoir to adjacent formations, and those from by production of hot fluids.

15.3.1 Heat Loss from Surface Pipes

The heat loss from unit length of surface pipe (q_{ls}) is calculated from the temperature difference between the air (T_a) and the fluids in the pipe (T_f) and the overall specific thermal resistance (R_h):

$$q_{ls} = \frac{T_f - T_a}{R_h} \tag{15.14}$$

It is assumed that heat loss rate is steady state, because the transient period is generally short. However, transient state is important during short steam-soak operation. If the temperature of surface is low, radiation is negligible. If there is wind, convection is important. The heat loss from surface pipes is about 1–3% of the injected heat for 1000 m.

15.3.2 Heat Loss from a Wellbore

The heat loss from a wellbore never reaches a steady state. It attains a quasi-steady state in which the rate of heat loss is a monotonically decreasing function of time (Ramey, 1962; Willhite, 1967). Generally, convection is neglected and the pressure is assumed unchanged. Thus, only steam quality is reduced from the wellhead to the sand face. The calculation equation is similar in form to that for the heat loss from surface pipes, but the thermal resistance R_h is a variable and has to be determined by iteration. The heat loss from wellbores is about 10–20% of the injected heat for 1000 m.

15.3.3 Heat Loss to Over- and Underburden Rocks

The heat flow through the overburden is usually modeled by the equation of linear flow of heat in a semi-infinite medium:

$$\frac{\partial^2 T}{\partial z^2} = \frac{1}{\alpha}\frac{\partial T}{\partial t} \tag{15.15}$$

where α is the thermal diffusivity of the overburden rock. The initial and boundary conditions may be expressed as

$$T(z,0) = T_R \tag{15.16}$$

$$T(0,t) = T_s(t) \tag{15.17}$$

$$T(\infty,t) = T_R \tag{15.18}$$

where T_R is the initial reservoir temperature and $T_s(t)$ is the variable steam temperature at the reservoir-overburden interface. The similar equations can be written for the underburden rock. Note that heat conduction in the x- and y-directions are neglected.

If we assume constant thermal properties and a constant reservoir boundary temperature, the analytical solution of the boundary value problem described above is given as

$$T(z,t) = T_R + (T_s - T_R)\text{erfc}\left(\frac{z}{2\sqrt{\alpha t}}\right) \tag{15.19}$$

where erfc is the complementary error function. The heat loss to the overburden per unit area, u_{lb}, is calculated by (Marx and Langenheim, 1959):

$$u_{lb} = -\lambda\frac{\partial T}{\partial z}\bigg|_{z=0} = \frac{\lambda(T_s - T_R)}{\sqrt{\pi\alpha t}} \tag{15.20}$$

However, the reservoir boundary temperature is not a constant and varies with time. An analytical solution for the problem outlined above requires the use of superposition, as done by Grabowski and Aziz (1977), and Abou-Kassem (1981). Chase and O'Dell (1973) applied variational principles. However, the problem can be also solved numerically, as done by Coats et al. (1974). A simple method was proposed by Vinsome and Westerveld (1980).

15.3.4 Heat Loss from Produced Fluids

To predict the heat loss from produced fluids, we need to know the variation with time in the temperature of produced fluids. To do that, we need to predict how the average temperature in the heated zone changes. One of the theories for such calculation is the Boberg and Lantz (1966) model which is described in the chapter of cyclic steam stimulation.

15.4 ESTIMATION OF THE HEATED AREA

When reservoirs are heated by hot-fluid injection, a significant fraction of the injected heat is lost to the surrounding formations. In the Marx and Langenheim model, the reservoir is considered to have uniform thickness (h)

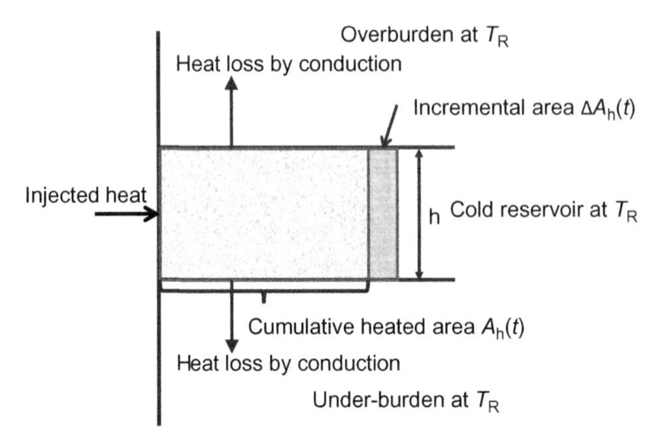

FIGURE 15.1 Schematic description of the Marx and Langenheim heating model.

and fluid and rock properties. The temperature is uniform throughout the vertical cross section. Steam and condensate do not segregate under the influence of gravity.

The steam zone forms as soon as steam is injected to increase the reservoir temperature from T_R to T_s. Heat is lost to the over- and underburden, but not to the cold reservoir zone ahead of the steam front by conduction. However, the heated area increases in the direction of displacement. See Figure 15.1 for the schematic description.

The heat balance equation in the Marx and Langenheim model is

$$m_s h_s = M_R(T_s - T_R)\frac{\Delta V}{\Delta t} + 2\int_0^{A_h} u_{lb}(t-\tau)\mathrm{d}A_h \tag{15.21}$$

where m_s is the steam injection rate, h_s is the heat content of the injected steam relative to the reservoir temperature T_R, M_R is the volumetric heat capacity of the heated zone, A_h is the heated area, ΔV is the volume of the heated zone within Δt, the reservoir thickness is h, u_{lb} is the rate of heat loss to the overburden or the underburden of unit area, t is the total time since the start of injection, τ is the time of arrival of heated zone at a specific location, and A_h is the heated area at time τ.

According to Eq. (15.20), the above equation becomes

$$m_s h_s = M_R h(T_s - T_R)\frac{\mathrm{d}A_h}{\mathrm{d}t} + 2\int_0^{t} \frac{k_h(T_s - T_R)}{\sqrt{\pi\alpha(t-\tau)}}\frac{\mathrm{d}A_h}{\mathrm{d}\tau}\mathrm{d}\tau \tag{15.22}$$

The solution of the above equation gives an expression for the heated area A_h as a function of dimensionless time as

$$A_h = \frac{m_s h_s M_R h}{4(T_s - T_R)\alpha M_b^2}\left[e^{t_D}\mathrm{erfc}(\sqrt{t_D}) + 2\sqrt{\frac{t_D}{\pi}} - 1\right] \tag{15.23}$$

where

$$t_D = 4\left(\frac{M_b}{M_R}\right)^2\left(\frac{\alpha}{h^2}\right)t \tag{15.24}$$

M_b is the volumetric heat capacity of the over- and underburden.

The rate of growth of the heated zone is given by

$$\frac{dA_h}{dt} = \frac{m_s h_s}{(T_s - T_R)M_R h}e^{t_D}\text{erfc}(\sqrt{t_D}) \tag{15.25}$$

We have so far described the estimation of the steam zone assuming enough steam (heat) is injected. As the steam zone expands, more heat is lost to the over- and underburden rocks. By sometime, steam is condensed into hot water. Now we present when this happens (Green and Willhite, 1998).

From Eq. (15.22), the rate of heat loss to the over- and underburden rocks is

$$q_{ls} = 2\int_0^t \frac{k_h(T_s - T_R)}{\sqrt{\pi\alpha(t-\tau)}}\frac{dA_h}{d\tau}d\tau = m_s h_s - M_R h(T_s - T_R)\frac{dA_h}{dt} \tag{15.26}$$

Substituting Eq. (15.25) into the above equation yields

$$q_{ls} = 2\int_0^t \frac{k_h(T_s - T_R)}{\sqrt{\pi\alpha(t-\tau)}}\frac{dA_h}{d\tau}d\tau = m_s h_s\left[1 - e^{t_D} erfc\left(\sqrt{t_D}\right)\right] \tag{15.27}$$

The entire heated region is filled with steam as long as

$$q_{ls} = m_s h_s\left[1 - e^{t_D}\text{erfc}\left(\sqrt{t_D}\right)\right] \le m_s(f_s L_v)_{\text{downhole}} \tag{15.28}$$

We may use this equation to estimate the critical time t_{cD} at which a hot-water zone forms:

$$e^{t_{cD}}\text{erfc}\left(\sqrt{t_{cD}}\right) = 1 - \frac{(f_s L_v)_{\text{downhole}}}{h_s} \tag{15.29}$$

After this critical time, the heat loss to the over- and underburden rocks (q_{lb}) will be the sum of the heat loss from the steam zone and that from the hot-water zone. Mathematically,

$$\begin{aligned} q_{ls} &= 2\int_0^t \frac{k_h(T_s - T_R)}{\sqrt{\pi\alpha(t-\tau)}}\frac{dA_h}{d\tau}d\tau = m_s h_s\int_0^{t_D}\frac{\exp(\tau_D)\text{erfc}(\sqrt{\tau_D})}{\sqrt{\pi(t_D-\tau_D)}}d\tau_D \\ &= m_s h_s\int_0^{t_{Ds}}\frac{\exp(\tau_D)\text{erfc}(\sqrt{\tau_D})}{\sqrt{\pi(t_D-\tau_D)}}d\tau_D + m_s h_s\int_{t_{Ds}}^{t_D}\frac{\exp(\tau_D)\text{erfc}(\sqrt{\tau_D})}{\sqrt{\pi(t_D-\tau_D)}}d\tau_D \end{aligned} \tag{15.30}$$

where the dimensionless time t_{Ds} is the time that the steam front travels; the first term is the rate of heat loss from the steam zone, and the second term is that from the hot-water zone.

Because the heat loss from the steam zone is supplied totally by condensation of steam,

$$m_s h_s \int_0^{t_{Ds}} \frac{\exp(\tau_D)\operatorname{erfc}(\sqrt{\tau_D})}{\sqrt{\pi(t_D - \tau_D)}} d\tau_D = m_s (f_s L_v)_{downhole} \tag{15.31}$$

The above equation is used to estimate t_{Ds}. Once t_{Ds} is known, we can estimate the steam zone:

$$A_s = \frac{m_s h_s M_R h}{4(T_s - T_R)\alpha M_b^2} \left[e^{t_{Ds}} \operatorname{erfc}(\sqrt{t_{Ds}}) + 2\sqrt{\frac{t_{Ds}}{\pi}} - 1 \right] \tag{15.32}$$

The preceding methods based on the Marx and Langenheim model assume that the injection heat content (injection rate) is constant. If not constant, use Ramey's (1959) extended method which basically uses the principle of superposition about the injected heat.

When the injected heat breaks through production wells, some heat is removed. To estimate the heated zones, we can still use the preceding methods, but the total injected heat must be minus the produced heat. Then we must consider the variation of injected heat following Ramey's (1959) method.

15.5 ESTIMATION OF OIL RECOVERY PERFORMANCE

After the heated zones are estimated as presented in the preceding section, we can estimate the oil recovery performance. Several models were presented in the literature, for example, by Myhill and Stegemeier (1978). The main idea is to assume the piston-like displacement of oil (frontal advance models) by steam and hot water. For example, the oil rate by steam drive is

$$q_o = \phi \frac{dV_s}{dt} \left(\frac{S_{oi}}{B_{oi}} - \frac{S_{ors}}{B_{os}} \right) \tag{15.33}$$

where the subscript s denotes steam drive, the subscript i denotes an initial condition, V_s is the steam swept volume, S_{ors} is the residual oil saturation by steam drive, and B is the oil formation factor.

Alternatively, in some injection patterns (e.g., a five-spot pattern), the oil rate is estimated from the equations derived for waterflooding (WF) by using steam and reduced oil viscosities. By doing so, the heated zones (radius of each zone) must be estimated first.

One economic parameter in thermal recovery is oil–steam ratio (OSR) which is defined as the barrels of oil produced by injecting one barrel of steam in cold water equivalent. In a typical condition, about 0.07 barrel of oil is burned to generate one barrel of steam (Green and Willhite, 1998). This ratio gives some reference for economic evaluation, because steam cost is a major cost in a thermal project. One OSR economic limit is 0.15

(Liu, 1997). In practice, the economic cutoff of OSR is higher than that (e.g., 0.3). Generally, OSR in steam soak is higher than that in steam flooding (SF) (also called steam drive).

15.6 MECHANISMS

One obvious mechanism of SF is the reduction of oil viscosity by steam injection. Another important mechanism is the increased reservoir pressure (energy) owing to steam injection. For light oil steam flood, viscosity reduction is not important; instead, steam distillation is the most important recovery mechanism (Konopnicki et al., 1979; Volek and Pryor, 1972; Wu, 1977).

In reservoirs containing volatile oils, very low residual oil can be obtained by a combination of steam displacement and steam distillation (William et al., 1961). In a steam distillation process, hydrocarbons are more readily vaporized because of a lowering of their partial pressures in the presence of steam vapor. The light components are distilled from the residual oil and transported to the steam front where they recondense and mix with the oil bank to form a solvent slug. This is solvent extraction mechanism or steam-strip drive (Hagoort et al., 1976). As the steam zone advances, the solvent slug is displaced and redistilled to further increase oil recovery.

Other mechanisms may include (1) thermal swelling, (2) gas drive, (3) gravity drainage, (4) relative permeability modification and wettability alteration, and (5) emulsification by forming oil/water emulsions (Doscher, 1967; Grathoffner, 1979). As the temperature is increased, connate water saturation increases, residual oil saturation decreases, water relative permeability decreases and oil relative permeability increases (Nakornthap and Evans, 1986). However, the absolution permeability and effective oil relative permeability decrease. Hong (1994) split the mechanisms according to the primary, horizontal, and vertical processes.

By the way, steam drive is a stable process (Harmsen, 1971; Miller, 1975; Prats, 1982). This is because the steam front is condensed owing to heat transfer between the steam fingers and the cold fluids and rock at the displacement front.

15.7 SCREENING CRITERIA

Taber et al. (1997), and Green and Willhite (1998) summarized the general screening criteria for SF. The criteria from these two presentations are a little bit different for some parameters. Table 15.1 summarizes our criteria by updating their presented values with some actual field practices. These parameters are for SF only and steam is generated at surface. The design criteria summarized by Farouq Ali (1974) under which some of actual field projects

TABLE 15.1 Screening Criteria for SF

Parameters	Criteria Values	Design Criteria	Average Field Data
Oil gravity (°API)	9–25	12–25	14.6
In situ oil viscosity (cP)	20–20,000	<1000	3000
Oil saturation (fraction)	>0.4		>0.45
Oil content (bbls/acre-ft)	500–780		
Net thickness (ft)	>20	>30	70.5
Porosity (ϕ) (fraction)	>0.2	≥0.3	0.31
Permeability (mD)	>200	~1000	2300
Transmissibility (mD-ft/cP)	>5		
Depth (ft)	300–5000	<3000	1182
Reservoir pressure (psi)	<1670		300
Reservoir temperature (°C)			29
Gas cap	Not desirable		
Aquifer	Not desirable		
Fracture	No		
Clay	Low		
Water–oil ratio	<10		
Steam quality (%)		80–85	60
Steam pressure (psig)		<2500	595
Spacing (acres)		2–10	4.5

have been successful are also listed in this table. The average field data are also listed in this table. The field data are from Farouq Ali (1974), Farouq Ali and Meldau (1979) and a Chinese project survey (unpublished).

For steam-soak projects, the ranges of parameters can be wider. In practice, for more viscous heavy oil reservoirs, steam soak is conducted before SF.

The API limit is related to *in situ* oil viscosity. Generally, WF may work better than steam flood if the *in situ* viscosity is lower than 20 cP (Taber and Martin, 1983). The average field data is 3000 cP. A high saturation and a high porosity are required for the economic reason because of high steam injection cost. Chu (1985) added that the product of saturation and porosity (both in fraction) should be greater than 0.08. A high permeability is needed so that steam can transport fast enough to beat heat loss. The heat loss to the over- and underburden rocks also limits the reservoir thickness. Generally, the reservoir thickness should be greater than 30 ft (10 m), but could be as low as 15 ft (5 m) if the reservoir depth is below 500 m. The average field data is 70.5 ft (21.5 m).

The limit of reservoir depth is related to heat loss to formation through the vertical wellbore. It is also related to injection pressure. For a too shallow formation, the injection pressure required may exceed the fracture pressure.

From this point, the reservoir depth is also related to well spacing. Generally, if the depth is less than 500 m, the injector−producer distance should be about 100 m; if the depth is 800−1600 m, the distance is 140−150 m and should be less than 200 m (Liu, 1997). The actual field data is 4.5 acre which is equivalent to a well distance of 190 m.

The fraction of heat injected as latent heat decreases as pressure increases. And the temperature of steam increases with pressure, which leads to a higher heat loss to the surrounding formation. Therefore, the equivalent OSR decreases with steam pressure (Myhill and Stegemeier, 1978). These facts limit the maximum reservoir pressure for SF. Another consideration is that for a given mass of water, a greater volume of steam can be converted to steam at a lower pressure and thus a greater volumetric sweep of the reservoir can be achieved. The recovery efficiency increases with the volume of displacing steam. From this point, a lower reservoir pressure is preferred.

Clay content must be low. Some clays are water sensitive, for example, montmorillonite clay swells when contacting with steam or water. The formations with such clays may be treated with clay-stabilizing chemicals like potassium and zirconyl chlorides, hydroxyaluminium, and organic polymers (Young et al., 1980).

The screening criteria are used in the first step in developing a steam injection projects to screen candidate reservoirs for the potential application of the method using available reservoir and fluid property data. Because the criteria are vague and have a wide range, full commitment cannot be made based on the criteria. If sufficient reservoir and fluid data are available, a reservoir model may be constructed and used for a scoping study (Hong, 1986). This kind of study is to define the economic potential of the project under various operating scenarios.

15.8 PRACTICE IN STEAM FLOODING PROJECTS

Some of the design criteria are summarized by Farouq Ali (1974) and presented in Table 15.1. The practices presented in this section may also serve as references for design criteria.

15.8.1 Formation

Almost all steam flood projects were conducted in sandstone reservoirs, except a few cases in carbonate or naturally fractured reservoirs. One was a steam-drive pilot in a fractured carbonate reservoir, Lacq Superieur field (Sahuquet and Ferrier, 1982). Another one was in the Garland field in Big Horn County, Wyoming (Enhanced Recovery Week, 1987). The third one was in the Teapot Dome field, Wyoming (Olsen et al., 1993).

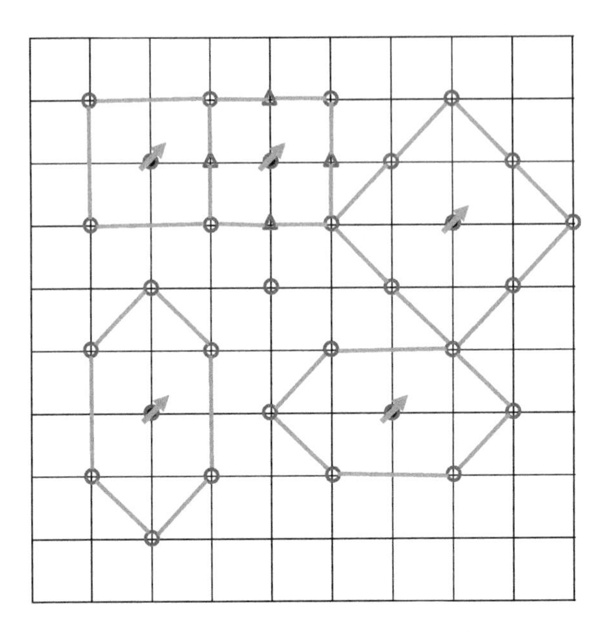

FIGURE 15.2 Five-spot patterns converted to inverted seven- and nine-spot patterns.

15.8.2 Injection Pattern and Well Spacing

The rule is to choose a flood pattern that can provide good sweep efficiency and more producers to achieve high rates. Five-spot, inverted seven-spot, and inverted nine-spot patterns were used in SF. In very viscous oil reservoirs, inverted nine-spot pattern was more often used in Chinese projects. Inverted five-spot pattern was most often used.

One reason inverted five-spot pattern was often used is that it can be converted to inverted nine-spot pattern or inverted seven-spot pattern, as shown in Figure 15.2. Then the ratio of producer to injector can be changed from one for inverted five-spot patterns to two for inverted seven-spot patterns, and three for inverted nine-spot patterns. For example, the original five-spot patterns in the Kern River field were converted to inverted nine-spot patterns in the later infill phase (Hong, 1994).

The distances between injectors and producers from Chinese projects are shown in Figure 15.3. At the 50% probability, the distance was about 100 m. The data from Farouq Ali (1974) and Farouq Ali and Meldau (1979) showed about 190 m. Generally, if the reservoir depth is less than 500 m, the well distance is about 100 m. If the reservoir depth is 800–1600 m, the distance is 140–150 m but not greater than 200 m (Liu, 1997). Many SF projects in the United States, the smallest well spacing is from 2.5 ac (equivalent to the well distance of 71 m for a five-spot pattern or the single-well drainage area of 1.25 ac); the largest well spacing is 20 ac (the well distance of 200 m for

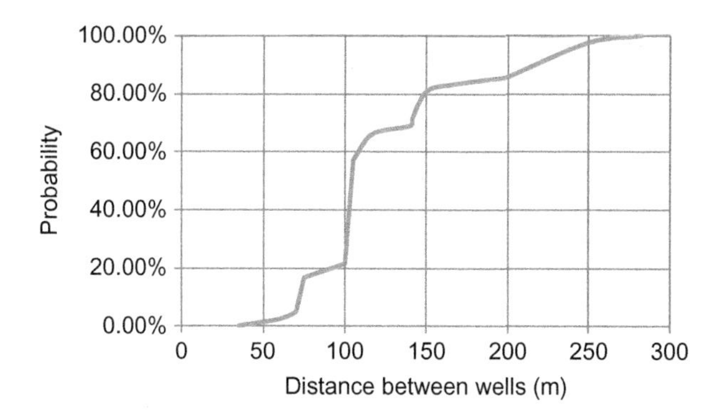

FIGURE 15.3 Distances between wells in Chinese projects.

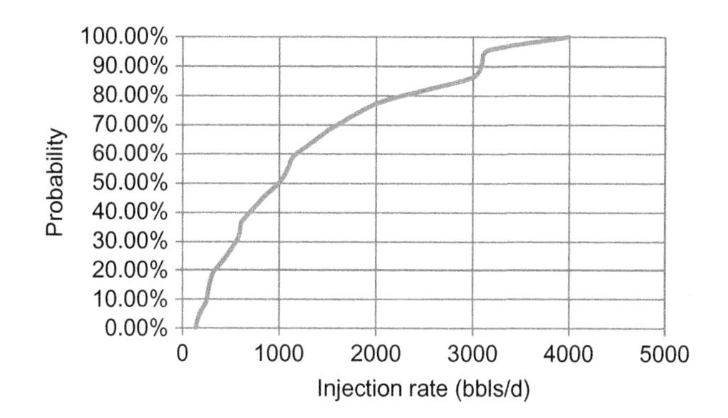

FIGURE 15.4 Injection rates in actual field projects.

a five-spot pattern). The reservoir depths are less than 800 m. In the M-6 SF project in Venezuela, inverted seven-spot patterns were used. The well distance was 231 m. This large well distance might be a reason for its lower-than-expected recovery.

15.8.3 Injection and Production Rates

Slow injection velocity will result in higher heat loss. Therefore, the steam injection rate should be as high as possible and is limited by fracture pressure. The injection strength is 1.24–1.4 bbls/(d. ac. ft) in successful field projects (Zhang, 2006). The steam injection rate should be reduced after heat through to minimize steam projection. The actual field data showed an average well injection rate of 1000 bbls/d, and the injection pressure is 959 psi, as shown in Figures 15.4 and 15.5. In a thicker reservoir, the injection rate and steam quality could be lower. In a thin reservoir, the injection rate and steam quality must be high.

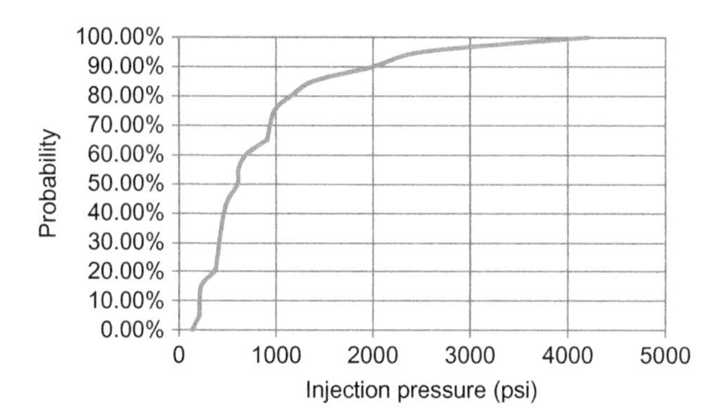

FIGURE 15.5 Injection pressure in actual field projects.

The production should be higher than the injection rate in a reservoir, because the reduced pressure leads to higher steam-specific volume, and water flashed to vapor. The resulting benefits include: (1) to establish a pressure gradient from an injector to a producer to increase liquid and oil production rate, (2) to facilitate steam movement, (3) to reduce heat loss, (4) to reduce production time, and (5) to increase simultaneous OSR. Large pumps should be used to handle high water rates and achieve low fluid levels even when steam flashing reduces pump efficiency (Farouq Ali and Meldau, 1979).

15.8.4 Injection Schemes

In many cases, for example, in the case of very high oil viscosity, steam soak is performed before SF. The steam soak not only solves the problems of high injection pressure and late response in SF, but also provides performance data which help to design the subsequent SF. For a typical steam injection project, soak period is about 3–5 years, while SF is about 6–10 years, with total 10–15 years; the cumulative steam injection during steam drive is about 1.2–1.5 PV (Liu, 1997).

15.8.5 Time to Convert Steam Soak to Steam Flood

To convert steam soak to steam flood, the following conditions need to be considered.

1. The higher oil saturation would lead to higher steam flood OSR and recovery. It is required that the oil saturation be greater than 50%, or at least greater than 45%. The movable oil saturation in the steam flood period should be greater than 27% (Liu, 1997).
2. The reservoir pressure should be reduced to a reasonably low value before converting to steam flood to have a high steam-specific volume.

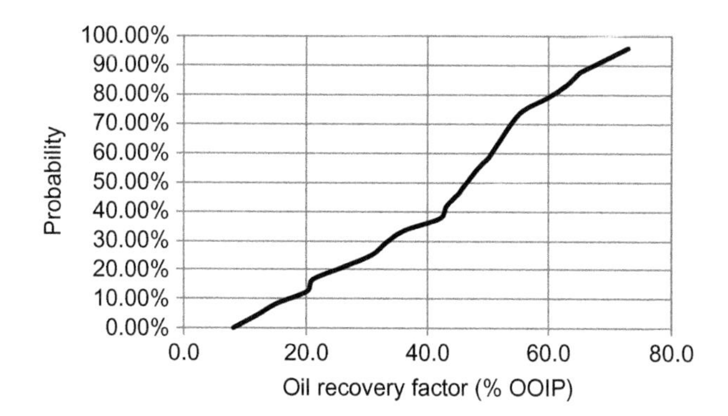

FIGURE 15.6 Oil recovery factor from actual data.

3. The pressure near producers should be lower than that at the corresponding injectors so that a pressure gradient can be established. To achieve that, sometime producers may still have to continue cyclic steam injection after steam flood starts. This will also help to establish the connection between injectors and producers.
4. In the existence of gas cap or edge or bottom water, the time for conversion will be more important and should be studied.

At a later phase of SF, several measures can be taken to improve heat efficiency: lowered steam injection rate (Messner, 1990), reduced steam quality, intermittent steam injection (Hong, 1988), water-alternating-steam process (WASP) (Hong, 1990), conversion of steam to water injection (Hong, 1987, 1988), injection profile modification, and so on.

15.8.6 Oil Recovery and OSR

Figures 15.6 and 15.7 show the statistical analysis of actual field data for oil recovery factor and OSR. The recovery factor and OSR at the 50% probability are 46.5% and 0.195, respectively. The data are from Farouq Ali (1974), Farouq Ali and Meldau (1979), and Chinese steam flood projects.

15.8.7 Completion Interval

Injection wells are usually completed in the lower one-third to one-half of the interval to retard steam gravity override. If there is a large-scale horizontal shale layer, completion of top interval may be needed. Production wells are normally completed over the entire production interval. These are conventional practices without detailed analysis and optimization. Obviously, the exact completion intervals should depend on detailed reservoir and flow conditions, especially reservoir thickness, ratio of vertical to horizontal permeability, well spacing, injection and production rates, whether there is a

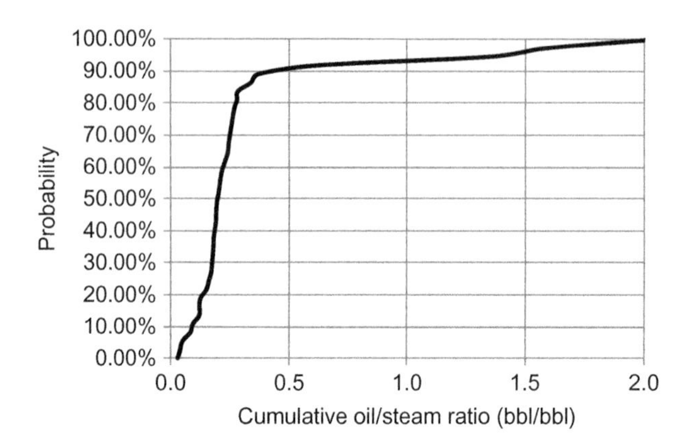

FIGURE 15.7 Cumulative OSR from actual data.

horizontal shale barrier, etc. Chu (1993) did a systematic simulation study on the subject. In his base models, the single-well spacing was 1.25 ac, and the injection rate was 2 bbls/(d. ac. ft). His results showed that without deterministic shales, the optimal completion intervals for injectors and producers are:

Patterns (inverted)	Injectors	Producers
Five-spot	$^1/_6$–$^1/_3$	$^1/_3$–$^1/_2$
Seven-spot	$^1/_3$	$^1/_3$
Nine-spot	$^2/_3$	$^5/_6$ (corners), $^1/_2$ (sides)

15.8.8 Production Facilities

To reduce thermal expansion effect, prestressed casings should be used. Casing sizes are larger than 7 in. Most used casings are N80 or P110 casing of 9.19 and 10.36 mm. To improve thermal stability 30–40% silica flour can be added in the cement (Zhang, 2006).

Thermal-resistant tubings could be type III tubing, hydrogen-resistant tubing, and vacuum tubing. Annulus should be filled with helium or nitrogen to reduce wellbore heat loss. Pumps used in lifting heavy oils are generally rod pumps.

15.8.9 Water Treatment

A lot of water is needed in a steam injection project. Generally, the ratio of water needed to the oil produced is 4 to 5 (Liu, 1997). Water treatment includes sodium exchanger to reduce hardness, and equipment to add chemicals and to remove oxygen. Water quality for boilers is given in Table 15.2.

TABLE 15.2 Water Quality for Boilers

Suspended solids (ppm)	<2
Hardness (ppm)	<0.1
pH	7.5–11
Oil (ppm)	<2
Dissolved oxygen (ppm)	<0.05
Iron (ppm)	<0.05
Copper (ppm)	<5
Silicon dioxide (ppm)	<50
Sodium (ppm)	<10

15.8.10 Monitoring and Surveillance

Injection wellhead temperature, pressure and injection rate are measured daily. The downhole steam quality is measured half yearly. For production wells, liquid rate and oil rate are measured continuously for 4–8 h once a week. The tubing head pressure, casing pressure, and the temperature of produced oil are measured once a week. The dynamic liquid level is measured once a month (Zhang, 2006).

15.9 FIELD CASES

Five field cases are presented in this section: Kern River in California, Duri steam flood (DSF) in Indonesia, West Coalinga Field in California, Karamay Field and the Qi-40 block in Laohe, China.

15.9.1 Kern River in California

Kern River was a large heavy oil field 5 miles northeast of Bakersfield, CA. The original oil in place (OOIP) was higher than 4 billion bbls. The reservoir was shallow and consisted of an alternating sequence of unconsolidated sands with interbedded silts and clays. The porosity and permeability ranged 28–35% and 1–5 D (4 D on average), respectively. The oil viscosities were on the order of 4000 mPa·s at the reservoir temperature of 70°F. The reservoir pressure was low (100 psig). At 250°F, the viscosity was reduced to 15 mPa·s. The Kern River formation represented a continental–alluvial fan deposit derived largely from the westward-flowing Kern River.

The heat had been used in Kern River since 1950s. First downhole heaters were used, and then hot water was injected. In 1964, the hot-water project was converted to a steam-drive project and expanded to 47 injectors. This project is known as the Kern project. After 1964, four other projects were started: San Joaquin, Kern “A”, G & W “A”, and Reed. Five-spot

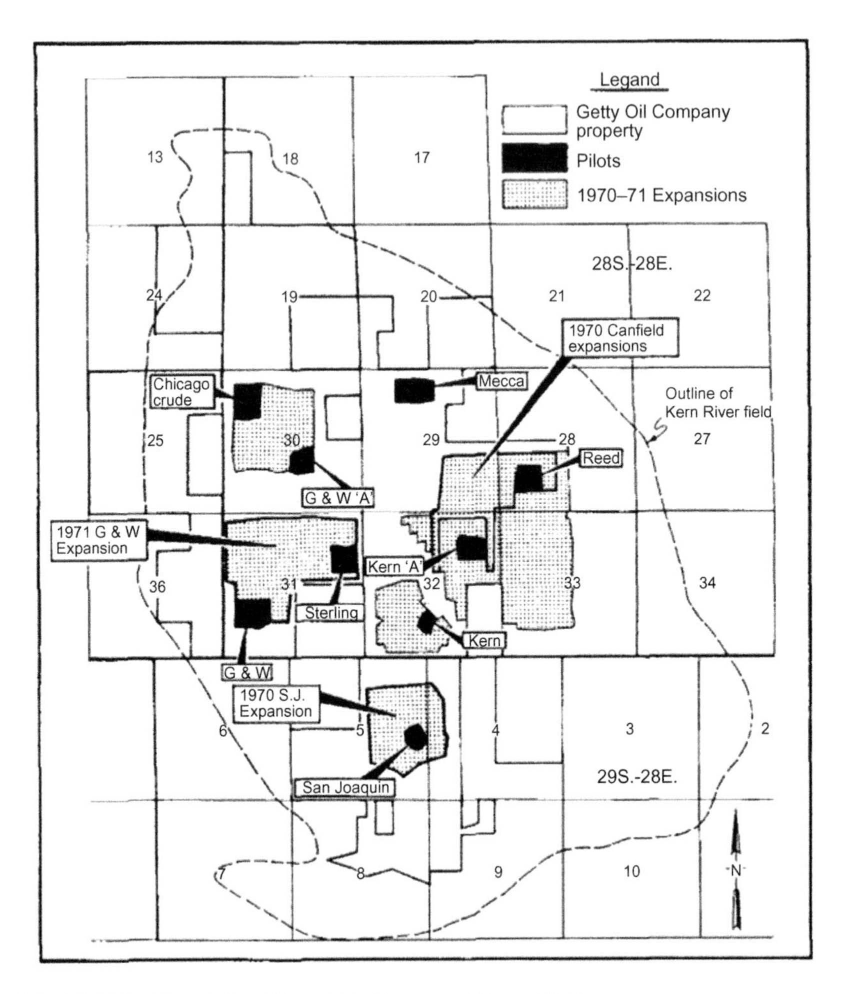

FIGURE 15.8 Map of Kern River field (Greaser and Shore, 1980).

pattern with $2^1/_2$ spacing were used (Bursell, 1970). Figure 15.8 is the map of Kern River field with steam flood pilots and 1970–71 expansions (Greaser and Shore, 1980). A more detailed summary report of the steam injection projects in different parts of the Kern River field was provided by Green and Willhite (1998).

Five basic types of completions were used in these steam projects. They were punched liners, slotted liners, selectively perforated cemented casing, inner liner completions and gravel-packed liners. Since 1966, producing wells had been completed by cementing casing through the oil zone and selectively jet-perforating 50–60 ft of interval near the bottom of the zone. Where sand production became a problem, inner liners were run. Although this helped to limit the sand production, in many cases it caused plugging.

Steam flood performance showed poor vertical sweep efficiency. Two methods were used to improve steam profile: mechanical restriction of the production interval and adding foam diverters in the injected system (Greaser and Shore, 1980).

Kern River sands were separated by silt and clay interbeds. Individual sands were believed to be isolated. Steam injection was started from the deepest sand and moved upward to the next zone after the current zone was depleted. This was done by selectively perforating injection wells in the desired zone. Production wells were generally completed with all zones open. Because steam was injected for several years, some of the heat lost to the overburden preheated the reservoirs above the injection layer (Restine, 1983).

15.9.2 Duri Steam Flood (DSF) Project in Indonesia

The DSF project was the largest steam flood project in the world. The field was located in the Riau Province, on the island of Sumatra, in Indonesia. Duri was the second largest field in the country, producing about 200,000 BOPD by steam flood. The DSF project was planned to develop over 15,000 ac of reservoir utilizing over 4000 producing wells (Pearce and Megginson, 1991).

The reservoir depth was 600 ft with a net pay of 109 ft. The porosity and permeability were 36% and 1550 mD, respectively. The reservoir temperature was 100°F. The oil viscosity at the reservoir temperature was 157 cP. The API gravity of oil was 23.

The field began production in 1958. Cyclic steam stimulation began in 1967. By 1977, 339 steam-stimulation jobs had been performed. A steam flood pilot was initiated in 1975. With the success of the pilot, steam flood was expanded in Area 1. Area 1 consisted of 95 inverted seven-spot patterns, each 11.625 ac in size, which was the main pattern used in the field. There were more than 420 producers.

The earlier design was a concentric injection string in which one string of tubing was inside another. The problem was that heat transfer between the flow streams resulted in high-quality steam entering one interval, while low-quality stream or even hot water entering into another. That caused one sand being preferentially heated over the other. The solutions employed were: (1) to utilize a downhole choke configuration to inject steam down a single string and use the principle of critical flow to achieve a proper split between sands; and (2) to use "twin" existing injectors for patterns in which the total rate per sand was greater than that achievable with the choke design.

It was discovered that "hybrid" development scheme with a combination of 15.5-acre five- and nine-spot patterns maximized oil recovery and improved economics. Additionally, by changing to nine-spot patterns in the thickest net pay areas, the producer-to-injector ratio improved from 2.4 for seven-spot patterns to 3.5 for nine-spot patterns.

The DSF project is unique in that it simultaneously involved the management of existing steam flood areas, the development of new steam flood areas, and the design of future areas to maximize both oil recovery and production efficiency.

15.9.3 WASP in West Coalinga Field, CA

A Water-Alternating-Steam Process (WASP) was applied to vertical expansion (VE) sands in the pilot area of Section 13D, the West Coalinga field to stop wasteful steam production and to improve vertical conformance of injected steam. The section 13D steam flood began in 1973 with 32-acre, six patterns. Steam was injected into the three lowermost sands (H, J, and J_v). In 1984, after 11 years of steam injection, the steam flood in these sands showed high steam–oil ratios and high flow line temperatures. The temperature survey of the sand G indicated 240°F as a result of hot-plate heating from the immediately underlying bottom layer.

In April 1984, these sands were converted to waterflood. In 1985, twin injectors were drilled and the pilot was vertically expanded into the above sands E and G with hot-water injection continuing in the lower sands. In early 1988, five pilot wells produced vapor-phase steam. Observation wells showed an extensive steam chest in the sand G with temperatures in excess of 400°F during steam injection. This steam chest breakthrough caused well sanding, cutting of downhole tubulars, and high-temperature-fluid handling problems. To alleviate these problems, pumps had to be raised in five wells and one well had to be shut in, reducing oil production from the VE sands and the lower waterflooded zones.

A WASP field test, based on a numerical simulation study, was implemented in July 1988 with alternating slugs of water and steam, each injected over 4 months. The WASP eliminated steam production, allowing the pumps to be lowered and the shut-in well to return to production. Oil production remained constant through the first WASP cycle and increased during the second cycle. Sales oil (total production minus oil used to generate steam) increased as a result of saving generator fuel during the water leg of each WASP cycle, resulting in improved project economics (Hong and Stevens, 1992).

15.9.4 Karamay Field, China

There were four pilot areas with different oil viscosities: 9_1^1, 9_1^2, 9_3, and 9_6. Cyclic steam stimulation was started in these pilots in 1984, 1985, 1987, and 1988, respectively. And they were subsequently converted into SF in 1987–1990. These pilots are the first batch of SF pilots in China. Some reservoir and fluid data, and well pattern information are listed in Table 15.3 (Liu, 1997).

TABLE 15.3 Data for the Four Pilots in the Karamay field

	9_1^1	9_1^2	9_3	9_6
Reservoir depth (m)	105	170–109	230–250	175.5
Net thickness (m)	15	16.5	8.5	9.5
Oil saturation (%)	69.3	70	61	64
Porosity (%)	32.7	34	32	29
Permeability (mD)	4160	6170	3496	1260
Reservoir pressure (MPa)	2.12	2.16	2.0	1.8
Reservoir temperature (°C)	10	17	19	22
Oil viscosity *in situ* (mPa·s)	1100	1600		
Gas-free oil viscosity at 20°C (mPa·s)	3100	2640	4000	20,000
Pattern (inverted)	Seven	Nine	Five	Five
Number of patterns	3	4	9	9
Well distance (m)	100	70 × 50	100	50
Soak time (months)	30	56	36	23
Cycles per well	3.06	3.7	4.5	2.9
Soak OSR (bbl/bbl)	1.31	0.54	0.4	0.317
Oil prod. velocity (%)	7.5	6	6	
Soak recovery (%)	24.2	27	20.3	23.7
Production started in	May 84	Oct. 85	January 97	August 88
Converted to SF in	July 87	June 90	January 90	November 90
Flood OSR (bbl/bbl)	1.34	0.5	0.39	
Flood recovery (%)	27	34	20.3	
Flood oil production velocity (%)	8.5	7	6	

The following are some lessons from these pilots.

1. A high steam quality and reasonable steam injection velocity are necessary conditions for success.
2. The ratio of production to injection should be higher than 1.
3. The steam breakthrough was problematic, especially in the blocks 9_1 and 9_6 where the well distances were short.
4. If steam breakthrough occurred in the soak period, it would become more severe during the flood period. Therefore, measures should be taken to stop or reduce breakthrough in the soak period.
5. Time to convert soak to steam flood is important. For a better steam flood performance, the conversion should be made before the end of the third cycle.

15.9.5 Qi-40 Block in Laohe, China

An SF pilot was conducted in the Qi-40 block in the Huang Xi Ling field, Laohe Field, China. The pilot was in the Lian (Hua) II layer. The average porosity was 25% and the permeability was 1.49 D. The average oil column was 60.5 m and the net-to-gross ratio was 0.484. The reservoir depth was 910–1045 m. The original reservoir pressure was 8–10 MPa, and the original reservoir temperature was 36.8°C at 850 m. The dead oil viscosity at 50°C was 2639 mPa·s (the *in situ* oil viscosity was 3750 mPa·s). The block had been under steam soak since June 1987. The initial development plan was to use 200 m square patterns in the areas of greater than 15 m reservoir thickness. Two development patterns were used in the areas of greater than 20 m reservoir thickness. Before SF, oil saturation was 0.57, and the recovery factor was 24%.

A simulation study was conducted to compare continuous steam soak, WF and SF in 1997. The results for the Lian II are shown in Table 15.4 (Ma et al., 2005).

The results showed that if steam soak was continued, the recovery factor would be low. If converted to hot WF, the recovery factor was not significantly increased. The recovery factor was the highest if converted to SF.

The conversion of steam soak to hot WF was tried starting in April 1996. Three inverted five-spot patterns with well distance of 141 m were used. After conversion, the water cut increased from 31% before the conversion to 85% within 2 months. The oil rate decreased. The WF was stopped in May 1997. And it was decided to convert steam soak to SF.

Four inverted nine-spot patterns of a well distance of 70 m were initially steam soaked in January 1998, and converted to SF in October 1998. There were 4 injectors, 21 producers and 2 observation wells. The performance may be divided into three phases.

Phase I, from the start of conversion to March 1999, the group liquid rate increased from 154 to 330 t/d, and water cut increased from 63.3% to 90%. The oil rate decreased from 56 to 30 t/d following the trend from steam-soak period.

TABLE 15.4 Simulated Results of Different Schemes After Steam Soak Was Converted

Scheme	Production Time (d)	Cum. Steam Injected (10^3 tons)	Cum. Oil Produced (10^3 tons)	Cum. Water Produced (10^3 tons)	Recovery Factor in Period (%)	OSR	Net Oil Produced (10^3 tons)
Soak to end	1210	101.8	43.8	87	13.9	0.43	37
Soak to hot WF	2975	49.8	63.8	67.6	20.2		60
Soak to SF	1555	466.8	108.5	449	34.4	0.23	75

Phase II, March–July 1999, six low-liquid-rate wells were steam soaked. The group liquid rate increased to 440 t/d, the water cut decreased slightly, and the oil rate increased to 70 t/d.

Phase III, from July 1999 on, the group liquid rate increased to 578 t/d by March 2000, but decreased to 470 t/d during August–September 2000 owing to well sand production and low pumping efficiency. The oil rate fluctuated within 35–125 t/d.

Until the end of 2003, the cumulative OSR was 0.21, the SF recovery factor was 40.3%, and the total recovery factor including steam soak was 64.3%. The pilot was extended to include seven injectors in July 2003. The SF was continued until the end of 2004. By that time, 815 tons/d of steam was injected, and the average production rate was 7.8 tons/d per well in the expanded pilot area. The water cut was 83.3% and the OSR was 0.2. Compared with the steam-soak recovery factor, 17.65% incremental oil recovery was obtained. It was predicted that the incremental oil recovery factor would be 24.66%, the production time would be 6 years, the sweep efficiency by steam was 45%, and the heat efficiency would be 38.8% (Ma et al., 2005).

The monitoring and surveillance program included (Cheng et al., 2005):

1. periodically measure wellhead steam quality and flow rates,
2. take downhole samples to measure downhole steam quality quarterly,
3. measure steam profile half yearly,
4. extended tests to monitor production performance for key wells,
5. measure temperature profiles in observation wells,
6. install permanent pressure gauges at key observation wells.

The measured steam profiles in injection wells showed 23.4–64.9%, indicating overall low profiles. High-temperature profile modification was conducted. Before modification, steam injection was stopped to convert to hot-water injection for 10–15 days. Then high-temperature profile-modification agent was injected. After that steam injection was resumed.

Downhole gauges were installed below production pumps to monitor pressure and temperature. The success rate was 100%. The longest monitoring time was 6 months. The data from 5 wells showed 230°C during the initial steam-soak period, 80–90°C in the middle of production period, and 60–70°C in the late production period. The data showed a decline trend in temperature. The flow pressure was 5–6 MPa in the initial production period, but 0.2–0.3 MPa at the lowest during the late production period. The monitoring data helped to adjust well production parameters. For low-rate wells, more perforations were added or some chemical stimulation measures were taken. For high-potential wells, larger pumps were installed. For steam-channeling wells, profile-modification measures were taken.

The measure steam qualities from downhole samples at 7 injection wells for 30 times were 65–73% at 30 m depth, 56–66% in a middle depth, and 50–56% at a lower depth. The steam quality from the boiler was 75–76%.

REFERENCES

Abou-Kassem, J.K., 1981. Investigation of Fluid Orientation in a Two-Dimensional, Compositional, Three-Phase Steam Model. The University of Calgary, Calgary, AB, Canada (Ph.D. dissertation).

Andrade, E.N.DA C., 1930. The viscosity of liquids. Nature March (1), 309–310.

Boberg, T.C., Lantz Jr., R.B., 1966. Calculation of the production rate of a thermally stimulated well. JPT December, 1613–1623 (*Trans.* AIME, **237**).

Bursell, C.G., 1970. Steam displacement—Kern River field. JPT 22910, 1225–1231.

Chase, C.A., O'Dell, P.M., 1973. Application of variational principles to cap and base rock heat losses. SPEJ August, 200–210.

Cheng, Z.-P., Zhao, Y.-W., Deng, Z.-X., Wang, P.-H., 2005. Monitoring and surveillance techniques and application in steam flooding. In: Yan, C.-Z, Li, Y. (Eds.), Tertiary Oil Recovery Symposium. Petroleum Industry Press, Beijing, pp. 156–160.

Chu, C., 1985. State-of-the-art review of steamflood field projects. JPT 37 (1), 1887–1902.

Chu, C., 1993. Optimal choice of completion intervals for injectors and producers in steamfloods. Paper SPE 25787 Presented at the SPE International Thermal Operations Symposium, 8–10 February 1993, Bakersfield, CA.

Coats, K.H., George, W.D., Chu, C., Marcum, B.E., 1974. Three-dimensional simulation of steamflooding. SPE J. 14 (6), 573–592.

Doscher, T.M., 1967. Technical Problems in in situ Methods for Recovery of Bitumen from Tar Sands, Proceedings of Seventh World Petroleum Congress, Mexico City.

Enhanced Recovery Week, 1987. Marathon to expand carbonate steamflood at Garland, 24 August.

Farouq Ali, S.M., 1974. Current status of steam injection as a heavy oil recovery method. J. Can. Petrol. Technol. 34 (1), 54–68.

Farouq Ali, S.M., Meldau, R.F., 1979. Current steamflood technology. JPT 31 (10), 1332–1342.

Grabowski, J.W., Aziz, K., 1977. A preliminary investigation of in situ combustion by the method of collocation. In: Canada–Venezuela Oil Sands Symposium, Edmonton, AB, Canada.

Grathoffner, E.H., 1979. The role of oil in water emulsions in thermal oil recovery processes. Paper SPE 7952 Presented at the SPE California Regional Meeting, 18–20 April, Ventura, CA.

Greaser, G.R., Shore, R.A., 1980. Steamflood performance in the Kern River field. Paper SPE 8834 Presented at the SPE/DOE Enhanced Oil Recovery Symposium, 20–23 April, Tulsa, OK.

Green, D.W., Willhite, D.P., 1998. Enhanced Oil Recovery. SPE Textbook Series, vol. 6. The Society of Petroleum Engineers.

Hagoort, J., Leijinse, A., van Poelgeest, F., 1976. Steam-strip drive: a potential tertiary recovery process. JPT December, 1409–1419.

Harmsen, G.J., 1971. Oil recovery by hot-water and steam injection, Proceedings of the Eighth World Petroleum Congress, **3**. Applied Science Publishers, London, 243–251.

Hong, K.C., 1986. Numerical simulation of light oil steamflooding in the Buena Vista hills field, California. Paper SPE 14104 Presented at the International Meeting on Petroleum Engineering, 17–20 March, Beijing, China.

Hong, K.C., 1987. Guidelines for converting steamflood to waterflood. SPERE 2 (1), 67–76.

Hong, K.C., 1988. Effects of shutting in injectors on steamflood performance. SPERE 3 (3), 945–952.

Hong, K.C., 1990. Water-alternating-steam process improves project economics at West Coalinga field. Paper CIM 90-84 Presented at the Annual Technical Meeting of CIM, 10–13 June, Calgary, AB, Canada.

Hong, K.C., 1994. Steamflood Reservoir Management—Thermal Enhanced Oil Recovery. PennWell, Tulsa, OK, pp. 258–259, 299–301.

Hong, K.C., Stevens, C.E., 1992. Water-alternating-steam process improves project economics at west coalinga field. SPERE 7 (4), 407–413.

Konopnicki, D.T., Traverse, E.F., Brown, A., Deibert, A.D., 1979. Design and evaluation of the Shiells Canyon field steam-distillation drive pilot project. JPT 31 (5), 546–552.

Liu, W.-Z., 1997. Steam Injection Technology to Produce Heavy Oils. Petroleum Industry Press, Beijing, China.

Ma, D.-S., Zhao, H.-Y., Hu, C.-H., 2005. Practice and learning of the steam flooding in the block 40 of Liauhe field. In: Yan, C.-Z, Li, Y. (Eds.), Tertiary Oil Recovery Symposium. Petroleum Industry Press, pp. 300–305.

Marx, J.W., Langenheim, R.H., 1959. Reservoir heating by hot fluid injection. Trans. AIME 216, 312.

Messner, G.L., 1990. A comparison of mass rate and steam quality reductions to optimize steamflood performance. Paper SPE 20761 Presented at the SPE Annual Technical Conference and Exhibition, 23–26 September 1990, New Orleans, LA.

Miller, C.A., 1975. Stability of moving surfaces in fluid systems with heat and mass transport, III. Stability of displacement fronts in porous media. AIChE J. 21 (3), 474–479.

Myhill, N.A., Stegemeier, G.L., 1978. Steam-drive correlation and prediction. JPT February, 173–182.

Nakornthap, K., Evans, R.D., 1986. Temperature-dependent relative permeability and its effect on oil displacement by thermal methods. SPERE 1 (3), 230–242.

Olsen, D.K., Sarathi, P.S., NIPER; Hendricks, M.L., Schulte, R.K., and Giangiacomo, L.A.. 1993. Case history of steam injection operations at naval petroleum reserve no. 3, teapot dome field, Wyoming: a shallow heterogeneous light-oil reservoir. Paper SPE 25786 Presented at the SPE International Thermal Operations Symposium, 8–10 February, Bakersfield, CA.

Pearce, J.C., Megginson, E.A., 1991. Current status of the Duri steam flood project, Sumatra, Indonesia. Paper SPE 21527 Presented at the SPE International Thermal Operations Symposium, 7–8 February, Bakersfield, CA.

Prats, M., 1982. Thermal Recovery. SPE Monograph. The Society of Petroleum Engineers, Richardson, TX.

Ramey Jr., H.J., 1959. Discussion of reservoir heating by hot fluid injection. Trans. AIME 216, 364.

Ramey Jr., H.J., 1962. Wellbore heat transmission. JPT April, 427–440 (Trans. AIME, 225).

Restine, J.L., 1983. Effect of preheating on kern river field steam drive. JPT 35 (3), 523–529.

Sahuquet, B.C., Ferrier, J.J., 1982. Steam-drive pilot in a fractured carbonated reservoir: Lacq Superieur field. JPT 34 (4), 873–880.

Taber, J.J., Martin. 1983. Technical screening guides for the enhanced recovery of oil. Paper SPE 12069 Presented at the SPE Annual Technical Conference and Exhibition, 5–8 October, San Francisco, CA.

Taber, J.J., Martin, F.D., Seright, R.S., 1997. EOR screening criteria revisited—part 2: applications and impact of oil prices. SPERE 12 (3), 199–206.

Vinsome, P.K.W., Westerveld, J., 1980. A simple method for predicting cap and base rock heat losses in thermal reservoir simulators. JCPT 19 (3), 87–90.

Volek, C.W., Pryor, J.A., 1972. Steam distillation drive—Brea Field, California. JPT 24 (8), 899–906.

Willhite, G.P., 1967. Over-all heat transfer coefficients in steam and hot water injection wells. JPT May, 607–615.

William, B.T., Valleroy, V.V., Runberg, G.W., Cornelius, A.J., Powers, L.W., 1961. Laboratory studies of oil recovery by steam injection. JPT July, 681–690.

Wu, C.H., 1977. A critical review of steamflood mechanisms. Paper SPE 6550 Presented at the SPE California Regional Meeting, 13–15 April, Bakersfield, CA.

Young, B.M., McLaughlin, H.C., Borchardt, J.K., 1980. Clay stabilization agents—their effectiveness in high-temperature steam. JPT 32 (12), 2121–2131.

Zhang, Y.T., 2006. Thermal recovery. In: Shen, P.P. (Ed.), Technological Developments in Enhanced Oil Recovery. Petroleum Industry Press, Beijing, China, pp. 189–234.

Chapter 16

Cyclic Steam Stimulation

James J. Sheng
Bob L. Herd Department of Petroleum Engineering, Texas Tech University, Lubbock, TX 79409, USA

16.1 INTRODUCTION

In cyclic steam stimulation (CSS), steam is injected into a production well for a period. Then the well is shut in and allowed to soak by steam for some period before it returns to production. The initial oil rate is high because of high initial oil saturation, high increased reservoir pressure, and lowered oil viscosity. As the oil saturation becomes lower, the reservoir pressure becomes lower and the oil viscosity becomes higher due to heat losses to the surrounding rock and fluids, oil rate declines. At some point, another cycle of steam injection is initiated. Such cycle may be repeated several times or many times.

The terms of steam soak and steam huff-and-puff (huff-n-puff, huff 'n' puff) are also used to describe CSS.

In this chapter, we will first briefly discuss CSS mechanisms, theories to estimate production performance, and screening criteria. After that we will focus on practice and field cases of CSS projects.

16.2 MECHANISMS

The first mechanism of CSS is the reduced oil viscosity owing to the steam injection. Steam injection increased the reservoir pressure. Thus the pressure drop is high. According to the Darcy equation, the oil rate is increased.

Figure 16.1 is a schematic of a radial flow model after steam stimulation. Let us use the steady-state Darcy flow equation. The production rate at the downhole conditions after steam stimulation, q_{oh}, is

$$\begin{aligned} q_{\mathrm{oh}} &= \frac{2\pi h(p_{\mathrm{e}} - p_{\mathrm{w}})}{\dfrac{\mu_{\mathrm{oc}}\ln(r_{\mathrm{e}}/r_{\mathrm{h}})}{k} + \dfrac{\mu_{\mathrm{oh}}\ln(r_{\mathrm{h}}/r_{\mathrm{d}})}{k} + \dfrac{\mu_{\mathrm{oh}}\ln(r_{\mathrm{d}}/r_{\mathrm{w}})}{k_{\mathrm{d}}}} \\ &= \frac{2\pi k k_{\mathrm{d}} h(p_{\mathrm{e}} - p_{\mathrm{w}})}{k_{\mathrm{d}}\mu_{\mathrm{oc}}\ln(r_{\mathrm{e}}/r_{\mathrm{h}}) + k_{\mathrm{d}}\mu_{\mathrm{oh}}\ln(r_{\mathrm{h}}/r_{\mathrm{d}}) + k\mu_{\mathrm{oh}}\ln(r_{\mathrm{d}}/r_{\mathrm{w}})} \end{aligned} \tag{16.1}$$

Enhanced Oil Recovery Field Case Studies.

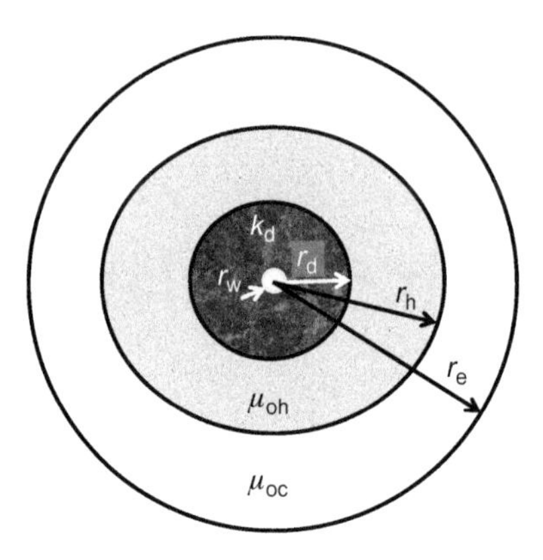

FIGURE 16.1 Schematic of a radial flow model after steam stimulation.

The production rate before steam stimulation, q_{oc}, is

$$q_{oc} = \frac{2\pi h(p_e - p_w)}{\frac{\mu_{oc}\ln(r_e/r_d)}{k} + \frac{\mu_{oc}\ln(r_d/r_w)}{k_d}} = \frac{2\pi k k_d h(p_e - p_w)}{k_d\mu_{oc}\ln(r_e/r_d) + k\mu_{oc}\ln(r_d/r_w)} \tag{16.2}$$

The ratio of production index after steam stimulation (J_h) to that before (J_c) is

$$\frac{J_h}{J_c} = \frac{k_d\mu_{oc}\ln(r_e/r_d) + k\mu_{oc}\ln(r_d/r_w)}{k_d\mu_{oc}\ln(r_e/r_h) + k_d\mu_{oh}\ln(r_h/r_d) + k\mu_{oh}\ln(r_d/r_w)}$$
$$= \frac{(k_d/k)\ln(r_e/r_d) + \ln(r_d/r_w)}{(k_d/k)\ln(r_e/r_h) + (k_d/k)(\mu_{oh}/\mu_{oc})\ln(r_h/r_d) + (\mu_{oh}/\mu_{oc})\ln(r_d/r_w)} \tag{16.3}$$

Assume a practical case: $r_e = 500$ ft, $r_h = 50$ ft, $r_d = 5$ ft, $r_w = 0.25$ ft, $k_d/k = 0.1$, and $\mu_{oh}/\mu_{oc} = 0.01$. Then

$$\frac{J_h}{J_c} = \frac{(k_d/k)\log(r_e/r_d) + \log(r_d/r_w)}{(k_d/k)\log(r_e/r_h) + (k_d/k)(\mu_{oh}/\mu_{oc})\log(r_h/r_d) + (\mu_{oh}/\mu_{oc})\log(r_d/r_w)}$$
$$= \frac{(0.1)\log(500/5) + \log(5/0.25)}{(0.1)\log(500/50) + (0.1)(0.01)\log(50/5) + (0.01)\log(5/0.25)} = 12.2$$

In other words, the productivity is increased by 12.2 times after stimulation when the damage is not removed by steam (k_d is unchanged). When the damage is removed by steam (k_d is equal to k *after* stimulation), the productivity is increased by

$$\begin{aligned}\frac{J_h}{J_c} &= \frac{\log(r_e/r_d) + k/k_d \log(r_d/r_w)}{\log(r_e/r_h) + (\mu_{oh}/\mu_{oc})\log(r_h/r_d) + (\mu_{oh}/\mu_{oc})\log(r_d/r_w)} \\ &= \frac{\log(500/5) + (10)\log(5/0.25)}{\log(500/50) + (0.01)\log(50/5) + (0.01)\log(5/0.25)} \\ &= 14.7\end{aligned}$$

This example calculation shows that the productivity is increased by a similar magnitude whether the formation damage is removed or not by steam injection. It is implied that the main mechanism of cyclic steam injection is the reduction in oil viscosity. Although removing damage does improve productivity, the improvement is much less (by 20% in this example) than that by viscosity reduction.

However, if the steam injection has removed the formation damage and the reservoir has been cooled, the improvement in the productivity is significantly improved. For this example, the improvement is

$$\begin{aligned}\frac{J_{\text{after CCS}}}{J_{\text{before CCS}}} &= \frac{\log(r_e/r_d)/k + \log(r_d/r_w)/k_d}{\log(r_e/r_d)/k + \log(r_d/r_w)/k} \\ &= \frac{\log(500/5) + \log(5/0.25)(10)}{\log(500/5) + \log(5/0.25)} \\ &= 4.55\end{aligned} \tag{16.4}$$

This example calculation shows that the mechanism of steam stimulation to remove formation damage near the production well will be in effect after the reservoir cools down. This is the second mechanism. The formation damage could be caused by deposition of solids, paraffin, or asphaltene near the production well.

The third mechanism may be explained by this situation. When the oil viscosity is very high, or the well spacing is too large, an unrealistically high injection pressure is needed in steam flooding. In such situation, CSS may work by heating a small zone and using a lower injection pressure. In fact, CSS is a precursor to steam flooding in most reservoirs.

Other mechanisms may include fluid expansion and rock compaction (de Haan and van Lookeren, 1969), gravity drainage, relative permeability modification, wettability alteration, distillation, and interfacial tension reduction. Rock compaction was observed to be significant in some cases.

16.3 ESTIMATING PRODUCTION RESPONSE FROM CSS—BOBERG AND LANTZ MODEL

As shown in Chapter 15, we use the Marx and Langenheim (1959) model to compute the radius of heated zone as the steam is injected into a reservoir.

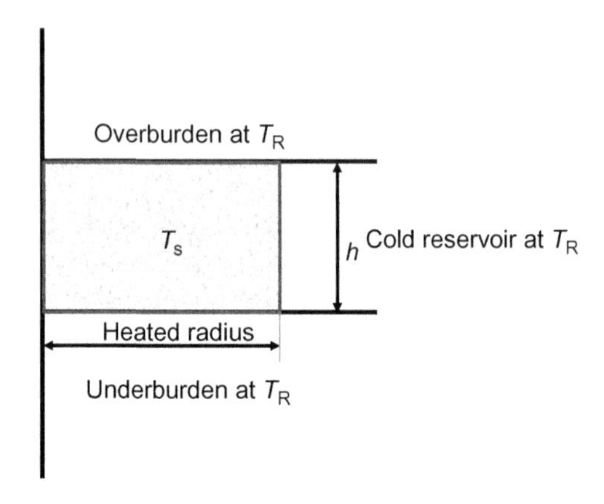

FIGURE 16.2 Initial reservoir temperature distribution in the Boberg and Lantz model.

When the steam injection stopped and the well is put in production, the reservoir temperature will decrease owing to the heat loss to the overburden and underburden rocks and the heat loss by produced fluids. This decrease in the average temperature of the reservoir is computed in the Boberg and Lantz (1966) model.

Although the heat loss to the overburden and underburden is considered to compute the radius of the heated reservoir in the Marx and Langenheim model, the temperatures outside the heated zone are treated to maintain at the initial reservoir temperature T_R before steam injection. Therefore, the initial temperatures outside the heated zone in the Boberg and Lantz model will also be at T_R, as shown in Figure 16.2.

During the shut-in and production periods, the heat losses by conduction in the vertical and horizontal direction are considered. Although the colder fluids enter the heated zone, the average temperature model does not explicitly account for this effect.

The differential equation used in the mode is

$$\frac{k_h}{r}\frac{\partial}{\partial r}\left(r\frac{\partial T}{\partial r}\right)+k_h\frac{\partial^2 T}{\partial z^2}=M_R\frac{\partial T}{\partial t} \tag{16.5}$$

where M_R is the average heat capacity of the overburden or underburden rock and the reservoir, and k_h is the thermal conductivity coefficient. The initial and boundary conditions are shown in Figures 16.2 and 16.3, respectively. This model considers the heat loss to the overburden and underburden rocks by conduction only.

The solution is presented in dimensionless quantities:

$$\overline{T}_D=\overline{T}_{Dr}\overline{T}_{Dz} \tag{16.6}$$

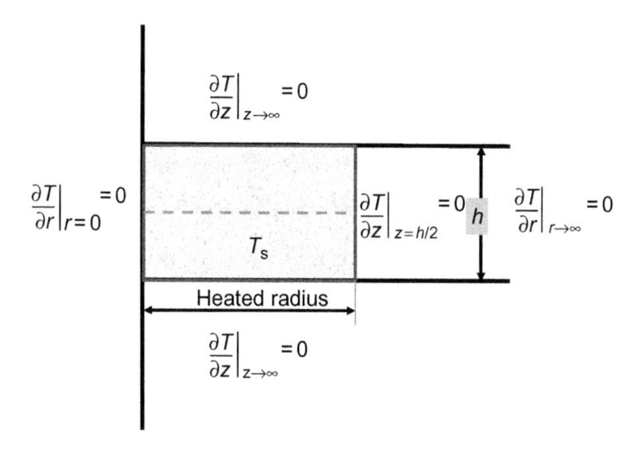

FIGURE 16.3 Boundary conditions in the Boberg and Lantz model.

where

$$\overline{T}_{\mathrm{D}} = \frac{\overline{T} - T_{\mathrm{R}}}{T_{\mathrm{s}} - T_{\mathrm{R}}} \tag{16.7}$$

$\overline{T}_{\mathrm{D}r}$ and $\overline{T}_{\mathrm{D}z}$ are the components of $\overline{T}_{\mathrm{D}}$in the r and z directions, respectively. They may be found from a chart presented by Boberg and Lantz (1966) or calculated using the following equations.

For $0.01 < t_{\mathrm{D}r} \leq 0.1$ (G. Paul Willhite, January 9, 2012, personal communication),

$$\overline{T}_{\mathrm{D}r} = 1 - \sqrt{\frac{t_{\mathrm{D}r}}{\pi}}\left(2 - \frac{t_{\mathrm{D}r}}{2} - \frac{3}{16}t_{\mathrm{D}r}^{2} - \frac{15}{64}t_{\mathrm{D}r}^{3} - \frac{525}{1024}t_{\mathrm{D}r}^{4} - \cdots\right) \tag{16.8}$$

The above expansion only gives approximate values as $t_{\mathrm{D}r}$ becomes smaller.

For $t_{\mathrm{D}r} > 0.05$,

$$\overline{T}_{\mathrm{D}r} = \frac{1}{4t_{\mathrm{D}r}}\left[1 - \frac{1}{4t_{\mathrm{D}r}} + \sum_{k=2}^{\infty} s_k\right] \tag{16.9}$$

where

$$s_1 = 1/4$$

$$s_k = \left[\left(-\frac{1}{t_{\mathrm{D}r}}\right)\frac{(0.5+k)}{(1+k)(2+k)}\right]s_{k-1} \tag{16.10}$$

The expanded form of Eq. (16.9) is

$$\overline{T}_{\mathrm{D}r} = \left(\frac{1}{4t_{\mathrm{D}r}} - \frac{1}{16t_{\mathrm{D}r}^{2}} + \frac{5}{384t_{\mathrm{D}r}^{3}} - \frac{1}{439t_{\mathrm{D}r}^{4}} + \frac{7}{20480t_{\mathrm{D}r}^{5}} + \cdots\right) \tag{16.11}$$

It turns out that the series expansion is a fairly accurate representation of the average radial temperature for $t_{Dr} > 0.05$ but requires a number of additional terms for small t_{Dr}. It is necessary to continue the series until the last value is on the order of 10^{-5}.

Although Boberg and Lantz did not place a restriction on this solution, Bentsen and Donohue (1969) added the restriction of $t_{Dr} > 10$.

$\overline{T}_{Dz}$ is calculated from (Bentsen and Donohue, 1969)

$$\overline{T}_{Dz} = \mathrm{erf}(1/\sqrt{t_{Dz}}) - (\sqrt{t_{Dz}/\pi})(1 - \exp(-1/t_{Dz})) \tag{16.12}$$

In these equations, t_{Dr} and t_{Dz} are defined as

$$t_{Dr} = \frac{\alpha(t - t_i)}{r_h^2} \tag{16.13}$$

$$t_{Dz} = \frac{\alpha(t - t_i)}{(h/2)^2} \tag{16.14}$$

where t_i is the initial time and α ($=k_h/M_R$) is the thermal diffusivity.

In the above formulation, the initial temperature outside the heated zone is assumed to be the cold reservoir temperature T_R. In reality, there is a temperature gradient from the heated zone to the outside. To consider this temperature gradient, a hypothetical thickness Δh is added in the actual reservoir thickness h in Eq. (16.14). Δh can be estimated from the overall energy balance:

$$m_s h_s = \pi r_h^2 M_R (T_s - T_R)(h + \Delta h) \tag{16.15}$$

In the above formulation, the heat loss from the produced fluids is not included. To include this heat loss, a correction factor δ is added in Eq. (16.6). Thus the average pressure in the heated zone is calculated from

$$\overline{T}_D = \overline{T}_{Dr}\overline{T}_{Dz}(1 - \delta) - \delta \tag{16.16}$$

where δ is defined by

$$\delta = \frac{1}{2}\frac{\int_{t_i}^{t_p} \dot{Q}_p \, dt}{Q_i} = \frac{1}{2}\int_{t_i}^{t_p} \frac{\dot{Q}_p \, dt}{\pi r_h^2 h M_R (T_s - T_R)} \tag{16.17}$$

where Q_i is the total heat at the initial time t_i, t_p is the production time which is measured from the termination of steam injection, and $\dot{Q}_p$ is the rate of produced heat. Equation (16.16) is an empirical equation proposed by Boberg and Lantz (1966). Note when δ is equal to ½, $\overline{T}_D$ may be negative according to Eq. (16.16). In this case, $\overline{T}_D$ is forced to be zero.

The rate of produced heat $\dot{Q}_p$ at the downhole condition is related to the production rates and average temperature of the heated zone by

$$\dot{Q}_p = \left(q_o M_o + q_w M_w + q_g M_g + q_s M_w + \frac{q_s \rho_w L_v}{T_s - T_R}\right)(T_s - T_R) \tag{16.18}$$

Note that all these rates are at downhole conditions, q_s is the steam injection rate in the cold water equivalent (CWE). Downhole oil and gas rates are not difficult to estimate from surface one. But the water and steam production rates must be corrected from that measured at the surface to account for heat losses. If the calculation is made in time steps, these rates are the average rates in the time intervals. Be careful that consistent units must be used.

Several other analytical models are available for the CSS process (Clossmann et al., 1970; de Haan and van Lookeren, 1969; Martin, 1967; Seba and Perry, 1969). From the description of the Boberg and Lantz model, we can see that the hand calculation is tedious and many assumptions are made in the model. In modern days, numerical simulators are much easier to handle such calculation.

16.4 SCREENING CRITERIA

Taber et al. (1997) and Green and Willhite (1998) summarized the general screening criteria without differentiating steam flooding and steam soak. In fact, the ranges of parameters used in actual field steam soak projects are wider than the ranges presented by these two groups. In other words, the applicable conditions for steam soak are less restrictive than for steam flooding. Table 16.1 summarizes our modified criteria for the steam soak process

TABLE 16.1 Screening Criteria and Average Field Data

Parameters	Criteria Values	Design Criteria	Average Field Data
Oil gravity, °API	8–35	<15	14.4
In situ oil viscosity, cP	50–350,000	4000	5247
Oil saturation, fraction	>0.4		
Net thickness, m	>6	>9	24.2
Net/gross ratio, fraction	>0.4		
Porosity (ϕ), fraction	>0.18	≥0.35	0.32
Permeability, mD	>50	≥1000	1736
Transmissibility, mD-ft/cP	>5		
Depth, m	<1525	<915	518
Gas cap	Not desirable		
Aquifer	Not desirable		
Fracture	No		
Clay	Low		
Steam quality, %		80–85	
Steam pressure, psig		~1500	900
Injection time, days		14–21	11
Soak time, days		1–4	6.25
Number of cycles		3–5	3
Cycle length, months		~6	~6

by including the parameters used in field practices. The design criteria summarized by Farouq Ali (1974) are also listed in this table. Under these design criteria, some of field projects have been successful.

Based on our survey data on Chinese field projects (unpublished) and the survey data reported by Farouq Ali (1974) and Farouq Ali and Meldau (1979), we did statistical data analysis (rank and percentile) for some of parameters. Some of the average data for the surveyed field projects are presented in Table 16.1. The field data were at 50% probability for the survey data. Performance data will be presented in figures in the next section.

When the measured temperature was not the same as the reservoir temperature, the oil viscosity is interpolated or extrapolated assuming that 10°C increase would result in the decrease in oil viscosity by half. When a range of viscosities were reported, the middle point viscosity is picked. For any other parameter, when a range of values are provided, a simple arithmetic average value is used in the statistical analysis.

The gross thickness is also an important parameter. Unfortunately, enough data were not collected to do the analysis.

Generally, a gas cap or a bottom aquifer is not desirable because the former will promote the gravity override of steam, and a large aquifer will serve as a heat sink. In such a reservoir, optimized well placement and development plan are needed, as was done in the Gaosheng field case presented later in this chapter. The values of other screening parameters are listed but not discussed here.

16.5 PRACTICE IN CSS PROJECTS

Some of the design criteria are summarized by Farouq Ali (1974) and presented in Table 16.1. The practice presented in this section may also serve as references for design criteria.

16.5.1 General Producing Methods

If reservoir oil viscosity is 50–150 mPa·s, waterflooding is carried out first followed by steam flooding. If the viscosity is 150–10,000 mPa·s, steam flooding is directly applied because waterflooding may not be effective. CSS followed by steam flooding will be more effective.

When the reservoir oil viscosity is 10,000–50,000 mPa·s, CSS is needed. Subsequent steam flooding is carried out if favorable reservoir conditions are met. When the oil viscosity is above 50,000 mPa·s, special production techniques are needed, such as fracturing, horizontal wells, and adding chemicals.

For a multilayer reservoir, the steam injection should be started from the bottom layer and moved up so that the top layers are preheated. At a proper time, steam soak is converted to steam flooding.

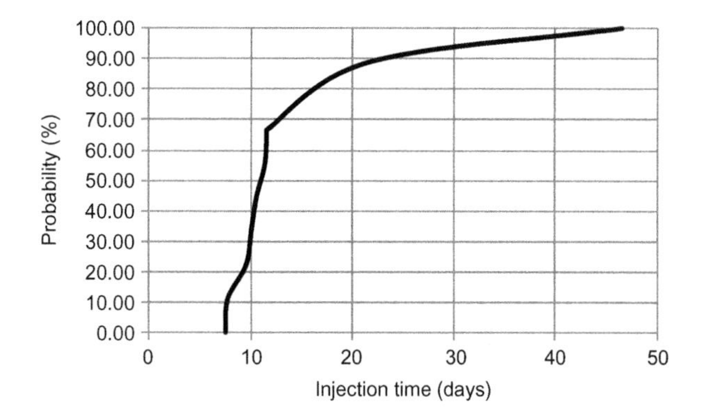

FIGURE 16.4 Injection time in actual CSS projects.

For a reservoir with gas cap or edge or bottom aquifer, the pressure balance is controlled between the oil zone and water or gas zone. Well completion intervals need to be optimized. Wells should be drilled in the oil zone first and then expanded toward the edge water zone. In the existence of bottom aquifer, perforation should be above the water zone, e.g., 15 m in the Shu 175 block.

Liu (1997) investigated the conditions under which a heavy oil reservoir can be economically developed by CSS using simulation approach. He assumed that oil viscosity, reservoir thickness, and depth are the main parameters which determine the steam soak performance and did sensitivity studies on these parameters.

16.5.2 Injection and Production Parameters

Steam injection period could be a few days to a few weeks. Figure 16.4 shows the injection time from actual field projects. The data sources are the same as those in Table 16.1. The average injection time at the 50% probability is 11 days.

If soak time is too short, more heat is accumulated near the wellbore and will be produced when the well is open. If soak time is too long, heat loss to overburden and underburden will be high and the production time becomes longer. However, if the reservoir has sufficient pressure, a long soak period may be desirable in order to increase the thermal efficiency of the process (Farouq Ali, 1974). Adams and Khan (1969) found an optimum soak time of 9 days based on a correlation of six months' cumulative oil production versus soak time. The field data show an average soak time of 6.25 days as shown in Figure 16.5. Liu (1997) observed that 2–3 days of soak time should be enough.

The average production time is 180 days (about half a year) as shown in Figure 16.6.

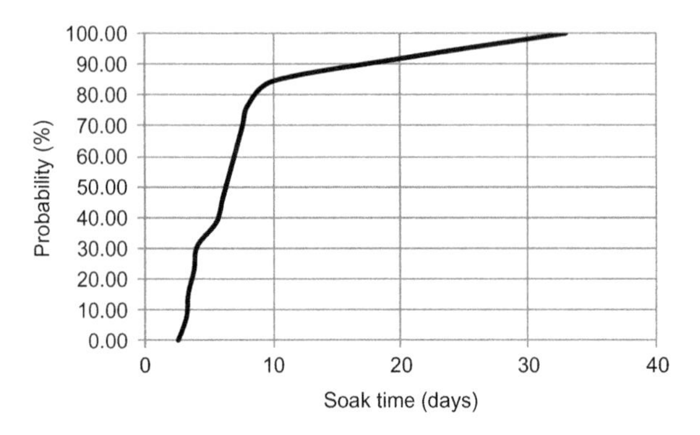

FIGURE 16.5 Soak time in actual CSS projects.

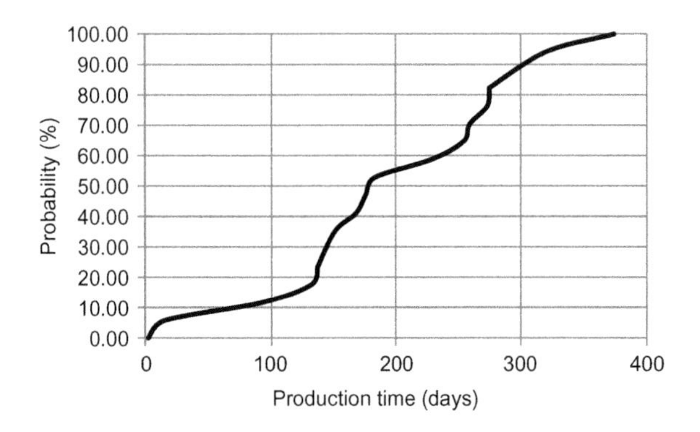

FIGURE 16.6 Production time in actual CSS projects.

Figure 16.7 shows the well total injection per cycle (CWE) from actual field projects with the average of 10,800 bbls of CWE. The average injection pressure is 900 psi, as shown in Figure 16.8. The amount of steam injection is typically 80–160 tons/m of oil column, with the higher side for thinner reservoir and the lower side for a thicker reservoir. The amount of steam injected increases with the cycle number by 10–15% (Liu, 1997).

When we discussed the CSS mechanism, we mentioned that one mechanism is to reduce formation damage. This is achieved by cleanup during backflow period. From this point of view, the amount of steam injected in the first cycle should not be too high, because a high volume of steam may displace the plugging materials far away from the wellbore, and then it will be more difficult for the plugging materials to be flushed back.

In a very high viscous reservoir, generally the performance in the second and third cycles is better than that in the first cycle. From this point of view as well, the amount of steam injected should not be too high in the first cycle, but the steam should have a high quality.

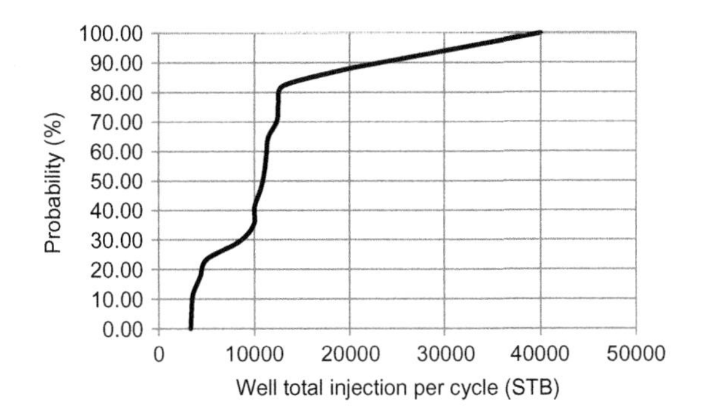

FIGURE 16.7 Well total injection rate per cycle (CWE).

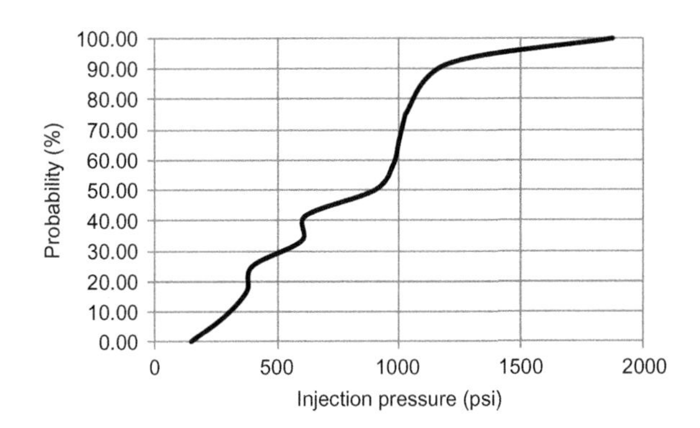

FIGURE 16.8 Well injection pressure.

CSS is mature even at deep reservoirs (e.g., 1700 m). The number of steam stimulation cycles which are economical and effective is 6–7 and should not be greater than 10 (Liu, 1997). Generally, the peak oil rates are in the second and third cycles and decrease sharply during the fourth to sixth cycles. After the seventh cycle, oil rate decreases slowly. On average, three stimulation cycles have been used, as shown in Figure 16.9. The data are from Farouq Ali (1974).

To make the heat more efficient, development should be made area by area for steam soak, although steam soak is conducted in single wells. When the oil rate is about one-third of the rate at the beginning of the cycle, the next cycle of steam injection should be started. In other words, switching to next cycle should be made when the pressure is high and rate is high. Otherwise, the performance of the subsequent cycles will be deteriorated.

Liu (1997) suggested that all the wells should be drilled based on a designed well spacing, instead of drilling infill wells at a later time. This will avoid formation damage and sand production caused by infill drilling when the reservoir pressure is low. It will also help the subsequent steam flooding.

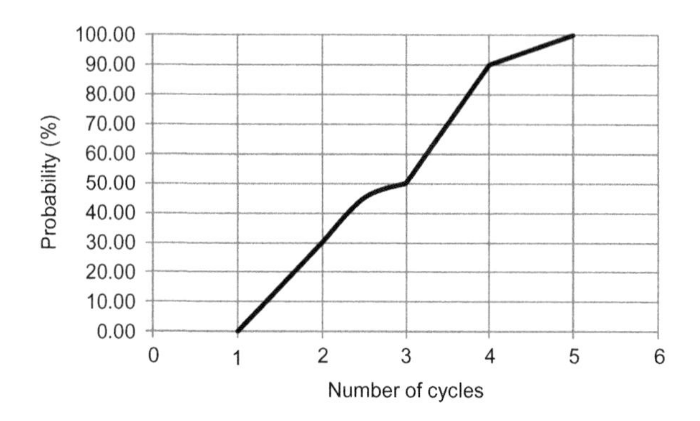

FIGURE 16.9 Number of cycles in actual CSS projects.

16.5.3 Completion Interval

Similar to steam flooding process, wells should generally be completed in the bottom half of the oil layer. If there is a bottom aquifer, the completion should be away from the aquifer for some distance. In a steam soak process, completion in the bottom part may help improve the subsequent steam flood performance.

16.5.4 Wellbore Heat Insulation

For the reservoir depth of 300–400 m, ordinary tubing can be used with packers and nitrogen filled in the annulus. If the depth is above 400 m, insulated tubing and heat-resistant packers are needed. If the depth is 800–1600 m, high-quality insulated tubing and packers must be used with nitrogen filled in the annulus.

16.5.5 Incremental Oil Recovery and OSR

Figures 16.10 and 16.11 show the statistical analysis of actual field data for incremental oil recovered per well and oil-steam-ratio (OSR). The incremental oil recovery and OSR at the 50% probability are 8775 bbls and 0.43, respectively.

16.5.6 Monitoring and Surveillance

During steam injection, injection wellhead temperature, pressure, steam quality, and injection rate are measured. The steam quality at the exit of a boiler and at wellbore should be higher than 75% and 40%, respectively.

During soak period, pressure and temperature are monitored. During production, production rate, wellhead pressure, casing pressure, and the temperature of produced fluid are measured. The water cut and temperature are monitored. The dynamic liquid level is measured once a week in the beginning.

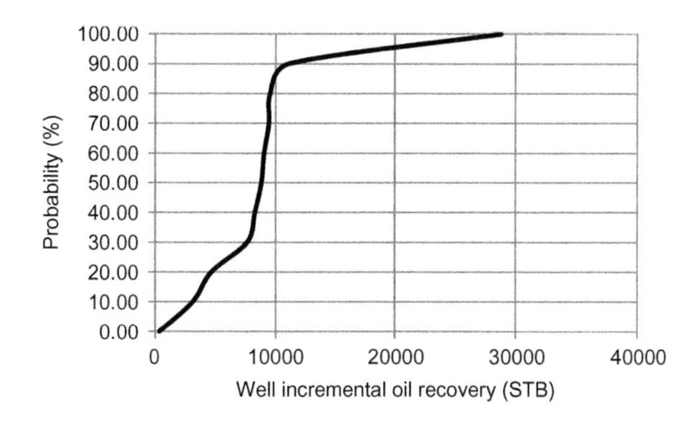

FIGURE 16.10 Incremental oil recovery per well in actual CSS projects.

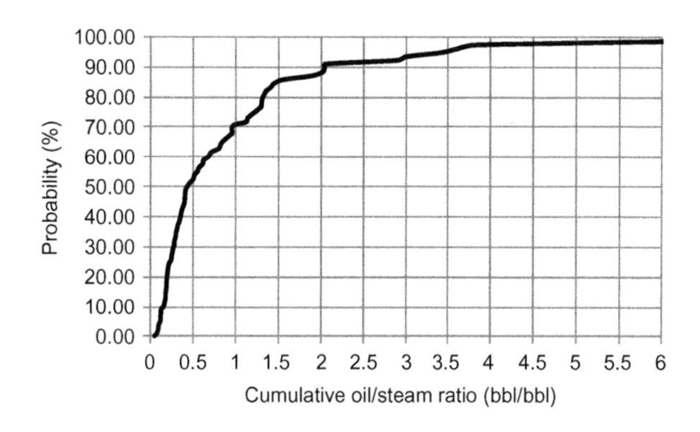

FIGURE 16.11 Cumulative OSR in actual CSS projects.

Fluid samples are taken and analyzed for 30% wells in the first cycle and 15% wells in the second cycle. Water cut, sand content, and chloride ion content are monitored (Zhang, 2006).

16.6 FIELD CASES

Seven field cases are presented which include Cold Lake in Alberta, Canada, Midway Sunset in California, Du 66 block in the Liaohe Shuguang field, Jin 45 Block in the Liaohe Huanxiling field, Gudao Field, Blocks 97 and 98 in the Karamay field, and Gaosheng Field in China.

16.6.1 Cold Lake in Alberta, Canada

This was the largest CSS project in oil sands. Cold Lake was one of the four major Alberta oil sands deposits. It contained an estimated 160 billion barrels of low gravity (10.2°API) and highly viscous oil (100,000 mPa·s at 13°C reservoir temperature). The reservoir depth was from 300 to 600 m. Therefore, the oil was too deep to be produced by surface mining or too viscous to be pumped at a reasonable rate at original conditions. The formation porosity was 37% and the permeability was 3000 mD. The reservoir thicknesses were on the order of 33 m.

Esso Resources Canada began laboratory and engineering studies in the early 1960s, progressing to small-scale field pilots in 1964. In order to provide a sound planning base for future operations, an evaluation program of drilling and coring was started in 1973 (Buckles, 1979).

Because Cold Lake oil (bitumen) at reservoir conditions (450 psi and 13°C) was practically immobile, it was necessary to stress the formation to the point of yielding for steam injection. It was found that the Clearwater formation of main interest would yield to a downhole pressure of 1300 psi. The initial breakdown pressure might be 30–50% higher. With high injection pressure, vertical and horizontal fractures were generated to accommodate large volumes of hot fluids.

The Ethel pilots were initiated in late 1964 and operated until 1970. The stimulation wells were completed in the Clearwater bitumen zone and were stimulated through eight cycles. The size of steam treatments ranged from 3000 to 5000 bbl. Gas was injected with steam in seven of the cycles, and air and water with steam were injected in two cycles. These additives were not convinced to be beneficial.

A soak period of about 5 days was unusually allowed for heat dissipation in the reservoir. The well was then opened for production for a few weeks which might continue for 5–8 months, depending on the fluid temperature and observed decline in oil rate.

In October 1969, a bottom water 5-spot steam flood was initiated. The flood contained one central producer, four steam injectors, and four confining producers, all of which were open to the bottom water. The objective was to determine whether heating conformance in the oil zone could be improved by injecting steam into the more mobile lower water zone. The rate of vertical heating was found to be slow and the experiment was terminated in April 1970.

Steam generation and fluid handling facilities were upscaled to the commercial scale, based on the pilot tests and engineering studies: 20% bitumen recovery, well production rate of 80 bbl/day over an average of 6 year life, and 0.4 OSR.

16.6.2 Midway Sunset in California

This case is about sequential steaming process in the Midway-Sunset field. According to the sequential steaming process, wells would be steamed in rows, on strike, sequencing from down- to updip. To take advantage of the theoretical and observed benefits of asynchronous steaming, a row would be steamed in two stages of alternate, adjacent wells (Jones and Cawthon, 1990; McBean, 1972), as shown in Figure 16.12.

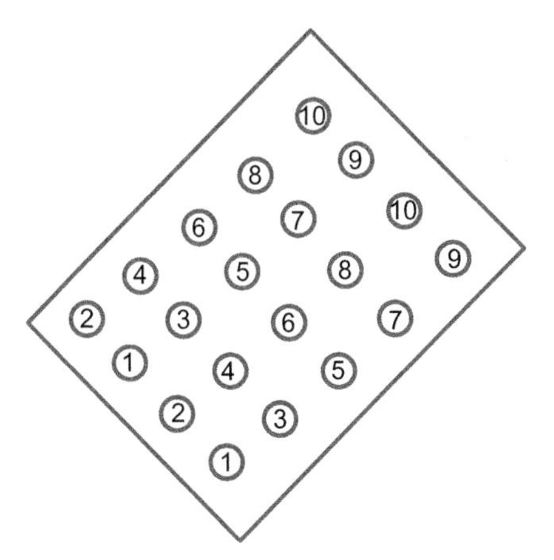

FIGURE 16.12 Schematic of sequential well steaming process (the values of numbers indicate the sequence).

The Potter sands were the primary producing zones in the northern end of the Midway-Sunset field. The sands were predominantly a series of fan-channel complexes formed during rapid subsidence of the basin. Coarse granitic debris from the coastal range flowed down marine canyons into a deep-water fan. The depositional character of the Potter is extremely variable, ranging from massive, conglomeritic, debris-flow-filled, narrow channels to thinly bedded, laterally extensive, very-fine-grained, distal-fan turbidites. The subsidence produced a generally transgressive sequence; i.e., the deeper-water, thinly bedded deposits tended to overlie stratigraphically the coarser, shallow-water channels. Subsequent uplift and tilting produced an erosional surface. The producing sands outcropped at the western limit of Potter and were covered by a rapidly thickening wedge of Tulare silts and sands toward the east. Dips ranged from about 40° at the west to 20° at the east and trended north to east.

Common reservoir properties in these leases were low-gravity crude (11.5–13°API), fairly steep dips, and high lateral permeability between well locations.

Cyclic steam operations in the Midway-Sunset field began in early 1964. The first cycle experienced peak oil rates of nearly 200 B/D and convinced that steam was a very attractive EOR method in this reservoir.

The first steam drive began in August 1967 in an up-structure attic location. Unfortunately, the project performed poorly. A post-audit of this drive showed conclusive evidence of uncontrolled steam loss to the up-structure air zone. Several steam drive projects in later years were not successful.

CSS was tested in this field. More than 19,000 steam cycles were performed in 1500 wells in the Midway-Sunset field. Most of wells were Potter wells. More than 75 wells received 30 or more cycles, and more than 350 wells received 20 or more. The first cyclic well produced 10 bbl/day at the

39th cycle and with cyclic peaks in the 100 bbl/day range. With the much cyclic activity, a number of field tests were performed to optimize the cyclic steam process. Finally, it was found that the sequential steaming process to take advantage of gravity drainage is the process in this dip reservoir.

Here is an example performance of sequential steaming process. In the 27 USL lease, a steady decline of 2% per year was arrested and reversed after implementing an infill drilling and sequential-steam program in 1980 and 1981. Well spacing was decreased from about in 1¼ acre and ⅝ acre. A steam injection schedule was set up in the pattern shown in Figure 16.12 with steam rates increased from about 8000 to 12,000 bbl/year per well. As a result, oil production increased from 15 to 24 bbl/day per well. The key thermal-efficiency indicator, OSR, remained between 0.53 and 0.83 bbl/bbl.

16.6.3 Du 66 Block in the Liao Shuguang Field, China

The Du 66 block was in the Liao Shuguang field. The block had many thin layers. It had an oil-bearing area of 4.9 km^2 with original oil in place (OOIP) of 39.4 million tonnes. The reservoir depth was 800–1200 m. The average reservoir thickness was 42.1 m, average permeability 780 mD, and average porosity 25%. The reservoir had many think layers and about 30 clay interbeds. The average thickness of these interbeds was thicker than 3 mm. The net-to-gross ratio was low (<0.5). The oil viscosity was 300–2000 mPa·s. The reservoir temperature was in the range of 47–54°C. The measured pressure was 9.69–11.04 mPa.

Cyclic steam injection was initiated at the well Shu-1-37-35 in March 1985. Steam was injected from March 16 to March 27 for 12 days. A total of 2302 tons of steam was injected. The well was under natural flow for 3 days. The average oil rate was 104 tons/day. Steam was injected again from April 5 to April 16, 1985. The total steam injection was 2554 tons. Then there was a steam soak for 2 days followed by 8 days of natural flow and 237.2 days of pumping. The cumulative OSR was 2.9.

The steam injection was expanded in 1986. The injection patterns were in 200 m square patterns (5-spot patterns). By October 1989, a total of 187 wells had been drilled according to the development plan. Afterward, new wells were drilled to adjust the injection patterns. By February 1990, a total of 358 wells had been drilled including 200 producers, 138 injectors, and 20 observation wells. At a later time, infill wells were drilled.

In September 1991, a steam flooding pilot was initiated. By December 1993, 343 wells had been drilled with 325 wells open. The total oil rate was 1496.4 tons/day, the water cut was 46.7%. The recovery factor was 9.64%. The cumulative OSR was 1.01. The cumulative oil production was 1.3–2.3 times that of the analog block, Du 84 block, where waterflooding was implemented.

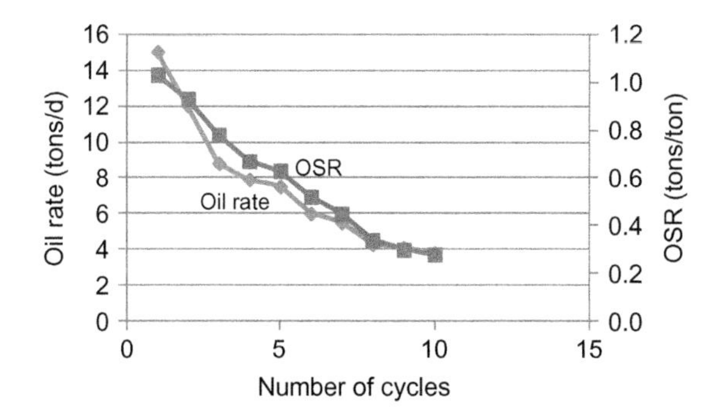

FIGURE 16.13 Oil rates and OSR at different cycles.

During 1994–1998, more infill wells were drilled. After 1998, no new wells were drilled. In the Du 166 and Du 97 well patterns, hot water was added in the steam stream. In the well Shu-1-45-31 well pattern, water-alternate-steam injection and hot water injection were tested.

In June 2003, the Du 66 block had 538 wells with 428 wells open, 36 hot water or steam flooding wells with 24 wells open, and 10 observation wells. The water cut was 62.6%, the recovery factor was 19.76%, and the cumulative OSR was 0.64. The average oil rate was 1.5 tons/day. The average reservoir pressure was 1.2 mPa, significantly reduced compared with the initial reservoir pressure. The average cycles were 8. The initial oil rates per well and OSR in each cycle are shown in Figure 16.13. The oil rates and OSR decreased with the cycle.

The Du 66 block was in the late stage of cyclic steam injection. The question was whether it should be converted to steam flooding. A simulation study of a pilot zone of four patterns showed that 55.14% recovery factor for steam flooding, 49.4% for intermittent steam injection (2-month injection and 1-month pause), and 49.8% for steam flooding for 4 years followed by cold waterflooding. The simulation results showed steam flooding should be continued. The subsequent pilot showed that injection profile needed to be improved before steam flooding. At the end, the main layers were under steam flooding.

Several production techniques have been implemented. One was to use prestressed casing. Prestressed casings were installed in more than 250 wells. Casing was found damaged only at 1 well. Rods were heated electrically so that the oil in the wellbore was heated and the oil viscosity was reduced. Mixing produced oil with hot oil also reduced the oil viscosity.

Several lessons from this block are (Liu, 1997):

1. Combine thin layers into several thick layers and selectively perforate thick layers.
2. Steam quality should be high in such reservoir with many thin layers.

3. Take measures to prevent clay swelling.
4. Use packers to achieve steam injection in separate layers to reduce steam crossflow between layers or between wells.

16.6.4 Jin 45 Block in Liaohe Huanxiling Field, China

The Jin 45 block had an active edge and bottom aquifer. Its area was 9.05 km^2. CSS was tested from May 1985 to July 1986. Edge aquifer broke in the test well 18–24 during the third cycle. Starting in 1986, steam soak was implemented in the entire block using four developing layers. Square patterns of 167 m well distance were used. By June 1991, a total of 295 wells were drilled with 232 well open. The average well oil rate was 9 tons/day and the water cut was 67.2%. Overall, steam soak performed well.

The reservoir depth was 890–1180 m. The average porosity and permeability were 29% and 800 mD, respectively. There were two groups of layers which had two separate water–oil contacts: 1020–1060 and 1120–1160 m. Both layers had edge and bottom aquifers. The reservoir temperature was 44.6–50°C, and the initial reservoir pressure was 10 mPa. The oil viscosity at 50°C was 486–7696 mPa·s.

During the first cycle, 89% of surveyed 171 wells could flow naturally in the beginning. For the surveyed 111 wells, the percentages of production from natural flow in the entire cycle were 23.3 in the first cycle, 13.3 in the second cycle, and 3.6 in the third cycle. This was because of strong edge and bottom aquifer to provide pressure support. The performance in the first two or three cycles was good. After that, aquifer broke in and the water cut was above 50%. And the oil rate decreased significantly. However, for the wells near edge and bottom aquifer, oil rate decreased more slowly, especially in the first and second cycles. For the wells in the top center area, oil rate and pressure decreased faster; and when water broke in, pressure built up, water cut rose, and steam soak performance became poorer. On the average, the maximum cycles of a single well were 6–7, and the CSS lasted about 5 years.

Several lessons from the CSS in this block are as follows (Liu, 1997):

1. Although edge and bottom aquifer provided pressure support and increased oil rate in the first two cycles, the water breakthrough reduced the number of cycles and deteriorated the steam soak performance. To control the aquifer breakthrough, larger pumps were used in edge wells. For some water coning wells, water shutoff workover was performed.
2. Steam was injected separately into layers so that the steam injection rates in different layers were controlled.
3. It was observed that steam injected broke through neighbor wells. This was because injection pressure was too high, some fractures were formed;

and steam broke through along faults. Therefore, steam injection rate, injection pressure, and injection strength should be controlled.

4. Sand production was a problem. Measures must be taken to control sand production.

16.6.5 Gudao Field, China

We focus on production techniques in developing heavy oil reservoirs using this field case, the Gudao field in China. The Ng_5-Ng_6 sand groups in this field were unconsolidated. The clay content was 7.5–12%. The porosity was 30–35%, the permeability was 770–2000 mD, and the initial oil saturation was 56–65%. The oil viscosity was 5000–24,562 mPa·s. The oil viscosity was sensitive to temperature. If the temperature was raised each 10°C above 50°C, the oil viscosity was reduced by half. The reservoir depth was about 1300 m.

A single well CSS was started in the well Zhong 25–420 from August 4 to 27, 1991. The injection pressure was 10.5–13.5 mPa, and the injection temperature was 270–310°C with the steam quality of 40–75%. The injection rate was 168 tons/day and total of 2206 tons of steam were injected. The initial oil rate was 23.5 tons/day and the production lasted 191 days. A subsequent test well had similar performance. The CSS was extended to larger scales and finally to a commercial scale. Some of production techniques practiced in this field are discussed below.

Well Completion

Low-solid drilling fluid was used (<5% solid). The hydraulic pressure of drilling fluid was not 5–8% greater than the reservoir pressure. The cement had 30–40% silica flour and the cement filled up to the surface. The casing was prestressed with 1.2×10^6 N (including the casing gravity).

Sand Control

Owing to unconsolidated sand, several sand control techniques were developed and implemented. In Zhong Er Bei Unit 5, coating sand was used in 78 wells. The treatment was successful in 59 wells and was effective for 171 days (75.6% success rate). However, when applied to 33 wells (times) in another unit, only 19 wells (times) was successful (57.6% success rate).

Another technique for sand control was wiring wrapped screen. The wiring wrapped screens were applied at 292 wells (times); 271 treatments were successful (93.9% success rate) and were effective for 250 days. However, such treatment was expensive. The implementation took a long time resulting in more heat loss. Other tools were being developed to improve the technology.

Tests were conducted to combine coating sand and wiring wrapped screens. Wells were filled with coating sand first under high pressure to form a strong stable borehole. Then metal wiring wrapped screens were installed with gravel packing; 26 wells (times) were tested and 24 cases were successful and effective for 198 days.

Prevent Clay Swelling

Owing to high content of clays (7.5%), clay stabilizing agents were developed. One product was FGW-1, which was a combined system of organic cationic polymer and inorganic compounds. A survey of 35 wells (times) showed 21.8 days longer for cyclic production.

Application of Detergents

The original reservoir pressure was about 12 mPa. The injection pressure in some wells reached 15 mPa. Detergents like BN-5 were used to clean up plugging near wellbores. Nitric acid was also used for wellbore cleanup.

Application of Thin Film Spreading Agents

Thin film spreading agents like HCS were used to break oil films, to demulsify W/O emulsion to O/W emulsion, and to generate emulsions for improved sweep efficiency.

16.6.6 Blocks 97 and 98 in Karamay Field, China

This case is about an ultrahigh viscous oil case. The top of the reservoir was at 70–150 m. The average thicknesses of oil zones were 10.5 m in the block 97 and 12.2 m in the 98 block. The formation was the Qi-Gu group. The average porosity and permeability in the oil-bearing layers were 30.6% and 1287 mD, respectively. The vertical permeability was 597 mD on the average. The formation of total dissolved solids was about 3127 ppm. The initial oil saturation was 70.68%. The initial reservoir pressure and temperature at the middle depth of 145 m (155 m subsea) were 1.63 mPa and 17.1°C, respectively. The average oil viscosity at 20°C was 350,000 mPa·s. The oil viscosity in the block 98 reduced by 50 times from 20°C to 50°C. At 80°C, the viscosity could reduce to 1000 mPa·s. The oil viscosity in the block 97 reduced by 100 times from 20°C to 50°C. At 90°C, the viscosity could reduce to 1000 mPa·s so that the oil could flow in the wellbore and underground.

Field trials and development may be divided into two phases. In the first phase from September 1989 to December 2004, several methods were tested and implemented in these blocks before CSS, such as horizontal wells in the 98 block, and large-diameter wellbores and downhole heating in the 97 and 98 blocks. During the period, reducing oil viscosity using chemicals, microbes, and mixing with light oils was tested. Some of tests in the first phase are detailed below.

1. The first CSS was started in September 1986; $100 \times 140\ m^2$ well patterns were used. The oil viscosity in the pilot area was 81,016 mPa·s at 20°C. Total of 18 wells were drilled. At the end of the pilot in December 1993, the water cut was 71% and the OSR was 0.24. The total cycles were 3.9, and cumulative oil production per well was 2195 tons on the average.
2. Another pilot area was started in 1988 and started production in 1989; $100 \times 140\ m^2$ well patterns were also used. The oil viscosity in the pilot area was 213,954 mPa·s at 20°C. Total of 48 wells were drilled. By August 2004, the water cut was 69% and the OSR was 0.19. The cumulative oil production per well was 2613 tons and well oil rate was 2.7 tons/day on the average. In the first two cycles, the production time was about 110 days and the oil production was above 500 tons. The well oil rate was above 4 tons/day and the OSR was 0.22, demonstrating good performance. However, during later cycles, performance became worse. The main reasons were low steam quality due to long steam lines, lowered steam injection strength in the third cycle, and sand production.
3. Total of 141 wells including five large-diameter wells were drilled in another area in the 98 block with relatively lower oil viscosity (172,567 mPa·s at 20°C) in 1993–1996. By August 2004, the well oil rate was 2.5 tons/day, water cut 70.5%, and OSR 0.25 on the average. For the five large-diameter wells, 4008 tons of oil was produced by 1 well and the OSR was 0.28–0.39 in the first four cycles. However, starting in the fifth cycle, oil rate and liquid significantly decreased because of low steam injection and low steam quality.
4. Because of the success in using large-diameter wells in the 98 block, a $70 \times 100\ m^2$ inverted 9-spot pattern was drilled (total 9 wells) in the 97 block to further test large-diameter wells. The oil viscosity was 566,869 mPa·s at 20°C. The pattern started production in September 1998. By August 2004, the cumulative oil production was 58,000 tons, cumulative water production 89,400 tons, and cumulative OSR 0.31. The well production time was 1872 days with oil rate 3.44 tons/day and water cut 60.7%. During the test, high steam quality was maintained, and measures like using chemicals to reduce oil viscosity were taken. In 2001, additional eight patterns were drilled. Because some of wells were not perforated in the main producing layers, low steam quality, and higher oil viscosity, the performance in these patterns was poor. Also considering high cost of drilling, therefore, the method using large-diameter wells was not further expanded.
5. A pilot of 15 vertical injectors and 4 horizontal producers in the 98 block was executed in 1997. The distance between vertical wells or between a

vertical well and a horizontal well was about 50 m. The middle reservoir depth was 180 m. The reservoir thickness was 12.5 m and the porosity was 30%. The oil viscosity at 20°C was 121,800 mPa·s. By October 1999, one cycle of steam soak and one cycle of intermittent steam injection had been executed in the four horizontal wells. The oil rates were 8.9–14.4 tons/day with an initial rate 45 tons/day. Compared with the nearby old vertical wells, the liquid production was 2.7 times, oil production 1.5 times, and oil rate 2.8 times those of vertical wells. However, because these horizontal wells were infill wells (the nearby old vertical wells had been through more than four cycles before drilling horizontal wells), these horizontal wells had high water cut (70%) and low OSR (0.21). In the second cycle, steam broke through and the wells were buried by sand. Owing to high cost of workover, the test was abandoned.

6. Downhole heating was tested in 5 wells in the 97 block in 2004. The oil viscosity at 20°C was 1,500,000 mPa·s or so. By October 2005, the production time was 364 days, and the oil rate was 3.6 tons/day on the average. In the first cycle, the wellhead temperature was 40°C, the production time was short (46 days), and the oil production in the cycle was 214 tons per well. In the second cycle, the wellhead temperature was above 70°C, the production time was 118–211 days, and the well oil production in the cycle was 751 tons. Two more wells were tested and were successful. These tests showed that oil with more than 1,000,000 mPa·s could be produced using downhole heating.

In the second phase (from January 2005 onward), it was designed to use 70 m square patterns to develop the blocks. According to the development plan, the pipeline to transport steam was shortened from 1.5–2 to 0.5 km, reducing the heat loss rate from 14.7 to less than 5%. As a result, the steam quality at the wellheads was raised from 60% to above 70%. More horizontal wells were drilled. Screening completion was used to prevent sand production. A large-scale steam injection was implemented in these blocks. At the end of 2005, the oil-bearing area was 5.44 km^2, and the producing oil in place was 19.73 million tons. Total number of wells including abandoned 59 wells was 814. The water cut was 74%, OSR 0.22, recovery factor 8.8%, and well oil rate 2 tons/day.

The performance from these blocks may be summarized as follows.

1. Because of high oil viscosity (50,000–1,000,000 mPa·s at 20°C), oil could not flow without heat injection.
2. Steam soak made oil flow naturally. But the production time was 1–32 days with an average of 7.5 days. Initial rate was high (>6 tons/day) but declined quickly (very low rate in 10 days).
3. The oil production and OSR in the second and third cycles were higher than those in the first cycle.
4. Steam breakthrough and sand production were the problems.

16.6.7 Gaosheng Field, China

This field in Liaohe, China, had gas cap. Although it also had a bottom water, there was a barrier so that water coning was not observed. The gas–oil level was 1510 m, the oil–water level was 1690 m, and the reservoir depth was 1500–1800 m. The developed area was 14.5 km^2. In the horizontal direction, there were seven blocks among which the blocks 3, 246, and 3618 were the mainly oil-bearing blocks. In the vertical direction, there were eight layers. Among these layers, Layers L1–L4 were gas-bearing layers, L5, L6, and L7 were the main oil layers (88% oil in place), and L8 was the aquifer layer. The reservoir thickness was 67.7 m on average. The porosity was 22–26% and the air permeability was 1000–2300 mD. The reservoir temperature at 1600 m was 60°C and the initial reservoir pressure was 16.1 mPa. The oil viscosity *in situ* was 74–605 mPa·s. The oil viscosity decreased to 6 mPa·s when the temperature was raised to 200–220°C (Liu, 1987).

Initially the field was produced by mixing light oil and heating rod pumps. Starting in September 1982, CSS was tested and found successful. In 1984, a development plan was designed which included:

1. Five-spot patterns of 210 m later infilled to 150 m.
2. Separately developing L5, L6, and L7 because of existence of gas cap and bottom water.
3. Four phases: initial depletion by mixing light oil and heating rod pumps, CSS, steam flooding, and cold waterflooding.
4. Completion included gravel packing, wiring wrapped screen, and perforated prestressed casing.
5. Wells were drilled along the gas–oil ring in L5 to make use of gas cap energy and control pressure.

Because the reservoir was deep, it was important to reduce heat loss through wellbores. Measures to reduce heat loss included tubing insulation, high-temperature metal packer, and filling nitrogen in the annulus. The heat loss was controlled to be less than 12%.

It was observed that the back-produced water was only 7.8% of the injected. Such low flow back was caused by high content of clay (7–10%), especially montmorillonite (90%). Clay swelling adsorbed a lot of water and reduced permeability. The cumulated water slowed down the heat dissipation into the reservoir during injection. To solve this problem, surfactants and chemicals to prevent clay swelling were added in the steam. Adding nitrogen in the steam also helped water production. Adding thin film spreading agents also helped.

To stop gas cap breakthrough, several wells were drilled to produce gas under a controlled mode. The pressure of gas cap was controlled not lower than 8 mPa, and the pressure difference between gas cap and oil layer was controlled.

REFERENCES

Adams, R.H., Khan, A.M., 1969. Cyclic steam injection project performance analysis and some results of a continuous steam displacement pilot. JPT 21 (1), 95–100.

Bentsen, R.G., Donohue, D.A.T., 1969. A dynamic programming model of the cyclic steam injection process. JPT December, 1582–1596 (Trans., AIME, 246).

Boberg, T.C., Lantz Jr., R.B., 1966. Calculation of the production rate of a thermally stimulated well. JPT December, 1613–1623 (Trans., AIME, 237).

Buckles, R.S., 1979. Steam stimulation heavy oil recovery at Cold Lake, Alberta, Paper SPE 7994 Presented at the SPE California Regional Meeting, 18–20 April, Ventura, CA.

Clossmann, P.J., Ratliff, N.W., Truitt, N.E., 1970. A steam-soak model for depletion-type reservoirs. JPT 22 (6), 757–770 (Trans., AIME, 249).

de Haan, H.J., van Lookeren, J., 1969. Early results of the first large-scale steam soak project in the Tia Juana field, Western Venezuela. JPT 21 (1), 101–110.

Farouq Ali, S.M., 1974. Current status of steam injection as a heavy oil recovery method. J. Can. Petrol. Technol. 34 (1), 54–68.

Farouq Ali, S.M., Meldau, R.F., 1979. Current steamflood technology. JPT 31 (10), 1332–1342.

Green, D.W., Willhite, D.P., 1998. Enhanced oil recovery. SPE Text Book Series. vol. 6. The Society of Petroleum Engineers, Richardson, TX.

Jones, J., Cawthon, G.J., 1990. Sequential steam: an engineered cyclic steaming method. JPT 42 (7), 848–853, 901.

Liu, W.-Z., 1987. Pilot steam soak operations in deep wells in China. JPT 39 (11), 1441–1448.

Liu, W.-Z., 1997. Steam Injection Technology to Produce Heavy Oils. Petroleum Industry Press, Beijing, China.

Martin, J.C., 1967. A theoretical analysis of steam stimulation. JPT 19 (3), 411–418.

Marx, J.W., Langenheim, R.H., 1959. Reservoir heating by hot fluid injection, trans. AIME 216, 312.

McBean, W.N., 1972. Attic oil recovery by steam displacement, Paper SPE 4170 Presented at the SPE California Regional Meeting, 8–10 November, Bakersfield.

Seba, R.D., Perry, G.E., 1969. A mathematical model of repeated steam soaks of thick gravity drainage reservoirs,. JPT, 21 (1), 87–94 (Trans., AIME, 246).

Taber, J.J., Martin, F.D., Seright, R.S., 1997. EOR screening criteria revisited—part 2: applications and impact of oil prices. SPERE 12 (3), 199–206.

Zhang, Y.T., 2006. Thermal recovery. In: Shen, P.P. (Ed.), Technological Developments in Enhanced Oil Recovery. Petroleum Industry Press, Beijing, pp. 189–234.

Chapter 17

SAGD for Heavy Oil Recovery

Chonghui Shen
Shell Canada Limited, 400, 4th Ave., SW, Calgary, Alberta, Canada, T2P 2H5

17.1 INTRODUCTION

The oil sands of northern Alberta, Canada, contain about 1.7 trillion barrels of bitumen, making it the second largest hydrocarbon resource on earth. Only recently has technology developed, allowing for efficient recovery of the bitumen in the subsurface. The steam assisted gravity drainage (SAGD) is one of the *in situ* recovery processes for Alberta oil sands, and it is the focus of this chapter.

SAGD is a thermal oil recovery process, originally conceived by Dr. Roger Butler (1982), for recovering viscous heavy oil and bitumen from a reservoir by continuously injecting steam to lower the viscosity of the oil and concurrently producing the mobilized oil and steam condensate. Figure 17.1 shows the schematics of the SAGD process mechanism. To achieve continuous injection and production at the same time, a pair of parallel wells is used in the process. Steam is injected into the formation via an (upper) injection well and forms a steam saturated zone which is usually called the "steam chamber." Steam flows toward the edge of the steam chamber and releases its latent heat to the formation and condenses, the viscous hydrocarbons are mobilized and drain by gravity toward the (lower) production well. The steam chamber grows vertically and spreads laterally in the formation with time under gravitational force.

Butler's gravity-drainage theory yields an analytical equation to predict the rate of oil drainage of the steam chamber, with the assumptions that only steam flows in the steam chamber, oil drains along the vertical steam chamber, oil saturation is residual, and heat transfer ahead of the steam chamber to cold oil is only by steady-state conduction (Butler, 1994):

$$q = 2L\sqrt{\frac{1.3\phi\Delta S_o kg\alpha h}{m\nu_s}} \tag{17.1}$$

where L is the length of the horizontal well, ϕ is the porosity of the formation, ΔS_o is the difference between initial oil saturation and residual oil

Enhanced Oil Recovery Field Case Studies.

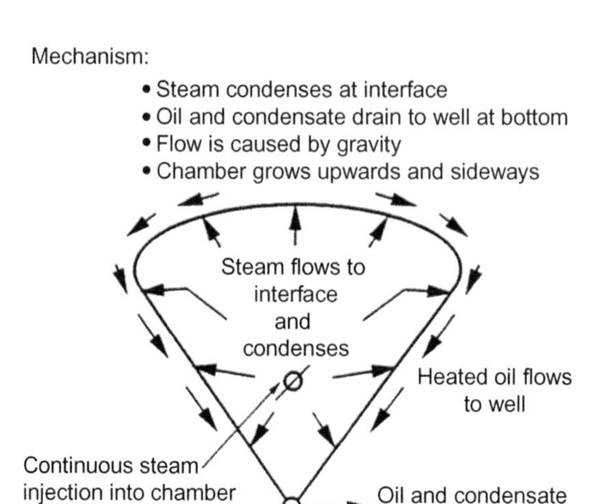

FIGURE 17.1 The schematic of SAGD process mechanism (Butler, 1994).

saturation to steam, k is the effective (vertical) permeability for the flow of oil, g is the acceleration due to gravity, α is the thermal diffusivity, h is the steam chamber height, m is a dimensionless parameter (typically 3 to 4) depending on the oil viscosity versus temperature relation, and ν_s is the kinematic viscosity of oil at steam temperature. It has been realized that the derivation of the above SAGD drainage equation relies on some crude assumptions; e.g., it neglects (1) liquid buildup toward the bottom of the steam chamber and (2) film drainage of oil in the SAGD chamber. For details of assumptions, one may refer to Butler's original derivation.

The SAGD process is dominated by the gravitational force, as shown in Eq. (17.1). With gravity drive, the process is relatively stable and self-correcting if the operation is properly managed. During the operation, one maintains the balance of injection and production to achieve a targeted, stable, operating pressure. On the other hand, the gravitational drive force is relatively weak compared to the viscous force which usually dominates in flooding processes, and the drainage process is slow. Therefore, it requires sufficiently high fluid mobility to achieve reasonable well productivity. This can be achieved by the combination of lowering the oil viscosity via heating and selecting a formation with high permeability, especially in the vertical.

The typical well configuration for SAGD is a pair of vertically aligned horizontal wells drilled into the oil reservoir, one a few meters above the other, with the upper wellbore used for steam injection and the lower wellbore for fluid production, as shown in Figure 17.2. It has been estimated that the oil production rate for Alberta oil sands reservoir is in the range of 0.1–0.3 m^3/day per meter of wellbore, according to Eq. (17.1), as shown in Figure 17.3. The utilization of

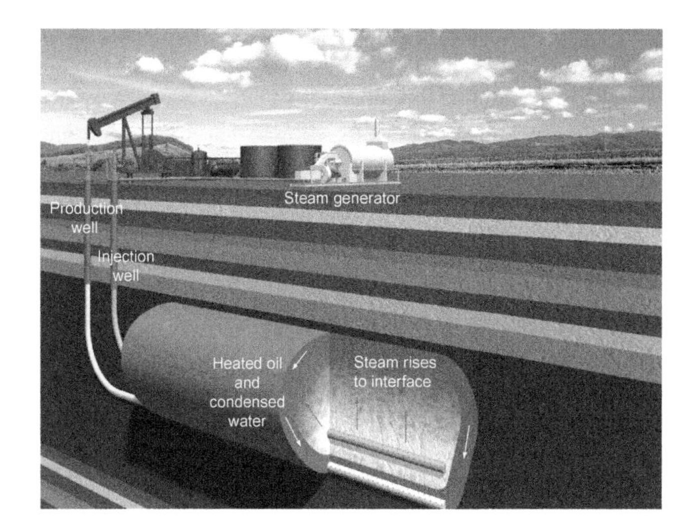

FIGURE 17.2 The schematic of typical SAGD well pair configuration.

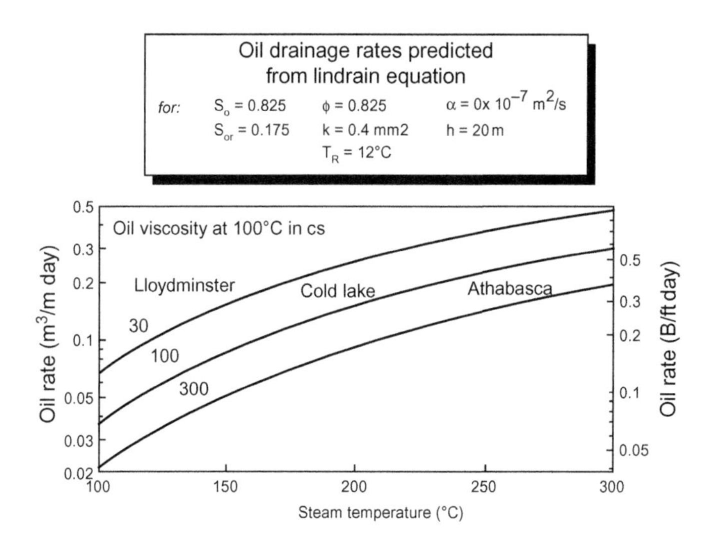

FIGURE 17.3 Drainage rates for three Canadian heavy crudes (Butler, 1994).

horizontal wellbores in SAGD greatly improves reservoir contact and well productivity. To achieve a reasonable oil production rate, one would choose a pair of horizontal well 500–1000 m long. The horizontal producer is placed close to the base of reservoir pay to maximize the reservoir drainage volume, and the injector is drilled above and parallel to the producer. The vertical separation of the wellbores is 4–6 m for Alberta oil sands formation. The SAGD wellbores are directionally drilled to ensure precise placement of the wellbores in the formation and with respect to each other.

As in all thermal recovery processes, the SAGD process is also energy intensive. The energy balance analysis shows that in general the energy injected by the steam can be roughly divided into three equal streams: one-third retained in the steam chamber, one-third dissipated to formation rock outside the steam chamber, and one-third produced to surface (Yee and Stroich, 2004). For the process to be economical, the energy efficiency measured in terms of cumulative steam to oil ratio (cSOR) is generally in the range of 2–4 tonnes/m^3.

SAGD for heavy oil recovery can achieve a recovery factor of 60% in general, or even as high as 70–80% in some favorable reservoirs.

17.2 EVALUATION OF SAGD RESOURCE

17.2.1 Importance of Resource Quality

It is generally agreed that resource quality is the most critical factor in SAGD project performance, because SAGD economics are more sensitive to resource quality than to most other operating parameters. The key indicators of reservoir quality for SAGD are (1) the thickness of continuous pay, (2) the oil concentration (which is usually defined as weight percent of bitumen), and (3) formation (vertical) permeability. It has been found that the SAGD process performs best in thick, highly permeable, oil wet reservoirs with no or discontinuous nonpermeable layers, while initial oil viscosity plays a less significant role. Based on the cross-plot of oil concentration versus formation permeability, as shown in Figure 17.4, the general ranking of reservoir quality of three Canadian oil sands deposits/formations are in the order of Athabasca/McMurray, Cold Lake/Clearwater, and Peace River/Bluesky. Shin and Polikar (2005) conducted an economic evaluation of the three Canadian oil sands resources and revealed that SAGD economics depends strongly on the product of permeability and thickness, as shown in Figure 17.5. This is the main drive behind the much higher concentration of SAGD projects in the Athabasca/McMurray formation than in others.

The pay thickness determines the maximum drainage head. The thicker the pay, the greater the drainage head and the higher the production rate will be, as shown in Eq. (17.1). A thick oil pay also means less heat loss to over- and underburden formation rocks. For commercially successful SAGD operations using current technology and proven production, a reasonable pay thickness cutoff is in the range of 10–15 m.

Oil concentration is directly related to thermal efficiency. In the SAGD process, the gross rock volume in the entire steam chamber needs to be heated to the steam temperature regardless of reservoir quality. Higher oil content means that more oil can be recovered with the same amount of

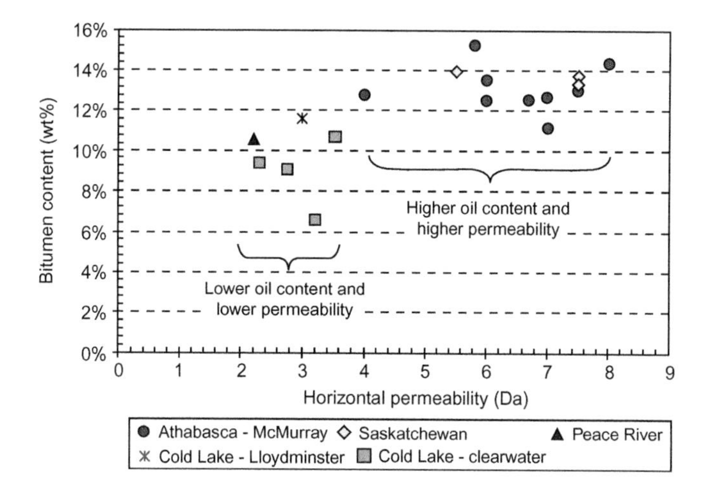

FIGURE 17.4 Cross-plot of weight percent bitumen versus permeability of Canadian SAGD projects (Scott, 2002).

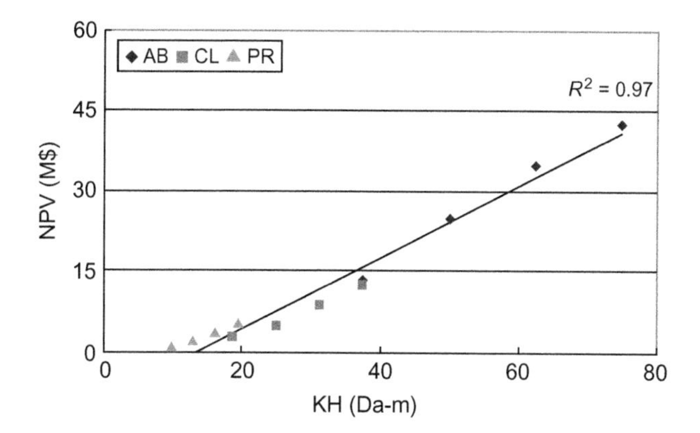

FIGURE 17.5 Effects of reservoir productivity on SAGD process economics (Shin and Polikar, 2005). AB, Athabasca; CL, Clearwater; PR, Peace River; NPV, net present value; KH, product of permeability and thickness.

thermal energy consumption, and this translates to a high thermal efficiency (or a lower cSOR). Using current technology and proven production, SAGD operations would be commercially viable at 10% weight of bitumen. It can be reasonably expected that in SAGD, the ultimate recovery factor will be close to 60%, or as high as 80% for some top quality reservoirs.

The founding principle of SAGD is that the injected steam rises and steam condensate and bitumen flow down to the production well via gravity. Hence, the need for uninterrupted vertical permeability becomes paramount. In Alberta oil sands, the vertical permeability is strongly controlled by shale streaks in the depositional bodies or the boundaries in between. Understanding the distribution of resource sedimentary facies is critical to the success of development. For example, the presence of shale layers, or breccias, within the sandstone formation, if located between the injector and producer wells, can hamper the rise of the steam chamber and/or the producer/injector communication, or even limit the rise of the steam chamber and effectively reduce the drainage head.

Geological heterogeneities will cause variations in steam chamber rise and liquid drawdown along wellbores. The effects of reservoir heterogeneity on the SAGD process has been numerically evaluated by Chen et al. (2008) using a stochastic model of shale distribution. They found that the drainage and flow of hot fluid within the near wellbore region are very sensitive to the presence and distribution of shale. The presence of shale features above the well region will adversely affect (vertical and horizontal) expansion of the steam chamber, but only when the above well region contains long, continuous shale bodies, or a high fraction of shale. The findings of this study point to the importance of determining the base of SAGD interval and the requirement of placing the producer in clean sands. This has been confirmed by some SAGD projects with producers placed into shaly (low-quality) reservoir zones (Figure 17.6), leading to serious performance issues, and by the performance improvement when wells were redrilled and placed into better reservoir zones in the Tucker Lake project (Husky Energy, 2011).

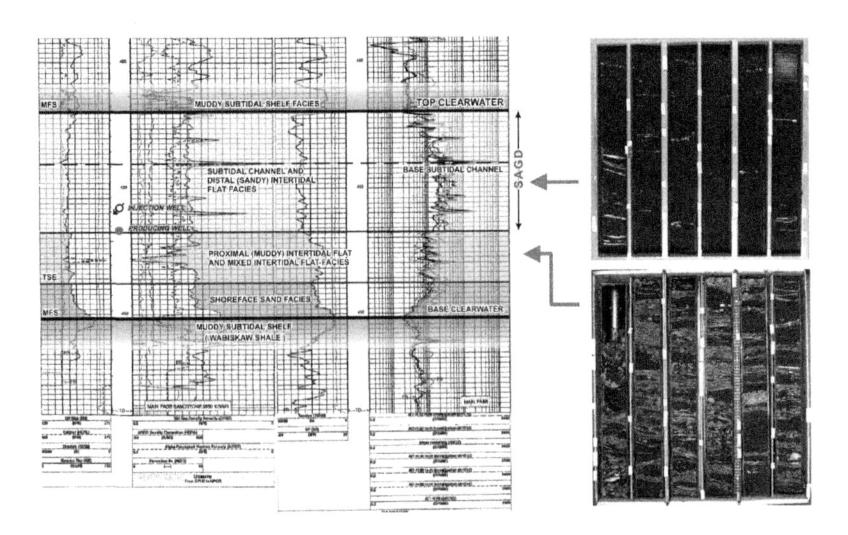

FIGURE 17.6 Composite well log and representative core photos of Clearwater formation in the area of Shell's Orion project (Shell Canada, 2011). The lower reservoir facies is much shalier than the upper reservoir facies and hence low vertical permeability.

17.2.2 Focus of Delineation

Alberta oil sand deposits are generally comprised of multiple episodes of incision, valley creation, and subsequent infill with fluvio-estuarine sediments. One of the examples is shown in Figure 17.7. Fluvio-estuarine deposits are, by their nature, heterogeneous. Significant reservoir heterogeneity will lead to nonuniform steam chamber development, a low wellbore utilization factor, and low production rate. For a 500–1000 m long horizontal well, it can be reasonably expected that the wellbore penetration will cover more than one geological body or facies. To maximize wellbore utilization, it is critical to have a good understanding of reservoir geology and to optimize the placement of horizontal wellbores with respect to it.

The drilling of delineation wells provides critical information about reservoir tops and quality. To achieve a proper level of geological understanding, the industry is moving toward drilling closely spaced delineation wells to 400 m or less apart. For example, Figure 17.8 shows the delineation well density in the Phase 2 area of MEG Energy Christina Lake Regional Project. The development area has a density of delineation wells up to 30 per square mile.

Delineation wells can be logged and cored or only logged. Cored wells through the anticipated pay zone contribute more to the understanding of geological details by allowing identification of thin shale streaks and direct determination of oil concentration. Well logs can provide specific information but are usually not good enough by themselves to fully evaluate formation quality; especially in the case of small-scale shale streaks which are below the resolution of the logging tools but important to vertical permeability and relevant to SAGD performance. The advancement of formation microimaging (FMI) logging technology has lead to it being used as an alternative to core evaluation.

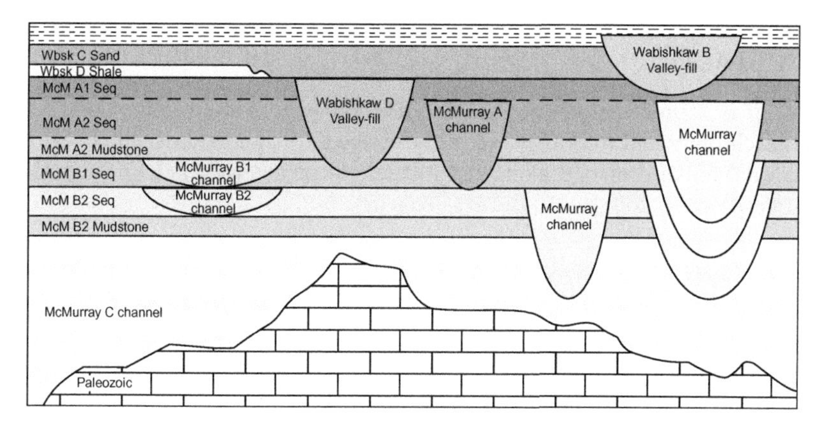

FIGURE 17.7 Schematic of McMurray formation geology in Athabasca deposit. Source*: From Hein and Cotterill (2006).*

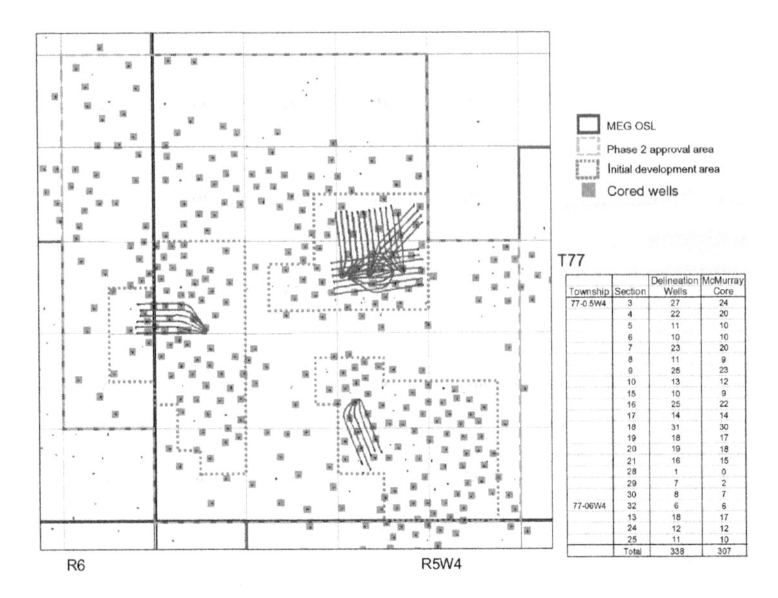

Township	Section	Delineation Wells	McMurray Core
77-0.5W4	3	27	24
	4	22	20
	5	11	10
	6	10	10
	7	23	20
	8	11	9
	9	25	23
	10	13	12
	15	10	9
	16	25	22
	17	14	14
	18	31	30
	19	18	17
	20	19	18
	21	16	15
	28	1	0
	29	7	2
	30	8	7
77-06W4	32	6	6
	13	18	17
	24	12	12
	25	11	10
	Total	338	307

FIGURE 17.8 Delineation well density in Phase 2 Area of MEG Energy Christina Lake Regional Project (MEG Energy, 2011). One section covers one square mile.

The application of high-resolution 3D seismic allows one to map the general structure of the reservoir deposits, as shown in Figure 17.9 (Hubbard et al., 2011). Seismic can help to provide subsurface information on the depositional environment, helping to identify zones that are best suited for the SAGD process because of pay thickness, sand quality, and permeability. Once the better reservoir intervals are mapped out, one can design the specific patterns for the SAGD development.

17.3 START-UP

Start-up is a process to bring the SAGD well pair in fluid communication with each other and to initiate a steam chamber. It is aimed at achieving uniform and active drainage along the full length of the well pair through the start-up process.

17.3.1 Circulation Heating and Inter-Well Communication Initialization

Due to the very nature of high bitumen viscosity, the initial mobility of oil in the formation is very low or near zero. To promote inter-well fluid communication and to lower the oil viscosity, one needs to preheat the formation between the wells.

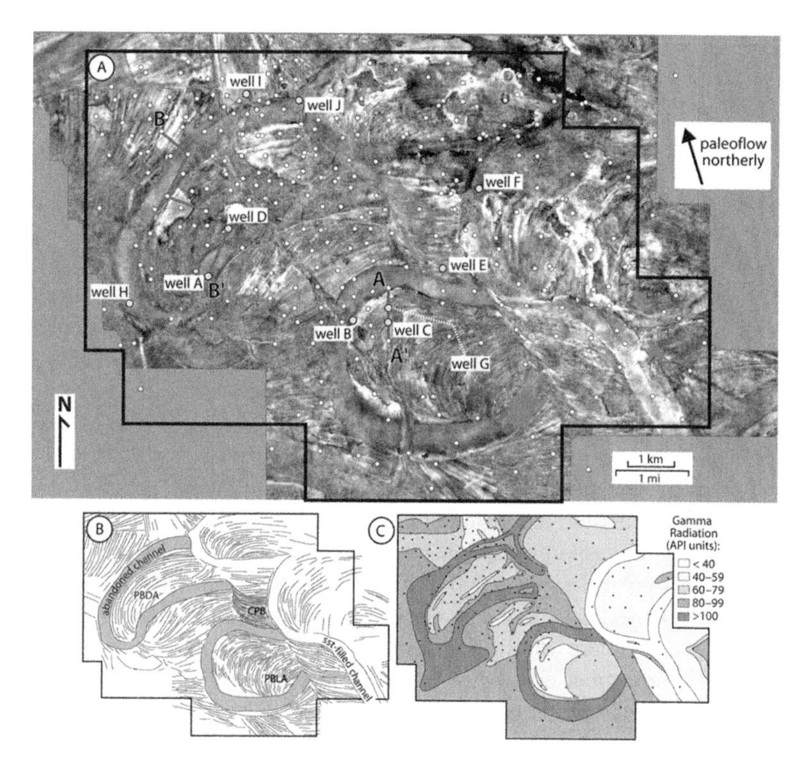

FIGURE 17.9 A geological model developed using an integrated approach combining seismic, log, core, etc. (Hubbard et al., 2011).

Steam circulation in both the injector and the producer at the same time is a common practice in the industry. During the circulation period, the process is dominated by thermal conduction with heat transmitted from wellbore to formation rock, which is usually called a "hot-pipe" effect. When the rock between the wellbores is sufficiently heated, the oil mobility becomes high enough to allow inter-well fluid communication. To ensure full-length wellbore heating, the steam is usually injected at the toe end through a tubing string and condensed steam (fluid) returns taken a second, shorter, tubing string or the wellbore annulus. For a well pair separation of 4–6 m in an oil sands reservoir, the circulation period ranges from 3 to 5 months to bring the inter-well formation to between 50°C and 100°C and to establish sufficient fluid mobility and communication. As noted in more detail below, it is generally best during the circulation phase to minimize net fluid production from the reservoir in order to promote heat absorption into the formation surrounding each wellbore. After the circulation period, the operation is transitioned to normal SAGD operation by gradually switching the injector to net steam injection and the producer to net production.

The steam circulation rate is determined based on the wellbore length and heat loss in the overburden and along the horizontal. The rate should be high enough to be able to deliver live steam to the toe end of the wellbore and is usually in the range of 70–100 tonnes/day. Vanegas Prada et al. (2005) have evaluated the impact of operational parameters and reservoir variables during the start-up phase of an SAGD process. Yuan and McFarlane (2011) in their simulation study found that (1) for the given tubing and liner sizes and reservoir properties, relatively low circulation rates at high steam quality are more favorable for faster initialization and development of uniform temperature between a horizontal well pair and (2) a small pressure difference, offsetting the natural hydraulic pressure, appears to be more favorable for faster and more uniform initialization.

End of the start-up period or the time to convert to normal SAGD is usually determined by either a targeted heating conformity (Duong et al., 2008) or an indication of sufficient pressure communication between the well pair (monitoring the producer pressure response to injector pressure variation). The approach of using production response to judge the well communication has been numerically assessed by Parmar et al. (2009). If wellbore temperature monitoring is in place, one can periodically assess the heat conformance. This is usually done by examining the wellbore shut-in temperature profiles. Figure 17.10 shows an example of several fiber distributed temperature sensing (DTS) profiles in one of the wells in Shell's Orion project (Shell Canada, 2011).

Circulation start-up can be operated in either a pressure-balanced or unbalanced process. In the pressure-balanced start-up, the pressures in both of the injector and producer are maintained to be equal or nearly equal. This

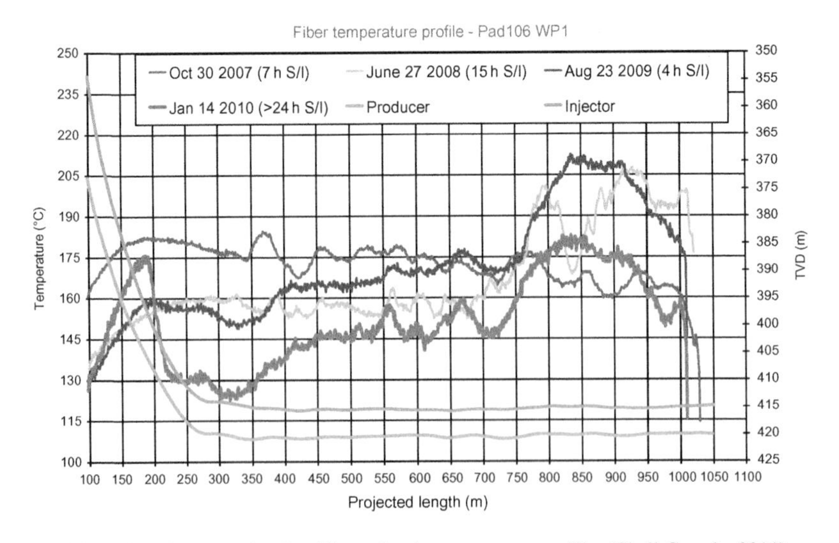

FIGURE 17.10 An example of wellbore shut-in temperature profiles (Shell Canada, 2011).

will result in minimal net fluid injection or production into the formation during the circulation period. The benefits of pressure-balanced operation is more uniform heating of the inter-well formation and a lower risk of localized steam breakthrough potential after switching to normal SAGD operation (Edmunds and Gittins, 1992). In the unbalanced start-up, however, a small pressure differential (in the order of 100–200 kPa) is applied to the well pair, usually with a positive pressure drop from injector to producer. The benefit of unbalanced operation is the enhanced heating from convection when the fluid mobility becomes high enough and shortens the start-up period. On the other hand, the drawback of this approach is a greater risk of getting localized steam breakthrough from nonuniform heating due to reservoir heterogeneity and the variations in well separation along the wellbores.

17.3.2 Well Separation and Start-Up Period

The vertical separation of an SAGD well pair is determined by competing demands of shortening the start-up time for inter-well fluid communication and minimizing steam coning (or breakthrough) potential. Because the preheating time required through thermal conduction heating is proportional to the distance squared (Edmunds and Gittins, 1992), one prefers to place the well pair close together in order to shorten the preheating period and to achieve normal SAGD operation sooner. On the other hand, the steam coning potential into the production well increases with decreasing well separation and liquid level above the producer. To minimize the coning potential, one prefers to place the SAGD wells further vertically apart. In SAGD, one intends to operate the process with the liquid level somewhere between the well pair. The balance of these two contradicting demands leads to a compromised SAGD well pair separation of 4–6 m and 3–5 months of circulation time for Canadian oil sands reservoirs.

17.3.3 Wellbore Effects

As mentioned previously, one aspires to achieve uniform and active drainage along the full length of the well pair through the start-up process. In reality it is difficult to achieve such uniform formation heating by the SAGD wells during the preheating period due to (1) nonconstant well separation due to wellbore undulation (Edmunds and Gittins, 1992; Shen, 2011), (2) heterogeneity of formation properties (thermal conductivity or thermal diffusivity) (Duong et al., 2008), and (3) differing pressure differential along the wellbores resulting from wellbore hydraulics (Edmunds and Gittins, 1992; Ong and Butler, 1990; Thorne and Zhao, 2009).

Wellbore undulation, due to limitations in well drilling control and accuracy, results in varying well separation along the SAGD well pair. The typical variation of inter-well distance is in the range of 1–2 m along the target interval with the latest horizontal drilling technology. The well undulation results in nonuniform heating of inter-well formation and the development of localized hot spots, and eventually uneven initiation and growth of the steam chamber (Edmunds and Gittins, 1992; Shen, 2011). This also leads to slower oil production ramp-up and a lower peak oil rate than that with a pair of perfectly parallel wellbores (Shen, 2011).

SAGD performance can be detrimentally impacted by high localized pressure gradients in injection wells. Wellbore hydraulics also affects wellbore subcool control (which will be discussed later) with high frictional pressure drop and leads to the need of aggressive subcool control, appropriate artificial lift selection, and operational controls (van der Valk and Yang, 2007). If the well capacity is too small, it will cause hydraulic losses in the well and skew the liquid interface parallel to the well pair (Ong and Butler, 1990), and impair well productivity (Parappilly and Zhao, 2009).

17.4 WELL COMPLETION AND WORK-OVER

17.4.1 Steam Circulation for Start-Up

The start-up of an SAGD process in Canadian oil sands requires steam circulation inside the wellbores due to near zero initial oil mobility. The well completion should accommodate wellbore circulation. Dual string completions are a common practice for this purpose. A tubing string is inserted into the toe end of the liner, and a short string is placed at the heel of the wellbore, as shown in (Figure 17.11). This also facilitates the flexibility of operating the SAGD well pair with multiple choices of injection and production position and combinations (e.g., injection/production at heel or toe or both with different ratios), in addition to wellbore circulation.

17.4.2 Thermal Wellbore Insulation

Design of the well completion for SAGD should be aimed to minimize heat loss to the overburden section. This will benefit the process by (1) maximizing heat delivery to the sand face and (2) avoiding significant overburden heating for environment protection.

Wellbore insulation is an effective way to achieve this and can be accomplished by either an annulus gas blanket or insulated tubing. A gas blanket is formed by injecting an inert gas (usually nitrogen) into the casing/tubing annulus to push the liquid level down to the reservoir and form a gas-filled section with low thermal conductivity. The concept of either mineral insulation or vacuum insulation has advanced and insulation systems with high thermal performance tubular is available for steam injection wells (Lombard et al., 2008; Luft et al., 1997). In general, insulated tubing has higher cost than gas blanketing. The utilization of insulated tubulars should be evaluated

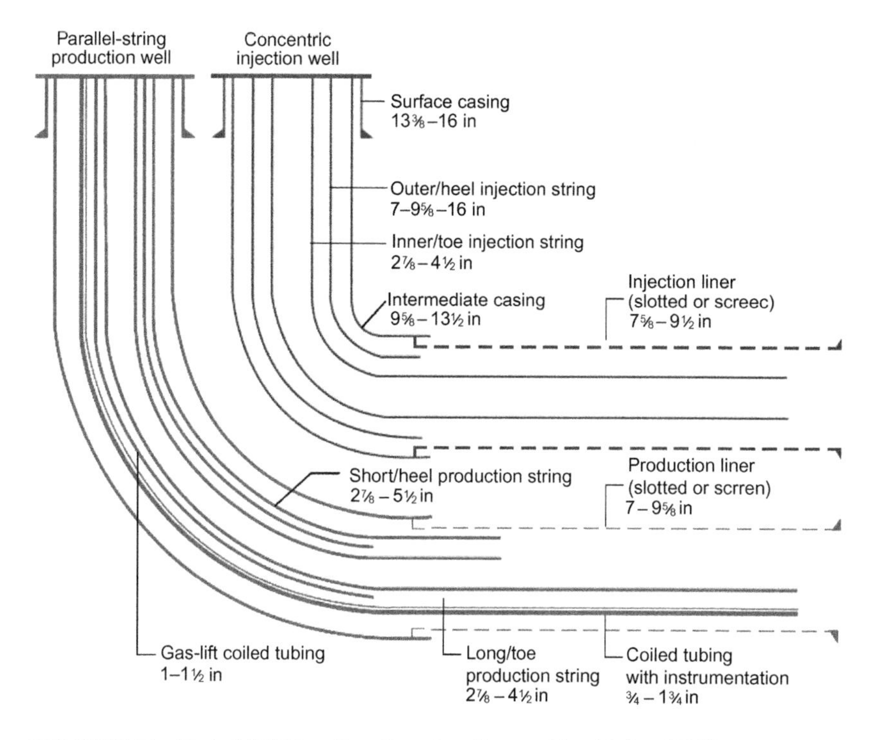

FIGURE 17.11 Typical SAGD well configuration. Source: *After Medina (2010).*

based on (1) the requirement of high-performance insulation for effective heat delivery, such as for deep reservoirs or long wellbores, (2) the cost savings associated with treatment of less produced water because of overall high thermal efficiency, and (3) the need for environmental protection, such as minimizing arsenic desorption in prone areas and contamination of shallow aquifers caused by overburden heating.

ConocoPhillips used vacuum insulated tubing (VIT) in some of their Surmont SAGD injection and production wells, for the potentials of minimizing heat transfer to permafrost zones to prevent subsidence, minimizing annular pressure buildup, hydrate prevention, paraffin prevention, and heat retention for heavy oil and steam flooding projects (ConocoPhillips, 2011).

17.4.3 Sand Control Liner

The majority of oil sands formations in which SAGD applications are conducted are generally poorly- or unconsolidated, and sand control is usually required. Slotted liner or wire-wrap screen are two of the most popular choices. Slotted liners have been used in most SAGD applications to date because of their superior mechanical strength and integrity in contrast to other mechanical sand control devices for long horizontal well completions.

The slot design needs to be adequate for sand control. The slotted liners can have a variety of configurations with varying slot density, slotting patterns, slot apertures, and slot internal geometries. If available, it is good practice to use actual core samples from nearby wells and the approximate target well placement interval, for slot or screen sizing. Once cleaned of bitumen, the core sand can be sieved and tested for best sand control sizing. The overall objective of a successful slotted liner design is to ensure that the liner allows minimal pressure drop, while retaining the majority of the formation sand and preventing the filling of horizontal section of the well with solids, erosion and failure of downhole pumps, and surface equipment. Both laboratory and field experience demonstrate that keystone and rolled-top slot design are superior to straight-cut slot because of less plugging potential (Bennion et al., 2009).

Wire-wrap screen results in less pressure drop across the liner than slotted liner because of its larger open area or high liner permeability. Its mechanical strength, however, is traditionally lower than slotted liner. Advancements in new wire-wrap screen designs have greatly improved their mechanical integrity, and the industry has renewed its interest in field application because of the much greater flow area that wire-wrapped screens offer over that of a slotted liner.

The slotted liner tends to experience higher plugging potential than wire-wrap screen because of difference in the area of opening. As demonstrated by Japan Canada Oil Sands Limited (JACOS, 2011) in their Hangingstone SAGD demonstration project, shown in Figure 17.12, there is a clear difference in using slotted liner versus wire-wrap screen. It is observed that with the wire-wrap screen, there is lower pressure drop from injector to producer (<500 kPa) which decrease with time, but a greater likelihood of sand production. In contrast, the slotted liner provides better sand control, but with a large bottom-hole pressure drop between injector and producer (>500 kPa and up to 1500 kPa) which tends to increase with time. The larger pressure drop with slotted liners may result from conservative slot sizes and mineral plugging.

17.4.4 Liner Plugging Issue and Treatment

Liner plugging has been experienced in many SAGD wells and leads to increased pressure drop and impaired well productivity. The mechanisms of liner plugging may range from fines migration, clay swelling, and inorganic scales to asphaltene precipitation and deposition, but in general are still not fully understood. Many SAGD wells experience liner plugging issues attributed to calcium carbonate scale formation. Acid stimulation has been tried

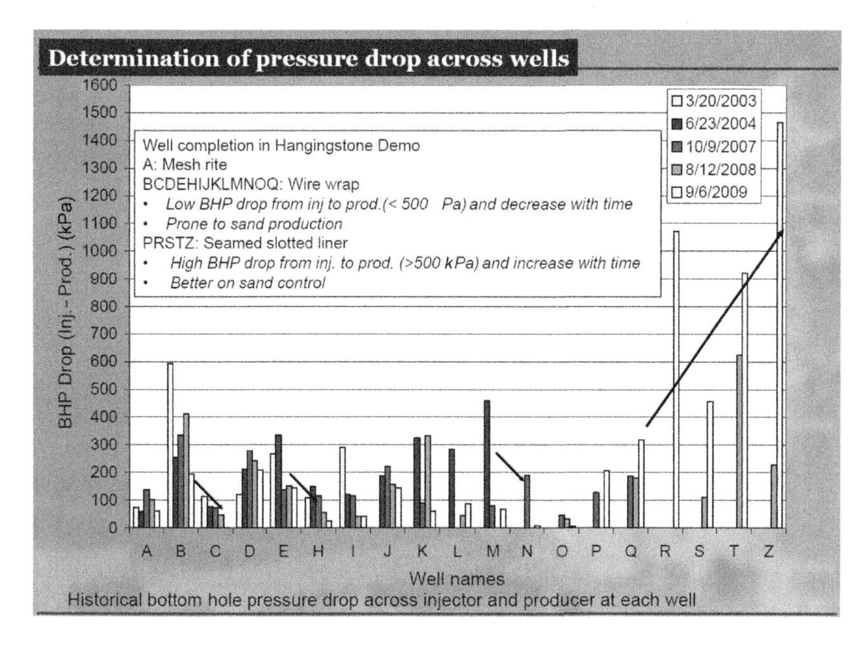

FIGURE 17.12 The comparison of bottom-hole pressure drop with slotted liner versus wire-wrap screen in JACOS' Hangingstone SAGD Demonstration Project (JACOS, 2011).

with varying degrees of success, but in general with diminishing effectiveness with successive treatments.

Bennion et al. (2009) have conducted systematic evaluation of slotted liner plugging mechanisms. They found that many factors can cause slot plugging: varying from grain size distribution of the sand, slot geometry, clay content of the formation, flow velocity, wettability, and fluid pH. Clay plugging at the top (external) portion of the slots has been found to be the dominant damage mechanism in laboratory experiments. Low pH environments tend to reduce the plugging potential, and the use of dilute acids can be effective at removing existing restrictions or plugs, though this will depend on composition of the scale and the way the acid is introduced into the wellbore and the treatment performed.

Carbonate scale can cause liner plugging. Erno et al. (1991) stated that "scaling due to pressure and temperature changes is most prevalent in or around production wells. Radial flow into small diameter wellbores results in high linear fluid velocities in the immediate wellbore regions, and thus large pressure decreases. The drop in pressure causes dissolved gas to evolve from the fluids. Loss of carbon dioxide (and other acid gases) from a brine results in increased pH and causing a shift in the bicarbonate−carbonate equilibrium toward carbonate. This can lead to super-saturation with calcite, and subsequent precipitation produces a scale deposit. Once a deposit begins to form,

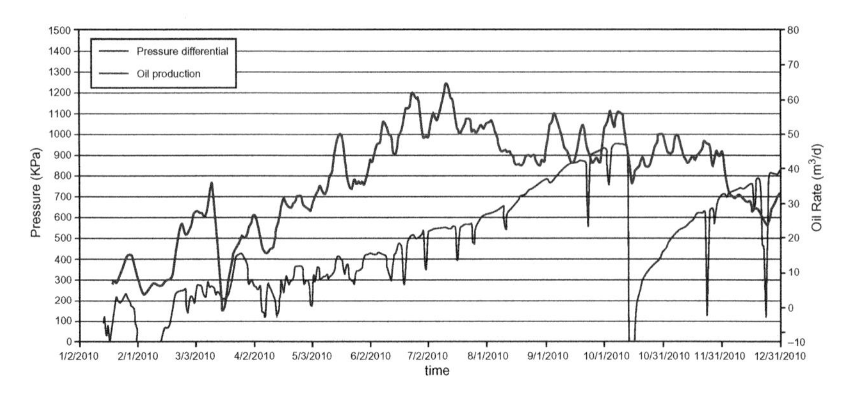

FIGURE 17.13 The trend of bottom-hole pressure differential and oil rate in one of well pairs of Shell's Orion SAGD project (Shell Canada, 2011).

it often restricts flow, which increases the pressure drop, enhances gas separation, and increase precipitation. This self-aggravating effect can lead to a very rapid decline in fluid flow."

The general symptoms of liner plugging are the gradual increase of bottom-hole pressure differential between the SAGD wells of the pair and the decrease of well productivity, as shown in Figure 17.13. Acid treatment to remove calcium carbonate scale plugging has been successfully tried in different SAGD wells (Brand, 2010; Nexen, 2006; Shell Canada, 2011). Brand (2010) has reported that the repeated acid treatments of the plugging liners tend to have diminishing returns with time (Figures 17.14 and 17.15). This indicates that the treatment does not address the root cause of the plugging issue and only alleviates the problem temporarily.

17.4.5 Recompletion to Fix Local Steam Breakthrough

In field operation, localized steam breakthrough may occur occasionally due to a variety of reasons. To improve effective wellbore utilization and thermal efficiency, it is desirable to reduce and eliminate local steam breakthrough. The installation of a scab liner is one of the effective ways to fix local steam breakthrough. Following is one example from Shell's Orion SAGD project (Shell Canada, 2011). As shown in Figure 17.16, the flowing wellbore temperature profile of July 19, 2010, indicates that there is a clear point of hot fluid entry at 300 m of projected length. A scab liner, as shown in Figure 17.17, was installed in the heel section of the horizontal wellbore and the local breakthrough point was effectively eliminated as indicated by a postinstallation wellbore temperature profile of December 10, 2011.

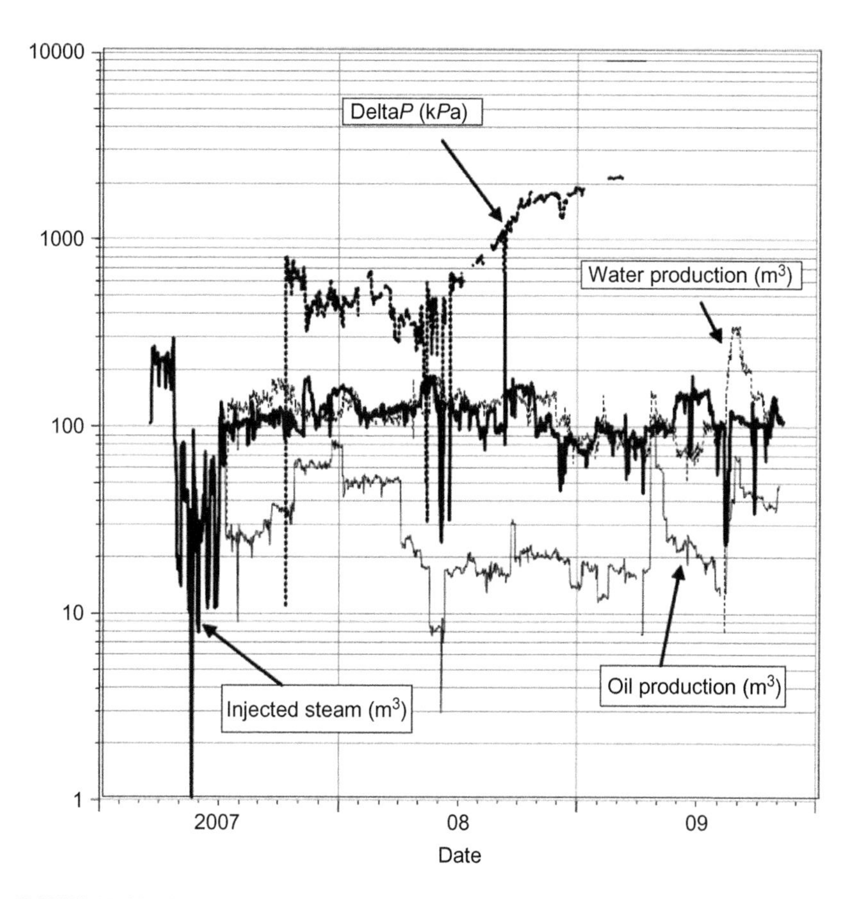

FIGURE 17.14 An example of increasing pressure differential in Husky's Lloydminster SAGD project (Brand, 2010).

17.4.6 Intelligent Well Completion

One of the biggest challenges of the SAGD process is achieving even steam conformance along the horizontal injection wellbore and even liquid drainage along the horizontal production wellbore. This is a result of both the heterogeneity associated with any reservoir and the limited control of fluid inflow or outflow along the wellbore. Consequently, uneven steam injection and heating can occur in the well and lead to development of a nonuniform steam chamber and impairment of well performance. This effect will be amplified by increasing well length and heterogeneity.

Intelligent completion technologies have been developed for improving both steam injection and production conformance in the SAGD process. An intelligent well completion can be either active or passive in flow distribution control.

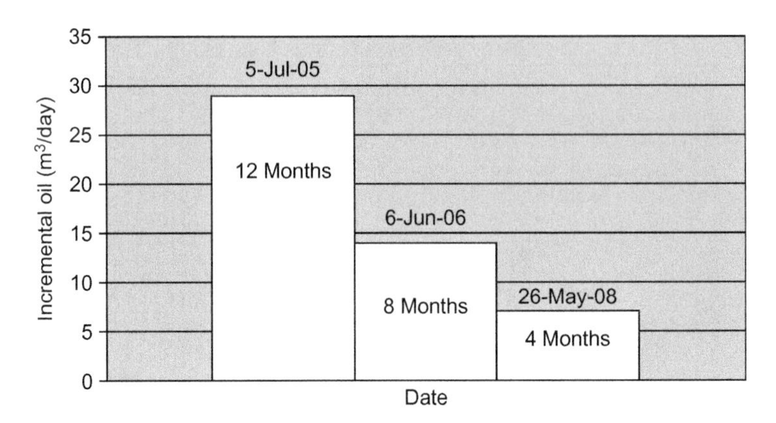

FIGURE 17.15 An example of diminishing return of acid treatment (of well 12C1-32) in Husky's Lloydminster SAGD project (Brand, 2010).

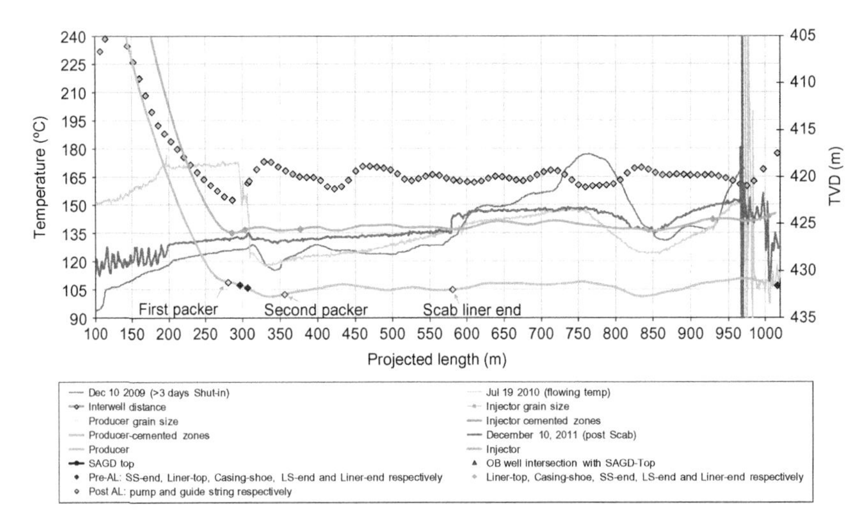

FIGURE 17.16 An example of wellbore temperature profiles before and after installation of a scab liner to fix the local steam breakthrough (Shell Canada, 2011).

Clark et al. (2010) have reported a successful field trial of an active completion technology by incorporating interval control valves (ICVs), well segmentation, and associated downhole instrumentation. The completion provides the ability to selectively open and close any one of four segmented sections of the wellbore and monitor the key parameters of temperature and pressure from surface. The trial proved operability of the system in high-temperature thermal applications, and demonstrated the feasibility of modifying steam distribution.

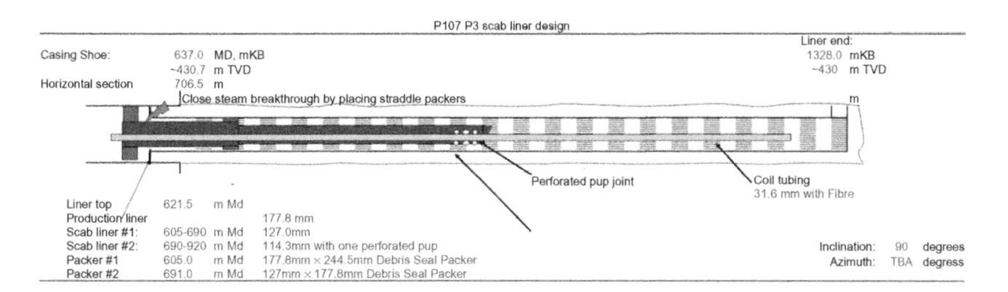

FIGURE 17.17 Schematic of scab line installation to fix local steam breakthrough (Shell Canada, 2011).

One type of passive control completion has been tested by ConocoPhillips (Stalder, 2012) in their Surmont SAGD project. The passive fluid distribution control system consists of multiple Baker Equalizer (combination of orifice and screen) sections periodically placed along a liner. This enables uniform steam distribution along the injector liner, and minimizes steam ingress along the production liner, while not hindering production capacity. This flow distribution liner system has been deployed in a pair of SAGD wells and achieved superior steam conformance and SAGD performance without a toe string in either well after a preheat circulation period. The elimination of toe string allows one to (1) reduce liner size in both injector and producer and/or (2) increase completion length beyond the current 800–1000 m typical range because of reduced drag of smaller liners.

17.5 PRODUCTION CONTROL

17.5.1 Steam Trap

Steam trap control is used as an operational means to reduce or prevent steam withdrawal from the steam zone in the reservoir. Prevention of live steam production will benefit the process by (1) conserving energy and lowering SOR, (2) reducing high vapor flow which affects negatively the lifting capacity of the well and surface facilities, and (3) reducing fine-sand and other solids movement through the liner which may cause erosion of the liner and massive sand production.

There are mainly two types of steam traps: mechanic (liquid level) and thermodynamic (subcool) (Edmunds, 2000). The mechanic type directly controls the condensate–steam interface, letting liquid out but retaining vapor via keeping the interface above a drainage point. The thermodynamic type merely infers nearness of the interface by comparing the local temperature and pressure with the steam saturation curve. An adequate temperature setting is critical for proper and stable operation of a thermodynamic trap.

Numerical investigation for a prototype Athabasca reservoir has found that subcool dynamics are rather complex (Gates and Leskiw, 2008). In three dimensions, inflow temperature can vary significantly along the well according to local conditions. Maintaining uniform subcool control along the length of the well pair is difficult because of the heterogeneity of the reservoir along the length of a well pair. When a long well pair is operated so that the mixed subcool is in a presumed optimum range, some sections of the producer wellbore may tend to produce small amounts of steam, while others may tend to be much cooler than the target, optimum, temperature.

van der Valk and Yang (2007) pointed out that it is critical to ensure that the subcool control point is not in the open horizontal liner section (adjacent to inflow) to adequately capture the notion of mixed, or "effective" subcool. Both numerical and field experiences indicate that an effective subcool should be in the range of 20–30°C (Edmunds, 2000).

Steam breakthrough will lead to high-velocity flow through the liner and may cause erosion and liner failure. The high-velocity flow results mainly from the significantly different volume ratio of steam to condensate (e.g., ~55 at 3000 kPa). Stable producer bottom-hole pressure control should be pursued to minimize pressure fluctuation and localized steam breakthrough.

17.5.2 Wellbore Lift

A good wellbore lift should be able to facilitate both an efficient lift of fluid to surface and stable control of well bottom-hole pressure. SAGD by nature should be operated under stable bottom-hole pressure control in order to maximize fluid drainage and to minimize the potential of steam breakthrough. The following are popular wellbore lift options for SAGD:

- *Natural lift*: It can be operated when the reservoir pressure is high enough to overcome both the hydrostatic and dynamic pressure heads of the produced fluids.
- *Gas lift*: Gas is injected through the tubing–casing annulus to gasify the fluid and reduce its density, when the reservoir pressure is not high enough to produce the well by itself.
- *Electric submersible pump (ESP)*: The availability of wide range lift capacities makes it the preferred choice for SAGD operation, if it can be operated within the temperature limitation. The current, maximum, downhole fluid temperature limit for ESP is approximately 250°C (Burleigh, 2009).
- *All metal progressing cavity pump (AMPCP)*: AMPCP has a wide operating range in terms of production and the capability of handling varying downhole operating conditions such as bottom-hole pressures, temperatures, and fluid viscosity. The all metal rotor and stator are able to withstand a broad range of static and dynamic temperatures up to 350°C. The application of AMPCP in Shell's Orion SAGD project demonstrated that it can handle a wide range of operating temperature (with bottom-hole

temperatures ranged from 125°C to 260°C), from low to high intake and differential pressures (Rae et al., 2011).

Beam pump is a well-known lift method in other thermal operations, but it is not popular in SAGD. The reason behind this is that a beam pump system cannot generate a stable-enough bottom-hole pressure, the likes of which an ESP and PCP can. The reciprocating stroke action can result in significant changes in fluid fillage of the pump with each stroke (due to gas or steam interference) and cause variations in flowing bottom-hole pressure. This works against maintaining a constant subcool or at least as constant as possible.

The wellbore lifting systems can be classified into two main categories from reservoir and surface pressure coupling point of view:

1. *Pressure coupled:* natural or gas lift.
2. *Pressure decoupled:* downhole pumps (ESP or PCP).

The pressure decoupled lift option is preferred for SAGD. The use of downhole pumps will allow one to directly control the well bottom-hole pressure and to minimize the steam breakthrough potential. The pressure coupled option is usually operated with surface pressure control and tends to produce less stable downhole pressure, and in extreme cases can lead to a geysering phenomenon which will be discussed next.

17.5.3 Geysering Phenomenon Under Natural Lift

SAGD well production with surface pressure control could experience severe unstable flow under certain conditions. Figure 17.18 shows an example of production rate fluctuation during a 12 h period, from Blackrock Venture

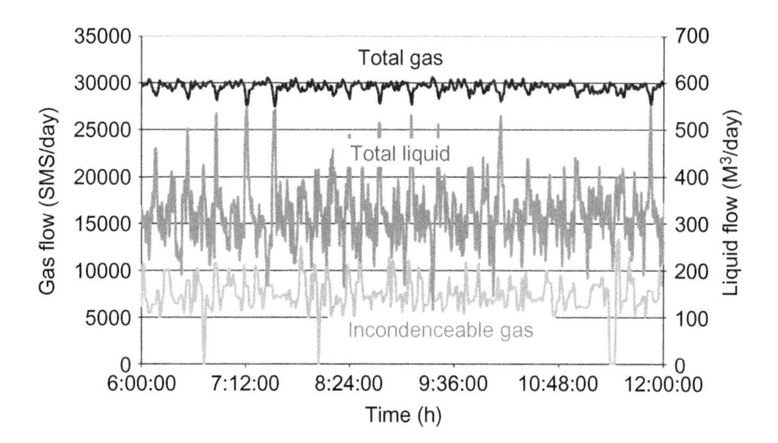

FIGURE 17.18 Produced fluid production rate, January 1, 1999, Hilda Lake Pilot (Donnelly, 1999).

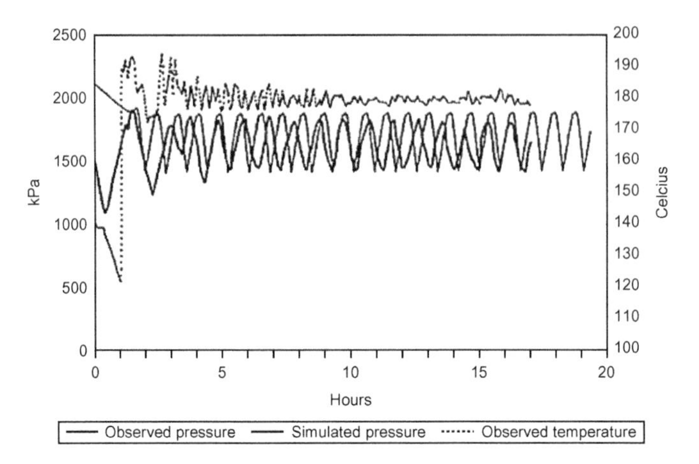

FIGURE 17.19 UTF riser bottom conditions during geysering (December 1–2, 1988) (Edmunds and Good, 2006).

Inc's Hilda Lake SAGD Pilot project (Donnelly, 1999). The well produced into a gas–liquid separator which was maintained at a constant pressure, with no artificial lift system in use. The fluid pulses into the separator are reminiscent of the pulses one observes in the coffee percolator, which is commonly known as geysering phenomenon. The well geysering can be expected to simultaneously generate significant pressure pulses downhole, as observed in UTF Phase and shown in Figure 17.19 (Edmunds and Good, 2006). This unstable flow behavior is very destructive to smooth operations both underground for steam trap control and on surface for fluid handling, as well as imposing a greater risk of a condensate induced water hammer effect (Carlson, 2010).

As pointed out by Edmunds and Good (2006), the geysering can be minimized by back-pressuring the top of the production string so that flashing will not occur, and lifting the fluids against this pressure by a downhole pump.

17.6 WELL, RESERVOIR, AND FACILITY MANAGEMENT

Well, reservoir, and facility management (WRFM) is a critical part of SAGD operation, which will enable one to monitor the progress of start-up, steam chamber development and steam conformity, well inflow behavior and steam trap conditions, surface movement and caprock integrity, and surface equipment performance/stability. The data or information collected in WRFM is critical to the diagnosis potential operational issues, the optimization of the recovery process, and to maintain operational integrity of an asset including the surface facilities.

An integrated WRFM program for SAGD covers the following key aspects:

- SAGD wells
- Reservoir
- Overburden (or underburden)
- Surface equipment.

17.6.1 Wellbore Pressure and Temperature

Most of the SAGD wells operated in various projects have temperature monitoring using either a thermal-couple string to measure temperature at discrete positions or optical fibers for DTS. Knowledge of the temperature distribution along the production well can be used to estimate the conformance of the steam chamber, wellbore utilization, and steam trap control. Figure 17.20A shows an example of using temperature falloff profiles to assess the formation temperature along the wellbore and indicate that the toe section of the wellbore is much hotter than the heel end.

Subcool control with only downhole temperature measurement has its limitations. Krawchuk et al. (2006) noticed from downhole data gathering that during warm-up, significant thermal gradients can exist across the horizontal producing well's diameter. This makes the knowledge of temperature measurement location in relation to the wellbore critical for accurate interpretation and utilization for subcool control. The differences across a wellbore are attributed to the laminar nature of fluid and gas flow in horizontal sections, the differing heat capacities of each phase, and thus the effective heat loss to the surrounding formation in the lower versus the upper portion of a wellbore.

Downhole pressure monitoring is usually done by bubble-tubes or high-temperature pressure sensors. The monitoring of downhole pressure can be supplemental to the downhole temperature and will provide direct bottom-hole pressure measurement which can be important when downhole conditions are significantly away from the steam saturation curve. With downhole pressure, one can fine-tune well operations to achieve a stable SAGD pressure and more efficient steam trap control (minimizing live steam production downhole).

17.6.2 Reservoir Monitoring

Time-lapse seismic has evolved into a mature technology and is being used in many of the operational projects. The differences in seismic response between two consecutive surveys is the most effective means to estimate the aerial or volumetric extent of steam chamber development and oil depletion. Figure 17.20C shows an example of time-lapse seismic response from an SAGD operation; the area with hotter color indicates a high degree of alteration.

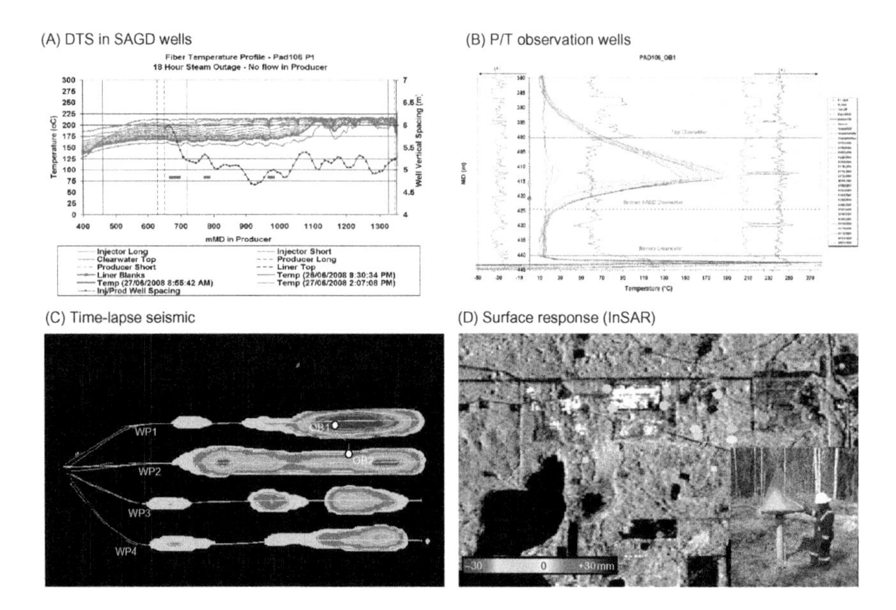

FIGURE 17.20 An example of surveillance employed in Shell's Orion SAGD project.

Observation wells placed at various locations and equipped with pressure and temperature sensors enable one to effectively evaluate the status of the depletion process and vertical steam chamber growth. The temperature response at the observation well locations can indicate the evolution of steam chamber boundary development (Figure 17.20B) and flags potential flow barriers at the location.

Temperature monitoring for the steam chamber development may not be sufficient on its own, due to the fact that noncondensable gas (NCG) accumulates at the top of the heated chamber and drains oil from beyond (above) the steam zone. This means that the extent of the steam chamber may not equal the extent of the oil depletion zone in the presence of significant amount of NCGs in the depleted volume. Figure 17.21 presents a comparison of the simulated steam chamber with the oil depletion zone and illustrates the difference. The NCG at the top of the reservoir provides thermal isolation and reduces heat loss to the overburden, thus potentially improving SOR.

Formation pressure monitoring is aimed at the evaluation of vertical continuity and *in situ* fluid mobility, as well as compliance with the operational reservoir pressure limit.

17.6.3 Rock Deformation Evaluation and Surface Monitoring

Microseismicity and tiltmeters have been used to monitor SAGD steam injection (Maxwell et al., 2009) and demonstrated that they can detect

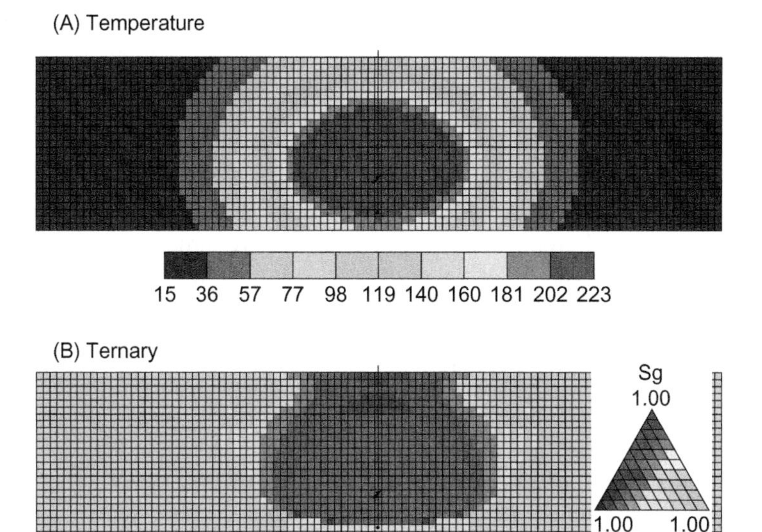

FIGURE 17.21 A comparison of the simulated (A) temperature and (B) ternary at a given time with the presence of NCG.

(1) significant seismic events resulting from formation deformation during SAGD steam injection, (2) two distinct types of microseismic events associated with thermal expansion of the uncemented liner and induced shear fracturing around the steam chamber, and (3) differential responses of surface tiltmeters along the SAGD well.

Satellite Interferometric Synthetic Aperture Radar (InSAR) techniques are capable of remotely detecting millimeters to meters of vertical motion spanning days, months, years, and decades, across specific sites or wide areas (1000 s km^2). The application of InSAR to monitor production activity at the Cold Lake heavy oil field has been reported by Stancliffe and van der Kooij (2001). Figure 6.1D shows an example of InSAR application for surface deformation monitoring in an SAGD project, with many corner reflectors installed across the area of interest. Surface deformation (heave or subsidence) information can be used for (1) assessing potential impacts on environment and the integrity of operational surface facilities, (2) monitoring steam chamber development for optimization, (3) predicting the maximum gradient at the edge of patterns for future pipeline and plant facilities design, and (4) calibrating geomechanical simulation model. The surface heave, however, may only provide a crude indication of steam zone position if the reservoir is shallow, i.e., steam zone size is in the order of or greater than the reservoir depth below surface.

The importance of maintaining overburden rock integrity during SAGD has been exemplified by the catastrophic containment failure of Total's Joslyn Creek SAGD project in northeastern Alberta in May 18, 2006, when pressurized steam burst up through the thin caprock, blasting out a crater 20 m wide and 5 m deep. Fortunately, no one was injured in the blast. The postaccident investigation concluded that the most likely cause of the steam release was the injection of steam at excessively high pressures which lead to the initiation and propagation of a vertical fracture through the water sand directly underlying the Clearwater caprock and subsequently caused the Clearwater caprock to fail under shear to surface (ERCB, 2010).

17.7 SAGD WIND-DOWN

At late stage of SAGD operation, as instantaneous SOR increases with time, it is no longer economic to continue pure steam injection; however, the reservoir is still hot, and the energy in place can still be utilized. The wind-down process is aimed to gradually lower the formation temperature to salvage the heat remaining in the formation for improved energy efficiency and to prolong oil production for better ultimate oil recovery.

One of the wind-down processes is to inject NCG or a mixture of NCG and steam to sustain formation pressure while the formation temperature declines. The results of a laboratory experiment and corresponding numerical history matching showed that (1) the hot chamber continued its expansion after steam injection was stopped and gas injection was initiated and (2) co-injection of steam and NCG gave the best result. The continuous, expanding, period represented the most productive period in the gas injection wind-down process (Zhao et al., 2005).

The results of co-injection of steam with natural gas and then flue gas into UTF Phase B as a wind-down process were reported by Yee and Stroich (2004). Figures 17.22 and 17.23 show that the SAGD wind-down with NCG and steam co-injection performed better than expected. They concluded the following from the numerical evaluation and field experience:

- The addition of NCG to steam injection has proved to be a technically viable method to effectively wind-down a mature SAGD chamber.
- The gas/steam wind-down process has performed better than expected. It has achieved good oil rates and steam–oil ratios.
- The heat transfer rates into the colder regions of the reservoir are not affected significantly by the presence of NCG. This is different from the results predicted by numerical simulation.
- NCG travels ahead of the apparent "steam" front, which results in higher depletion levels than previously thought.
- Flue gas has been used successfully to replace natural gas in the wind-down operation. The injection system was designed and operated to ensure the flue gas temperature is always above the dew point of the gas,

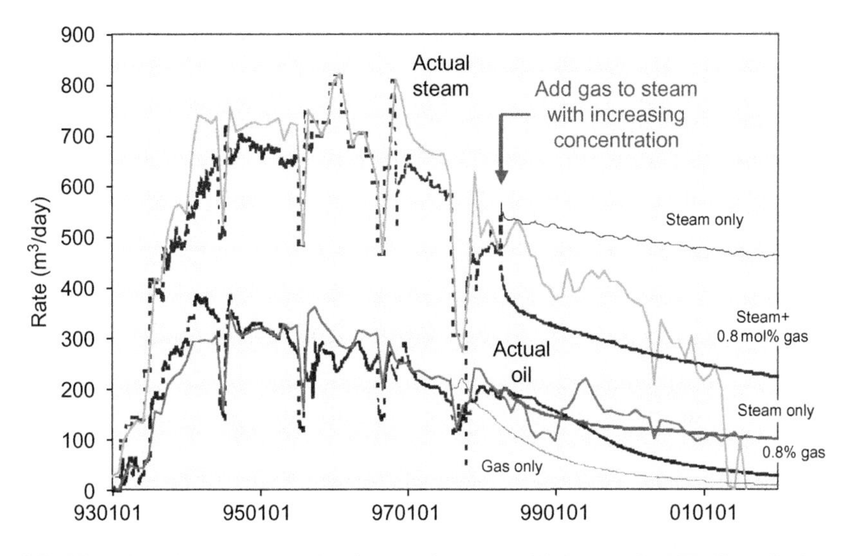

FIGURE 17.22 Comparison of wind-down performance with forecast for UTF Phase B (Yee and Stroich, 2004).

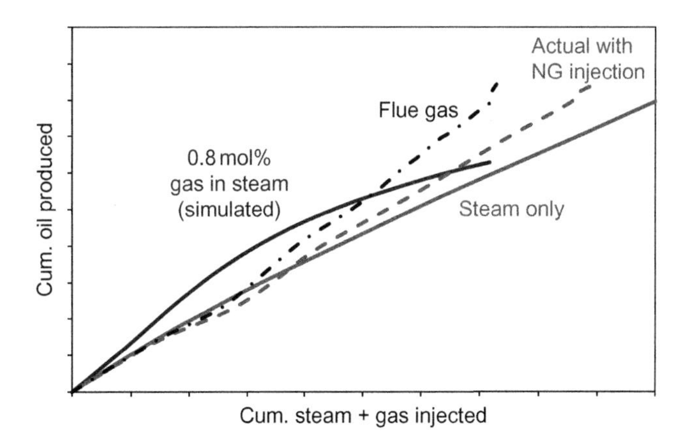

FIGURE 17.23 UTF Phase B performance since April 1, 1998. Gas injection converted to steam equivalent (Yee and Stroich, 2004).

so condensation will not occur. This technology has the potential to reduce the cost of injection substantially and to provide a means to sequester some of the exhaust gas from steam generation.

Belgrave et al. (2007) proposed the use of air injection as a follow-up process to SAGD operations. Laboratory work has demonstrated the feasibility to maintain a burning front in a mature steam chamber. Simulation studies indicate the potential to significantly increase the recovery factor over methane blow-down. The process has been successfully tested in Cenovus' Christina Lake Thermal Project (CLTP). Currently the CLTP operates with an air gas cap overlying the bitumen zone, with no negative impacts to SAGD operations (Cenovus Energy, 2011). The oxygen is consumed in an exothermic reaction with the residual oil, producing carbon dioxide as a by-product. There has been no oxygen produced in the producer wells, and no observed change in oil quality once the SAGD chambers intersect the air gas cap.

17.8 INTEGRATION OF SUBSURFACE AND SURFACE

SAGD is a continuous injection and production process and does not like downtime. The interruption of stable operation due to downtime, for a variety of surface or subsurface reasons, leads to well cooling and load-up. This will cause operational difficulties when trying to bring the wells back on.

To achieve smooth operation, the surface and subsurface should be fully integrated, balanced, and properly monitored through an effective WRFM process as described earlier. The reliability of the whole system is the aggregation of subsystem reliabilities. The key interface between the subsurface and surface is the water balance: the subsurface calls for sufficient and stable steam supply, while the surface demands adequate supply of makeup water and reuse of produced water.

In case of downtime, the operational procedure should call for continued circulation of fluid in the producer and injector, allowing for a more efficient start-up after a shutdown or upset.

17.9 SOLVENT-ENHANCED SAGD

Solvent and steam co-injection for enhanced SAGD has been proposed and developed by Nasr and Isaacs (2001). The patented process is called expanding solvent-SAGD "ES-SAGD." The main mechanism of the process is that the injected solvent will condense with steam around the steam chamber/cold formation interface causing oil dilution and viscosity reduction. This process has been successfully tested in fields (Gupta et al., 2005; Gupta and Gittins, 2006) and resulted in improved oil rate and oil-to-steam ratio (OSR), as well as lower energy and water requirements as compared to SAGD. The field application of solvent-enhanced SAGD has achieved solvent recoveries greater than 70%, and improvements in both oil rate and OSR are greater than 30%.

Solvent-enhanced SAGD processes can be operated in a thermodynamic window where they are uniquely effective for the benefits of driving down the cost for maximum profit, for energy efficiency, and for minimizing the environmental impact. Figure 17.24 shows the diagram of solvent–steam

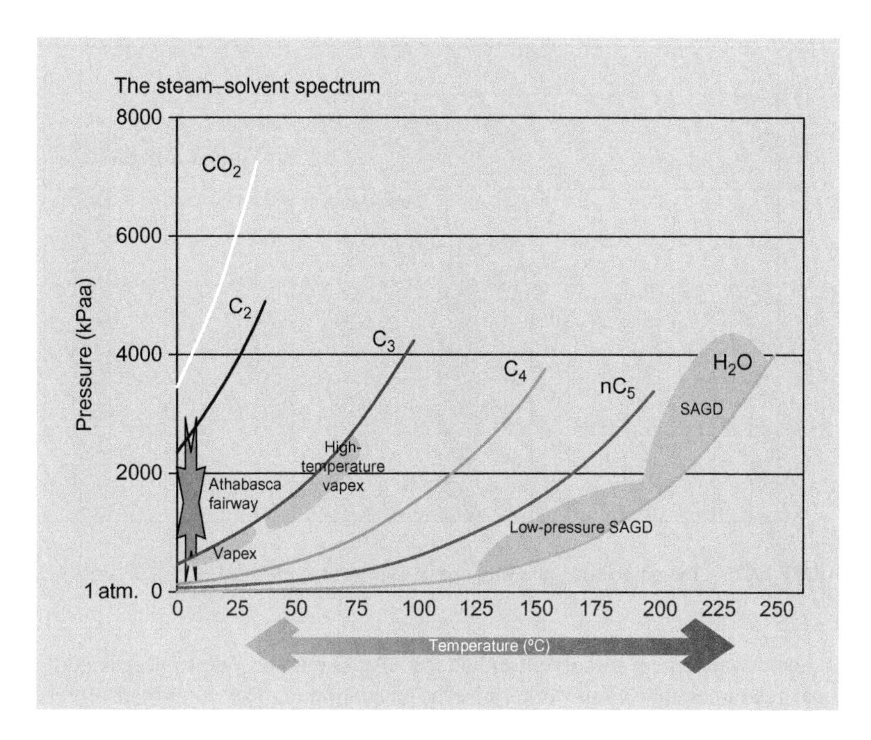

FIGURE 17.24 The diagram of steam–solvent spectrum (Schmidt, 2010).

spectrum which indicates the operation windows of different solvents. The selection of a solvent or a solvent mixture is aimed in such a way that the solvent can vaporize and condense at the same temperature/pressure conditions as the water phase, and the phase change of solvent is expected to be the same as steam along the vapor/bitumen interface.

Nasr et al. (2003) found that the maximum oil drainage rate under ES-SAGD occurred when the steam temperature matched the solvent vaporization temperature. Figure 17.25 shows the variation of oil drainage rate with different steam/solvent pairs in a laboratory test study. The tests were conducted under the conditions of injected steam (215°C at the operating pressure of 2.1 MPa).The drainage rate improves with the increase of carbon number of the injected hydrocarbon and as the vaporization temperature of the hydrocarbon additive became closer to the injected steam temperature. The drainage rates peaked with hexane, which has the closest vaporization temperature to steam temperature in the test, and then started to decline with octane, which has a higher vaporization temperature as compared to injected steam temperature. Using a diluent (mainly C_4–C_{10}) with steam resulted in a drainage rate comparable to that from steam/hexane and much higher (three times higher) than that from steam-only injection.

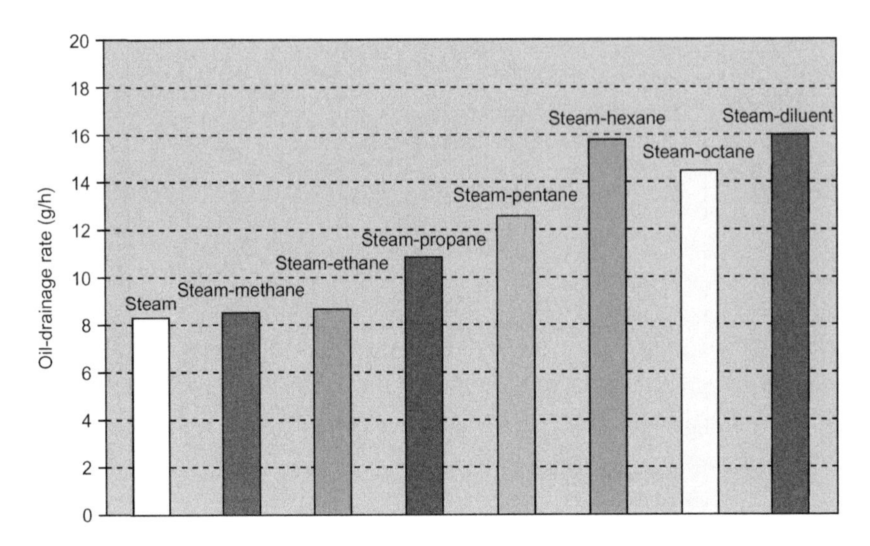

FIGURE 17.25 The oil drainage rate with carbon number at 2.1 MPa and 215°C (Nasr et al., 2003).

The advancement of solvent-enhanced SAGD enables one to access additional resources that would otherwise be uneconomic. The combined effects of fluid heating and dilution can result in more effective viscosity reduction and allows one to operate the process at lower temperatures than with steam only. This translates to reduced SOR, increased plant productivity for a given steam capacity, and reduced capital intensity. The solvent SAGD process also has the potential for increased recovery via reduced residual oil saturation (possibly due to reduction of interfacial tension), and higher volumetric sweep.

REFERENCES

Belgrave, J.D.M., Nzekwu, B., Chhina, H.S., 2007. SAGD optimization with air injection, Paper SPE 106901 Presented at the 2007 SPE Latin American and Caribbean Petroleum Engineering Conference, 15–18 April, Buenos Aires, Argentina.

Bennion, D.B., Gupta, S., Gittins, S., Hollies, D., 2009. Protocols for slotted liner design for optimum SAGD operation. J. Can. Pet. Tech. 48 (11), 21–26 (SPE-130441).

Brand, S., 2010. Results from acid stimulation in Lloydminster SAGD applications, Paper SPE-126311 Presented at the 2010 SPE International Symposium and Exhibition on Formation Damage Control, 10–12 February, Lafayette, LA.

Burleigh, L., 2009. Pushing the limit: taking ESP systems to 250 C, Paper Presented in Middle East Artificial Lift Forum, Manama, Bahrain.

Butler, R.M., 1982. Method for continuously producing viscous hydrocarbons by gravity drainage while injecting heated fluids, US Patent No. 4,344,485.

Butler, R.M., 1994. Horizontal wells for the recovery of oil, gas and bitumen, Petroleum Society Monograph Number 2, Alberta, Canada.

Carlson, M., 2010. What every SAGD engineer should know about condensation induced water hammer, Presentation Presented at the SPE Calgary Section Technical Luncheon, 5 May, Calgary, Alberta, Canada.

Cenovus Energy, 2011. Annual Christina Lake SAGD performance review approvals 8591, In-situ Progress Report to ERCB, 15 June, Calgary, Alberta, Canada.

Chen, Q., Gerritsen, M.G., Kovscek, A.R., 2008. Effects of reservoir heterogeneities on the steam-assisted gravity drainage process. SPE Res. Eval. Eng., 921–932.

Clark, H.P., Ascanio, F.A., van Kruijsdijk, C., Chavarria, J.L., Zatka, M.J., Williams, W., et al., 2010. Method to improve thermal EOR performance using intelligent well technology: Orion SAGD field trial, Paper SPE 137133 Presented at the Canadian Unconventional Resources & International Petroleum Conference, 19–21 October Calgary, Alberta, Canada.

ConocoPhillips, 2011. Annual Surmont SAGD performance review approvals 9426F and 9460C, In-situ Progress Report to ERCB, April 6, Calgary, Alberta, Canada.

Donnelly, J.K., 1999. Hilda lake a gravity drainage success, Paper SPE-54093 Presented at the 1999 SPE International Thermal Operations and Heavy Oil Symposium, 17–19 March 1999, Bakersfield, CA.

Duong, A.N., Tomberlin, T.A., Cyrot, M., 2008. A new analytical model for conduction heating during the SAGD circulation phase, Paper SPE 117434 Presented at the International Thermal Operations and Heavy Oil Symposium, 20–23 October, Calgary, Alberta, Canada.

Edmunds, N.R., 2000. Investigation of SAGD steam trap control in two and three dimensions. J. Can. Pet. Tech. 39 (1), 30–40.

Edmunds, N.R., Gittins, S.D., 1992. Effective application of steam assisted gravity drainage of bitumen to long horizontal well pairs. J. Can. Pet. Tech. 32 (6), 49–55.

Edmunds, N.R., Good, W.K., 2006. The nature and control of geyser phenomena in thermal production risers. J. Can. Pet. Tech. 34 (4), 41–48.

ERCB, 2010. Staff review and analysis: total E&P Canada Ltd., Surface Steam Release of May 18, 2006 Joslyn Creek SAGD Thermal Operation, 11 February.

Erno, B.P., Chriest, J., Miller, K.A., 1991. Carbonate scale formation in thermally stimulated heavy-oil wells near Lloydminster, Saskatchewan, Paper SPE 21548 Presented at the International Thermal Operations Symposium, 7–8 February, Bakersfield, CA.

Gates, I.D., Leskiw, C., 2008. Impact of steam trap control on performance of steam-assisted gravity drainage, Paper PETSOC-2008-112 Presented at the Canadian International Petroleum Conference/SPE Gas Technology Symposium 2008 Joint Conference (the Petroleum Society's 59th Annual Technical Meeting) , 17–19 June, Calgary, Alberta, Canada.

Gupta, S.C., Gittins, S.D., 2006. Christina Lake solvent aided process pilot. J. Can. Pet. Tech. 45 (9), 15–18.

Gupta, S.C., Gittins, S.D., Picherack, P., 2005. Field implementation of solvent aided process. J. Can. Pet. Tech. 44 (11), 8–13.

Hein, F.J., Cotterill, D., 2006. The Athabasca oil sands—a regional geological perspective, Fort McMurray Area, Alberta, Canada. Nat. Resour. Res. 15 (2), 85–102.

Hubbard, S.M., Smith, D.G., Nielsen, H., Leckie, D.A., Fustic, M., Spencer, R.J., et al., 2011. Seismic geomorphology and sedimentology of a tidally influenced river deposit, Lower Cretaceous Athabasca oil sands, Alberta, Canada. AAPG Bull. 95 (7), 1123–1145.

Husky Energy, 2011. Tucker thermal project, commercial scheme approval no. 9835, In-situ Progress Report to ERCB, June 27, Calgary, Alberta, Canada.

JACOS, 2011. Hangingstone demonstration project 2010 thermal in-situ scheme progress report, approval no. 8788I, In-situ Progress Report to ERCB, February 9, Calgary, Alberta, Canada.

Krawchuk, P., Beshry, M.A., Brown, G.A., Brough, B., 2006. Predicting the flow distribution on total E&P Canada's Joslyn project horizontal SAGD producing wells using permanently installed fiber-optic monitoring, Paper SPE-102159 Presented at the 2006 SPE Annual Technical Conference and Exhibition, September 24–27, San Antonio, TX.

Lombard, M.S., Lee, Jr. R., Manini, P., Slusher, M.A., 2008. New advances and a historical review of insulated steam injection tubing, Paper SPE-113981 Presented at the 2008 SPE Western Regional and Pacific Section AAPG Joint Meeting, 31 March–2 April, Bakersfield, CA.

Luft, H.B., Bennion, D.B., Arthur, J., 1997. Thermo-fluid mechanic characteristics of insulated concentric coiled tubing (ICCT) and the SW-SAGD process, Paper PETSOC-97–98 Presented at the 48th Annual Technical Meeting of the Petroleum Society, 8–11 June, Calgary, Alberta, Canada.

Maxwell, S.C., Du, J., Shemeta, J., Zimmer, U., Boroumand, N., Griffin, L.G., 2009. Monitoring SAGD steam injection using microseismicity and tiltmeters. SPE Res. Eval. Eng. April, 311–317.

Medina, M., 2010. SAGD: R&D for unlocking unconventional heavy-oil resources. Way Ahead 6 (2), 6–9.

MEG Energy, 2011. Christina Lake Regional Project, In-situ Progress Report to ERCB, June, Calgary, Alberta, Canada.

Nasr, T.N., Isaacs, E.E., 2001. Process for enhancing hydrocarbon mobility using a steam additive, US Patent No. 6,230,814.

Nasr, T.N., Beaulieu, G., Golbeck, H., Heck, G., 2003. Novel expanding solvent-SAGD process ES-SAGD. J. Can. Pet. Tech. 42 (1), 13–16.

Nexen, 2006. Long lake project—pilot performance review, In-situ Progress Report to EUB, 20 June, Calgary, Alberta, Canada.

Ong, T.S., Butler, R.M., 1990. Wellbore flow resistance in steam-assisted gravity drainage. J. Can. Pet. Tech. 29 (6), 49–55.

Parappilly, R., Zhao, L., 2009. SAGD with a longer wellbore. J. Can. Pet. Tech. 49 (6), 71–77.

Parmar, G., Zhao, L., Graham, J., 2009. Start-up of SAGD wells—history match, wellbore design and operation. J. Can. Pet. Tech. 48 (1), 42–48.

Rae, M., Seince, L., Mitskopolos, M., 2011. All metal progressing cavity pumps deployed in SAGD, Paper WHOC11-578 Presented at the 2011 World Heavy Oil Congress, Mach 14–17, Edmonton, Alberta, Canada.

Schmidt, G., 2010. A passion for solvents, A Presentation to SHARP Consortium Workshop, June 5, Calgary, Alberta, Canada.

Scott, G.R., 2002. Comparison of CSS and SAGD performance in the Clearwater formation at Cold Lake, Paper SPE 79020 Presented at SPE International Thermal Operations Symposium, 4–7 November, Calgary, Alberta, Canada.

Shell Canada, 2011. In Situ Oil Sands Progress Presentation, Hilda Lake Pilot 8093, Orion 10103. In-situ Progress Report to ERCB, April 21, 2011, Calgary, Alberta, Canada.

Shen, C., 2011. Evaluation of wellbore effects on SAGD startup, Paper SPE-148819 Presented at the Canadian Unconventional Resources Conference, 15–17 November, Calgary, Alberta, Canada.

Shin, H., Polikar, M., 2005. Optimizing the SAGD process in three major Canadian oil-sands areas, Paper SPE 95754 Presented at the 2005 SPE Annual Technical Conference and Exhibition, 9–12 October, Dallas, TX.

Stalder, J.L., 2012. Test of SAGD flow distribution control liner system, Surmont field, Alberta, Canada, Paper SPE 153706 Presented at the SPE Western Regional Meeting, 19–23 March, Bakersfield, CA.

Stancliffe, R.P.W., van der Kooij, M.W.A., 2001. The use of satellite-based radar interferometry to monitor production activity at the cold lake heavy oil field, Alberta, Canada. AAPG Bull. 85 (5), 781–793.

Thorne, T., Zhao, L., 2009. The impact of pressure drop on SAGD process performance. J. Can. Pet. Tech. 48 (9), 41–46.

van der Valk, P.A., Yang, P., 2007. Investigation of key parameters in SAGD wellbore design and operation. J. Can. Pet. Tech. 44 (6), 49–56.

Vanegas Prada, J.W., Cunha, L.B., Alhanati, F.J.S., 2005. Impact of operational parameters and reservoir variables during the startup phase of a SAGD process, Paper SPE-97918 Presented at the 2005 SPE International Thermal Operations Symposium, November 1–3, Calgary, Alberta, Canada.

Yee, C.-T., Stroich, A., 2004. Flue gas injection into a mature SAGD steam chamber at the Dover project (Formerly UTF). J. Can. Pet. Tech. 43 (1), 55–61.

Yuan, J.Y., McFarlane, R., 2011. Evaluation of steam circulation strategies for SAGD startup. J. Can. Pet. Tech. January, 20–32.

Zhao, L., Law, D.H.-S., Nasr, T.N., Coates, R., Golbeck, H., Beaulieu, G., et al., 2005. SAGD wind-down: lab test and simulation. J. Can. Pet. Tech. January, 49–53.

Chapter 18

In Situ Combustion

Alex Turta
Alberta Innovates Technology Futures, Calgary, Canada

The technical aspects of the *in situ* combustion (ISC) are thoroughly reviewed based on published information on the ISC experience during six decades of field application and author's experience. Compared to all other books treating this subject, there are three completely new areas, namely:

- Comprehensive treatise of line drive versus patterns.
- Application of ISC to light and very light oils.
- Novel ISC processes, involving the use of horizontal wells, including the toe-to-heel air injection (THAI) process.

The most interesting ISC pilots (dry and wet forward combustion and reverse combustion)—which are both well instrumented and well documented—are discussed with field examples from both conventional heavy oil and extra-heavy oil reservoirs. Current status of commercial ISC application for both heavy oil reservoirs and deep, light oil reservoirs is provided; this is based on ample field examples; also, operational problems are briefly presented.

18.1 FUNDAMENTALS

18.1.1 Introduction and Qualitative Description of *In Situ* Combustion Techniques

In a porous rock of an oil reservoir, the oil can be ignited around the wellbore by means of an artificial igniter, by spontaneous ignition (SI) or other means and then the ISC front generated is slowly propagated through the reservoirs; displaced oil is collected by production wells. ISC process is sustained/(ISC front is moved) due to continuous injection of air (oxygen-enriched air) or air/water. Therefore, the ISC is basically a gas injection process with heat as an adjuvant to improve recovery. A small portion of the oil (5−10%) is burned out furnishing heat to the rock and its fluids; the ISC front surface can be vertical or very tilted, almost horizontal. The main consequences of the ISC front movement are:

- Direct displacement of the oil from the burned volume, which represents a relatively small percentage of the total volume (15%–30%).
- Reduction of oil viscosity for some of the oil ahead and adjacent to the ISC front due to:
  - Generation/displacement with steam and hot gases under pressure.
  - Generation of a limited (light) oil bank by condensation of the light components of the vaporized oil.
  - Limited light oil generation by thermal cracking.

The idea of ISC was patented in 1923 by Wolcott and Howard in the United States. The first field test attempts to ignite oil in a very shallow reservoir were conducted in 1930s by Sheinman and others in Soviet Union, but there was no continuation with a more extended field pilot or semicommercial operations. The field tests were started again in the United States during 1952 and extended after the first favorable laboratory results were published by Kuhn and Koch (1953) and by Grant and Szasz (1954). However, it seems that the ISC process was practiced *unknowingly* by many operators in the United States who injected air into oil sands, as a gas injection method, for sustained periods of time (Wilson, 1979); in some cases, when the produced gas was analyzed, 10%–15%.

CO_2 was measured. In retrospect, it clearly confirms that a spontaneous ignition did lead to the generation of an ISC front (Chu and Hanzlik, 1983).

Despite the theoretically very efficient heat utilization of ISC, steam injection methods (mainly steam flooding) have produced by far much more incremental oil commercially. This has been due to the following factors:

1. Poor ISC process control as evidenced by poor sweep efficiency and well productivity impairment.
2. Damage to well completions owing to severity of downhole conditions, especially after heat breakthrough.
3. ISC is a highly complex process; there is a clear necessity of more initial knowledge and well ahead of time preparation of specialized, dedicated personnel.
4. The labor intensive character of the process as compared to steam injection methods.

The negative effect of the first and second factors was to some extent ameliorated by switching from pattern operation to the line drive operations starting updip; however, in cases of pronounced overriding and/or channeling, the premature heat breakthrough still can happen and this is equivalent to early project termination. A total of more than 270 ISC field pilots (more than 200 in the United States) have been conducted in 60 years; probably only 10–20% of them have reached the commercial phase. In general, most failed pilots were implemented in poor prospects by unknowledgeable operators that compounded odds against success; projects undertaken by larger companies generally tended to be more successful than those initiated by small companies or smaller independents (Sarathi, 1999). Evaluation of an ISC pilot has been one of the most challenging aspects in the ISC field testing.

Based on direction of the ISC front propagation in relation to the air flow, the process can be classified as forward ISC and reverse ISC. The forward ISC is also named *co-current ISC* as the ISC front advances in the general direction of air flow; the reverse ISC is also called *countercurrent ISC* as the front moves in a direction opposite to that of air flow. So far only the forward ISC has reached the commercial application status. The forward ISC is further categorized into "dry" ISC and "wet" ISC according to the injection fluid used to sustain the process: air for dry ISC and air and water for wet ISC.

Both dry and wet ISC process can be further categorized into *frontal ISC process* and *segregated ISC process*; for relatively thin layers ($h < 5-6$ m) containing a not very heavy oil a quasi-frontal ISC process can be expected, while for thick layers a segregated forward ISC behavior will occur. There is a lot more knowledge about frontal ISC process although in the field it was clearly seen that both of them were a reality. However, there are less methods for the design of a segregated ISC process.

A typical temperature profile for dry ISC with frontal advancement is shown in Figure 18.1; the main saturation zones generated by the process are also shown. From Figure 18.1, two points are highlighted; point A, which represents the position of the ISC front and point B, which is the so-called convection point, showing the advancement of the convective wave. Most of the heat generated during ISC process is stored/comprised between these two points, therefore in the burned zone. The incentive to switch to wet ISC was related to the desire to reduce this amount of heat and transfer as much heat as possible, ahead of the ISC front where the oil is.

Typical comparative temperature profiles for dry and wet ISC with frontal advancement are shown in Figure 18.2. It can be seen that for dry ISC (having a water–air ratio (WAR) zero) most of the heat is in the burned out zone, while for moderate wet ISC processes, having moderate air–oil ratio (AOR) values, the high peak temperature still exists, but a higher fraction of the generated heat is brought ahead of the ISC front, practically in the water vaporization–condensation zone (steam plateau). When the WAR is further increased, the peak temperature disappears, but the ISC is still sustainable due to a very long "combustion zone" at the steam temperature; the *super-wet ISC* process is obtained. Typical temperature profiles correlated with saturation profiles both for moderate ISC and super-wet ISC are shown in Figures 18.3 and 18.4; it can be seen that in super-wet process some residual fuel remains in the burned zone.

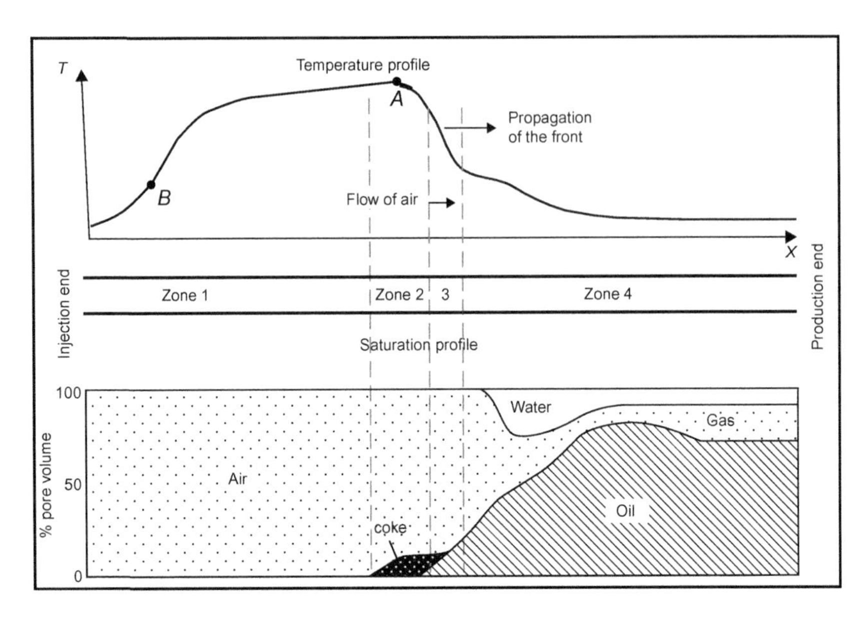

FIGURE 18.1 Temperature and saturation profiles during dry ISC process.
Source*: Modified from Burger et al. (1985)*

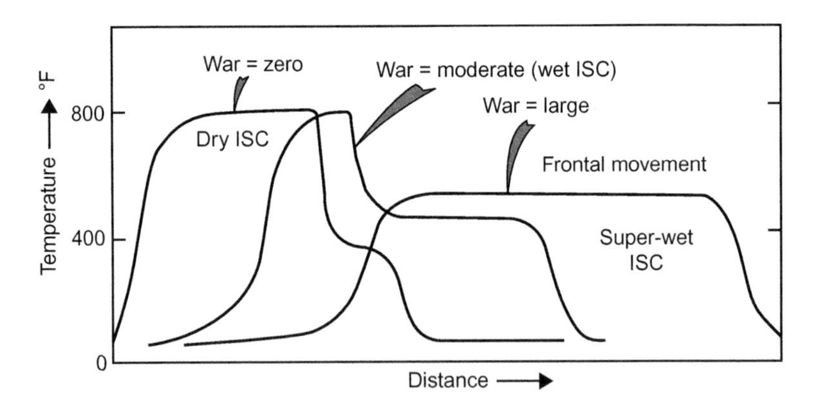

FIGURE 18.2 Comparative temperature profiles for dry, moderate wet, and super-wet ISC process (Improved Oil Recovery Handbook, 1983, Chapter ISC by Chu and Crawford).

A typical temperature profile for segregated ISC is difficult to present as this is extremely complex; a 3D combustion surface is formed. A very simplified schematics of the various zones formed during the segregated ISC are shown in Figure 18.5. Usually, the injection wells are perforated only in the lower half of the pay zone, while the production wells avoid the top of formation.

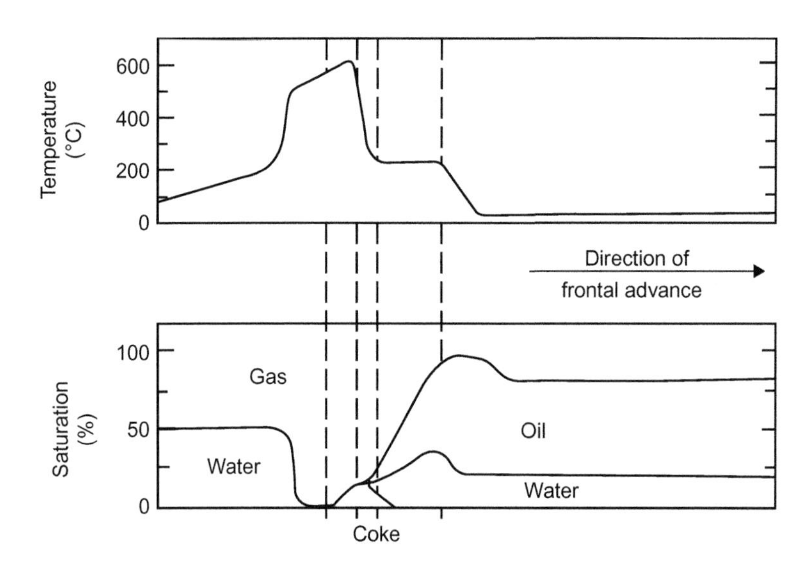

FIGURE 18.3 Temperature and saturation profiles for moderate wet ISC process.
Source: *Courtesy of UNITAR Centre, Mehta, and Moore (1996)*

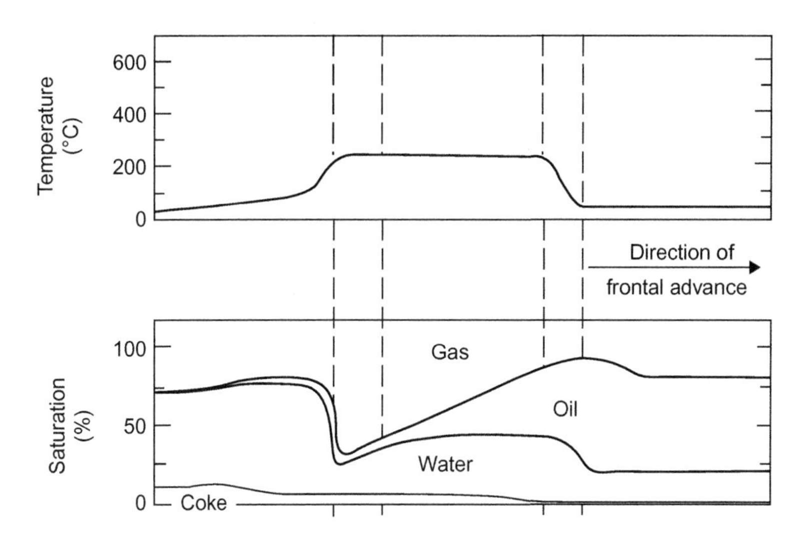

FIGURE 18.4 Temperature and saturation profiles for super-wet ISC process.
Source: *Courtesy of UNITAR Centre, Mehta, and Moore (1996)*

For forward ISC where an O_2-enriched air is injected, both temperature profiles and saturation profiles are similar to those described for normal/air ISC, both for frontal advancement and segregated ISC (Figures 18.1–18.4).

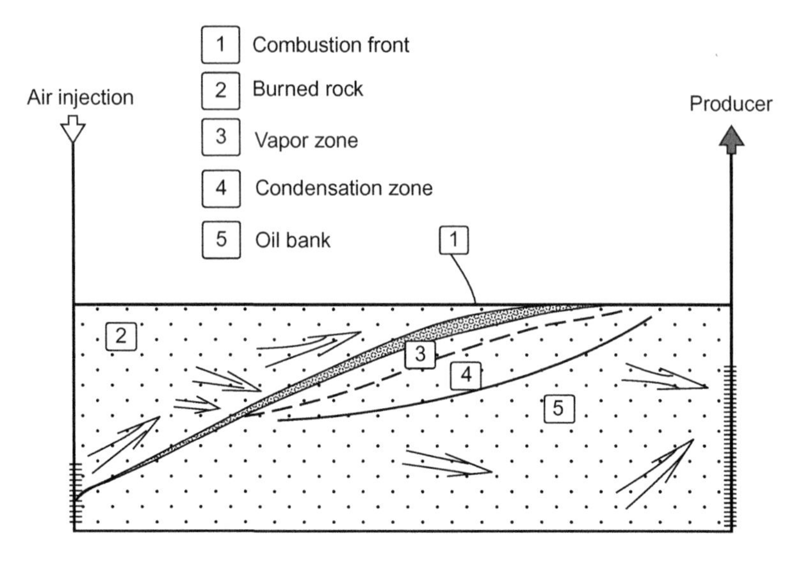

FIGURE 18.5 Schematics of various zones formed during segregated ISC process.

When the oil is too viscous to flow under reservoir conditions, but the reservoir has an adequate air permeability, the reverse ISC application can be contemplated. As mentioned, in this process, the ISC front moves counter to the air flow (from low pressure to high pressure!), while the displaced oil flows through the hot burned zone. Unlike forward ISC, the production well is ignited; after ignition the production well is put into production and another well is used for air injection (Figure 18.6); the production well remains in the hot zone for the whole life of the project. The front moves in the same way in which a cigar is consumed by expelling the air instead of inhaling it. The process has seen limited use in the field; in some situations, it may be used to correct the advancement of a forward ISC front. The temperature of reservoir should be low enough in order to allow the oxygen supply of the ISC front.

In the field, only forward ISC has the status of commercial application; 70–80% dry and only 20–30% wet; in terms of ISC pilots it was a bit different, but generally wet ISC pilots such as East Tia Huana, Schoonebeek, and Sloss were not developed up to a commercial status. Only three commercial wet ISC projects are known, namely Bellevue (USA) and Balol and Santhal (India).

For heavy oil reservoirs, the most successful ISC projects have been operated as peripheral line drive projects (started updip) as opposed to patterns projects. This is so in spite of the fact that for patterns projects, there are more developed design methods; putting gravity to work for operator is crucial and cannot be stressed enough.

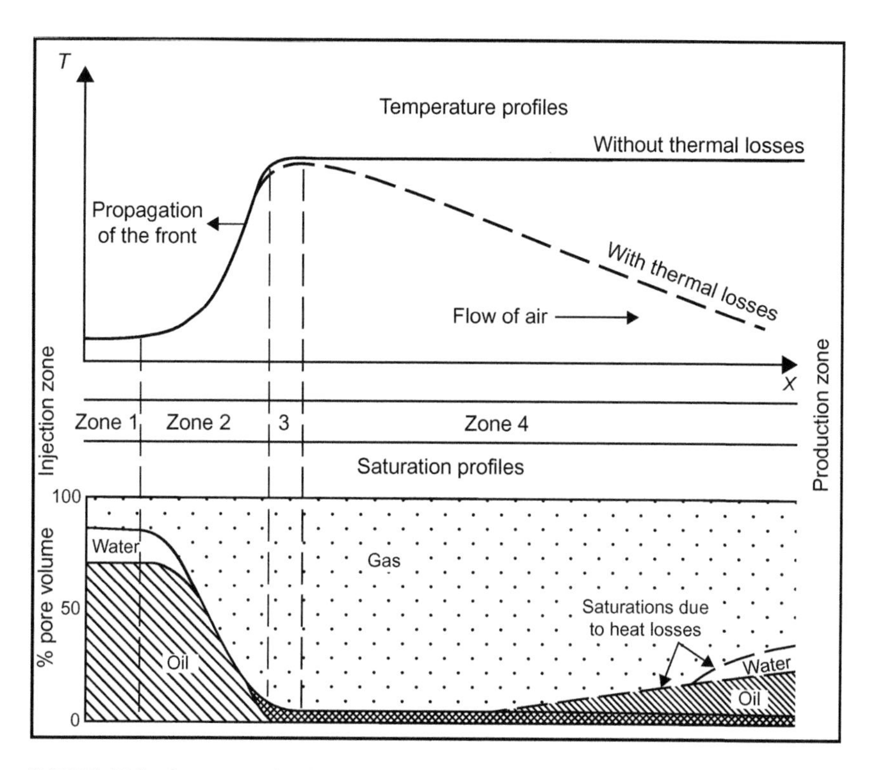

FIGURE 18.6 Reverse combustion: temperature and saturation profiles.
Source*: After Burger et al. (1985)*

To determine feasibility of ISC applicability, there are three main categories of routine ISC laboratory tests, namely:

- Ramped Temperature Oxidation (RTO)
- Accelerated Rate calorimeter (ARC) technique
- Combustion Tube (CT) tests.

The laboratory tests (RTO and ARC) will give information about the chemical reactivity of oil–rock couple, and some kinetic data to be used in mathematical simulation, while the CT tests will provide a base figure for the amount of fuel deposited and/or burned during combustion and for air requirement (per unit volume). This last figure cannot be obtained from pilot data, whatever detailed analyses and measurements are carried out in the field. Also, from the pilot, there is no reliable means of determining or checking the value of air requirement, which, also, has to be determined in laboratory.

Details on the equipment and procedures used for RTO and CT tests can be found in different manuals, among them the most important being Burger et al. (1985) and White and Moss (1983). Essential details on ARC tests are provided by Yanimaras and Tiffin (1995).

18.1.2 Design, Operation, and Evaluation of an ISC Field Project

The focus of this section is to highlight how the geometry of the structure/reservoir (as a whole) must be accounted for and incorporated in the management of a future ISC process to be applied in that reservoir. This would make a great difference in the operation and the evaluation of the pilot and future commercial development and, generally, in achieving a successful project, i.e., an operation with a lower air–oil ratio and leading to a higher ultimate oil recovery (UOR).

Ignition Operation

Unlike any other enhanced oil recovery (EOR) processes, application of ISC requires an additional step, namely initiation of ISC, which is known as ignition operation. During this step, the full ISC front is generated by the increase of the temperature around the future air injection well, up to a temperature close to the peak temperature necessary for self-supporting ability of the process.

Ignition operation is a key-phase and it is of crucial importance for the process; many ISC pilot failures were due to the failure of ignition. In many ISC pilots, the disappointment started with ignition and very often due to impossibility to recognize its achievement (or its failure) in a reasonable period of time.

Four factors are important for the ignition: cumulative of heat generated, maximum temperature locally reached, time interval in which this is done, and chemical modifications of the oil during the ignition period. Taking into account these parameters, the operator can choose from a large variety of methods/devices. Generally, ignition can be performed by involving one of the following actions:

- Preheat injected air.
- Preheat the oil formation around the well.
- Improve reactivity of fluid–rock couple around the well.

Function of what action is taken, in practice, there are two categories of methods:

1. Artificial methods/artificial devices (ADs)
2. SI and enhanced SI

ADs refer to the downhole electrical heaters or downhole gas/liquid fuel burners; SI refers to the pure SI, while enhanced SI refers either to chemical means or via preliminary steam injection. Combination of these methods is also possible. Generally, for depths higher than 1000 m, the SI methods should be considered first.

Our focus will not be on mechanical details but on how to choose and evaluate an ignition operation conducted using a certain method. Use of AD involves more preparatory work (including mechanical/electrical devices), front investment (for specialized equipment), and continuous supervision during a relatively short-period operation (several days), but the assessment of ignition is very straightforward. On the contrary, SI methods involve a lot less preparatory work and equipment/instrumentation, but evaluation of ignition (recognizing achievement or failure) is more complex while the period of ignition significantly longer, up to a few months. In other words, ignition based on ADs is equipment/labor intensive in the field, while SI methods are labor intensive in the office, for evaluation of ignition.

Artificial ignition: For the mechanical details of the AD, the reader is directed to the Baibakov book (1989), while both mechanical details and ignition procedures/results are provided in White and Moss's book (1983). The only detail given here is that for any artificial ignition, a thermocouple should be located in the perforations in order to allow the control of the bottomhole temperature (BHT) during the ignition.

Spontaneous ignition (SI): As a pure SI (just air injection), it can be applied to oil reservoirs having a reservoir temperature higher than 60–70°C. The period needed for the generation of the ISC front is called ignition delay. The ignition delay for a series of specific reservoir conditions was given by Strange (1964). The graph shown in Figure 18.7 (Dietz and Weijdema, 1970) provides an estimate of the ignition delay, function of reservoir temperature, and pressure, as estimated with an analytical expression (Tadema and Weijdema, 1970), using the oxidation characteristics determined in the laboratory. SI is worthwhile pursuing only in cases the ignition delay is shorter than 1–2 months. Otherwise, the enhanced SI should be considered. Very often the enhanced SI by chemical means is adopted and this involves the injection of a slug of linseed oil before starting air injection. Linseed oil is a very effective ignition enhancer as its oxidation starts at a temperature as low as 33°C. There is extensive experience using linseed oil; in the Videle–Balaria fields (Turta and Pantazi, 1986), this method was commercially used for more than 60 ignition operations; the ignition delay was determined based on the variation of apparent atomic hydrogen/carbon ratio and it was generally shorter than 2–3 weeks. This kind of ignition can be taken into consideration for oil reservoirs with a temperature in the range of 35–70°C. A chart indicating how to choose an ignition method is shown in Figure 18.8 (Turta, 2011).

Indications of ignition: When choosing a certain ignition method, it should be kept in mind that increase of oil production, therefore recording of incremental oil, is supposed to appear only after the generation of the full ISC front, therefore after a period of time longer than the ignition delay; during the ignition period, only low-temperature oxidation (LTO) reactions occur and they are not conducive to significant mobilization of the oil.

Artificial ignition is achieved very quickly. The ignition achievement is marked by the continuous decrease of O_2 percentage; eventually, O_2

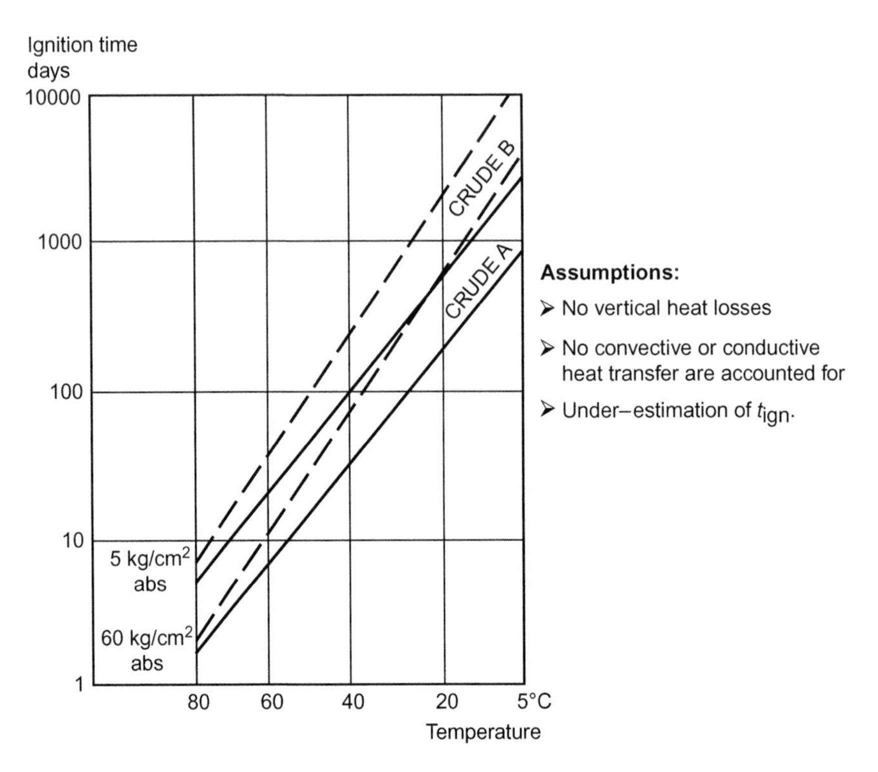

FIGURE 18.7 Ignition delay (t_{ign}), as function of reservoir T and P, calculated with an analytical expression, using the oxidation characteristics determined in the laboratory (Dietz and Weijdema, 1970).

percentage gets close to zero (from the quasi-stabilized value obtained before ignition), while the CO_2 percentage stabilizes in the range of 11–16%.

For all kinds of ignition methods, another indication of ignition is the increase of injection pressure two to three times when the full ISC front is generated (White and Moss, 1983); this seems to be due primarily to buildup of the oil bank displaced by the burning front. However, this symptom is missing when initially the well was blocked and its de-blocking is related to the ignition operation.

The most complex is the rigorous determination of ignition delay for an SI or enhanced SI. This can be done by analyzing the gas composition evolution, but very frequently this is not very reliable due to different parasite effects such as CO_2 solubility effects and other effects. That is the reason that this determination is better done based on a synthetic indicator of the gas composition, namely the apparent hydrogen–carbon (H/C) ratio. Such a determination is shown in Figure 18.9 for a linseed oil ignition at Videle–Balaria field (Turta and Pantazi, 1986); the ignition delay was approximately 3 weeks.

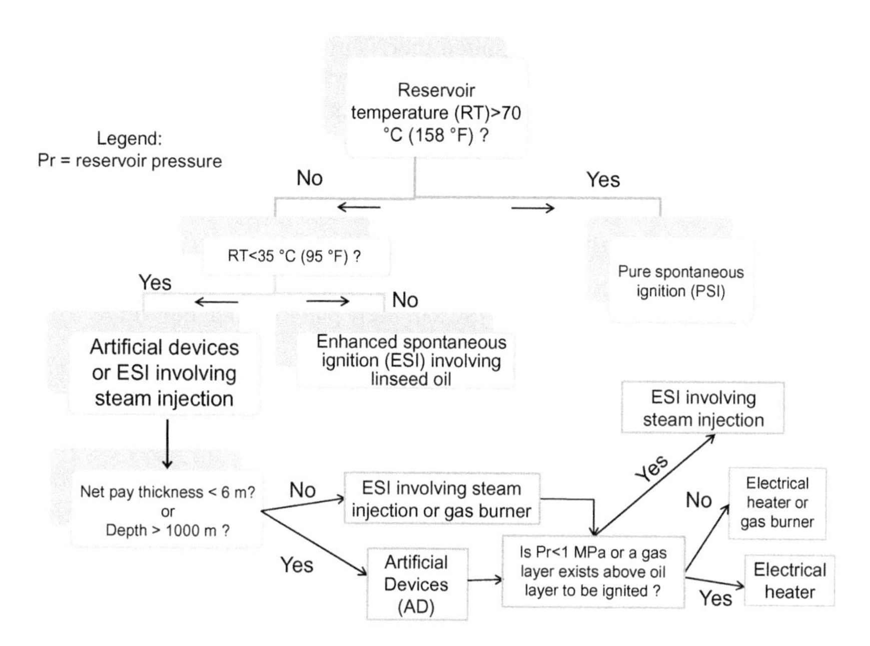

FIGURE 18.8 How to choose the most appropriate ignition operation.

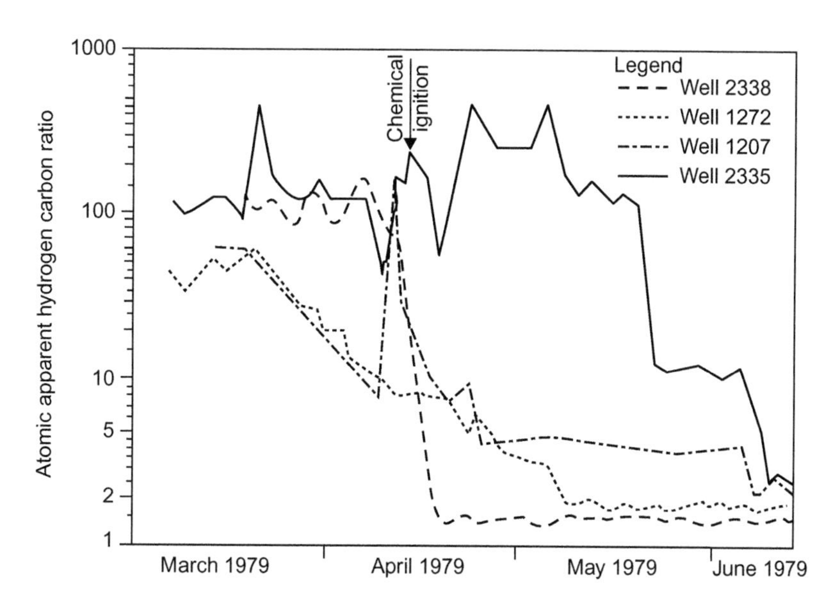

FIGURE 18.9 Apparent atomic H/C ratio variation during the chemical ignition (linseed oil) of the pattern A_1, Videle East, Romania (Turta and Pantazi, 1986).

Irrespective of the way the ignition time is estimated, reaching the stabilized burning-front conditions will take longer as the reservoir pressure increases and as the oil is lighter, therefore it is more difficult. For these stabilized conditions, the H/C value, calculated from the gas composition, has to be less than 2.5–3 both for heavy oil and light oil ignition operations.

Line Drive Versus Pattern Application: Choosing the Best Location of the Pilot

There are two ways of applying ISC: in well patterns and line drive well configuration. The patterns can be contiguous or isolated. The location of patterns may be upstructure or downstructure (downdip). So far all three configurations were tried, but most of applications used contiguous patterns and peripheral line drive configurations. The isolated patterns were only used in one commercial operation (West Newport).

The options for the expansion from the pilot to a field scale commercial application are displayed in Figure 18.10. As shown, the line drive is possible to be applied only starting from the upper part of the reservoir. For this reason, it is extremely important to place the pilot upstructure; this way, after the test is finished, one can have both options of commercial, i.e., either line drive or patterns.

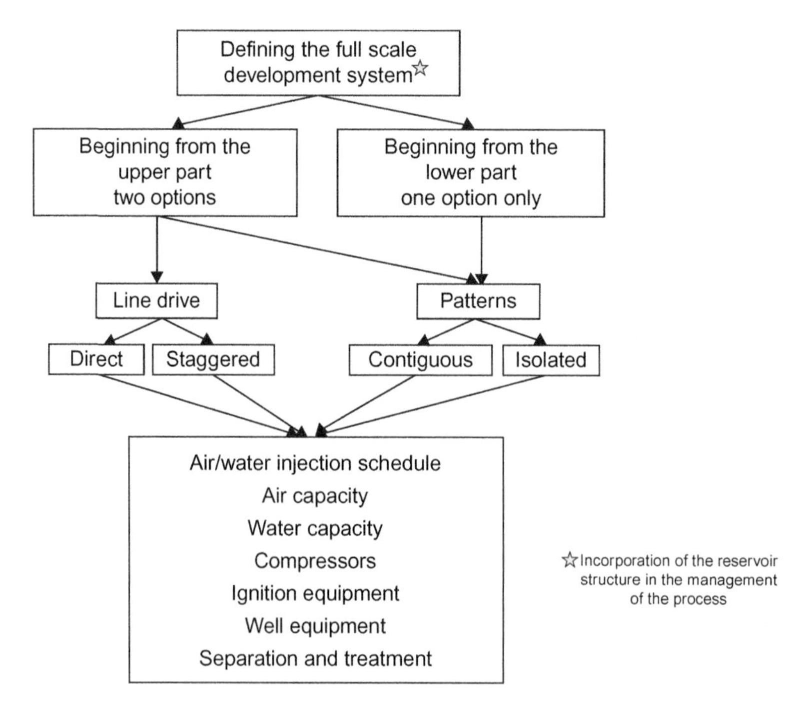

FIGURE 18.10 The options for expansion of ISC pilot to field scale operation.

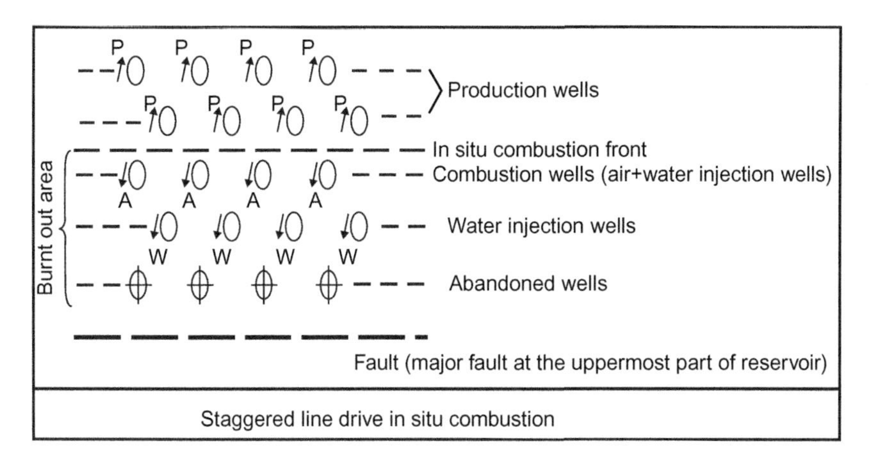

FIGURE 18.11 A generic peripheral line drive application.

The peripheral line drive well configuration: A schematic of line drive well configuration is shown in Figure 18.11. Essentially, the first row of wells at the uppermost part of the structure consists of air (air/water) injection wells (combustion wells), while the displaced oil is produced by the nearby two to three production well rows, which are isobathically below the injector row's isobath. Once the closest production row is intercepted by the combustion front, this row is converted into air (air/water) injection, while behind, the former air injection row is used for water injection or simply shut off. Therefore, except the first uppermost well row, all other wells are utilized first as producers and afterward as combustion wells. One more exception is the last lowest row on the structure which is used only as production well row. In this system, a secondary air gas cap is formed and this is bigger and bigger as the ISC front is advancing downdip. A schematic of a real line drive application, that of the Suplacu de Barcau, after 20 years of commercial operation is shown in Figure 18.17. As a rule, the front has been propagated as much as possible parallel to the isobaths. So, the front was moved toward the oil–water contact (Condrachi and Tabara, 1997).

To decide between line drive and patterns application is one of the most important responsibilities of the designer. All the design parameters shown at the bottom of the diagram in Figure 18.10 will depend on the above-mentioned choice. For the line drive, the rate of oil recovery is limited by the length of isobath. Therefore, one cannot use any oil production rate, being limited on the higher side of the air rate values; there are some time constraints as far as the total life of exploitation is concerned.

ISC, in principle, is a gas injection which has additional beneficial effects associated with the propagation of the heat wave generated by the ISC front. Like a conventional peripheral gas displacement, it is just normal to start the process upstructure. Actually, only the existence of an extended primary gas cap should prevent locating the pilot upstructure.

When the pilot is located upstructure, there is a good possibility of more rigorous evaluation of oil recovery. There is a rigorous way to delineate a volume of reservoir located at the upper part of the reservoir and which is under the influence of the combustion and for which both AOR and incremental oil can be reliably calculated; this will be addressed later on in this chapter.

Another positive side of locating the pattern at the upper part of the structure is that in case the test is inconclusive or deemed uneconomical—although the burning was OK or almost OK—the operator can just stop the air (air/water) injection and walk away. Oil resaturation of the burning zone is not going to take place or it should be minimal.

If the pilot is not located updip, when stopping air injection, the burned zone may be unintentionally transformed into a cracking reactor with the formation of large quantity of coke. For this reason, at the abandonment of the ISC pilot, it is recommended to inject a volume of water equal to at least 0.8 burned pore volume for the dry combustion and less for wet combustion.

All in all, the main advantages of the line drive over the well pattern configuration are:

- Full advantage of the gravity (higher oil recovery) because the oil displacement is gravity stable.
- Full control regarding the avoiding of oil resaturation of the burned area, which can lead to an important increase of air requirement and decrease of efficiency of process.
- Easier evaluation of the process (mainly incremental oil evaluation).
- Easier operation due to the following reasons:
  - Each producer is intercepted by the ISC front only once. For the pattern system, as many as four ISC fronts may intercept the producer, and the risks of damaging the wells are higher.
  - The area of combustion gas distribution is much smaller (less gas analyses for the same oil production).
  - Less artificial ignition operations, given the possibility just to make an air transfer to the new row of recently intercepted (by the ISC front) producer row.
  - Easier and more reliable tracking of ISC front. After the first row of producers is intercepted by the front, the tracking of the front position is a lot easier, just by assessing the interception for each producer.

On the other hand, the main advantages of the pattern configuration are the use of different completions for injectors and producers (including perforating different interval in injectors and producers), and the liberty to select any rate of oil production, by operating simultaneously as many patterns as the operator wants (Machedon et al., 1993, 1994). The first advantage is important mainly for the application in a multilayer oil formation, when the separation between layers is not very well defined. When several companies are operating on the same reservoir, the application of the line drive requires the unitization of the pool, while the patterns system does not.

Some operators reported increase of oil production when reducing the air rates or even stopping the air injection in the patterns. Sometimes this was presented as an advantage of the patterns system, *which seems to be incorrect*. In reality, the advantages of this increase in oil production may be misleading as in the air injection stoppage period, the oil outside the swept zone is flowing into the high temperature, burned zone, *and the cracking of the oil produces large amounts of coke*. This phenomenon is even more harmful as compared with the pushing of the oil into the gas cap during conventional waterflood. The increase of air requirement and of the air–oil ratio is the direct consequences; significant volumes of "coked rock" will remain in the reservoir. Evidence of this phenomenon was revealed by the coring wells, in the burned area of the first ISC pilot at Suplacu de Barcau (Carcoana et al., 1975). As shown in Figure 18.17, the coke deposits as high as 170 kg/m^3 rock were measured along a pay thickness portion of 3.5 m at the boundary between the burned and unburned zones (as compared with 35 kg/m^3 rock, the normal, coke deposit for this reservoir). These very high deposits were caused by frequent fluctuations in the injection pressure and frequent stoppages of air injection, which were inherent for the very beginning of the process, when the running of the air compressors was not a routine operation for the field personnel.

Design and Evaluation of an ISC Pilot Project (the Integrated Approach)

In order to apply the ISC commercially, one needs to conduct the following four phases: laboratory tests, field pilot test design, evaluation of the field pilot test, and expansion to field scale operation.

As mentioned, the geometry (architecture) of the structure/reservoir (as a whole) must be accounted for and incorporated in the management of a future ISC process to be applied in that reservoir. This is one of the most crucial tasks of people designing the ISC application and will have an effect on the entire life of the project, including the easiness of ISC operations and its results. That is why the subtitle of this section is the "integrated approach."

Is the field pilot test a necessity? Usually it is. More precisely, to carry out a pilot is a rule, and the skipping of the pilot is rather an exception. The field pilot test will show if the ISC has the self-supporting capacity, if it has the ability to give incremental oil and what operational difficulties are to be expected in a future full-scale process. The pilot evaluation is totally dependent of the way the experimental pattern is located on the structure.

The most important performance indices for the evaluation of an ISC process are injected air/incremental produced oil ratio, AOR, and the incremental oil recovery factor (incremental oil/ original oil in place (OOIP))—IORF. The value of incremental oil is not necessarily the same figure for AOR and IORF calculations, because for the patterns system, the offset wells show an enhancement in oil rates, as well. The incremental oil from the offset wells is taken into consideration only when calculating AOR. Therefore, usually, the incremental oil for the calculation of AOR is associated with a larger area than in the case of determining the incremental oil for the calculation of IORF. The calculations of these performance indices are made easier if the pattern is located upstructure. This recommendation is valid also for the cases where the dip is very small, even 3–4°.

Given the very high value of front-end investment (mainly for compressors), a gradual development of ISC is recommended. The experience showed that it is a good idea to expand the pilot into a semi-industrial process.

Field examples of pilot location: One of the most important difficulties associated with an ISC pilot is to establish the incremental oil obtained by ISC. Locating the pilot updip can substantially reduce the frustrations associated with the assessment of this parameter.

In the following, three field cases will be presented, illustrating the main aspects associated with the three possible approaches of locating the pilot.

Suplacu de Barcau Field (line drive): The most in-depth exploration of the effect of location of pilot on the reservoir was performed at the Suplacu de Barcau Field (Carcoana et al., 1975).

The Suplacu de Barcau reservoir is a monocline. There is a major fault limiting the reservoir to the South, while to the North the oil reservoir borders upon an aquifer. Both the depth and the thickness increase from south to north. The reservoir properties are given in Table 18.6.

To select the best location for starting ISC process, three different experimental patterns were located: up-, middle-, and downstructure, as shown in Figure 18.12. These patterns were operated for more than 5 years. The upper pilot gave the best result in terms of AOR value. Thus, for the ones in the middle and in the lower part of the reservoir, the AOR was in the range of 16,800–22,800 scf/bbl (3000–4000 sm^3/m^3), which are unfavorable values as compared to the value of 8400 scf/bbl (1500 sm^3/m^3) for the pilot at the upper part of the field. These unfavorable AOR were due to the intensive channeling of gases toward the upper part of the field, without causing increases in the oil rate for the updip wells located in that regions.

After 5 years of operation, even for the pattern located at the upper part, the IORF was not yet possible to be calculated and the decision to go ahead with a semiindustrial phase—consisting of six contiguous patterns located updip—was made (Figure 18.13).

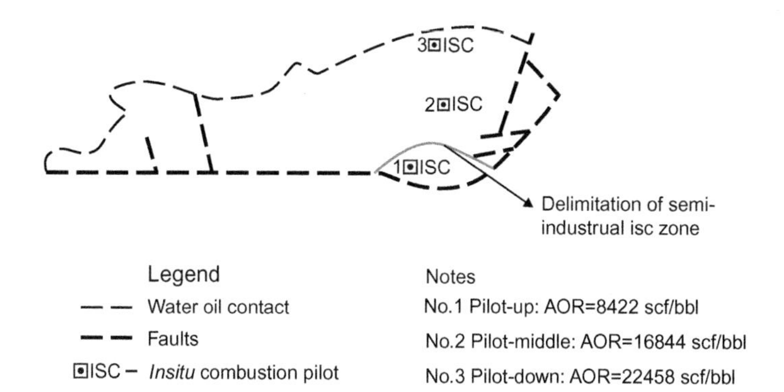

FIGURE 18.12 Three different experimental pattern locations at Suplacu de Barcau: up-, middle- and downstructure.

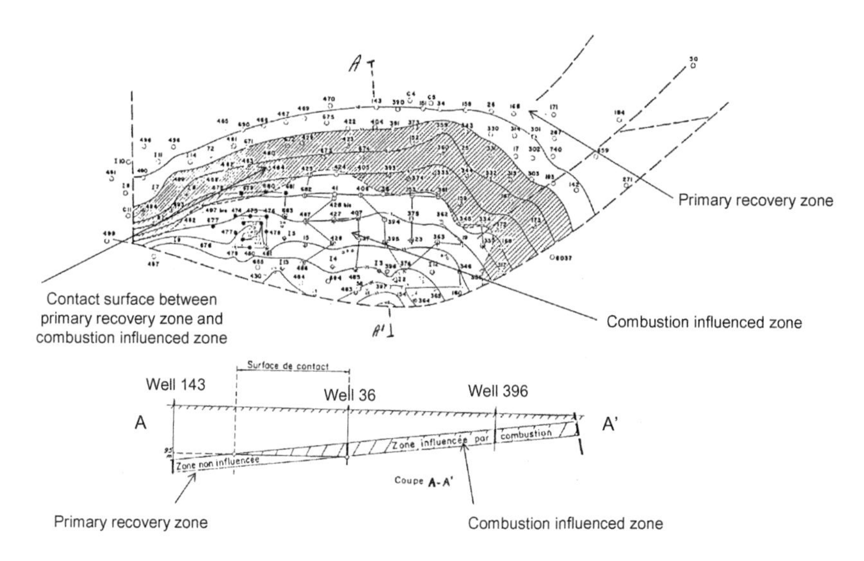

FIGURE 18.13 Suplacu de Barcau commercial ISC as of April 1975. ISC drainage area, as confined by the "contact zone".

During semiindustrial phase, a procedure to calculate the incremental oil production was established (Carcoana et al., 1975). For this procedure, the delimitation of the zone considered for the calculation of IORF was essential. The procedure consists of the following: The most downstructure producer, which recovered combustion gas, was picked up and through the base of its perforations a horizontal plane was drawn, representing "the contact surface" between ISC-affected and nonaffected zone. For this delimited region, both current IORF and the UOR factor were calculated. For the inside producers

(producers on the affected area), the entire oil production was taken into consideration, while for all the producers located on the "contact surface," only half of the oil production was considered. The sum of these two figures divided by the OOIP in the "reservoir delimited by the contact surface" gave the current oil recovery factor. Because the majority of wells were still producing oil, production extrapolation allowed the calculation of the UOR. This calculation was made every 2–3 years. A conservative oil recovery factor of 35% was obtained for the first calculation, and this figure increased in the subsequent years.

An important point to be made here is that by applying this procedure, the value for the incremental oil is the same figure for both AOR calculations and oil recovery factor calculations.

West Balaria Field (contiguous patterns): The pilot patterns were located somewhere in the middle of the block (Petcovici, 1982). However, to have a possibility to calculate the oil recovery factor, the four inverted 5-spot patterns were arranged such that they formed a confined pattern (inside the four injectors) with only one producer for this confined area (Figure 18.14). The incremental recovery factor is easily obtained by dividing the incremental accumulated oil production of the central production well (1605 in the figure) to the OOIP existent in the confined area. The procedure is risky because only the producer could be damaged during the process. Actually,

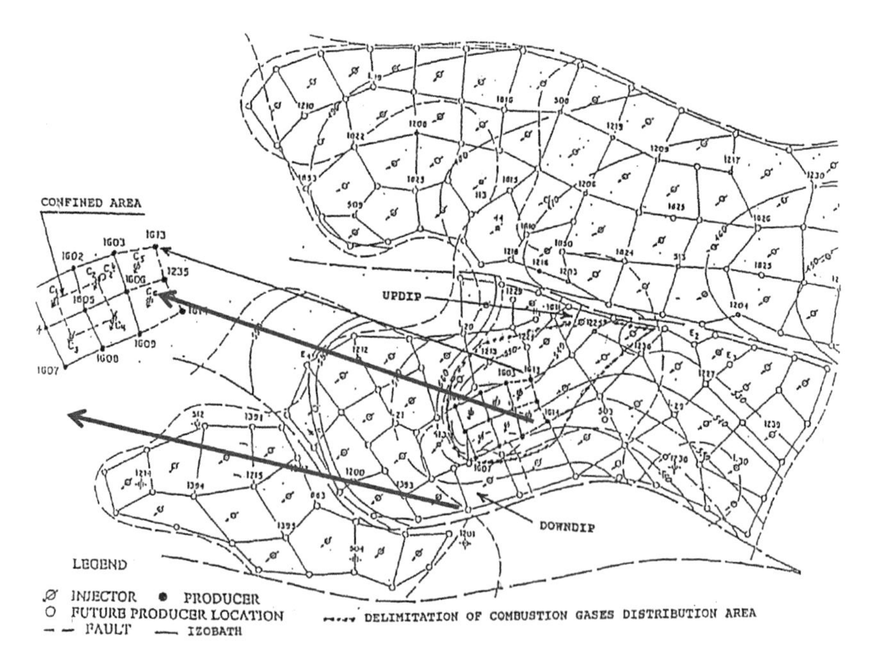

FIGURE 18.14 Balaria reservoir, contiguous six patterns (confined drainage area). Determination of oil recovery potential Petcovici, 1982.

later on, two more contiguous patterns were added and in this way the second confined pattern (area) was formed.

On the other hand, the incremental oil used in the AOR calculation is determined for an enlarged zone, larger than the pattern area, as many of the offset wells experienced increased oil rates. This situation happened in many pilots worldwide, such as Golden Lake Sparky and Aberfeldy (Miller, 1987), to cite just two pilots. There are means of calculating this parameter, based on extra oil recovered (over the primary performance curve), for the enlarged zone. However, this incremental oil figure cannot be reliably used for oil recovery calculations; *an attempt to do this resulted in very different estimates for the patterns area and "enlarged" area* (Miller and Staniford, 1988). Actually, even the allotment of fireflood related oil production to a particular injector (pattern) is practically impossible, although it is sometimes claimed.

South Belridge Field (isolated pattern): The pilot was located somewhere in the middle of the structure; the pilot consists of an isolated pattern unconfined in space and time! Why that? Because this is an example of a pattern which was operated way beyond its 5-spot boundary; it was operated more than 20 years between 1956 and 1978, at least (Gates and Ramey, 1958; Gates et al., 1978). The pilot was very well instrumented and evaluated and brought important information; also it is well documented. First of all, this was a clear example of a segregated ISC process. This can be easily understood from Figure 18.15, which shows the burned thickness after 3 years of operation and the vertical downward advancement of the tilted ISC surface. The burned zone was confined at the top of the layer, and there was a clear tendency for the ISC front to go preferentially upstructure; the area encompassed by the ISC front contour was three times the area of the original pattern only in 3 years of operation; then it increased even more. Obviously in this case, it is impossible to calculate any incremental oil obtained by ISC, as the value of OOIP to refer to, cannot be reliably established.

The information from this section shows that failure to locate the pilot on the crest or very close to the crest will negatively impact the project during its whole life; generally, attempts to correct afterward will not be very successful!

Performance Prediction Methods and Mathematical Modeling

The main prediction methods were developed for dry ISC conducted in patterns and are to be used for the "pattern" exploitation only. The most important methods are:

- Nelson and McNiel (1961)
- Brigham et al. (1980)
- Gates and Ramey (1980).

The Nelson and McNiel method is the oldest one and still remains the most complete and reliable method; it provides both an air injection profile

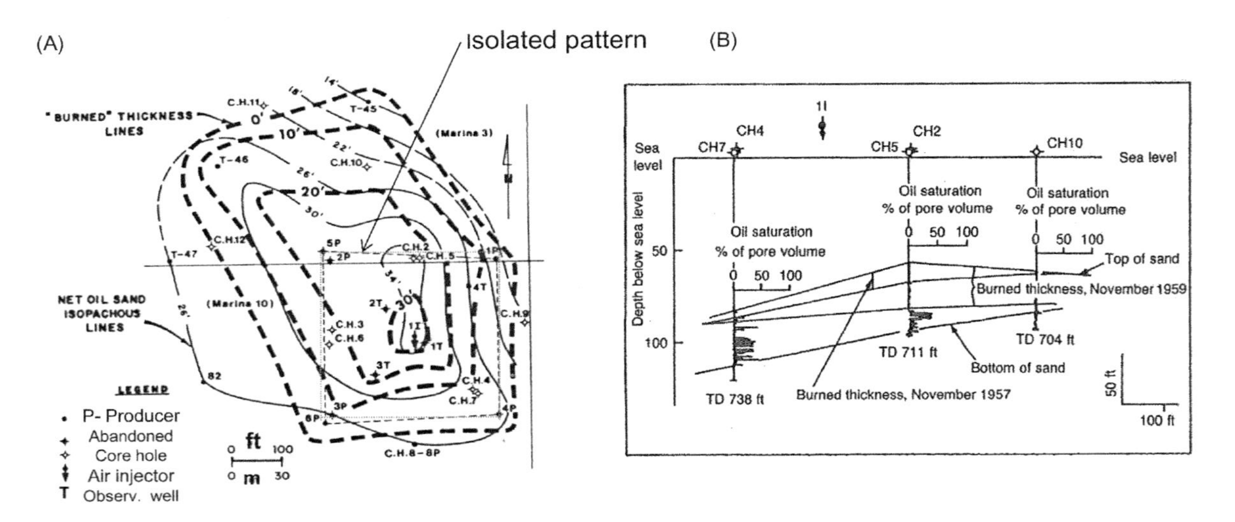

Core hole (CH) number	Sand thickness	Burned thickness (m)	
	(m)	November 1957	November 1959
CH2 and CH5	11	3.4	7.9
CH3 and CH6	9.4	2.7	4.3
CH4 and CH7	8.8	0.9	2.1
CH10	7.0	-	5.5

FIGURE 18.15 South Belridge ISC Project. Unconfined isolated pattern (Gates et al., 1978). (A) Burned thickness as of November 1959. (B) cross section showing the growth of the burn-out reservoir volume with vertical thickness values (in the table below).

and a way to estimate air injection pressure, oil recovery, and AOR. Brigham, Satman, and Soliman is empirical, based on some US pilot performance, while Gates and Ramey is formulated based on South Belridge pilot performance.

Neither of them incorporates any results from the mathematical simulation of the process (either numerical or analytical).

According to their capacity to describe either thermal or displacement aspects or both, there are two categories of analytical models:

1. Heat transfer models (HTM)
2. Heat transfer and displacement models (complete simulation).

While the second category is used mainly for the simulation of laboratory tests, the first category can be used for field simulation and generally assumes a moving vertical ISC front; in this case, only the temperature distribution, both in the oil formation and the adjacent formations, is determined; fluid saturations are not determined as the thermal aspects are totally separated from the hydrodynamic phenomena.

The HTM offering the most complete analysis of a radially expanding vertical ISC front of infinitesimal width are the model of Chu and that of Thomas, both published in 1963. While Chu model is built for a thin-sand situation, Thomas model assumes a burned interval located somewhere in the middle of a thick oil formation; more details on Thomas model are given elsewhere (Thomas, 1963). Neither of these models is applicable to segregated ISC (overriding combustion front).

Typical results obtained by Chu, giving the temperature as a function of position around the combustion front (radial profiles for different horizontal planes and vertical profiles for different radial distances and isotherms), are shown in Figure 18.16. From these profiles, one can easily notice that most of the generated heat remains behind the ISC front. Using this model, it was found that stability in time of ISC process is very good. After a few months of vigorous ISC, the process can be interrupted for a few weeks without any risk of losing the process; this finding was confirmed by ISC field tests, where interruptions of up to 18 days were possible (Turta, 2012).

In many field applications, however, the combustion is segregated and in this case other analytical models can be considered: Gottfried Model (1965), Prats et al. (1968), and Khelil Model (1969).

18.2 FIELD APPLICATIONS

18.2.1 Screening Guide

At this stage of development, the screening criteria will be presented separately for heavy oil reservoirs and deep, very light oil reservoirs. The thinking behind this separate presentation is related to the fact that so far the

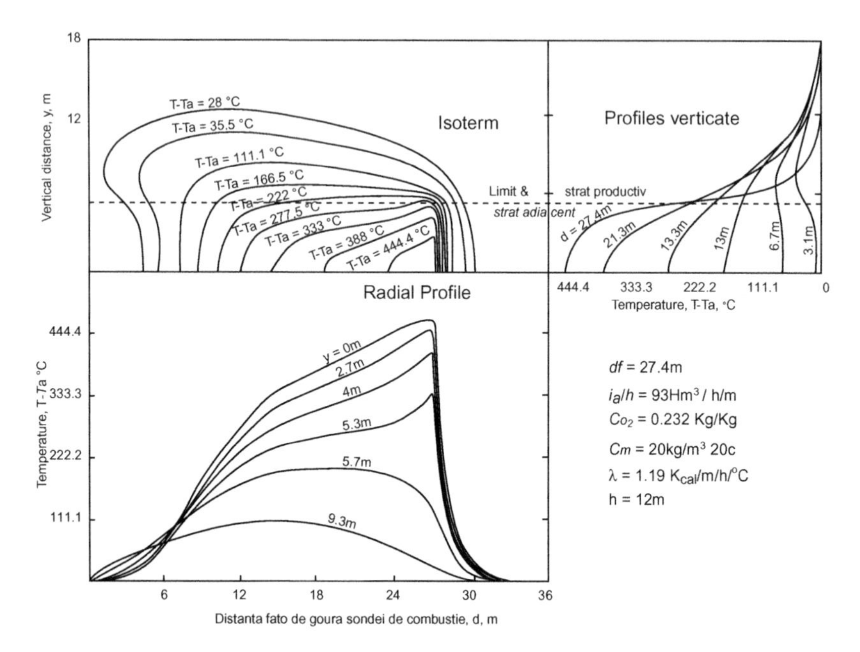

FIGURE 18.16 Typical isotherms and radial/vertical temperature profiles for dry ISC with frontal advancement (Chu, 1963).

commercial application of the ISC to reservoirs having an oil viscosity between 2 and 60 mPa·s has not been made. For both heavy and very light oils, there has been an incentive in applying it due to a low oil recovery by primary methods and easiness of application. That easiness was related to two aspects:

1. Low-pressure application for the heavy oils and the simplification of application by skipping of the ignition operations for the deep light oil reservoirs.
2. Fewer operational problems due to low corrosion in heavy oils and integral oxygen consumption in case of very light oils.

The screening criteria are presented in Table 18.1.

Although the criteria above do not distinguish between dry and wet ISC process, the wet process should be contemplated mainly in case of relatively thin formations of good permeability, containing a relatively medium–heavy oil (not very heavy); although only moderate wet combustion is considered here, an operating pressure of at least 1.8 MPa is recommended (to allow a continuous sustainment of the process even in some areas where WAR would exceedingly increase (due to some local conditions). In case oil viscosity (reservoir conditions) is higher than 1000–1500 mPa·s, the local preheating may be necessary before the ISC application; this can be

TABLE 18.1 Screening Criteria for Application of ISC

Criterion	Heavy Oil Reservoirs	Very Light Oil Reservoirs
Fracturing	No	No
Presence of gas cap	No	No
Net pay thickness, m	>3	>3
Permeability, mD	>100	>5
Oil transmissibility	16 mD m/mPa·s	–
Oil viscosity, cp	In the range of 60–10,000	<2
Porosity, fraction	>0.18	>0.10
Oil content (ΦS_o)	>0.07	>0.07
Current oil recovery, %	<20	<10
Reservoir temperature, °C	–	>90
Presence of bottom water	No	–

S_o, oil saturation, fraction.

accomplished by cyclic steam stimulation (CSS). Otherwise, serious difficulties may be experienced with the injectivity and the process may be compromised.

For the light oil reservoirs, the above criteria are based entirely on the Williston Basin, USA, high-pressure air injection (HPAI) commercial operations (Turta, 2001—with some modifications), as applications elsewhere, at different conditions, met with limited success.

Actually, it is important to mention that for both these two categories of oils (heavy and very light), application of ISC as a commercial tertiary EOR method after waterflooding has not been proven yet. There have been some commercial and semicommercial operation, respectively (ex: Videle, Romania, Machedon et al., 1993) for heavy oils and (ex: Madison, USA, Erickson et al., 1993) for very light oils, but the economic results were marginal, with air–oil ratios in the range of 3000–6000 sm^3/m^3. Also, the commercial application of ISC for exploitation of oil sands has not been proven yet.

18.2.2 Monitoring and Evaluation of an ISC Pilot/Project

The question most frequently asked about an ISC project is where the ISC front is located, and if the ISC front arrival to a certain production well is somehow heralded by a change in a property of a fluid, or other means, before the temperature gets high values and the production well can be damaged. Unfortunately, there are some partial responses and a proper analysis of field operations depends heavily on the engineering interpretation of routine project data and the previous ISC experience, which is extremely important for that. Some responses will be provided in this section; the equipment used in the ISC project, the procedures used for the treatment of the effluent, and the safety measures are outside the scope of this chapter/book. Details on these topics can be found in White and Moss textbook (1983).

Before starting an ISC pilot, a gas injectivity test is useful. Gas injection tests are easy to conduct and simple to interpret; operating time is generally short and portable compressors can be used to this effect. At the end of the injectivity test, the operator has information on wells in communication, directional gas flow, air permeability, expected injection pressure, etc. When conducted in a pattern, generally, it is desirable to have at least half of the production wells in communication; otherwise the pilot may not have a lot of chances.

Ignition is a key-phase and it is crucial for the success of the project; actually, knowingly or unknowingly, numerous ISC pilots failed due to improper ignition operations. In the first period of ISC development—probably decades of 60 and 70—this was so for some cases where the ADs have been used for ignition, but later on the ADs were significantly improved and a clear/crisp ignition was obtained, as a rule. However, later on, this problem has lingered/remained in case of SI, chemical ignition, and steam-based ignition methods. In these last cases, the ignition times can be longer and the estimate of ignition delay can be more subtle; in some cases, actually, the operators did not rigorously determine if they got the ISC front or not, and they injected a very large cumulative of air just to finally see that they had a simple gas injection; the oil mobilizing effect of an ISC front was missing. *In other words, the incremental oil due to ISC appears only after the ISC front (with a high peak temperature) is established.* In these cases, the evaluation of ignition should be made based on the variation of the apparent atomic hydrogen–carbon (H/C) ratio, as determined from the complete combustion gas analysis.

Burning efficiency (E_B) refers specifically to the percentage of the oxygen taking place in high-temperature oxidation (HTO) reactions in the ISC front, and as a rule is calculated from the produced gas analyses when a stabilized composition was obtained (steady-state conditions). The burning efficiency is different from the oxygen utilization efficiency (EO_2), which is calculated using the equation

$$E_{O_2} = (O_{2\,\mathrm{inj}} - O_{2\,\mathrm{prod}})/O_{2\,\mathrm{inj}}$$

$O_{2\,\mathrm{inj}}$ and $O_{2\,\mathrm{prod}}$ are cumulative amounts (sm^3 or scf) of injected and produced oxygen. The EO_2 parameter is *unanimously* accepted for evaluating the ISC efficiency in the conventional ISC, but for some types of ISC processes, it may not correctly reflect the efficiency of process. In other words, the burning efficiency *may no longer be synonymous to oxygen utilization efficiency*. It has been proposed (White and Moss, 1983) to use a more comprehensive approach, which takes into account the oxygen consumed in the LTO reactions. The approach is to calculate the ISC burning efficiency (E_B) as follows:

$$E_B = E_{O_2} \times F_B$$

F_B is the fraction of air *reacting in the reservoir* that is consumed in the HTO zone (fraction of air taking part in the burning process).

$$F_B = 1 - [(m+1)(H/C - (H/C)_{CT})]/[2 + 4m + H/C(m+1)]$$

where m is the CO_2/CO ratio, H/C is the apparent H/C ratio in the field (steady conditions), $(H/C)_{CT}$ is the apparent H/C ratio in the laboratory CT test (steady conditions), conducted using the specific reservoir rock and oil.

Similarly, the fraction of air reacting in the reservoir that is consumed in LTO reactions, or (fraction of air taking part in the LTO process) (F_{LTO}) is defined as

$$F_{LTO} = (1 - F_B)$$

Experience shows that for an ISC project where the burning is very vigorous, we may use just the old EO_2 parameter, but when there are some questions about burning efficiency, or too much LTO activity, the use of the parameter E_B is an absolute must.

Generally, the arrival of the ISC front at a production well is not correlated with a change in gas composition (for instance, the increase of nonsaturates in the combustion gas). A production well is considered intercepted by the ISC front (located in the burned out zone) only when the BHT increases significantly (an increment of at least 50–60°C) and the percentage of O_2 in the produced gas increases significantly (more than a few percentages). For deep projects, only the BHT measurements may be useful; the surface temperatures may not sense anything. The routine analysis of the water and oil is important mainly in the second half of the project; sometimes, a specific change in pH of the water may happen (Chattopadhyay et al., 2002, 2003, 2004).

Methods for the location/tracking the ISC front are different for the projects operating in patterns and those operating in a line drive mode. In a pattern operation, frequently it is difficult to determine the position of the front inside the pattern, while in line drive, the method is based on finding "intercepted production wells."

For an isolated ISC pattern, the following methods can be used:

1. Based on routine analyses (gas analyses and BHT)
2. Backflow of the injection well
3. Buildup pressure analysis
4. Seismic methods.

All these method can be applied in contiguous patterns, but their results are less reliable. The backflow of the injection well is based on calculation of the gas volume in the burned zone but is not recommended as it can lead to explosions. The buildup pressure analysis in fact is a falloff pressure test analyzing the profile of pressure decrease during the shutoff of the air injection; this method can have good results when it is known that the ISC surface is close to a frontal front (quasi-vertical), in situations of thin formations and/or not so viscous oils, etc; it gives an estimate of the permeability in the burned zone, as well. More details are provided elsewhere (White and Moss, 1983).

In a line drive operation, such as in a large commercial operation, the method based on routine analyses (gas analyses and BHT constitute an easier and more reliable method). An example of ISC front contour is given later on, in Figure 18.27 for Suplacu de Barcau project. The seismic methods can be applied both in patterns and in a line drive; the results are more reliable for dry ISC, as compared to wet ISC.

Once the position of the ISC front is approximately known, remedies for correcting the ISC front advancement are thought, as follows:

- Reducing the produced gas rate (choking the well) and/or oil production rate (including the shut in) for an offending producer—with too much gas channeling, for instance—works but only to some extent
- For reservoirs having an oil viscosity >2000 mPa·s, preheating of the region around producers using extensive CSS or applying cyclic ISC (CISC) sometimes succeeds in connecting the well with the main ISC front.
- Attracting the ISC front by reverse ISC, sometimes works but this can be applied only in reservoirs of low temperature.

As the methods listed are only partially effective. it is important to emphasize the importance of the injectivity test and its indications on directional communications; the pilot should be started only when there is enough communications in many directions; preventative measures are more important than the corrective ones.

Finally, the geometrical conformance of the burned zone is based on all the information collected during the process. In addition, the information acquired at the end of the process, from coring wells drilled in the burned zone can be very useful; not only the extracted cores analysis is important, but the neutron logs and induction logs also can indicate the thickness of the burned portion, generally located toward the top of the layer.

Location of the coring wells has to be done with maximum of attention to details. Coring wells are important for determination of the gravity segregated character of the process-conformance factor. Determination of the volume burned or even the fuel deposit/m^3 rock are auxiliary. Figure 18.17 shows the results of Suplacu de Barcau coring well K2 and K3; sometimes the coke band between burned zone and unburned zone can be very thick (up to 2.5 m in well K3).

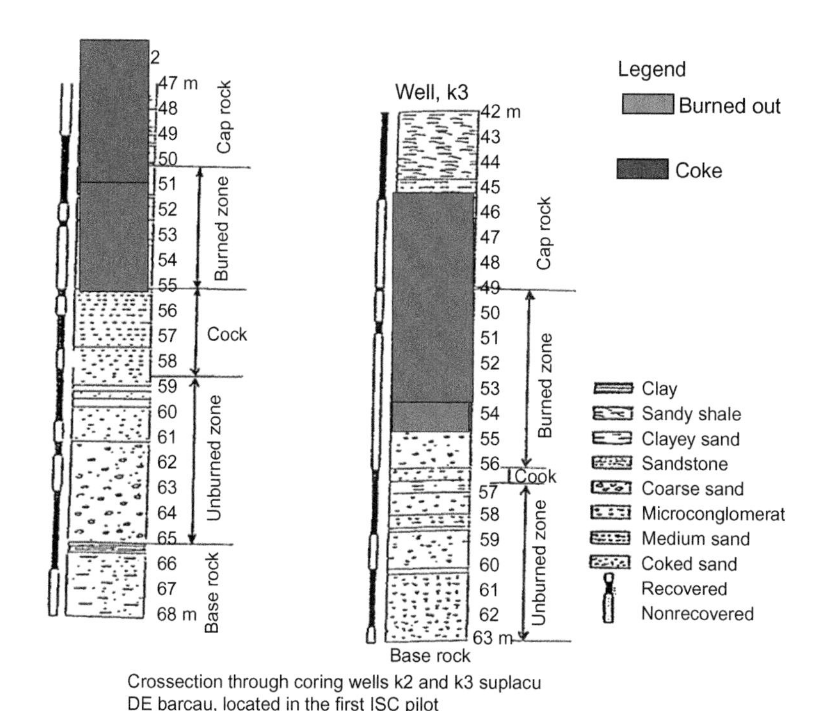

FIGURE 18.17 Burned thickness in the coring wells K2 and K3 Suplacu de Barcau, drilled in the experimental pattern.

The most important challenges in the assessment of an ISC test are evaluation of AOR and incremental oil recovery by ISC (confined and unconfined testing area). The evaluation of incremental oil recovered due to ISC is the hardest one and many pilots were not converted to commercial operation due to the reliability matters related to this evaluation. Examples of evaluation of incremental oil were provided in the previous section, both for line drive, contiguous patterns, and isolated patterns.

18.2.3 ISC Pilots

Information from the Most Instrumented Conventional ISC Pilots

The analysis in this section will be focused not only on the most instrumented pilots, but also on those which are well documented in the technical literature, as well.

Heavy Oil Reservoirs

Dry ISC: Most information came from this category of pilots. Next we will analyze a few pilots in which at least two coring wells and one to two observations wells were drilled in order to get direct information relative to the maximum temperature recorded and/or information related to the conformance factor/volumetric sweep efficiency.

For Suplacu de Barcau segregated ISC process (Tables 18.6 and 18.7), a maximum temperature of 620°C was recorded in a production well at the top of formation. This was confirmed by a Bayou State Oil Corporation (BSOC) project in Bellevue. The coring wells have revealed sometime a very thick band coke (2.5 m in K3 well Suplacu, see previous section). However, in tighter rock of lower thickness, Balaria project (Table 18.4), this cock band has not been identified, when two coring wells were drilled in the burned out zone. It seems that other projects did not find this band, either. Probably, the coke band may be an indication of segregated ISC, which presumably did not occur in the last two cases; however, this is not fully understood.

Another high maximum temperature (649°C) was recorded in S.E. Kansas project (Emery, 1962), although this was conducted in a thin layer (3 m) containing a relatively medium oil (70 cp) in which oxygen utilization efficiency was only 80%. In this project, a vertical-conformance factor of around 80% was revealed by the numerous coring wells drilled in the burned zone.

Another classic segregated ISC process was South Belridge (Gates et al., 1978). Here, as a unique information—from the coring wells—the downward advancement of the tilted ISC surface was followed in time (Figure 18.15); position of the coring wells are also shown in Figure 18.15, where isopach of the burned thickness (at the top of layer) are presented. The maximum burned thickness is 10 m around the injection well of this isolated pattern. A high volumetric sweep efficiency is reported in this project: 25–50%. In all other heavy oil projects, generally, the volumetric sweep efficiency has been less than 25%.

Finally in the North Tisdale ISC test (Martin et al., 1971), conducted in the presence of a bottom water, the observation and coring wells showed that even at an approximately 72% oxygen utilization efficiency, a sustained ISC process was a reality; this pilot was a technical success although economically was at the limit.

Wet ISC: Most reliable information is provided by the Getty and Cities Service companies in Bellevue pilots (Joseph and Pusch, 1980; Long and Nuar, 1982). A direct comparison of dry and wet ISC in two adjacent group of patterns found slightly better results for the wet process (Burger et al., 1985); for the same cumulative of air injected, at least 10–25% more oil production was obtained by wet ISC; therefore, the AOR should have been lower. It is also inferred that the volumetric sweep efficiency by wet process is slightly higher; however, it has never been proved that this directly resulted in a higher UOR.

At Suplacu de Barcau, the wet process—tested in an alternative air–water regime at low pressure—had slightly better results than the dry process; however, the operators did not adopt it due to additional operating problems, mainly the rebellious emulsions generated (Turta, 1977). The composition of the produced gases during the wet process showed a smaller percentage of CO, leading to less environmental pollution; during air injection period CO_2 percentage in the produced gases was in the range of 12–15% and increased slowly up to 18–24% during the water injection period, therefore a slight cyclical variation was noticed. Actually, during the water injection period, the percentage of hydrocarbon gases in the produced gases increased considerably, as well, up to 40–50%. The cyclical variation of gas composition was not so apparent in the high-pressure wet ISC process applied using alternative air–water injection in Balol and Santhal (Turta, 1996).

Light Oil Reservoirs

There is even less information in this area as generally Williston Basin processes have not been benefited of any instrumentation as pilots; the main objective was oil production and only based on production performance the process was developed to field scale.

The most useful information is associated with Sloss, Nebraska light oil (0.8 cp) ISC semicommercial tertiary application in a hot (93°C), sandstone thin formation (5 m) at an extremely low oil saturation (approximately 30%) caused by a thorough previous waterflooding. Simultaneous air–water injection was conducted at an average WAR of 5.6 L/sm^3, but the instantaneous values reached as high as 14 L/sm^3. Five coring wells drilled at the end of the experiment showed that the burned portion was at the top, and indirectly (inferred from mineralogical modifications) a maximum temperature of at least 510°C was reported.

More information can be extracted from the pilots in Delaware Childers, Fry and MayLibby, Delhi Field (Barnes, 1965; Bleakley, 1971; Grant and Szasz, 1954; Hardy et al., 1970). The most important conclusion was that the vertical-conformance factor was very high, close to 100%; consequently, the production wells were generally abandoned once they were intercepted by the ISC front.

Testing of ISC Process for Extra-Heavy Oils with Some Mobility at Reservoir Conditions; PC-ISC (Morgan)

This test was conducted by Amoco at the Morgan Field in the Lloydminster area of Alberta, from 1985 to 1992 (Margerrison and Fassihi, 1994).

The reservoir is unconsolidated sand with a high porosity and permeability, located at a depth of 600 m. Oil viscosity is 6800 mPa·s at the reservoir temperature of 21°C. Net pay thickness is 10 m and the sand is relatively homogeneous. The reservoir was put into production in 1980 and produced significant amounts of sand (1000 m^3/year) indicating the cold heavy oil production with sand (CHOPS) process occurred and suggesting therefore that the reservoir was wormholed. Clear signs of foamy oil mechanisms were the high foaminess of oil at surface and rapid communication between wells during some injection tests.

ISC was tested in nine contiguous inverted 7-spot patterns, occupying a whole section. Initially, in the period 1981–1985, the wells were extensively stimulated using CSS to an average of 5 cycles/well. The steam/oil ratio (SOR) gradually increased from 1 to 4 m^3/m^3. Initially, pure steam was used in CSS operations. However, later on, air was co-injected with steam and the results (in terms of SOR) were better.

Dry ISC was applied by using large periods of (air) injection and noninjection (average injection/noninjection cycle of 1.5 years). In the period from 1985 to 1992, a total of five such ISC cycles were conducted; production wells were always open to production although some producers with too much gas production were shut in toward the air injection period and this caused the injection pressure to vary between 2.8 and 7 MPa, during the air injection period. There are few details on combustion gas composition, but it was stated that the oxygen utilization was 100% (Jensen, 1990a,b).

All in all, the performance was exceptional; the oil rate was in the range of 5–20 m^3/day and the total cumulative oil recovered per well was in the range of 5000–29,000 m^3 at an air–oil ratio of 355 std. m^3/m^3. By 1992, an oil recovery of 23% OOIP was obtained. The reasons for stopping the project are not known.

The most important feature of this project was the production of upgraded oil. Out of 21 production wells (where measurements of density were made), only three or four did not produce upgraded oil. The upgrading intensity was cyclical in nature, following the air–water cycles (Jensen, 1990a,b) with a maximum upgrading during the middle of the noninjection period. The average upgrading was 10 points from 12° to 22°API.

Testing of ISC Process for Oil Sands Reservoirs: Forward and Reverse ISC Tests

For oil sands, both forward and reverse ISC were tested; four tests used the forward process, while two tests used the reverse ISC process.

Special Applications of Forward ISC

Both wet and dry ISC processes were used, but most of the tests utilized moderate wet combustion in applications with injectivity created due to a fracturing process or a pronounced oscillation/pulsing of the injection pressure. The main data for all the pilots is provided in Table 18.2.

Gregoire Lake, Athabasca: A very comprehensive testing of ISC was conducted by Amoco at Gregoire Lake, south of the city of McMurray in Athabasca, during a period of 13 years (1958–1971). This testing involved

TABLE 18.2 Main Data for the Most Significant ISC Pilots in Oil Sands

Field/ Region/ Company	Period of Testing	Type of ISC: Dry (D), Wet (W) or Reverse ISC (R)	Type of Preheating: CSS, or No Preheating (N)	Communication Type: Fracturing (F); Horizontal Fractures (HF); Vertical Fractures (VF); Gas inj. (GI)	Net Pay (m)	Depth (m)/ Res. Temp. T_R (°C)	Permeability (mD)	Oil Viscosity. (cP)/API of Oil	Injection Wells/ Prod. Wells	Injection Pressure (MPa)	Air/ Oil Ratio (Sm^3/m^3)	Oil Recovery %	Observations
Gregoire Lake, Athabasca, Ca/Amoco and Aostra	1958–1971 (6 phases)	R	Air	–		?/15??	High	>1,000,000	1	1		50	Highly unconfined (93% of gas lost)
		W with PC	CSS	F					1	4			VIU (8–12 API)
Marguerite Lake, Alberta Ca/ British Petroleum	1977–1988 (2 phases)	W[1] with PC	CSS	VF	23 (3 layers)	450/15	Coarsening upward	100,000 @ $T_{R/8}$	1	2	2,000	20	$T_{peak} = 600°C$
			CSS	VF					4	9	?	≥60	$T_{peak} = 980°C$ Pronounced override
Wabasca, Jolie Fou, Wabasca, Ca	1981–1983	W	CSS	F		360/22		40,000–100,000		≤10.3		2.7	Highly unconfined (85% of air lost) premature O_2 breakthrough

(Continued)

TABLE 18.2 (Continued)

Field/ Region/ Company	Period of Testing	Type of ISC: Dry (D), Wet (W) or Reverse ISC (R)	Type of Preheating: CSS, or No Preheating (N)	Communication Type: Fracturing (F); Horizontal Fractures (HF); Vertical Fractures (VF); Gas inj. (GI)	Net Pay (m)	Depth (m)/ Res. Temp. T_R (°C)	Permeability (mD)	Oil Viscosity. (cP)/API of Oil	Injection Wells/ Prod. Wells	Injection Pressure (MPa)	Air/ Oil Ratio (Sm^3/ m^3)	Oil Recovery %	Observations
Kyrock, Kentucky, USA[a]	1959–1960	D	–	HF (confirmed after the operation)	30	90??/13	2,000	100,000 @ T_R	1/4	Low	7,500	54	Highly unconfined (36% of air lost) VIU (10–14 API)
Bellamy Field, Montana, USA[b]/	1955–1958	R	–	AI (2 weeks)	2–4	20/13	800	500,000	15/8	0.35	7,400	67	T_{peak} = 454–871°C MU (10–26 API)
N Asphalt Ridge, Utah, USA[b]/LETC	1977–1978	R first, then D	–	AI (several days)	4–6	107/11	85[2]	>1,000,000	6/3	2.5–3.5	25,000	25	Highly unconfined (50% of air lost) T_{peak} = 400–1, 100°C VIU (14–20 API); inconsistent!

CSS, cyclic Steam Stimulation; PC, pressure cycling; VIU, very intensive upgrading; MU, maximum upgrading.

Observations:

[a]*Most instrumented test in the world (35 observation, coring, fracturing wells, and other auxiliary wells); reservoir in communication with an outcrop*

1. Moderate wet ISC with alternative air–water injection having very long cycles (1–3 months air injection and 0.5–1 month water injection.)
2. For the oil-saturated core.

[b]*Line drive operation (one production row in-between two injection rows); very small distances between wells (2–5 m for Bellamy and 7–20 m for N Asphalt Ridge) and very short tests: 9 and 180 days, respectively.*

six phases of increasing complexity (Giguerre, 1977). The bitumen is totally immobile at reservoir conditions.

Initially in 1958 and 1959 (Phases 1 and 2), a pair of wells, one production and one injection, were located at an offset distance of 33 m. First the production well was ignited and reverse combustion was sustained for a limited period. This was followed by direct forward combustion. The oil rate reached 4 tons/day and the total oil produced was 70 tons. No fracturing was necessary in this case, as low air fluxes were used.

In Phase 3 (1960–1961), the distance between the two wells was increased to 80 m in an area where the bitumen saturation was very high. Air injection was made after fracturing. In Phase 4 (1963–1965), several wells were utilized for wet combustion testing, where the formation temperature was always maintained above 93°C around each production well probably by using CSS. The maximum oil production from this test was 21 tons/day. No other details are available.

The most important was Phase 5 (1966–1968), in which wet combustion was tested in a very small 0.4 ha inverted 5-spot pattern with observation wells used to monitor the process. Air injection was made at a pressure higher than the fracturing pressure. The most significant event was that ignition was adapted to the fractured rocks by heating the perforations for a very long, 7 month period. At the end of this period, a temperature increase was seen in some production wells. This extensive preheating was followed by a reservoir depressurization (blowdown) period of 4 months. Then wet combustion was continued for 6 months. It was observed that the produced oil was upgraded and had a viscosity 100 times lower than that of the original bitumen and the API of the oil increased from 8 to 10–14°API. In this test, an amount of oil equivalent to approximately 50% of the pattern oil reserve was recovered.

Phase 6 (1969–1971) consisted of an inverted 4 ha 9-spot pattern with some observation wells for monitoring. The results were not encouraging due to the large pattern area.

Based on knowledge obtained in the period from 1958 to 1971, Amoco envisaged the use of wet combustion in small patterns (1 ha/pattern) in future applications with the use of the following phases:

- Forced forward combustion (with air injection above fracturing pressure).
- Blowdown (no injection, only oil production).
- Wet combustion displacement phase.
- Blowdown (no injection, only oil production), etc.

In fact, in 1976 Amoco and AOSTRA agreed to test this technology on a large scale in nine contiguous-spot patterns of 1 ha/pattern. However, this application was canceled prematurely because of some difficulties with casing damage; therefore, the possibilities for evaluating oil recovery and global efficiency of the process were eliminated once the scale of piloting was significantly reduced (Marchesin, 1982).

Marguerite Lake, Canada: The most comprehensive testing was conducted at Marguerite Lake, Wolf Lake region of Alberta, from 1979 to 1988 (Hallam and Donelly, 1988).

The reservoir is a very fine to fine-grained sand, being divided into three layers C_1, C_2, and C_3 with C_3 at the bottom; there is a general coarsening upward trend. The net pay thickness is 23 m. Oil viscosity is 100,000 mPa·s at the reservoir temperature of 15°C (Table 18.2).

From 1979 to 1982, a three well pattern of two producers, 100 m apart, and a middle injector was initially stimulated using CSS; during the steam injection fracturing of the formation occurred. This was followed by moderate wet combustion. During the first test (north area—3 wells) 3700 m^3 of oil corresponding to 20% OOIP was produced at an AOR of 1400–2000 std. m^3/m^3. Oil rates were in the range of 1–3.5 m^3/day/well. The 20% oil recovery should be considered cautiously as the pattern was unconfined. A peak temperature of 600°C was recorded in an observation well.

In the second period (1983–1985), an extended testing of moderate wet ISC was made in four contiguous inverted 5-spot 4 ha patterns located close to the initial testing area. All the wells were first stimulated using CSS but it is not known for how long and how many cycles were performed. Steam injection was conducted above the fracturing pressure of the formation. The maximum oil production from this test was 21 tons/day. No other details are available. Some communication paths and heated zones were created before wet ISC was started in the main area (Hallam, 1991).

In the third period (1987–1988), testing of the moderate wet ISC with oxygen was made in the same four contiguous inverted 5-spot patterns. The process applied was very close to Amoco's process developed at Gregoire Lake. The authors indicate that vertical fractures are formed during CSS (Hallam and Donelly, 1988; Mehra, 1991).

Moderate wet ISC was practiced in a WAG mode, using very long periods of air and water injection such as 1–3 months for the air injection and 0.5–1 month for the water injection. WAR was in the range of 2–2.7 L/m^3. During the air injection period, the observation wells located on-trend showed a maximum temperature of 980°C, which decreased to 200°C during the water injection period. The temperature profiles showed a pronounced override, both for CSS and for ISC. The ISC front started in the C_3 layer and migrated to the upper C_2 layer.

The main shortcoming of the test was the pronounced performance decline of the test area (Mehra, 1991), which was linked directly to a very low degree of control of the ISC process.

Joli Fou, Wabasca, Canada: The reservoir contains a bitumen with a viscosity of 40,000–100,000 mPa·s at reservoir temperature (Table 18.2).

Wet combustion was tested in a large inverted 7-spot 8 ha pattern with one observation well (Alderman et al., 1983) starting in 1981. Air and water injection were conducted in parallel with a preheating process carried out with CSS, which was neither very intensive nor very effective. The injection pressure, 10.3 MPa, during CSS was close to the fracturing pressure and it is believed that formation fracturing occurred.

Probably, due to fracturing of the formation, the test finished prematurely within 2 years because of oxygen breakthrough. Moreover, a massive loss of the injected air occurred as only 15% of it was recovered. The unconfined nature of the pattern was a serious problem.

KYROCK Project, Kentucky, USA: The reservoir is in communication with an outcrop. The oil viscosity is 100,000 mPa·s at reservoir temperature (Table 18.2).

Dry ISC was tested in a very small 0.14 ha inverted 5-spot pattern (Terwilliger, 1975). The test pattern is shown in Figure 18.18. *This has been the most instrumented ISC test in the world, as 35 auxiliary wells (fracturing wells, observation wells, temperature wells, and coring wells) were drilled besides the five normal wells of the pattern.* The main feature of the test was that before starting ISC, the region surrounding the production wells was fractured using the specially designed horizontal fracturing wells located close to the producers. In reality, the producers were drilled after the fracture was executed and this way *the correctness of the horizontal fracturing was verified.*

This low-pressure ISC process began in September 1959 with a strong artificial ignition (gas burner). The whole test lasted 5 months (October 1959 to February 1960); approximately 500 m^3 of oil was recovered, equivalent to 54% of OOIP inside the pattern. This oil was obtained at an AOR of 7500 std. m^3/m^3. The produced oil was substantially upgraded, from 100,000 to 2000 mPa·s, corresponding to 4° of upgrading (10.5–14.5°API).

The main difficulties of the process were:

- Pronounced asymmetrical advance of the ISC front in the NW–SE direction.
- Relatively low oxygen utilization efficiency due to the fractured nature of the pattern.
- Loss of an approximately 36% of the injected gas through the outcrop.

The first problem was partially solved by applying reverse ISC to the offending producers (isolated, noncommunicating producers) and by restricting the flow to the prolific producers. The second problem was not solvable and the producers were shut in when oxygen percentage in the produced combustion gases reached 6%.

A typical temperature profile is shown in Figure 18.19 and it demonstrates a relatively well spread zone of at least 2–4 m at a temperature higher than 538°C.

The final volumetric sweep (isopach map) is presented in Figure 18.20. After the completion of the test, 12 coring wells were drilled in the pattern.

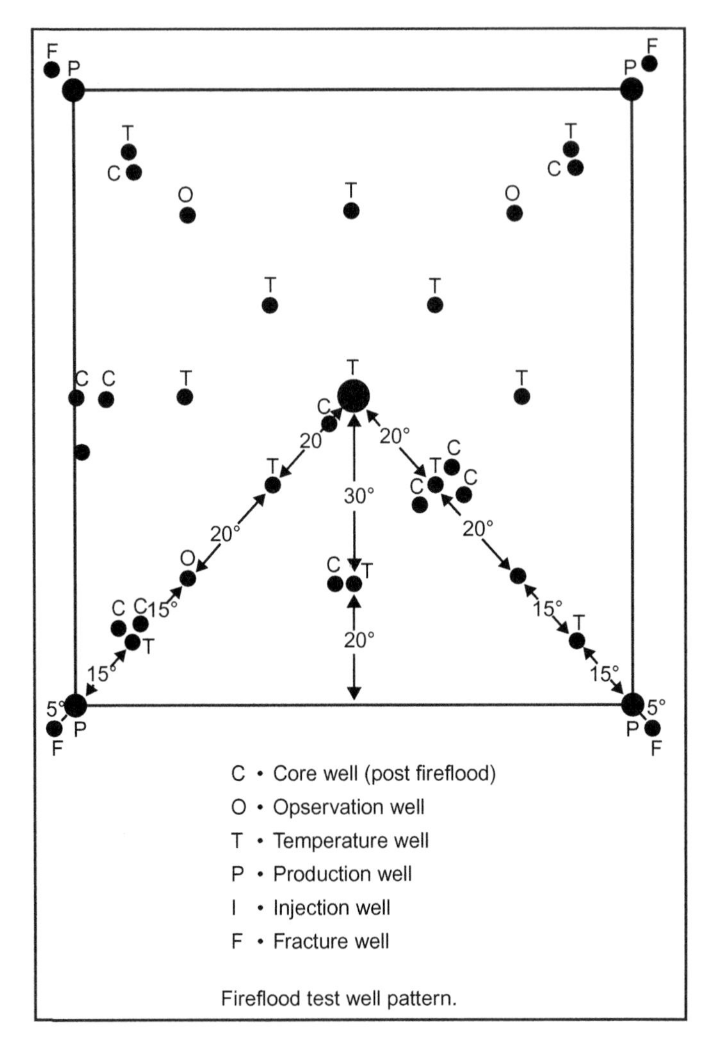

FIGURE 18.18 KYROCK Oil Sand ISC Project: Arrangement of wells (Terwilliger, 1975) (testing of fracture-assisted ISC).

It was found that around the injection well (5–6 m, distance) almost the entire pay thickness was completely burned (Figure 18.31). Also, it was found that *even in the presence of a horizontal fracture, the ISC front had a tendency to migrate toward the top of the formation.*

Reverse ISC

The two tests presented here are the only complete reverse ISC tests ever conducted. Therefore, there is extremely limited experience with the field

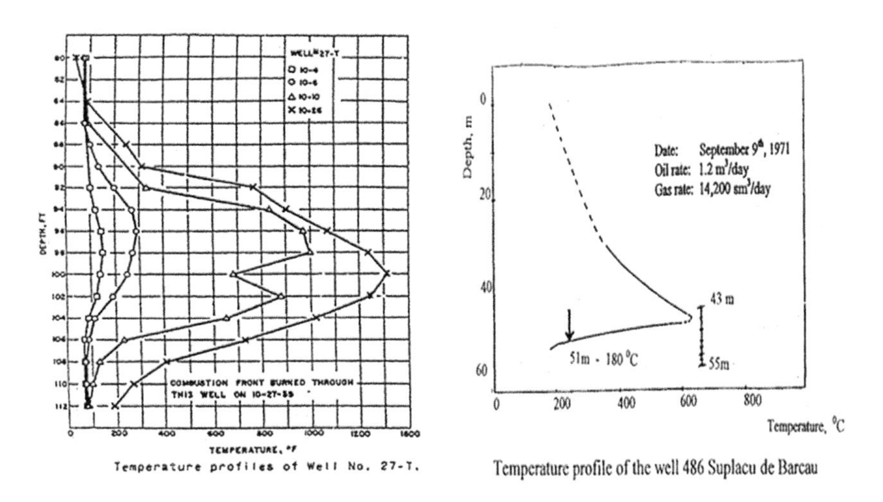

FIGURE 18.19 Temperature profiles in an observation well in Kyrock Project and in a production well at Suplacu de Barcau.

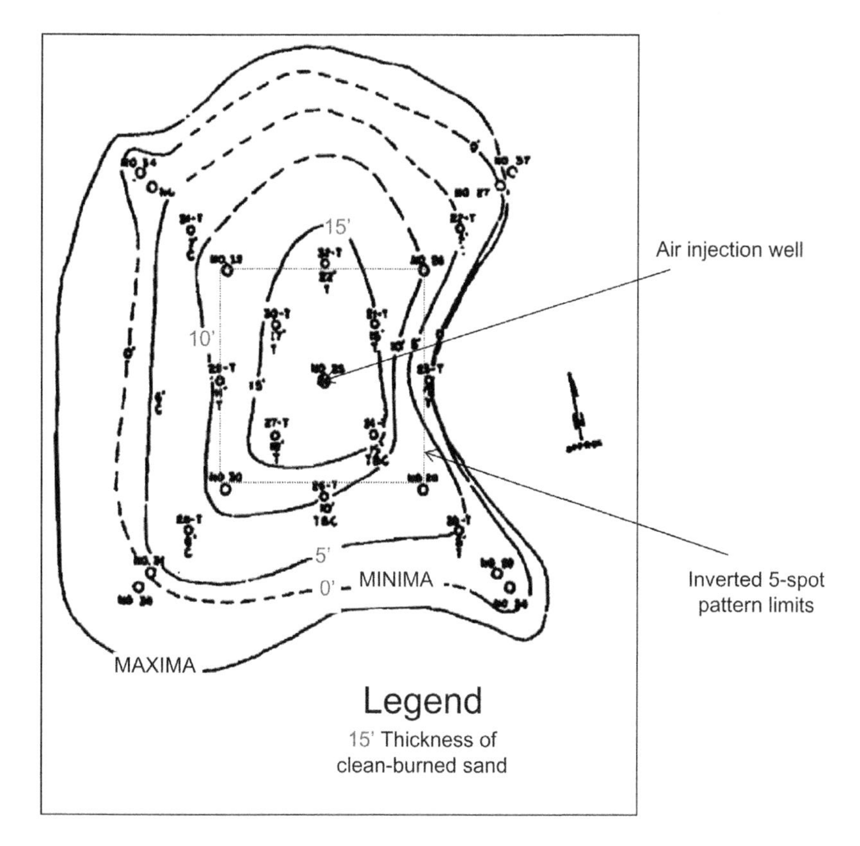

FIGURE 18.20 KYROCK ISC Project (fracture-assisted ISC) (Terwilliger, 1975). Isopachous map of the burned zone.

application of this process. These two tests were conducted in a thin layer using a line drive configuration, both of them having a production row between two injection rows. *The use of a greater number of injectors compared to the number of producers was an essential characteristic of both reverse ISC pilots.*

Bellamy Field, Montana, USA: This is a low-pressure test (0.35 MPa) in a thin formation (2–4 m). Bitumen has a viscosity of 500,000 mPa·s at the reservoir temperature of 13°C (Table 18.2).

A line drive was formed by 15 injection wells arranged in two parallel rows and eight production wells in a row in-between the injection rows (Trantham and Marx, 1966). The distance between rows was 5 m while the distance between two consecutive wells in a row was 2 m. This was a well instrumented test as 20 observation and coring wells were drilled.

Before starting ISC, the region surrounding the production wells was dried out in 2 weeks by injecting air into the central wells with all other wells shut in. This procedure significantly reduced the amount of water within the confinement of the patterns.

The reverse ISC process began with artificial ignition of the five production wells; details on the special ignition can be found elsewhere (Trantham and Marx, 1966). The whole test lasted 9 days, finishing when the thermal fractures from the advancing ISC front reached the injection wells leading to oxygen breakthrough. The composition of the combustion gas produced was steady and normal and showed hydrogen percentages of 0.4–0.7%.

During this pilot, an amount of oil equivalent to 67% of OOIP (from in-between the two injection rows) was recovered, at an average AOR of 7400 std. m^3/m^3. The produced oil showed maximum upgrading from 500,000 to 10 mPa·s, corresponding to 16 points of upgrading (10–26°API). A light oil was produced, with a viscosity of 5–15 mPa·s.

During the test the maximum temperatures recorded in the observation wells were in the range of 454–871°C. As oil is produced through the burned zone, which has a very high temperature, the temperature of the production wells was maintained in the range of 200–450°C, by injecting water intermittently.

After the completion of the test, coring wells drilled in the burned out zone of the pattern found that the entire pay thickness was completely burned; due to the very high temperatures, local fracturing occurred, causing a substantial increase in average permeability (20 times).

Northwest Asphalt Ridge, Utah, USA: This is a relatively high pressure test (2.5–3.5 MPa). A two-phase ISC process was tested with a Phase 1 reverse ISC and a Phase 2 forward ISC, started when the reverse ISC front reached the air injection wells. The authors call this process echoing ISC. The entire design of this test was based on a previous Energy Research and Development Administration (ERDA) test (Arscott and David, 1977).

The reservoir is located at a depth of 107 m with a permeability of 85 mD for the bitumen-saturated cores. It contains bitumen with a viscosity

$>$1,000,000 mPa·s at the reservoir temperature of 11°C. Pay thickness is 4–6 m (Tables 18.3 and 18.4).

The process was tested in the period 1977–1978; a line drive was formed by six injection wells arranged in two parallel rows and three production wells arranged in a row in-between the injection rows (Johnson et al., 1980). The distance between rows was 20 m, while the distance between two consecutive wells in a row was 7 m (pattern area = 0.04 ha). Thirteen observation wells and seven coring wells were drilled. Well layout is shown in Figure 18.21.

Before starting ISC, a preliminary air injection was conducted for several days in all six future injection wells without any fracturing required. This preliminary injection roughly established well-to-well communications.

The reverse ISC process started with artificial ignition at the central production well (well 2P2 in Figure 18.21). When, after 20 and 47 days, respectively, the reverse ISC front intercepted the other two production wells (2P1 and 2P3), these wells were considered ignited and converted to production wells. This was considered the start of the whole reverse ISC front with propagation toward the six lateral injectors where the air injection was initiated. Approximately 90 days after ignition, the test was converted to forward ISC when the front reached the injection wells. The whole test lasted 180 days, therefore, half the test was reverse ISC and half was forward ISC.

During the reverse ISC phase, there were periods of out of control operation, with the process reverting from reverse to forward ISC. This may explain why the composition of the combustion gas produced fluctuated wildly. Hydrogen and monoxide carbon reached values of up to 14%, methane reached values of up to 34%, and CO_2 increased up to 28%. Oil production also had a wide fluctuation.

During this test, 92 m^3 of oil, equivalent to 25% of OOIP was recovered. The oil was obtained with a water cut of approximately 85% at an average AOR of 25,000 std. m^3/m^3; half of the injected air was lost out of the pattern. Oil was produced both during reverse and forward ISC phases and showed some upgrading but was inconsistent. On average it was upgraded from 14 to 20°API. In practicality, a light oil and a heavy oil were produced but details are missing. This average reflects a combination upgraded light oil and slightly downgraded heavy oil (higher asphaltene content compared to the original oil).

Maximum temperatures recorded in the observation wells were in the range of 150–400°C for the reverse ISC phase and 540–1100°C for the forward ISC phase.

After test completion, coring wells drilled in the burned out zone found an average coke band 0.5–0.7 m thick; in a certain region a large 4 m thick coke band was formed. Based on data from coring wells and observation well (temperature profiles), it was determined that the volumetric sweep efficiency was considerably better for reverse ISC (86%) than forward ISC (33%).

TABLE 18.3 US Past Commercial ISC Processes (Situation as of 1994) Turta, 1994

Field, Company	Depth (ft)	Permeability (mD)	Oil Viscosity (cp)	Injection Wells	Prod. Wells	Daily Oil Production by ISC (BOPD)	Air/Oil Ratio (scf/bbl)	Injection Pressure (psi)
West Newport, Mobil[a]	1,600	750	750	36	139	980	10,700	200
Lost Hills, Mobil[c]	300	1790	410	7	45	520	6,200	–
Midway Sunset, Mobil[a]	2,700	1,500	110	3/up	31	900	6,700	800
Midway Sunset S FE Energy[a]	1,700	1,300	5,000	10	40	700	–	–
South Belridge, Section 12,M.[c]	1100	3000	1600	2	?	900	6,000	>500
Bellevue, Texaco[c]	400	650	660	15	85	420??????	16,300	250
Forest Hill, GO[c]	5,000	950	1,060	21	100	400	–	2,000
Brea Olinda[c]	3,300	300	20	2/up	20	650	7,700	1,000

a, active; c, completed.
GO = Greenwich Oil.
up—ISC process started from the upper part of reservoir.
LD—ISC process using the line drive system.
u-L—ISC process started from the upper part of reservoir, using the line drive system.

TABLE 18.4 Past Commercial ISC Projects Outside the United States (Updated Situation as of 2012) Machedon, 1994, 1999 and Turta, 1994

Field, Company, Country	Depth (ft)	Permeability (mD)	Oil Viscosity (cp)	Injection Wells	Prod. Wells	Daily Oil Production by ISC (BOPD)	Air/Oil Ratio (scf/bbl)	Injection Pressure (psi)
Karazhanbas, Kaz.	1,300	800	300	72/LD	261	7,000[a]	5,600	–
Balahani, Azer.	910	500	140	6/up	35	600	6,700	500
Battrum, Mobil, Can.	2,900	930	70	19	101	3,000?	10,000	400
Morgan, Amoco, Can.	1,940	4,400	8,100	9	35	940	2,000	400–1,000[b]
W. Videle, Sa 3c, Rom.	2500	900	100	19/u-L	50	610	17,000	600
E. Videle, Rom.	2,100	1,200	100	33/u-L	89	660	21,000	700
West Balaria, Rom.	2200	500	116	22	60	820	24,500	850
East Balaria, Rom	1,500	500	416	15/u-L	47	550	22,500	850

Kaz., Kazakhstan; Azer., Azerdbaidjan (former Soviet Union).
up—ISC process started from the upper part of reservoir.
LD—ISC process using the line drive system.
u-L—ISC process started from the upper part of reservoir, using the line drive system.
Note: All presented processes were completed before or by 1999; Battrum and all Videle–Balaria projects were terminated by 1999.
[a] *Estimated.*
[b] *Pressure-cycling ISC.*

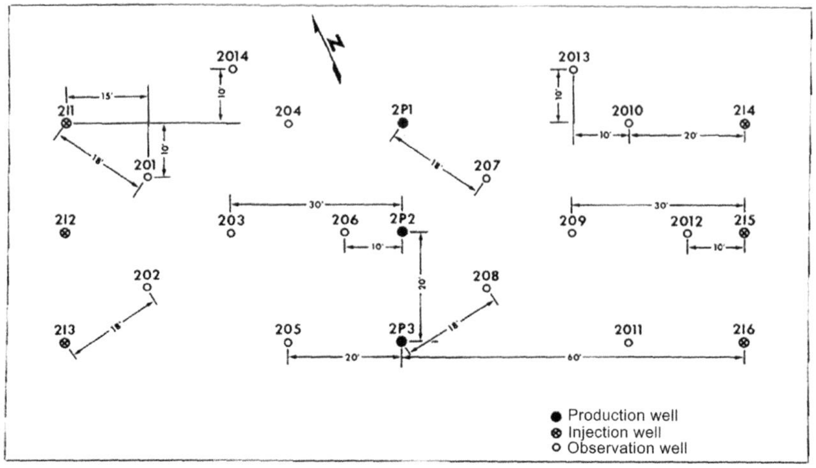

FIGURE 18.21 Northwest Asphalt Ridge Oil Sand Reverse ISC test: arrangement of wells (Arscott and David, 1977).

The most important challenges of this test were the lack of confinement and of control over the ISC front propagation. These caused an extremely high AOR and led to a lack of consistency in oil upgrading.

Commercially Unproved Applications of ISC; Bottom Water Reservoirs and Naturally Fractured Reservoirs

In both these cases, ISC has not attained commercial operation status, but the experience from these field tests is still very valuable, either for ISC application in other areas or in the development of new, more successful procedures, for these two cases.

Bottom Water ISC (BW-ISC): So far, two procedures have been tested in oil reservoirs with bottom water. These are the conventional ISC-pattern application, or the so-called BW-ISC and the basal combustion (BC). The BC process was applied using just vertical wells, for both injection and production. The difference between these two methods is in how the ignition of the formation is conducted. While the ignition is normally done in the upper part of layer in BW-ISC, in basal ISC, this ignition is intentionally done at the water–oil contact with the intent to use the high mobility water zone for oil flow toward the producers.

Two tables presenting the main data and results for nine BW-ISC tests and three BC tests are provided elsewhere (Turta et al., 2009). From the testing of BW-ISC tests, the following conclusions can be drawn:

- Out of nine tests mentioned, the most complete and the most successful ISC project was North Tisdale, Wyoming (four injectors, 15 producers, and operation for more than 9 years), where an amount of oil corresponding to an oil recovery increase from 10% to 50% was obtained at an AOR of 3500 sm^3/m^3. The relatively high AOR is mainly due to a low oxygen utilization of around 72%; however, for BW-ISC operations, this AOR is still the best ever obtained (Martin et al., 1971). It is probable that the low oxygen utilization was due to some injected air entering and flowing through the bottom water zone. The fact that the thickness of the bottom water was only 12 ft (4 m) compared to 50 ft (15.3 m) for the oil zone represented relatively favorable reservoir conditions.

- The Cado Pine test, although incomplete in information, revealed some interesting features. In this test, there was no clear evidence of air bypassing through the bottom water zone (Horne et al., 1981). The pay zone was only 9 m while the bottom water had a thickness of 30 m.
- Evidence of air bypassing through the bottom water zone also did not exist in the third test at Pauls Valley which had very viscous oil (8000 mPa·s) and where the cold heavy oil production with sand (CHOPS) process had been applied unintentionally (Elkins and Morton, 1972). However, the test was stopped after 16 months due to serious operational problems caused by the existence of the wormholes causing mechanical failures of producers.
- Test #4, Zerotin, has incomplete information. In this test, although the reservoir temperature is very low, good ignition and good self-supporting features resulted in good oxygen utilization (88.5%). This may imply that little air bypassed through the bottom water zone. Good oxygen utilization may also be linked to an oil recovery of 25% at the inception of the test, and therefore, probably good air flux. AOR was 2600 std. m^3/m^3 for this single-pattern test which operated for 4 years (Juranek, 1959).
- Test #5, Eyehill, involved nine, 5-spot patterns, with nine injectors and 16 producers, and has been operated for 10 years. The water cut was around 80% at the inception of the ISC test and has receded to 50% but only temporarily for 2–3 years. An attempt to apply CSS was not successful; however, it is not known if this is directly related to the existence of bottom water (Farquharson and Thornton, 1986). Although burning was satisfactory, the performance was mediocre, probably related to the high heterogeneity of this channel type reservoir and numerous operational problems.
- The remainder of the tests do not provide any additional interesting information. Only of note is that based on the Suffield test (test #2) where the antiwater coning technology (AWACT) was developed (Wells, 1980). This technology, although not spectacular, has been proven useful in more than 60 field operations.

Based on the tests analyzed, a schematic of BW-ISC is shown in Figure 18.22. The partition of the injected air between the main ISC front and the secondary one will probably decide the efficiency of the process.

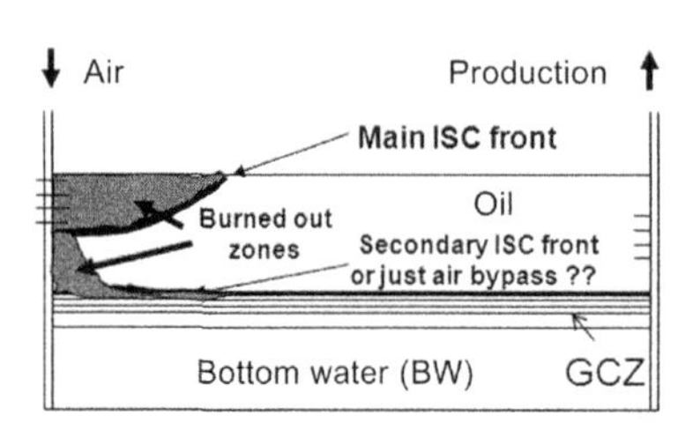

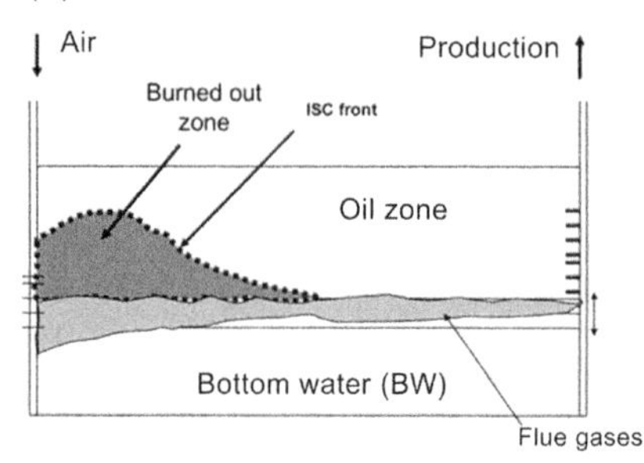

FIGURE 18.22 Simplified schematics for the application of ISC in a bottom water situation: (A) conventional ISC and (B) BC.

For testing of BC, the following conclusions can be drawn (Turta et al., 2009):

- In the Carlyle pool (depth 260 m, oil viscosity 700 mPa·s, and reservoir temperature 24°C), three different pilots were conducted. Only information for two pilots, Wiggins B pilot (dry ISC) and Riggs Pilot (wet ISC), is available. Above the oil/water contact, a streak of lime exists (Elkins et al., 1974). In the Wiggins B pilot, the injection well was perforated above the oil/water contact, but it was drilled through the lime streak into the water zone. After 7 years of operation, three coring wells were drilled into the burned out zone. Although ignition was achieved in the oil zone which is 34 m thick, ISC took place up to 7 m into the water zone. Incremental oil recovery was 31%, while incremental production was obtained at a very high water cut of approximately 90%. The main stumbling block was the very low productivity of the producers, which practically impeded the expansion to a commercial scale. In a reverse ISC operation, SI at the oil–water contact occurred after 8–10 months.
- In the Riggs pilot, the injection well did not go through the lime streak; it stopped in the oil zone so that the ignition was performed by a gas burner at the upper part of layer. In this case, burning was confined to a 3 m interval of high permeability in the oil zone so it was above the oil/water contact. Oil recovery was just 8.5% while the AOR was high at 7600 std. m^3/m^3.
- An attempt to use SI for the creation of the ISC front at the oil/water interface was made in SE Pauls Valley Field, where 10 months of air injection did not succeed at a reservoir temperature of 43°C.
- In the third test, Wabasca (depth 25 m, oil viscosity >1,000,000 mPa·s, and reservoir temperature 13°C), the ISC front was successfully initiated just above the oil/water interface (Thornton et al., 1996). However, in this test an extensive preheating by CSS preceded the ignition, which was

obtained using a gas burner. An upward extending burned zone was formed, while all fluids were produced through the vertical producer. No official performance exists, as the test lasted just 18 days in order to check the ignition. Unofficial information, however, shows that for an additional 4–5 months, the pilot continued with unsatisfactory results.

Based on the tests analyzed, a simplified schematic of the BC is shown in Figure 18.22; in the lower zone, denoted as flue gases, there may or may not be an ISC front depending on the air flux and oil saturation in the transition zone.

It should be pointed out that BW-ISC and BC have been minimally investigated in the laboratory. Greaves et al. (1994) were the only ones to investigate the use of horizontal well assisted ISC for heavy oil reservoirs with extensive bottom water.

ISC in naturally fractured reservoirs (ISC-NFR): ISC-NFR has been investigated very little in the laboratory and in the field.

The first laboratory study on ISC-NFR, involving heavy oil, was conducted in the Netherlands (Schulte and de Vries, 1985). This study was continued in England (Greaves et al., 1991). By processing the results of these two laboratory studies, the following basic conclusions were made:

- To sustain the ISC-NFR process, a preliminary extensive preheating of the air injection end (inlet well), in addition to the local preheating for ignition has to be conducted. In Netherlands the investigators preheated to 473°C, 14% of the length CT (Schulte and de Vries, 1985).
- During air injection after ignition, it is believed that two combustion fronts develop. A rapid LTO quasi-front due to the oxidation taking place on the lateral (and frontal wall in the field) blocks, and a slow HTO front (the real ISC front) inside the block due to penetration of the air inside the matrix.
- Most of the oil was recovered in the first half of the run as the test was usually stopped well before the HTO front reached the outlet. The termination was due to the fact that the oxygen content in the produced gas increased significantly.
- Increased fuel deposit/air consumption, coupled with relatively high hydrogen–carbon (H/C) ratios, with simultaneous high AOR values of 3400 to 11,000 std. m^3/m^3 were measured.

The increased value of air consumption was confirmed by two other studies of ISC-NFR processes done in Russia (Baibakov and Garusev, 1989) and Canada (Bennion, 1986) as routine work before the field testing. Baibakov reported this information in relation to the testing of ISC-NFR in the very heterogeneous rock of Zybza Glubokii Yar reservoir, while Bennion reported this in relation to the testing of ISC-NFR in the Grosmont Carbonate formation at Buffalo Creek reservoir in Alberta.

The ISC-NFR test in the fractured-vuggy Grosmont 2 carbonate formation at Buffalo Creek reservoir was conducted in the period 1978–1979 (Bennion, 1986). Oil viscosity is 100,000 mPa·s under reservoir temperature of 17°C. It is not recorded if the formation was karsted or nonkarsted (Roche, 2006). Two very short-term ISC tests were conducted: one in 1978 and the second one in 1979. The 1978 test was a 3-month test in which it was seen that the most difficult aspect was the massive loss of air injected to the Grossmont 3 formation positioned above Grosmont 2 formation in which the ISC was tested.

The 1979 test was a 6-month test in which two main periods of air injection were recorded. The ignition was done again using a gas burner and all six involved wells were preheated using CSS operations leading to a BHT in the range of 80–120°C. After ignition, oxygen breakthrough occurred very rapidly (in 10 days) in one production well located at 150–200 m, causing it to burn out. A second production well experienced temperature increase and was shut. The most difficult and intractable problem was the massive loss of air injected to the upper Grosmont 3 formation. It was estimated that approximately 50% of the air injected was lost from the area of interest. Another reason for failure was the use of an initial air injection rate of 38,000 std. m^3/day, which was considered far too high for the reservoir conditions. However, all aspects considered, the test showed that oil can be produced from bitumen carbonates. The five wells involved in the second ISC test produced 3500 m^3 oil due to both CSS and ISC operations.

The most comprehensive testing of ISC-NFR in a very heterogeneous rock comprising microfractures was conducted in Russia at Zybza Glubokii Yar reservoir in the period 1972–1973 (Baibakov and Garusev, 1989). Oil viscosity was 2000 mPa·s at a reservoir temperature of 29°C. Eighteen wells were involved, including two observation wells located close to a line of three air injection wells. Many difficulties were recorded during gas-burner ignition operations as no preheating was conducted and injectivity consistently decreased after ignition. Finally, it was decided to change to an electrical heater and extend the ignition operation up to 37 days in a form of preheating during ignition operation. This succeeded in creating a vigorous ISC front which propagated normally for the first 5 months. There were no BHT measurements in the production wells or observation wells. As shown in Figure 18.23, the oxygen concentration in the produced gas increased slightly from 0% to 5%, while the CO_2 content decreased from 15% to 10% under a constant injection pressure and air injection rate. When the test was stopped 12 months after ignition, the oxygen concentration was over 10% for all four production wells. There was literally no oil production, although the ISC process ran normally for about 5 months. The results were considered *totally negative* and the authors concluded that the application of ISC for recovering of oil from fractured heavy oil reservoirs was not possible.

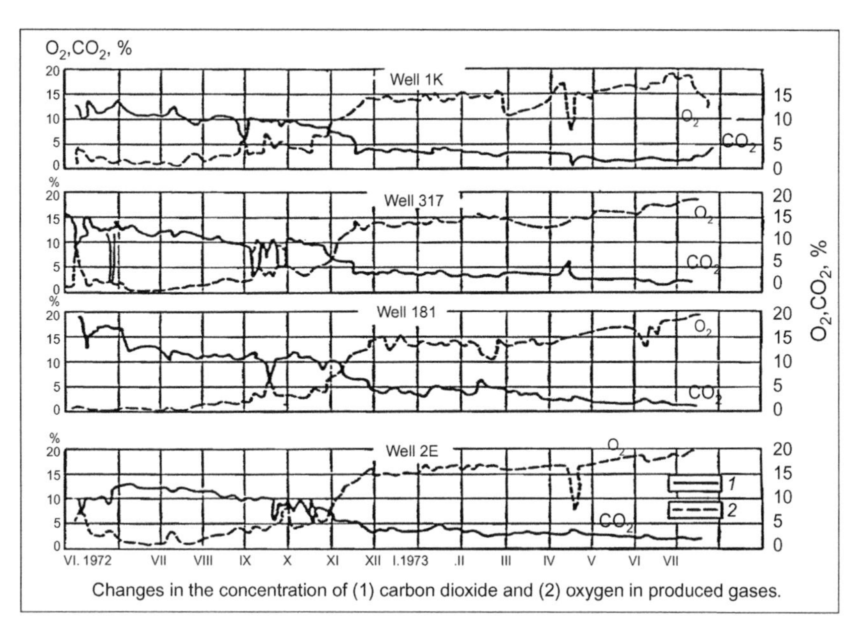

FIGURE 18.23 Combustion gas composition for a fractured formation ISC process: Zybza Glubokii Yar Reservoir (former Soviet Union) Baibakov and Garusev, 1989.

Based on the experience acquired so far, it can be concluded that application of a conventional ISC process has little potential in naturally fractured reservoirs.

18.2.4 Commercial ISC Projects in Heavy Oil Reservoirs

The World's Most Significant Past *ISC Projects*

Limited information is available on the commercial past ISC projects. Although not published at all, a relatively large commercial ISC process had been in progress for a long period of time in Albania, in Drisa reservoir; this process involved more than 40 air injection wells (Gjini et al., 1999).

For the listed commercial processes, the viscosity is in the range of 5–8000 mPa·s and the conclusions of the analysis in its entirety refer to this range. What is stated here applies to the ISC applications in the case of recovery for oils with some natural mobility at reservoir conditions. The main characteristics of the project and their results are provided in Table 18.4. The process can be applied in a wide range of depths, from shallow to deep reservoirs, and to a wide range of permeability.

Out of the listed projects, nine were started from the uppermost part of the structure, with seven of them operated using a line drive well configuration.

The most important parameters, indicative of economic efficiency, are AOR and injection pressure. For the same value of AOR, lower injection pressure means better economics. The AOR was in the range of 1000–4500 sm^3/m^3 (6000–25,000 scf/bbl) for injection pressures of 1.3–6.4 MPa (200–900 psi).

Dry ISC Processes

Moco Zone in Midway Sunset Field: This is one of the earliest examples where the air injection wells were positioned upstructure *to take full benefit of the effect of gravity*. The reservoir is an anticline with a dip in the range of 20–45° at a depth of 700–900 m; there are six major sands which have been exploited comingled (Table 18.5). The ISC was started when primary recovery was 17%. ISC initiation was by enhanced SI through five wells and air injection was conducted in all six sands; 67 producers were used in this process. The process was operated more than 20 years with an AOR of 2900 scf/bbl (550 sm^3/m^3) and UOR predicted at 45% (Gates and Sklar, 1971).

Very Thin Layer ISC Projects (Trix-Liz, Glenn Hummel, Gloriana and Casa Blanca, Texas, USA)

There were four projects conducted, in very thin layers, as low as 2.8 m; although they have only essential documentation, all of them lasted for a period of more than 5 years, generally with good results (Buchwald et al., 1972; OGJ February 17, 1969). Some reservoir properties and essential results are provided in Table 18.5. The following conclusions can be drawn from these tests:

- Oil was medium to heavy, with a viscosity in the range of 26–174 mPa·s; reservoirs are located at a depth of 500–1000 m.
- The tests were all of small size (2–3 injection wells, located at the upper part of structure); they were started in 1968–1969 by the same company; the duration was short, in the range of 3–5 years, at the time of evaluation.
- The ignition operations were done using very different techniques (electric, gas burner, spontaneous, and chemical), but in all cases were successful, as no problems were reported and oxygen efficiency during the process was in the range of 87–98% for all these projects. The reservoir temperature is in the range of 38–59°C, therefore being a favorable factor.
- In all cases, the process was applied in small stratigraphic traps (sand lenses) of low dip, with a high degree of confidence in the evaluation of

TABLE 18.5 Texas, USA: ISC Projects Within Thin Pay Layers

Field Company	Net Pay (m)/Gross Pay (m)	Depth (m)	Reservoir Temperature (°C)	Permeability (mD)	Oil Viscosity (cP)	Injection Wells	Prod. Wells	Daily Oil Production by ISC (m^3/day)	Air/Oil Ratio (Sm^3/m^3)	Injection Pressure (kPa)
Glenn Hummel Sun[a]	2.5/3.7	741	45	1000	72	2	31	100	820	14,000
Gloriana Sun[a]	1.3/3.1	488	45	1000	174	1	12	40	1700	7000
Trix-Liz Sun[a]	?/2.8	1112	59	500	26	3	11	32	1420	4200
Casa Blanca Mobil[a]	3.3/5	314	38	600	36					2800

[a]*ISC process started from the upper part of reservoir.*

oil recovery; at the time of evaluation (oil recovery at start was probably in the range of 10–20%, by primary exploitation) by ISC, this oil recovery was 30–31%, with a predicted UOR of 56–60% in 10–14 years.

- Air–oil ratio was in the range of 800–2200 sm^3/m^3, for these four projects.

Videle East Project: Also represents a thin-layer ISC project but it is unique, as it is conducted as a post-waterflood process (Machedon, 1994; Machedon et al., 1993; Turta and Pantazi, 1986). This is the longest project (more than 16 years) and constitutes a full-scale secondary ISC application after a commercial waterflood recovered approximately 10–12%. The oil viscosity is approximately 100 cp (Table 18.4). It should be noted that the ignition of this waterflooded pool (original temperature 54°C) was easily obtained by applying a linseed oil-based chemical ignition; a 3-week ignition delay was estimated for most ignition operations (Figures 18.9 and 18.24). During normal operation, the percentage of CO_2 in the produced gas was in the range of 11–14%, while the percentage of O_2 was less than 1–2%. Due to ISC application, the water cut decreased from 70% to 60%. The air–oil ratio was approximately 4000 sm^3/m^3 for an injection pressure of 3000 kPa; it was only marginally economic, as the full energy balance was

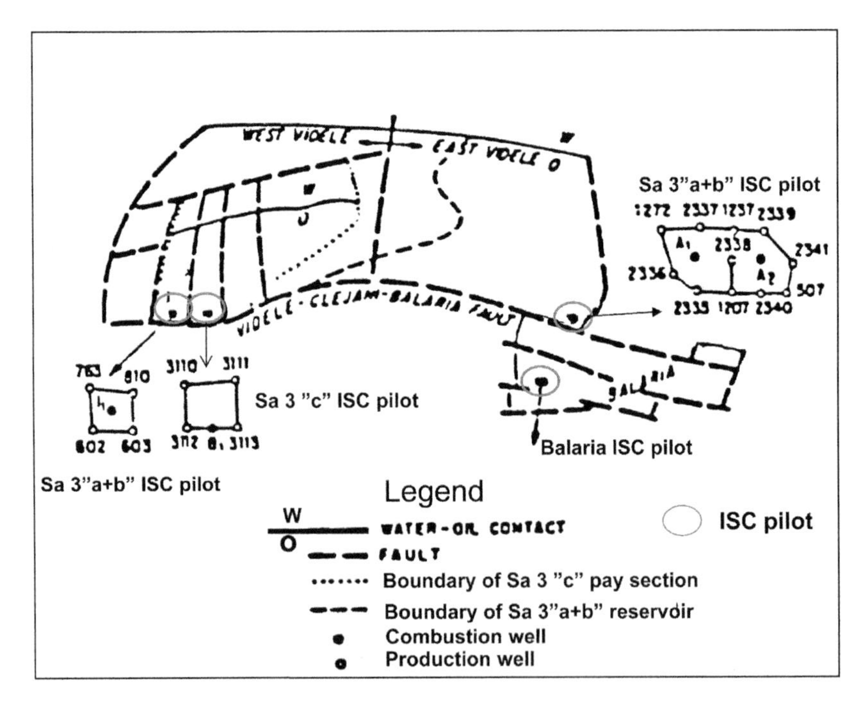

FIGURE 18.24 Videle/Balaria fields—schematic map showing the location of the experimental combustion patterns.

approximately 30%. The UOR for different areas was up to 30–37%. The main operational problems were related to coke bridge formation in the hot production well perforations, reduction of air injectivity in time, and emulsion generation.

18.2.5 Wet ISC Projects

Karazhanbas Field, Kazakhstan: This field became a commercial oil producer in 1980. It has been exploited by commercial ISC from the very beginning (1980) and by commercial steam flooding since 1982 (Bocserman et al., 1991; Mamedov and Bocserman, 1992; Turta, 2003).

The reservoir is a major one having a length of approximately 30 km and width of 5–6 km. The major areas of the reservoir are the western area, central area, and eastern area.

Lithologically, the reservoir is made up of interbedded sandstone and shales in a deltaic-predelta sequence. However, each sand is continuous over large areas. Sandstones are very fine grained, unconsolidated to semiconsolidated, and silty. The clay content seems to be medium.

The reservoir is tectonically simple: a giant monocline having a dip of about 2–4°. A lateral aquifer is found in the south and southeast sides of the structure; there is no information on how big the aquifer is. Toward the north, the trap is formed probably by a pinch out. The net pay thickness maps (isopachs maps) show a net pay thickness in the range of 14–16 m. However, the net pay is higher, up to 25 m, in the eastern area.

ISC was applied on a commercial scale as a moderate wet combustion mainly in the upper part of the western and central areas (Turta, 2003).

The ISC commercial application, which took place from 1980 to 1996, consisted of a line drive system involving, at its maximum development, 72 air/water injectors, and 261 producers. In 1996, continuous water injection in the former air injection wells started and was still in operation by 2003. This was a successful project. The estimated AOR was 500–1000 std. m^3/m^3 for the period 1986–1989 when both oil production and air injection were constant and the number of water injectors was relatively small. This is a good value compared to other commercial ISC operations worldwide.

As of 2002, the oil recovery was 28%. Given the actual oil rate and the water cut (87%) in 2003, it was estimated that a final oil recovery of at least 33–34% could be obtained.

Co-currently, a commercial steam drive operation was conducted in the central area, starting in 1983. The number of injectors increased continuously up to 60 between 1991 and 1998 and then decreased to 42 in 2002. The steam–oil ratio varied between 3 and 5 for the period between 1991 and 1998, when both oil production and the number of steam injectors were almost constant. As of December 2002, the oil recovery of the area submitted to steam injection was 23%. Given the current (2003) oil rate and the water cut (87%), it was estimated that a final oil recovery of at least 30% can be obtained.

Battrum Field, Saskatchewan, Canada: There is very little detailed information about this commercial ISC project which was operated by Mobil Canada starting in 1965 (Table 18.4). No history case paper treating the details of this project has been found. This has been the largest Canadian ISC project and it was operated in patterns (8 ha/pattern) as wet ISC (WAR up to 2.8 L/m^3). This constituted a commercial ISC operation conducted in a reservoir with the lowest oil viscosity, 70 mPa·s. No other ISC pilots in reservoirs of low temperature with oil viscosity less than 70 mPa·s—except very deep, light oil, high-temperature reservoirs—succeeded in attaining the commercial stage.

The process was continuously expanded until 1990, when the highest total air injection rate (approximately 720,000 std. m^3/day) into 19 air injectors was attained. In 1999, the process was discontinued (OGJ August 12, 1968 and OGJ EOR surveys 1994–2000).

The project was relatively successful, but the operational problems were very troublesome. With this relatively light oil, by far the most challenging problem was the generation of very problematic emulsions. These emulsions were promoted by the high solids content and iron oxide in the liquids and added significantly to the treating costs. Corrosion, sand production, and asphaltene deposition also constituted important problems that, too, added to costs. It can be concluded that the most important feature of this commercial operation was the existence of very serious operational problems, perhaps greater than in any other commercial ISC worldwide.

Generally, very high peak temperatures were not recorded in the observation wells and production wells, or indicated by the coring wells.

Bellevue Field Louisiana: Three separate ISC projects were started by Getty Oil, Cities Service Oil (CSO), Cities along with DOE (Bodcau Zone) in 1963, 1971, and 1976, respectively; the fourth one Bayou State Oil Corp. (BOSC) process—to be presented in the next section—was started in 1971, as well (Turta et al., 2007). A schematic map of Bellevue field showing the areas operated by these operators is shown in Figure 18.25; the properties of reservoir are very similar for all areas and are given in Tables 18.3 and 18.6. Essentially, this was an ISC project applied in a very shallow (130 m), relatively thick reservoir containing an oil of 450–700 cp, at an extremely low reservoir pressure (0.3 MPa); very small 5-spot pattern areas (up to 1–2 acres) were the norm.

The reservoir is located in the northwest corner of Louisiana, USA, in Bossier Parish, it is a dome structure; it was discovered in 1921 and the initial oil rates were up to 1000 bbl/day/well (159 m^3/day/well). In 1963, Getty Oil initiated an ISC pilot in a 2.5 acre (1 ha) inverted 9-spot pattern, and based on its successful operation four more experimental patterns were added. In 1971, CSO began ISC piloting within several patterns. Both Getty

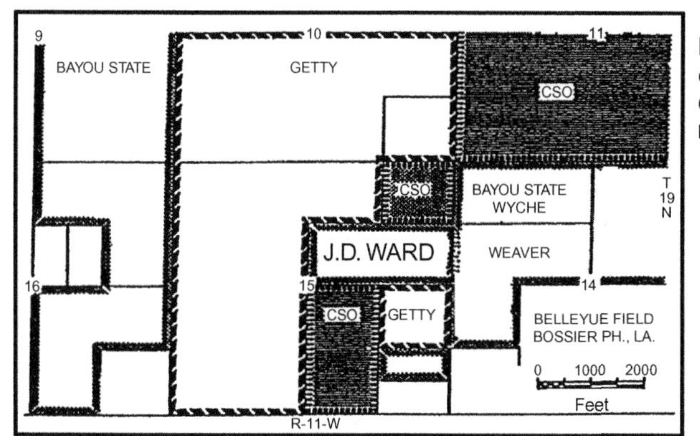

FIGURE 18.25 Schematic of Bellevue field map showing the ISC areas, operated by different companies (Turta, 2005).

and CSO projects were operated as wet combustion processes and they were expanded several times up to 1978. Getty and CSO discontinued their ISC projects between 1984 and 1990.

Performance of Getty's Bellevue project is shown in Figure 18.26A. In 1982, 223 wells were involved in the project, with an oil production of approximately 438 m^3/day (2750 bbl/day) at an air–oil ratio of 3500 std. m^3/m^3. This value represented 25% of the roughly 10,000 bbl/day oil production from ISC in the United States, at that time (Boberg, 1988), when actually this was the biggest ISC project in the United States.

Generally, in this field, after wet ISC, the wrapping of the process was done by a continuous water injection in order to scavenge additional amounts of heat from the burned zone in a kind of hot waterflooding. In all three projects, UOR of about 60% is expected.

This field also was the first one in which a comparative investigation of dry and moderate wet ISC was performed. Initially, Getty Oil used dry ISC followed by water injection. However, later on they tested moderate wet ISC with simultaneous or alternate air–water injection in two groups of patterns and they showed that wet ISC was superior (Burger et al., 1985); for all 3 years of parallel operation, the oil recovery was estimated to be higher for wet ISC patterns.

Subsequently, more investigations were done by Cities Services on the "Bodcau" lease in cooperation with US Department of Energy by comparing the performance of four patterns operated by dry ISC with four patterns operated by moderate wet ISC (WAR of 1.2 L/m^3) for a period of 20 months (Figure 18.27). At the end of this period, for the same cumulative of air injected, the volume heated at 315°C (or more) was almost two times higher

TABLE 18.6 Active Commercial ISC Projects: Reservoir Properties (Turta, 2007)

Field, Company, Country	Formation	Dip Degrees	Depth (ft)	Res. Temp (T_r °F)	Gross pay/ Net pay (ft)	Porosity (%)	Connate Water Saturation (%)	Oil Saturation at Start (%)	Permeability (mD)	Oil Viscosity at (T_r cp)	Oil Gravity (°API)	Initial Res. Pressure/ Reservation Pressure at Start of ISC (psi)	OOIP MMbbl
Suplacu de Barcau, Romania	S[a]	5–8	115–720	65	27–290/ 20–89	32	15	<85	5,000–7,000	2,000	16	140/80	310
Balol, India	S[b]	4–7	3280	158	10–95/ 9–50	28	30	70	3,000–8,000	100–450	16[c]	1450/1450	128
Santhal, India	SS	3–5	3280	158	16–195/ 9–50	28	30	70	3,000–5,000	50–200	18[c]	1450/1450	300
Bellevue, Louisiana, USA	SS	0–5	400	75	70[d];30[d]/ _ ; _	32	27	73	650	676	19	/40	4.6[d]; 10.6[d]

S, sand; SS, sandstone.
[a]*Unconsolidated.*
[b]*Coal and carbonaceous material are present (as streaks) within the sand formation (approximately 10% vol).*
[c]*Sulfur content 0.14%*
[d]*Lower sand and upper sand, respectively.*

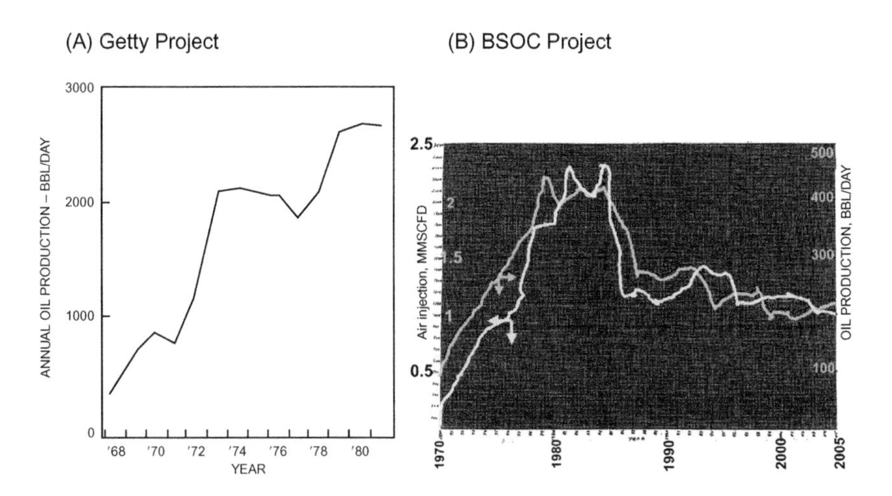

FIGURE 18.26 Bellevue ISC Project: performance of getty oil (A) and BSOC (B) projects.

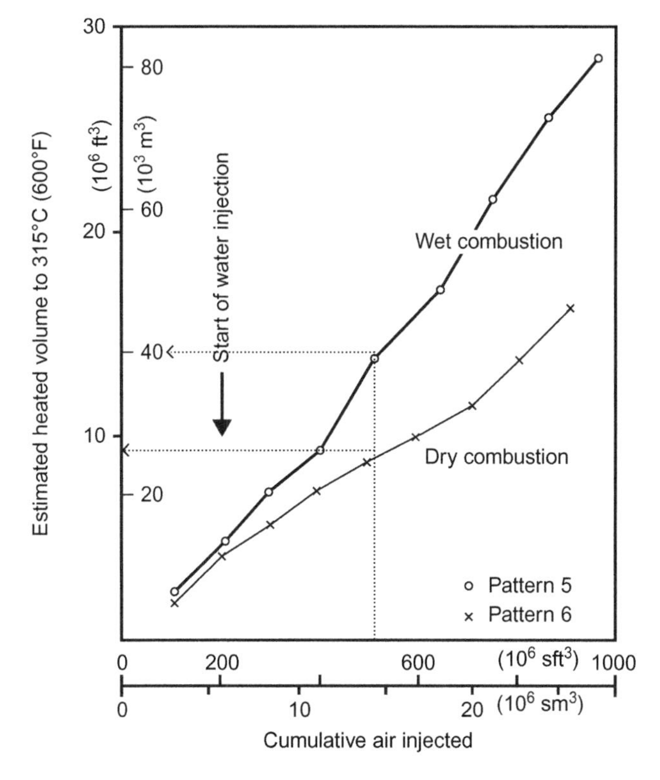

FIGURE 18.27 Bellevue Field. Comparison of heated volume in dry and moderate wet combustion; for the same of cumulative air injected, the heated volume is higher for wet combustion (Joseph and Pusch, 1980).

in the wet ISC patterns, mainly due to additional heating at the lower part of layer. This finding was confirmed by the Texaco field investigation, as well (Boberg, 1988).

The World's Currently Active Commercial ISC Projects: Bellevue, Suplacu de Barcau, Balol, and Santhal

Essential information on the four largest projects of the world, two in India, one in Romania, and one in the United States, is provided. The commercial ISC project at Suplacu de Barcau, Romania, is the largest project of this kind and it has been in operation for 41 years. The Balol and Santhal projects in India have been in operation for more than 15 years. Currently, each of these projects produces more than 500 m^3/day (3450 bbl/day). The fourth project, a dry ICS process operated by BSOC in Bellevue, Louisiana, USA, has been in operation for more than 40 years. Currently, it has 15 air injectors and 90 production wells and produces 48 m^3/day (300 bbl/day).

The Suplacu de Barcau, Balol, and Santhal projects are operated using a peripheral line drive starting from the uppermost part of the reservoir and going downdip, while the Bellevue project has been operated in patterns. Suplacu de Barcau project occurs in a shallow heavy oil reservoir with typical solution gas drive, while Balol and Santhal projects are applied successfully in deep reservoirs, having strong lateral water drive.

In the presentation of these projects, the focus is on the main characteristics of the process and the major operational problems. The properties of the reservoirs exploited by ISC are presented in Table 18.6, while Table 18.7 provides the results.

Dry ISC: Bellevue BSOC Project: The BSOC Bellevue project is a dry ISC process conducted in a shallow reservoir (400 ft) of relatively low permeability (700 mD) and high heterogeneity. It contains two layers which are operated separately (see more details in the previous section).

The producing zone is the upper Cretaceous Nacatosh, which consists of unconsolidated sand with considerable permeability variations (Table 18.6). The sand is crossed by faults and streaked with layers of sandy shale and fossilized lime. The formation is divided into two main sands (lower and upper one) (Turta et al., 2007).

ISC was started when oil recovery was 10%. It was initiated in the lower sand with three inverted 7-spot patterns located in the middle of the structure (first phase). More patterns were added such that in 1978, there were 10 air injection wells, all of them operated in the lower sand. In 1983, three patterns located at the eastern edge (downdip) were ignited in the upper zone, commencing simultaneous ISC operation in both lower and upper sands. Generally, the patterns have a small area, in the range of 1.5–3 acres/pattern.

As of 2004, there were 15 active air injection patterns and 53 wells are producing combustion gases. However, only 62.5% of the produced gases are

TABLE 18.7 Active Commercial ISC Projects: Results (Turta, 2007)

Field, Company, Country	Start Date (of Commercial Operation)[a]	Injection Pressure (psi)	No. of Injection Wells	No. of Production Wells	Daily Oil Production by ISC (bbl/day)	Current Water Cut (%)	O_2 Utilization (%)	Air/Oil Ratio (scf/bbl)	Expected Oil Recover (%)
Suplacu de Barcau, Romania	1971	150–200	111[b]	736[b]	9,000[c]	82	95	14,000	52
Balol, India	1997	1300–1600	30	75	4400	60	>95	5,600	38
Santhal, India	1997	1200–1500	30	105	4000	60	>95	5,600	36
Bellevue, Louisiana, USA	1970	60	15	90	300	90	80	15,000	60

[a] *The ISC piloting starting 3–7 years earlier than this date (7 years for Suplacu, Balol and Santhal and 4 years for Bellevue).*
[b] *At any time, there are also 24 production wells under CSS.*
[c] *It includes the contribution of CSS, estimated between 18% and 25% of the daily oil production.*

measured because at several production wells gas is not measured and a considerable amount of gas migrates across the lease boundaries. The oil production and air injection are shown in Figure 18.26B. For all 15 injectors, a total of 45,000 std. m^3/day (1,600,000 scf/day) of air is injected, producing 50 m^3/day (320 bbl/day) of oil at an AOR of 2700 std. m^3/m^3 (15,000 scf/bbl). The average air injection rate per well is 9000 std. m^3/day (300,000 scf/day), for an average pattern area of 1 ha (2.5 acres). BSOC has been able to separately operate (including ignition) the upper and lower layers. The 5.5 m (18 ft) thickness of lime (separation) is enough to ensure separate and simultaneous operation.

The main results of the project are provided in Table 18.7. The water cut is quite variable but in general, higher at the lower zone, between 95% and 98%, and lower in the upper zone, between 90% and 96%.

As mentioned, ignition is carried out using electrical heaters. Generally, for this process, using low air injection rates, 1.3–2.5% oxygen percentages are observed in the combustion gases, while the CO_2 concentration is 15–17%.

From the BHT profiles, a maximum temperature of 400°F (204°C) was recorded. Some coring wells were drilled in the burned area and showed a 10 ft (3 m) burned out zone at the top, with a 40 ft (12 m) unburned heated zone below.

Some production well burn-outs were experienced; approximately 10% of the producers have been replaced with new wells.

The combustion gases, which contain some hydrogen sulfide, sulfur dioxide, oxygen, and water vapor, when combined with temperatures of 100–150°F (38–65°C), create a very corrosive environment. Corrosion is not eliminated, but is somewhat controlled by downhole treating with corrosion inhibitors and biocides.

Emulsion problems have been troublesome. The combination of viscous crude with brine, coke from the combustion, fines, and iron oxides produces emulsions that are sometimes difficult to treat. Several brands of emulsion breakers have been tried over the years. Occasionally, the nature of the crude changes as the ISC front gets closer to the wellbore and a new treatment has to be developed.

A full development of ISC for the upper sand across the lease is contemplated. This is to be conducted by using injectors perforated in the lower sand; when ISC is to be initiated in the upper sand, the lower sand will be plugged off and the wells reperforated in the upper zone.

Dry ISC: Suplacu de Barcau Field: This is a dry ISC project conducted at low pressure (less than 200 psi), in a very shallow reservoir (less than 180 m) using a very small well spacing (50–100 m distance between wells). The oil viscosity is relatively high, around 2000 mPa·s and the project is operated in a peripheral direct line drive (Condrachi and Tabara, 1997; Gadelle et al., 1981; Machedon, 1994; Machedon et al., 1993; Panait-Patica et al., 2006; Turta et al., 2007).

The reservoir is located in the northwestern part of Romania, close to the Oradea town. It represents an east-west oriented anticline upfold, axially faulted by the major fault of Suplacu de Barcau, which limits the field to the south and east (Figure 18.12). The length of the monocline is approximately 15 km. To the north and west, the field is bordered by a weak aquifer. Both depth and thickness increase from east to west and north to south. Depth ranges from 115 ft (35 m) to 660 ft (200 m) and the thickness ranges from 14 ft (4 m) to 80 ft (24 m).

The reservoir was put into production in 1960, with solution gas drive being the main mechanism. An UOR of 9% was predicted. Initial oil rates were in the range of 12–36 bbl/day/well (2–5 m^3/day/well), but decreased quickly to 2–6 bbl/day/well (0.3–1 m^3/day/well).

Both ISC and steam drive methods were tested at the upper part of the structure in the period between 1963 and 1970 (Figure 18.12). Both methods were tested in a 0.5 ha pattern. A semicommercial operation consisting of six contiguous patterns of 2–4 ha was used for both methods and based on the performance the decision to use ISC for the commercial exploitation was made in 1970. At the same time, it was decided that steam injection would be used permanently in CSS mode for preheating the production wells located close to the ISC front. Also, the decision to convert the pattern exploitation to line drive exploitation was made. The decision to sweep the reservoir starting from the uppermost part of reservoir was supported by two separately operated experimental ISC patterns located at the middle and at the lowest parts (close to the water–oil contact) of the structure. These pilots were operated for more than 5 years and they showed that the control and the efficiency of the process are lower when pilots are not located updip.

For more than 40 years, a linear ISC front has been propagated continuously downstructure, parallel to isobaths. Starting in 1986, the process was expanded in new areas of the western part of the reservoir. The position of the ISC front, as of 2004, is shown in Figure 18.28; the air injection wells are included in an east-west line for more than 10 km. The distance between two adjacent wells within a row ranges between 152.5 and 229 ft (50 and 75 m). Based on the performance of the wells in the zone processed by ISC, an UOR of 55% was calculated.

In 1983, a second linear ISC front, parallel to the main one, was started in the middle of the structure in the wider eastern part of the reservoir. The operation of two parallel ISC fronts has been challenging, primarily due to the reduction of the main updip ISC front propagation velocity and the downgrading of the whole performance. In 1996, the operation of the second ISC front was discontinued.

The performance of the commercial ISC project is shown in Figure 18.29. Water cut has increased to its current values of 82%. This is explained by the extended water–oil contact and the proximity of

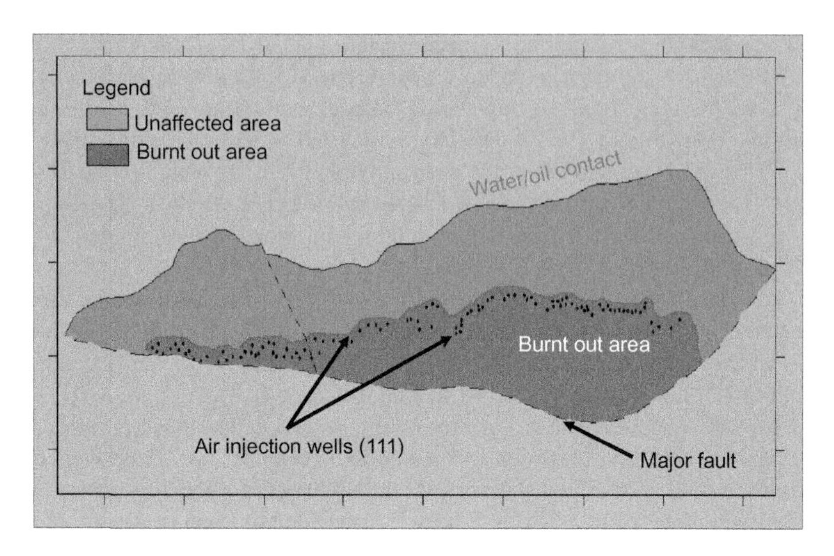

FIGURE 18.28 Suplacu de Barcau. Position of the ISC front as of July 1, 2004. Source: *After 33 years of commercial operation.*

producers to the current oil/water contact. Maximum oil production was seen from 1985 to 1991 when the total air injection rate was at a maximum AOR has increased to a present value of 3000 std. m^3/m^3. As CSS has been applied continuously in parallel with the main ISC process, it is estimated that its contribution to oil production was as much as 18% for the eastern part and 25% for the western part, where the net pay thickness is significantly higher.

This has been one of the best instrumented ISC projects in the world. Hundreds of BHT profiles have been taken in the observation and production wells and some of them recorded very high peak temperatures (around 600°C) in the upper part of the layer indicating the segregated nature of the ISC process. Additionally, some burning of the producers have been experienced, as approximately 15% of the producers have been replaced with new wells; 8 coring wells in the burned area have been drilled. They recorded 5–7 m burned out at the top of the formation and 7–10 m of unburned but heated rock below. This is shown in the lithologic column for two of the coring wells (Figure 18.17).

For reasons related to the safety of workover operations at the hot producers, a special drilling mud was developed to kill these "hot wells," with temperatures in the range of 80–250°C. This fluid ensures equilibrium in the well, avoids blockage of the formation, and presents rheological–colloidal characteristics appropriate to the specified temperature (Aldea et al., 1988).

ISC application has led to the increase in concentration of natural emulsifiers in the produced oil such as asphaltene, resins, naphthenic acids, and

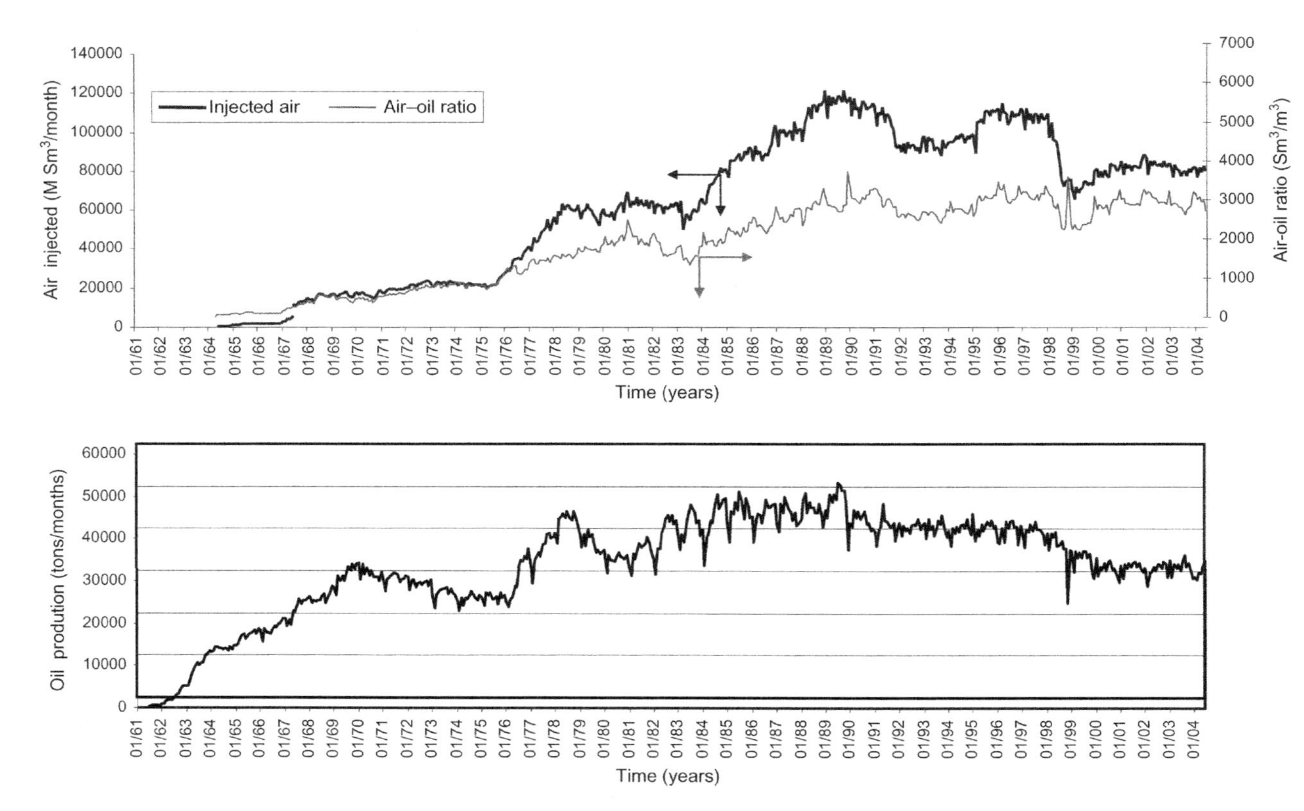

FIGURE 18.29 Performance of the Suplacu de Barcau ISC Project.

finely dispersed solid particles which lead to the formation of difficult emulsions. For dehydration and desalting of the crude oil, a special technology for the thermal–chemical treatment with a final stripping step was developed.

A major challenge was the leakage to the surface of some combustion gases through certain kinds of mud and steam volcanoes, which appeared high on the structure and accompanied the ISC commercial exploitation since its beginning. This gas escape can be related to the very shallow depth of the reservoir and to improper sealing of some old producers (Carcoana, 1990).

Wet ISC: Balol and Santhal Projects: These projects are conducted at high pressure (greater than 10.3 MPa), in relatively deep reservoirs (approximately 1000 m) producing under a strong lateral water drive (Tables 18.6 and 18.7). The oil viscosity is medium, between 50 and 200 mPa·s for Santhal and between 200 and 1000 mPa·s for Balol (Bhatia and Singh, 1998; Roychaudhury et al., 1995). Both reservoirs are operated using a peripheral direct line drive, sometimes direct, sometimes staggered.

Santhal and Balol reservoirs belong to the heavy oil belt located in the northern area of the Gujarat State in West India. Oil production began in 1974 in Santhal and in 1985 in Balol. The structure consists of a combined structural–stratigraphic trap. The oil trap is limited updip by a pinch-out line and downdip by a water–oil contact (Figure 18.29). The pay section is unconsolidated sand with interbedded shale, carbonaceous shale, and coal. The unique feature of the reservoir is that coal and carbonaceous material are present in the oil formation both as dispersed coal (black particles and centimetric laminations) and as coal and carbonaceous material streaks with a horizontal extension from a few meters up to the distance between two wells. The thickness of the coal streak varies from 0.2 m inside the pay zone to a few meters outside the pay zone.

The oil recovery mechanism is the natural edge water drive under conditions of unfavorable mobility ratio. For both Santhal and Balol, the static pressure has been almost constant up to the present indicating a strong aquifer and that the oil still remained above the bubble point pressure. Originally, oil was produced through natural flow and then sucker rod pumps were installed in the wells. During the ISC application, the majority of wells are on pump although some wells are self-flowing and helped by the heat effect.

No significant sand influx problems in the production wells have been encountered. In general, the gravel packing of the producers has worked properly.

The ISC was tested in Balol beginning in 1990. Initially, one 2.2 ha inverted 5-spot pattern was used. The pattern was enlarged to 9 ha by drilling four new peripheral wells. This was followed by a second ISC pattern (Figure 18.30). In 1996, based on the favorable performance of these two patterns, a decision was made to use ISC for commercial exploitation. The

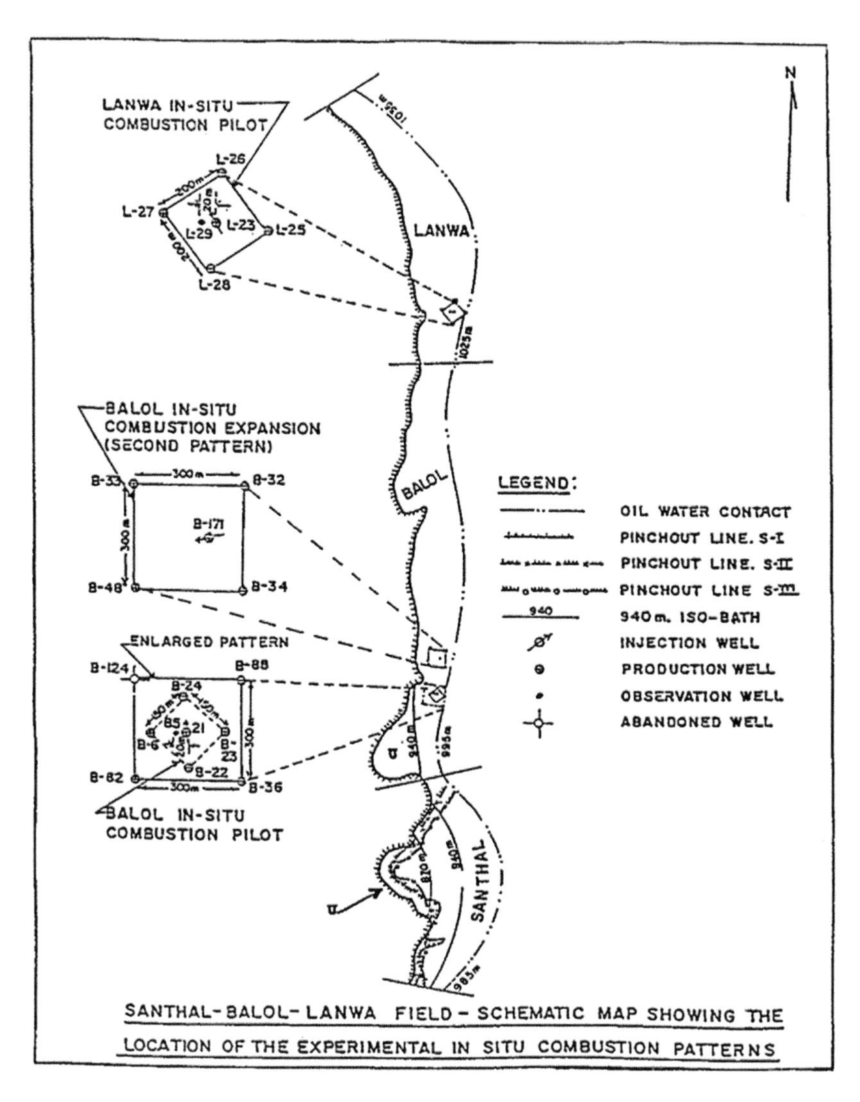

FIGURE 18.30 Schematic map of Balol–Santhal Field showing the experimental ISC patterns.

commercial exploitation in Balol started in 1997 and from the beginning the peripheral updip line drive was adopted (Turta, 1996).

In Santhal, the commercial ISC process was initially applied in inverted 5-spot patterns. Although the initial design recommended an exploitation in patterns, based on Santhal initial experience and the experience from the adjacent Balol ISC exploitation, the operation was changed to a peripheral updip line drive, which is now in operation (Chattopadhyay et al., 2003, 2004).

Both in Balol and Santhal, wet ISC was tested and then applied commercially. The WAR has been between 1 and 2 L/std. m^3, and the air and water injection has been done in an alternating way (nonsimultaneous injection). The wet combustion might have helped in moderating the high peak temperatures in the ISC front, leading to the reduction of H_2S in the combustion gases. The percentage of H_2S in the combustion gases has been in the range of 100–1500 ppm, with spikes of up to 4000 ppm (Turta, 1996).

The injection wells are spread over a line of more than 12 km in Balol and more than 4.6 km in Santhal. Given the relatively high reservoir temperature (70°C) and pressure, SI is used for the initiation of the ISC process.

There was a continuous increase in oil production due to expansion of the Balol and Santhal ISC projects. Figure 18.31 shows a typical performance of a production well in Balol, located close to the oil–water contact, which displayed a dramatic reduction of the water cut, from 75% to less than 5% as the ISC displaced oil and pushed it downdip toward the well; a similar effect was noticed in Santhal. This shows that ISC is a very effective method for the exploitation of heavy oil in the presence of a strong lateral water drive as the water is pushed back into the aquifer. The main results of these projects are provided in Table 18.7. The air/oil ratio has a favorable value of 1000 std. m^3/m^3 (5600 scf/bbl) for both projects.

To solve the pollution problem caused by H_2S, SO_2, and hydrocarbon gases, the produced gases are vented through tall flare stacks equipped with outside makeup gas and electronic igniters. Unlike almost all other commercial ISC worldwide, which typically have up to 2% hydrocarbon gases in the produced combustion gases, this percentage for Balol and Santhal is higher, approximately 6%, helping in reducing the amount of outside makeup gas required for flaring.

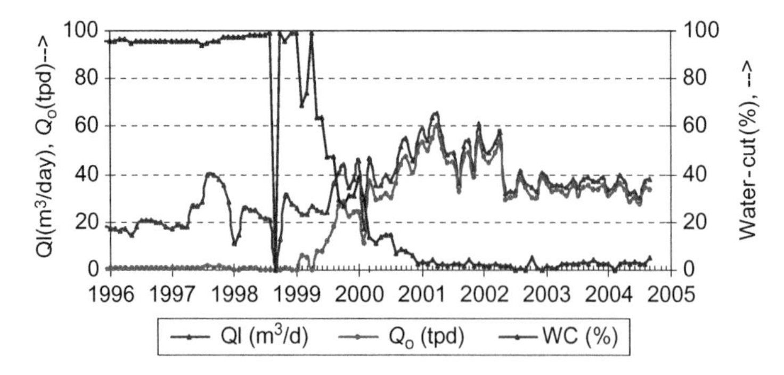

FIGURE 18.31 Balol ISC project, conducted in the presence of edge water drive; production performance of a typical ISC-affected production well in Balol Legend: Ql, total liquid; Q_o, oil production; WC, water cut.

Brief Considerations on Application of Wet ISC Process

Theoretically, at least for the frontal ISC process, the wet ISC is superior to dry ISC. Actually, the wet process has proved net superior to the dry process in the laboratory testing, due to:

- Its lower fuel burned per volumetric unit of rock, leading to lower air consumption per volumetric unit of rock;
- Its higher velocity of propagation leading to a shorter life of the project;
- Its lower AOR.

Unfortunately, the field tests (both pilots and commercial operations) could not prove with accuracy all these advantages, one by one, mainly for the first two advantages; it rather expressed that by obtaining a lower AOR and even based on AOR considerations, this was not long enough and done on a large scale. In exchange, some operational field problems, such as the more pronounced emulsion problems, and some injectivity problems impeded the widespread utilization of wet ISC process. Additionally, some problems with the frequent changes of the production systems (natural flow, lift, etc.) contributed to the reluctance of operators to replace dry ISC with wet ISC; this was the case of the commercial operations in Suplacu de Barcau and BSOC Bellevue projects.

In Situ *Upgrading Considerations for Conventional ISC Process*

Normally, *in situ* upgrading should be expected as a result of the thermal cracking and utilization/consumption of the heaviest oil fractions as a fuel. However, although this happens right ahead of the ISC front, the upgrading is largely lost due to the mixing of the small amount of upgraded oil with the original (not upgraded oil) while flowing through the cold region.

In Suplacu de Barcau project, no upgrading was observed. In Balol and Santhal projects, analysis of the produced oil's properties over a long period of time did not show consistent viscosity reduction although some reductions, in some periods, were recorded (Chattopadhyay, 2002).

A little bit more consistent was the upgrading observed in the West Newport project; API of produced oil increased from 15.2 to 17 API (sometimes to 20 API), while the sulfur decreased from 2.2% to 1.8% (wt%). The maximum corresponding viscosity reduction (for the 20 API case) was from 4600 to 270 cp (at 60°F) (Chu and Crawford, 1983).

In South Belridge project, API of produced oil increased from 12.9 to 14.2 API with a viscosity reduction from 2700 to 900 cp (at 87°F) (Chu and Crawford, 1983).

Generally, there is no consistency as to whether the *in situ* upgrading occurs all the time and even when it occurs it is not very significant (2–3°API). Therefore, for the conventional forward ISC process, this *in situ* upgrading is not a feature to be accounted for in the design of the process.

18.3 ISC PROJECTS IN LIGHT OIL RESERVOIRS

There are two categories of projects of this kind: (1) those from Williston basin, which generally have been conducted as a secondary recovery method applied at a very low primary recovery factor and have been proved successful in this situation, therefore they reached the commercial status and (2) those (generally from outside Williston Basin) which tried to expand the process after a natural water drive or a waterflood, preferentially for dipping reservoirs, which have not been fully proved and these processes are still under development.

18.3.1 Commercial HPAI Projects in Very Light, Deep, Williston Basin Oil Reservoirs

An important milestone in the advance of air injection processes was the implementation of commercial air injection projects in the Williston Basin of North and South Dakota, USA, starting in 1979 (Erickson et al., 1993; Fassihi et al., 1994; Kumar et al., 1994). The process was applied in dolomite reservoirs with low porosity (11–19%) and permeability (less than 20 mD), containing very light oils (viscosity of less than 2 mPa·s under reservoir conditions), where water injection encountered significant problems due to extremely low injectivity. The dolomite contains some microfractures, but extensive fracturing or faulting is not known to exist.

The main data on the Williston Basin projects (and other similar projects) are summarized in Table 18.8. The most important features of these projects are:

- The *high reservoir temperature (103–112°C)*, which facilitates the initiation of the process by SI.
- The very low oil recovery at the start of ISC (less than 10%), due to extremely undersaturated oil, stored in a very compact rock.

The first paper on these projects appeared in 1993, and it introduced the term HPAI (Erickson et al., 1993). The fact that the projects seemed to involve a miscible process was indirectly supported by the fact that the operator tried hard to inject air at sufficiently high air rates in order to maintain a high-pressure level in the reservoir (Miller, 1994). For these projects, it seems that the reservoir and operating conditions were conducive to the generation of a self-sustaining ISC front, although this was not confirmed by BHT in the producers or in the observation wells, as temperature measurements were not made. However, the fact that CO_2 percentage in the produced gases has been around 12% seems to suggest that a self-sustaining process has been accomplished. Some corrosion problems were experienced in these projects, but they were solved by routine methods.

A schematic map of the Cedar Creek Anticline with five of its commercial operations is shown in Figure 18.32; the biggest exploitations are those

TABLE 18.8 Data on the *Very* Light Oil Reservoir Projects (HPAI Projects) in the United States (as of 2005)

Field, Company, Country	Ref.	Rock Type	Pay Thickness (m)	Depth (ft)	Temperature (°C)	Injection Pressure (psi)	Porosity (%)	Permeability (mD)	Oil Viscosity (mPa·s)	Daily Oil Prod. by ISC (bbl/day)	Air/Oil Ratio (scf/bbl)	Oil Recovery, % (primary/HPAI)	Observations
West Hackberry, Amoco, Louisiana	T	S	10[a]	3,000–12,000	94	–	27	300	0.9	–	–	–	
Sloss[b], Amoco, Nebrasca	T		6	6,200	94	3,600	11	190	0.8	480	16,900	–	
Mphu, Continental, Williston Basin (WB)	T	D&L	6	9,500	104	4,400	17	5	0.5	600	12,000	15/29.2	
Buffalo, Continental, WB	T	D	4.6	8,500	102	4,400	19	18	0.5	2,500	10,000	6.5/15.6	
Madison Capa[b], Koch Expl., WB	T	L	?	8,600	99	4,400	11	10	0.5	–	20,000		
Cedar Hills (North and South Units),	OGJ 2010			9,000	102		18	10	2	6,000/ 13,300(as of 2012)	?		Using only horizontal wells

(*Continued*)

TABLE 18.8 (Continued)

Field, Company, Country	Ref.	Rock Type	Pay Thickness (m)	Depth (ft)	Temperature (°C)	Injection Pressure (psi)	Porosity (%)	Permeability (mD)	Oil Viscosity (mPa·s)	Daily Oil Prod. by ISC (bbl/day)	Air/Oil Ratio (scf/bbl)	Oil Recovery, % (primary/HPAI)	Observations
Continental, WB													
Pennel (2 Phases), Encore Acq., WB	OGJ 2010			8,800	93		17	10	1.4	1980/260 incremental	?		Vertical and horizontal wells
Little Beaver, Encore Acq., WB	OGJ 2010			8,300	93		17	10	1.4	1,650 (as of 2012); 750 incremental	?		Vertical and horizontal wells

MPHU, Medicine Pole Hills Unit; WB, Williston Basin, N and S Dakota; Continental, Continental Resources; Encore Acq., Encore Acquisition Company; D, dolomite; L, limestone; T, Turta and Singhal, 2001; OGJ, Oil and Gas Journal, April 2010.

[a] *DIP = 23–35°.*

[b] *Fractured rock.*

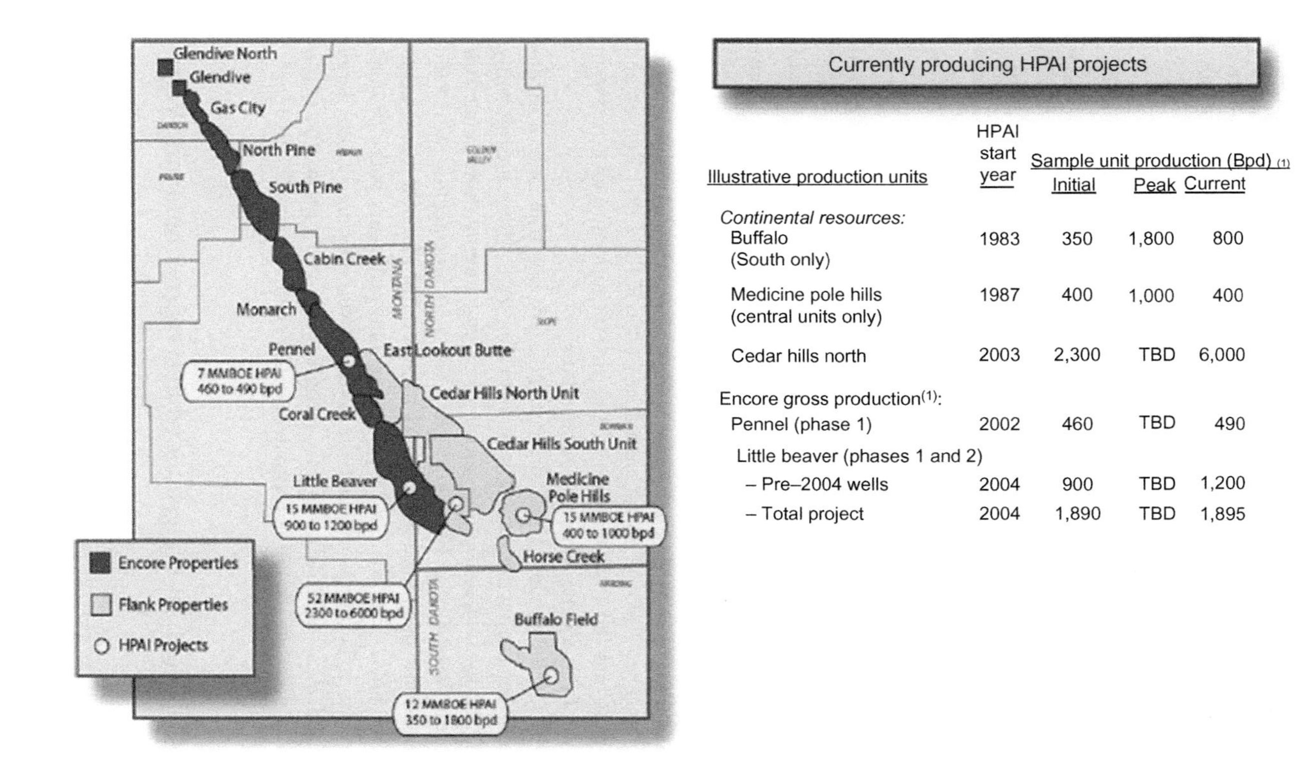

Illustrative production units	HPAI start year	Sample unit production (Bpd) (1) Initial	Peak	Current
Continental resources:				
Buffalo (South only)	1983	350	1,800	800
Medicine pole hills (central units only)	1987	400	1,000	400
Cedar hills north	2003	2,300	TBD	6,000
Encore gross production(1):				
Pennel (phase 1)	2002	460	TBD	490
Little beaver (phases 1 and 2)				
– Pre–2004 wells	2004	900	TBD	1,200
– Total project	2004	1,890	TBD	1,895

FIGURE 18.32 Schematic map of Cedar Creek Anticline (Williston Basin) and main data on ENCORE Acquisition's HPAI projects as of 2005 (Encore's web site).

from Cedar Hills, Medicine Pole Hills, and Buffalo; Little Beaver and Pennel projects were started recently (around 2004), and they are applied after waterflooding.

Figure 18.33 shows the performance of the Medicine Pole Hills Project, conducted in patterns using vertical wells. The increase of oil production is dramatic; based on this performance, an increase of UOR from 15% to 29% is predicted (Kumar et al., 1994), and this represents a maximum increase in UOR recorded in this region.

Figure 18.34 shows the performance of the South Buffalo HPAI project, also conducted in patterns using vertical wells. A very thorough study, comparing 18 years of side-by-side exploitation by water injection and HPAI on this reservoir found that UOR increased from 10% to 15% for waterflooding and from 10% to 15–17% or a little bit higher, for HPAI (Kumar and Gutierez, 2007), for which the main advantage is faster oil recovery (three times faster).

In this region, another project – applied in patterns – was started in Horse Creek reservoir in 1996 (Watts et al., 1997). The project involved three injection wells and 11 producing wells. The injection pressure was 4700 psi (32.4 MPa); oil production increased from 300 to 700 bbl/day; AOR was 3200 sm^3/m^3. However, the performance period was too short to allow clear conclusions.

After year 2000 operators started to drill horizontal wells, including them in the HPAI projects, at the beginning only as producers and subsequently as injectors (Belgrave, 2006). This happened in Cedar Hills, Little Beaver and Pennel projects and led to a decrease of the air–oil ratio due to increase in oil production from a better pressurized reservoir.

Cedar Hills is the biggest HPAI project and, currently uses only horizontal wells both for production and injection (Table 18.8); they are probably arranged in side-by-side configuration. By 2005, considering both Cedar Hill North Unit and West Cedar Hill Unit, there were 30 air injection wells and the oil production was around 6,000 bbl/day; this oil production increased to around 13,300 bbl/day in 2012, with 127 horizontal producers and 126 horizontal injectors. It is interesting that the utilized well spacing is very large (1 mile), (OGJ, April 2, 2012). For this secondary recovery mode, the maximum potential incremental oil is estimated at 12% OOIP (from 6% to 18%).

18.3.2 ISC Projects in Waterflooded Reservoirs Containing Very Light Oil

As shown in Table 18.8 air injection was tested in the extensively fractured CAPA Madison reservoir, North Dakota (Erickson et al., 1993), where air was injected at the end of a waterflood process. After 1.5 years of pattern flooding in this watered out reservoir, the AOR was twice that of other

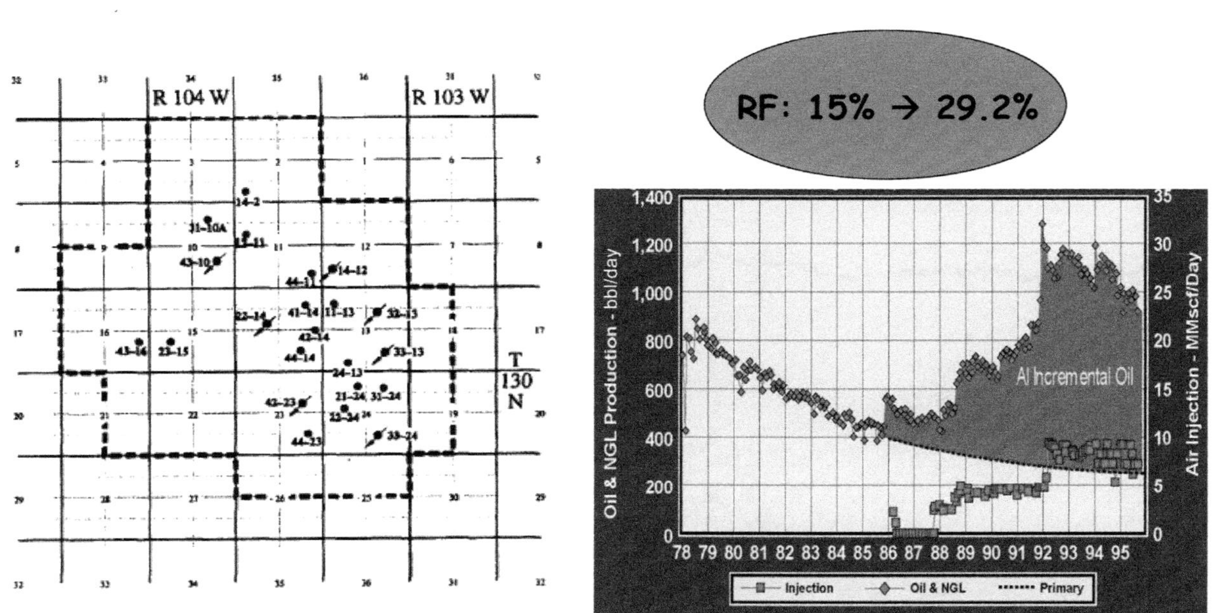

FIGURE 18.33 Performance of Continental Resources' Medicine Pole Hills HPAI project. Vertical wells; pattern flood; air injection after primary recovery (Belgrave, 2006).

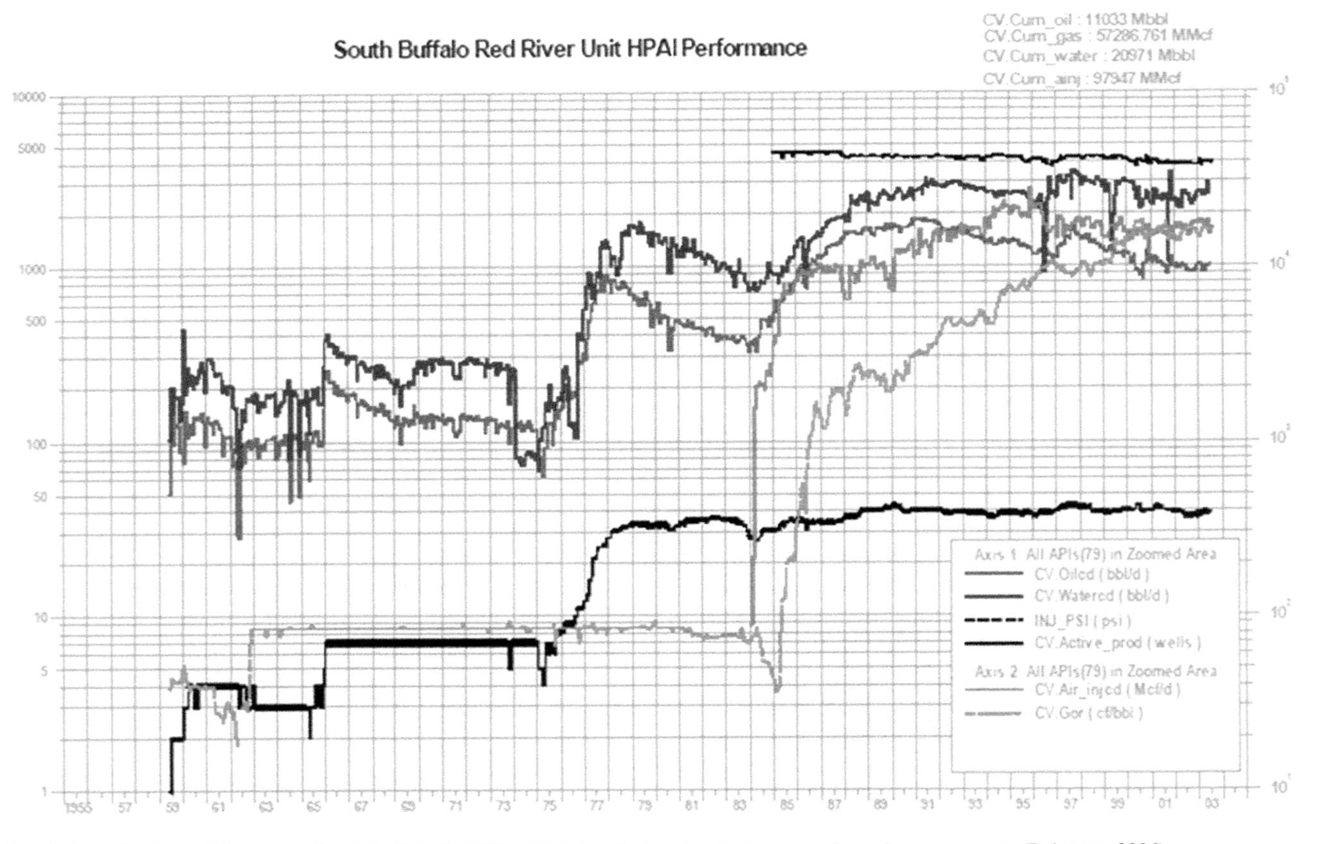

FIGURE 18.34 Continental Resources South Buffalo Red River Unit (vertical wells; air injection after primary recovery (Belgrave, 2006).

Williston Basin projects (around 20,000 scf/bbl or 3500 sm^3/m^3), and the process was discontinued.

After a natural water drive/waterflood, two projects were recently initiated. The first project began at West Hackberry (Table 18.8), USA, in 1994 and consisted of two pilots: first one, a high-pressure one and second one a low-pressure project. Both these projects took full advantage of gravity by locating the injection wells at the *uppermost part of the structure*. In the first pilot, the reservoir has a dip of 23–35°, while in the second, the dip is 60°. The oil production increased by 65% for the second test and 30% for the first test, probably indicating that the formation dip is more important than miscibility. Here, the process was applied at a late stage when a natural edge water drive already achieved an oil recovery of 50%; the first results are encouraging but the final results are not available yet (Turta and Singhal, 2001); the pilot showed that all oxygen was consumed in reservoir as it never appeared in the production wells. A similarly applied project was tested by TOTAL S.A. in Handil reservoir, Indonesia, but the test was in an early stage at the date of results reporting (Clara et al., 2000).

As mentioned, after year 2000 operators started to drill horizontal injectors; they included them in the HPAI processes applied after waterflooding, as well. This happened in Little Beaver and Pennel projects. In Little Beaver in 2006, after 4 years of application, the oil production was 25% higher than that corresponding to an extrapolated waterflooding performance, while after 10 years, in 2012 (OGJ April 2, 2012), the total oil production was 1650 bbl/day with 750 bbl/day (45%) as incremental oil, by utilizing 29 injectors and 57 producers. Pennel Project was developed in two areas: Phase 1 and Phase 2 areas, starting in 2002; in 2012 the total oil production was 2000 bbl/day with 260 bbl/day (13%) as incremental oil, utilizing 32 injectors and 78 producers (OGJ April 2, 2012). After 10 years of operation, both projects were considered promising, and the expansion was considered likely. Therefore, as a tertiary application after waterflood, the final proof (of HPAI as a commercial, successful operation) may be available very soon.

18.3.3 ISC Failures in Reservoirs with Light-Medium Oils

In reservoirs with oil viscosity in the range of 2–60 cp, there were many ISC failures (Turta and Singhal, 2001). The conclusion of a failure is indirect, as indicated by the lack of expansion of the project to semicommercial or commercial application. ISC failures in Romanian fields, such as Ochiuri, Babeni, etc. are presented elsewhere (Carcoana et al., 1983; Machedon et al., 1993). Main ISC failures in the US fields are listed by Chu (1977, 1983), while some ISC failures in Canadian fields are listed by More et al. (1994).

Many lessons were learned from these failures. For instance, in case of Ochiuri field, the initiation of the ISC was possible even in a secondary gas cap after many decades of secondary gas injection. It seems that not too many failures were due to the lack of sufficient fuel available for the process. The most frequent causes for failures were:

- Very troublesome operational problems (corrosion, erosion, frequent change of pumping equipment, etc.)
- Existence of very old wells, which did not ensure the confinement for the test area.

Two exceptions constituting at least technical successes are the West Heidelberg (WH) project and another project in China, which is under way at this time. Although not applied after waterflooding it is important to mention that one of the most efficient air injection project in a light oil reservoir (oil of 6 cp viscosity)—from the point of the value of its performance, mainly UOR obtained—was the WH project (Huffman et al., 1983), which was applied in a gravity stable mode (top-down displacement). WH reservoir has a depth of 3500 m with a reservoir temperature of 105°C. Permeability is low (85 mD), but the dip is relatively high (5–15°); it contains a strongly undersaturated oil (bubble point pressure is five times less than initial pressure). The air injection started when oil recovery was 6% and in a period of 11 years, with three injectors updip, this recovery increased to 30%, at an AOR of 1800 sm^3/m^3. A maximum BHT of 242°C was recorded in a production well.

An air + foam field test was ongoing in China (as of 2008) in a light oil reservoir; the results are encouraging (Yu et al., 2008). In a small, extensively waterflooded 5-spot pattern (reservoir temperature 90°C and oil recovery 21%) of extremely high heterogeneity (nitrogen brokethrough in 3 days) after 6 months of air–foam injection (0.45 PV) no N_2/O_2 breakthrough occurred; oxygen consumption *in situ* was complete (oxygen utilization efficiency was 100%) even in the presence of foam. Oil production increased to 20%, while the water cut slightly decreased (97–92%). However, more performance history is needed to assess the results. Two more similar tests are planned. It appears to be a promising direction mainly for formations with relatively good permeability.

The main factors, leading to a successful ISC project in light oil reservoirs, seem to be high reservoir temperature, high dip, and relatively high oil saturation.

18.4 CISC APPLICATIONS

CISC or "burn and turn" ISC is, to some extent, similar to CSS in the sense that an ISC front is initiated and propagated a few meters around a well and then, after a heat soak period, the same well is put into production. Oil then flows through the burnt-out zone, absorbs some of the heat stored there, and is produced by the well. Oil flow through the burnt-out zone is similar to that taking place during a reverse ISC process; the only difference is that during CISC, there is no continuous pressure support as the air injection is already stopped.

While CSS has been widely used, CISC has seen limited utilization, primarily due to field and personnel limitations, directly related to the complexity of the ISC process. The main effects of CISC are related to oil production stimulation and sand consolidation. Another application is sand consolidation by injecting hot air (controlled coking), in which the ignition of the formation is not intended; a controlled coking of the oil around the well—using LTO to this effect—resulting in sand consolidation by hot gas injection is targeted.

18.4.1 CISC Application for Heavy Oil Production Stimulation

In Canada, the CISC was applied by BP Canada Resources at Marguerite Lake, but no results have been presented in the open literature (Moore et al., 1993). In the United States, CISC was applied in a sandstone reservoir containing a medium–heavy oil of 800 mPa·s viscosity. The oil rate increased four times, and the incremental oil production due to the thermal effect lasted for 5 months; the well produced hot fluids in the first 3 weeks (White, 1965). Also, CISC was applied in Balol, India, in a production well relatively close to the edge water/oil contact. Oil rate increased threefold and the effect lasted more than 1 year. Water cut decreased from 95% to approximately 40%, but increased gradually up to 80% within 1 year (Rao et al., 1997).

A more systematic application was conducted in Albania and Romania. In Albania, in the unconsolidated sand heavy oil reservoir Patos-Marinza, with an oil viscosity of 9000 mPa·s, the operation was conducted mainly to stop the sand influx. It was carried out for eight producers which were already shut in due to sand problems. Typically, oil rates were 0.3 tons/day and after CISC, they increased up to 1.5 tons/day, with incremental oil obtained for a period of up to 1.5 years (cumulative oil up to 650 tons/well). The sand content of the produced oil decreased from 8% to 2%, after the CISC (Gjini et al., 1999).

In Romania, during the commercial ISC process conducted in heavy oil reservoirs of Suplacu de Barcau and Videle–Balaria, the CISC operation was conducted for both production stimulation and stopping the sand influx (Turta et al., 1985). For Suplacu de Barcau, the following conclusions were made. For Suplacu de Barcau, the following conclusions were made:

- Out of 14 operations, nine were a success both as sand consolidation and as oil production stimulation. After the CISC operations, the earliest sand influx appeared after 5–6 months of production and in some wells the sand influx never appeared.
- The average daily oil rate increased 3 to 15 times and for the successful operations the additional oil was between 200 and 6000 m^3/well. The thermal stimulating oil production effect continued for 0.5–2 years.

- For the successful operations, the AOR was less than 1000 std. m^3/m^3.
- In the case of wells 43 and 2537, the BHT was not kept under control and cracking (coking) took place in the producing interval. Coke was deposited in the wellbore and around the perforations making cleaning by normal means extremely difficult.
- Wells 800, 801, and 802 did not produce because of flowing clay. The CISC operations converted this clay from hydratable into nonhydratable stabilized clay so that the clay no longer migrated from the formation into the wellbore. This had a direct result the increase of the oil rates from 0.8 to 6 m^3/day.

For a CISC to be successful, during the ignition, the operator should keep the peak temperature at less than 400°C to avoid damaging the casing. This damage can make exploitation difficult and make it impossible to take advantage of the thermal effect as was the case with wells 402 and 802. The performance of well 516, before and after the CISC, is shown in Figure 18.35. The peak oil production was around 29 m^3/day (average monthly rate around 24 m^3/day) in the first period (after CISC). There is an increase in the water cut toward the end of production period, but it was still maintained less than 10%, similar to that before the CISC implementation. The additional oil for 1.5 years after CISC equals 50% of 4 years oil production prior to CISC operation.

An interesting case is that of well 801, located close to initial oil–water contact, which CISC experienced frequent sand and clay influx and had very low production (the amount of oil produced in 4 years was 400 tons) before

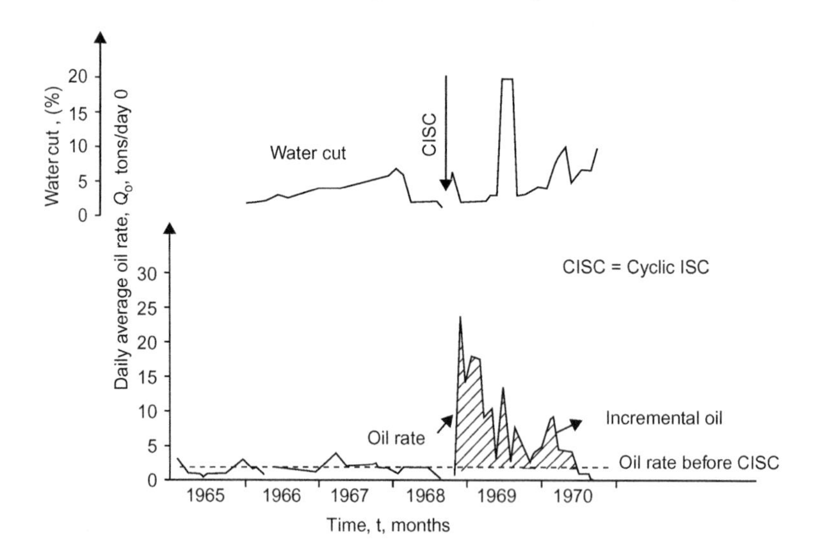

FIGURE 18.35 Performance of Well 516 Suplacu de Barcau before and after CISC operation (Turta et al., 1985).

CISC operation. The first 70 days after CISC the well was self-flowing and the oil rate was around 30 m^3/day. The CISC did not totally solve the sand influx and required three sand cleanings in the first year after CISC implementation.

For Videle–Balaria, the data and results of CISC operations imply the following conclusions:

- Out of seven operations, three were a success both as sand consolidation and as oil production stimulation effect. Two CISC operations were a success only as sand consolidation effect. After the CISC operation, the sand influx disappeared completely at three wells and for the rest there were some sand influx problems but much less frequently than before.
- On average, the daily oil production increased two to three times and for the successful operations, the additional oil was between 2000 and 18,000 tons/well. The thermal stimulating effect on oil production lasted for 2–3.5 years.
- The most efficient operation took place at the well 1215 where the front propagation period was a month and the cumulative air injected was relatively low, at 220,000 m^3. This well was put in production only due to the CISC operation. The cumulative oil produced before CISC operation was 400 m^3 during a 2-year period. After CISC operation, the well produced continuously for 12 years with peak oil production of around 11 m^3/day.
- For successful operations, the AOR was less than 1500 std. m^3/m^3. Generally, the performance was better at the wells where the front propagation periods and the corresponding cumulative air injected were lower.
- The decrease of water cut occurred in all successful CISC operations. Generally, after CISC, the wells began flowing for a certain period.
- The shut-in time was based on the burning period length and it was between 3 and 18 days. For sand consolidation cases, better results were obtained, when the shut-in periods were greater than 10 days.

More details on the operations are given elsewhere (Trasca and Paduraru, 1993; Turta et al., 1985).

At both Suplacu and Videle–Balaria, after the well was put back into production, some sand influx was still present at the beginning. However, after a sand cleaning, sand influx generally disappeared. In reality, after the burn period, clean unconsolidated sand exists. Only when hot oil is flowing over, does the consolidation occur.

Both in Albania and Romania operations, in those wells for which the ignition operation was doubtful, the oil stimulation effect was low. Nevertheless, sand influx was completely eliminated at some wells.

All CISC operations described so far were designed *a priori* as CISC operations. However, in Romania, on several occasions, the normal ISC projects (continuous ISC in a pattern) were converted into CISC operations when air injection was stopped and the well was put into production. *This forced conversion* was related with too low compressor pressure, as compared to the air injection pressure required; it was an undertaking of last resort. For nine cases of this kind in different reservoirs, it was clearly concluded that the CISC operations were unsuccessful if the injection wells were converted into production after very long periods of active combustion. In such cases, the wells were put into production only after long workover periods (3–12 months) and the oil rates achieved by the wells were still lower than the neighboring primary wells. For instance, at Balaria, the wells C1, C2, and C4, which were converted in CISC operations after 1.5–4 years of continuous ISC, had production rates of 0.5–1 m^3/day/well, versus the average reservoir rate of 2 m^3/day/well, for primary production wells (Turta et al., 1985).

Most of the CISC operations reviewed so far were not optimized. Based on all technological advances made so far, it can be envisaged that their efficiency can be significantly improved. Therefore, it is believed that, to some extent, these operations could replace the CSS operations in those places where water for steam generation is at premium.

18.4.2 Increase of Injectivity for Water Injection Wells

Six water injection wells in a sandstone reservoir (Oliver reservoir, Indiana) of low permeability (15 mD), and containing a 9 mPa·s viscosity oil, were stimulated by CISC. On average, water injection rates increased from 10 to 50 m^3/day, leading to a total oil production increase of 85% or a cumulative oil increase of approximately 450 tons/stimulated injector (Stallings, 1965). It seems that the formation of small thermal fractures is the mechanism for the significant increase of permeability.

18.4.3 Sand Consolidation by Hot Air Injection ("Controlled Coking")

The first results were reported for six applications in the United States (Fitzgerald, 1966), where good results were obtained for sand consolidation. The oil production stimulation was very limited, for less than 1–2 months. Hot air (up to 200°C) was injected for 7 days and then, for another 7 days only cold air was injected. After this 14 day period, the well was put into production and it produced slightly upgraded oil in the first 2 weeks. This could be connected to the fact that real ignition actually took place after the first 7 days, and an ISC front was propagated for the remaining 7 days generating a burnt-out region immediately around the well. This interpretation is supported by the fact that some sand influx occurred immediately after putting the well into production, but completely disappeared afterward.

Controlled coking operations were carried out in the former Soviet Union, where 17 operations of this kind were conducted for nine wells in the Pavlov Gor reservoir (Baibakov and Garusev, 1989). In eight of them, the operation was repeated two to three times. The reservoir is composed of unconsolidated sands and sandstones and contains oil with a viscosity of 173 mPa·s. Many of wells were shut in due to sand influx problems. The sand influx was eliminated for a period ranging from 3 to 26 months, with an average of 9 months. The incremental oil was greater when the temperature during the heating of the well (using electrical heaters) was higher, ideally 350°C, and a higher cumulative of heat was injected.

The consolidation by hot air has the disadvantage of a low control on the operation as it seems that it can result either in a straight ignition or in an SI after a prolonged period of injection; this disadvantage increases significantly when the reservoir temperature is higher.

18.5 NEW APPROACHES TO APPLY ISC IN COMBINATION WITH HORIZONTAL WELLS

Currently, there is little information in the area of horizontal well assisted ISC, for both laboratory research and field testing.

18.5.1 Horizontal Wells Drilled in Old Conventional ISC Projects

So far, horizontal wells in conjunction with conventional ISC processes were drilled in two Canadian projects (Turta, 1994). In both cases, they were used as producers.

In the Eyehill Project, Saskatchewan, three wells with horizontal legs of 1000–1200 m were drilled after 2 years from the complete termination of the dry combustion, which was active for about 10 years; it was operated in adjacent patterns with relatively low air rates, leading to an oil recovery of 10%. The reservoir is underlined by bottom water column height as high as 15 m. Oil viscosity at reservoir conditions is around 2000 mPa·s and the net pay thickness is 5–8 m.

Of the three wells, one had very good production performance. This well produced for a long time with oil rates of 55–60 m^3/day. The good behavior is explained by the fact that it was located very close to the boundary of the project area (*just in the oil bank created by the former ISC process*) but did not intercept any burned zones. The other two wells did not perform very well as they were located either too far from the project area or in the burned zone.

In the second project, at Battrum, Saskatchewan, one horizontal well was drilled in conjunction with the commercial wet combustion process which has been in progress in this reservoir since 1965 (Ac et al., 1993). This process takes place in a reservoir having a relatively low oil viscosity (70 mPa·s) and a net pay thickness of 9–18 m (see Table 18.4). The horizontal well had a horizontal leg of 610 m and it was positioned between the gas tongue and the water tongue of an exploitation using patterns system. The performance of this well was very good as the oil rate increased 5–10 times (from 3 to 15 m^3/day for a vertical well, to 35 to 75 m^3/day for the horizontal well) with a decrease in water cut from 90% to 20%. Another very important advantage of the horizontal well was the substantial reduction of operating problems like sand influx and emulsions, to the extremely low drawdown during oil flow toward the horizontal well.

18.5.2 Long-Distance Versus Short-Distance Displacement

With the advent of horizontal wells, an alternate approach to improved recovery of heavy oil is being used. Instead of attempting to move mobilized oil over hundreds of meters, long-distance oil displacement (LDOD), in a pattern or line drive flood, short-distance oil displacement (SDOD), typically over a few meters or tens of meters, is carried out. In many situations with heavy oil, due to the oil high viscosity, the displacement to producers located long distances away from injectors is not practical because of the high injection pressures required to sustain reasonable injection rates. Often, the exceedingly high injectant/oil mobility ratio leads to gravity override/underride or extensive channeling resulting in low volumetric sweep efficiency, poor sustained oil rates, low recovery, and marginal or poor economics. Either way, in most heavy oil situations, LDOD processes do not work.

When an LDOD process takes place in a heterogeneous layered formation, the tendency of preferential advancement of the displacement front is a function of the overall (integrated) flow resistance in every layer, from the injection well to the production well. The mobility ratio is crucial in intensifying or reducing this tendency (frontal instability), initiated by variable permeability. Thus, in LDOD processes, performance depends on distribution of properties (mainly permeability and viscosity of the injectant and oil) along the flow path, between injection and production wells.

Generally, the deterioration of performance of any LDOD process may be due to:

- Rock heterogeneity, leading to channeling of the injectant.
- Unfavorable mobility ratio (Mr) between the displacing fluid/oil. $Mr = (Kr/\mu)d/(Kr/\mu)oil$.
- Gravity segregation leading to overriding/underriding of the injected fluid.

While heterogeneities and gravity segregation can have a combined positive or negative effect, the unfavorable mobility ratio always has an adverse effect and disproportionably increases the negative effect of the other two parameters. Therefore, the negative effects of rock heterogeneity and gravity segregation are aggravated when the viscosity of oil increases, as mobility ratio becomes more unfavorable. This aspect is very important for light oil reservoirs and it is critical for heavy oil reservoirs.

Almost all conventional floods are LDOD processes. In spite of being relatively inefficient, these are still economically acceptable when oil viscosity is low (<10 mPa·s), but the necessity to switch to SDOD becomes acute in heavy oil situations, especially when combined with horizontal wells.

In SDOD processes, mobility (viscosity) of injectant remains important but is not so dominate as in LDOD processes. The more significant feature is the short travel distance for any oil particle before it is produced. It must be pointed out that SDOD processes are specifically designed for unfavorable mobility ratios between injectant and oil. *Instead of looking for solutions to make the mobility ratio more favorable (like in polymer flooding), the SDOD processes reduce its importance.* This approach is, by far, more practical. For most heavy oil pools, even if a favorable mobility ratio of one is attained, injection pressures required to sustain an economically acceptable oil rate may be impractical or lead to fracturing, which is generally undesirable in displacement processes.

There are two types of SDOD processes:

1. With the swept zone surrounding the horizontal producer and forming an ever-expanding chamber (displacement front is quasi-parallel to horizontal producer); SAGD type.
2. With the displacement front quasi-perpendicular to the horizontal producer. Swept zone starts from the toe and moves toward the heel; toe-to-heel (TTH) displacement type.

The first type (SAGD process) uses two parallel horizontal wells, one for injection and the other for production: the second type is characteristic for THAI process and uses a vertical injector and a horizontal producer with the toe of the producer located in proximity to the toe of the injector (Figure 18.36). In the first type, streamlines are perpendicular to the horizontal section of the producer and the well produces through the entire horizontal section during the producing life. In the second type, the streamlines bend toward the producer, characterizing a distribution of flow, which results from a combined effect of drive (parallel to the horizontal well) and injectant/oil gravity segregation. Successively smaller sections of the horizontal well are thus utilized for production. A mobile oil zone (MOZ) forms slightly ahead of the displacement front. In fact, this zone is a "double" MOZ, as mobile oil is flowing down to the horizontal wellbore and the zone itself moves from toe to heel. In principle, for very viscous oils, there is very little flow in the region ahead of the MOZ and displacement performance is not affected by that region.

An important feature of SDOD processes is the mitigation of heterogeneity effect. The negative effects of heterogeneity are abated, mainly due to use of

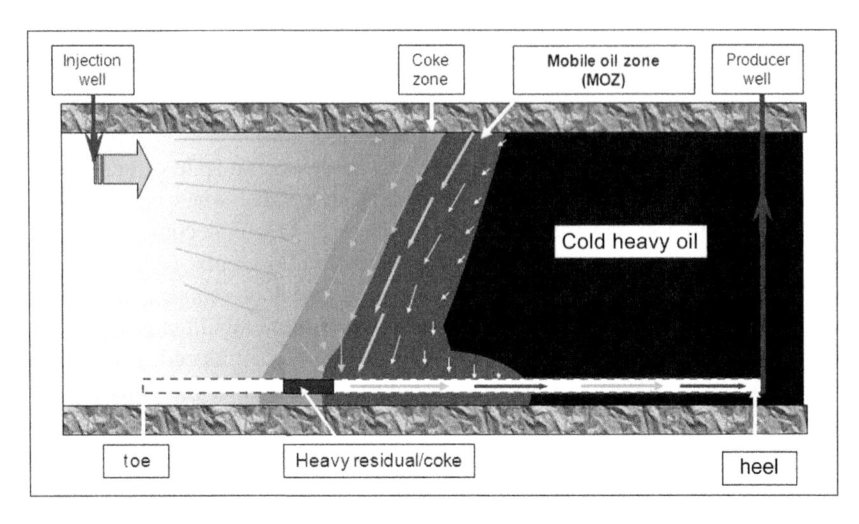

FIGURE 18.36 Schematics of THAI process. Fluid Banks.

the horizontal well as a linear sink. Any formed finger will not be allowed to become the large dominant finger as the streamlines would not follow the bedding planes. Also, these processes fully utilize gravity segregation.

The concept of SAGD was introduced in the late 1970s and the idea of TTH displacement processes appeared in 1992. TTH displacement processes can be applied as THAI™, with its variant catalytic THAI™ or CAPRI™. First THAI was investigated and based on THAI, the CAPRI was developed by using the horizontal leg of the horizontal well not only as an oil producer but also as a catalytic reactor.

Figure 18.35 shows the schematic of a THAI process, which can also be used to describe the CAPRI process; the horizontal leg of the producer is located in the lower part of the pay and a vertical well is used for injection. The toe of the horizontal well is close to the vertical well, but offset at a distance. This is a critical characteristic of a TTH process. Gravity contributes both to the stabilization of the displacement front and to a reduction in injection pressure by providing part of the hydraulic head necessary for liquid flow. Heated fluids flow downward toward the horizontal leg of the producer mostly within the MOZ.

The following section describes the current understanding of THAI process derived mainly from laboratory and field testing, and numerical simulation.

18.5.3 THAI Process

Laboratory testing: THAI was extensively tested in a low-pressure (60 psi) laboratory 3D model using Wolf Lake heavy oil ($\approx$50,000 mPa·s) and Athabasca bitumen.

More than 100 THAI tests were carried out in both dry and wet combustion modes, and results showed that THAI performed better than the classical ISC using vertical wells. The propagation of the front was not associated with gas override (near total stability). Also, a significant amount of thermal upgrading took place. The viscosity of oil was reduced by a factor of 5 to values as low as 10,000 mPa·s, for Wolf Lake oil. Unlike the conventional ISC process, THAI yielded consistently upgraded oil. In conventional ISC, although an upgrading of oil occurs (Freitag and Exelby, 1998) the produced oil in the field still does not show any significant upgrading because it is mixed with nonupgraded oil. THAI laboratory tests yielded even more promising results for recovery of Athabasca oil sand where thermal upgrading resulted in a viscosity reduction from 1,000,000 to 500 mPa·s (Turta, 1994). The propagation of the front was also very stable.

THAI produces thermally upgraded oil within the reservoir. This ability was catalytically enhanced by a new process, catalytic THAI or CAPRI. The CAPRI process is a variant of THAI in which oil and entrained gases at temperatures of over 400°C pass over a catalyst placed downhole, around the horizontal leg of the producer using a conventional gravel pack placement, while the required high temperature is generated by ISC; the tests using 50,000 mPa·s Wolf Lake oil and nickel and molybdenum catalysts showed a final upgrading to 30 mPa·s. Tests using Athabasca bitumen and the same catalysts showed an upgrading to an oil viscosity of 40 mPa·s at even steadier rates. *An additional upgrading by 3°API over straight THAI is obtained by using CAPRI.* In tests conducted with both oils, the heavy metals and sulfur content of the produced oil were significantly reduced. For Wolf Lake oil tests, sulfur was reduced from 43,000to 5100 ppm and vanadium from 195 to 8 ppm; these reductions have major environmental implications.

The most remarkable feature of this process is the ease of control compared to any conventional ISC operations. THAI can be applied to formations as thin as 6 m. Unlike SAGD, THAI does not require large amounts of water and does not need to burn natural gases. These considerations are very important in certain situations.

Field testing: Field testing of a THAI process in Athabasca oil sands at Conklin, near Fort McMurray, Alberta (Whitesands Project), began in July 2006. The layout of the experimental pilot is shown in Figure 18.37. For the first of the three THAI modules/pairs, preheating by steam injection took place in the period March–July 2006 and ignition occurred in July 2006. Following the preheating, air injection began in January 2007, in the second THAI module, and in June 2007, in the third module.

The burning in the ISC front was normal and the propagation of the ISC front seemed to be normal as peak temperatures of 700°C were recorded and the oxygen utilization efficiency was almost 100%. There was no air short circuit through the horizontal producer. The oil cut increased slowly from 15% to 50%; generally, the oil rate was lower than in an SAGD producer of

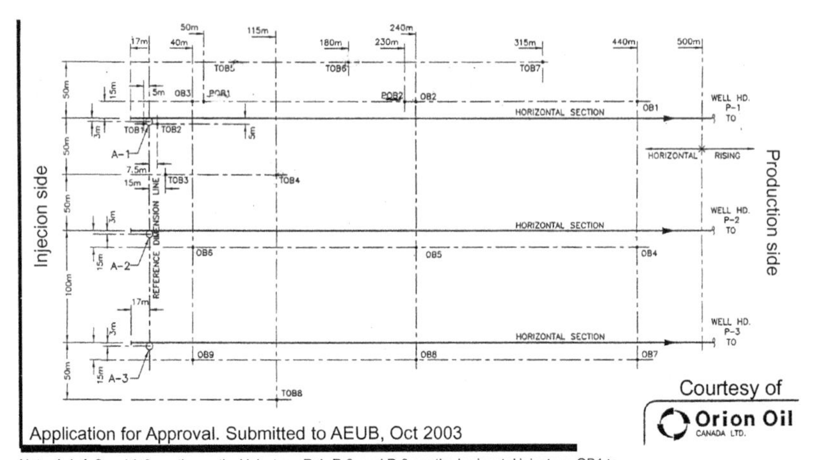

Note: A-1, A-2 and A-3 are the vertical injectors. P-1, P-2, and P-3 are the horizontal injectors. OB1 to OB9 — observation wells. TOB1 to TOB5 — temperature obs. Wells POB1 to POB2 — pressure observation wells

FIGURE 18.37 The layout of the WhiteSands THAI pilot.

the same length. The production wells have exhibited high sand production volumes and facilities (sand knockout vessels) to handle the sand were designed and installed. Two horizontal producers were redrilled.

Analysis of the produced oil has shown a consistent *in situ* upgrading effect; at times, the upgrading was up to 8 points (from 8 to 16 API), while viscosity was decreased up to 100 mPa·s, as compared to 500,000 mPa·s, for the original bitumen; on the average the upgrading was approximately 4°API (www.Petrobank.com, August 2007). The test was wrapped up in September 2011. Results of this initial pilot resulted in a large-scaled commercial design.

A second THAI test started in Kerrobert field, Saskatchewan, Canada, in 2010 in a single module. Then, by the end of 2011 to start of 2012, extension to 12 pairs took place. This process takes place in a conventional heavy oil reservoir. As of July 2012, the process is under way.

Fundamental considerations on THAI application: The TTH process, in its fundamental design, places the initial point of production very close to the injection. In principle, if the horizontal section offers an extremely large conductivity, significant bypassing of the reservoir occurs. At the other extreme, if the well offers extremely low conductivity compared to that of the reservoir itself, TTH displacement may not occur, rather overriding of the air may be accentuated.

The air will have a tendency to override the oil, while heterogeneities may cause some small frontal instability with a tendency to distort the displacement front. Due to the short-distance displacement nature, any developed finger will not grow significantly. The main design objective in completion of the horizontal section will be to provide the capability to effectively drain the oil by achieving the most favorable pressure distribution along the horizontal section. Pressure distribution around the well is determined, among other factors, by lateral pressure drop within the horizontal section of the well.

Recent investigations through laboratory tests and numerical simulation, showed that in THAI, there is an important mechanism for stability, i.e., the existence of a moving coke deposit over a limited portion ahead of the ISC front (local blocking). This coke deposit was proved to generate a gas seal (Greaves et al., 2007). However, the process may be stable even without this feature (Greaves et al., 2011).

In application of SDOD processes to heavy oil and bitumen recovery, the most important aspect is the creation of an initial communication link between injectors and producers. This is known to be an important step for SAGD. However, this is *even more important* for THAI process, where it is intended to create not only the communication, but also the initial quasi-vertical ISC front and later on to anchor it properly at the toe of the horizontal well. Because the distance over which the communication is needed may be larger than in SAGD, this is a greater challenge. *The communication step is a crucial step for the THAI process because subsequent development of the processes will depend to a great extent on the quality of this initial front.*

For oil sands, in order to create the initial communication, there are only two options: either heat the inter-well region until the oil contained in the region achieves a certain minimum mobility or mechanically develop some artificial paths (fractures). In the latter case, this may hurt performance due to a lack of control on creation of such an intense heterogeneity. The first approach seems preferable and different procedures such as steam circulation in both the injector and producer can be used. Certain minimum oil mobility and some initial reservoir energy may greatly help in creating communication by an initial CSS of the injection and production wells.

18.5.4 Other ISC Approaches (COSH and Top-Down ISC)

The schematic of the combustion override split-production horizontal well (COSH) process is provided in Figure 18.38. Several ISC fronts are initiated through the vertical injectors (perforated high in the pay zone) and then propagated downward, toward a unique horizontal producer located low in the pay zone (Kisman and Lau, 1994). The process is more complex as it requires three categories of wells that must be pressure balanced for it to work; special wells just for gas venting are set up.

The COSH concept was tested through extensive numerical simulation, but it seems that subsequent laboratory tests have not conclusively proved the concept.

The schematic of the top-down ISC (TD-ISC) process is to some degree similar to that of COSH (Coates et al., 1995). Several ISC fronts are initiated

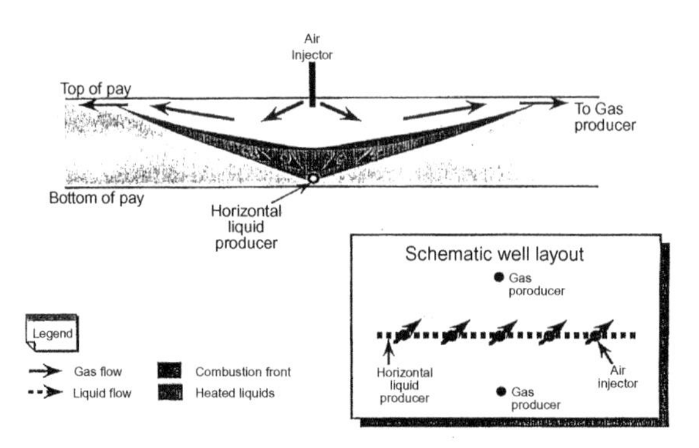

FIGURE 18.38 Schematics of COSH process (Kisman and Lau, 1994).

through vertical injectors (perforated high) and then propagated downward toward a horizontal producer located low in the pay zone, which, unlike COSH, produces both liquids and all gases; no venting gas wells exist.

18.6 OPERATION PROBLEMS AND THEIR REMEDIES

An in-depth analysis of most of the operational problems can be found elsewhere (Chu, 1977). Here the accent will be on gravity of the problem and the consequences for each problem, as far as the future of the ISC project is concerned.

The main operational ISC problems belong to two categories:

1. Critical
2. Noncritical

Generally, the critical problems are those that are serious in nature, as they can bring an end to the process, cause substantial permanent reduction of the efficiency of the process, or cause general discouragement/disappointment related to the continuation of the project.

On the other hand, noncritical problems usually lead to the increase of operational expenses. Generally, they can be solved in a certain time frame. Even if they are not totally solved, they do not entirely compromise the project; operators can often live with them to the end of the process.

Among the critical problems, the most important are:

- Escape of combustion gases or injected air to the surface.
- Escape of combustion gases/injected air to another oil/gas layer located at a lower depth.
- Explosions.

As far as the noncritical problems are concerned, the following ones will be discussed:

- Poor injectivity or a pronounced reduction of injectivity over time
- Severe gas production or pumping problems
- Sand production at reservoir temperature (low-temperature sand influx)
- Hot wells with or without a coke bridge in perforations
- Corrosion and/or erosion problems
- Channeling of the ISC front or uneven advancement of the ISC front.

18.6.1 Critical Problems

Escape of combustion gases or injected air to the surface represents the most critical problem. As a rule, if this occurs during an ISC pilot, testing is discontinued and further application of ISC on that reservoir is ruled out. If this happens in a commercial operation, the process may continue but environmental and other problems may remain permanent (Carcoana, 1990). This actually happened in the world's biggest ISC project at Suplacu de Barcau. Generally, the escape occurs either behind the casing of a poorly cemented well or by fracturing the overburden layers. The latter case seemed to be what happened at Suplacu de Barcau where the reservoir depth at the place of failure was only 35 m and there were no very compact/impermeable layers in the overburden.

Today this problem has to be handled during the design stage through an in-depth geologic analysis of overburden layers. The existence of a compact limestone layer immediately above the pay section is desirable. Otherwise, depths greater than 150–200 m should be contemplated for ISC application. In parallel, the possibility of gas migration to existent, shallow freshwater layers should also be analyzed.

The second problem, escape of injected air or combustion gases to another oil/gas layer located at a lower depth, is less critical, but can still seriously compromise the process; there is a risk of explosion during the ignition operation using a gas burner if there is a gas layer in the overburden layers. Later on in the process, the problem of confinement of injected gas in the target formation is related to either the gas escape behind the casing of a poorly cemented well or the development of new escape pathways in the overburden layers, in conjunction with the high temperature developed by the ISC front at top of the layer. For the last situation, it is now well established that a limestone layer of at least 4–5 m is enough to confine the process in the subjacent layer and to allow a separate ISC operation for the shallower oil layer.

The third critical problem, explosions, may be a serious constraint for the continuation of an ISC pilot. Explosions may occur anywhere in the system: air compressor station, air pipelines, and injection, production or observation wells.

With today's technology, the compressor station explosion is a thing of the past. The advent of nonlubricated compressors has totally solved this problem. Explosions in the air conduits are also almost totally eliminated. In the air injection wells, most the explosions took place during the ignition operations mainly when using artificial igniters, especially gas burners. This problem can be eliminated by:

- Using wells in good mechanical condition. (new wells if the old ones are not reliable)
- Using automated, advanced equipment for ignition.

During the ignition, free-of-oil perforations are a must. A standby air compressor is also a necessity. For a few months after ignition, there is a higher risk for explosion in case of an air interruption. If a standby compressor is not present, a slug of water or nitrogen should be injected to ensure the safety of the operation.

Generally, any time it is possible, wet ISC should be used rather than dry ISC in order to ensure long-term safety conditions. In production wells, explosions occur less frequently. Measuring compositions of produced gasses and ensuring component concentrations are within nonexplosive limits, by and large assures the safety of operations. As far as the observation wells are concerned, the explosions could be related to the failure/collapse of wells during their interception by the ISC front. Generally, observation wells are completed as blind wells.

18.7 NONCRITICAL PROBLEMS

At any time during an ISC process, the air injection rate should be higher than the value corresponding to the minimum air flux. If this condition is not met, the peak temperature within the combustion front will decrease continuously with time until eventually the ISC process is lost. Repeated tries to theorize the optimum (meaning "small") air injection rates have not had the appropriate fundamental background and have caused more harm than good; therefore, it is better to err on the higher side of the air injection rate than on the lower side.

Even if air injectivity is high enough in the preignition operations, after ignition and establishment of the ISC front, hot oil displacement in the cold regions generates a blockage (a pronounced reduction of gas permeability). If there are no natural channeling pathways, the increase of injectivity by man-made methods is practically impossible and the problem may not have any solution once it happens. That is why it should be anticipated during the design phase. In such cases, an extensive preheating, either by using CSS or other means, should be designed. Also, in such cases, use of large well spacing/patterns should be avoided.

In cases where the injectivity decreases *over time*, due to the reduction of permeability around the injection well, the problem can be solved by applying periodic acidifications. Simultaneously, fines or other deposits in the perforations can be eradicated by reverse circulation or by applying a chemical treatment.

Pumping problems due to emulsion generation and other causes (such as too high gas production rates) can be solved by using special pumps or producing the gases separately through the annulus. Sometimes, the most difficult problems occur during the transition from pumping to the self-flowing regime.

Sand production can be a very challenging problem. When this happens at reservoir temperature, it can be solved by conventional methods of gravel packing. However, this is not a universal solution, as in some field ISC cases gravel packing did not give complete satisfaction and its repair is extremely difficult and time consuming. In the field where the conventional ISC method is planned, massive sand production such as in the CHOPS process should be avoided as it can create pathways for ISC channeling.

Sand influx at high temperature imposed by the ISC process should be regarded seriously as it can be associated with coke bridge formation in the perforations of the production wells. Once formed in the production well, this bridge cannot be solved in any way. Therefore, the solution is to prevent its formation by keeping the BHT under 110–120°C by measuring the BHT and designing a cooling system for production wells.

While an erosion problem is treated in conjunction with high gas production rates and sand production, the corrosion can be solved after a few different chemical recipes are tested. Corrosion activity decreases as the oil viscosity increases; higher viscosity oils generate a film for tubing protection.

Even if there is no sand influx and no bridge formation, once a high temperature is recorded in a production well, the hot well constitutes a problem for any workover operations. In such cases, a specially designed drilling mud has to be used in order to kill the well but not block the formation around the well.

Uneven global advancement of the ISC front (in different directions) has been recorded in almost all ISC projects and its extent is dictated by permeability heterogeneity in different directions (permeability trends). Generally, a total remedy does not exist and preventative measures are the best solution. This assumes that the permeability trends are known before the start of the ISC process and they are accounted for in the design of well configuration. Otherwise, in the absence of preventative measures, a partial solution would be the "go with the ISC front" procedure, which means that new production wells should be placed in the regions of preferential advancement of the ISC front in order to intercept the oil banks. However, there is little experience with this procedure.

Premature arrival of the ISC front at certain production wells is caused by intensive channeling due to high permeability streaks and this is even more aggravated when these streaks are located at the top of the formation. As a direct influence of the action of gravity, the overriding phenomenon itself causes premature arrival of the ISC front, first at the production wells located updip, and then to those located on the same isobath. These negative phenomena can be significantly attenuated by applying the conventional ISC in a gravity or quasi-gravity stable system. This means a peripheral line drive started from the uppermost part of the structure and advancing downhill. In this case, the stabilizing effect works almost as in vertical gas miscible flooding.

REFERENCES

1968. First commercial Canada fireflood okayed for mobil, Oil Gas J. August 12.

1969. Texas fireflood looks like a winner, Oil Gas J., February 17.

2012. EOR Survey. Oil Gas J. April 2.

Ac, M., Ames, B., Grams, R., Pebdani, F., Sandes, R., 1993. Horizontal well economics in a mature combustion project Battrum field, Saskatchewan, Canada, SPE/CIM Third Annual One Day Conference on Horizontal Well Technology and Economics, 15 November, Calgary, Canada.

Aldea, G., Turta, A., Zamfir, M., 1988. The *in situ* combustion industrial exploitation of Suplacu de Barcau field, Romania, The Fourth International Conference on Heavy Crude and Tar Sands, 7–12 August, Edmonton.

Alderman, J.H., Fox, R.L., Antonation, R.G., 1983. *In situ* combustion pilot operations in the Wabasca Heavy Oil Sands Deposit of North Central Alberta, Canada, 58th SPE Annual Technical Conference and Exhibition, 5–8 October, San Francisco.

Arscott, R.L., David, A., 1977. An evaluation of *in situ* recovery of tar sands, *In Situ*, No. 3. March

Baibakov, N.K., Garusev, A.R., 1989. Thermal Methods of Petroleum Production. Elsevier, Amsterdam, Oxford, New York, Tokyo.

Barnes,, A.L., 1965. Results from a thermal recovery test in a watered-out reservoir. JPT November.

Belgrave, J.D.M., 2006. Air injection: timely and widely applicable!, Paper 2006-432. World Heavy Oil Conference, November Beijing China.

Bennion, D.W., 1986. Evaluation of fireflooding in the Grosmont carbonate reservoir, Report by Hycal Energy Research Laboratory, December 18.

Bhatia, A.K., Singh, D., 1998. Reservoir management of heavy oil reservoirs of north Gujarat, India. Petrotech-1998, New Delhi, India.

Bleakley, B.W., 1971. Evolution of the fry in-situ combustion project. Oil & Gas. J. 69 (18).

Bocserman, A.A., Mamedov, Y.G., Antoniady, D.G., 1991. Diverse methods spread thermal EOR in U.S.S.R. Oil Gas J. October 7.

Boberg, T.C., 1988. Thermal Methods of Oil Recovery. An Exxon Monography. John Willey and Sons Inc.

Brigham, W.E., Satman, A., Soliman, M., 1980. Recovery correlations for *in situ* combustion field projects and application to combustion pilots. J. Petroleum Technol. 32 (December).

Buchwald, R.W., Hardy, W.C., Neinast, G.S., 1972. Case history of three *in situ* combustion projects. JPT July.

Burger, J., Sourieau, P., Combarnous, M., 1985. Thermal Methods of Oil Recovery. Editions Technip, Paris.

Carcoana, A., 1990. Results and difficulties of the world's largest *in situ* combustion process: Suplacu de Barcau Field, Romania, SPE/DOE 20248. Seventh SPE/DOE Symposium on EOR, 22–25 April Tulsa, OK.

Carcoana, A., Turta, A., Burger, J., Bardon, C., 1975. Balayage et Recuperation d'un Gisement d'Huile Lourde par Combustion *In Situ*, Colloque Internat. sur les Techniques d'Exploitation et d'Exploration des Hydrocarbures, 10–12 December Paris.

Carcoana, A., Machedon, V., Pantazi, I., Petcovici, V., Turta, A., 1983. *In-Situ* combustion. An effective method to enhance oil recovery in Romania, Eleventh World Petroleum Congress, August 28–September 2, London.

Chattopadhyay, S.K., 2002. Enhanced oil recovery by *in situ* combustion process in Santhal and Balol Fields of Cambay Basin, Mehsana, Gujarat, India—A case study, Presentation to Shell Holland.

Chattopadhyay, S.K., Ram, B., Maviya, C., Das, B.K., Mittal, V.K., Meena, H.L., et al., 2003. Enhanced oil recovery by *in situ* combustion process in Balol Field of Cambay Basin, India—a case study, Indian Oil and Gas Review Symposium IORS-2003, 8–9 September, Mumbai, India.

Chattopadhyay, S.K., Ram, B., Das, B.K., Bhattacharya, R.N., 2004. Enhanced oil recovery by *in situ* combustion process in Santhal Field of Cambay Basin, Mehsana, Gujarat, India—a case study, Paper SPE 89451 SPE/DOE Symposium on Improved Oil Recovery, 17–21 April, Tulsa, OK.

Chu Chieh, 1963. Two-dimensional Analysis of a Radial Heat Wave. JPT October.

Chu, C., 1977. A study of fireflood field projects. JPT February.

Chu, C., Crawford, P.B., 1983. Chapter VI—*in situ* combustion from the book Improved Oil Recovery. Interstate Oil Compact Commission, Oklahoma City, OK, USA.

Chu, C., Hanzlik, E.J., 1983. Case histories—thermal recovery (West Newport Case), Southern Methodist University Institute for the Study of Earth & Man Enhanced Oil Recovery for the Independent Production Symposium, 9–10 October, Dallas, TX.

Clara, C., Durandeau, M., Quenault, G., Nguyen, T.H., 2000. Laboratory studies for light-oil air injection projects: potential application in Handil Field, SPE Reservoir Evaluation and Engineering, June.

Coates, R., Lorimer, S., Ivory, J., 1995. Experimental and numerical simulations of a novel top-down *in situ* combustion process, Canadian International Petroleum Conference, 19–21 June, Calgary, Alta, Canada.

Condrachi, A., Tabara, G., 1997. Review of the performance of the Suplacu de Barcau Field, exploited by thermal methods (*in situ* combustion and cyclic steam stimulation), Report ICPT, Campina, Romania.

Dietz, D.N., Weijdema, 1970. Combustion: state of the art. JPT May.

Elkins, F.E., Morton, D., 1972. Experimental fireflood in a very viscous oil—unconsolidated sand reservoir, S.E. Pauls Valley Field, Oklahoma. SPE 4086.

Elkins, L.E., Skov, A.M., Martin, P.J., Lutton, D.R., 1974. Experimental fireflood—Carlyle Field, Kansas, SPE Paper 5104, 49th Annual Fall Meeting, 6–9 October, Houston, TX.

Emery, L.W., 1962. Results from a multi-well thermal recovery test in southeastern kansas. J. Petrol. Technol.

Erickson, A., Legerski, J.R., Steece, F.V., 1993. Appraisal of high pressure air injection (HPAI) or *in situ* combustion results from deep, high-temperature, high gravity oil reservoirs, Fifteenth Anniversary Field Conference, August, Casper, Wyoming.

Farquharson, R.G., Thornton, R.W., 1986. Lessons from Eyehill. JCPT March–April.

Fassihi, M.R., Yanimaras, D.V., Kumar, V.K., 1994. Estimation of recovery factor in light oil air injection projects, SPE International Petroleum Conference & Exhibition of Mexico, 10–13 October Veracruz, Mexico.

Fitzgerald, E.M., 1966. Warm air coking—a new completion method for unconsolidated sands. JPT January.

Freitag, N.P., Exelby, D.R., 1998. Heavy oil production by *in situ* combustion – distinguishing the effects of the steam and fire fronts. JCPT April.

Gadelle, C.P., Burger, J.G., Bardon, C.P., Machedon, V., Carcoana, A., Petcovici, V., 1981. Heavy-oil recovery by *in situ* combustion—two field cases in Rumania. J. Petr. Techn. 33 (11).

Gates, C.F., Ramey, H.J., 1958. Field results of South Belrige thermal recovery experiment. In Trans. AIME 213.

Gates, C.F., Ramey, H.J., 1980. A method for engineering *in situ* combustion oil recovery projects. JPT February.

Gates, C.F., Sklar, I., 1971. Combustion—a primary recovery process, Moco Zone Reservoir-Midway Sunset Field, California. JPT August.

Gates, C.F., Jung, K.D., Surface, R.A., 1978. *In-situ* combustion in Tulare Formation, South Belrige Field, California. JPT May.

Giguerre, R.J., 1977. An *in situ* recovery process for the oil sands of Alberta, In Energy Processing, September–October, Canada.

Gjini, D., et al, 1999. Experience with cyclic *in situ* combustion in Albania, Paper 99-51 Presented at the CSPG and Petroleum Society Joint Convention, 14–18 June, Calgary.

Gottfried, B.S., 1965. A mathematical model of thermal oil recovery in linear systems. Soc. Petrol. Eng. J. 5, 196–210.

Grant, B., Szasz, J., 1954. Development of an underground heat wave for oil recovery. Trans. AIME.

Greaves, M., Javanmardi, G., Field, R.W., 1991. *In-situ* combustion in fractured heavy oil reservoirs, The Sixth European IOR Symposium, 21–23 May, Stavanger, Norway.

Greaves, M., Wang, Y.D., Al-Shamali, O., 1994. *In situ* combustion process: 3-D studies of vertical and horizontal wells, European Symposium on Heavy Oil Technologies in a Wider Europe, 7–8 June, Berlin, Germany.

Greaves, M., Xia, T.X., Turta, T.A., 2007. THAI process—theoretical and experimental observations, Eighth Canadian International Petroleum Conference, 12–14 June, Calgary.

Greaves, M., Dong, L.L., Rigby, S., 2011. Upscaling THAI: Experiment to Pilot' Canadian Unconventional Resources Conference, 15–17, November, Calgary.

Hallam, R.J., 1991. Operational techniques to improve performance of *in situ* combustion in heavy-oil and oil-sand reservoirs, SPE Paper 21773, Western Meeting of SPE, 20–22 March, Long Beach, CA.

Hallam, R.J., Donelly, J.K., 1988. Pressure-up blowdown combustion: a channelled reservoir recovery process, SPE 18071. 63rd Annual Technical Conference and Exhibition, 2–5 October, Houston, TX.

Hardy, W.C., Fletcher, P.B., Shepard, J.C., Dittman, E.W., Zadow, D.W., 1970. *In-situ* combustion performance in a thin high gravity oil reservoir, Delhi Field. SPE 3053.

Horne, J.S., Bousaid, I.S., Dore, T.L., Smith, L.B., 1981. Initiation of an *in situ* combustion project in a thin oil column underlain by water—Cado Pine Project. Paper SPE 10248.

Huffman, G.A., Benton, J.P., El-Messidi, A.E., Riley, K.M., 1983. Pressure maintenance by *in situ* combustion, West Heidelberg, Jasper County, Mississippi. JPT October.

Jensen, E., 1990a. Morgan pressure cycling *in situ* combustion project, Specialists Forum on *In Situ* Combustion, 25 October, Calgary.

Jensen, E., 1990b. Morgan *in situ* Combustion Project, Unpublished Talk from the Calgary Section CIM Heavy Oil Special Group's Specialists Forum on *In-Situ* Combustion, 25 October Calgary.

Johnson, L.A., et al., 1980. An echoing *in situ* combustion oil recovery project in a Utah Tar Sand. JPT February.

Joseph, C., Pusch, W.H., 1980. A field comparison of wet and dry combustion. JPT September.

Juranek, J., 1959. Erdolforderung durch Teilverbrennung *in situ*, Forschung und practische Anwandung in der CSSR, August, Erdol und Kohle.

Khelil, C., McCord, D.R., 1969. A study of *in situ* combustion in a segregated system. SPE 2519.

Kisman, K.E., Lau, E.C., 1994. A new combustion process utilizing horizontal wells and gravity drainage. JCPT March.

Kuhn, C.C., Koch, R.I., 1953. *In situ* combustion—newest method of increasing oil recovery. Oil Gas J. August.

Kumar, V.K., Gutierez, D., 2007. High-pressure air injection and waterflood performance comparison of the two adjacent units in Buffalo Field, Canadian International Petroleum Conference, 12–14 June, Calgary.

Kumar, V.K., Fassihi, M.R., Yanimaras, D.V., 1994. Case history and appraisal of the Medicine Pole Hills Unit Air Injection Project, SPE/DOE Ninth Symposium on IOR, 17–20 April, Tulsa, OK.

Long, R.E., Nuar, M.F., 1982. A study of Getty Oil Co.'s successful *in situ* combustion project in the Bellevue field, SPE/DOE Paper 10,708.

Machedon, V., 1994. Romania—30 years of experience in *in situ* combustion, DOE/NIPER Symposium *In Situ* Combustion Practices—Past, Present, and Future Application, 21–22 April, Tulsa, OK.

Machedon, V., Popescu, T., Paduraru, R., 1993. Application of *in situ* combustion in Romania, Joint Canada–Romania Heavy Oil Symposium, 7–13 March, Sinaia, Romania.

Mamedov, Y.G., Bocserman, A.A., 1992. Application of improved oil recovery in U.S.S.R., SPE/DOE 24162, SPE/DOE Eight Symposium on EOR, 22–24 April, Tulsa, OK.

Marchesin, L.A., 1982. Gregoire Lake Block 1 Pilot, 1976–1981, 33rd Annual Technical Meeting of CIM, 6–9 June, Calgary.

Margerrison, D.M., Fassihi, M.R., 1994. Performance of Morgan pressure cycling *in situ* combustion project, SPE Paper 27793, SPE/DOE Symposium on IOR, 17–20 April, Tulsa, USA.

Martin, W.L., Alexander, J.D., Dew, J.W., Tynan, J.W., 1971. Thermal recovery at North Tisdale Field, Wyoming. SPE 3595.

Mehra, R.K., 1991. Performance analysis of *in situ* combustion pilot project, SPE Paper 21537, International Thermal Operations Symposium of SPE, 7–8 February, Bakersfield, CA.

Mehta, S.A., Moore, R.G., 1996. Heavy Crude: Energy Alterantives for Development – Heavy Oil Workshop Sponsored by UNITAR Centre for Heavy Crude and Tar Sands, vol.2. Campina, Romania, 3–6 June, pp 3–9.

Miller, K., 1987. EOR pilot review—husky experience, First Petroleum Conference of the South Saskatchewan Section of CIM, 6–8 October, Regina.

Miller, K., Staniford, K.R., 1988. Recent observations at the Golden Lake Sparky fireflood pilot. JCPT January–February.

Miller, R.J., 1994. Koch's experience with deep *in situ* combustion in Williston Basin, Key Note Address at the Forum on Field Applications of *In-Situ* Combustion—Past Performance/ Future Application, 21–22 April, Tulsa.

Moore, R.G., Belgrave, J.D.M., Ursenbach, M.G., Mehta, S.A., Laureshen, C.J., 1993. A Canadian perspective on *in situ* combustion, Joint Canada–Romania Heavy Oil Symposium, Sinaia, 7–13 March, Romania.

Nelson, T.W., McNiel, J.S., 1961. How to engineer an *in-situ* combustion project. Oil Gas J. 5.

Panait-Patica, A., Serban, D., Ilie, N., 2006. Suplacu de Barcau Field–a case history of a successful *in situ* combustion exploitation, SPE Europec/EAGE Annual Conference and Exhibition, 12–15 June, Vienna.

Petcovici, V., 1982. La Combustion *In-Situ* Assure un Taux de Recuperation Eleve dans un Important Gisement D'Huile Lourde de Roumanie. Le Sarmatien de Balaria, Second European IOR Symposium, 10–12 November, Paris, France.

Rao, N.S., et al., 1997. Results of Spontaneous Ignition Test in Balol Heavy Oil Field, SPE paper 38067, Presented at the Asia Pacific Oil and Gas Conference, Kuala Lumpur, April 14–16.

Prats, M., Jones, R.F., Truitt, N.E., 1968. *In-situ* combustion away from thin, horizontal gas channels. Soc. Petrol. Eng. J.

Roche, P., 2006. Carbonate Klondike: the next Oilsands? New Technol. Mag. Summer.

Roychaudhury, S., Rao, N.S., Saluja, J.S., 1995. Experience with *in situ* combustion pilot in presence of edge water, Paper 154, UNITAR International Conference on Heavy Oils and Tar Sands, 12–17 February, Houston, TX.

Sarathi, P.S., 1999. In-Situ Combustion Handbook–Principles and Practices. DOE, Tulsa, Oklahoma.

Schulte, W.M., de Vries, A.S., 1985. *In-situ* combustion in naturally fractured heavy oil reservoirs. Soc. Petroleum Eng. J. February.

Stallings, E.W., 1965. Thermal stimulation ups injection rate 800% in Indiana Field. Oil Gas J. July 12.

Strange, L.K., 1964. Ignition: key phase in combustion recovery. Petroleum Eng. 36 (12).

Tadema, H.J., Weijdema, J., 1970. Spontaneous ignition in oil reservoirs. Oil Gas J. December 14.

Terwilliger, P.L., 1975. Fireflooding shallow tar sands—a case history, SPE Paper 5568, Presented at 50th Annual Fall Meeting of SPE, September 28–October 1, Dallas, TX.

Thomas, G.W., 1963. A study of forward combustion in a radial system bounded by permeable media. In J. Petrol. Technol. October.

Thornton, B., Hassan, D., Eubank, J., 1996. Horizontal well cyclic combustion Wabasca air injection pilot. JPT November.

Trantham, J.C., Marx, J.W., 1966. Bellamy Field Test: Oil From Tar by Counter-flow Underground Burning SPE paper 1269.

Trasca, N., Paduraru, R., 1993. Stimulation of oil inflow and consolidation of the productive formation by cyclic combustion, In Joint Canada/Romania Heavy Oil Symposium, 7–13 March, Sinaia, Romania.

Turta, A., 1994. *In situ* combustion—from pilot to commercial application, Paper No. ISC 3 Presented at the DOE/NIPER Symposium *In Situ* Combustion Practices—Past, Present, and Future Application, 21–22 April, Tulsa, OK.

Turta, A., 1996. Review of the Balol, Santhal and Lanwa *in situ* combustion field tests, Four reports (1991, 1992, 1994 and 1996) Prepared for UNDP of UN, New York and ONGC, India, Ahmadabad.

Turta, A., 2011. Review of steam-based ignition operations for initiation of *in situ* combustion process, World Heavy Oil Congress, March, Edmonton.

Turta, A., 2012. Stability of *in situ* combustion process to the air injection stoppage, SPE Heavy Oil Conference Canada, 12–14 June, Calgary.

Turta, A., Pantazi, I., 1986. Development of *in situ* combustion on an industrial scale at Videle field. SPE Reservoir Eng. November.

Turta, A., Singhal, A., 2001. Reservoir engineering aspects of light-oil recovery by air injection. SPE Reservoir Eval. Eng. August.

Turta, A., Socol, S., Trasca, N., Ilie, N., 1985. Application of cyclic combustion in Romania, Mine Petrol si Gaze, July (in Romanian).

Turta, A., Coates, R., Greaves, M., 2009. *In-situ* combustion in the oil reservoirs underlain by bottom water, Review of the Field and Lab. Tests, Canadian International Petroleum Conference, 16–18 June, Calgary.

Turta, T.A., 2003. Assessment of thermal projects in Karazhanbas oil field, Kazakhstan, Contract Work for Petronas, March, Malaysia.

Watts, B.C., Hall, T.F., Petri, D.J., 1997. The Horse Creek Air-Injection Project: an overview, SPE Rocky Mountain Regional Meeting, 18–21 May, Casper, Wyoming.

Wells, B., 1980. Suffield Heavy Oil Pilot Project, AOSTRA's Non-conventional Oil Techn., 29–30 May, Calgary.

White, P.D., Moss, J.T., 1965. High-temperature thermal techniques for stimulating oil recovery. J. Petrol. Technol.

White, P.D., Moss, J.T., 1983. Thermal Recovery Methods. PennWellBooks, Pennwell Publishing Co., Tulsa, OK.

Wilson, Q.T., 1979. Progress report—Willow Draw Field, Attic air injection project, Park County, Wyoming, Prepared for the Energy Research and Development Administration Under Contract No ET-76-C-O2-1810.

Yu, H., Yang, B., Xu, G., Wang, J., Ren, S.R., 2008. Air foam injection for IOR: from laboratory to field implementation in Zhong Yuan Oilfield China, SPE/DOE Improved Oil Recovery Symposium, 19–23 April, Tulsa, OK.

Chapter 19

Introduction to MEOR and Its Field Applications in China

James J. Sheng
Bob L. Herd Department of Petroleum Engineering, Texas Tech University, Lubbock, TX 79409, USA

19.1 INTRODUCTION

Microbial-enhanced oil recovery (MEOR) is based on biological technology to enhance oil recovery in reservoirs. There are two essential components: microbes and nutrients. The sources of microbes could be exogenous and indigenous. Exogenous bacteria (microbes) are cultivated at the surface, while indigenous bacteria exist natively in reservoirs. The sources of nutrients could be *ex situ* and *in situ*. *Ex situ* nutrients such as nitrate and molasses are added in water and injected into reservoirs. We use molasses or any organic sugar source as a carbon nutrient because it is easily available, and nitrogen and phosphorous fertilizers as inorganic nutrients which are well understood and are routinely adopted in most trials (Maudgalya et al., 2007). *In situ* nutrients such as residual oil exist in reservoirs. If bacteria and nutrients are mixed at the surface, and microbial reaction products such as biopolymer and biosurfactants are generated at the surface. Such a process is similar to chemical injection, and probably nothing different from the reservoir point of view. Therefore, we will not discuss such process in this chapter. If bacteria or nutrients, or both in most cases, are injected, the bioreaction occurs in the large reservoir, and microbial reaction products are generated *in situ* to enhance oil recovery. Such process (*in situ* generation) is similar to alkaline injection in chemical flooding and *in situ* combustion. *In situ* generation presents a potential advantage in logistics, especially if residual oil can be used as an *in situ* carbon source. This is likely to be the most important, and possibly the only advantage over other processes. But *in situ* generation introduces a new set of technical difficulties beyond those facing other EOR processes, as discussed later in this chapter. Compared with other EOR methods, other advantages are that MEOR methods are low cost, which is similar to waterflooding cost, able to use existing waterflooding facilities with slight modification, and environmentally friendly because MEOR products are biodegradable. But overall environmental impact is unknown. Analysis of a few projects showed that a price of US$2.3–6.6 per incremental barrel of oil and this can be reduced if the scale of operation is larger (Maudgalya et al., 2007). Another advantage is that microbial activity increases with microbial growth. This is opposite to the case of other EOR additives in time and distance. The disadvantages are that the oxygen deployed in aerobic MEOR can act as a corrosive agent on nonresistant topside equipment and downhole piping; anaerobic MEOR requires large amounts of sugar limiting its applicability in offshore platforms owing to logistical problems; exogenous microbes require facilities for their cultivation; indigenous microbes need a standardized framework for evaluating microbial activity, e.g., specialized coring and sampling techniques (Awan et al., 2008).

In this chapter, we will present MEOR mechanisms, and screening criteria. Subsequently, we will present different MEOR applications.

19.2 MEOR MECHANISMS

Table 19.1 lists the oil recovery mechanisms related to microbial reaction products which are summarized from Momeni and Yen (1990), Sen (2008), and Yu (2006).

Table 19.1 lists the beneficial effects of MEOR. Some detrimental effects are that biologically produced hydrogen sulfide, i.e., souring, causes corrosion of piping and machinery; consumption of hydrocarbons by bacteria reduces the production of desired chemicals (Van Hamme et al., 2003). However, some field application showed that MEOR reduced reservoir souring (Zahner et al., 2011). Some effects could be beneficial or detrimental. For example, permeability reduction can be beneficial in some cases but detrimental in others. Negatively, microbial metabolites or the microbes themselves may reduce permeability by depositing biomass (biological clogging), minerals (chemical clogging), or other suspended particles (physical clogging). Positively, attachment of bacteria and development of slime, i.e., extracellular polymeric substances (EPS), favor the plugging of highly permeable zones (thieves zones) leading to increased sweep efficiency. Field tests showed a high percentage of success for the mechanism of permeability profile modification (Maudgalya et al., 2007). MEOR mechanisms have not been well understood and interpreted. Many questions have been raised (e.g., by Bryant and Lockhart, 2002) regarding whether these mechanisms are workable in practical reservoirs. One puzzle is the amount of microbial products generated is little compared with the chemicals injected in conventional chemical injection projects. However, a survey (in 1995) of MEOR projects (322) in the United States showed that 81% of the projects successfully increased oil production, and there was not a single case of reduced oil production (Lazar et al., 2007). We offer more discussions about MEOR mechanisms and the feasibility of these mechanisms below.

TABLE 19.1 Microbes, Microbial Products and their Roles in EOR

Microbial Products	Microbes	Roles in EOR
Gases (H_2, N_2, CH_4, CO_2)	*Clostridium, Enterobacter, Methanobacterium, Desulfovibrio.*	Increase reservoir pressure, swell oil volume, displace immobile, reduce oil viscosity, increase permeability by dissolving carbonate rocks
Acids (low MW fatty acids, formic acid, propionic acid, (iso) butyric acid, etc.)	*Clostridium*, mixed acidogens, *Desulfovibrio, Bacillus.*	Increase porosity and permeability by dissolving carbonate precipitates, reduce permeability due to clay movement, help emulsification, produce CO_2 through reaction with carbonate minerals to reduce oil viscosity and to swell oil droplets
Solvents (propanol, butanol, acetone, propan-2-diol, etc.)	*Clostridium, Zymomonas, Klebsiella, Arthrobacter.*	Reduce oil viscosity by dissolution of asphaltene and heavy components in oil, increase oil permeability by dissolving heavy components from pore throats, have cosurfactant effects
Biosurfactants (emulsan and alasan, surfactin, rhamnolipid, lichenysin, glycolipids, viscosin, trehaloselipids, etc.)	*Acinetobacter, Bacillus, Pseudomonas, Rhodococcus, Arthrobacter, Corynebacterium, Clostridium, Mycobacterium, Nocardia.*	Have various surfactant effects, e.g., reduce IFT, reduce residual oil saturation, change wettability, emulsify crude, etc.
Biopolymers (xanthan gum, pullulan, levan, curdlan, dextran, scleroglucan, etc.)	*Xanthomonas,* Aureobasidium, *Bacillus, Alcaligeness, Leuconostoc,* Sclerotium, *Brevibacterium, Enterobacter.*	Have various polymer effect, e.g., increase drive water viscosity, plug high-perm channels
Biomass (i.e., flocks or biofilms) (cells and EPS (mainly exopolysaccharides))	*Bacillus, Leuconostoc, Xanthomonas*	Act as plugging agents, displace oil by its growth, reverse wettability, reduce oil viscosity and pour point, emulsify and desulfurize oil

Bryant and Lockhart (2002) raised the constraint that the residence time, the time fluid spends within the microbial incubation zone, must be longer than the time required for a microbial product to reach a desired concentration. In other words, the metabolic rate must be large enough compared with the fluid injection rate. But they did not present whether the required metabolic rate is high enough in a typical MEOR process. This kind of constraint applies to a finite metabolic zone, either fixed or growing. If indigenous or mobile microbes are used, the residence time is the travel time between injector and producer. In these cases, even slow kinetics could eventually yield the desired concentration. Nevertheless, there remains incentive to achieve the required product concentration as rapidly as possible so as to contact as much of the reservoir as possible. From the residence point of view, a smaller well spacing requires a higher metabolic rate. Or, higher concentrations of microbes and nutrients are required so that a required concentration of microbial product can be generated before the fluid reaches the producer. These issues should be considered in designing an MEOR process.

Table 19.1 lists the microbial products of biopolymers and biosurfactants. For their mechanisms to work, they must satisfy their minimum concentrations (e.g., at least higher than a critical micellar concentration for a biosurfactant) after satisfying their adsorption requirements. Maudgalya et al. (2007) stated that the effectiveness of biosurfactant production and wettability changes is questionable because no biosurfactant, alcohols or polymers were measured in the production streams in a few waterfloods and single-well tests. The effectiveness was claimed based on pretrial laboratory core experiments and improvements in oil flow rate in these trials. Bryant and Lockhart (2002) raised a stoichiometric constraint by a fixed amount of *in situ* carbon source within a fixed microbial zone. If the microbes are mobile, all oil in the reservoir is a potential carbon source, thus such a constraint will practically not be imposed. However, microbial systems generally require several nutrients. For aerobic microbes, one nutrient required is oxygen. The solubility of oxygen in water is limited (approximate 5.6 g/m^3 at 50°C and atmosphere, http://www.engineeringtoolbox.com/). Bryant and Lockhart (2002) calculation showed that the process is limited by other nutrients, but not by oxygen in the case of a small stationary (fixed) microbial zone. For the generated biopolymers and biosurfactants to work effectively, one important parameter is the salinity in the injection fluid and in the formation water. The salinity in the biosurfactant system should be at the optimum for biosurfactants to work most effectively.

In generating a viscous biopolymer fluid *in situ*, Bryant and Lockhart (2002) argued that any heterogeneity in permeability or microbial reactivity would make mobility control less efficient, in contrast with the conventional polymer flooding. In the cases of permeability heterogeneity, their argument was that when the injected fluid would be directed toward a lower-permeability layer from a higher-permeability layer because initially high viscous biopolymer is generated in the high-permeability layer, resulting in shorter residence time in the lower-permeability layer. With the shorter residence time, the time required to generate enough microbial products may not be enough. Then high viscous fluid cannot be generated in the lower-permeability layer, and the injected fluid will continue flowing through the lower-permeability layer without generating viscous fluid. Thus mobility control would fail. Our argument is that the injected fluid contains microbes and nutrients; if more fluid is directed toward to the lower-permeability layer, the microbial reactions would be higher, or fewer constraints could impose on the microbial reactions, resulting in viscous biopolymer fluid generated in the lower-permeability layer; then the injected fluid cannot continue flowing into the lower-permeability layer, or less injected fluid will flow through the lower-permeability layer. Thus mobility control is improved like in the conventional polymer flooding. Another argument of ours is that when the fluid is diverted to the lower-permeability layer, it is true that the residence time becomes shorter due to a higher fluid velocity. However, this shorter residence time should be still longer than that in the higher-permeability layer. If viscous fluid can be generated in the higher-permeability layer, then it can be generated in the lower-permeability layer as well. Thus, it will not happen that the injected fluid continues flowing through the lower-permeability layer without generating microbial products. Our argument was supported by some published experiments. Li et al. (2005) reported an experiment where two cores were set up in parallel in the same flow system. One gravel-packed core had the initial permeability of 11,000 mD, and the other core had the initial permeability of 900 mD. The microbe used was *Enterobacter* sp. (sp. means species) (code CJF-002). The two cores were initially saturated with formation water. Two PV of liquid with 10^5/mL microbe concentration and 5% molasses was injected through the cores. The cores were closed at their ends and the microbes were cultured for 4 days. Since the 5th day, the formation water was flooded through the two cores once a day. On the 6th day, the permeability of the gravel-packed core decreased from 11,000 to 4000 mD, 64% lower, whereas the permeability of the other core decreased from 900 to 600 mD, 33% lower. The permeability reduction in the gravel-packed core was higher, because it had higher initial permeability, and more microbial liquid was injected in this core with more biopolymer generated. A separate experiment using the same microbe also showed that permeability reduction is higher in higher-permeability channels than in lower-permeability channels (Yu, 2006).

In the cases of heterogeneous reactivity, Bryant and Lockhart (2002) argued that in a lower-reactivity region, the lower extent of reaction and the resulting lower fluid viscosity attract more fluid into that region. The higher flux reduces the residence time, lowering the extent of reaction even further and reinforcing the tendency of fluid to be diverted into the lower-reactivity region. Thus, variations in reactivity are self-amplifying. Similarly, our argument is that if more fluid is diverted toward to the lower-reactivity region, the microbial reactions would be enhanced because of more microbes and/or nutrients, or fewer constraints could impose on the microbial reactions, resulting in more viscous biopolymer fluid generated. As a result, the fluid diversion to the lower-reactivity region will be mitigated.

Immobile biomass accumulates initially near wellbore along high-permeability channels, thus it acts as plugging agents and diverts the water flow into lower-permeability zones. However, if biomass accumulates too fast, the well injectivity could lose.

Because the oxygen solubility in water is much lower (about 100 times lower) than CO_2 solubility, generating CO_2 in the underground requires a large amount of water injected, if the oxygen required is supplied by the injected water only. Bryant and Lockhart (2002) calculation showed that 100 m^3 PV water needs to be injected per cubic meter of PV, if free CO_2 saturation reaches 0.1. To have enough CO_2 generated for its role in EOR, we need an alternative source of oxygen. Or oxygen is injected separately from water. If less oxygen is available, less CO_2 will be generated. Then CO_2 will play fewer roles in enhancing oil recovery. In contrast, *in situ* generation of CH_4 does not require an external source like oxygen. But for a small fixed microbial zone, the generated CH_4 is limited by the crude available. Bryant and Lockhart's (2002) calculation showed that miscible gas displacement is unlikely to contribute to MEOR because limited volumes of gases can be generated. We expect that generated gases play some roles in enhancing oil recovery, but not in the mode of miscible displacement.

From the above discussions, understanding MEOR mechanism is still far from being clear. Although many laboratory experiments have been performed and several mechanisms have been proposed, how feasible these mechanisms could work in real reservoirs is far from explored. Obviously, more fundamental research needs to be conducted to understand MEOR mechanisms.

19.3 MICROBES AND NUTRIENTS USED IN MEOR

Table 19.1 lists some of the microbes which generate different microbial products. Among these microbes, *Clostridium*, *Desulfovibrio*, *Acinetobacter*, and *Nocardia* are anaerobic microbes; *Pseudomonas*, *Xanthomonas*, *Corynebacterium*, and *Mycobacterium* are aerobic microbes; and *Bacillus*, *Leuconostoc*, *Arthrobacter*, *Enterobacter*, *Nocardia*, *Acinetobacter*, *Clostridium*, and *Pseudomonas* are falcutative microbes. The microbes used in field trials so far are generally mixed anaerobic or facultative anaerobic: *Clostridium*, *Bacillus*, Arthrobacterium, *Micrococcus*, *Peptococcus*, *Mycobacterium*, etc. (Lazar et al., 2007). *Clostridium* and *Bacillus* were the most commonly used microbes. Field trials showed a higher percentage of success when *Bacillus* and *Clostridium* were used. Spores of *Clostridium* were used more frequently. Most of the successful experiments used anaerobic microbes. Oil degrading microbes were reported in most trials but they were never specific about the types of microbes that were actually used. Microbial behavior was generally inconsistent. They could behave differently in laboratory experiments and field trials. A specific microbe could lead to successful tests at times but unsuccessful tests at other times (Maudgalya et al., 2007). Most experiments used a combination of bacteria to take advantage of their different abilities. Nitrate reducers could modify permeability and had the benefit to reduce souring. Sometimes the word "bacteria" is used indistinctively with microbes. The word "bacteria" refers to uncharacterized microbes (Daims et al., 2006).

The injected microbes must have these characteristics for them to work effectively:

- Anaerobic or facultative anaerobic, so that they can grow at the surface as well as in the underground, or aerobic which needs oxygen injected. It is better to inject a mixture of anaerobic and aerobic so that aerobic grow at the surface, while anaerobic grow in the underground.
- They can survive and grow in high temperature, high pressure, and high salinity, and they grow faster than indigenous microbes (otherwise, nutrients are taken by indigenous microbes).
- The nutrients are residual oil and inorganic salts in the formation.
- Environmentally friendly.

The number of microbes should be in the range of 10^5-10^6/mL injection water (Wang, 2005).

Nutrients are the largest expenses in the MEOR processes, and it is important that the right combination and quantity are available. Sugar or crude oil is a carbon source needed by microorganisms. The most common carbon source is molasses because it is easily available as slurry and can be pumped down a well. Other nutrients used are nitrates and phosphorous salts which are provided by fertilizers (ammonium phosphate, superphosphate, ammonium nitrate, and sodium nitrate) (Donaldson et al., 1989).

19.4 SCREENING CRITERIA

General MEOR screening criteria are summarized in Table 19.2 based on the literature information (Bryant and Lindsey, 1996; Ohno et al., 1999). Criteria proposed by different authors are slightly different. Table 19.2 includes a wider range of the data presented in the literature. These are general criteria only. Some projects were carried out outside the range defined by the criteria. The most critical parameter is formation temperature. Temperature affects enzymes (biological catalysts) function. Some microorganisms can only survive at pressures up to 20 mPa and temperature up to 80°C. Some can survive at 115°C (Brown, 2011) or 121°C (Kashefi and Lovely, 2003). Zahner et al.

TABLE 19.2 MEOR Screening Criteria

Formation temperature	<98°C, preferably <80°C
Pressure	10.5–20 mPa
Formation depth	<2400–3500 m
Porosity	>0.15
Permeability	>50 mD
Formation water TDS	NaCl <10–15%
pH	4–9
Oil density	<0.966 g/cm^3
Oil viscosity	5–50 mPa·s
Residual oil saturation	>0.25
Element	Arsenic, mercury <15 mg/L
Well spacing	40 acres

(2011) survey shows that MEOR has been successfully applied to reservoirs with oil gravity as high as 0.96 g/cm^3 and as low as 0.82 g/cm^3, at reservoir temperatures as high as 93°C, and at formation salinities as high as 140,000 ppm total dissolved solids (TDS). Formation depth is related to the temperature and pressure. For the 40 National Institute for Petroleum Energy Research (NIPER)collected worldwide MEOR projects, the average depth is 1800 ft (the deepest is 2600 ft), and the API oil gravity is 34–40 (Pautz and Thomas, 1991). One study has concluded that substantial bacterial activity is achieved when there are interconnections of pores having at least 0.2 μm diameter (Fredrickson et al., 1997). It is expected that pore size and geometry may affect chemotaxis. However, this has not been proven at oil reservoir conditions. One study (Wang, 2005) showed that microbes can flow though a formation of 30 mD. The formation water salinity and pH will affect enzymatic activity and change cellular surface and membrane thickness (Fujiwara et al., 2004). Reasonably high residual oil saturation is to ensure an economic MEOR project. Well spacing is important for MEOR success because microbes consume nutrients and grow as they move forward, thus limiting the distance microbes can survive. Table 19.2 lists only general screening criteria. For a specific mode of application, more criteria should be used.

19.5 FIELD APPLICATIONS

In this section, we will present several microbial field applications: single-well microbial huff-and-puff, microbial waterflooding, wellbore stimulation to remove wellbore or formation damage, and MEOR using indigenous microbes.

TABLE 19.3 Leng-43 Reservoir and Fluid Data

Average porosity (%)	20.5
Peremability (mD)	725
Formation depth (subsea) (m)	1410–1650
Reservoir temperature (°C)	48
Oil viscosity (mpa · s)	9620–43,000
Asphaltene content (%)	37–42
Pour point (°C)	−3–16
Formation water TDS (mg/L)	5434.7
pH	7

19.5.1 Single-Well Microbial Huff-and-Puff

During a single-well microbial huff-and-puff process, microbes and nutrients are injected in a single well; the well is shut in for several days or weeks for microbial inoculation and growth, and for the generation of microbial products; finally the well is produced. Such process is used to degrade heavy crude components (reduce oil viscosity) and to remove plugging near the wellbore. In an alternate single-well treatment, microbes and nutrients are added into a producer periodically, but the well is not shut in. The following is an example of such process to recover heavy oil (Yu, 2006).

Reservoir and Fluid Description

The pilot area, Block Leng-43, was in the Liaohe field. Some of the reservoir and fluid data are shown in Table 19.3. The block was at the late stage of steam injection. The formation water was $NaHCO_3$ type.

Microbes and Effects of Microbial Products

The selected microbes were *Pseudomonas* (code LH-18) and *Brevibacterium* (code LH-21). Their sizes were 0.4×0.8 μm and 0.5×0.6 μm. They were cultured in the medium: 40 g liquid paraffin, 2 g urea, 5 g KH_2PO_4, 0.6 g Na_2HPO_4, 0.015 g fermentation power, and 1000 mL formation water. After the culture at 48°C for 48 h, their concentrations became 6.9×10^8/mL for LH-18 and 3.9×10^8/mL for LH-21.

The generated microbial products reduced oil viscosity by 26–65%, and reduced water/oil interfacial tension by 10.7–19%, from 60 to 50 s mN/m. The water pH became slightly less than 7. Note that the interfacial tension was still very high (50 mN/m) in contrast with the ultralow interfacial tension in a typical synthetic surfactant system. Zhao et al. (2005) reported the interfacial tension reduced from 0.138 to 0.0126 mN/m at the laboratory using mixed microbes coded FM. Their oil viscosity reduction was only 16–18.5%.

Pilot Test

The two microbes, LH-18 and LH-21, were mixed at the ratio of 1:1 in the injected water. The nutrients of 0.2 kg urea, 0.6 kg KH_2PO_4, and 0.03 kg Na_2HPO_4 were added in 1 m^3 of water. During the test, first 10 m^3 nutrient liquid was injected through the annulus between tubing and casing, then 10 m^3 of the mixture of microbes and nutrient liquids were injected followed by injection of another 25 m^3 of the nutrient liquid. About 1 ton of microbe liquid was injected for each test well. Test wells were shut in for 4–7 days for microbial inoculation and growth.

Test Results

For three test wells, water cuts decreased from 50–67% to 33.3–57.1%, oil rates increased from 2–3 tons/day to 4.5–5 tons/day. Total incremental oil of 491 tons was produced from these three wells. One well produced only 7 days in 1 month with the oil rate being 12 tons/month before the treatment. After the treatment, the well produced every day with the oil rate being 1.1 tons/day. Total incremental oil of 38.5 tons was produced from this well. Among the five test wells, only one well did not respond to the treatment.

19.5.2 Microbial Waterflooding

In a microbial flooding process, microbes and nutrients are injected into a target reservoir where microbial products are generated. These microbial products listed in Table 19.1 react with the rock and fluids *in situ* to produce some residual oil. A field application of this process is presented next (Li et al., 2005).

Selection of Microbes

The microbial flooding was conducted in the Fuyu field, Jilin. MEOR research was started in 1996 for this field. First, the microbe coded #48 was selected for this field. Later, another microbe coded CJF-002 (*Enterobacter* sp.) was identified. It is anaerobic or facultative. The exponential phase (log phase) is 10–18 h. One of its microbial products is linear and long-chain biopolymer. The accumulation phase of the biopolymer is 16–24 h. The linear and long-chain polymer molecules tangle each other to form a 3-D network structure so that large-size gels can be formed. The gel can form a kind of tight biofilm of very low-permeability when its connate water is squeezed out under high hydraulic pressure. The biopolymer can be stable under oxidization, biodegradation, and at different temperatures, salinities and pH. When pH is less than 5, the biopolymer can be under enzymatic degradation (Yu, 2006). The compatibility of this microbe with the reservoir fluids and rocks was studied. Because there were other bacteria in the reservoir, these bacteria would compete with the selected microbe for nutrients and undesired microbial products could be generated. It was found that when the CJF-002 concentration was higher than 10^5/mL, the microbe could compete with the other bacteria even the bacteria concentration was higher than 10^7/mL, and desirable microbial products could be generated. To reduce the cost, corn starch replaced molasses and it was workable.

Injection Facility

To prevent undesired bacteria to grow, nutrients were supplied to the injection well heads using supply tanks, and the microbe liquid was transported through pipeline in the early phase of the test. Later, when changing test locations, it was realized that the injection pipelines had to be changed. So it was decided to use skid-mounted equipment to supply microbes and nutrients to the injection well heads. For a larger scale test, such skid-mounted equipment had its limitation. Therefore, changes were made so that microbe liquid was supplied to the injection well heads by pump tracks, while the nutrient liquid was supplied through pipelines.

Field Tests and Test Results

In the Fuyu field, five pilot tests were conducted in four blocks. We only present the test in Block Dong 24–26. Some of reservoir and fluid data are shown in Table 19.4. The formation water was $NaHCO_3$ type.

Two injectors were used. The injection rate of each well was 25 m^3/d. The microbial injection rate was 0.5 m^3/d, and the molasses injection rate was 2.5 m^3/d (10% water injection rate). Fifty-four percent sugar molasses cost 750 Chinese dollars (about US$100) per m^3 (Yu, 2006). The injection time was 60 days. After injection, the microbe concentrations at the 13 neighbor producers were 10^5–10^6/mL. The biopolymer concentrations at

TABLE 19.4 Block Dong 24–26 Reservoir and Fluid Data

Area (km^2)	0.203
Original oil in place (tons)	50×10^4
Average porosity (%)	26.9
Permeability (mD)	241
Medium grain size (μm)	0.13
Medium pore radius (μm)	3.8
Formation depth (subsea) (m)	320–450
Reservoir temperature (°C)	32
Formation water TDS (mg/L)	3617
pH	7–8
Recovery before microbial injection (%)	10.6
Water cut before test (%)	84
Average well oil rate before test (tons/day)	6.7

these producers were 40–60 mg/L. The heavy components of produced oil decreased. The average oil rate increased from 6.7 to 19.7 tons/day, and the water decreased from 84% to 70%. From this test, the total incremental oil was 2916 tons, and the resulting input–output ratio was 1:4.3. The main mechanism for this pilot test is mainly biopolymer effect.

Some Conclusions and Lessons

Some conclusions and lessons from the pilot tests in the Fuyu filed are summarized or discussed here.

- Microbial effect depended on the injection modes of microbe and nutrients. Slug injection mode was intermittent injection of (high concentration) microbes during continuous injection of nutrients. Mixed injection mode was continuous injection of mixed microbes and nutrients. It was observed that the former was more effective than the latter (Yu, 2006).
- Using fresh water was better than using produced water because produced water had some undesired bacteria.
- Inadequate injected volumes and/or concentrations of microbes and nutrients were the main reason for some poor performance (Yu, 2006).
- Before biopolymer was generated, CJF-002 microbes and nutrients could be penetrated deep into formation. Therefore, we do not have the injectivity issue.
- Good microbes like CJF-002 would be attacked by other bacteria or degraded, when they stayed in reservoirs for a long time. Their effectiveness lost in less than 5 years (Wei et al., 2005).
- Because desired microbes could be attacked by other undesired bacteria, higher concentrations of desired microbes should be injected.
- When the system was not closed, CJF-002 microbes in a culture tank disappeared in 18–24 h, or disappeared in a transportation tank in 8–12 h (Wei et al., 2005). Therefore, microbes should not be stored for a long time because any system could not be an absolutely closed system.

19.5.3 Well Stimulation to Remove Wellbore or Formation Damage

Well stimulation to remove wellbore or formation damage is to inject microbes through the annulus between tubing and casing, and to use microbial products such as biosurfactants to remove the deposits of heavy components such as wax and asphaltene in the wellbore or in the formation near wellbore. Such well stimulation is generally applied to wells of lower oil rates and with severe scales, once for every month or every 3 months. An example is presented next (Di and Lü, 2005).

A lot of well stimulation work was done in the Qianda field belonging to Jilin Oil Company. Some of the reservoir and fluid data are shown in

TABLE 19.5 Qianda Field Reservoir and Fluid Data

Average porosity (%)	16.5
Permeability (mD)	5.5
Formation depth (subsea) (m)	1250
Reservoir temperature (°C)	62
Reservoir pressure (mPa)	8.4
Formation water TDS (mg/L)	12,000–16,000
pH	6–8
Oil viscosity *in situ* (mPa · s)	48.7
Asphaltene content (%)	23.1
Water cut before test (%)	87.1

Table 19.5. Note the permeability was very low (5.5 mD) which is outside the range in Table 19.2.

In the Qianda field, typically 150–300 kg microbial liquid is injected into a well followed by injection of water or nutrient liquid. After injection the well is shut in for 5–10 days. Total 374 wellsxtimes well stimulation jobs were carried out in this field from 1997 to 2001. Two hundred fifty-four wellsxtimes were effective with 7714 tons of cumulatively incremental oil. Such amount of incremental oil was less than what was expected from such stimulation. Analysis of produced liquid samples in 2001 showed that the liquid had 3–5 types of microbes which took up more than 90% of the total microbes present in the liquid. And these types of microbes did not have strong functions to degrade heavy components or generate biosurfactants. Low concentration of microbes and lower dosage of injection were believed to be the main causes for the less-than-expected oil production.

Well stimulation was applied to other fields in Jilin from 1998 to 2001. Even using the same microbes, the performance at different wells were quite different. Some wells produced more oil, some wells did not, and some wells even lost oil because of shut-in time. Because the performance was not exciting, the well stimulation gradually lost its attraction. Some argued that to recover a large amount of oil in practice, a very large number of microbes were needed. Some laboratory tests showed that the rate of microbial reaction on oil was very slow. People started to have some doubt about this method.

19.5.4 MEOR Using Indigenous Microbes

Indigenous microbes are generally carried into a reservoir during water injection. They remain relatively stable during water injection for some period of time. In deep reservoirs, relatively less microbes may exist because of high temperature and high pressure. In shallower reservoirs, a variety of microbes can survive. Indigenous microbes rely on residual oil as the carbon source. Add air and inorganic salts such as nitrate and phosphate in injection water, so that the indigenous microbes can be created. There are two phases. In the first phase, aerobic and facultative microbes are fermented. Microbial products like acids, biosurfactants, solvents, CO_2, etc. are generated. Some of these microbial products can work as nutrients for anaerobic microbes. In the second phase, anaerobic microbes are fermented. Microbial products like CH_4 are generated. The following is a field application (Feng et al., 2005; Yu, 2006).

Reservoir and Fluid Data

The pilot test was conducted in the Kongdian field belonging to Dagang Oil Company. Some of the reservoir and fluid data in the test area, the Kong-Er-Bei block, are shown in Table 19.6. The formation water was $NaHCO_3$ type. The identified indigenous microbes were *Pseudomonas* which can generate biosurfactants, fermentative bacteria which can generate gas, acid, solvent, biopolymer, and Methanobacteriaceae.

Pilot Test and Results

From 2001 to 2002, air and nutrients were injected into reservoir five times. Each time, 6000 m^3 air and 2800 kg nutrients were injected. Results showed that although the indigenous microbes were incubated, the microbes did not grow in a large enough quantity. Limited producers were seen to have the benefits. In 2003, the amount of nutrients was doubled compared with 2001 and 2002 tests, and 5.5 m^3 fermented indigenous microbe liquid was added each time. The amount of air injected and other injection parameters were

TABLE 19.6 Block Kong-Er-Bei Reservoir and Fluid Data

Area (km^2)	1.6
Original oil in place (tons)	674×10^4
Average porosity (%)	33
Permeability (mD)	1878
Formation depth (subsea) (m)	1206.8–1412
Reservoir temperature (°C)	60.7
Formation water TDS (mg/L)	5518–6300
H_2S in formation water (mg/L)	<0.35
Oil viscosity *in situ* (mPa · s)	69.4–73
Asphaltene content (%)	25.8
Pour point (°C)	−11.8
Water cut before test (%)	94.4
Average well oil rate before test (tons/day)	10.8

unchanged. After injection of air and nutrients, water injection continued for 1 day followed by 1 day of shut in before normal water injection.

Test results showed that no phosphate or dissolved oxygen was produced; the produced water viscosity increased by 0.11 mPa · s; oil density, viscosity and heavy contents decreased by some levels. Near injectors and producers, the concentration of sulfate-reducing bacteria decreased (rate up to 3364 μg S^{2-}/(L d)), and *Pseudomonas* increased. Methanobactericeae increased up to 10^5/mL near injectors, three orders higher than before testing and the growth rate up to 97 μg CH_4/(L d). Saprophytic bacteria reached 10^7/mL.

In the produced water, fermentative bacteria increased in the 70% affected producers by up to 10^9/mL, five to seven orders of magnitude higher. *Pseudomonas* increased in the 30% affected producers by up to 10^6/mL, three orders of magnitude higher. Saprophytic bacteria increased in the 50% affected producers by up to 10^3/mL, one order of magnitude higher. Methanobactericeae increased in the 50% affected producers by up to 10^4/mL, one to three orders of magnitude higher. The rate to produce CH_4 increased only in the 40% producers by 25 times, with the maximum rate of 26.3 μg CH_4/(L d). SRB reduction rate increased in the 50% producers by 50 times, with the maximum rate of 205.8 μg S^{2-}/(L d). Because the formation water had low concentration of SO_4^{2-} (80 mg/L), no H_2S was detected.

The amount of methane in the produced water was not high. The produced water was observed to have acetate in 70% producers. The maximum acetate concentration reached 172.6 mg/L; it increased by 43.3 times. Near injectors or in injected water, acetate concentration reached 30 mg/L; it increased by 11.6 times. Isobutyrate was also observed in producers. Its concentration increased then decreased, which indicated that the injected nutrients were not enough.

The injected components were monitored. Almost no phosphate, dissolved oxygen, nitrate, etc. were detected in the affected producers. The pH varied from 7.5 to 9.0. Carbonate ions increased by 100−350 mg/L. Some wells had increased TDS. The interfacial tension was 8−25 mN/m with less than 20 mN/m in 85% of the affected producers. CH_4 content increased by 3−5%. Oil viscosity reduced by 7.7%.

For the affected 22 producers, oil increased and water cut decreased in 2 producers; decline rate reduced in 7 producers but these 7 producers had other well treatments to boost production, the decline rates unchanged in the other 13 producers.

For the 11 injectors, 3 injectors had lower startup pressure, indicating microbes removed some damage near the wells. The other eight injectors had higher resistance factor, indicating microbes generated polysaccharide and emulsions to increase flow resistance.

Lü et al. (2003) pointed out that because nutrients like carbon, nitrate, and phosphate are common nutrients to a target microbe and other undesired microbes as well, it becomes difficult to inject nutrients specifically for the target indigenous microbe. It would be more practical to culture the target microbe in a large quantity, and to inject the microbe together with nutrients into a reservoir.

ACKNOWLEDGMENTS

The review comments from Dr. Lewis R. Brown are appreciated.

REFERENCES

Awan, A.R., Teigland, R., Kleppe, J., 2008. A survey of North Sea enhanced-oil-recovery projects initiated during the years 1975 to 2005. SPE Reservoir Eval. Eng. 11 (3), 497–512.

Brown, L.R., 2011. Personal communication, July 15.

Bryant, R.S., Lindsey, R.P., 1996. World-wide applications of microbial technology for improved oil recovery. Paper SPE 35356 Presented at the SPE/DOE Improved Oil Recovery Symposium, Tulsa, OK, 21 – 24 April.

Bryant, S.L., Lockhart, T.P., 2002. Reservoir engineering analysis of microbial enhanced oil recovery. Reservoir Eval. Eng. October, 365–373.

Daims, H., Taylor, M.W., Wagner, M., 2006. Wastewater treatment: a model system for microbial ecology. Trends Biotechnol. 24 (11), 483–489.

Di, S.-J., Lü, Z.-S., 2005. Research of application of microbial technologies in the Jilin field. Nanfang Oil Gas 18 (3), 54–59.

Donaldson, E.C., Chilingarian, G.V., Yen, T.F. (Eds.), 1989. Developments in Petroleum Science: Microbial Enhanced Oil Recovery. Elsevier.

Feng, Q.-X., Ma, L.-J., Ni, F.-T., Zhou, L.-H., 2005. MEOR pilot tests in Daqang. In: Yan, C.-Z., Li, Y. (Eds.), Tertiary Oil Recovery Symposium. Petroleum Industry Press, Beijing, China, pp. 143–149.

Fredrickson, J.K., McKinley, J.P., Bjornstad, B.N., Long, P.E., Ringelberg, D.B., White, D.C., et al., 1997. Pore-size constraints on the activity and survival of subsurface bacteria in a late Cretaceous shale-sandstone sequence, northwestern New Mexico. Geomicrobiol. J. 14, 183–202.

Fujiwara, K., Sugai, Y., Yazawa, N., Ohno, K., Hong, C.X., Enomoto, H., 2004. Biotechnological approach for development of microbial enhanced oil recovery technique. Pet. Biotechnol. Dev. Perspect. 151, 405–445.

Kashefi, K., Lovely, D.R., 2003. Extending the upper temperature limit for life. Science 301 (5635), 934. 10.1126/SCIENCE.1086823.

Lazar, I., Petrisor, I.G., Yen, T.F., 2007. Microbial enhanced oil recovery (MEOR). Pet. Sci. Technol. 25 (11), 1353–1366.

Li, X.-C., Cui, J., Hong, C.-X., 2005. Research and application results of microbial technologies in the Fuyu field, Jielin. In: Yan, C.-Z., Li, Y. (Eds.), Tertiary Oil Recovery Symposium. Petroleum Industry Press, Beijing, China, pp. 279–286.

Lü, Z.-S., Di, S.-J., Wang, L.-F., 2003. Applications of DNA detecting technology in MEOR. Xingjiang Pet. Geol. 24 (2), 164–166.

Maudgalya, S., Knapp, R.M., McInerney, M.J., 2007. Microbially enhanced oil recovery technologies: a review of the past, present and future. Paper SPE 106978 Presented at the SPE Production and Operations Symposium, Oklahoma City, OK, 31 March–3 April.

Momeni, D., Yen, T., 1990. Introduction to microbial enhanced oil recovery. In: Yen, T. (Ed.), Microbial Enhanced Oil Recovery: Principle and Practice. CRC Press, Boca Raton, FL.

Ohno. K., Maezumi, S., Sarma, H.K., Enomoto H., Hong, C., Zhou, S.C., et al., 1999. Implementation and performance of a microbial enhanced oil recovery field pilot in Fuyu Oilfield, China. Paper SPE 54328 Presented at the SPE Asia Pacific Oil and Gas Conference and Exhibition, Jakarta, Indonesia, 20–22 April 1999.

Pautz, J.F., Thomas, R.D., 1991. Applications of EOR Technology in Field projects—1990 update (NIPER—513), January.

Sen, R., 2008. Biotechnology in petroleum recovery: the microbial EOR. Prog. Energy Combust. Sci. 34 (6), 714–724.

Van Hamme, J.D., Singh, A., Ward, O.P., 2003. Recent advances in petroleum microbiology. Microbiol. Mol. Biol. Rev. 67 (4), 503–549.

Wang, W.-D., 2005. MEOR studies and pilot tests in the Shengli oilfield. In: Yan, C.-Z., Li, Y. (Eds.), Tertiary Oil Recovery Symposium. Petroleum Industry Press, Beijing, China, pp. 123–128.

Wei, Z.-S., Wang, F.-F., Lü, Z.-S., Di, S.-J., 2005. Field practice and understanding of MEOR in the Jilin field. In: Yan, C.-Z., Li, Y. (Eds.), Tertiary Oil Recovery Symposium. Petroleum Industry Press, Beijing, China, pp. 129–137.

Yu, L., 2006. Microbial technologies to enhance oil recovery. In: Shen, P.-P. (Ed.), Technical Advances in Enhanced Oil Recovery. Petroleum Industry Press, Beijing, China, pp. 276–312.

Zahner, R.L., Tapper, S.J., Marcotte, B.W.G., Govreau, B.R., 2011. What has been learned from a hundred MEOR applications. Paper SPE 145054 Presented at the SPE Enhanced Oil Recovery Conference, Kuala Lumpur, Malaysia, 19–21 July.

Zhao, H.-T., Chen, H., Wang, C.-L., Zhang, S.-H., 2005. Field testing of microbial profile modification in the Wenmingzhai field. In: Yan, C.-Z., Li, Y. (Eds.), Tertiary Oil Recovery Symposium. Petroleum Industry Press, Beijing, China, pp. 138–142.

Chapter 20

The Use of Microorganisms to Enhance Oil Recovery

Lewis Brown

Mississippi State University Biological Sciences, 449 Hardy Road, Room 131 Etheredge Hall, P.O. Box GY, Mississippi State, MS 39762, USA

20.1 ORIGIN OF THE MEOR CONCEPT

The use of microorganisms to increase oil recovery from petroleum reservoirs is commonly referred to as microbial-enhanced oil recovery (MEOR). At times microorganisms are also employed to remove a buildup of hydrocarbons around the wellbore and while this will help increase the flow of oil from the reservoir, it is not truly MEOR. Generally speaking, only 10% of the original oil in a reservoir is recovered during primary production (Ollivier and Magot, 2005). Even after secondary recovery processes, such as waterflooding, nearly two-thirds of the original oil is left in the ground (Brown, 2010).

The ever-increasing demand for oil is not going to be satisfied by finding new oil fields, but rather by finding ways to recover a greater percentage of oil from the existing oil deposits. A multiplicity of tertiary methods of recovering oil have been tried, including polymer flooding, surfactant flooding, alkaline flooding, injection of steam, or even *in situ* combustion, but none of these methods have been adopted by the oil industry. In studying the natural disappearance of oil from the environment, Beckman pointed out as early as 1926 that the world's supply of oil is limited and that a large percentage of the oil remains in the earth (Beckman, 1926). He questioned as to whether bacteria might be employed to cause the oil to flow again. Nothing was ever done with his suggestion until ZoBell (1946) received a patent on his process to get more oil out of the ground. ZoBell's patent involved injecting *Desulfovibrio hydrocarbonoclasticus* and nutrients into a well. The nutrients consisted of a carbon source and oxidized sulfur compounds upon which the microorganism would grow and produce products, such as gas, surfactants, etc., that would increase the recovery of oil. In his patent ZoBell cites five mechanisms by which microorganisms can liberate more oil from the petroliferous formation. These processes are (1) dissolution of limestone and other calcareous materials, (2) production of gases, such as carbon dioxide, methane, and hydrogen, (3) production of detergents which help remove hydrocarbons, (4) attachment to solid surfaces which tends to displace oil, and (5) reduce the viscosity of the oil. However, Beck (1947) carried out extensive experiments with ZoBell's culture, but got only inconsistent results. He concluded that ZoBell's cultures would be useless in the field. ZoBell (1953) later received another patent in which he employed organisms in the genus *Clostridium* and other hydrogen-producing microorganisms. Both patents were based only on laboratory results—no field trials were ever performed.

Updegraff and Wren also obtained a patent in 1953, wherein they injected microorganisms of the genus *Desulfovibrio* and possibly a symbiont bacterium. They found that utilization of crude oil by the bacteria was extremely slow and their patent included the injection of molasses along with the bacteria to accelerate growth. Updegraff obtained another patent in 1957 wherein he suggested injecting a gas-producing facultative or obligate anaerobe along with a water-soluble carbohydrate (sugar). He also specified a number of bacteria that could be employed as the bacterium. Once again, the patent was based on laboratory work, not field experiments.

20.2 EARLY WORK ON MEOR

There is no question that aerobic microorganisms have the ability to utilize hydrocarbons to produce substances that will enhance oil recovery. Unfortunately, the anaerobic metabolism of hydrocarbons by microorganisms was unknown until the late 1980s (Heider et al., 1999). Later it was shown that microorganisms were able to anaerobically degrade oil in the subsurface (Aiken et al., 2004; Knopp et al., 2000). While the process was anaerobic, it was slow and therefore useless in MEOR. However, the procedure of adding microorganisms to reservoirs to help recover more oil continued in spite of the concern for plugging. For example, Beck (1947) as well as O'Bryan and Ling (1949) had plugging problems in their laboratory studies. Furthermore, Updegraff (1983) pointed out that the by-products of microbial metabolism alone could cause plugging. Hitzman (1962) suggested using spores rather than vegetative cells because of their smaller size. Lappin-Scott et al. (1988) argued that spores could cause plugging and suggested using ultramicrobacteria (UMB) because of their smaller size. On the basis of calculations, Jack et al. (1991) stated that injected microorganisms needed to be small, spherical, and less than 20% of the size of the pore throat in the formation. Davis and Updegraff (1954) stated that the diameter of the injected cells must be one-half the pore entry diameter. Chang and Yen (1984) even suggested using a lysogenic strain of bacteria since they will disintegrate on their own.

In spite of the problems of plugging the well, patents continue to be issued on the subject of adding microorganisms to wells as attested to by US

Patent No. 7,776,795 issued on August 17, 2010, wherein a reservoir is inoculated with a specific bacterial isolate (Keeler et al., 2010).

In addition to the size, another problem confronting the practitioner of MEOR is that of temperature. Oil degradation at temperatures exceeding 82°C had never been observed in the 1970s (Philippi, 1977). Further, Wilhelms et al. (2001) suggested that oil reservoirs were uplifted from hotter regions where the hydrocarbon-degrading bacteria would have been inactivated at temperatures greater than 80–90°C. Bernard et al. (1992) failed to cultivate microbes at temperatures higher than 82°C and Grassia et al. (1996) failed to find microbes that grew above 85°C. Roling et al. (2003) stated that anaerobic biological hydrocarbon degradation is apparently inhibited at temperatures above 80–90°C. However, Stetter (2006) reported on microorganisms that grow above 80°C up to 113°C, but he did not try to employ these microorganisms in MEOR. Azadpour et al. (1996) was able to demonstrate that one of the microbes she isolated from cores from oil reservoirs grew at temperatures up to 116°C. More recently, as reported later in this paper, there is good evidence that microorganisms were present in a core taken from a petroliferous formation at 115°C and that they apparently grew in the petroliferous formation when supplied with nitrate and phosphate (Schmitz et al., 2005).

20.3 PATENTS ON MEOR

Actually, the use of microorganisms to enhance oil recovery is not a single mechanism but a collection of mechanisms. Further, it should be pointed out that most of the reports and patents on MEOR are based on laboratory results rather than field work, and many of these studies were carried out on microorganisms isolated from production fluid, not from cores from petroliferous formations. While some of the methods advocate adding microorganisms to the wells, a few of the methods are designed to employ the microbes present in the petroliferous formations. This distinction between using laboratory data versus field data are not apparent in the name MEOR and that is one of the reasons many professional petroleum engineers are skeptical of MEOR. In the laboratory the cores from the formation (if indeed the cores came from the petroliferous formation) are only inches to a few feet in length and cannot reflect the true characteristics of the petroliferous formation. Furthermore, the only experience many petroleum engineers have with microorganisms is that they can plug wells and consequently, they are opposed to adding microorganisms to their oil reservoirs. Therefore, since many of the patents are based on laboratory results rather then field data; they do not prove or disprove MEOR. The only way to test MEOR prior to going to the field will be to use cores from the petroliferous formation and use conditions as close to those that exist in the field. The field trials must be of sufficient duration and incorporate enough scientific proof that indeed the process will enhance oil production without damaging the well or the petroliferous formation. In other words, for MEOR to be a successful technology it must be repeatable and proven under field conditions.

Hitzman (1965) points out that the problem of injecting microorganisms into the production zone of oil wells tends to plug the formation. The microorganisms tend to lodge in the area immediately adjacent to the wellbore, and thus few organisms penetrate very deeply into the formation. He proposed a unique method of getting around the problem. Instead of only drilling the well into the petroliferous formation, he suggested continuing to drill into the water-bearing formation below the petroliferous formation. The water formation is then inoculated with a species of the family *Bacteriaceae* and a species of the family *Pseudomonadaceae*. After enough water (approximately 1000 barrels) enters the oil-bearing strata, the water is shut off. The bacteria which grow at the oil–water interface will assist oil production in a variety of ways including producing gases or helping dissolve carbonates, dolomite, limestones, and other stratal materials.

A process is described by Lindblom et al. (1967) wherein a heteropolysaccharide-thickening agent, produced by a bacterium from the genus *Xanthomonas*, is injected to increase oil production. Interestingly enough they suggest adding a bactericide to the solution containing the thickening agent to prevent microbial growth.

In spite of arguments against injecting microorganisms into oil reservoirs, there are a number of investigators who still advocate this approach to recover more oil from petroliferous formations. Because of the potential economic value of a process to recover more oil from the ground, processes are still being patented. For example, in 1984 Thompson and Jack (1984) proposed injecting microorganisms capable of producing insoluble exopolymers into the petroliferous formation. The introduction of a compound that triggers the formation of the exopolymer is also injected. The exopolymer reduces the permeability of the more permeable zones and both the microorganism and the trigger compound (sucrose) are injected into the formation in an aqueous medium. Theoretically, the exopolymer will plug the more porous zones in the formation. The patent was elaborated on further by Thompson and Jack (1985).

McInerney et al. (1985) proposed injecting a specific microorganism (*Bacillus licheniformis* strain JF-2, ATCC No. 39307) into the petroliferous formation, nutrients from which the microorganism will produce the surfactant lichenysin, and sealing off the waterflooded formation. The nutrients injected were molasses, grain malts or grain worts, and one of the following sources of nitrogen—alkali metal nitrates, ammonia, alkali metal ammonium salts, protein digests, protein hydrolysates, protein peptones, or corn steep liquor.

Sometimes during the course of waterflooding as a secondary method of oil recovery, thickening agents are added to the water employed in the process. Hitzman (1984) developed a process whereby a well that had previously been injected with a polymeric viscosifier-thickening agent was subsequently injected with a microorganism capable of metabolizing the polymeric viscosifier. Obviously, the thickening agent would have to be biodegradable. A list of common genera of microorganisms is listed in his claims. Interestingly, no mention of temperature is included in the claims and the genera of microorganisms listed do not contain hyperthermophiles. Therefore, the temperature of the petroliferous formation in which this process would work will be a limiting factor.

Clark (1986) realized that it was difficult to prevent microbial cells from sticking to the material in the formation so he devised a way of getting around the problem. Microbial cells have a strong surface charge and are, therefore, adsorbed to the surface of stratal material. His method involved injecting an agent to adsorb onto the active sites in the formation thereby diminishing the number of microbial cells adsorbed. For example, when the injected sample contained 3.43×10^9 cells the untreated Berea sandstone core allowed the passage of 8.6×10^4 cells while the treated core permitted 7.9×10^5 cells to pass through the core. Obviously, more of the injected microbial cells will move further into the formation. The microbial cells will produce metabolic products, like carbon dioxide, that will enhance oil production.

The characteristics of the microorganism to be injected into a reservoir to increase oil production are delineated by Silver et al. (1989). These characteristics of the microorganism are that it is motile, a facultative anaerobe, halotolerant, thermotolerant, produces an exopolymer, and sporulates. A nutrient solution containing phosphate is injected prior to, along with, or after injection of the microorganism into the formation and hopefully into the highly permeable zones of the formation.

Bryant's (1990) patent specifies that the field in which the described process is used has a minimum of one injection well and one producing well. The microorganisms employed in her patent are a surfactant-producing microorganism (*Bacillus licheniformis*) and a species from the genus *Clostridium* which secretes a solvent. An aqueous solution of molasses is injected, and the injection well is shut in to give the microorganisms time to grow and produce the desired products.

Another patent by Silver and Bunting (1990) only has one claim but proposes to use a phosphate compound which will not precipitate out of solution, chelate alkaline soil, rare earth, and heavy metal ions and simultaneously will serve as a nutrient source for bacteria.

Sperl and Sperl (1991) reported that their methodology is for a carbonate-containing rock formation wherein a denitrifying microorganism is introduced in the presence of a sulfur-containing compound and produces sulfuric acid. The sulfuric acid dissolves the carbonate, thereby releasing oil which then can be produced. The microorganisms specified in the patent are species of the denitrifying thiobacilli.

Sheehy's (1990) patent uses the indigenous microorganisms in the petroliferous formation to help recover oil. Basically, they limit at least one of the nutrients required for growth of the culture, thereby causing the cells to be reduced in volume up to 70%. The reduced cells have increased surface-active properties which help release oil from the stratal material.

Rather than injecting microorganisms into the petroliferous formation, Clark and Jenneman (1992) patented a procedure whereby they injected only microbial nutrients into the formation. However, they also included in their patent injecting both microorganisms that are indigenous to the formation and other microorganisms that were sometimes injected into the formation. Their procedure called for sequentially injecting individual nutrients into the formation until a complete nutrient medium was generated in the formation for the purpose of enhancing oil recovery.

It is interesting to note that Sunde (1992) uses aerobic oil-degrading bacteria in his enhanced oil recovery process unless there are already aerobic oil-degrading bacteria present in the formation. He injects the bacteria and/or the mineral solution plus vitamins into the injection well and produce the oil from a nearby production well. The microorganisms employed are members of the genera *Pseudomomas, Corynebacterium, Mycobacterium, Acinetobacter*, and *Nocardia*. The injection water contains at least 5 mg/L of oxygen. Oil is produced from a nearby production well.

In a later patent, Sundae and Torsvik (2004) used facultative or anaerobic sulfate-reducing, nitrate-reducing, iron-reducing, or acetogenic bacteria that are either present in the formation or introduced into the formation. They supplied them with vitamins, phosphates, and an electron acceptor. The microorganisms grew on the oil and produced substances that released oil from the formation.

Lal et al. (2009) employed acidogenic, barophilic and, hyperthermophilic anaerobic bacterial strains for enhanced oil recovery in oil reservoirs with temperatures in the range of 70–90°C. These bacteria produce carbon dioxide, methane, biosurfactants, volatile fatty acids, and alcohols from specifically designed nutrient medium. The bacteria are species of *Thermoanaerobacterium, Thermotoga*, and *Thermococcus*. The injection water contained a carbon source, mineral nutrients, nitrogenous substrates, reducing agents, trace minerals, and vitamins. The compounds produced by the bacteria dissociate oil from the strata thereby releasing it for production.

Another recent patent by Brigmon and Berry (2009) is directed toward wells with strata in the temperature range of 20–65°C using a group of specific microorganisms (ATCC Nos. PTA-5570 through PTA-5581). The oil field must have at least one injection well and one production well. The microorganisms are injected either simultaneously or after the injection of a nutrient mixture capable of supporting the microbial population. At least one of the microorganisms should produce a surfactant while another microorganism must produce an enzyme that dislodges oil from the oil-bearing substrate. The authors also suggest using recycled production water because it will contain the desired microorganism.

The microorganism, *Thauera* strain AL 9:8 (ATCC No. PTA 9497), is the subject of a recent patent by Hendrickson et al. (2010). It is claimed that this microorganism with one or more additional microorganisms could alter the permeability of the subterranean formation to improve water sweep efficiency in a waterflood operation. This is accomplished by producing biosurfactants, mediating changes in wettability, producing polymers that facilitate mobility of petroleum, generating gases that increase formation pressure, and reducing the viscosity of oil. These activities of the microorganisms enhance the recovery of oil.

While there is no question as to the ability of microorganisms to release oil from oil sands, the introduction of microorganisms into single wells will only yield a small fraction of the oil remaining in the formation. Therefore, it seems obvious that if MEOR is to be a tertiary method of recovering more oil from oil reservoirs, the microorganisms indigenous to the petroliferous formation will have to be the agents responsible for increasing oil recovery because of their presence throughout the petroliferous formation and it has been shown repeatedly that the injection of microorganisms into the petroliferous formation does not allow them to penetrate very far into the formation.

20.4 OUR PROJECTS ON MEOR

Many of the reports in the literature claiming success of MEOR are viewed with skepticism due to lack of scientific validation of the processes. This helped lead the Department of Energy (DOE) in 1992 to initiate a program entitled "Class I Oil Program-Mid Term Activities" (Stephens et al., 2000). This led an executive of an oil company (Jim Stevens), a petroleum engineer (Alex Vadie), and a microbiologist (Lewis Brown) to submit a proposal to DOE to pursue a contract to try to extend the life of a particular oil field scheduled to be abandoned in 2004 due to dwindling oil production. Specifically, the objective of this cooperative project was to demonstrate the effectiveness of a microbial permeability profile modification (MPPM) technology to enhance oil recovery from a fluvial-dominated deltaic reservoir and to document the scientific basis of the technology. The process to be employed was developed by Mississippi State University with funds provided in part by industry and improved under a DOE grant in the early 1990s (Stephens et al., 2000; Brown, 1984). The oil field for this study was the North Blowhorn Creek Oil Unit (NBCU) situated in Lamar County, AL. At the beginning of the project, the field consisted of 20 injection wells and 32 producing wells. The producing formation was the Carter Sandstone of Mississippian Age at a depth of about 2300 ft. Analysis of a typical core sample from the petroliferous formation indicated that it was 90% quartz containing dolomite (4%). The clay fraction was mixed layer clay (3%) with 2% kaolinite and less than a trace of siderite ($FeCO_3$). Permeability of the core sample varied widely from 1 to 198 mD and porosity varied from 7% to 19%. Oil saturation varied between 34% and 45% with connate water saturation about 17%. The field was discovered in 1979 and waterflooding began in 1983.

At the outset of the project, two wells were drilled outside of the area being swept by the waterflood, specifically to obtain a core for laboratory studies. Unfortunately, a core was not obtained from the first well drilled, but a core was obtained from the second well. Special precautions were taken to insure that the inside of the core was not contaminated with exogenous microorganisms. Immediately upon withdrawal of the core barrel from the well, the core barrel was opened, the core broken into 1-ft lengths, each piece was wiped with 70% ethanol, and placed in an anaerobic jar containing a Becton Dickinson Gas Pak®, transported to the laboratory, and the anaerobic jar placed in a refrigerator at 4°C. The inner 3.5 in of the 4-in core was tested for barium (barite used in the drilling mud), but none was found indicating that drilling fluid had not penetrated the core. Samples of the core were tested and found to contain microorganisms, although few in number.

Two core plugs were drilled from a core using a sterile coring device under an atmosphere of nitrogen. Each plug was 3–4 in (76.2–101.6 cm) in length by 1.5 in (38.1 cm) in diameter. An entry and exit port were placed on opposite ends of the cores and allowed for entry and exit of liquids into and out of the core. The plugs were then inserted into special heat shrink plastic tubes and the entire assemblies inserted into thick rubber sleeves and employed in coreflood experiments in a specially constructed coreflood facility shown in Figure 20.1.

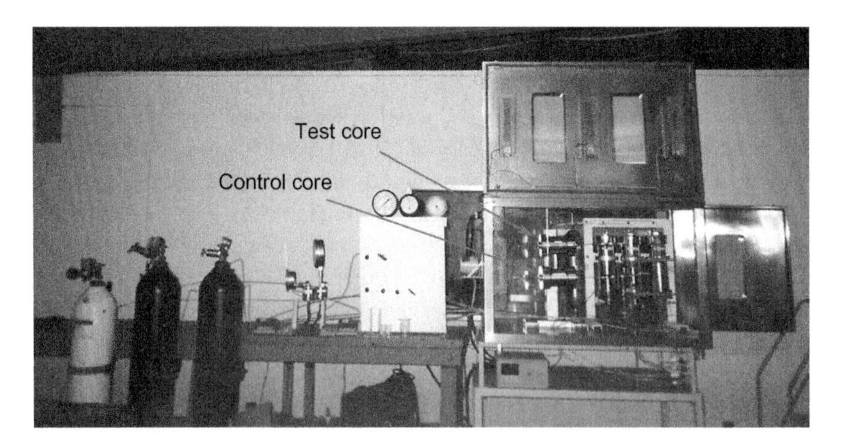

FIGURE 20.1 Coreflood facility.

Initially, sterile simulated production water was allowed to flow through both plugs for 48 h. Thereafter, the control plug received only simulated production water while the test plug received simulated production water containing potassium nitrate and disodium hydrogen phosphate. The control plug showed a steady increase in flow rate, while the flow rate from the test plug decreased with time. Oil was found in the effluent from the control plug only once while oil was found in the effluent from the test plug multiple times. This experiment was repeated three times, each time with a new core, with the same results. The test was repeated using 0.1% sterile molasses added to two of the test plugs and the results indicated that the MPPM could be accelerated by the addition of small amounts of molasses to the feeding regime. The experiment was repeated using actual injection water instead of simulated injection water with the same results.

Prior to commencing the field study, a tracer study was conducted wherein tritiated water (containing a radioactive isotope of hydrogen) was added to the injection water injected into one of the injection wells. The distance between the injector and the nearest production well was approximately 0.7 miles. Months later, tritium was detected in three of the surrounding production wells. The results indicated that it would be 7–12 months before nutrients could be expected to reach the surrounding production wells.

The field trial employed 4 of the 20 injection wells as test injectors to which KNO_3 and NaH_2PO_4 were added as shown in Table 20.1. Molasses was added to two of the test injectors. The producing wells surrounding the test injection wells were monitored for oil production. Four other injection wells were employed as control injectors. The producing wells surrounding each control well were monitored for oil production and served as control producing wells. In some cases producer wells served in more than one test or control pattern. The locations of the test and control patterns are shown in Figure 20.2.

TABLE 20.1 Nutrient Additions to the Four Test Injectors from November 1994 to April 1996

Nutrient and Concentration	Day of the Week	Test Injector Well			
		No. 1	No. 2	No. 3	No. 4
KNO_3 0.12% (w/v)	Monday	+	+	+	+
NaH_2PO_4 0.034 (w/v)	Wednesday	+	–	+	–
NaH_2PO_4 0.034 (w/v)	Friday	+	+	+	+
Molasses 0.1% (v/v)	Wednesday	–	+	–	+

+ = addition to test injector; – = no addition to test injector.

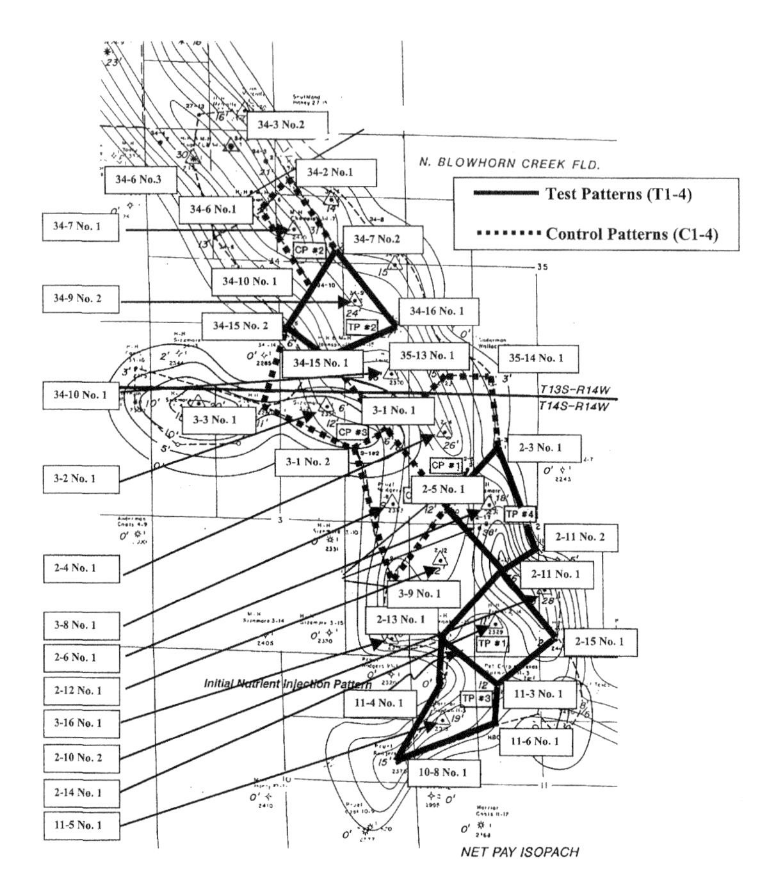

FIGURE 20.2 Isopach of NBCU Oil Field showing locations of wells in the four test and control patterns (Stephens et al., 2000).

Skids, where the nutrients were mixed prior to injection into the test injection wells, enabled the chemicals to be dissolved in 100–300 gal of water prior to being monitored into the injection water being pumped into each injection test well (see Figure 20.3).

After 12 months, 8 of the 15 wells in the test areas demonstrated an increase in oil production while none of the seven wells in the control areas showed a positive response. The nutrient additions to the injection wells were adjusted after 16 months as shown in Table 20.2. After a total of 30 months of adding nutrients to the four test injection wells, an additional six field injection wells began to receive nutrients as shown in Table 20.3. Overall, 11 of the producer wells in the test patterns gave a positive response as well as two of the control producer well. Two of the test producer wells also gave a questionable response as shown in Table 20.4.

FIGURE 20.3 Picture of skid in which chemicals are mixed for injection into petroliferous formation (Stephens et al., 2000).

TABLE 20.2 Nutrient Additions to the Four Test Injectors from April 1996 to June 1997

Nutrient and Concentration	Day of the Week	Test Injector Well			
		No. 1	No. 2	No. 3	No. 4
KNO_3 0.12% (w/v)	Monday	+	+	+	–
KNO_3 0.06% (w/v)	Monday	–	–	–	+
NaH_2PO_4 0.034 (w/v)	Wednesday	+	+	+	–
NaH_2PO_4 0.017 (w/v)	Wednesday	–	–	–	+
Molasses 0.2% (v/v)	Friday	+	+	+	–
Molasses 0.3% (v/v)	Friday	–	–	–	+

+ = addition to test injector; – = no addition to test injector.

In order to determine if the nutrients were indeed being distributed in the petroliferous formation, an additional three wells were drilled in the field. Chemical analyses showed the presence of nitrate and phosphate in the oil-producing strata in all three of the new wells, demonstrating that the nutrients were being widely distributed throughout the oil-bearing formation. Electron micrographs of cores from these three wells showed a large number of microbial cells (see Figures 20.4 and 20.5).

Originally the field was scheduled to be shut down in 2004 due to production dropping below 1500 bbl of oil per month, but the field is still producing nearly 3000 bbl per month even though the injection of nutrients had been stopped after 42 months. Currently, the field is not expected to reach

TABLE 20.3 Feed and Feeding Regime for all 10 Injectors from July 1997 to June 1998

Well No.	Monday	Tuesday	Wednesday	Thursday	Friday
34-16 No. 1		0.16N 0.04P	–	0.28M	–
2-4 No. 1	0.10N 0.03P	–	0.20M	–	–
2-6 No. 1	0.05N	–	0.30M	–	0.02P
34-9 No. 2	0.11N	–	0.18M	–	0.05P
3-16 No. 1	–	0.19N 0.05P	–	0.32M	–
34-7 No. 1	–	0.17N 0.04P	–	0.21M	–
2-10 No. 2	–	0.12N 0.02P	–	0.19M	–
11-5 No. 1	0.15N	–	0.29M	–	0.04P
2-12 No. 1	–	0.26N 0.07P	–	0.43M	–
2-14 No. 1	0.08N	–	0.47M	–	0.02P

N = percent potassium nitrate (w/v); P = percent sodium dihydrogen phosphate (w/v); M = percent molasses (v/v).

1500 bbl/month until 2018. Thus far, MPPM has been responsible for over 360,000 bbl of extra oil production and the field is expected to produce another 230,000 bbl before being shut down in 2018.

It should be pointed out that certain activities in a reservoir, such as drilling a new well, shutting-in a well, increasing the water injection rate in a new injection well, etc., can alter the performance of other wells in the field. Thus, an increase in oil production in a well is not necessarily proof that the treatment of the field was responsible for that increase.

However, in the above study, our treatment, including the addition of nitrate and phosphate to feed indigenous microorganisms, did indeed cause the increased production of oil as attested to by the following.

1. The sulfide content of the produced fluids from the wells was absent after 6 months of nutrient injection. This was not unexpected since both nitrate and nitrate-reducing bacteria inhibit sulfate-reducing bacteria.
2. Gas chromatographic profiles of oil from 10 of the producing wells in the test areas indicated the presence of new oil in the produced oil, i.e., the appearance of new smaller hydrocarbon compounds in the produced oil.
3. There was an increase in the amount of oil produced in 11 of the 15 test producers being influenced by the addition of nutrients.

TABLE 20.4 Oil Production Response from all Producer Wells Included in the Project

Well No.	Pattern(s)	After 12 months	June 1998
2-11 No. 1	T1, T4	Positive	Positive
2-15 No. 1	T1	None	Questionable
11-3 No. 1	T1, T3	None	Positive
2-13 No. 1	T1, T3	Positive	Positive
34-7 No. 2	T2, C2	None	Positive
34-16 No. 2	T2	Positive	Questionable
34-15 No. 1	T2, C3	Positive	Positive
34-15 No. 2	T2, C3	Positive	Positive
34-10 No. 1	T2, C2	None	Positive
10-8 No. 1	T3	None	Positive
11-6 No. 1	T3	Positive	Positive
11-4 No. 1	T3	None	None
2-11 No. 2	T4	Positive	Positive
2-3 No. 1	T4, C1	Positive	Positive
2-5 No. 1	T4, C1, C4	None	None
35-13 No. 1	C1	Natural Decline	Natural Decline
35-14 No. 1	C1	Shut-in	–
3-1 No. 1	C1, C3, C4	Natural Decline	Natural Decline
34-2 No. 1	C2	Natural Decline	Positive
34-6 No. 1	C2	Shut-in	–
3-3 No. 1	C3	Natural Decline	Natural Decline
3-1 No. 2	C3, C4	[a]	[a]
3-9 No. 1	C4	Natural Decline	Positive

[a] *Oil production increased due to an increase in the volume of injection water in control injector well 3-2 No. 1.*

4. The presence of an increased amount of propane in the produced gas was more like the original gas produced when the field was new.
5. The slope of decline in oil production had lessened from 18% prior to MPPM to 7.5% when 10 injection wells received nutrients for 12 months.

In light of the fact that MPPM was apparently effective at a moderate temperature, the question arises as to whether it will work at higher temperatures. It should be pointed out that the temperature of the Carter sand in the NBCU field is only 32°C, and the question arises as to the highest temperature at which MPPM would be expected to work. Oil degradation had been reported in many reservoirs but never in reservoirs at temperatures exceeding 82°C according to Philippi (1977), although Kashefi and Lovley (2003) had described a thermophilic microorganism that has a maximum growth temperature of 121°C.

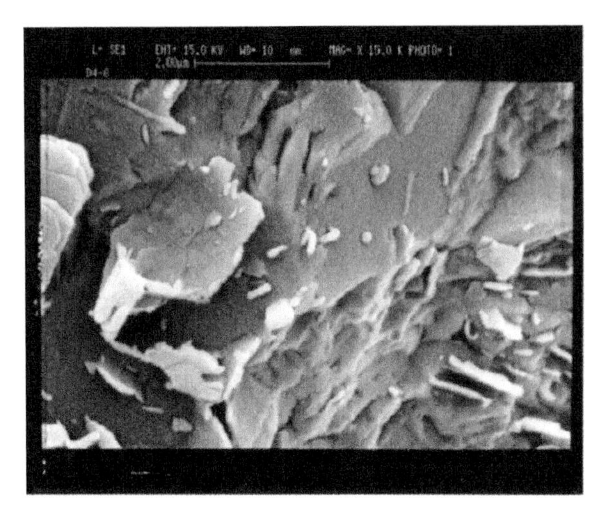

FIGURE 20.4 Electron micrograph of a sample of core from a test production well (note the scattered microbial cells) (Stephens et al., 2000).

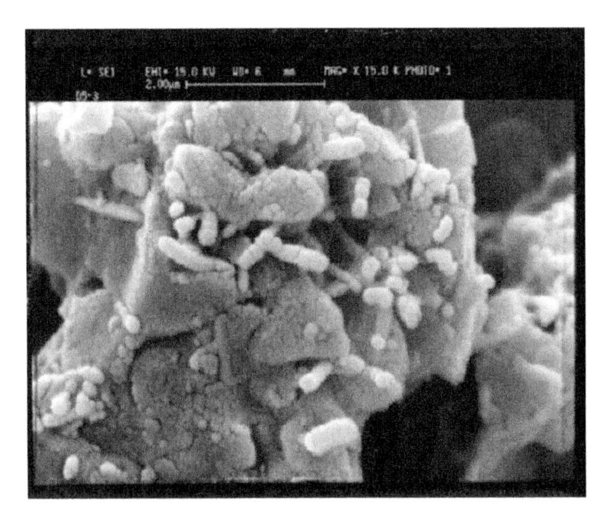

FIGURE 20.5 Electron micrograph of a sample of core from a test production well (note the large number of microbial cells) (Stephens et al., 2000).

To help answer the question in regard to temperature, an opportunity presented itself wherein we were able to evaluate MPPM in an oil field with a petroliferous formation at a temperature of 115°C. The experiment was conducted in an oil field that was currently undergoing CO_2 flooding as the secondary method of oil recovery. While no cores were available from the field, a core was obtained from a nearby field from the same petroliferous

FIGURE 20.6 Device designed to act as a growth chamber that can be incubated in an oven at a temperature of 115°C.

formation. The temperature of the petroliferous formation from which the core was obtained was 115°C. In order for MPPM to function, microorganisms must be present and be able to grow. Since the temperature of the petroliferous formation was greater than the boiling point of water (100°C), a special device was constructed to test for microorganisms capable of growing at 115°C (see Figure 20.6). Microorganisms must have water in the liquid state in order to grow and, therefore, by completely filling the device, the water will remain in the liquid state even though the temperature is greater than 100°C. Samples from the inner portion of the core were placed in these specially constructed containers along with simulated injection water containing potassium nitrate and disodium hydrogen phosphate. After 50 days of incubation at 115°C, stains were made of the contents of the containers but, microbial cells could not be distinguished from the particulate matter from the core. Electron microscopic examination also failed to differentiate between particulate material from the core and microbial cells. Therefore, the material was stained with propidium iodide, a DNA stain and DNA was shown to be present in the samples. The material was also stained with a second DNA stain (DAPI, 4′-6-diamino-2-phenylindole) and it likewise showed the presence of DNA in the samples. Thus, not only were there microorganisms present in the core material at 115°C, but they were able to proliferate when given nitrate and phosphate. Since microorganisms were present in the core from the same formation as the petroliferous formation in the oil field, it was decided to go forward with the field trial.

While the project is not complete at this time, the results thus far are certainly encouraging. One of the producer wells has shown an increase of 100 bbl a day, another well has increased production by 50 bbl, and three other wells have shown 5–10 extra barrels a day. Chemical analyses showed that there were also changes in the composition of the produced oil indicating the presence of new oil. More importantly, it was shown that microorganisms were indeed present in petroliferous formations at a temperature of 115°C and will grow when supplied with the nutrients necessary for growth. This greatly expands the number of oil fields where MPPM can potentially be employed. It also has been shown that fields employing CO_2 flooding as the secondary recovery method are candidates for MPPM. The upper temperature limit at which MPPM can be employed is not known, but according to Setter, Hoffmann, and Huber the highest theoretical limit for life is 150°C (Ollivier and Magot, 2005).

20.5 FUTURE STUDIES

A review of the literature reveals that many of the patents relating to MEOR were conducted in the laboratory, not in the field and, therefore, do not take into account the problems associated with the characteristics of the petroliferous formation themselves. As stated earlier, it is virtually impossible to inject microorganisms any appreciable distance into the petroliferous formation as shown by many workers (Beck, 1947; Chang and Yen, 1984; Davis and Updegraff, 1954; Hitzman, 1962; Jack et al., 1991; Lappin-Scott et al., 1988; O'Bryan and Ling, 1949; Updegraff, 1983). Thus only a small percentage of the oil will be recovered even if the microorganisms recover 50% of the remaining oil. For example, if an oil reservoir had a thickness of 20 ft, with a porosity of 18%, and an oil saturation of 65%, and the injected microorganisms recovered 50% of the oil within a radius of 100 ft area, it would only recover 4572 bbl of oil. However, if the injection well was in the center of four producing wells in a 40-acre tract with indigenous microorganisms in the formation, even if it only recovered 25% of the oil, it would recover 238,858 bbl of oil. Of course it would require more nutrients, but the reward would be 13 times more oil recovered than if four single wells were treated as above.

Another point worth considering is the purpose for which the microbial growth is being stimulated. If the purpose of adding microorganisms to the formation or stimulating them or those indigenous to the formation to grow and produce solvents, detergents, etc. will require large numbers of microorganisms to produce the quantities of solvents or detergents needed. Contrariwise, if the purpose of feeding the indigenous microorganisms in the formation is to alter the direction of the flow of water in a waterflood, it will require considerably fewer microorganisms to accomplish the goal. Furthermore, it has been pointed out that microorganisms behave one way in the laboratory but behave differently in the field (Kashefi and Lovley, 2003).

Obviously, in order to study the microflora of the petroleum formation, obtaining a core must be done at the time the well is being drilled. While sidewall samples can be obtained later, their value is limited. Some investigators have attempted to retrieve microorganisms from the production fluids but there is a serious question as to whether they represent microorganisms indigenous to the petroliferous formation. Most of the microorganisms in the petroliferous formations are attached to the stratal material and therefore not free-floating. Also, many of the microorganisms are in the UMB form, and our experience has been that many UMBs cannot be cultured in normal media (many of them require diluted media).

Another consideration that must be taken into account is the temperature of the petroliferous formation. Temperatures of near the boiling point of water (100°C) require special facilities in order to grow the microorganisms. Similarly, some of the waters in the petroliferous formations contain higher than normal salinities as well as pH values distant from 7.0. While our experiences have shown that most microorganisms obtained from cores of the petroliferous formations have the ability to use oil as a food, other microorganisms from the formation may have other nutritional requirements.

Most of the procedures involving MEOR are directed toward the production of chemicals that assist in the recovery of more oil, such as biosurfactants. It is evident that this will require large numbers of microorganisms situated throughout the petroliferous formation to have any impact on the recovery of more crude oil from the formation. Therefore, the question is whether you can produce or add a sufficient number of microorganisms to accomplish the production of biosurfactant and other products to enhance oil recovery without plugging the petroliferous formation. Under these circumstances it is easy to see where the method of redirecting the flow of water from the injection well to the production wells requires fewer cells to be effective, as is the case in our examples. The question then becomes how many petroliferous formations have a resident population of microorganisms. Obviously, more cores from a variety of wells need to be examined to insure the presence of microorganisms in the cores. Thus far we have found microorganisms in all of the 17 cores we have examined from petroliferous formations

It has been pointed out that the MPPM has been the most successful method of recovering more oil because indigenous bacteria were employed (Maudgalya et al., 2007). Additionally, when nitrate is used as one of the compounds to feed the indigenous bacteria, it will inhibit the sulfate-reducing bacteria that produce hydrogen sulfide that is undesirable in the produced oil and lowers the value of the produced oil. It is obvious that in order for the use of microorganisms to gain acceptance by the oil industry, quantitative measures of microbial performance must be established and verified scientifically.

REFERENCES

Aiken, C.M., Jones, D.M., Larter, S.R., 2004. Anaerobic hydrocarbon biodegradation in deep subsurface oil reservoirs. Nature 431, 291.

Azadpour, A., Brown, L.R., Vadie, A.A., 1996. Examination of thirteen petroliferous formations for hydrocarbon-utilizing sulfate-reducing microorganisms. J. Ind. Microbiol. 16, 263.

Beck, J.V., 1947. Penn grade progress on use of bacteria for releasing oil from sands. Producers Mon. 11, 13.

Beckman, J.W., 1926. Action of bacteria on mineral oil. Ind. Eng. Chem. News 4, 3.

Bernard, F.P., Connan, J., Magot, M., 1992. Indigenous microorganisms in connate water of many oil fields: a new tool in exploration and production techniques (SPE24811). In: Proceedings of the SPE 67th Annual Technical Conference. Richardson, TX. Society of Petroleum Engineers Inc.

Brigmon, R.L., Berry, C.J., 2009. Biological enhancement of hydrocarbon extraction. US Patent No. 7,472,747 B1.

Brown, L.R., 1984. Method for increasing oil recovery. US Patent No. 4,475,590.

Brown, L.R., 2010. Microbial enhanced oil recovery (MEOR). Curr. Opin. Microbiol. 13 (3), 316.

Bryant, R.S., 1990. Microbial enhanced oil recovery and compositions therefor. US Patent No. 4,905,761.

Chang, P.L., Yen, T.F., 1984. Interaction of *Escherichia coli* B and B/4 and bacteriophage T4D with Berea sandstone rock in relation to enhanced oil recovery. Appl. Environ. Microbiol. 47, 544.

Clark, J.B., 1986. Oil recovery processes. US Patent No. 4,610,302.

Clark, J B., Jenneman, G.E., 1992. Nutrient injection method for subterranean microbial processes. US Patent No. 5,083,611.

Davis, J.B., Updegraff, D.M., 1954. Microbiology in the petroleum industry. Bacteriol Rev. 18 (4), 215–238.

Grassia, G.S., McLean, K.M., Glenat, P., Bauld, J., Sheehy, A.J., 1996. A systematic survey for thermophilic fermentative bacteria and archaea in high temperature petroleum reservoirs. FEMS Microb. Ecol. 21, 47.

Heider, J., Spormann, A.M., Beller, H.R., Widdel, F., 1999. Anaerobic bacterial metabolism of hydrocarbons. FEMS Microb. Rev. 22, 459.

Hendrickson, E.R., Jackson, R.E., Keeler, S.J., Luckring, A.K., Perry, M.P., Wolstenholme, S., 2010. Identification, characterization, and application of *Thauera* SP AL 9:8 useful in microbiologically enhanced oil recovery. US Patent No. 7,708,065 B2.

Hitzman, D.O., 1962. Microbiological secondary recovery. US Patent No. 3,032,472.

Hitzman, D.O., 1965. Use of bacteria in the recovery of petroleum from underground deposits. US Patent No. 3,185,216.

Hitzman, D.O., 1984. Enhanced oil recovery process using microorganisms. US Patent No. 4,450,908.

Jack, T., Steckmeier, L.G., Islam, M.R., Ferris, F.G., 1991. Microbial selective plugging to control water channeling, Microbial Enhancement Oil Recovery—Recent Advances, 433. Elsevier.

Kashefi, K., Lovley, D.R., 2003. Extending the upper temperature limit for life. Science 301, 934.

Keeler, S.J., Hendrickson, E.R., Hnatow, L.L., Jackson, S.C., 2010. Identification, characterization, and application of *Shewanella putrefaciens* (LH4:18), useful in microbially enhanced oil release. US Patent No. 7,776,795 B2.

Knopp, K.G., Davidova, I.A., Suflita, J.M., 2000. Anaerobic oxidation of *n*-dodecane by an addition reaction in a sulfate-reducing bacterial enrichment culture. Appl. Environ. Microbiol. 66, 5393.

Lal, B., Reddy, M.R.V., Agnihotri, A., Kumar, A., Sarbhai, M.P., Singh, N., et al., 2009. Process for enhanced recovery of crude oil from oil wells using novel microbial consortium. US Patent No. 7,484,560 B2.

Lappin-Scott, H.M., Cusack, F., Costerton, J.W., 1988. Nutrient resuscitaton and growth of starved cells in sandstone cores: a novel approach to enhanced oil recovery. Appl. Environ. Microbiol. 54, 1373.

Lindblom, G.P., Ortloff, G.D., Patton, J.T., 1967. Displacement of oil from partially depleted reservoirs. US Patent No. 3,305,016.

Maudgalya, S., Knapp, R.M., McInerney, M.J., 2007. Microbial enhanced-oil-recovery technologies: a review of the past, present, and future. SPE Production and Operations Symposium, Oklahoma City, OK, 31 March – 3 April.

McInerney, M.J., Knapp, R.M., Menzie, D.E., 1985. Biosurfactant and enhanced oil recovery. US Patent No. 4,522,261.

O'Bryan, O.D., Ling, T.D., 1949. The effect of the bacteria, *Vibrio desulfuricans* on the permeability of limestone cores. Texas J. Sci. 1, 117.

Ollivier, B.M., Magot, M., 2005. Petroleum Microbiology. ASM Press, Washington, DC.

Philippi, G.T., 1977. On the depth, time and mechanism of origin of the heavy to medium-gravity naphthenic crude oils. Geochim. Cosmochim. Acta 41, 33.

Roling, W.F.M., Head, I.M., Larter, S.R., 2003. The microbiology of hydrocarbon degradation in subsurface petroleum reservoirs: perspectives and prospects. Res. Microbiol. 154, 321.

Schmitz, D., Brown, L.R., Lynch, F., Kirkland, B.L., Collins, K., Funderburk, W., 2005. Improvement of carbon dioxide sweep efficiency by utilization of microbial permeability profile modification to reduce the amount of oil bypassed during carbon dioxide flood. DOE Contract Award No. DEFC2605NT15458-05090806.

Sheehy, A., 1990. Recovery of oil from oil reservoirs. US Patent No. 4,971,151.

Silver, R.S., Bunting, P.M., 1990. Phosphate compound that is used in a microbial profile modification process. US Patent No. 5,044,435.

Silver, R.S., Bunting, P.M., Moon, W.G., Acheson, W.R., 1989. Bacteria and its use in microbial profile modification process. US Patent No. 4,799,545.

Sperl, G.T., Sperl, P.L., 1991. Enhanced oil recovery using denitrifying microorganisms. US Patent No. 5,044,435.

Stephens, J.O., Brown, L.R., Vadie, A.A., 2000. The utilization of the microflora indigenous to and present in oil-bearing formations to selectively plug the more porous zones thereby increasing oil recovery during waterflooding. Work report under DOE Contract No. DE-FC22-94BC14962.

Stetter, K.O., 2006. Hyperthermophiles in the history of life. Philos Trans. R Soc. Lond B Biol. Sci. 361, 1474.

Sunde, E., 1992. Method of microbial enhanced oil recovery. US Patent No. 5,163,510.

Sunde, E., Torsvik, T., 2004. Method of microbial enhanced oil recovery. US Patent No. 6,758,270.

Thompson, B.G., Jack, T.R., 1984. Method of enhancing oil recovery by use of exopolymer producing microorganisms. US Patent No. 4,460,043.

Thompson, B.G., Jack, T.R., 1985. Method of enhancing oil recovery by use of exopolymer-producing microorganisms. US Patent No. 4,561,500.

Updegraff, D.M., 1957. Recovery of petroleum oil. US Patent No. 2,807,570.

Updegraff, D.M., 1983. Plugging and penetration of petroleum reservoir by microorganisms. Proceeding of the International Conference on Microbial Enhancement of Oil Recovery, 80–85.

Updegraff, D.M., Wren, G.B., 1953. Secondary recovery of petroleum oil by *Desulfovibrio*. US Patent No. 2,660,550.

Wilhelms, A., Larter, S.R., Head, I., Farrimond, P., Di-Primio, R., Zwach, C., 2001. Biodegradation of oil in uplifted basins prevented by deep-burial sterilization. Nature 411, 1034.

ZoBell, C.E., 1946. Bacteriological process for treatment of fluid-bearing earth formations. US Patent No. 2,413,278.

ZoBell, C.E., 1953. Recovery of hydrocarbons. US Patent No. 2,641,566.

Chapter 21

Field Applications of Organic Oil Recovery—A New MEOR Method

Bradley Govreau[1], Brian Marcotte[1], Alan Sheehy[1], Krista Town[2], Bob Zahner[3], Shane Tapper[2] and Folami Akintunji[4]

[1]*Titan Oil Recovery, Inc., 9595 Wilshire Blvd., Suite 303 Beverly Hills, CA 90212, USA,* [2]*Pengrowth Energy Corporation 222 Third Avenue SW, Suite 2100 Calgary, Alberta T2P 0B4, Canada,* [3]*Venoco, Inc., 6267 Carpinteria Avenue, Suite 100 Carpinteria, CA 93013, USA,* [4]*Atinum E&P, 333 Clay Street, Suite 700Houston, TX 77002, USA*

21.1 INTRODUCTION

Microbial-enhanced oil recovery (MEOR) is a group of processes based on increasing oil recovery by use of microbes. Microbes are single-cell organisms that exist almost everywhere in nature. Oil reservoirs are no different and hold huge numbers of such organisms—each reservoir with its unique ecosystem. In general, the mechanisms can be grouped into those which alter oil, water, reservoir, or interfacial properties, usually through mimicry of chemical EOR processes and those that use the biological mass (biomass) for flow diversion (Gao et al., 2009). MEOR traditionally has involved the injection of particulate microbes and the appropriate nutrients needed to generate either the EOR chemicals or biomass.

There are very few documented applications of successful MEOR projects in waterfloods. Most successful MEOR applications are single producer treatments that would be better described as wellbore cleanup (Saikrishna et al., 2007). A few are attempting to create biomass for flow diversion, changing the sweep pattern in waterfloods (Bauer et al., 2011; Brown et al., 2000; Jackson et al., 2011).

Between July 2007 and the end of 2011, there have been 183 applications of organic oil recovery (OOR), which is defined in Section 21.2, to enhance recovery of North American waterfloods. The application of this process typically consists of five steps: (1) initial field screening, (2) well sampling and laboratory analysis, (3) application of the nutrient formula developed in the laboratory to a single producing well to assure the microbial response under

Enhanced Oil Recovery Field Case Studies.

TABLE 21.1 Over One Hundred and Eighty Applications Were Performed in 98 Wells Through 2011

Summary	Number of Wells	Number of Treatments	Number of Increases	Success Rate	Oil Increase (%)
Producers	41	44	32	78%	186
Pending	1	1			
Injectors	41	123	40	98%	35
Pending	15	15			
All wells					
Confirmed results	82	167	72	88%	102
Pending	16	16			
Total	98	183			

actual field conditions replicates lab results, (4) pilot testing (if applicable) in a representative portion of the waterflood, and (5) full-field application. Forty-four treatments have been applied to 41 producing wells and 123 treatments have been applied to 41 injection wells. From the results available to date, on average the wells and their adjacent producers have seen an oil-production increase in 88% of the applications. On average, these applications have resulted in a 102% increase from pretreatment rates to posttreatment maximum rates. Table 21.1 shows the results available as of January 1, 2012 (Akintunji et al., 2012).

21.2 OIL RELEASE MECHANISM

Unlike many previous attempts at MEOR, this OOR process does not attempt to introduce microbes into oil-producing reservoirs. Instead, through a sophisticated analysis of produced fluids, microbes that are resident in the oil reservoir are identified and quantified. Certain species of resident microbes have the capability of cellular change resulting in the interactivity with trapped or unproducible oil resources. Once becoming interactive with the oil, these resident microbes insert themselves into the oil/water interface around any trapped oil in the reservoir. The flow characteristics of the trapped oil are affected by the presence of microbes at the oil/water interface. The changes in the oil/water/rock interfaces result in the deforming of the residual oil, allowing small droplets to form and be released into the active flow channels of the reservoir. Creation of a microemulsion at the oil water interface changes the interfacial relationship of the oil and water. This microemulsion lowers the interfacial tension between the two liquids,

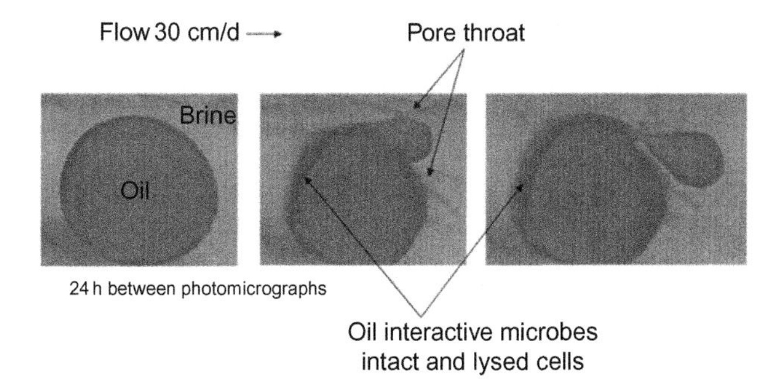

FIGURE 21.1 Oil Release Process—Microbes migrate to the oil water interface to help break up the oil.

allowing the oil to flow more freely. With the reduced interfacial tension micro oil droplets are released (Figure 21.1) (Town et al., 2010).

The approach needs to be customized to accommodate the different microbial ecologies in each reservoir. In the ideal application, the water-injection system becomes the transport medium for the nutrients, distributing the nutrients through the reservoir. In the higher permeability portions of the reservoir, newly released oil, water, and microbes may interact to form a transient (temporary) microemulsion that may alter the sweep efficiency of the injected water as it moves through the reservoir. Based on laboratory data, it is believed that in a waterflood project, this OOR process can recover up to an additional 10% of the original-oil-in-place (Davis et al., 2009).

Although the first application of this process was on a production well in the Alton field in Australia (Sheehy, 1990), this OOR process targets oil fields currently using conventional water-injection (waterflood) operations as a means of secondary recovery. The process is designed for crude-oil production and is not currently suitable for either natural gas or condensate fields; nor is poorly mobile oil currently a target of this process.

Typically in a mature waterflood, oil exists as trapped immovable droplets such that there is little or no relative permeability to oil. Residue hydrocarbons tend to bond and coat the reservoir grains and act as pore-filling material. In a very mature waterflood only water can flow toward the wellbore. The little bits of oil that are produced tend to be dragged to the producers as water moves through the pore channels. This OOR process releases oil that would normally be trapped in the reservoir. Substantial quantities of oil may be released near wellbore as a result of the nutrient treatment. This release of oil resaturates the main producing channels in the reservoir rock. This resaturation of the reservoir changes the relative permeability and results in lower water flow and increased oil flow (Akintunji et al., 2012).

21.3 DISCUSSION OF APPLICATIONS (ZAHNER ET AL., 2011)

Based on over 180 treatments over 4 years, the application of OOR can be applied to many more reservoirs than originally thought with little downside risk. The review of more than 180 OOR applications expands the types of reservoirs where OOR can be successfully applied. Low risk and economically attractive treatments can be accomplished when appropriate scientific analysis and laboratory screening is performed prior to treatments.

Observations and conclusions include the following:

1. Screening reservoirs is critical to success. Identifying reservoirs where appropriate microbes are present and oil is movable is the key.
2. OOR can be applied to a wide range of oil gravities. OOR has been successfully applied to reservoirs with oil gravity as high as 41°API and as low as 16°API.
3. When microbial growth is appropriately controlled, reservoir plugging or formation damage is no longer a risk.
4. Microbes reside in extreme conditions and can be manipulated to perform valuable *in situ* "work." OOR has been applied successfully at reservoir temperatures as high as 200°F and salinities as high as 140,000 ppm total dissolved solids (TDS).
5. OOR can be successfully applied in dual-porosity reservoirs.
6. A side benefit of applying OOR is that it can reduce reservoir souring.
7. OOR has been successfully applied in reservoirs with permeability as low as 7.5 mD (Akintunji et al., 2012).
8. An oil response is not always seen when treating producing wells, but this is generally believed to be related to localized, unusually low residual oil saturation in very mature waterflood projects.

21.3.1 Screening Reservoirs Is Critical to Success

Measured by increased oil production, the success rate is very high at approximately 90% (Table 21.1). It is believed that the screening process is the major factor in delivering this high success rate. Identifying reservoirs where microbes are present and oil is movable is the key to the success.

Microbes must be present in the reservoir since no microbes are injected. Typical initial screening criteria for the presence of microbes in oil reservoirs are reservoir temperature of less than 80°C (180°F) and water salinity below 10%, 100,000 ppm TDS. Also, pH should be neutral, 6–8. Since this process does not add any energy to the system or change the characteristics of the oil, it should be historically documented that the heavier oils are movable and there is sufficient pressure differential within the reservoir to allow flow to occur. The best documentation of movable oil is waterflood response. If the reservoir has a documented waterflood response and appropriate microbes are present, it should respond favorably to OOR treatments.

It is preferred to work in reservoirs with active water injection or water drive. In addition to providing energy, the water carries and distributes the nutrients throughout the reservoir. Reservoir permeability greater than 50 mD is preferred, although the main characteristics are injectivity and ability of the released oil to move to the producers. If a reservoir meets all of these conditions, it is an ideal candidate for an OOR application.

The second step in reservoir screening is well sampling of produced fluids and a rigorous laboratory analysis. Fluid samples are analyzed to determine if resident microbes can be manipulated. If appropriate microbes are present, the lab work results in a nutrient formulation specifically suited for the particular reservoir.

Reservoir oil saturation before treatment is unknown for all applications. Although it is very likely that most sandstone reservoirs are water wet, wetting conditions and interfacial tension between oil and water are often unknown. All documented applications, but one, are inactive waterfloods. In one application in an inactive waterflood, low reservoir energy in the system may have contributed to the lack of response.

21.3.2 Organic Oil Recovery Can Be Applied to a Wide Range of Oil Gravities

OOR can be applied to a wide range of oil gravities. Typically this means OOR applications are limited to reservoirs with oil gravity of 20°API and above. However, as noted, successful treatments have been conducted on reservoirs with oil gravity as low as 16°API. The bottom limit of oil gravities where this process can be successfully applied is not yet known.

The highest gravity oil to be treated is 41°API gravity oil in the Devonian sandstone in Alberta, Canada. The lowest gravity oil to be treated is 16°API gravity. This has occurred in two locations, the Sparky sandstone in Alberta, Canada and the Upper Topanga Sandstone (Miocene) offshore California. All three reservoirs have shown good waterflood response. Some of the reservoir parameters are shown in Table 21.2.

Results of applying OOR in these extreme gravity reservoirs have been impressive. In the Devonian reservoir (41-gravity oil), a test of nutrients was conducted in a producing well in April 2008. The well saw an increase in production from 2.5 m^3/d oil (16 BOPD) + 29.2 m^3/d water (184 BWPD), 92% water cut to 5.1 m^3/d oil (32 BOPD) + 29.1 m^3/d water (183 BWPD), 85% water cut (Figure 21.2). Subsequently, a pilot project was initiated in this reservoir.

The nutrient test in the 16-gravity Sparky reservoir conducted in August 2010 saw outstanding results. Production went from 1.4 m^3/d oil (9 BOPD) + 22.9 m^3/d water (144 BWPD), 94% water cut to 9.0 m^3/d oil (57 BOPD) + 51.0 m^3/d water (321 BWPD), 85% water cut (Figure 21.3).

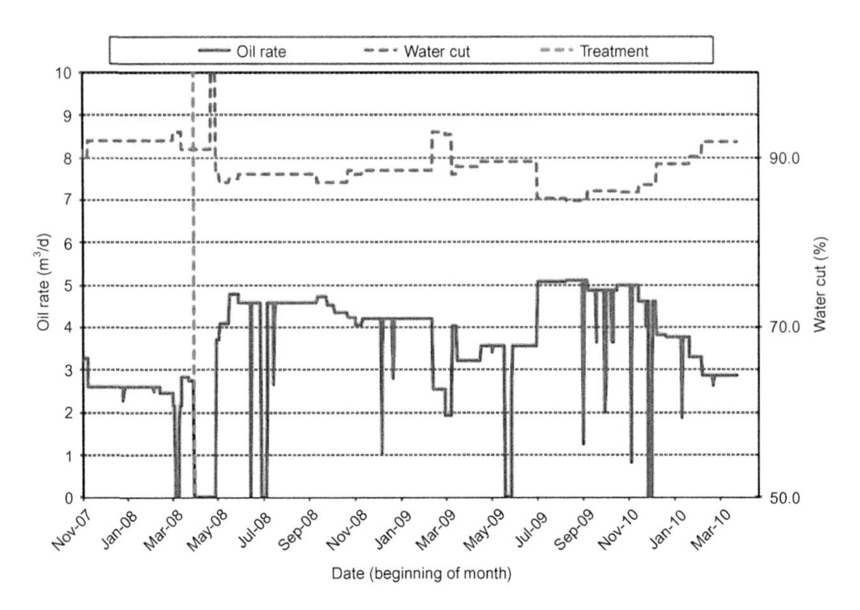

FIGURE 21.2 Devonian producer with 41°API oil shows positive results.

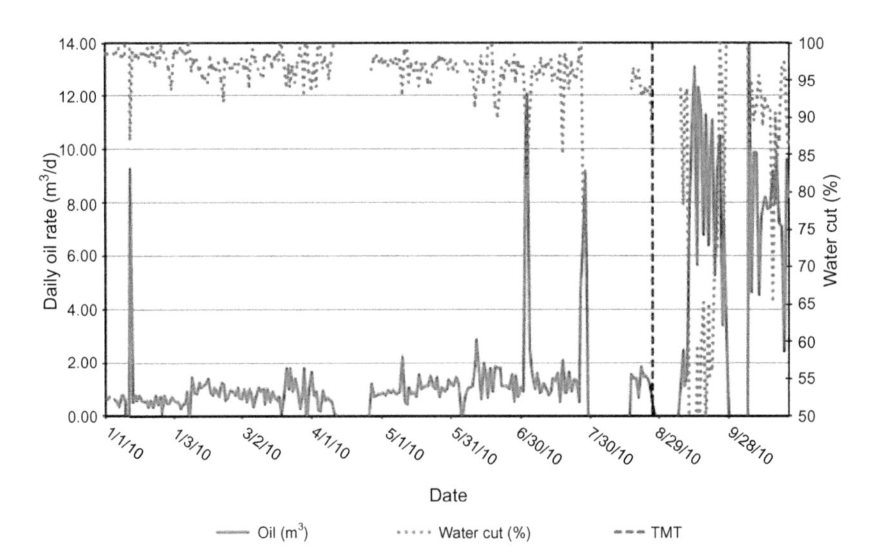

FIGURE 21.3 Sparky producer with 16°API oil shows positive response to treatment performed on August 10, 2010.

Since this field only has five injectors, a full-field application was conducted.

At another 16-gravity reservoir, three injectors have been treated in an Upper Topanga reservoir. The nutrient application in a producing well in this

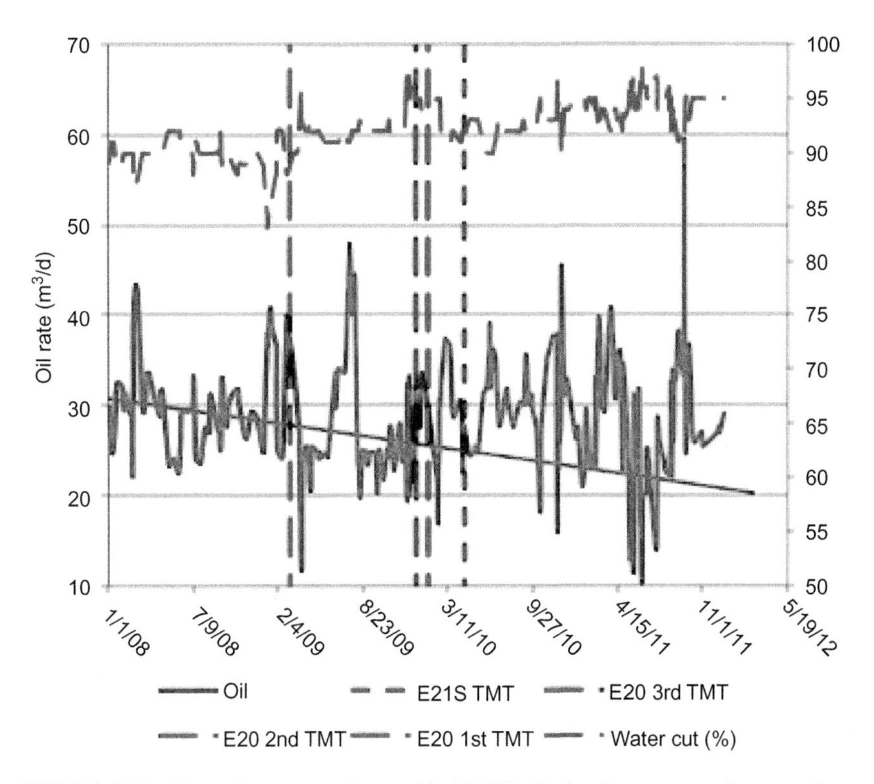

FIGURE 21.4 Upper Topanga producer with 16°API oil showing response from treating injectors.

reservoir saw a production increase in the treated producer from 28.7 m^3/d oil (181 BOPD) + 177 m^3/d water (1113 BWPD), 86% water cut to 42.2 m^3/d oil (266 BOPD) + 169 m^3/d water (1065 BWPD), 80% water cut. Although the response was short lived, a pilot was initiated and later expanded. The pilot injector was treated three times. After the third treatment, well tests of the offset producers increased 7% from 403 m^3/d oil (2539 BOPD) + 3660 m^3/d water (23,055 BWPD), 90% water cut to 430 m^3/d oil (2707 BOPD) + 3241 m^3/d water (20,419 BWPD), 88% water cut. The pilot was expanded with the treating of two additional injectors (Figure 21.4).

21.3.3 Reservoir Plugging or Formation Damage Is No Longer a Risk

Historically, MEOR has often been associated with reservoir plugging. No doubt biomass is created when microbial growth is stimulated. The formation of biomass can be a problem in injectors, even when not associated with MEOR treatments. An example is the documented plugging of injectors in the East Beverly Hills and San Vincente fields prior to this new generation of OOR technology (Cusack et al., 1985, 1987). Utilizing the nutrients in the water-injection system at that time, microbial growth was stimulated and biomass plugged wells. In general, the new approach to OOR eliminates the likelihood of such a circumstance.

Through 183 applications of OOR there have been no indications of any perforation or formation damage. This high rate of successful applications is due to two major changes between the current application practices of OOR and past practices of MEOR applications. First, microbes are not injected. It is believed that one of the reasons for reservoir plugging in the past was the practice of injecting microbes, which could plug small pore throats in the reservoir. Second, glucose nutrients are not used in this new OOR process. Injecting glucose nutrients stimulates the growth of a very wide variety of microbes. Uncontrolled biomass growth can result in excessive growth and lead to reservoir plugging. Today nonglucose environmentally benign nutrients are used. In a few instances, there has been research done in the area of controlled biomass growth to optimize waterflooding by changing the sweep (Brown et al., 2000).

21.3.4 Microbes Reside in Extreme Conditions and Can Be Manipulated to Perform Valuable *In Situ* "Work"

OOR has been applied successfully at reservoir temperatures as high as 93°C (200°F) and salinities as high as 142,000 ppm TDS. An application in the Hauser formation in California has demonstrated successful OOR application in a reservoir that ranges in temperature from 88°C to 93°C (190–200°F) (Zahner et al., 2010). Before treatment the target well was producing about 4.4 m^3/d (28 BOPD) + 30.3 m^3/d (91 BWPD). After peaking at 17.8 m^3/d oil (112 BOPD) + 11.4 m^3/d water (72 BWPD), the well tests averaged 12.5 m^3/d oil (79 BOPD) and 30.3 m^3/d water (91 BWPD) for the next 3 months. It is estimated that this single treatment yielded 4500 bbl of incremental oil. Because this field only has three injectors, a full-field OOR application was initiated. Other parameters of this reservoir are listed in Table 21.2.

The previously mentioned Devonian reservoir has a water salinity of 142,600 TDS. Reservoir parameters of this reservoir are listed in Table 21.2 and the previously mentioned successful nutrient application can be seen in Figure 21.2. The biological environment of this reservoir was limited (due to

TABLE 21.2 Parameters for Typical Reservoirs Where OOR Has Been Successfully Applied

Reservoir	Devonian Sandstone	Sparky	Upper Topanga	Hauser	Sparky C
Oil gravity, °API	41	16	16–18	22–26	20
Depth, meters (ft)	1056 (3466)	600 (1970)	1585 (5200)	1650–2636 (5413–8647)	661 (2169)
Temperature, °C (°F)	49 (120)	20–25 (68–77)	71–74 (160–165)	88–93 (190–200)	26 (79)
Pay thickness, meters (ft)	9 (29)	2–4 (6–13)	15–46 (50–150)	14–76 (45–250)	4 (13)
Permeability, mD	300	700	100–1000	10–100	600
Porosity, %	14–16	16	26	18–30	30
Salinity, ppm TDS	142,600	80,642	35,000	18,900	70,000
Cumulative recovery, % OOIP	22	6	20	[a]	34
Current water cut, %	98	95	85	85	88

[a]*Unknown, because zones commingled.*

limited species being present), but still provided a window of opportunity to successfully improve flow characteristics.

21.3.5 Organic Oil Recovery Can Be Successfully Applied in Dual-Porosity Reservoirs

There is concern as to whether OOR can be applied in dual-porosity reservoirs. The biggest concern in applying nutrients to dual-porosity systems is whether the nutrients are able to penetrate the matrix or if they remain in the high-permeability streaks and bypass the matrix thereby limiting the amount of oil to be released. An application in a Sparky reservoir, in Alberta, Canada has proven that dual-porosity reservoirs can be treated. One of the unique characteristics of this reservoir is that it contains "wormholes," high-permeability channels that form within the reservoir (Tremblay et al., 1999). The permeability, which has been reported as high as 13 Darcies, is so high that communication between some injector/producer pairs is measured in hours (Yuan et al., 1999). This compares to months in more homogeneous reservoirs. The nutrient test in this Sparky reservoir was conducted in January 2009. Production in this producer increased from 1.3 m^3/d oil (8 BOPD) + 14.4 m^3/d water (91 BWPD), 92% water cut to a peak of 6.2 m^3/d oil (39 BOPD) + 14.8 m^3/d water (93 BWPD), 71% water cut. As

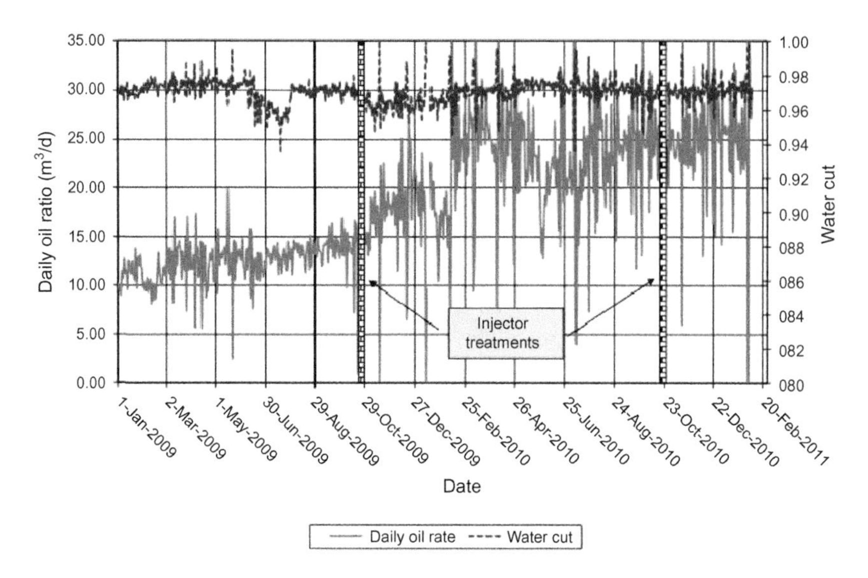

FIGURE 21.5 Twelve responding wells of 26-well pilot in dual-porosity reservoir showing positive results.

an indication of how permeable wormholes can be, it was noted that three 20-acre offset wells to the producing well responded to nutrient stimulation. Posttreatment while the well was shut in, it is believed that the nutrient material migrated through wormholes to the northeast stimulating microbial growth and oil response in the broader area. On average, the three offsets increased from 1.2 m^3/d oil (8 BOPD) + 20 m^3/d water (126 BWPD), 95% water cut to 2.6 m^3/d oil (16 BOPD) + 17 m^3/d (107 BWPD), 88% water cut. In total, the application of nutrients increased oil production in the well and offsets from an average of 4.8 m^3/d oil (40 BOPD) to 10.6 m^3/d oil (89 BOPD). A pilot application is currently underway and indications are that the offset producers are responding positively to injector treatments (Figure 21.5).

21.3.6 Applying Organic Oil Recovery Can Reduce Reservoir Souring

In March 2009 nutrient treatments began in the Upper Topanga Reservoir, offshore California. Injector E 20S was treated with nutrients on March 5, 2009, December 29, 2009, and January 25, 2010. In addition to seeing an increase in oil production in the frontline offsets to injector E 20S, it was noted that the produced hydrogen sulfide concentrations declined. All but well E 13L show this trend as noted in Table 21.3. There is no known operational change that contributed to this reduction. It is believed that the

TABLE 21.3 H_2S Concentrations Declined PostTreatment

H_2S Concentration (ppm)		
Well	**March 2009**	**August 2010**
E 5L	14,000	9000
E 10	24,000	20,000
E 11C	20,000	16,000
E 13L	2200	2600
E 23	5000	3500

microbes stimulated with nutrients outcompete the sulfate-reducing microbes (SRB) for nutrients and SRB growth is depressed.

21.3.7 Organic Oil Recovery Can Be Used in Tight Reservoirs (Akintunji et al., 2012)

In November 2010, two producing wells were treated in the Big Wells Field, Texas. This San Miguel sandstone reservoir has an average permeability of 7.5 mD, although sidewall cores indicated a range of up to 40 mD. When an oil-production increase and a water cut decrease were reported from the field, produced fluid samples were requested from both wells. This is when the field reported that well B-17 was not making any water and a water sample could not be taken. Prior to treatment, Well B-17 was producing 11 BOPD and 11 BWPD, 50% water cut. Two days after being returned to production, Well B-17 showed a quick production peak of 21 BOPD + 18 BWPD, 46% water cut on November 15, 2010. This was a little surprising since a 2-day shut-in during January 2010 did not show any production increase. Within a week production settled back to 11 BOPD + 11 BWPD, 50% water cut until March when it was verbally reported not to be making any water. On March 11, 2011, B-17 was making 13 BOPD + 2 BWPD, 13% water cut. Its water cut has slowly risen since then (Figure 21.6). This application confirms that OOR can be applied in permeability of less than 40 mD.

21.3.8 An Oil Response Is Not Always Seen When Treating Producing Wells

As seen in Table 21.1 9 of 41 producing wells did not show any oil increase after being treated, although all 9 showed good microbial response as a result of the treatments. Table 21.4 shows various reservoir parameters for five of the wells that did not show any response.

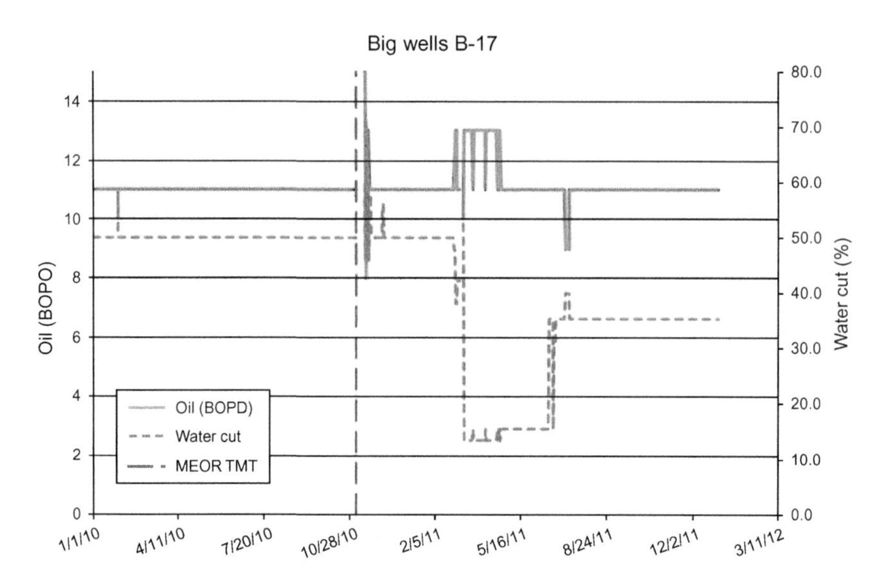

FIGURE 21.6 B-17 production.

TABLE 21.4 Parameters for Reservoirs Where Microbes Responded, but No Incremental Oil Was Measured

Reservoir	Upper Shaunavon	Pliocene Sandstone	Mannville	Mannville C	Mannville B
Oil Gravity, °API	22–24	21–24	22.2	15.5–21	15.5–21
Depth, meters (ft)	1200 (3927)	366 (1200)	1024 (3360)	957 (3140)	945 (3100)
Temperature, °C (°F)	47 (117)	49 (120)	35 (95)	31 (88)	31 (88)
Pay thickness, meters (ft)	3.4 (14)	285 (934)	6 (20)	6.3 (20)	8.5 (28)
Permeability, mD	567	850	400	1100	1500
Porosity, %	15–21	18–33	24	24	22
Salinity, ppm TDS	10,025	16,000	15,500	15,500	7000
Cumulative recovery, % OOIP	29	36	18	18	31
Current water cut, %	95	98	94	94	99

No correlations between reservoir parameters and unsuccessful applications have been identified. In fact, in comparing the parameters of this table to the parameters of successful applications in Table 21.2, it is difficult to see much difference. The mystery as to why no incremental oil is seen widens when looking deeper into the applications. For instance, the Upper Shaunavon reservoir had three successful producer applications, the original nutrient test and two additional producers (Town et al., 2010). Why the fourth application did not see any incremental oil is perplexing. The answer may lie in the degree of residual oil remaining in the portion of the formation receiving the nutrient materials—in short, very little oil present yields very little opportunity for increased production in such a producing well. Recent work has shown that residual oil saturation in water swept areas can be less than 15% (Romero et al., 2009).

The last two Mannville reservoirs are in different pools in the same field. This field consists of four separate pools. A nutrient test was conducted in each of the four pools. Two of the nutrient tests showed incremental oil; two of the nutrient tests did not. Before treatment, one of the successful wells was producing about 1.25 m^3/d (7.9 BOPD) + 16.9 m^3/d (106 BWPD), 93% water cut. After peaking at 4.4 m^3/d (27.7 BOPD) + 13.8 m^3/d (87 BWPD), 76% water cut, the well tests averaged 2.9 m^3/d (18 BOPD) + 19.4 m^3/d (122 BWPD), 87% water cut for the next month after which there were some mechanical issues, before stabilizing at 1.8 m^3/d (11.3 BOPD) + 13.9 m^3/d (87 BWPD), 89% water cut. It is estimated that this single treatment yielded 167 m^3 (1050 bbl) of incremental oil. Before treatment, one of the unsuccessful applications was producing about 0.78 m^3/d (4.9 BOPD) + 41.3 m^3/d (260 BWPD), 98% water cut. Posttreatment production dropped to an average of 0.31 m^3/d (2 BOPD) + 24.1 m^3/d (152 BWPD), 98% water cut for the next two and a half months after which production recovered to pretreatment volumes for the next 3 months, before eventually stabilizing at 0.4 m^3/d (2.5 BOPD) + 40 m^3/d (252 BWPD), 99% water cut (Figures 21.7 and 21.8).

It is worth mentioning that both of the nutrient test applications showing no incremental oil were in pools that had previously been under a tertiary alkaline polymer (AP) flood EOR scheme that ended 5 years prior to the nutrient treatments. Although the AP flood definitely increased the pH of the pool during treatment, it was thought that pH would have been restabilized at preflood levels by the time of the nutrient testing in May 2010. A direct correlation has not been proven between the lack of incremental oil success and the residual effects the AP flood in the reservoir, but it may be worth monitoring in similar future nutrient trials.

In all nine applications, the microbes responded as expected and as needed for oil release. It is believed that oil must not be in the presence of the microbes that have been stimulated with nutrients. No oil is being released because the area that is being treated is swept of any movable oil. There may be other parameters that effect results as well. One hypothesis is that very high water cut producers are less likely to see the release of oil by this organic recovery process. Note that all reservoirs in Table 21.4 have

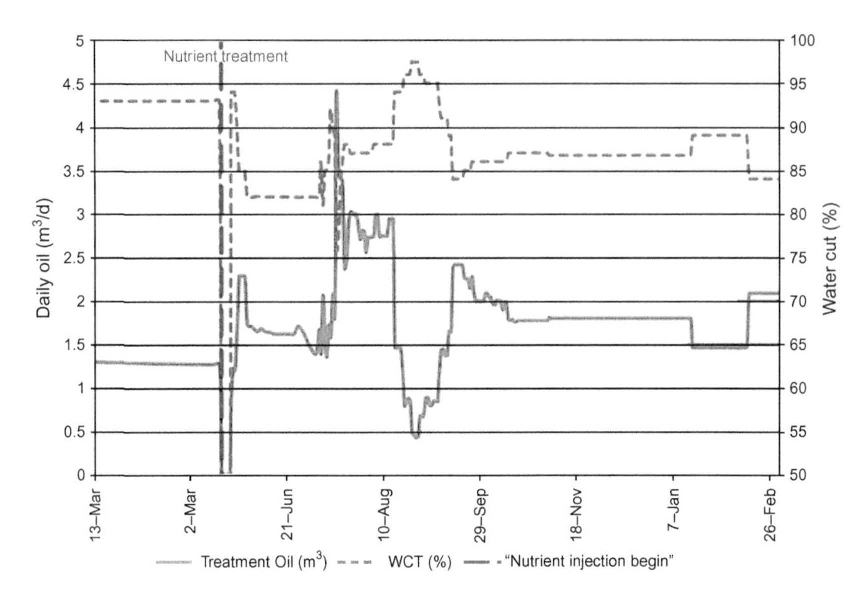

FIGURE 21.7 Mannville nutrient test shows positive results.

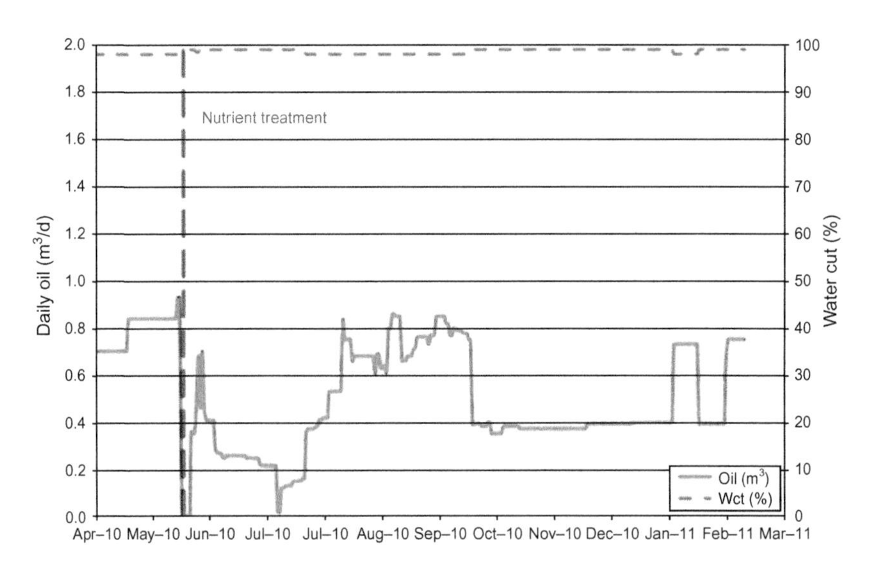

FIGURE 21.8 Mannville nutrient test shows unsuccessful results.

water cuts in the nineties. To date no correlation has been established to quantitatively support this hypothesis. Two reservoirs in Table 21.2 have water cuts in the nineties, but had successful applications. Another hypothesis is that reservoir energy is an issue, but as mentioned previously low energy has probably only been an issue with one application in an inactive waterflood.

21.4 CASE STUDY 1—TRIAL FIELD, SASKATCHEWAN (TOWN ET AL., 2010)

21.4.1 Background

The Trial field is located in the southwest corner of the province of Saskatchewan, Canada, southwest of Swift Current. The Trial field produces from the Upper Shaunavon sand. The field was discovered in 1952, and the waterflood was started in approximately 1967, initially set up as an inverted-five-spot pattern on 80-acre spacing.

The Upper Shaunavon sits on a structural high and has three members. The upper member is very high-quality sand and an excellent reservoir. The middle member, a poorer quality sand than the upper member, is isolated from the upper member. The lower member is a tight mixture of sands and shales. The average porosity ranges from 21.5% in the upper member to 15.2% in the lower member. The average permeability ranges from 567 mD in the upper member to 53 mD in the lower member. The average net pay is 2.6 m (8.5 ft) in the upper member, 1.8 m (5.9 ft) in the middle member, and 1.4 m (4.6 ft) in the lower member. Reservoir temperature is 47°C (117°F). Reservoir depth is 1200 m (3927 ft). TDS of the produced water are 10,025 mg/L.

Cumulative oil production is 3.3 million cubic metres (21 million barrels) by 2007, with average recovery of approximately 29% of the original-oil-in-place. Like most waterflooded reservoirs, low recovery makes the Upper Shaunavon an ideal EOR candidate. Oil gravity is 22–24°API. Current oil production is 62 m^3/d (391 B/D), with 1300 m^3/d of water (8190 BWPD) and 4250 m^3/d of gas. Current injection is 1700 m^3/d (10,700 BWPD).

21.4.2 Reservoir Screening and Laboratory Work

The reservoir parameters were reviewed to determine if this reservoir is a good candidate for OOR. There are two main criteria for a good candidate reservoir: mobile oil and the presence of specific species of microbes. In spite of relatively low oil gravity in the target field, the reservoir had good waterflood response, indicating that the oil is mobile. With the reservoir's moderate temperature of 47°C (117°F) and with produced water with only 9500 mg/L of chlorides, it was very likely that microbes were present in the reservoir. Laboratory analyses of microbial growth were conducted on the samples of produced water. Incubations were established with a range of nutrients and concentration of nutrients. The samples were examined by microscopy for evidence of cellular changes. Microbial growth patterns and replication rates consistent with the nutrients used as supplements were observed. Equally important, the nutrient manipulation resulted in the growth of a subpopulation of microbes capable of interaction at the oil/water interface. Specific nutrient combinations resulted in optimal potential for oil recovery and were recommended for use in this reservoir.

21.4.3 Field Application Process

The application of OOR to the field is performed in stages. First, the nutrients developed in the laboratory are used in treating a producing well. When the appropriate microbial response is observed, the second step is to treat an injection well. Since these were both successful, additional applications were administered in both producers and injectors. A description of each step and the result of each step follow.

21.4.4 Nutrient Test in Producer

Once laboratory work is complete, the formula devised specifically for this reservoir is applied to a producing well. This is done mainly to confirm that the appropriate microbes are stimulated. Pretreatment samples showed a low number of resident microbes. Very few of these were oil-interactive forms. After nutrient treatment, the number of microbes and number of oil-interactive forms increased dramatically.

The posttreatment samples showed the emergence in the field of oil-interactive forms. There was a substantial similarity between the microbial growth patterns observed in the laboratory and from the treated well's produced-water samples. Overall, posttreatment samples produced different population sizes compared to the laboratory, but the ratio of oil-interactive microbes to total microbes remained constant.

Often this step in the process also results in increased oil production. On December 6, 2007, Well A in the Trial field was treated with a 1.3-m^3 (8-bbl) tote of chemical nutrients solution, mixed with 13 m^3 (82 bbl) of injection water. The nutrient solution was injected into Well A through the tubing/casing annulus and displaced with 27 m^3 (170 bbl) of injection water. Well A was then shut in for 7 days to allow specific resident microbes to grow and multiply as a result of the nutrient stimulation. On 13 December, Well A was returned to production. Results were encouraging. The targeted species of microbes grew and reproduced exceptionally well. Also, oil production increased, with an associated decrease in water cut. Pretreatment daily production average for Well A was 1.2 m^3 of oil (8 BOPD) and 20.8 m^3 of water (131 BWPD), 94% water cut. Posttreatment daily production peaked at 4.1 m^3 oil (26 BOPD) and 19.0 m^3 of water (120 BWPD), 80% water cut. Well A was still seeing incremental production with daily

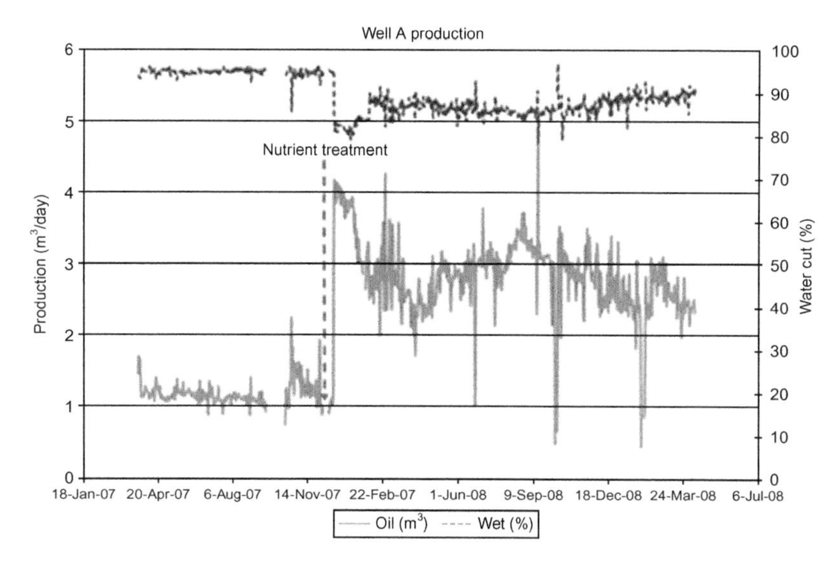

FIGURE 21.9 Producing Well A responds to cyclic treatment of nutrients.

production of 2.2 m^3 oil (14 BOPD) and 21.0 m^3 water (132 BWPD), a 91% water cut over a year after being treated. There was no change in the character of the produced fluid reported, and no treating problems were noted. This single-producing-well application result exceeded expectations by delivering approximately 500 m^3 (3150 bbl) of incremental oil. The water cut, percent water produced, also decreased significantly, which was another positive result of the treatment from an operating perspective (Figure 21.9).

21.4.5 Pilot

Now that the nutrients had been proven to be appropriate for this reservoir, a pilot project was initiated. Injection Well B was chosen for the pilot. It has three offset producers, Wells C, D, and E. The intent of this pilot test is to document the production response from the application of the OOR process under waterflood conditions. In addition to a production increase, it has been noted that a microemulsion may form in the reservoir, which will manifest itself at surface with lower injectivity in the pilot injector. The injection rate has been maintained on Injector B, and there has been no change in injectivity, which implies that no emulsion has formed. Also, there has been no indication of a microemulsion forming in the produced fluids (Figure 21.10).

The injector was batch treated with the nutrient solution, which was pumped down the injection well tubing and displaced into the reservoir with injection water. On April 24, 2008, Injector B was treated with a 1.3-m^3 (8-bbl) tote of chemical-nutrient solution, which was mixed with 16 m^3 of

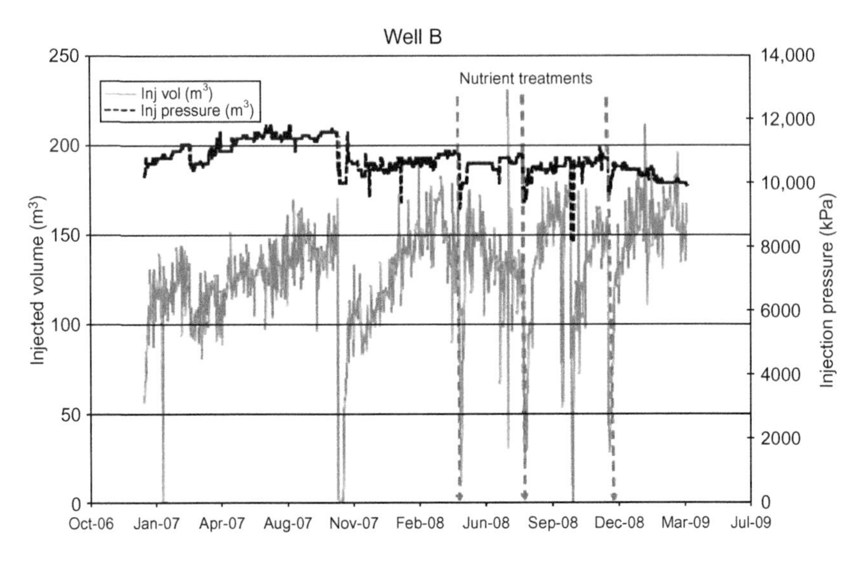

FIGURE 21.10 Injection rate into Well B, the pilot injector, is maintained.

injection water. After being injected, the nutrient solution was displaced with 32 m^3 (200 bbl) of water. Allowing the microbes time to incubate and populate, injection into Well B was limited for the next 8 days. Injected volumes were 10, 20, 50, and 75% of normal injection across the 8-day period.

After the nutrient injection, wellhead samples from the first offset production wells of the treated injector were taken and analyzed in the laboratory. Samples were tested by culturing and analyzing to determine changes in microbial composition and growth. The producers were monitored continually for rates, fluid levels, and produced-water chemistry. From this information, the coordination and scheduling of additional treatments were determined. Subsequent batch treatments were conducted on July 29 and December 3, 2008.

On 10 May, Well C increased from daily production of 1.5 m^3 of oil (9 BOPD) and 50.2 m^3 of water (316 BWPD), 97% water cut, to 4.6 m^3 of oil (29 BOPD) and 51.8 m^3 of water (326 BWPD), 92% water cut. Production continued to improve and the well peaked at 10.0 m^3/d of oil (63 BOPD) and 68.0 m^3 water (428 BWPD), 87% water cut. First response was expected in Well C because it is the nearest adjacent producer and it produces the most fluid. Production showed a 350% increase in oil production and an 8% decrease in average water cut. Laboratory analysis showed the targeted species of microbes grew and reproduced exceptionally well in Well C. Many microbes were in their oil-interactive state, in which they help to release additional oil (Figure 21.11).

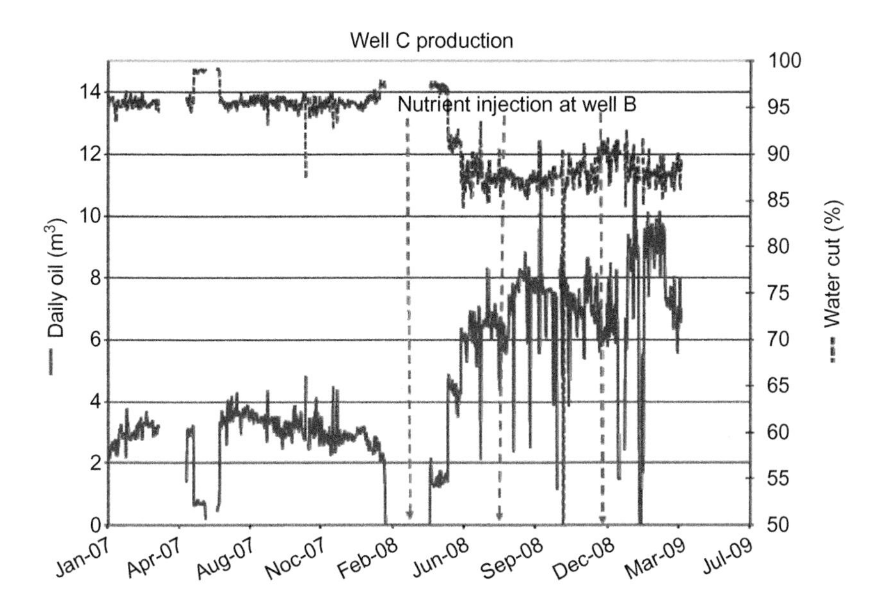

FIGURE 21.11 Producing Well C responds to treatments in offset injector Well B.

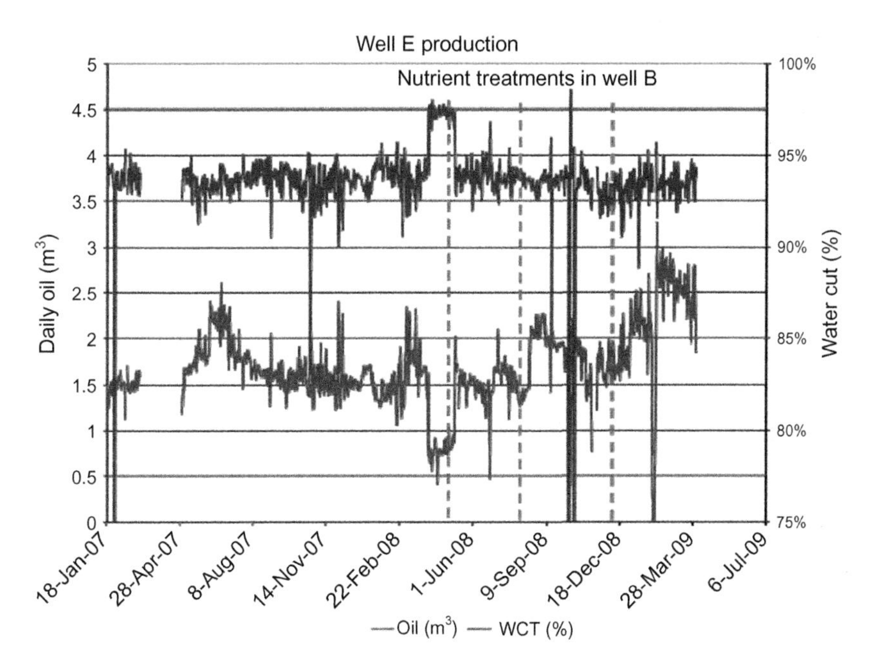

FIGURE 21.12 Producing Well E responds to treatments in offset injector Well B.

Gradually, positive response has been seen in the E offset producer. Starting at daily production of 1.5 m^3 oil (9 BOPD) and 25 m^3 of water (158 BWPD), 94% water cut, daily production peaked at 3.0 m^3 of oil (19 BOPD) and 38.3 m^3 of water (241 BWPD), 93% water cut (Figure 21.12). This behavior underscores that response is dependent on the movement of water through the reservoir as it carries nutrient material. Response time is volumetrically controlled by reservoir pore volume, water-injection rate and injection pattern conformance (both vertically and horizontally).

To date, no response has been seen in the other offset well, Producer D. This is not a surprise because transit time from Injector B is very likely to be longer, on the basis of well location, reservoir volume, and injection conformance. Well D daily oil production remains at 0.5 m^3/d of oil (3 BOPD) and 1.5 m^3/d of water (9 BWPD). Laboratory analysis of produced fluids from Well D indicates that only a small number of microbes are present. The low microbe concentrations in Well D indicate that the major impact of the nutrient effect had not yet reached this producing well.

21.4.6 Additional Producer Applications

As a result of the magnitude of oil response seen in the treatment on Well A, subsequent producer treatments were performed. On April 25, 2008, OOR treatments were performed on Wells F and G. Well F had been idle since 2005 and was reactivated to see what effect a nutrient treatment would have on a reactivated well. Because the microbes did not respond with the first treatment, Well F was retreated on 27 July. Then on December 4, 2008, a treatment was performed on Well H. In each case, a 1.3-m^3 (8-bbl) tote of chemical-nutrient solution was mixed with 13 m^3 (82 bbl) of injection water through the tubing/casing annulus and displaced with injection water. The test well was then shut in for 7–10 days to allow specific naturally occurring microbes to grow and multiply as a result of the nutrient stimulation.

Of the three production wells treated, Wells F and G have shown exceptional response. Well F increased from 0.6 m^3/d of oil (4 BOPD) and 3.2 m^3/d of water (20 BWPD), 84% water cut to 4.1 m^3/d of oil (26 BOPD) and 4.6 m^3/d of water (29 BWPD), 53% water cut. Well G averaged 0.5 m^3/d of oil (3 BOPD) and 30 m^3/d of water (189 BWPD), 98% water cut, before the second treatment, which was very similar to the 0.5 m^3/d of oil (3 BOPD) and 25 m^3/d of water (158 BWPD), 95% water cut that it was yielding in July 2005 when it last produced. After the second treatment, the well peaked at 3.0 m^3/d of oil (19 BOPD) and 20.8 m^3/d of water (131 BWPD), 87% water cut.

Even though initial oil production was disappointing, there was an excellent microbial response in Well H. It is believed that the lack of increased oil production is a result of other reservoir conditions (Figures 21.13–21.15).

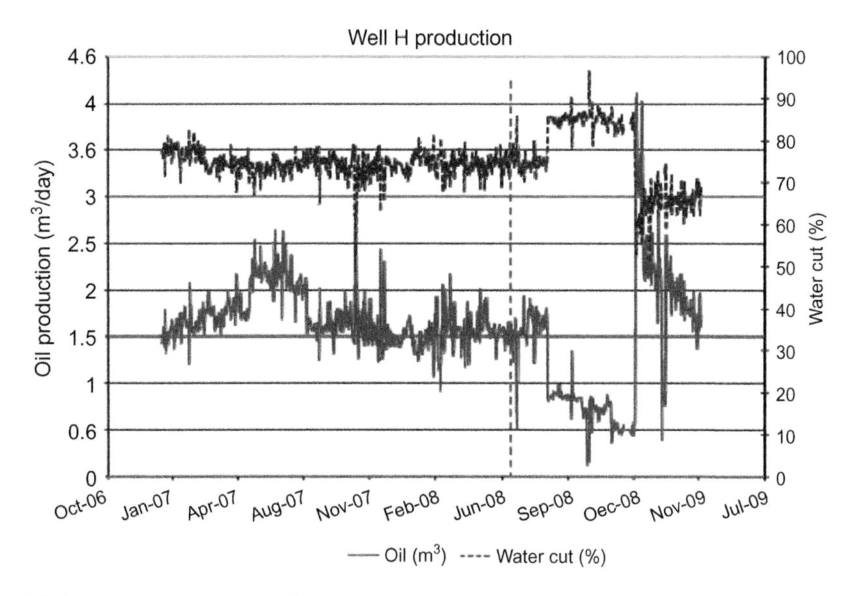

FIGURE 21.13 Producing Well H responds to cyclic treatment.

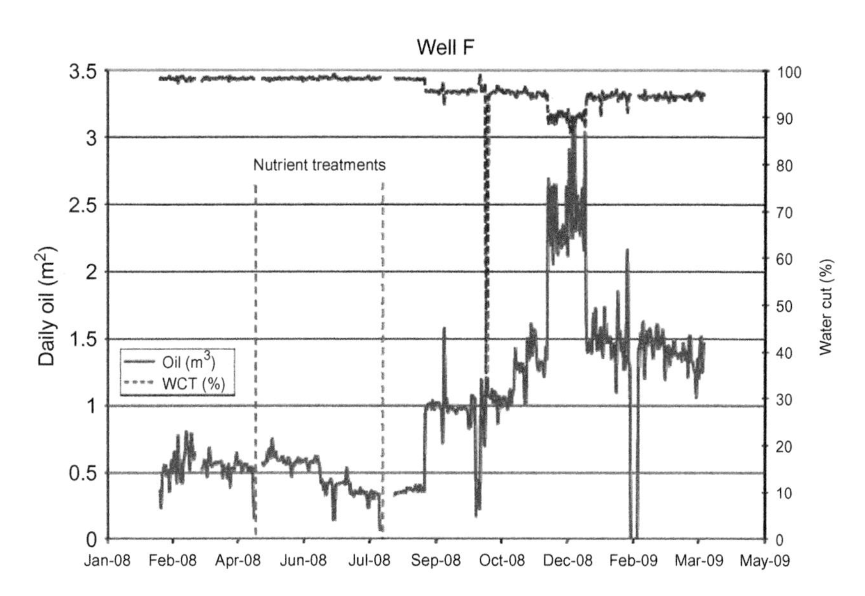

FIGURE 21.14 Idle producer Well F is reactivated and treated with nutrients.

21.4.7 Expanding the Pilot

After seeing the response in the pilot area, it was decided to apply the OOR process to a second injector. A batch treatment was pumped into Injector I

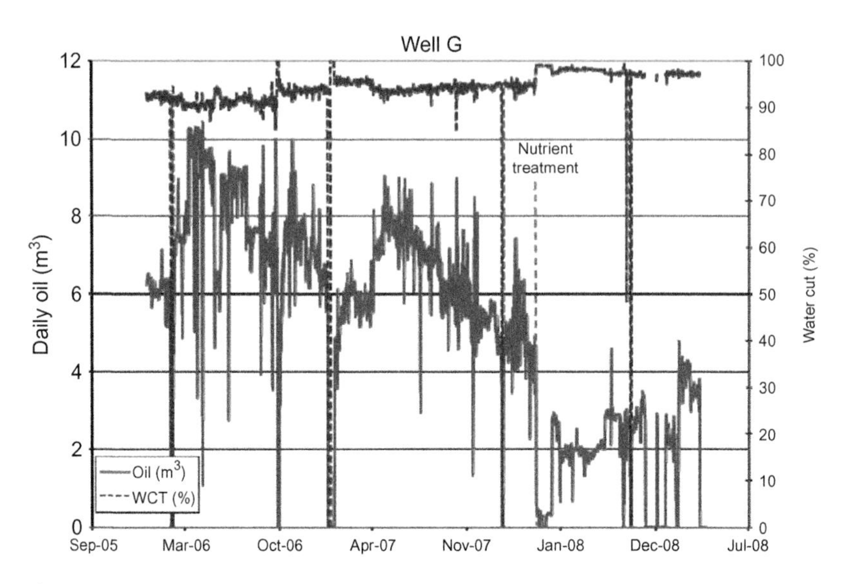

FIGURE 21.15 Producer Well G is treated with nutrients.

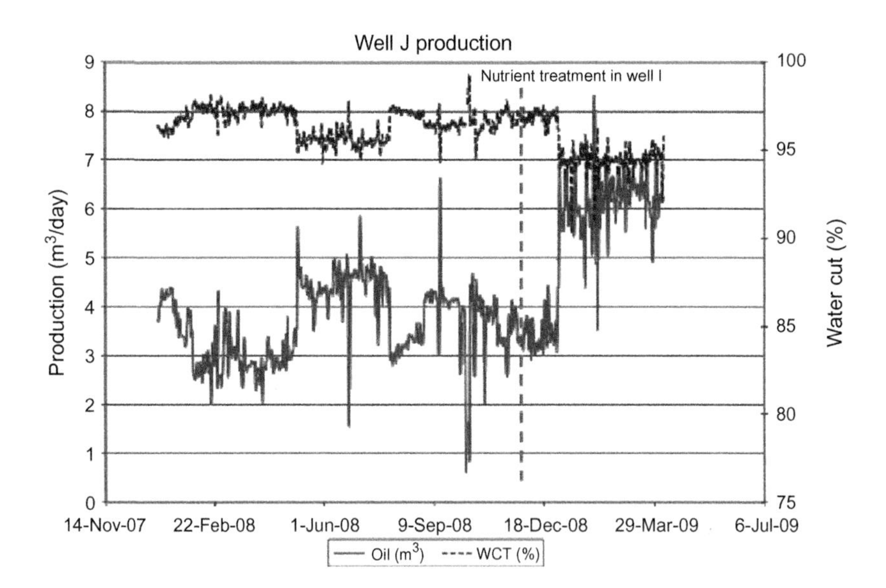

FIGURE 21.16 Producer Well J responds to treatment in offset injector Well I.

on December 4, 2008. As in the pilot, an oil-production increase was seen approximately 3 weeks after the first injection of nutrients. The three offset producers, Wells J, K, and L, responded. In total, they have increased

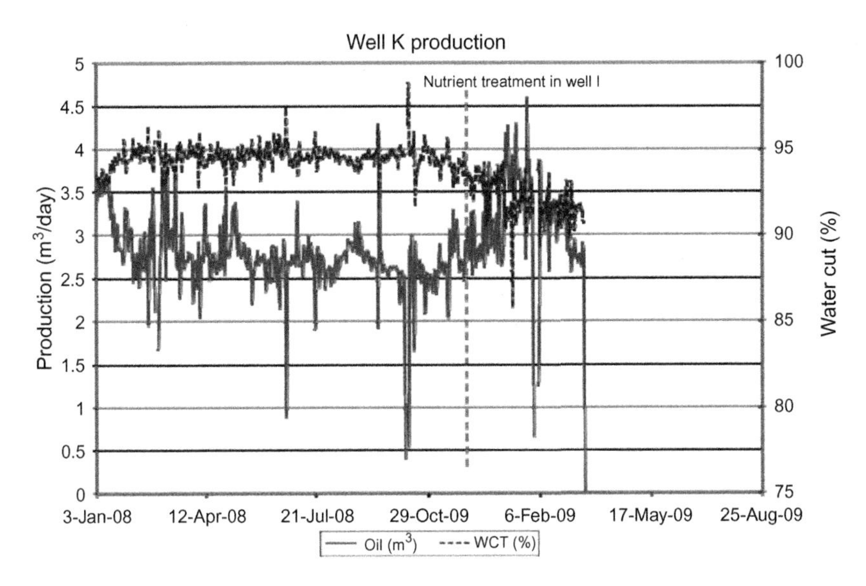

FIGURE 21.17 Producer Well K responds to treatment in offset injector Well I.

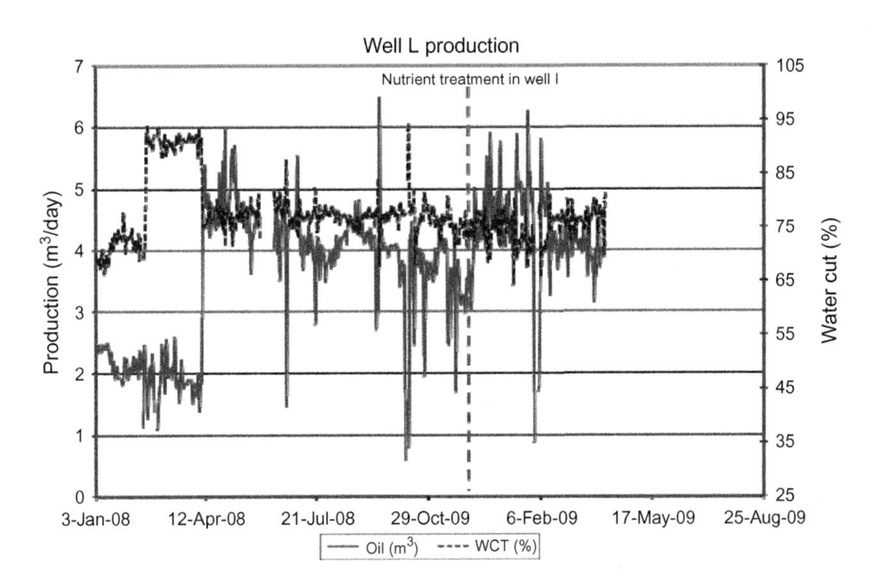

FIGURE 21.18 Producer Well L responds to treatment in offset injector Well I.

production from 10.2 m^3/d of oil (64 BOPD) and 157 m^3/d of water (989 BWPD), 94% water cut to a peak of 16.7 m^3/d of oil (105 BOPD) and 151 m^3/d of water (951 BWPD), 90% water cut (Figures 21.16–21.18). The process has now been applied to seven injectors within the field.

21.4.8 Discussion

The Trial field is experiencing several economic improvements. Not only is there an increase in oil production, but also there is an increase in oil recovery. With the increasing oil production and decreasing water cut, lifting costs are reduced. All these factors contribute to extending the life of the field. As in this application, a reduction in water production is often seen with these nutrient treatments. For instance, on Well A, water production dropped from 20.8 m^3/d (131 BWPD) to 19 m^3/d (120 BWPD). It is believed that the changes to the oil/water interface in the wellbore region change the relative permeability of water and oil.

There are several advantages of this process over other EOR processes, and even other MEOR processes. It is low cost to implement. Average incremental cost per barrel in the Trial field OOR application has been US$ 6.00 (Town et al., 2010). Some governments have programs in place to encourage EOR projects. This project is benefiting from provincial-government support to apply a new process. There is no capital outlay required to implement a project. Since the nutrients are batch treated even in injectors, permanent equipment is not required. There is little cost required to test the concept. Costs to conduct laboratory testing and to test nutrients in the field are minimal. Also, with batch treatments, the impact on field personnel is minimal. Another advantage is that it is low risk to implement. No microbes are injected, which minimizes the potential to cause reservoir plugging. The nutrient solutions that are injected are environmentally benign.

21.5 CASE STUDY 2—BEVERLY HILLS FIELD, CALIFORNIA (ZAHNER ET AL., 2010)

21.5.1 Background

The Beverly Hills field has two major producing horizons, the Hauser and the Ogden. The Hauser has been waterflooded since the mid-1980s, although producers in the field are generally commingled in both the Hauser and the Ogden formations. All water injection is into down-dip Hauser completions on the northeastern flank of the reservoir in the proximity of the original oil water contact. Oil gravity averages 22.5°API and ranges from 22–26°API. The field has 14 active producers and 3 active injectors with well spacing of approximately 10 acres. Field production is currently about 400 BOPD, 2000 BWPD, and 300 MCFD. All produced water is reinjected into Hauser.

The results reported are based on well tests. Oil production is metered at the field level and allocation to wells by their well tests. Usually, the sum of the well tests is within 10% of metered volumes. As in many fields, water cut data is limited as there is no provision for continuous sampling during well tests. Water cuts are based on wellhead samples taken by hand while the wells are being tested.

TABLE 21.5 OS-1 Well Test Data

	Date	Gross	Cut	Water	Oil
Well Tests before Treatment (Normal well production)	BFPD% BWPD BOPD				
	3/6/07	110	87	96	14
	3/11/07	105	82	86	19
	3/28/07	130	79	102	28
Average:		**115**	**83**	**95**	**20**
Well Tests after Treatment	7/9/07	177	56	99	78
	7/17/07	182	50	91	91
	7/24/07	162	57	92	70
	8/6/07	162	20	32	130
	8/14/07	156	61	95	61
	9/4/07	158	66	104	54
	9/26/07	134	34	44	88
Average:		**162**	**49**	**80**	**82**

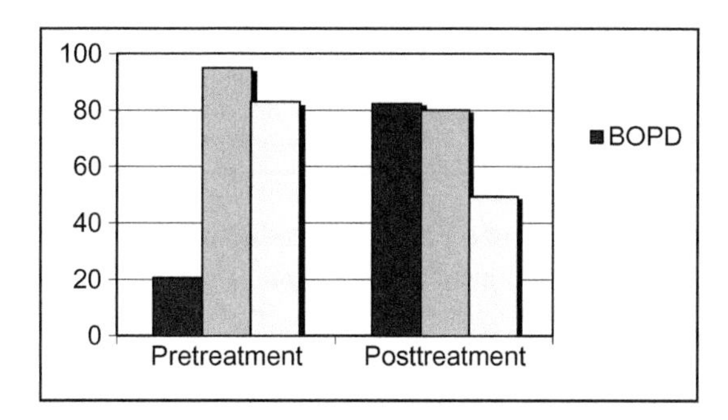

FIGURE 21.19 OS-1 well test results.

21.5.2 Nutrient Test in Producer

With a nutrient solution designed from the laboratory analysis of the field produced fluid samples, the nutrient test was conducted on producer OS-1 on July 2, 2007. A small volume (less than 8 bbl) of nutrients was injected into the well to check the reaction and behavior of resident microbes in the reservoir. The nutrient concentrate was mixed with 100 bbl of produced water and displaced with 350 bbl of injection water. The well was then shut in for 3 days to allow targeted resident microbes to grow and multiply as a result of nutrient stimulation.

Pretreatment production from OS-1 was 20 BOPD and 95 BWPD. After peaking at 130 BOPD and 32 BWPD, well tests average 82 BOPD and 80 BWPD for the first 3 months after treatment as shown in Table 21.5.

Over a year later, OS-1 was still producing 33 BOPD and 80 BWPD, although production was likely supported by treatment of offset injection wells as described later. This single-producing-well application yielded over 3000 bbl of incremental oil with a decrease in water produced (Figure 21.19).

21.5.3 Injection Well Treatments

Because this field only has three injectors, the pilot was skipped and full-field application immediately followed the successful producer test. The application was expanded to the full field by performing nine water-injection well treatments in the three active injection wells, OS-9, OS-10, and OS-14, and two additional producing well treatments, OS-8 and BH-15.

On November 29, 2007, an 8-bbl tote of highly concentrated chemical-nutrient solution was mixed with 250 bbl of injection water, injected into well OS-9 and displaced with 250 bbl of water. Giving the microbes time to incubate and populate, the water-injection rate into OS-9 was limited for the next 8 days. Each injector was given three similar treatments on the schedule listed below.

Well	Injector Batch Treatment Dates
OS-9	November 29, 2007, January 11 and March 20, 2008
OS-14	December 20, 2007, February 16 and April 15, 2008
OS-10	January 31, March 1 and May 1, 2008

Between July and September 2008 oil production increases were seen in the five active frontline producers, OS-1, OS-3, OS-4, OS-12, and OS-13. The targeted species of microbes grew and reproduced as nutrients migrated from injector to producer, freeing oil along the way.

Produced fluid samples taken on June 12 from the frontline producing wells indicated high concentrations of microbes were present in four of the five adjacent producers, OS-1, OS-3, OS-4, and OS-13. This was consistent with improved well tests seen on these four wells. The June 12 sample taken from OS-12 did not show any microbe activity, which was consistent with its well tests at the time. In July, OS-12 experienced a jump in oil production and another produced fluid sample was taken in August to determine if the oil-production increase was coincident with improved microbial activity. Laboratory results confirmed an increase microbial response with the increased oil seen in the well tests.

The frontline producers made more oil as a result of these treatments. From June through August 2008, the first 3 months of response, the frontline producers averaged 206 BOPD and 1480 BWPD. These wells averaged 179 BOPD and 1490 BWPD from March to May. Also, base production estimated in January 2008 for these wells was 179 BOPD. These wells produced

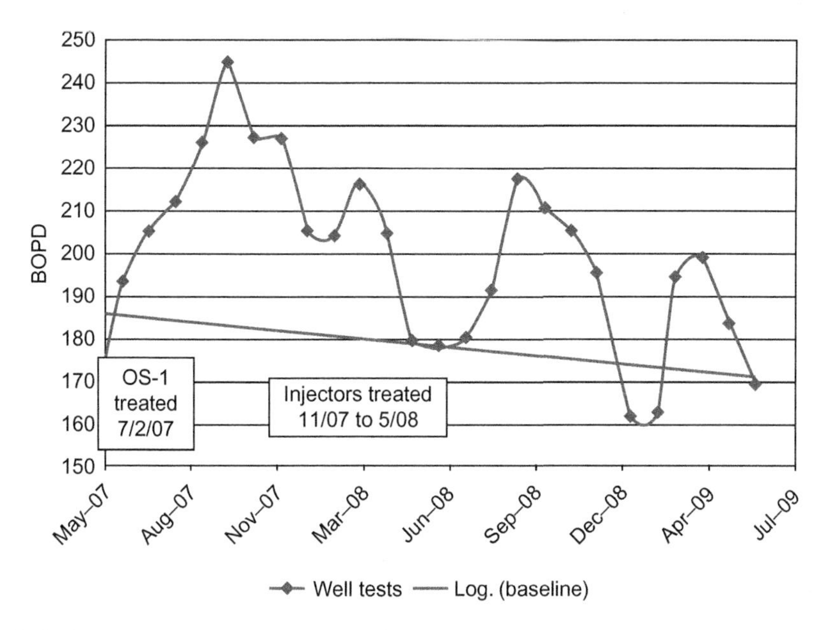

FIGURE 21.20 Well test versus baseline, frontline producers.

an average 27 BOPD over their base for the 3 months, June to August 2008 (See Figure 21.20). The frontline producers averaged 217 BOPD in July. The well tests peaked at 232 BOPD and 1669 BWPD. This is 53 BOPD over the base of 179 BOPD, a 30% increase overall. The frontline producers accumulated about 2500 bbl of incremental oil through August. As a result of these treatments, incremental oil continued to be produced above the baseline.

Based on the improved well tests and the advance seen in microbe activity in the other frontline producers, well OS-2 was returned to production on June 18, 2008. It had been shut in since April 2003 when it tested 2 BOPD and 217 BWPD. It tested no oil until November 2008 when it tested 11 BOPD and 142 BWPD. Well tests eventually peaked at 46 BOPD and 243 BWPD after the well's lift equipment was optimized (Table 21.6).

Hall Plots and derivative Hall Plots of the three injectors indicate that transmissibility has changed over time (Figures 21.21–21.23) (Izgec and Kabir, 2009). Well OS-10 showed a decrease in injectivity for a short time after the first treatment. This is a possible formation of a temporary emulsion. Sometimes an emulsion forms when oil, water and microbes are present; this emulsion tends to plug the higher permeability paths and improve the sweep efficiency of the waterflood. Other than this temporary decrease, all three injectors show a slight increase in injectivity with the nutrient treatments. These indications of increased injection match the field's

TABLE 21.6 OS-2 Well Tests

Well ID	Date	BOPD	BWPD	MSCFD	WC (%)
O.S. 2	4/17/03	2	217	0	99
	6/18/08	RTP			
	7/9/08	0	180	0	100
	7/21/08	0	166	0	100
	8/19/08	0	128	3	100
	9/22/08	0	99	1	100
	11/19/08	11	142	2	93
	11/21/08	13	132	2	91
	12/3/08	8	160	5	95
	1/10/09	10	191	5	95
	2/9/09	8	150	4	95
	3/28/09	20	158	5	89
	5/17/09	6	185	5	97
	5/20/09	46	243	5	84
	5/22/09	20	222	5	92
	5/23/09	42	222	5	84
	5/27/09	29	235	5	89
	5/28/09	15	238	5	94
	5/30/09	26	236	5	90
	7/9/09	33	224	5	87
	7/20/09	25	221	10	90
	8/13/09	23	203	10	90
	8/25/09	2	234	18	99
	9/24/09	31	228	18	88
	10/8/09	22	250	18	92
	11/6/09	22	226	10	91
	12/17/09	16	66	10	80
	1/15/10	31	224	12	88

increase in produced water from about 2100 BWPD to 2500 BWPD during the project.

21.5.4 Additional Producer Treatments

Based on the results of the treatment of OS-1, two producers, BH-15 and OS-8, were treated on April 18 and May 5, 2008, respectively. Each well was treated with an 8-bbl tote of chemical-nutrient solution mixed with 100 bbl of injection water. Displacement volume in the BH-15 was 400 bbl (200% of annular volume) and in the OS-8 the displacement volume was 700 bbl (150% of the annular volume). Giving the microbes time to incubate and populate, both wells were shut in for 4 days.

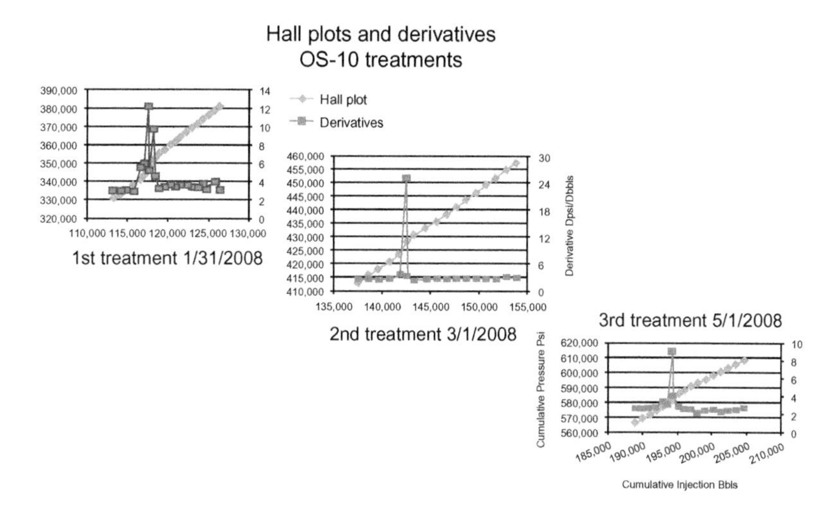

FIGURE 21.21 OS-10 Hall plots and derivatives.

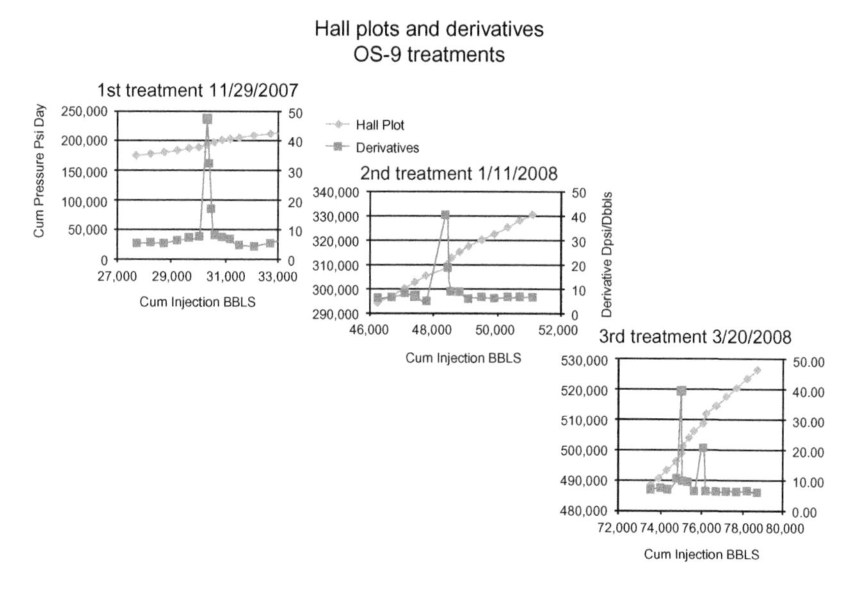

FIGURE 21.22 OS-9 Hall plots and derivatives.

21.5.5 OS-8

In both cases microbial populations increased, but neither well followed the normal pattern that was seen after treating OS-1 and other producing well applications. In the OS-8 well, the microbes showed the normal increase in population. However, the microbes did not move as rapidly into the oil-

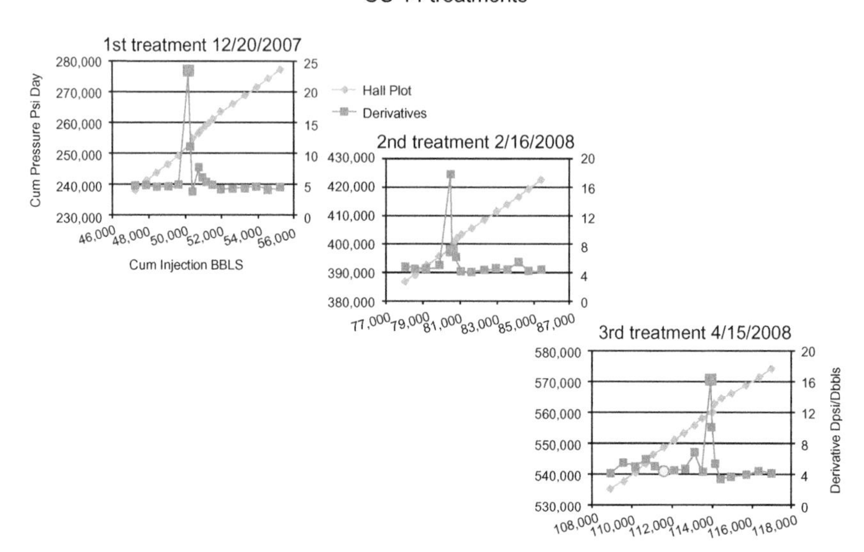

FIGURE 21.23 OS-14 Hall plots and derivatives.

interactive state as usual. The first month of produced fluid samples showed that the microbes were still in the growth stage, because nutrients remained plentiful. After seeing a production increase, additional samples taken on June 10, showed that the microbes were moving into an oil-interactive stage. This is a delayed transition to the oil-interactive stage as compared to OS-1. Not only did OS-8 see a delayed oil-production increase it saw no oil for some time. Initial well tests showed no oil. Oil was not seen until the May 22 well test, 13 days after the well had been returned to production. At this time the well had a cumulative production of about 850 bbl, which is about the treatment volume. No oil was seen until the entire treatment volume was recovered. As the microbes began to respond, the well started producing incremental oil. On June 9, it tested 37 BOPD and 28 BWPD, 43% water cut. In September OS-8 tested 33 BOPD and 30 BWPD, 48% water cut. Since the well averaged 29 BOPD and 36 BWPD, 55% water cut before it was treated, it made some incremental oil, but only a small amount (Figure 21.24).

21.5.6 BH-15

On the first day of production after being shut in for 4 days, the microbes in BH-15 showed an extraordinary increase in population as expected. On the second day of production, the microbe population decreased substantially and the high microbial count did not repeat. A similar varied oil-production response was

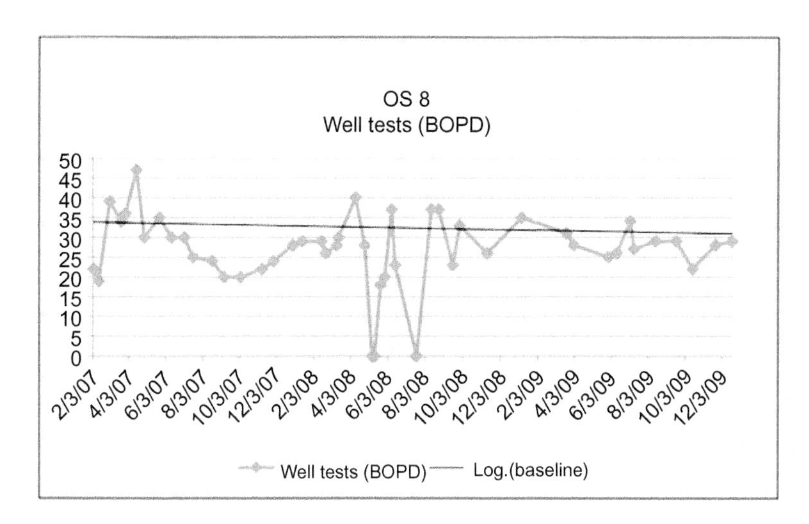

FIGURE 21.24 OS-8 well tests.

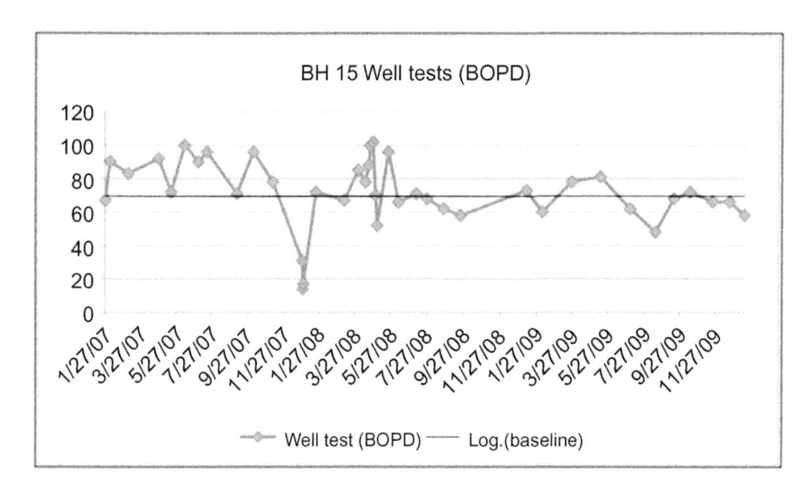

FIGURE 21.25 BH-15 well tests.

seen. BH-15 saw an early increase in oil production and decrease in water cut followed by a disappointing decline in oil production and increase in water cut back to its base production. During the first week, three well tests were taken and BH-15 was averaging 97 BOPD and 105 BWPD, 52% water cut. This initial high production was probably flush production from the well being shut in. Then, BH-15 dipped to 58 BOPD and 80 BWPD, 58% water cut in September 2008. Since the well averaged only 70 BOPD and 110 BWPD, 61% water cut, the well did not appear to make any incremental oil for the first few months. This is not surprising since microbes were not significantly stimulated (Figure 21.25).

21.5.7 Discussion of Results

Based on both biological indicators and production data, the field showed a positive response to nutrient treatments. As of the end of August 2008, adjacent producers appeared to be positively affected with a combined current production increase from five wells of over 30 bbl of oil per day—a production increase of as high as 30% over the base rate of the "frontline" of producing wells and an overall production increase of about 6% of total field production.

In general, producing well treatments had excellent microbial response, but only OS-1 showed significant incremental oil response. It appears that the large volume of displacement fluid in treating OS-8 temporarily hurt production from OS-8. The treatment may have temporarily changed the relative permeability near the wellbore. It took about 13 days to recover the treatment water, when first oil was reported on a well test. It took another 18 days before incremental oil was seen on June 9. This relative permeability problem was seen again following a tubing leak repair. The well was down from June 22 to July 15 due to a tubing leak. When the well was returned to production, its first well test on July 20 showed no oil. By August 8 the well returned to making incremental oil when it tested 37 BOPD and 41 BWPD, 53% water cut. To achieve a successful OOR treatment, the four components (oil, water, microbes, and nutrients) must make contact. In BH-15, since the microbes did not respond to the injection of nutrients, the nutrients may not have come in contact with the microbes. One possibility is that the nutrient didn't go in the oil zone. Reviewing BH-15, it is noticed that the gas oil ratio (GOR) on this well was much higher than the field average. Table 21.7 compares GORs among the key producing wells.

TABLE 21.7 Gas Oil Ratios

Well	Gas Oil Ratio, SCF/STB	Comment
OS-1	550	
OS-3	760	
OS-4	655	
OS-5	1000	
OS-7	1700	Structurally second highest producer
OS-8	857	
OS-11	915	
OS-12	400	
OS-13	160	Structurally lowest producer
BH-15	1700	Structurally highest producer

It is possible a secondary gas cap formed and the nutrients were injected into the gas cap. Another possibility is that the nutrients were injected into one zone and the well produced predominantly from another. In this case, there is no way to know where the nutrients are injected with multiple intervals open.

21.6 CONCLUSION

Over 180 nutrient applications have demonstrated that OOR can be successful in a variety of reservoirs. Application of OOR has been shown to complement secondary recovery projects using water injection since existing infrastructure can be used as the delivery system and avoid any significant new capital equipment expenditure requirements. Reservoirs with oil gravities between 16°API and 41°API have been successfully treated and yielded significant improvement in oil-production rate. Positive results have also been proven in higher temperature and higher salinity reservoirs and such reservoirs should still be considered for stimulation if other reservoir parameters are moderate. Even dual-porosity systems should be considered as potential candidates for OOR based on limited experience over the past 4 years. In addition to recovering incremental oil, side benefits include reduced H_2S levels. Knowing that there is essentially no risk of damaging the reservoir, there is little downside to trying OOR as long as due consideration is given to the microbial composition and other reservoir screening parameters. Rigorous scientific analysis prior to treatment is a requirement before initiating an OOR treatment to optimize potential benefits. OOR must be targeted and customized to each reservoir. When applied systematically, OOR can yield significant production and oil recovery increases.

REFERENCES

Akintunji, F., Atinum E&P, Inc., Marcotte, B., Sheehy, A., Govreau, B. Titan Oil Recovery, Inc., 2012. A Texas MEOR application shows outstanding production improvement due to oil release effects on relative permeability. Presented at Improved Oil Recovery Symposium, Tulsa, OK, 14–18 April.

Bauer, B.G., O'Dell, R.J., Merit Energy Company, Marinello, S.A., Babcock, J., Ishoey, T., et al., 2011. Field experience from a biotechnology approach to water flood improvement. Presented at SPE Enhanced Oil Recovery Conference, Kuala Lumpur, Malaysia, 19–21 July.

Brown, L.R., Vadie, A.A., Mississippi State University, Stephens, J.O. Hughes Eastern Corp. 2000. Slowing production decline and extending the economic life of an oil field: new MEOR technology. Presented at the IOR Symposium in Tulsa, OK, 3–5 April.

Cusack, F., U. of Calgary, Brown, D.R., Chevron Oil Field Res. Co., Costerton, J.W., U. of Calgary, et al., 1985. Field and laboratory studies of microbial/fines plugging of water injection wells: mechanism, diagnosis and removal.

Cusack, F., McKinley, V.L., Lappin-Scott, H.M., Microbios Ltd., Brown, D.R., Clementz, D.M., et al., 1987. Diagnosis and removal of microbial/fines plugging in water injection wells. SPE Annual Technical Conference and Exhibition, Dallas, TX, 27–30 September.

Davis, C.P., Marcotte, B., Govreau, B., 2009. MEOR finds oil where it has already been discovered. Hart Energy E&P November, 78–79.

Gao, C.H., Zekri, A., El-Tarbily, K., 2009. Microbes enhance oil recovery through various mechanisms. Oil Gas J. August 17 and 24.

Izgec, B., Kabir, C.S., 2009. Real-time performance analysis of water injection wells. SPE Reservoir Eval. Eng. J. February.

Jackson, S.C., Fisher, J., Alsop A., Fallon R., DuPont, 2011. Consideration for field implementation of microbial enhanced oil recovery. Presented at the SPE Annual Technical Conference and Exhibition, Denver, CO, October 30–November 2.

Romero, C., Aubertin, F., Cassou, E., Cheneviere, P., Tang, J.S., Odiorne, J., et al., 2009. Single-well chemical tracer tests (SWTT) experience in the mature Handil field: evaluating stakes before launching an EOR project. EAGA Fifteenth European Symposium on Improved Oil Recovery, 27–29 April.

Saikrishna, M., Anadarko, Knapp, R.M., McInerney, M.J., 2007. Microbial enhanced-oil-recovery technologies: a review of the past, present and future. Presented at Production and Operations Symposium, Oklahoma City, OK, March 31–April 3.

Sheehy, A.J., 1990. Field studies of microbial EOR. Presented at SPE/DOE Enhanced Oil Recovery Symposium, Tulsa, OK, 22–25 April.

Town, K., Sheehy, A.J., Govreau, B.R., 2010. MEOR success in south Saskatchewan. SPE Reservoir Eval. & Eng. J. 13 (5), 773–781.

Tremblay, B., Sedgwick, G., Vu, D., Alberta Research Council, 1999. CT imaging of wormhole growth under solution-gas drive. SPE Reservoir Eval. Eng. J. 2 (1), 37–45.

Yuan, J.Y., SPE, Tremblay, B., SPE, Babchin, A., Alberta Research Council, 1999. A wormhole network model of cold production in heavy oil. Presented at International Thermal Operations Symposium in Bakersfield, CA, 17–19 March.

Zahner, B., Venoco, Inc., Sheehy A., Govreau, B., Titan Oil Recovery, Inc., 2010. MEOR success in Southern California. Presented at IOR Symposium, Tulsa, OK, 24–28 April.

Zahner, R.L., Venoco, Inc., Tapper, S.J., Husky Energy, Marcotte, B.W.G., Govreau, B.R., et al., 2011. What has been learned from a hundred MEOR applications. Presented at SPE Enhanced Oil Recovery Conference, Kuala Lumpur, Malaysia, 19–21 July.

Chapter 22

Cold Production of Heavy Oil

Bernard Tremblay
Saskatchewan Research Council, EOR Field Development, Energy Division, Regina, Saskatchewan, Canada

Not all heavy oils show the same oil recovery factors. Pressure depletion experiments in combination with foamability, static and dynamic surface tension, and surface viscoelasticity measurements of the gas/oil interface have led to the following observations: (1) organic acids and bases assist in the formation of asphaltene networks at the interface between the gas and oil and in forming an elastic membrane which is more resistant to deformation and reduces gas diffusion, (2) resins reduce the surface activity of the asphaltenes, (3) greater pressure depletion rates increase gas bubble nucleation rates and decrease bubble size, and (4) gas bubble breakup at high capillary numbers reduces their coalescence.

Numerical modeling of the gas transport though porous media has shown that the nonequilibrium foamy oil approach, where gas transport is modeled as a series of chemical reactions, can history match the oil production in pressure depletion experiments in which a sand pack is saturated with live oil and depleted at one end. History matches of the oil production volumes can also be obtained by assuming thermodynamic equilibrium conditions but necessitates assuming very low gas relative permeability curves.

Cold production where sand production is encouraged is more commonly known as the CHOPS (cold heavy oil production with sand) process. Sand production experiments suggest that wormholes develop by the fluidization of the sand ahead of the wormhole tip without having to fail the formation. These wormholes would be sand-filled when they grow reaching diameters as large as 1 m. As they grow in length, an open channel would form, 6–10 cm in diameter, as suggested in tracer tests and in wormhole growth models. It is important to note that a wormhole is composed mostly of dilated sand, approximately 45% in porosity and can contain an open channel which develops eventually as the wormhole gets longer allowing more oil seepage. This seepage leads to a lower sand concentration in the wormhole and the development of an open channel by settling of the sand.

Large scale tracer tests in the field cannot be explained by the development of fractures since their width would be of the order of the sand grain diameters. Circulation loss observations at only a few locations in the field while drilling horizontal wells in cold produced reservoir indicate that the wormhole networks are not highly branched (if at all) and that dilated regions do not develop around CHOPS wells. The volumes of injected lost circulation material (LCM) during drilling indicate that the volume of the open channel network could reach more than 20 m^3. Assuming an open channel diameter of 10 cm and an average wormhole length of 250 m would lead to eight wormholes developing in the field.

Analyzing CHOPS field production data shows how variable the data can be since these reservoirs are usually quite thin 3–10 m and are discontinuous. Field data from a group of six CHOPS wells producing over a period of 3500 days was history matched for the first 2300 days. Using the same adjustable parameters, the CHOPS model was run to predict the oil, sand, water, and gas production for the period from 2300 to 3500 days. The reasonable predicted values indicate the potential for using multiwell CHOPS model to eventually be able to predict optimal well spacing, in-fill scheduling, and perhaps most importantly the design of post-CHOPS processes. For economic reasons, CHOPS wells are often not cored and sand and gas production are not recorded which makes history matching more difficult. Measuring this data would greatly help in predicting future post-CHOPS processes.

22.1 INTRODUCTION

Cold production is defined as a solution-gas drive process applied to heavy oils with viscosities in the range 200–40,000 cP. In reservoirs where the *in situ* viscosity is in the range 200–2000 cP, economical recovery factors are possible with little sand production. For example, de Mirabal et al. (1996) calculate a 10% original oil in place (OOIP) recovery factor due to solution-gas drive for the Hamaca area in Venezuela where the permeability can reach 10 D in the Orinoco Belt.

Due to the high viscosity of the heavy oil around Lloydminster (5000–30,000 cP), oil well operators have learnt through trial and error that the production of heavy oil could be increased by encouraging sand production. When sand is voluntarily produced, the process is called CHOPS, a term first used by Dusseault (2002). One advantage of the CHOPS process is that it can be applied to reservoirs as thin as 3 m where thermal processes are not economical.

In Canada, the CHOPS process has been applied in basically four regions in the Western Sedimentary Basin: Lloydminster, Lindbergh, Cold Lake, and Southwest Saskatchewan. The producing sands in the Lloydminster area are in the Lower Cretaceous Mannville Group (Colony, McLaren, Waseca, Sparky, General Petroleum, Rex, Lloydminster, Cummings, and Dina). The Sparky formation is the most prolific among these sands and also the cleanest in terms of clay (shale) content whereas the Colony has the lowest yield since it contains more clay as reported by Kimmel and Laviolette (1982). A CHOPS stratum normally ranges from 3 to 10 m in thickness with fine to medium grained uncemented sand with a D50 grain size diameter ranging from 80 to 150 μm. The permeabilility ranges from 1 to 5 D. The initial water saturation is fairly low since clean sand is produced and varies between 12% and 20%. These reservoirs range in depth from 400 to 800 m and the initial pressure varies between 3 and 7 MPa (Dusseault, 2002). Although CHOPS originated in western Canada, it is becoming more popular worldwide in Venezuela, Alaska, Albania, and China, for example, as conventional oil reserves decline.

It is interesting to note that sand production was considered a problem in cold production 30 years ago as expressed by Kimmel and Laviolette (1982), for example, who attributed the lower primary recoveries of 5.5% OOIP observed in the Loydminster heavy oil fields to adverse mobility due to high viscosity, the near absence of solution-gas drive, and severe sand production problems. In fact, their paper addressed the developments in artificial lift methods to improve the pumping rates of the sucker rod-beam pumping systems. The flow of the sand through the standing and floating valves led to considerable wear and sand could settle out in the dead zone between the moving plunger and the standing valve cage when high water cuts were produced.

A change in paradigm occurred in the early 1990s as advances in progressive cavity (PC) pump technology were implemented. More stages were added to the PC pumps and more wear resistant materials were used for the rubber stator and rotor. The back pressure was reduced by using larger production tubing and stronger pump rods allowed greater torques to be applied. It is striking to see the increase in oil production when PC pumps were installed in the 1990s as reported, for example, by Dusseault (2002) for Luseland cold production wells.

The objective of this chapter is to first review the literature on the two main mechanisms involved in the CHOPS process which are the enhanced solution-gas drive due to the depletion of heavy oil versus light oil and the enhanced reservoir access due to the production of sand. The different schools of thought regarding these mechanisms will be reviewed. The tracer field tests performed by industry to evaluate the effect of producing sand on the reservoir and bottom-hole pressure (BHP) data observed during the drilling of horizontal wells in CHOPS produced reservoirs will be analyzed. It will be shown that the development of high permeability channels (wormholes) explains better the tracer tests than the fracturing of a dilated zone around CHOPS wells. The oil, sand, water, and gas production data from a group of six CHOPS wells, which were put on production 9.6 years ago, were history matched using a multiwell CHOPS model for an initial production period of 6.3 years in a previous study (Tremblay, 2009b). The predictability of the multiwell model will be tested by using the same set of adjustable parameters to predict the oil production from year 6.3 to the present date (year 9.6) and by comparing these projections to the data.

22.2 MECHANISMS

22.2.1 Solution-Gas Drive

Background

The solution-gas drive process occurs naturally when a reservoir saturated with live oil (i.e., containing dissolved gas) is depleted below a certain pressure leading to the generation of gas bubbles which increases the compressibility of the reservoir fluid/gas mixture and slows down the pressure decline of the reservoir. If the pressure depletion rate is sufficiently slow, as would be performed in pressure–volume–temperature (PVT) measurements in the laboratory, to obtain quasi-equilibrium conditions, the pressure at which the bubbles appear would be defined as the bubble point.

Dissolved gas molecules are in constant random motion (thermal fluctuations) within the oil and statistically can form clusters momentarily simply from the collisions between the dissolved gas molecules which are more mobile than the heavy oil molecules. In the homogeneous thermodynamic model of bubble nucleation, the Gibbs functions of the gas and liquid are equal which leads to the following condition:

$$\mu_G = \mu_L \tag{22.1}$$

μ_G and μ_L are the chemical potentials for the gas and liquid, respectively. The mechanical equilibrium of a gas bubble in a liquid is given by the equation (Landau and Lifshitz, 1980):

$$P_g - P_L = 2\gamma/r \tag{22.2}$$

where P_G and P_L are the gas and liquid pressure, respectively, γ is the surface tension, and r is the radius of the gas bubble. As described by Bauget and Lenormand (2002), the equality of the chemical potentials between the gas and liquid leads to the conclusion that the pressure of the gas within the bubble is approximately equal to the pressure of the gas above a planar interface at equilibrium with the liquid. From Eq. (22.2), this implies that the pressure of the liquid containing a gas bubble would be lower than that of a liquid at equilibrium. The difference between the equilibrium pressure and the pressure of the liquid (i.e., the heavy oil and dissolved gas) mixture is defined as the supersaturation. In order for the bubble to grow in size, the liquid pressure has to decrease below the value given by Eq. (22.2). Gas will then diffuse into the bubble allowing it to grow until it reaches a radius given by Eq. (22.2) for the new liquid pressure. The supersaturation is then equal to the capillary pressure.

Bauget and Lenormand (2002) reviewed the literature on bubble nucleation in oil. According to these researchers, the homogeneous nucleation model cannot describe the generation of gas bubbles in heavy oils since supersaturation in the 1000 MPa range would be required for the clusters of gas molecules to aggregate with a diameter sufficiently large to resist redissolution from the capillary pressure. The model also does not predict a threshold supersaturation for bubble growth which is contrary to micromodel experiments. In the heterogeneous model, the nucleation of gas bubbles would occur on solid surfaces such as the surface of the sand grains. Bauget and Lenormand (2002) report studies which contradict the heterogeneous model since cavities of very specific angles would be required for nucleation to occur. Furthermore, the heterogeneous nucleation model also does not predict a supersaturation threshold. The authors argue that the only alternative would be the nucleation of gas bubbles on preexisting gas bubbles either in the bulk liquid or trapped in the roughness of the solid which would explain the observed threshold in supersaturation and the constant number of bubbles that develop for a given pressure drawdown. However, they do not mention how the presence of these preexisting gas bubbles in the reservoir could be confirmed. While preexisting gas bubbles can be entrained in hydraulic applications due to turbulence for example, it is more difficult to understand why gas bubbles would have resisted dissolution in an oil reservoir over geological timescales. Furthermore, the occurrence of homogeneous or heterogeneous nucleation in a porous medium cannot be excluded simply because the nucleation models do not agree with observations.

In reservoir engineering, the relative permeabilities to gas, k_{rg}, and oil, k_{ro}, often follow the Corey-type model given by

$$k_{ro} = k_{roi} S^{no} \tag{22.3a}$$

$$k_{rg} = k_{rgi}(1-S)^{ng} \tag{22.3b}$$

$$S = (S_o - S_{org})/(1 - S_{org}) \tag{22.3c}$$

where k_{roi}, k_{rgi}, no, and ng are constants, S_o is the oil saturation, and S_{org} is the residual oil saturation (Pooladi-Darvish and Firoozabadi, 1999).

Recovery Factor

As mentioned by Maini (1996), the recovery factor for light oil ranges from 1% to 5%, whereas for heavy oil it ranges from 5% to 15%. Smith (1988) suggested that micro-bubbles significantly smaller in diameter than the pore throats would form in heavy oil based on the thermodynamic analysis of Ward et al. (1982). The observation by field engineers that wellhead samples of live heavy oil resembled chocolate mousse led to the term foamy oil which was first used by Maini et al. (1993) to describe the oil in the reservoir. In fact, the use of the term gas dispersion or "bubbly oil" would be more accurate as suggested by Firoozabadi (2001) to describe the flow of the gas and oil in the reservoir.

According to Maini (1996), the greater oil recovery factor for heavy oil can be attributed to the higher production pressure drawdown imposed in cold production reservoirs, due to the high viscosity of the heavy oil, leading to the breakup and entrainment of pore throat-sized gas bubbles through the porous medium as observed by Bora et al. (1997) in visualization experiments using glass micro-models saturated with live heavy oil. The stability of dispersed gas bubbles would be governed mainly by the viscous drag forces, which tend to deform the bubbles and break them up, and the interfacial tension forces, which tend to resist deformation and give the bubble a stable spherical shape. The relative importance of these two forces can be better quantified by defining the capillary number as follows:

$$\mathrm{Ca} = U_{\mathrm{oil}}\mu_{\mathrm{oil}}/\sigma \tag{22.4}$$

where U_{oil} is the darcy oil velocity, μ_{oil} is the live oil viscosity, and σ is the surface tension between the oil and gas.

Solution-gas drive experiments were performed by Firoozabadi and Aronson (1999) at low capillary numbers and pressure depletion rates using a transparent core holder. Typically the first gas bubbles would appear within the sand pack at the surface of the core holder quite early in the experiment when only 1–2% of the oil was recovered; whereas the gas bubbles would start being produced out of the core only when 5–10% of the oil was recovered. They concluded that if gas flowed in the form of micro-bubbles with the oil, it would have been produced out of the sand pack at the onset of gas formation within the core. The same conclusion would apply to a gas-in-oil dispersion, since gas would flow along with the oil, although not necessarily at the same velocity, and should be produced out of the sand pack as soon as gas develops in the core, according to the foamy oil theory. They observed that the gas flowed in slugs accompanied by pressure fluctuations when the gas saturation was greater than the critical gas saturation.

Tang and Firoozabadi (2003) observed that approximately 10 times more gas bubbles were produced in sand packs saturated with heavy oil than with silicone oil using the same apparatus. Gas started to be produced out of the core at a much higher oil recovery factor for heavy oil (8%) than for silicone oil (2.5%) indicating that the depletion of the heavy oil was generating more bubbles which were more stable. The pressure drop across the length of the sand pack fluctuated sharply when the gas started to be produced for both types of oil which is consistent with the gas slugging observed in the experiments and with the low relative permeability to gas which they calculated (e.g., 2×10^{-5} at 15% gas saturation). Firoozabadi (2001) argued that high capillary numbers are not required to lead to higher oil recoveries for heavy oil.

Pressure Depletion Rate

Several researchers have observed in solution-gas drive experiments that oil recovery increases with increasing pressure depletion rates (Sahni et al., 2004; Sheng et al., 1999a) due to the increasing supersaturation of the oil leading to a greater number of gas bubbles and higher gas saturations. As explained by Moulu (1989), at high-pressure depletion rates, the gas diffusion is not as effective in countering the action of the pressure decline, whereas a smaller number of bubbles are necessary at lower rates since the gas has more time to diffuse into the bubbles.

Sheng et al. (1999a) modeled numerically three different pressure depletion experiments performed at the following depletion rates: 38, 115, and 221 kPa/h. They showed that the supersaturation in pressure in the sand pack increased with increasing pressure depletion rate leading to a greater volume fraction of dispersed gas and therefore a greater oil recovery. The dynamic model they developed history matched better the measured cumulative oil than an equilibrium model assuming a high critical saturation of 9%. They used a conventional gas relative permeability, k_{rg}, curve in their simulations with a k_{rgi}, of 0.61.

Sahni et al. (2004) caution against thinking that pressure depletion experiments performed at low depletion rates lead to higher gas saturation. They used a mechanistic model to suggest that a greater number of nucleation sites would be generated resulting in a greater gas saturation and higher critical gas saturation with increasing pressure supersaturation at the higher pressure depletion rates. They calculated the gas saturation in a sand pack at equilibrium from a material balance of the oil and gas using PVT data for the same live oil. The gas saturation which developed at the highest rate of 221 kPa/h of Sheng et al. (1999a) was in fact closest to the equilibrium gas saturation.

Sahni et al. (2004) showed that the gas saturation versus depletion pressure experimental results of Treinen et al. (1997) followed the equilibrium curve until a gas saturation of 9% was reached at which point the curve deviated sharply from the equilibrium curve since gas was suddenly produced. The solution-gas drive experiment of Treinen et al. (1997) was performed at the low capillary number of 1.6×10^{-7}. Their computed tomography (CT) observations of the sand pack indicate that gas bubbles grow in separate ganglia which do not move within the sand pack at low capillary numbers. When these ganglia connect, free gas is produced. The compressibilities of the gas and oil in the sand pack and of the produced gas and oil mixture at the outlet were measured. They observed that while the gas saturation and compressibility in the sand pack increased as the pressure decreased below the bubble point, the produced oil and gas compressibility did not change, indicating that only live oil was being produced without gas bubbles until a critical gas saturation of 9% was reached.

Bondino et al. (2003) suggest, based on network numerical modeling techniques, that the oil recovery depends on the length of the methane diffusion pathways, the local supersaturation gradients, and the gas cluster topology—all of which are related to the underlying connectivity of the pore system. According to their pore scale simulations, the greater the number of nucleated bubbles (as a result of faster depletion rate), the slower their individual growth rate. These authors showed, in pore scale network numerical simulations, that the connectivity of a porous medium influences the supersaturation. In their pore scale simulations, they assume that micro-bubbles do not form in any great numbers and do not flow with their associated oil, and that the gas remains trapped in the pore matrix under capillary dominated growth conditions.

In order to compare the capillary number, pressure depletion rate and pressure gradient measured in the previously discussed solution-gas experiments to the field case, a multiwell model developed at the Saskatchewan Research Council (Tremblay, 2009a,b) was used to first calculate the oil velocity and viscosity distribution in a typical CHOPS reservoir. The assumptions in these numerical simulations were: (1) a 40 acre spacing, (2) equilibrium methane solubility and k values, (3) a dead oil viscosity of 15,000 cP, (4) suppressed gas relative permeability (k_{rgi} of 10^{-3}, S_{gc} of 5%), (5) a pay thickness of 6 m and (6) an absolute permeability of 2 D. Previous history matching of oil, water, sand, and gas production from typical CHOPS wells suggested that the development of eight wormholes were sufficient to represent the field data (Tremblay, 2008a). One-quarter of the 40 acre drainage area was discretized into $80 \times 80 \times 10$ grid blocks $2.5\ m \times 2.5\ m \times 1\ m$ in size. The predicted oil velocity after 1200 days of production is shown in Figure 22.1 where the well is located at the top

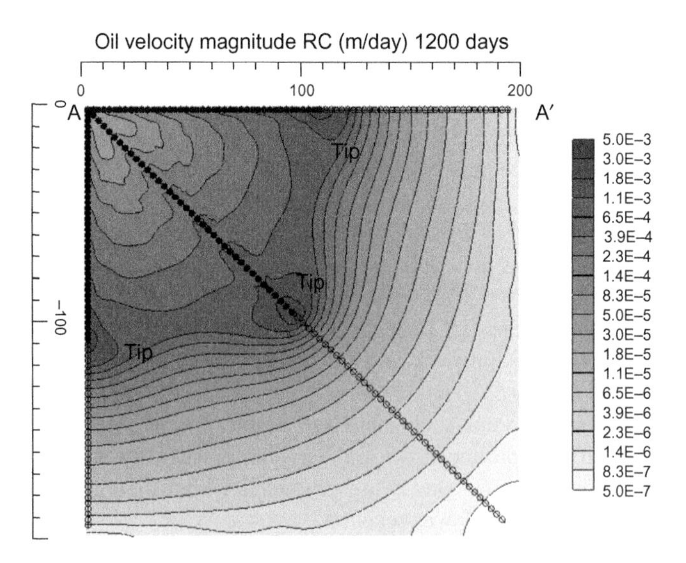

FIGURE 22.1 Areal distribution of oil velocity magnitude 1200 days after start of CHOPS production. Length scale is in meters. Well at location A (top left-hand corner).

left-hand corner of the area. Wormholes radiate along the X and Y axes and along the diagonal. In this simulation, the wormholes have grown out 100 m in 1200 days. As expected, the oil velocity is greatest at the tip of the wormholes reaching up to 5.0×10^{-3} m/day, whereas this velocity declines sharply in the reservoir. The capillary number along the X axis was calculated from Eq. (22.4) and is plotted in Figure 22.2 as a function of the radial distance from the well. The capillary number decreased sharply from its maximum value (2.7×10^{-5}) at the tip of the wormhole to values as low as 1×10^{-8} deeper into the reservoir. The minimum capillary numbers for some pressure depletion experiments are also plotted indicating that most experiments represent conditions within approximately 25 m of the wormhole tip. The field estimates of the pressure depletion rate and pressure gradient along the X axis in Figure 22.1 are shown in Figure 22.3 as a function of the distance from the CHOPS well. Again, the laboratory measurements are more representative of the conditions closer to the wormhole tip (within 25 m).

Critical Gas Saturation

The different definitions of the critical gas saturation, S_{gc}, are described in a review article by Sheng et al. (1999b). Normally S_{gc} is defined as the maximum gas saturation at which the gas relative permeability is zero (Firoozabadi, 2001). Other researchers define S_{gc} as the gas saturation at which the gas channels span the reservoir. Treinen et al. (1997) define the critical gas saturation as the gas saturation at which the produced gas–oil ratio (GOR) is greater than the initial dissolved GOR. These different

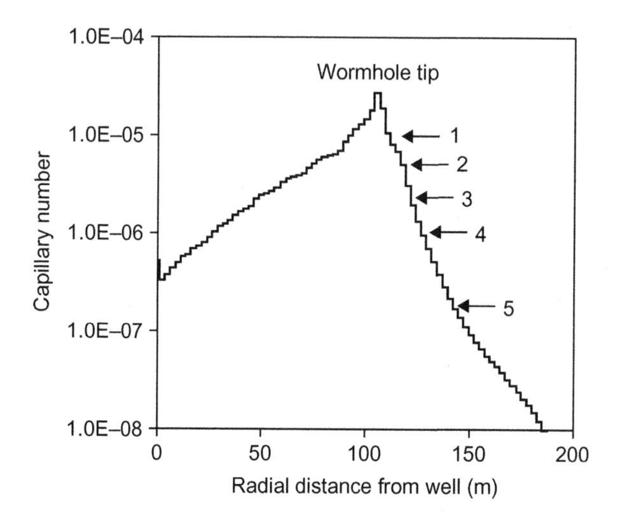

FIGURE 22.2 Capillary number versus radial distance from well after 1200 days. 1: Pooladi-Darvish and Firoozabadi (1999); 2: Ostos and Maini (2005); 3: Tang and Firoozabadi (2003); 4: Urgelli et al. (1999); 5: Treinen et al. (1997).

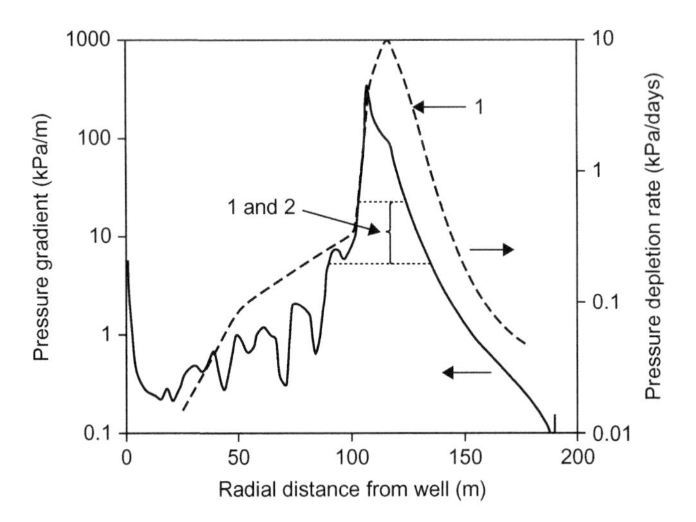

FIGURE 22.3 Radial pressure gradient versus distance from well after 1200 days. 1: Tang and Firoozabadi (2003); 2: Kumar and Pooladi-Darvish (2002).

definitions were developed because these authors model the gas flow in the reservoir differently. Sahni et al. (2004) obtained the following correlation between the critical gas saturation, S_{gc} (volume fraction) and the absolute permeability, K (mD), the live oil viscosity, μ (cP), the solution GOR (m^3/m^3), R_s (m^3/m^3), the pressure depletion rate dP/dt (kPa/h), and the initial water saturation S_{wi} (volume fraction):

$$S_{gc} = CK^{A1} \times \mu^{A2} \times R_S^{A3} \times (dP/dt)^{A4} \times (1-S_{wi})^{A5} \tag{22.5}$$

The coefficients C, $A1$, $A2$, $A3$, $A4$, and $A5$ are 2.62×10^{-3}, 0.076, 0.258, 0.17, 0.5, and 4.0, respectively. (Equation reproduced with permission of the copyright owner from Sahni et al. (2004). Further reproduction prohibited without permission.) This correlation was obtained using pressure depletion experiments which ranged in pressure depletion rate, dP/dt, from 41 to 38,400 kPa/day, in solution GOR, R_s, from 6.5 to 32 m^3/m^3, in initial water saturation, S_{wi}, from 0 to 32%, in live oil viscosity from 250 to 56,000 cP, and in permeability from 0.36 to 18 D.

Several researchers have observed a critical gas saturation of approximately 5% in low depletion rate experiments (Pooladi-Darvish and Firoozabadi, 1999; Tang and Firoozabadi, 2003; Urgelli et al., 1999). This value will be used in numerical simulations of the CHOPS process in this chapter. Firoozabadi (2001) suggests that critical gas saturations, S_{gc}, as high as 10% could be reached at very low-pressure depletion rates and high GOR. These latter observations appear to contradict Eq. (22.5) where S_{gc} is proportional to the pressure depletion rate to the power 0.5 and to the solution GOR, R_s, to the power 0.17. According to this latter correlation, S_{gc} should *decrease* with decreasing pressure depletion rate. If the field valuc for dP/dt is one order smaller than in the lab, S_{gc} would be reduced by a factor of 0.3. Sahni et al. (2004) defined the critical gas saturation as the saturation at which the GOR starts to increase. As the pressure depletion rates decrease, the physics of the gas flow is getting closer to the classical solution-gas drive conditions where few separate gas ganglia grow and eventually connect as observed by Treinen et al. (1997). The concept of critical gas saturation should be used ideally within the range of parameters given below Eq. (22.5) and only with care at pressure depletion rates below 41 kPa/day, which is at least one order of magnitude greater than the predicted values in the CHOPS process.

Oil Chemistry

Bauget et al. (2001) observed that asphaltenes enhanced foamability and oil film lifetime above a 10% weight concentration. The authors claim, based on a series of foamability, static and dynamic tension, viscosity, and surface viscoelasticity measurements using asphaltene and resin solutions dissolved in toluene, that asphaltenes can reorganize themselves at the interface, leading to prolonged surface tension and an increase in elastic modulus. They describe several studies, performed for the upgrading industry, where gas dispersions and water/oil emulsions need to be broken, which indicate that the adsorption of the asphaltene at the oil–water interface stabilizes crude oil emulsions due to the existence of a rigid protective films which limits coalescence. These studies are thought by Bauget et al. (2001) to also apply to gas dispersions in heavy oil. Their study also indicates that the resins have a role in reducing the surface activity of the asphaltenes since the resins partly dissolve the asphaltene aggregates. They report the work of Callaghan et al. (1985) who associate the stability of oil foam to the presence of short-chain carboxylic acids and phenols in the oil.

Tang et al. (2003) discuss the role of heavy, complex polar molecules in crude oil (such as asphaltenes, resins, and organic acids) and their attraction to oil–gas phase boundaries, where they form an elastic membrane that stabilizes the oil–gas interface which slows down bubble growth. They speculated that the formation of a skin adds resistance to the diffusion of gas through the bulk oil phase to growing bubbles and would reduce their growth rate.

Peng et al. (2009), in core scale depletion experiments, showed that oil composition plays a role in determining the metastability of dispersed gas bubbles in heavy oil. They observed that a high concentration of asphaltenes that exhibit acid and base functional groups tends to increase foamability and film lifetime of gas/crude oil dispersions. They conclude that acid and base groups within asphaltenes, and their interaction at the gas–oil interface, are a source of interfacial stability by allowing asphaltene molecules to form an interlinked network structure. This network would lead to higher surface tension of the gas bubbles, reducing their ability to be broken up.

Bora et al. (2003) also studied the role of asphaltenes in stabilizing gas bubbles in micro-model tests. They used four different oils: crude Lindbergh heavy oil, de-asphaltened Lindbergh heavy oil, mineral oil, and a light oil. The viscosity of the de-asphaltened Lindbergh oil was approximately the same as that of the mineral oil. In their visualization study, they counted the number of bubbles that developed in the micro-model during pressure depletion and observed that for all oils approximately the same number of bubbles was generated within the first 4 h. However, after 12 h, the number of bubbles was constant only for the crude Lindbergh oil, whereas the number decreased significantly for the other oils indicating that the asphaltenes have a role more in stabilizing the gas bubbles than in generating more bubbles.

Numerical Modeling

Foamy Oil

According to the foamy oil theory, the formation of a gas dispersion is a dynamic process which depends not only on pressure, temperature, and composition, but also on the flow conditions and on the history of the process (Sheng et al., 1999a). A dynamic model of dispersed gas flow was developed by Sheng et al. (1999a) which represents the transfer of dissolved gas to dispersed gas bubbles and then to free gas. In this model, the gas bubbles nucleate instantaneously below the bubble point. The radius of a gas bubble would be proportional to the square root of time as predicted by Moulu (1989).

Bayon and Cordelier (2002) history matched a series of long-core primary depletion experiments by using: (1) an equilibrium approach where the solution-gas is simply transferred directly to free gas; (2) a nonequilibrium approach where the transfer of dissolved gas to dispersed gas and finally to free gas was represented by two chemical reactions using the STARS® reservoir simulator that could represent the nonequilibrium gas transfer. In the first approach, they observed that a significantly lower gas relative permeability, k_{rg}, was required to history match the oil and gas at the higher pressure depletion experiment performed at 800 kPa/day compared to the lower depletion rate experiment at 80 kPa/day. In the nonequilibrium approach, the gas relative permeability was interpolated between a free gas relative permeability curve and a significantly lower k_{rg} curve. They were able to history match the oil and gas production for the two experiments with the same frequency factors. As seen in Figure 22.3, even the lowest pressure depletion rate of 80 kPa/day applied in the experiments by Bayon and Cordelier (2002) was one to two orders of magnitude greater than the field depletion rates away from the wormhole tip.

Uddin (2005) used a similar nonequilibrium approach except the gas nucleation and bubble growth steps were each represented by two chemical reactions. The reaction parameters were first determined by history matching constant volume withdrawal rate experiments (Lillico et al., 2001). Using these same parameters, the oil and gas production volumes for four separate constant pressure depletion experiments, performed at rates varying from 84 to 167 kPa/day, were predicted and found to compare well to the measured values. These pressure depletion rates are still at least one order of magnitude greater than estimated in the field case as shown in Figure 22.3. No details were given by Uddin (2005) regarding the gas and oil relative permeability curves used in the numerical simulations.

Low Gas Relative Permeability

Pooladi-Dravish and Firoozabadi (1999) first suggested using the conventional two-phase (oil and gas) relative permeability model and the absence of supersaturation in the reservoir, while assuming a very low gas relative permeability. They history matched quite well the gas saturation which developed in a sand pack saturated with live heavy oil by simply assuming that the gas, k_{rg}, and oil, k_{ro}, relative permeabilities followed the Corey-type model given by Eqs. (22.3a–c). For conventional oils, the constant k_{rgi} is normally in the 0.5–0.8 range. However, in order to fit the produced heavy oil volumes, that constant had to be decreased to 0.000025.

Pooladi-Dravish and Firoozabadi (1999) and Kumar and Pooladi-Darvish (2002) are assuming that the unsteady flow of gas in the form of spurts, as observed in the form of large pressure drop fluctuations in the sand pack, can be represented by the relative permeability concept. In their approach, the gas flow is severely restricted. This engineering approach has the advantage of being simpler but also the disadvantage of not representing the physics well. As will be shown in a further section, using the Corey-type equation with constant k_{rgi} values can lead to overestimates of oil production rates toward the end of the production life of a CHOPS well. The use of their method would require that the gas relative permeability curves be changed at different times during the simulation.

22.2.2 Sand Production

As mentioned in Section 22.1, in order to produce heavy oils with dead oil viscosities in the 2000–40,000 cP range, well operators observed that it is necessary to produce large quantities of sand. Certain CHOPS wells can produce up to 1500 m^3 of (bulk) sand. To model numerically the CHOPS process, it is essential to know the effect of producing such large quantities on the reservoir.

Background

The main forces acting on a sand grain in an uncemented porous medium are: (1) the contact forces, both normal and tangential at the location where the sand grains are in contact, (2) the pore pressure which acts in the normal direction to the surface of the sand grain, and (3) the adhesive forces at the contact point between the sand grains due to the capillary pressure between the fluid in the meniscus and in the pores surrounding the sand grain.

It is impractical to calculate microscopically a stress tensor at every point in the medium. Continuum approaches were first used by Terzaghi (1951) who decomposed the macroscopic stresses in a porous medium into two components: a neutral stress due to the fluid pore pressure and an effective stress which is the difference between the stress and the pore pressure. According to Terzaghi (1951), the effective stress provides the deformation of the porous medium. It can be thought of as the stress acting in the porous skeleton formed by the contacts between sand grains. This important concept is still used in geomechanics (Azizi, 2000).

The friction between two sand grains sliding along one another is proportional to the normal contact forces acting between the grains as described by Coulombic friction. This means that the shear strength of granular materials increases with increasing normal effective stress (Azizi, 2000). The cohesive strength of a soil or oil sand is defined as the maximum shear stress which the material can support under zero normal stress such as would occur on the surface of a wellbore.

The geomechanical properties of oil sands are often measured in triaxial cell tests in which a cylinder of sand is compressed axially while being confined radially (Azizi, 2000). By plotting the maximum effective axial stress which the material can withstand as a function of the effective radial stress, a failure envelope is created which can be used in calculations where the stress field is more complicated. One important test in determining the stability of wellbores is the unconfined compression test in which the effective confining radial stress is zero as would occur at the surface of a wellbore.

When an open horizontal channel (wellbore) is first formed in a reservoir, the radial effective stress, σ_r, at the surface of the wellbore is zero, in the absence of a filter cake. The tangential effective stress, σ_t, at the surface of the wellbore, increases until the unconfined compressive strength of the oil sand is reached. A plastic (dilated sand) zone develops around the borehole. According to Wong et al. (1994), this dilated sand region around an open channel would be composed of a residual zone and a softening zone. When a densely packed sand is sheared, it will start to dilate (softening zone) until it reaches a residual (critical) state at which the shear strength becomes constant (residual zone). The nonbrittle behavior of oil sands implies that the oil sand still has a shear resistance even after yielding. The elastic modulus of the dilated region (softening and residual zones) is significantly lower than that of the formation which reduces the tangential stress at the surface of the wormhole. The plastic zone is thus shielding the surface of the wellbore from the *in situ* stresses.

Dilated Zone

One geometry that field engineers have hypothesized develops during CHOPS was that of a symmetrical dilated zone around the wells. This geometry was first discussed by Smith (1988) who suggested that a cavity first develops above a perforation as shown in Figure 22.4. This cavity would be surrounded by an un-stressed sand region which failed. This latter region would then in turn be surrounded by a stressed sand region. According to Smith (1988), the cavities around each perforation would merge into one large cavity surrounding the wellbore as shown in Figure 22.5.

Nouri et al. (2008) and Dusseault (2002) describe the remolded zone concept as the gradual yield, dilation, and remolding of the formation material around a well due to shear and tensile failure. They surmise the following succession of events: (1) sand failure around the well when the well is drilled, (2) additional shear failure and dilation at the perforation opening, (3) development of small cavities that form and grow in size, (4) collapse of cavities, (5) formation of high porosity loose zones which grow in size into a liquified zone. Their dilated sand geometry is slightly different than that in Figure 22.5 since they do not assume that a cavity develops around the CHOPS well but a liquified sand zone in which the sand grains are not in contact with each other and can move freely. In addition, the plastic zone shown in Figure 22.5 is subdivided into a fully yielded zone and a further transition zones (Dusseault et al., 1998). No indications as to the approximate typical dimensions for these zones are given by Nouri et al. (2008) nor Dusseault et al. (1998).

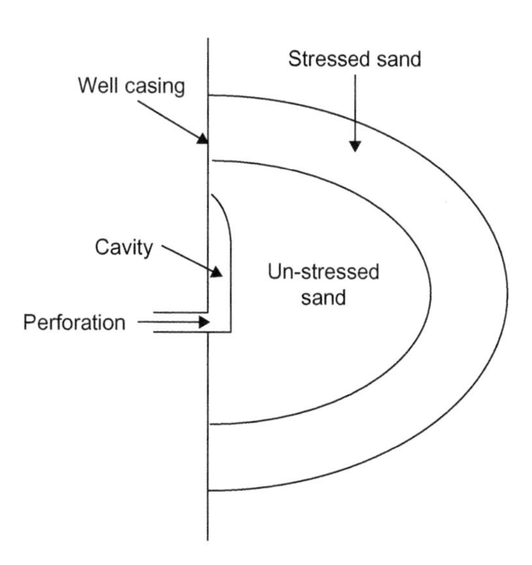

FIGURE 22.4 Sand arching around perforation.

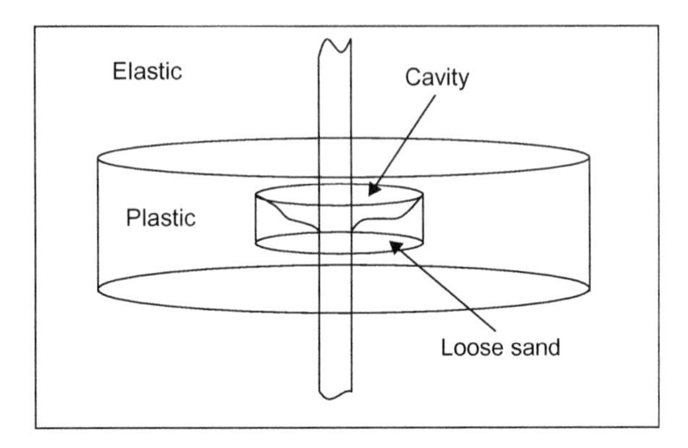

FIGURE 22.5 Dilated zone around wellbore.

If the dilated zone is assumed, it then would follow that a large part of the overburden stresses around the wellbore would be transmitted to the yielded and intact zones as suggested by Dusseault et al. (1998). They do not specify what the magnitude of the vertical stresses would be in the intact zone except to state that they would be greater than the initial vertical stress (overburden). This assumption poses a fundamental problem when analyzing injectivity tests as will be discussed in a further section since the injection pressure in tracer test studies (Squires, 1993) was no greater than the initial hydrostatic pressure and consequently significantly less than the initial vertical stress or horizontal stresses measured in the field (Bell et al., 1994). Consequently, a fracture could not propagate between CHOPS wells in a tracer test according to the scenario of Dusseault et al. (1998) or Nouri et al. (2008).

These authors do not in turn explain quantitatively how the overburden over a liquified zone would not collapse leading to the buckling of the well. Although well buckling due to excessive vertical stress compaction after large sand production has been observed in certain wells (Shute et al., 2004), the majority of CHOPS wells do not buckle; at least not enough to be detectable or to affect the oil production. It is not clear why the liquified zone would become cylindrical in shape as assumed by Nouri et al. (2008). Furthermore, they also did not calculate how much pressure gradient would be required to transport the sand through the liquified zone and whether or not this gradient would be large enough to move the sand to the well.

Wormhole Network

Wormhole Growth Experiments

One of the difficulties with the remoulded zone concept is in describing the initial mechanism by which cavities would hypothetically form around the perforations. In the experiments by Walton et al. (2002), uncemented water-wetted sand packs saturated with kerosene were initially perforated and then further flooded with kerosene until sand production occurred. They observed that perforating did not generate a sand-free tunnel in unconsolidated sands but instead a dilated sand region around the perforation. This hemispherical dilated region grew with increasing flow rate (pressure gradient) above a critical value below which the cohesion of the sand would have prevented it from being produced.

It is important to define what is meant by the term "wormhole." As will be explained in a further section when sand was produced in sand packs saturated with either dead or live oil, a cylindrical-shaped sand-filled region of higher porosity developed within the pack starting at the opening at the production end (Tremblay et al., 1999) as shown in Figure 22.6. This CT image shows the porosity within a sand pack saturated with live heavy oil where the pressure at the production end (outlet) was reduced at a constant rate of 205 kPa/day. It is important to note that the scanning length over which the CT image shown in Figure 22.6 was taken only covered 35.2 cm, whereas the full sand pack length was 85 cm. This means that the wormhole only

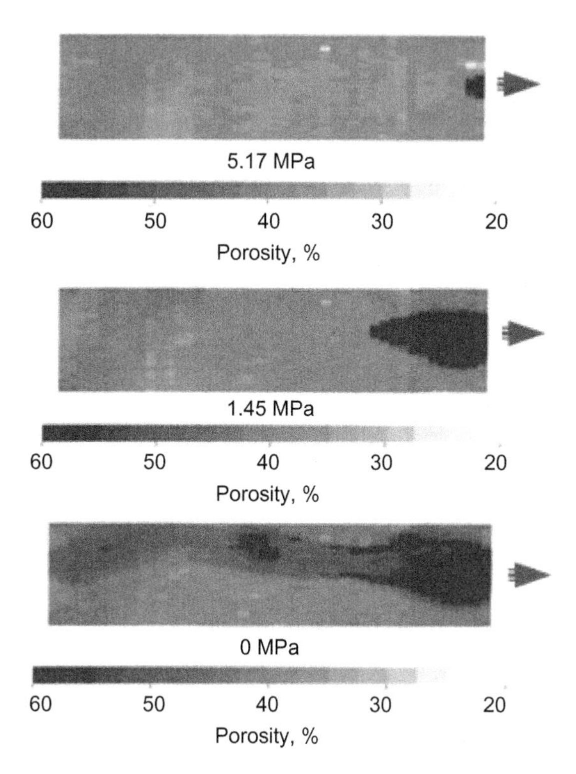

FIGURE 22.6 CT image of a sand pack after pressure at outlet reached atmospheric pressure (Tremblay et al., 1999). Source*: Copyright 1999 Society of Petroleum Engineers.*

grew out to 41% of the length of the sand pack under solution-gas drive. The top image shows the sand pack at the start of the experiment before the pressure at the production end was decreased from the initial saturation pressure of 5.17 mPa. Note that the porosity is greater at the top of the sand pack due to a packing artifact. In the middle image, the wormhole tip is pointed due to the concentration of the pressure gradient in the axial direction. The wormhole was completely sand-filled as it grew into the higher porosity region at the top of the sand pack.

Only in the experiments where a constant flow rate or pressure was maintained at the inlet of the sand pack did the wormhole reach the inlet (Tremblay et al., 1998a,b). The sand within the sand-filled region then started to empting out leading to an open sand-free channel. The sand was transported in a thin dilated layer at the interface between the oil at the top and the settled sand at the bottom (Tremblay et al., 1998a) by resuspension under laminar flow conditions. It is important to distinguish between a sand-filled cylindrical region and the open channel which forms within it. In the field, an open sand-free channel would likely develop when sufficient oil drainage occurs into the wormhole to decrease the sand concentration below approximately 45% for settling to occur. In addition, the open channel diameter (6–10 cm) would be much smaller than the sand-filled part of the wormhole which could reach 1 m in diameter as will be discussed further.

Sand production experiments performed by Meza-Diaz et al. (2011) showed that wormholes can grow under stress provided the rubber membrane which exerts a radial confining stress on the sand pack does not finger into the pack. Walton et al. (2002) attribute the complete radial collapse of their sand pack observed in their tests when the flow rate (pressure gradient) reached a critical value to the artifact of the constant pressure condition at the outer boundary of the sample and the finite sample size. Using the Bratli–Risnes stress analysis (1981), they calculated that the plastic envelope of the failed zone reached the boundary of their sample as the flow rate was increased leading to the complete radial collapse of the sand pack. In finite-sized test samples, the average (circumferential) hoop stress around the hole increases as the hole size grows, thereby expanding the plastic zone even further. As suggested by Walton et al. (2002), to simulate downhole conditions better, the confining pressure (stress) on the sample should be reduced as sand is produced and the hole expands. It is not likely that a cavity would form above the perforations in the field as shown in Figure 22.4 because the perforation and eventually the wormhole that can grow from it would be sand-filled at the start of production.

Bratli and Risnes (1981) developed an expression for the critical radial pressure gradient, ∇P_{crit}, at the surface of a spherical cavity of radius, R_c, at which sand production will occur:

$$\nabla P_{crit} = 2fC_u/R_c \tag{22.6}$$

where f is constant and C_u is the unconfined compressive strength. In their paper, f is equal to 1, however, Tremblay and Oldakowski (2003) found that this factor should be reduced to 0.1 for the dynamic case where the wormhole is growing. Since the unconfined compressive strength for an *in situ* oil sand (nondilated) is only approximately 90 kPa, this leads to a critical pressure gradient of only 60 kPa/m assuming a sand-filled wormhole tip radius, R_c, of 0.3 m. The pressure gradient between grid blocks along the X axis in Figure 22.1 increased sharply at the wormhole tip as shown in Figure 22.3 to a maximum value of 340 kPa/m. The pressure gradients were calculated at the level of the 2.5 m × 2.5 m × 1 m grid blocks which contained the wormhole and are understandably significantly lower than at the tip of the wormhole, roughly by a factor of 25 from continuity, assuming a 0.3 m radius at the tip. This implies that the pressure gradient at the wormhole tip in the field could be as large as 10 mPa/m and that uncemented oil sand can be easily produced given the large pressure gradients that develop in cold production without having to fail the formation before producing the sand contrary to the assumption by Rivero et al. (2010). It is important to note that the pressure gradient decreases sharply as a function of the radial distance from the tip (i.e., roughly to the inverse power of 2).

In a series of sand production experiments (Tremblay and Oldakowski, 2003; Tremblay et al., 1998a,b, 1999), the porosity within the wormholes was observed to be approximately constant, varying between 52% and 55%, while the wormholes were growing. The porosity in the wormhole did not depend on the oil velocity at the wormhole tip which in these experiments ranged from 0.001 to 0.04 cm/min. Multiplying the oil velocity at the wormhole at grid level (Figure 22.1) by a factor of 25, as argued previously, leads to a tip oil velocity of approximately 0.01 cm/min, which was within this experimental range. A typical longitudinal porosity profile as measured using a CT scanner, along a sand pack saturated with a heavy oil is shown in Figure 22.7. In this experiment, live oil was injected at one end of the pack at a constant pressure (Tremblay et al., 1998b) and produced at the other end (outlet). As shown in the figure, the sand was fluidized over a very thin zone within the length of one pixel (8 mm). From these observations, the following fluidization equation was developed (Tremblay, 2005), based on the continuity equation, to describe the advancement of the wormholes:

$$dr/dt = (1-\phi_f)/(\phi_f-\phi_i)\nabla P K_o/\mu_o \tag{22.7}$$

where φ_f is the porosity of the fluidized sand, φ_i is the porosity of the uneroded sand, ∇P is the pressure gradient at the fluidization surface, K_o is the permeability to oil, and μ_o is the live oil viscosity. This equation would explain why the wormhole developed in the higher porosity region at the top of the sand pack shown in Figure 22.6 since this region was more permeable. In the field, the viscosity can also vary as a function of depth indicating that

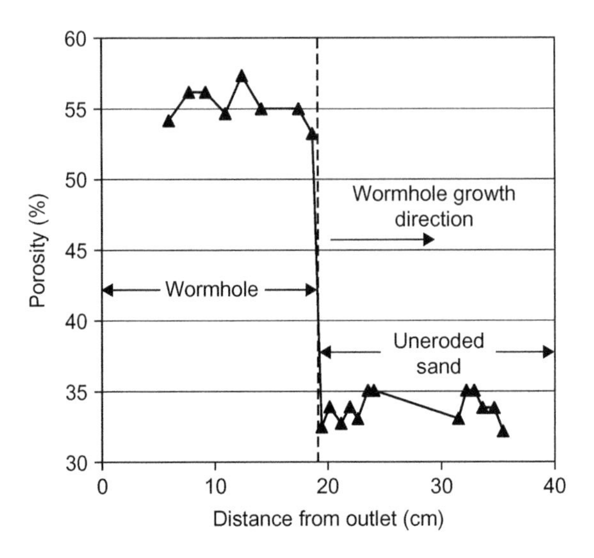

FIGURE 22.7 Longitudinal porosity profile.

the wormholes would grow in the higher mobility (permeability/viscosity) layers of uncemented sand.

One important criticism by geomechanics experts of the wormhole hypothesis is that the open channels within the wormholes would not be stable and would collapse. As mentioned earlier, when wormholes first grow they are sand-filled and can grow in the laboratory even if a stress is applied at the boundary of the sand pack provided the stress is not too large. Drillers of horizontal wellbores know that boreholes can remain stable for a few weeks. The cohesive strength of oil sands, typically 10 kPa in magnitude, is caused by the capillary pressure between the oil in the pores and the water in the pendicular rings at the contact area between the sand grains. When oil sand, comprised of sand grains and pore fluids, is sheared, it will dilate because of the interlocking between the sand grains (Tremblay and Oldakowski, 2003) which in turn creates a pore suction due to the low hydraulic diffusivity of the viscous oil leading to an increase in the effective confining stress and additional temporary cohesive strength. After a certain time, possibly 2–3 weeks, however, this effect would dissipate leaving only the cohesive strength due to the pendicular rings.

The stability of an open channel to solvent injection was investigated by Tremblay et al. (2008b) using the apparatus shown in Figure 22.8, in which a 10 cm diameter hole simulated an open channel, shown by the dashed lines, within a wormhole in the field. The absolute permeability of the sand was 20 D. The sand pack was first saturated with brine followed by two pore volumes of heavy oil (12,100 cP viscosity). A 10 cm diameter plug at the location of the open channel was pulled leaving an open hole in the sand

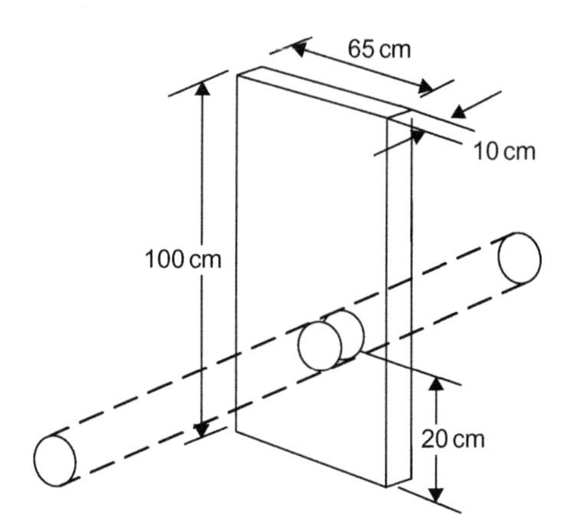

FIGURE 22.8 Wormhole stability apparatus.

pack which was stable for 2 weeks until butane was injected at the dew point leading to the collapse of the hole.

The yield stress of the oil sand was measured by Tremblay et al. (2008b) using a shear vane in separate tests for the cases where butane was not injected and where butane was injected at the dew point. In these tests, an oil sand sample within a cell was sheared using a vane by applying a uniform torque on the vane shaft until the material yielded completely. This vane was simply composed of four perpendicular blades welded to a central shaft. The peak and residual yield stress for the case where no butane was injected was 19.9 and 6.4 kPa, respectively. The corresponding values for the case where butane was injected at the dew point were almost the same at 19.2 kPa for the peak yield stress and 5.4 kPa for the residual yield stress. The shear vane tests were performed over sufficiently long times to avoid the previously mentioned transient strengthening effect due to the shearing and subsequent dilation of the oil sand saturated with a viscous oil leading to a pore suction and an additional confining stress on the sand.

The decrease in stability of the wormhole when butane was injected was most likely due to the significant reduction in viscosity of the heavy oil as the condensed solvent dissolved into the oil which decreased the transient strength of the oil sand. At the end of the experiment, the covers of the sand pack were alternately removed and the sand pack was photographed. Since large volumes of butane were injected, the sand was clean as shown in Figure 22.9. During the experiment, a separate chamber was in contact with the initially open hole such that the sand which collapsed in the hole could be produced. The tensile failure zone above the open channel rose as more sand was produced. The failure by gravity of the "roof" of the hole is consistent with the predictions of Wong et al. (1994) who identified this type of failure as being the most critical for wellbore failure.

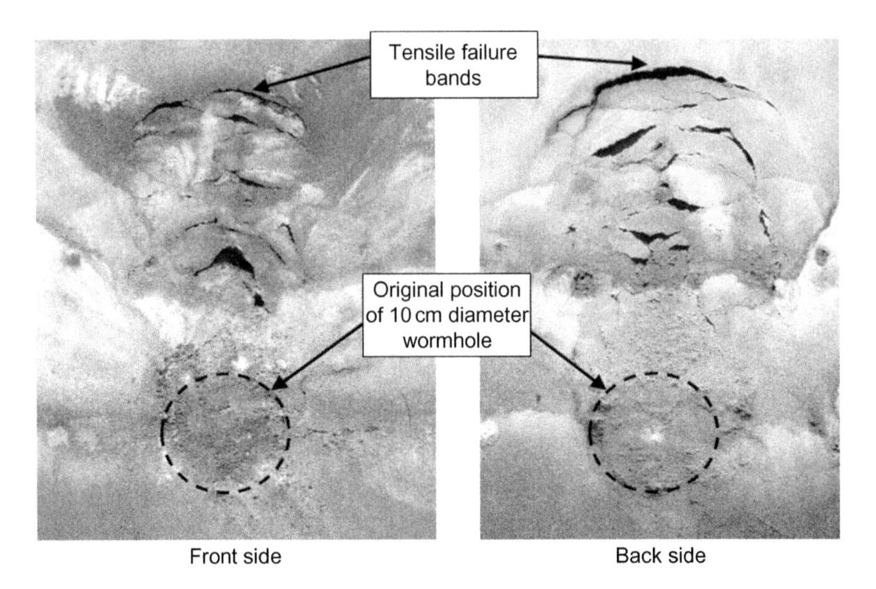

FIGURE 22.9 Front and back view of sand pack at end of experiment.

These images suggest that open channels within wormholes may collapse after a sufficiently long period as the transient cohesive strength dissipates. As sand is produced through the wormholes, the tensile failure zone may rise at the top of the wormholes until a more resistant layer of either cemented sand or shale is reached. A sand-free open channel would then be created since the sand would no longer collapse at the top of the wormhole. Vaziri et al. (2003) were the first to suggest that surficial erosion channels may develop beneath more resistant barriers such as shale layers based on centrifuge experiments simulating the production of sand into a central well.

Field Observations

The most convincing evidence of the presence of open sand-free channels in oil fields which have produced sand is the tracer injection tests performed by Amoco in 1993 at Elk Point, Alberta (Squires, 1993). When Amoco (Squires, 1993) performed a tracer test, the fluorescent dye which was injected in well #1 (Figure 22.10) was not diluted when produced from a neighboring well #2 and from further wells along the direction indicated by the solid arrows shown in Figure 22.10. In a separate experiment performed by Amoco, in which a fluorescent dye called fluorescein was injected and produced through a sand pack, the dye was completely adsorbed at the surface of the sand grains indicating that the dye had traveled through an open channel in the field. The complete path of the tracer dye is shown in Figure 22.10 by the solid arrows starting at well #1. The flow path was repeatable when the tests were

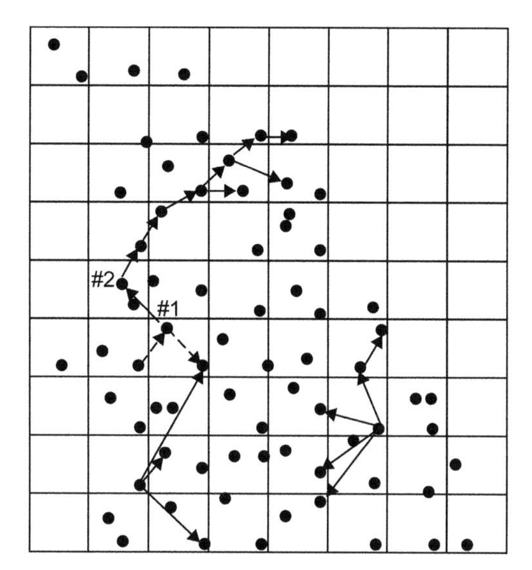

FIGURE 22.10 Fluorescent dye tracer path (solid line). Gel injection flow path (dotted line) (Squires, 1993).

performed again after 2 and 4 weeks. An estimate of 6 cm was obtained for the open channel diameter between wells #1 and 2 knowing the distance between the wells (400 m), the transit time between wells at the perforation level (1 h), and the dye solution injection rate (30 m^3/day). A chunk of dye 3 mm in diameter was recovered from well #2.

Nouri et al. (2008) explain the tracer tests by the development of fractures between wells but do not provide details as to the width or orientation of the fracture, whether vertical or horizontal. Since they assume that a dilated (remolded) region exists around the CHOPS well a potential fracture would have to form in a weaker sand. Controversy still exists whether dilated sand can be fractured. Di Lullo et al. (2004) injected a borate cross-linked gel into a sand pack saturated with water and observed that instead of a fracture, a wide cavity was observed. They claimed that the injection of a proppant/gel slurry into a well for stimulation purposes may lead to the squeezing or forcing by viscous fingering of the slurry into the formation rather than fracturing.

An upper bound for the width of a hypothetical fracture is obtained by assuming that the injected tracer solution did not leakoff and that the fracture was oriented vertically across the pay zone thickness of 15 m. From the injection rate (30 m^3/day) and tracer velocity along the flow path (7 m/min), as discussed previously, the maximum width would be equal to the unrealistic value of 0.2 mm which is the diameter of a sand grain. At such a small fracture width, the fluorescent dye would have adsorbed on the surface of the sand contrary to the observed concentration at the producer. Assuming a fracture width of 0.2 mm would lead to unrealistically high-pressure gradients of approximately 4 MPa/m. Therefore, fractures cannot explain the tracer tests of Squires (1993).

The SW/NE orientation of the flow path in Figure 22.10 was interpreted by some petroleum engineers as being indicative of the effect of the NW/SE direction of the minimum horizontal stress (S_{Hmin}) on the growth of the wormholes. It is important to note however that since the rate of growth of wormholes in an uncemented formation is proportional to the oil velocity, there is no reason to believe that wormholes are only growing from well #1 to well #2 for example since oil is drained radially from all directions. The oil pool where the tracer tests were performed has an ellipsoidal shape with the principle axis in the NW/SE direction which is also the direction of the oil/water contact line at the edge of the pool. Furthermore, the formation sands run updip in the NE direction. This means that the maximum pressure gradient due to the strong aquifer at the western edge of the reservoir was in the SW/NE direction which would explain why the wells watered out in sequence starting from the well closest to the edge. Because of the high-pressure at the western edge of the reservoir, wormholes would tend to grow more in the SW/NE direction until they connected. The tracer study itself would have likely enhanced the pressure gradients in the SW/NE directions leading to further wormhole connections. Multiwell wormhole growth simulations would be required to estimate the extent to which the aquifer modified the radial wormhole network growing from each CHOPS well.

As will be discussed in a further section, oil producers have started field pilot tests to recover oil after the CHOPS process. In one pilot test performed in western Canada, the possibility of using the vertical wells as solvent injectors and installing horizontal wells to serve as producers was tried. The circulation loss problems while drilling two horizontal wells for this pilot were analyzed in order to better describe the effect of producing sand on the CHOPS reservoirs. When the oil company drilled the horizontal well, shown as H1 in Figure 22.11, through a CHOPS reservoir, it encountered lost circulation episodes twice at the locations a and b indicated in the figure. When horizontal well H2 was drilled, circulation loss was encountered thrice at the locations c, d, and e.

The vertical well V1 had produced a total 71,000 m^3 of oil, which is large for CHOPS wells. The produced volume of sand was not recorded. However, the average cumulative sand concentration measured by the oil company for other CHOPS wells was 1% leading to a bulk sand production of approximately 700 m^3 for well V1. The average oil production from the surrounding 40 acre spacing wells (not shown in the figure) over their production life was 48,000 m^3 leading to an estimated cumulative sand production of approximately 500 m^3 per well. These wells were drilled at an average depth of 810 m.

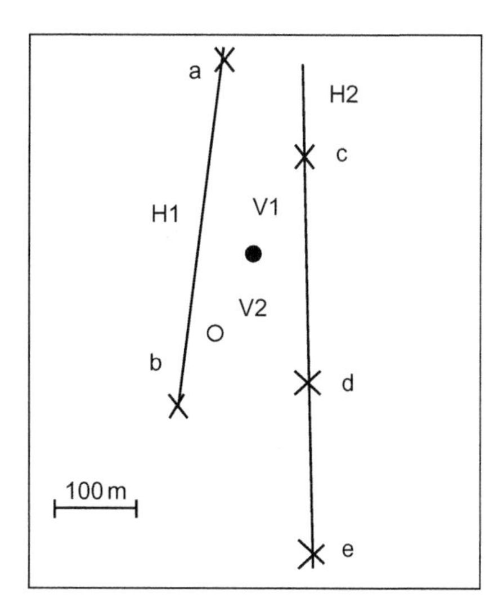

FIGURE 22.11 Lost circulation events (x-marks) while drilling horizontal wells (H1 and H2).

The drilling of the horizontal wells for the pilot was started when the CHOPS wells were depleted approximately 20 years after they were put on production. In order to observe the pressure and temperature in the reservoir during the pilot, an observation well V2 was first drilled at a distance of 107 m from well V1 (Figure 22.11). Interestingly, circulation was not lost while drilling well V2 although as mentioned earlier well V1 had produced 700 m^3 of sand.

The drilling of horizontal well H1 in Figure 22.11 was started at the same time as for well V2. Seven days after starting to drill well H1, circulation was lost at the location indicated by the letter a. The BHP at well V1, located approximately 185 m away from location a, showed a sudden jump to 300 kPaa from 225 kPa at the same time. When drilling mud circulation is lost in fractured and/or carbonate reservoirs with vugs, solids are added to the drilling mud to try to fill and block the cavities/fractures at the drill bit allowing the drilling mud to return to the wellhead where the cuttings are screened out in shakers. Blocking wormholes is a unique challenge because of the dimensions of the open channels which could reach 6–10 cm in diameter as estimated in the injectivity tests of Squires (1993) and in numerical simulations.

LCM, comprised of rubber crumb 2 cm in diameter (5% by volume), various commercial mineral fiber-based LCM systems (5% by volume), and calcium carbonate (7% by volume), was added to the drilling mud and drilling was reattempted. The total average solid volume injected was therefore approximately 17% by volume. During injection of the LCM, the pressure at V1 increased from 325 kPaa to a maximum of 1150 kPaa indicating that some communication occurred between the horizontal well and V1. The open channel was blocked allowing drilling to resume.

Circulation was again lost 18 days later at the location indicated by the letter b in Figure 22.11 which was located 322 m further down the horizontal wellbore from position a. Even after injecting 44 m^3 of lost circulation slurry followed by 18 m^3 of hydroxyethylcellulose (HEC) gel, drilling could not be resumed because of excessive drilling fluid losses as suggested by the elevated BHPs at well V1 reaching the maximum value of 3400 kPaa of the transducers. The LCM and HEC gel were injected in slugs, called pills in the drilling industry, of 10 and 5 m^3, respectively. At the start/end of injection of each pill, the pressure increased/decreased at a rate of approximately 100 and 50 kPa/min, respectively. In contrast, the pressure at observation well V2 showed a very different behavior where the pressure at the well increased to significantly lower values (maximum of 500 kPaa). The pressure signal at well V2 was very different in that it did not show clear increases and decreases in pressure as for well V1 when the pills were injected in slugs. Only 0.34 days after the start of drilling well H1 did the pressure at the observation well jump significantly from 125 to 400 kPaa at a rate of 23 kPa/min. Otherwise the pressure would increase at the low rate of 0.2 kPa/min. The BHP at well V2 then decreased suddenly at the rate of 30 kPa/min back down to a pressure of 125 kPaa, 1.2 days after the last slug of LCM was injected.

When the drilling mud entered well V1, it increased the hydrostatic pressure at the bottom of the well. The change in volume of drilling fluid, $\Delta V(t)$, within the wellbore at V1 during the injection of a pill was calculated from the change in BHP relative to the pressure P_i at the start of the injection using the following equation:

$$\Delta V(t) = \pi(r_c^2 - r_t^2)\bullet(P_{BH}(t) - P_i)/(\rho_m g) \tag{22.8}$$

where r_c is the casing radius (0.0889 m), r_t is the tubing radius (0.0385 m), ρ_m is the density of the drilling mud including the LCM (1400 kg/m^3), and g is the gravitational acceleration constant.

Substituting the average rate of increase of BHP, dP_{BH}/dt, at the start of injection of each pill, 100 kPa/min, into Eq. (22.8) yields a rate of filling of the annulus of 0.15 m^3/min which was lower than the fill rate observed during the drilling of well H2 as will be explained shortly indicating that the communication between the horizontal well and the wormhole network which developed from V1 may not have been as direct as in the drilling of well H2.

The drilling of a second horizontal well (H2 in Figure 22.11) was started 73 days after the drilling of horizontal well H1 was stopped. Circulation was lost at location c in Figure 22.11, approximately 125 m away from well V1, 2 days after spudding well H2. LCM, including calcium carbonate, was again added to the drilling mud in order to block the open channels in communication with well H2. The BHP at wells V1, V2, and H1 (location b) plotted in

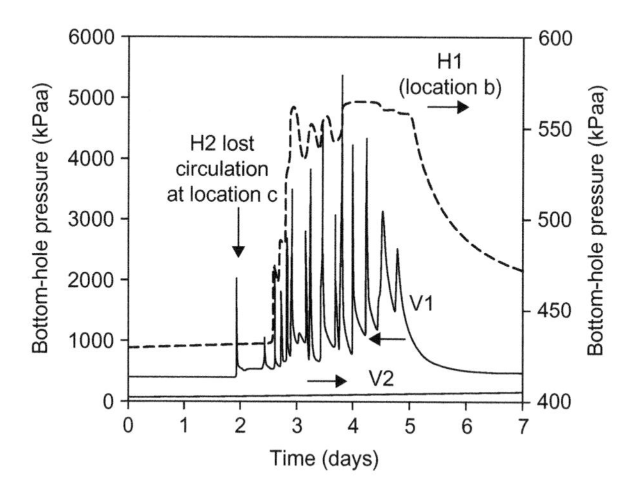

FIGURE 22.12 BHP at wells V1, V2, and H1.

Figure 22.12 responded differently to the lost circulation events. The BHP at well V1 increased suddenly as soon as circulation was lost at well H2 (location c), whereas well V2 did not respond at all to the circulation loss episodes. The BHP transducers at well V1 had a higher range than during the drilling of well H1 and could detect the maximum pressure of 5400 kPaa reached during the injection of the LCM pills which was still lower than the initial overburden pressure of the formation (10,000 kPaa) suggesting that the formation was not fractured during the injection. Well H1 (location b) showed an intermediate type of behavior in terms of the magnitude of the pressure response which was one order of magnitude lower than for well V1 as shown in Figure 22.12. Also, the fluctuations in pressure were not as pronounced at well H1 (location b) indicating that the communication between wells H2 and H1 was not as direct as between H2 and V1 since this latter well had produced approximately 700 m^3 of sand, whereas well H1 had been recently drilled and was not put on production yet. Communication between wells H1 and H2 would then likely have to go through the wormhole network generated by well V1.

Equation (22.8) was again used to estimate the volume change of the LCM within the wellbore of well V1. The pressure $P_{BH}(t)$ for well V1 in Figure 22.12 during the injection of the pill was substituted into Eq. (22.8) and the calculated volume change is shown in Figure 22.13 for each slug (pill). The average slope of the different curves (0.22 m^3/min) is the average fill rate within the well. As observed in the figure, the volume change increased to a maximum of 3.7 m^3 which was lower than the volume of the injected pills varying between 5 and 10 m^3 most likely due to the leakoff of the base fluid in the LCM pills to the formation from the extensive wormhole network. By taking the sum of the drilling fluid volumes of each pill

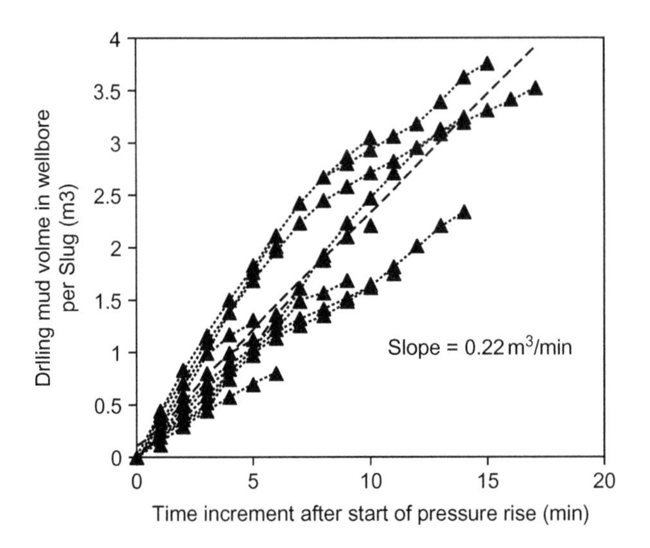

FIGURE 22.13 Calculated LCM volume in well V1.

injection, using Eq. (22.8), a total injected volume of 23 m^3 was obtained which is approximately half of the injected volume of drilling fluid.

The sudden change in pressure when the circulation loss occurred while drilling wells H1 and H2 can only be explained by the presence of an open channel between the horizontal well H2 and the vertical well V1. Because of the large diameter of the rubber crumb, 2 cm, the drilling fluid could not have gone through a porous medium indicating that the channel was open nor through a fracture.

An estimate of the volume of the open channel network was calculated from the volume of injected solids contained in the pill at a volumetric concentration of 17%. From the approximate volume of LCM injected (100 m^3) and assuming a range of open channel diameter of 6–10 cm, an estimate was obtained for the wormhole network length of 6000–2000 m, respectively. The multiwell CHOPS model developed by Tremblay (2009a,b) indicates that wormholes would most likely not be branched and could reach 250 m in length leading to an estimate of 24–8 wormholes respectively growing from the CHOPS well V1. The fact that circulation was lost only five times while drilling wells H1 and H2 indicates that the lower estimate of 8 wormholes may be more realistic.

Numerical Modeling

Dilated Sand Field Models

Denbina et al. (2001) assumed that the sand production, observed during cold production, leads to a radially symmetric, higher permeability zone, which would grow outward starting at a vertical cold production well. The transmissibility of the higher permeability zone would increase with decreasing pressure. They used the cumulative field sand production to determine the radial extent of the higher permeability zone. Kumar and Pooladi-Darvish (2002) in their cold production model used the sand production data from a typical well to calculate the increase in transmissibility of the higher permeability zone with increasing time (decreasing pressure). Tan et al. (2004) used time-dependent transmissibility multiplier factors to increase the transmissibility of the reservoir grid cells around a series of cold production wells. They history matched the cumulative oil and sand for a typical cold production well. All of the above models use the sand production data to either calibrate the transmissibility function or calculate the extent of the enhanced permeability function. Therefore, these models are nonpredictive in nature.

More recently, a series of multiwell CHOPS models has been developed (Aghabarati et al., 2008; Rivero et al., 2010; Vanderheyden et al., 2011), which treat the wormholed region as an equivalent permeability zone. Aghabarati et al. (2008) mention briefly that their model uses a hydro-erosion method to model the sand production. They did not history match the sand production field data. Rivero et al. (2010) model the wormholed region around CHOPS wells as an equivalent damaged zone, the size of which is determined solely by changes in the reservoir's stress field and distribution of the mechanical properties. They do not describe how the mechanical deformation of a wormhole network is related to an equivalent damage zone. They assumed that the disturbed zone due to the wormholes is homogeneous. This assumption is difficult to justify with a wormhole network of only 8 wormholes, for example. The model requires triaxial tests of core samples and acoustic log measurements of the elastic properties of the oil sand. Vanderheyden et al. (2011) modeled the wormholes as a series of horizontal wellbores that grow into the reservoir as modeled by Vittoratos et al. (2008). These models require several adjustable parameters such as the plugging length, variability of plugging, maximum extent, rate of advancement, and diameter of the open channels (wormholes). They used a particle and cell method to calculate the failure and fluidization of the sand in the sand pack shown in Figure 22.6 (Tremblay et al., 1999). Their porosity predictions compare qualitatively in terms of the length of the sand-filled wormhole; however, the diameter was significantly underestimated.

Wormhole Network Field Models

Loughead and Saltuklaroglu (1992) could match the pressure buildup curves for two cold production wells by assuming the presence of an elongated high permeability channel 50 m long. Lau (2001) developed a cold production model based on the assumption that wormholes act as growing horizontal wells (without branching). He assumed the diameter, number, and rate of advance of the wormholes.

A radial drainage cold production model was developed at Alberta Innovates (formerly Alberta Research Council) (Sawatzky et al., 2002) that assumes that the enhanced permeability region caused by sand production is composed of a highly branched wormhole network. This model requires several assumptions such as the wormhole diameter (sand-filled and open channels), number and branching of wormholes, and fraction of filled wormholes. In addition, the finite number of locations (5) at which circulation was suddenly lost while drilling horizontal wells in a previously described pilot project precluded the existence of a highly branched wormhole network. These observations also contradict models based on diffusion-limited aggregation (DLA) models (Liu and Zhao, 2005) which describe irreversible colloidal aggregation.

The multiwell CHOPS model by Tremblay (2005, 2009b) does not require *a priori* assumptions about the open channel diameter within the wormholes, nor their length or growth rate. These parameters are calculated by the model. The only adjustable parameters are the maximum diameter, D_w, of the sand-filled part of the wormhole (i.e., excluding the open channel diameter), the number of initial active perforations, N_{perf}, and the relative permeability to gas at residual oil saturation k_{rgi}. In this model, an active perforation is defined as one leading to the growth of a wormhole one grid block or more in length. In a parametric study (Tremblay, 2008a), it was shown that a unique combination of adjustable parameters is obtained when history matching the oil and sand production from CHOPS wells.

In the multiwell CHOPS model, the wormholes were assumed to grow horizontally in layers as straight wells. As mentioned earlier, the rate of advance of the wormholes in laboratory experiments was found to be proportional to the oil velocity at their tip (Eq. (22.7)). The wormholes were assumed to be un-branched for three reasons. Firstly, in order for a branch to develop at the surface of a wormhole, the oil velocity would have to be greater at that location than at the tip. However, numerical simulations indicate that the oil velocity is greatest at the tip of the wormhole (Tremblay, 2005). Secondly, as discussed previously, when an oil company tried to drill two horizontal wells in a CHOPS produced reservoir, they lost circulation at a finite number of locations (5) as shown in Figure 22.11. If the reservoir was dilated or if it contained several highly branched wormholes, it would have been impossible to drill the wells due to continuous fluid loss. Thirdly, the diameter of the sand-filled part of the wormholes (not open channel) could be as large as 1 m (Tremblay, 2008a), as suggested in numerical simulations using Eq. (22.7) to describe the growth of the sand-filled wormholes radially. From the volumes of sand produced, there would be very little branching.

CHOPS wells are typically perforated at 26 shots/m. It would not be practical to model the growth of each wormhole throughout the pay zone for several wells producing simultaneously over a 10-year period as in a typical simulation. The wormholes were assumed to grow instead along the x and y axis and in the diagonal directions within a plane, leading to a total of eight wormholes per layer. The number of layers was not restrained, however. The wormholes (sand-filled part; not open channel) start to grow as sand-filled perforations 20 cm in diameter, according to previous sand production experiments (Tremblay and Oldakowski, 2003). The model by Tremblay (2009a,b) simulates the increase in wormhole diameter until a maximum wormhole diameter is reached. The maximum diameter was adjusted depending on the volumes of sand produced. The diameters of the open sand-free channels which develop within the wormholes are calculated by the CHOPS model on the basis of slurry transport equations (Tremblay, 2005). The Saskatchewan Research Council (SRC) multiwell model (Tremblay, 2009a,b) is first run in order to generate a STARS data file in which the wormholes are modeled as superposed multilateral wells, with each adjacent well one block length longer than its immediate neighbor. The diameter of the open channel within the wormholes is equal to the diameter of the multilateral wells. The times at which the multilateral wells open and close, the diameters of the multilateral wells (open channels), and the negative skin factors due to the dilated region around the open channel are specified by the multiwell model. The STARS simulation is then run with the embedded growing wormhole network within the data file. A multi-well CHOPS model has been developed more recently by Istchenko and Gates (2012) where the growth of the wormholes is governed by a sand fluidization velocity. In their model, if the fluid velocity in a given direction within a block is greater than the sand fluidization velocity, the wormhole is advanced by one grid block in that direction. In their history matching of the oil, water and sand production from a CHOPS well, they assumed a very high fluidization velocity of 0.026 m/day. From the cross-sectional area of their 4 m wide by 0.5 m high blocks, a corresponding critical pressure gradient of 4 mPa/m was calculated at the tip surface of a 0.3 m radius wormhole (sand-filled). However, a critical pressure gradient of only 60 kPa/m was measured by Tremblay and Oldakowski (2003) for a sand pack prepared to the same cohesive strength as in the field as discussed in the section on wormhole growth experiments (page 628). By assuming unrealistically high fluidization velocity values, the authors will under-predict the extent of the wormhole networks.

22.3 FIELD CASE

22.3.1 Heterogeneity of Reservoirs

An example of some of the variability observed in CHOPS reservoirs will be given in this section. Seven neighboring 40 acre spacing wells, with very different oil productivities, were selected for this comparison. Their locations within a drainage boundary, assumed for numerical simulations, are shown

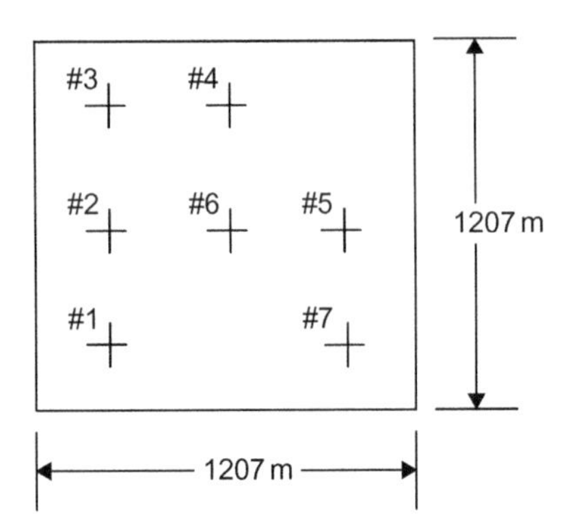

FIGURE 22.14 Flow boundary (within inner square).

in Figure 22.14. The oil, sand, water, and gas production field data for these wells are given in Tables 22.1 and 22.2. The production data as of 2225 days after the start of well #6 in Table 22.1 was used previously by the author to history match the production data at that time (Tremblay, 2009b). The data given in Table 22.2 was used more recently on the other hand to test the CHOPS model predictions by extending the numerical simulations to the present date (3488 days) using the same adjustable parameters and comparing the production data to the predictions. The OOIP per 40 acre spacing varied for each well since the pay zone thickness and porosity was not the same as given in Table 22.3. These wells were perforated in the Sparky formation which is described by Chugh et al. (2000).

As given in Tables 22.1 and 22.2, the recovery factors varied significantly between wells ranging from 0.019% for well #7 which only produced for 30 days to 6.6% for well #4 which is still producing at a rate of 3.5 m^3/day after 3488 days. As expected, the sand cut was higher in the period before 2225 days than after since the oil drainage of the reservoir through the wormholes increases with time leading to a lower sand concentration within the open channels and therefore to lower sand cuts. The water cut was lower in the latter production period (Table 22.2) which indicates that water encroachment was not occurring. The cumulative gas–oil ratio (GOR) did not change in the second period indicating that gas breakthrough was not occurring.

To reduce costs, the wells were not initially cored when drilled; therefore, the permeability distribution of the formation could not be measured directly. However, oil companies will often estimate the permeability distribution from well logs when cores are not available. In our example, the oil company developed a correlation between the vertical distributions of oil resistivity and absolute permeability. The assumption in this approach is that the oil would have originally migrated through the higher permeability part of the reservoir. In Figure 22.15, the absolute permeability is plotted as a function of the vertical distance from the bottom of the pay zone for wells #1, 2, and 3. Geologists

TABLE 22.1 Oil, Water, Sand, and Gas Production Field Data (2225 Days After Start of Well #6)

Well #	Start (day)	Stop (day)	Cumulative Oil (m^3)	Oil Recovery Factor (%)	Cumulative Bulk Sand (m^3)	Bulk Sand Cut (vol.%)	Cumulative Water (m^3)	Water Cut (vol.%)	Cumulative GOR (m^3/m^3)
1	249	1,825	1,560	0.45	70	3.3	475	22.6	8.1
2	255	Ongoing	14,800	4.5	597	3.2	3,280	17.6	7.0
3	618	Ongoing	4,161	1.4	172	2.7	2,008	32.0	25.5
4	254	Ongoing	17,393	4.8	598	2.9	2248	11.1	6.8
5	254	Ongoing	9,312	3.0	191	1.7	1,510	13.7	7.6
6	0	Ongoing	12,032	4.3	327	2.3	2,027	14.1	7.6
7	450	480	48.9	0.019	NA	NA	0	0	NA

TABLE 22.2 Oil, Water, Sand, and Gas Production Field Data (3488 Days After Start of Well #6)

Well #	Cumulative Oil (m^3)	Oil Recovery Factor (%)	Cumulative Bulk Sand (m^3)	Bulk Sand Cut (vol%)	Cumulative Water (m^3)	Water Cut (vol.%)	Cumulative GOR (m^3/m^3)
2	22,668	6.8	718	2.7	3,280	12.3	7.4
3	6,814	2.3	216	2.2	2,672	27.5	28.7
4	23,780	6.6	853	3.1	3,194	11.5	7.1
5	14,432	4.6	229	1.3	2,263	13.4	7.0
6	17,827	6.4	385	1.8	2,868	13.6	7.6

TABLE 22.3 Reservoir Properties

Well #	Net Pay (m)	Porosity (%)	OOIP (m^3)	Oil Viscosity (cP)	Initial Oil Saturation (%)	Average Gamma Ray Reading (API)	Average Pump Efficiency (%)
1	8	33.4	346,000	30,750	80	50	40
2	8	32.0	331,500	27,970	80	60	60
3	7	33.2	300,900		80	48	60
4	8.5	32.6	358,800	25,480	80	65	60
5	7.5	32.0	310,800	23,800	80	50	70
6	6	35.8	278,150		80	55	65
7	4						

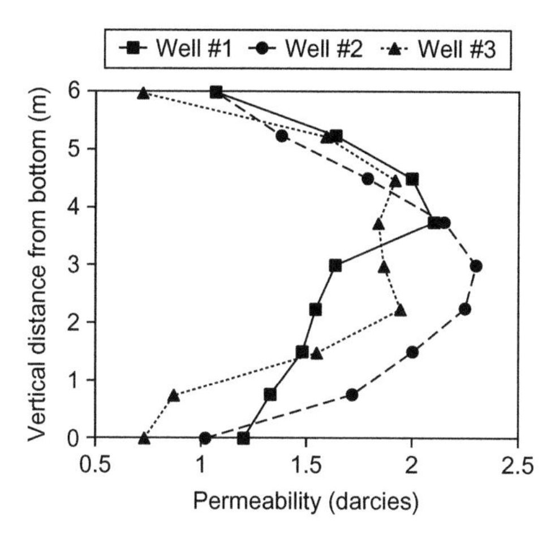

FIGURE 22.15 Vertical distribution of permeability for wells #1–3.

at the oil company however realized that the resistivity—permeability correlation does have its limitations, as the zones outside of the Sparky interval would indicate high permeability from the correlation; however, geologists know that these zones have low permeabilities. As a result, this correlation was suggested by the geologists to apply only for the Sparky zone which included the pay zone in (Figure 22.15). The uncemented sand was well sorted in the fine to very fine range as observed from sieve analysis from cores drilled one section away from the wells. The viscosity of the produced oil varied from well to well as shown in Table 22.3.

Tables 22.1 and 22.2 give that well #2 was significantly more productive than well #3 and even more so than well #1. When the permeability profiles for all wells were substituted into the CHOPS model, the oil production from wells #1, 3, 4, 5, and 6 was too high by a factor ranging from ten (well #1) to two (well #6). Some explanations for the large differences in measured oil production volumes between the wells may be that: (1) the actual permeability around the wells is different than that given by the correlation between the resistivity well logs and the permeability with the wells with lower oil production being drilled in a lower permeability sand; (2) the perforations were partially blocked with pieces of shale or other debris, leading to a large positive skin factor at the wellbore for the wells which produced less oil; and (3) the sand at these latter wells was partially cemented, which limited the growth of the wormholes.

The first and third explanations are somewhat similar since a partially cemented sand would likely have a lower permeability leading to limited wormhole growth, the rate of growth of the wormholes being proportional to the oil velocity at its tip, which in turn is proportional to the permeability, as described in Eq. (22.7). The average gamma ray values, which indicate the presence of natural radioactive elements typically present in shale, did not correlate well with the well productivity as seen by comparing these values given in Table 22.3 to the oil production data given in Table 22.2. The log for well #2, which was significantly more productive than well #1, showed in fact a higher gamma ray reading (60 API units versus 50 API units for well #1). It would appear that well #1 would not have produced more shale than well #2 based on the well logs. The well operators did not report sudden decreases in pump efficiency during the production of these wells which would have hypothetically been caused by pieces of shale entering the PC between the stator and rotor and increasing the friction on the rotor leading to high torques.

The torque on the rotor is measured indirectly from the pressure of the pump at the wellhead which circulates oil through a hydraulic drive head (motor) connected to the PC rotor. In general, well operators try to keep the BHP as low as possible without pumping off the wells, since high-pressure drawdowns lead to higher pressure depletion rates in the reservoir, which in turn lead to higher oil recovery factors as suggested by Eq. (22.5). To achieve these low BHPs, the operators increase the rotation rate of their pumps until the pump efficiency drops off and then decrease the rotation rate slightly and leave it turn at that rate. The average efficiency for the pumps (Table 22.3) did not correlate with oil production (Table 22.2) although well #1 stands out somewhat in having both a lower efficiency and significantly lower oil production. This lower efficiency may be a consequence of the lower production rates which require further adjustments to the rotational speed of the pump.

Well operators initially like to keep higher BHPs (typically 500 kPa) at the start of cold production due to the higher sand cuts which lead to higher torques requiring the rotation rate to be decreased. The higher fluid levels give a further safety margin in case the torque increases suddenly. The friction of the sand grains against the rotor as they are carried through the PC pump generates heat which is dissipated by the oil convection (flow) through the pump. The friction is also reduced by the oil film in the gap between the stator and rotor. After a few years, the sand cuts will decrease to less than 2%, reducing the danger of pumping the well dry which allows the reduction of the BHP to approximately 250 kPa.

For these wells, the sand production volume was approximately 3% by volume. Note that the sand volume in CHOPS is measured from the height of the settled sand within the trucks used to clean out the sand from the collection tanks at the surface of each well. Shale production was not noted by the producers. As given in Table 22.1, well #7 hardly produced any oil most likely because the reservoir quality dropped off to the east.

22.3.2 History Matching Cold Production Wells

In order to illustrate the application of a CHOPS multiwell model developed at the Saskatchewan Research Council (Tremblay, 2009a,b) described previously, the field production data presented in the previous section will be history matched to determine the adjustable parameters in the model. The calibrated model will then be used to predict oil, water, sand, and gas production from these wells on primary production. In a previous study (Tremblay, 2009b), the produced oil, sand, water, and gas data given in Table 22.1 was used to determine the following adjustable parameters: the permeability distribution, the maximum wormhole (sand-filled part) diameter (D_w), the number of initial active perforations, N_{perf}, the mobile water saturation (S_{wm}), and the relative permeability to gas at residual oil saturation (k_{rgi}). Instead of using the resistivity–permeability correlation discussed in the previous section, the permeability was tuned by history matching the cumulative oil production volumes. The permeability in the reservoir between the wells was interpolated linearly between well locations.

These parameters are interrelated so that, for example, if D_w is set, N_{perf} influences the total volume of sand produced since having more initial active perforations favors more wormhole growth leading to more sand production. In addition, D_w also has an indirect role in controlling the oil velocity at the wormhole tip and therefore the rate of advance of the wormhole which in turn controls the rate of pressure depletion of the reservoir and consequently the oil production rate. By decreasing the relative permeability to gas in the numerical simulations, the predicted produced GOR is lowered. In these simulations, thermodynamic equilibrium gas evolution was assumed. The k_{rgi} value in the Corey model for gas relative permeability (Eq. (22.3)(back)) was obtained from the laboratory measurements of Tang and Firoozabadi (2003) at low-pressure depletion rates. The final average set of adjustable parameters D_w and S_{wm} were 1.2 m and 1%, respectively.

The history match proceeded by first assuming a set of absolute permeabilities, a mobile water saturation for each well, and interpolating linearly these properties within the reservoir. An initial value for wormhole diameter

D_w was then chosen for each well and a simulation was run. After the first iteration, these parameters were changed by a factor given by the ratio of the actual to predicted oil, sand and water production volumes. The procedure was repeated until the ratio approached one for all parameters.

22.3.3 Predicting CHOPS Production

The cumulative oil production volumes, both field and numerical, are shown in Figure 22.16 for wells #1–6. The vertical dotted line indicates the boundary between the history matched portion of the plots (before 2225 days) and the predicted portion of the numerical plots (between 2225 and 3488 days). As expected, the history matched portion of the numerical plots is closer to the field data than the projected part. In general, the model tended to overestimate the oil production. The maximum error in prediction was for well #1 which also produced significantly less oil (Table 22.1). The time lag of almost 1000 days for well #1 is most likely due to the slow rate at which the wormholes grew from the well due to the lower permeability at that well. Since the drainage boundary was quite large (Figure 22.14), a coarse grid was used in the simulations (40 m × 40 m × 1 m). The wormholes growing from the initial perforations would have to reach a length of 20 m before a multilateral well simulating a wormhole was opened which took more time for well #1. The start of production for the other wells on the other hand was better history matched. The error in the projected cumulative oil volumes at 3488 days for wells #2–6 was −2.7%, +8.4%, +14.3%, +13.5%, and +9.1%, respectively.

The overprediction of the produced oil after 2000 days may in part be explained by the thermodynamic equilibrium assumption where the same gas relative permeability curves are used throughout the simulations. This assumption did not account for potential dynamic coarsening of the gas ganglia in the porous medium leading to more gas production with time. The average predicted cumulative GOR for the wells was 25 m^3/m^3, which is significantly larger than the field values given in Table 22.2. The produced gas was measured only at the annulus and did not include the gas produced through the tubing and then through the production tank (as free and dissolved gas) which could explain the lower cumulative GOR values measured in the field. The produced GOR from the wells operated by the oil company was measured only every 6 months. Low gas relative permeabilities had to be assumed to try to history match the produced gas GOR. Stopping the simulation and restarting with a higher gas relative permeability data set led to a decrease in oil production and a better history match (not shown in figures).

As expected, the cumulative bulk sand volumes were more difficult to history match and to predict because of the difficulty in modeling the wormholes as shown in Figure 22.17. In three cases, the cumulative bulk sand volume was overpredicted (wells #2, 3, and 5), whereas in one case it was underpredicted (well #4). Overall, the cumulative bulk sand predictions were acceptable and

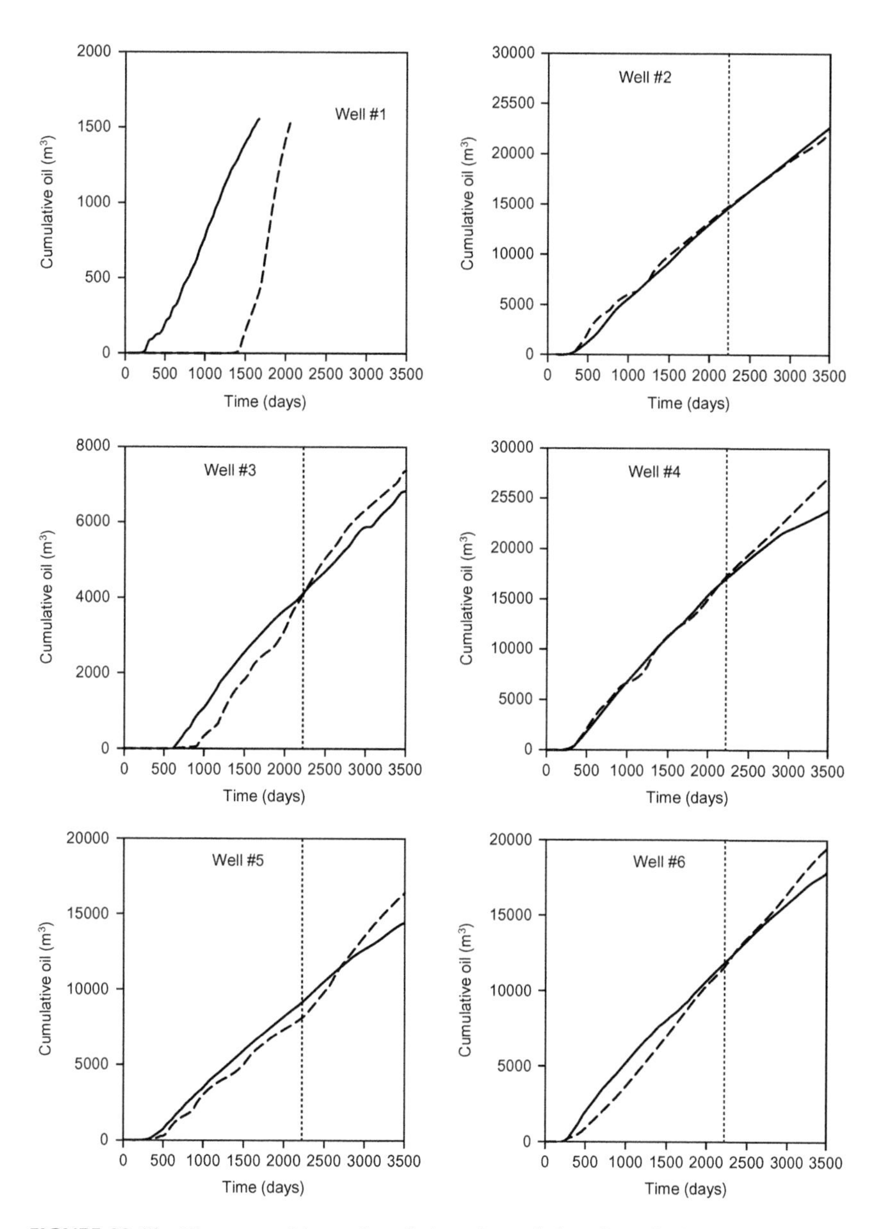

FIGURE 22.16 History matching and prediction of cumulative oil production: full line: data; dashed line: model.

predict that the wormhole network is still growing since the sand volumes are increasing. The error in the projected cumulative bulk sand volumes at 3488 days for wells #2–6 was +17%, +15.5%, −17.4%, +9.8%, and +2.4% respectively.

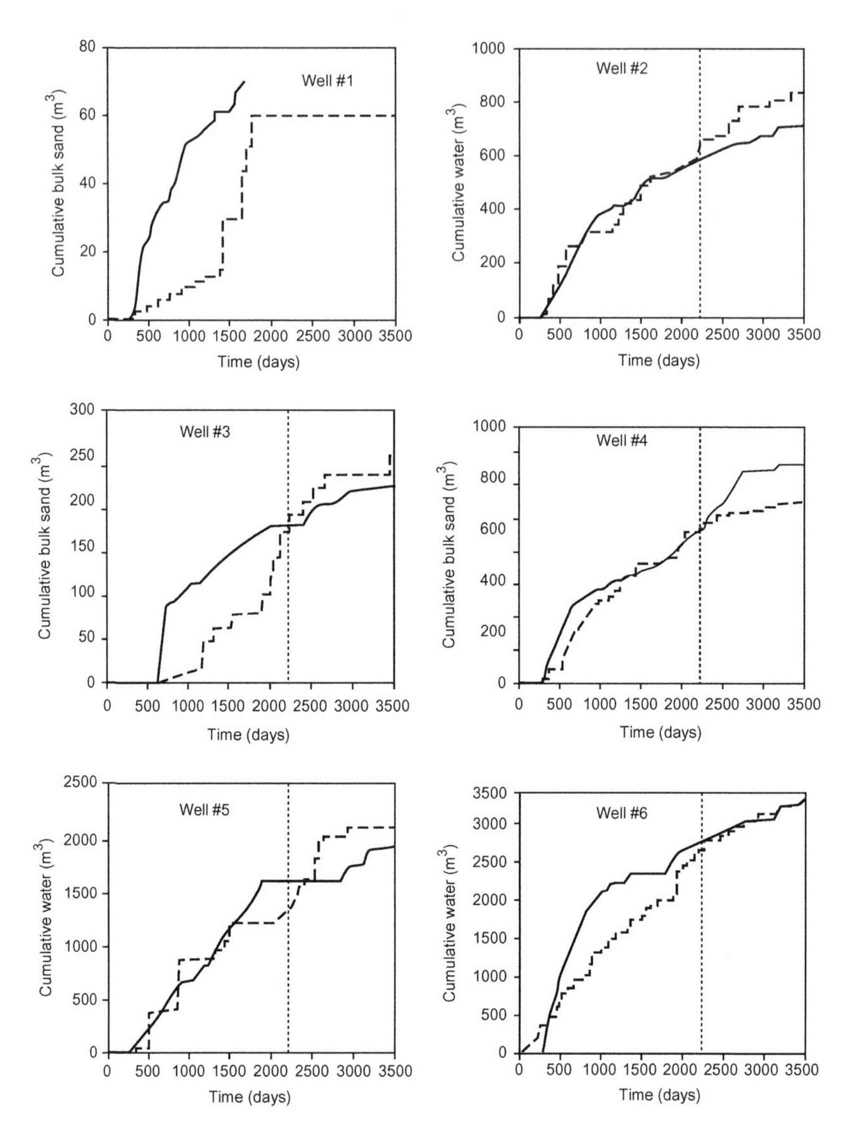

FIGURE 22.17 History matching and prediction of cumulative bulk sand production: full line: data; dashed line: model.

The best history match (i.e., prior to 2225 days) was for the cumulative water production volumes (Figure 22.18) which were obtained by assuming on average a mobile water saturation of only 1%. No aquifers were assumed in the numerical simulations. The water production predictions (2225–3488 days) were satisfactory except for wells #1 and #3. The reason for the lack of

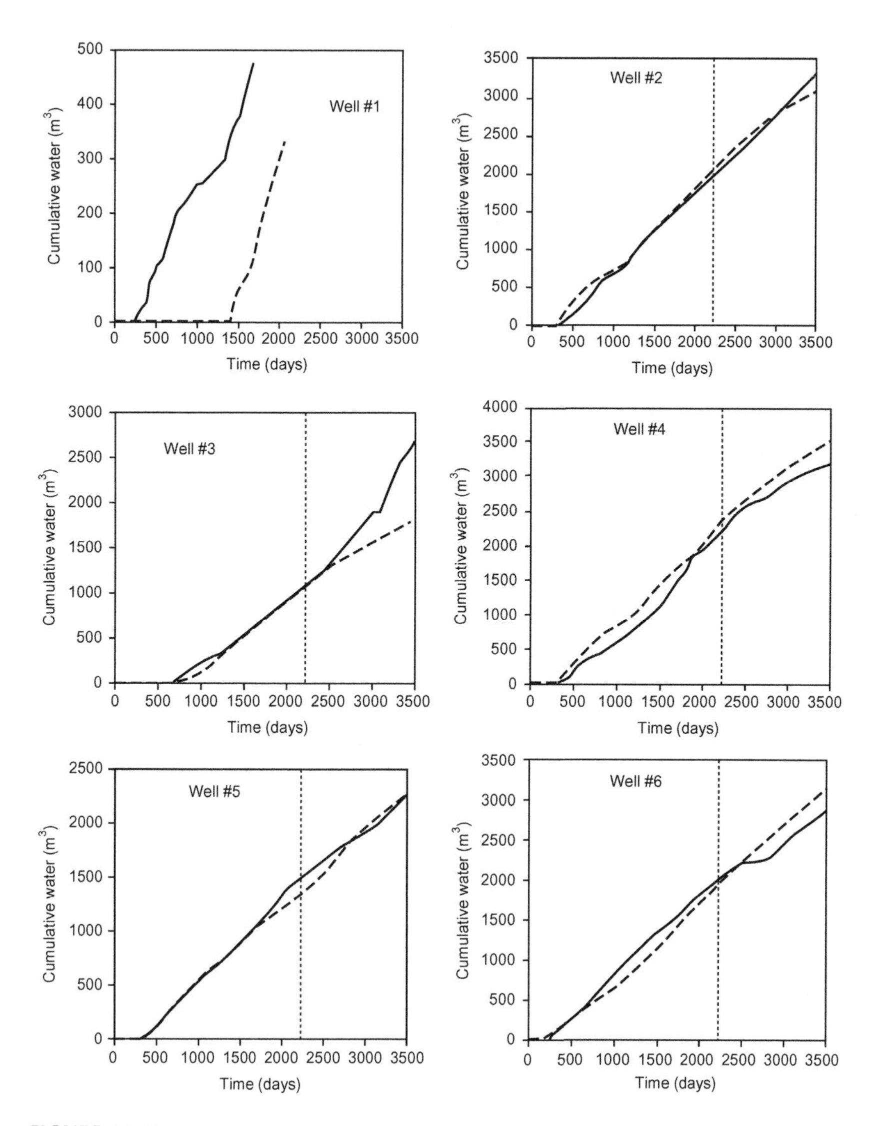

FIGURE 22.18 History matching and prediction of cumulative water production: full line: data; dashed line: model.

agreement in the case of well #1 would again be due to the low rate of advance of the wormholes around that well. The sudden rise in water production for well #3 may be due to a discrete region of higher water saturation called a water pocket in the industry. The error in the projected cumulative water volumes at 3488 days for wells #2–6 was −6.5%, −32.3%, +10.1%, −0.13%, and 8.9%, respectively.

22.3.4 Predicting Post-CHOPS Production

Using the multiwell model described in the previous section, the author previously performed numerical simulations of a potential post-CHOPS (waterflooding) process (Tremblay, 2009a) to better understand the reasons for the very moderate success of this process for heavy oils, as described by Miller (2006) and to investigate to what extent blocking wormholes would improve the process. Finally, an alternate process based on cyclic solvent injection was investigated.

In a first scenario, the open channels within the predominantly sand-filled wormholes were assumed to be open, whereas in a second scenario the channels closest to each other were assumed to be closed as would potentially be accomplished by injecting a blocking agent. In both scenarios, water was injected into well #6 at 100 m^3/day starting 3600 days after that well was put on primary production. The oil, water, and gas were produced through wells #1–5. The cumulative incremental oil production, defined as the total cumulative oil from wells #1–5 produced in waterflooding minus the total cumulative oil from the same wells produced in primary production, was plotted for both scenarios as a function of the start of the water injection shown in Figure 22.19. In both scenarios, as soon as the waterflooding started, less oil was produced than if water was not injected at all (i.e., if CHOPS had been continued). Surprisingly, blocking the open channels led to lower oil production since the oil viscosity was so high (Table 22.3) making it difficult for the injected water to mobilize the oil. The water saturation distribution around the production wells #2, #4, and #5 shown in Figure 22.20A and B shows that

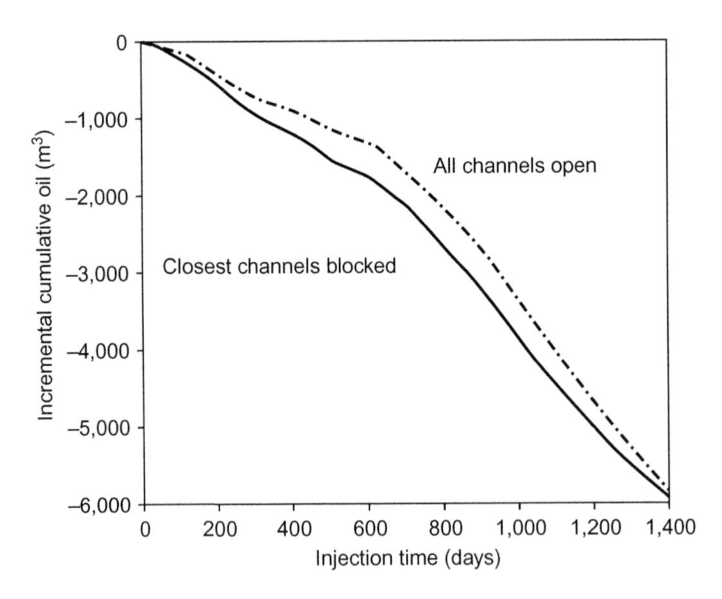

FIGURE 22.19 Incremental cumulative oil versus time after start of injection.

the blocked wormhole scenario led to a better sweep of the oil. However, due to the very unfavorable mobility ratio, the water cut rapidly reached 100% at wells #2 and #4 and latter at well #5. Since well #6 was converted from a producer to an injector, production is evidently lost from that well, which must be

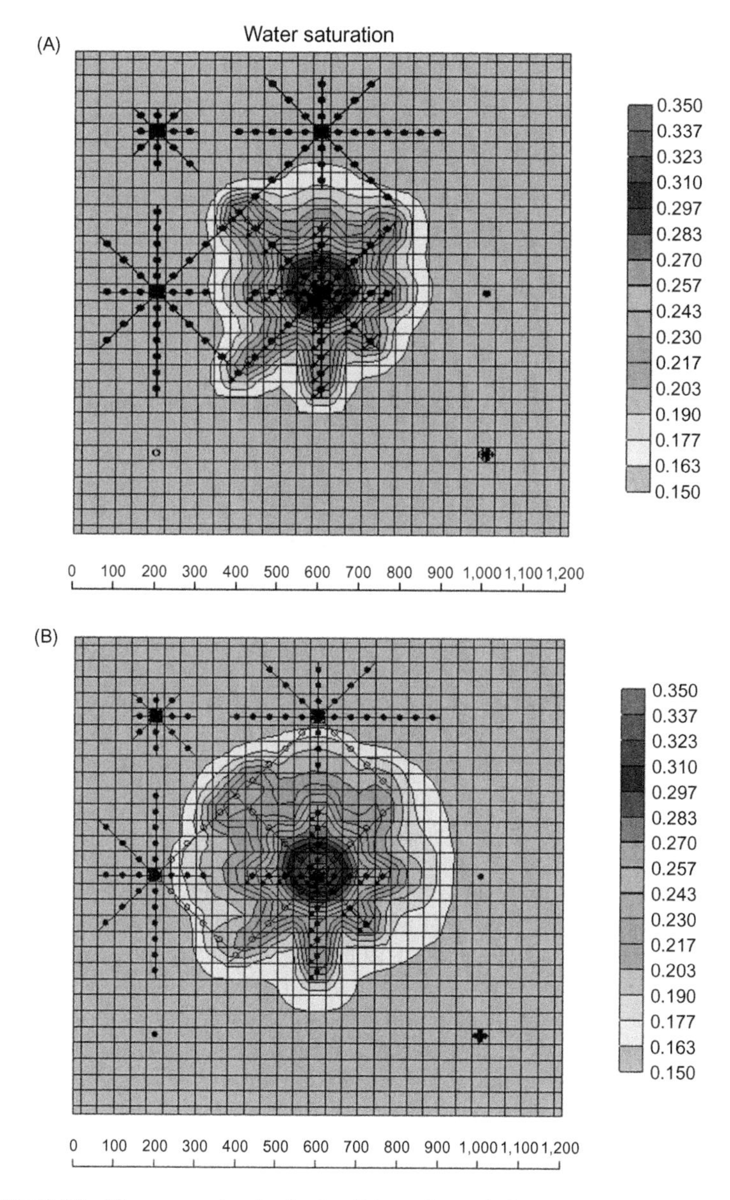

FIGURE 22.20 Water saturation distribution. Diagonal wormholes: (A) open and (B) closed.

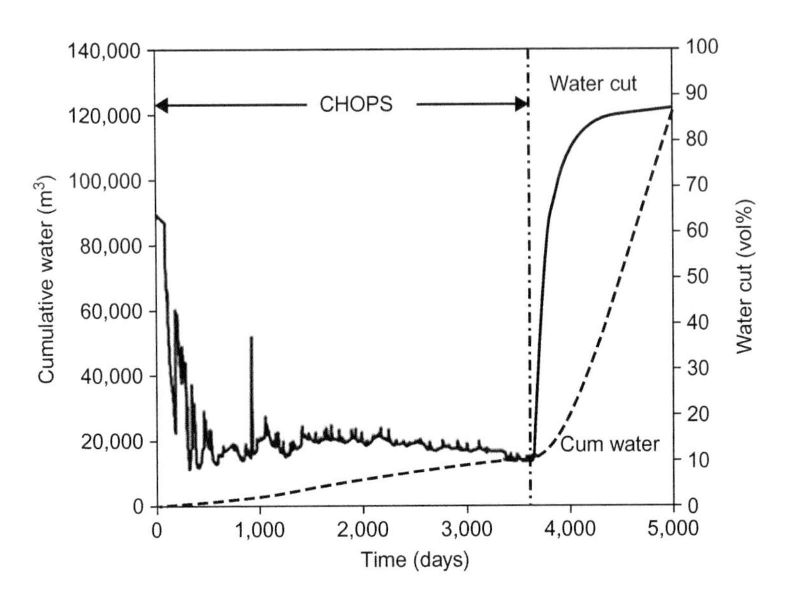

FIGURE 22.21 Cumulative water and water cut versus time after start of CHOPS production. Diagonal wormholes open.

considered in evaluating the waterflooding process economics. In addition to less oil being produced, the water cut was significantly higher as shown in Figure 22.21, increasing operating costs. In this figure, the water cut started to increase sharply at the start of waterflooding (3600 days).

Both Forth et al. (1996) and Smith (1992) stated that prior primary production was linked to a very poor performance when waterflooding heavy oil reservoirs. They suggested that the formation of a gas-saturated zone provided a channel between the wells for the injected water. This hypothesis was investigated by comparing the gas and water saturations at the end of the water injection period. As shown in Figure 22.22A, the gas saturation above production wells #2 and #5 was still high (43%) even after injecting water for 1400 days. The water saturation above these two producers was lower than at the location of the wormhole tips as shown in Figure 22.22B indicating that the water preferentially flowed through the wormholes closest to those which originally developed from well #6 during the CHOPS phase rather than through the higher gas saturation region at the top of the reservoir. Since the residual oil saturation was not reached at the top of wells #2 and #5, the injected water would have to push the viscous oil through the higher gas saturation region to reach the producers. Instead, the injected water followed the shortest path which was between the closest wormholes.

These simulations show that flooding processes are limited by the breakthrough of the displacing fluid between wormholes closest to each other. On the other hand, the contact surface between the open channels and the

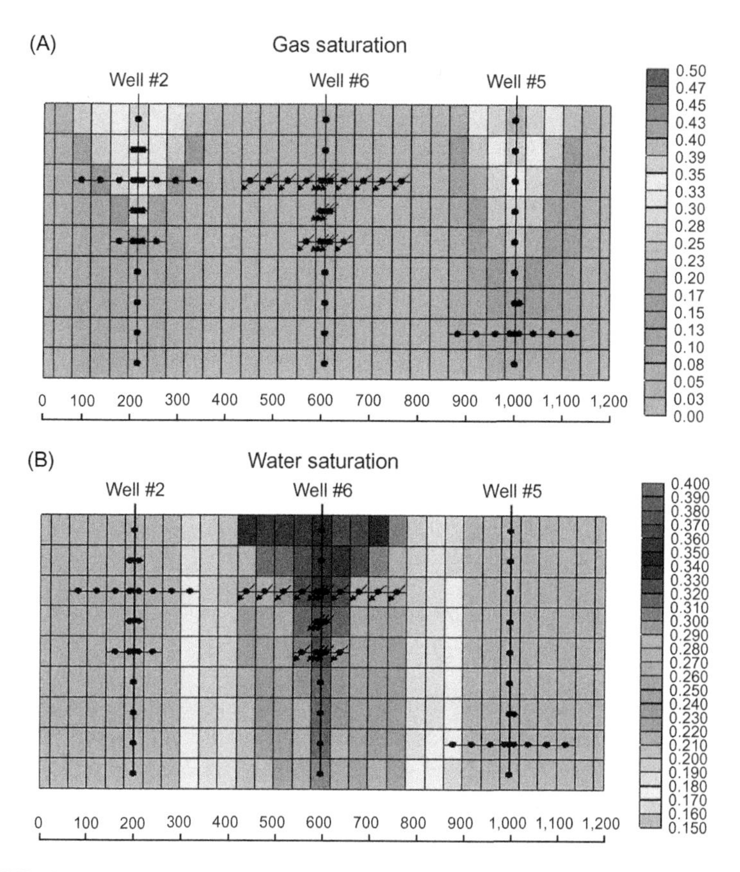

FIGURE 22.22 Cross-section of reservoir through wells #2, #6 (injector), and #5 at the end of water injection: (A) gas saturation and (B) water saturation.

reservoir at the end of the CHOPS process is quite large. For example, in these simulations, the average open channel diameter and total length of the wormhole network which developed from well #6 was predicted by the model to be 10 cm in diameter and 2000 m in length. The surface area in contact with the reservoir could then be approximately 630 m^2. A more promising process would then appear to be based on a cyclic process while keeping neighboring wells shut-in to prevent solvent breakthrough. In this approach, the reservoir is pressurized by solvent over the entire contact area.

A numerical simulation of a potential cyclic solvent process in which 6 cycles comprised of injecting a gaseous solvent into well #6 over a 4-month period up to 2500 kPaa and producing through the same well over an 8-month period for each cycle. The gaseous solvent was a mixture of propane (30 mole%) and methane (70 mole%) with a dew point of 3300 kPaa. The solvent injection cycles were started at 2900 days after CHOPS started

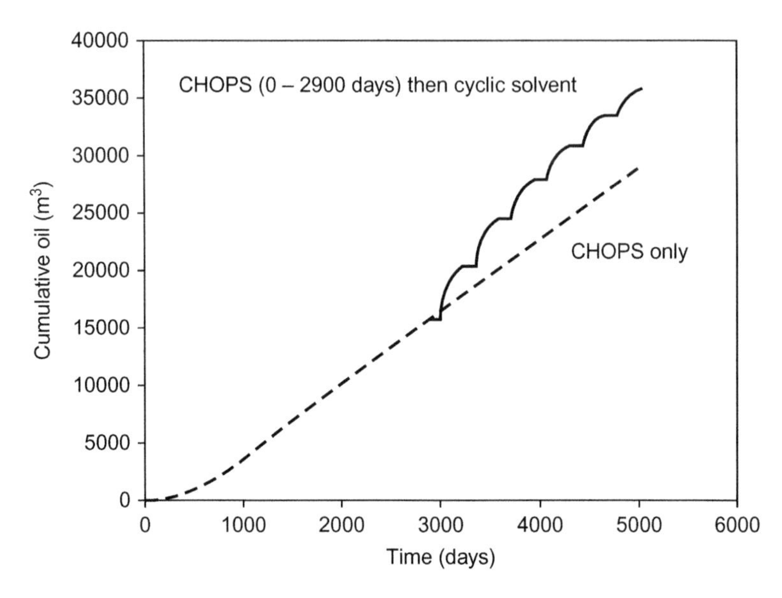

FIGURE 22.23 Comparison of cumulative oil (well #6) versus time. Effect of cyclic solvent injection.

as modeled numerically in the previous section. The predicted cumulative oil for the cyclic solvent process was greater than for the case where CHOPS was continued after 2900 days as shown in Figure 22.23. It is important to note that these predictions are for well #6 only; the other surrounding wells being shut in. The incremental produced cumulative oil was calculated from the difference between the curves and is plotted in Figure 22.24 as a function of the time increment after the start of the solvent injection. As expected, the incremental oil decreased during the injection part of the cycle since by definition oil was not produced during that period contrary to the case where well #6 would have been on CHOPS production only.

Several questions still remain to be answered before cyclic solvent injection reaches a commercial scale. The main question of course is that of the economics related to solvent, gas compression, and solvent loss costs. Other questions include the optimum solvent selection, the duration of the cycles, the maximum pressure at which the solvent should be injected, or the optimum pool development in terms of which wells should be shut in and which should be used as injectors.

22.4 CONCLUSIONS

The solution-gas drive mechanisms for both cold production with limited sand and the CHOPS process are becoming clearer after 20 years of research. The key role in the presence of asphaltenes at concentrations greater than

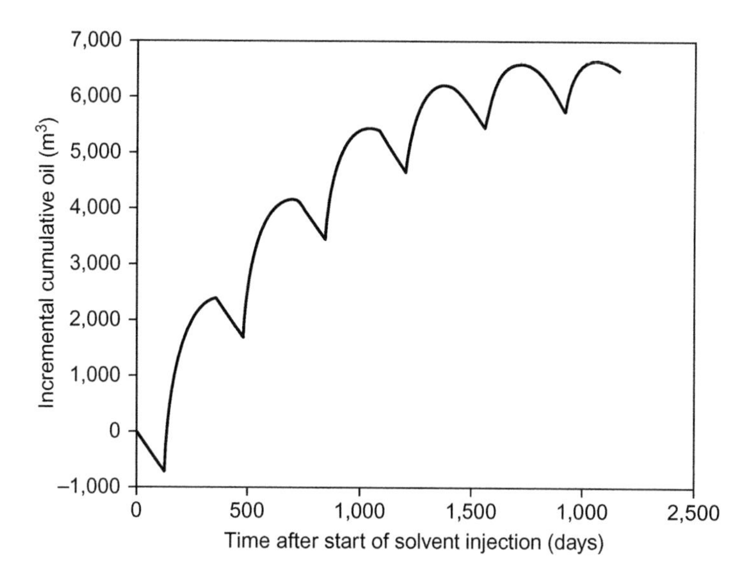

FIGURE 22.24 Incremental cumulative oil (well #6) versus time. Effect of cyclic solvent injection.

10% in stabilizing the gas bubbles has been confirmed in surface tension, surface viscoelastic moduli measurements, and solution-gas drive experiments. The additional role of resins in reducing the surface activity of the asphaltene is becoming clearer. The role of the organic acids and bases in the formation of asphaltene networks at the interface between the gas and oil and in forming an elastic membrane which is more resistant to deformation and reduces gas diffusion is important in stabilizing the gas bubbles.

The role of the depletion rate on nucleation rate and bubble size has been confirmed. The breakup of gas bubbles at high capillary numbers also helps in reducing bubble coalescence. However, numerical simulations of the capillary numbers and pressure gradients in the field indicate that the foamy oil effects are observed very close to the wormholes. The use of very low gas relative permeability curves to simulate cold production has been successful; however, further research is required to develop more predictive nonequilibrium solution-gas drive models which can better predict oil and gas production rates.

Sand production experiments suggest that wormholes can develop due to a fluidization of the sand in uncemented sand without having to fail the formation. These wormholes are sand-filled when they grow reaching diameters possibly as large as 1 m. As they grow in length, an open channel forms, of 6–10 cm, as suggested in tracer tests and in wormhole growth models. The tracer tests, showing rapid transit times and absence of surface adsorption of dyes, cannot be explained by the development of fractures since their width would be of the order of the sand grain diameters. Circulation loss observations while drilling horizontal wells in cold produced reservoir indicate that the wormhole networks are not highly branched (if at all) and that dilated regions do not develop around CHOPS wells. The volume of LCM used to try to block the wormholes in communication with the wellbore suggest that the volume of this open channel network within the significantly larger wormholes containing mostly dilated sand could be in the 20 m^3 or more range. Assuming an average open channel diameter of 10 cm and average wormhole length of 250 m would lead to eight wormholes per well.

Research is also progressing in the development of a multiwell CHOPS model to eventually be able to predict optimal well spacing, in-fill scheduling, and perhaps most importantly the design of post-CHOPS processes. The focus should be on trying to reduce the number of adjustable parameters in these models to make them more predictive in nature. One approach would be to investigate if all the adjustable parameters related, for example, to geomechanics effects are required as a first approach.

The field data for CHOPS production shows how variable the oil recovery can be since these reservoirs are thin and often heterogeneous. In general, in order to reduce costs since the pay zones are thin, wells are often not cored, and sand and gas are not monitored regularly which makes history matching a challenge. Fortunately, the availability of good bulk sand data was crucial in testing the predictability of a multiwell model developed by the author. Although more work is needed to further test the model with field data, the predictions look promising.

ACKNOWLEDGMENTS

Acknowledgement is gratefully extended to the Petroleum Research Centre (PTRC) and to participating industry members for their financial and technical support.

REFERENCES

Aghabarati, H., Dumitrescu, C., Lines, L., Settari, A., 2008. Combined reservoir simulation and seismic technology, a new approach for modelling CHOPS, Paper SPE/PS/CHOA 117581 (PS2008-317) Presented at the SPE Thermal Operations and Heavy Oil Symposium, Calgary, Alberta, October, 20–23.

Azizi, F., 2000. Applied Analyses in Geotechnics, first ed. E & FN Spon Inc, New York, NY.

Bauget, F., Lenormand, R., 2002. Mechanisms of bubble formation by pressure decline in porous media: a critical review, Paper Presented at the SPE Annual Technical Conference, San Antonio, TX, September 29.

Bauget, F., Langevin, D., Lenormand, R., 2001. Effects of asphaltenes and resins on foamability of heavy oil, Paper Presented at the SPE Annual Technical Conference and Exhibition, New Orleans, Louisiana, 30 September–3 October.

Bayon, Y.M., Cordelier, Ph.R., 2002. A new methodology to match heavy-oil long core primary depletion experiments, Paper Presented at the SPE/DOE Thirteenth Symposium on Improved Oil Recovery, Tulsa, OK, April 13–17.

Bell, J.S., Price, P.R., McLellan, P.J., 1994. *In situ* stress in the western Canada sedimentary basin, Chapter 29, Atlas of the Geology of the western Canada Sedimentary Basin, pp. 439–446.

Bondino, I., McDougall, S.R., Hamon, G., 2003. Interpretation of a long-core heavy oil depletion experiment using pore network modelling techniques, Paper SCA2003-11 Presented at the International Symposium of the Society of Core Analysts, Pau, France, September 21–24.

Bora, R., Maini, B.B., Chakma, A., 1997. Flow visualization studies of solution gas drive process in heavy oil reservoirs using a glass micro-model, Paper SPE 37519 Presented at the SPE Thermal Operation and Heavy Oil Symposium, Bakersfield, CA, February 10–12.

Bora, R., Chakma, A., Maini, B.B., 2003. Experimental investigation of foamy oil flow using a high pressure etched glass micromodel, SPE 84033 Paper Presented at the SPE Annual Technical Conference, Denver, CO, October 5–8.

Bratli, R.K., Risnes, R., 1981. Stability and failure of sand arches. SPE J. 21, 236–248.

Callaghan, I.C., McKechnie, A.L., Ray, J.E., Wainwright, J.C., 1985. Identification of crude oil components responsible for foaming. SPE J. 25 (2), 171–175.

Chugh, S., Baker, R., Telesford, A., Zhang, E., 2000. Mainstream options for heavy oil: part I—cold production. J. Can. Pet. Technol. 39 (4), 31–39.

Denbina, E.S., Baker, R.O., Gegunde, G.G., Klesken, A.J., Sodero, S.F., 2001. Modelling cold production for heavy oil reservoirs. J. Can. Pet. Technol. 40 (3), 23–29.

Dusseault, M.B., 2002. CHOPS—cold heavy oil production with sand in the Canadian heavy oil industry, Report prepared for the Alberta Department of Energy, March.

Dusseault, M.B., Geilikman, M.B., Spanos, T., 1998. Mechanisms of massive sand production in heavy oils, Paper Presented at the Seventh UNITAR Conference, Beijing, China.

Firoozabadi, A., 2001. Mechanisms of solution gas drive in heavy oil reservoirs. J. Can. Pet. Technol. 40 (3), 15–20.

Firoozabadi, A., Aronson, A., 1999. Visualization and measurement of gas evolution and flow of heavy and light oils in porous media. SPEREE December, 550–557.

Forth, R., Slevinsky, B., Lee, D., Fedenczuk, L., 1996. Application of statistical analysis to optimize reservoir performance. J. Can. Pet. Technol. October, 36–42.

Istchenko, C.M., Gates, I.D., 2012. The well-wormhole model of CHOPS: History match and validation, Paper SPE 157795 presented at the SPE Heavy Oil Conference, Calgary, Alberta, June 12–14.

Kimmel, T.B., Laviolette, D.J., 1982. Heavy oil production problems in the Lloydminster area, Paper presented at the Thirty-Third Annual Meeting of the CIM, Calgary, Alberta, June 6–9.

Kumar, R., Pooladi-Darvish, M., 2002. Solution–gas drive in heavy oil: field prediction and sensitivity studies using low gas relative permeability. J. Can. Pet. Technol. 41 (3), 26–32.

Landau, L.D., Lifshitz, E.M., 1980. Statistical Physics Part I, vol. 5. Pergamon Press, Oxford, England.

Lau, E., 2001. An integrated approach to understand cold production mechanisms of heavy oil reservoirs, Paper 2001-151 Presented at the 2001 CIPC Conference, Calgary, Alberta, June 12–14.

Lillico, D.A., Babchin, A.J., Jossy, W.E., Sawatzky, R.P., Yuan, J.-Y., 2001. Gas bubble nucleation kinetics in a live heavy oil. Colloids Surf. A 192, 25.

Liu, X., Zhao, G., 2005. A fractal wormhole model for cold heavy oil production. J. Can. Pet. Technol. 44 (9), 31–36.

Loughead, D.J., Saltuklaroglu, M., 1992. Lloydminster heavy oil – why so unusual?, Paper Presented at the Ninth Annual Heavy Oil and Oil Sand Technology Symposium, Calgary, Alberta, March 11.

Di Lullo, G., Curtis, J., Gomez, J., 2004. A fresh look at stimulating unconsolidated sands with proppant-laden fluids, Paper Presented at the SPE Annual Technical Conference and Exhibition, Houston, TX, September 26–29.

Maini, B.B., 1996. Foamy oil flow in heavy oil production. J. Can. Pet. Technol. 35 (6), 21–24.

Maini, B.B., Sarma, H.K., George, A.E., 1993. Significance of foamy oil behaviour in primary production of heavy oil. J. Can. Pet. Technol. 32 (9), 50–54.

Meza-Diaz, B., Sawatzky, R., Kuru, E., Oldakowski, K., 2011. Sand on demand: a laboratory investigation on improving productivity in horizontal wells under heavy-oil primary production. SPE Prod. Oper. 26 (3), 240–252.

Miller, K.A., 2006. Improving the state of the art of western Canadian heavy oil waterflood technology. J. Can. Pet. Technol. 45 (4), 7–11.

de Mirabal, M., Gordillo, R., Rojas, G., Rodriguez, H., Huerta, M., 1996. Impact of foamy oil mechanism on the Hamaca oil reserves, Orinoco Belt-Venezuela, Paper Presented at the Forth Latin American and Caribbean Petroleum Engineering Conference, Port-of-Spain, Trinidad and Tobago, April 23–26.

Moulu, J.C., 1989. Solution-gas drive: experiments and simulation. J. Pet. Sci. Eng. 2 (4), 379–386.

Nouri, A., Wang, X., Vaziri H., 2008. A review of sand production mechanisms in cold production heavy oil production, Paper Presented at the CIPC/SPE Gas Technology Symposium, Calgary, Alberta, June 17–19.

Ostos, A.N., Maini, B.B., 2005. An integrated experimental study of foamy oil during solution gas drive. J. Can. Pet. Technol. 44 (4), 43–50.

Peng, J., Tang, G.Q., Kovscek, A.R., 2009. Oil chemistry and its impact on heavy oil solution gas drive. J. Pet. Sci. Eng. 66 (1–2), 47–59.

Pooladi-Dravish, M., Firoozabadi, A., 1999. Solution-gas drive in heavy oil reservoirs. J. Can. Pet. Technol. 38 (4), 54–61.

Rivero, J.A., Coskuner, G., Asghari, K., Law, D.H.S., Pearce, A., Newman, R., et al., 2010. Paper Presented at the SPE Annual Technical Conference and Exhibition, Florence, Italy, September 19–22.

Sahni, A., Gadelle, F., Kumar, M., Tomutsa, L., Kovscek, A.R., 2004. Experiments and analysis of heavy-oil solution-gas drive. SPE Reservoir Eval. Eng. June, 217–229.

Sawatzky, R.P., Lillico, D.A., London, M.J., Tremblay, B.R., Coates, R.M., 2002. Tracking cold production footprints, Paper Presented at the Petroleum Society CIPC Conference, Calgary, Alberta, June 11–13.

Sheng, J.J., Maini, B.B., Hayes, R.E., Tortike, W.S., 1999a. Critical review of foamy oil flow. Trans. Porous Media 35 (2).

Sheng, J.J., Hayes, R.E., Maini, B.B., Tortike, W.S., 1999b. Modelling foamy oil in porous media. Trans. Porous Media 35, 227–258.

Shute, D.M., Kaiser, T.M.V., Munoz, H., 2004. Compaction resistant wellbore for sand-producing wells, Paper Presented at the Fifth Canadian International Petroleum Conference, Calgary, Alberta, June 8–10.

Smith, G.E., 1988. Fluid flow and sand production in heavy-oil reservoirs under solution-gas drive. SPE Prod. Eng. 3 (2), 169–180.

Smith, G.E., 1992. Waterflooding heavy oils, Paper SPE 24367 Presented at the SPE Rocky Mountain Regional Meeting, Casper, Wyoming, May 18–21, 473–480.

Squires, A., 1993. Inter-well tracer results and gel blocking program, Paper Presented at the 10th Annual Heavy Oil and Oil Sands Technical Symposium, Calgary, Alberta, March 9.

Tan, T., Slevinsky, R., Jonasson, H., 2004. A new methodology for modelling of sand wormholes in field scale thermal simulations. J. Can. Pet. Technol. 44 (4), 16–21.

Tang, G.Q., Firoozabadi, A., 2003. Gas and liquid-phase relative permeabilities for cold production from heavy-oil reservoirs. SPE Formation Eval. Eng. April, 70–80.

Tang, G.-Q., Leung, T., Castanier, L.M., Sahni, A., Gadelle, F., Kumar, M., et al., 2003. An investigation of the effect of oil composition on heavy oil solution-gas drive, Paper Presented at the SPE Annual Technical Conference, Denver, CO, October 5–8.

Terzaghi, K., 1951. The influence of modern soil studies on the design and construction of foundations, Proceedings, Building Research Congress, Division 1, Part III, London, England, pp. 139–145.

Treinen, R.J., Spence, A.P., de Mirabel, M., Huerta, M., 1997. Hamaca: solution gas drive recovery in a heavy oil reservoir, experimental results, Paper Presented at the Fifth Latin American and Caribbean Petroleum Engineering Conference and Exhibition, Rio de Janeiro, Brazil, 30 August–3 September.

Tremblay, B., 2005. Modelling of sand transport through wormholes. J. Can. Pet. Technol. 44 (4), 51–58.

Tremblay, B., 2008a. Wormhole reservoir characterization, Petroleum Technology Research Centre Report, PTRC Publication No. 001-00102-SRC, August.

Tremblay, B., 2009a. Multi-well cold flow numerical model: waterflooding cold produced reservoirs, PTRC Report No. 001-00149-SRC, September.

Tremblay, B., 2009b. Cold flow: a multi-well cold production (CHOPS) model. J. Can. Pet. Technol. 48 (2), 22–28.

Tremblay, B., Oldakowski, K., 2003. Modelling of wormhole growth in cold production. Trans. Porous Media 53 (2), 197–214.

Tremblay, B., Sedgwick, G., Forshner, K., 1998a. Modelling of sand production from wells on primary recovery. J. Can. Pet. Technol. 37 (3), 41–50.

Tremblay, B., Sedgwick, G., Forshner, K., 1998b. CT imaging of sand production in a horizontal sand pack using live oil, Paper 98–78 Presented at the Forty-Ninth Annual Technical Meeting of the Petroleum Society, Calgary, Alberta, June 8–10.

Tremblay, B., Sedgwick, G., Vu, D., 1999. CT imaging of wormhole growth under solution-gas drive. SPE Res. Eval. Eng. 2 (1), 37–45.

Tremblay, B., Smith, H., Exelby, R., 2008b. Wormhole stability to solvents, PTRC Report No. 00-00074-SRC, August.

Uddin, M., 2005. Numerical studies of gas ex-solution in a live heavy-oil reservoir, Paper Presented at the SPE International Thermal Operations and Heavy Oil Symposium, Calgary, Alberta, November 1–3.

Urgelli, D., Durandeau, M., Houcault, H., Besnier, J-F., 1999. Investigation of foamy oil effect from laboratory experiments, Paper SPE 54083 Presented at the SPE International Thermal Operations and Heavy Oil Symposium, Bakersfield, CA, March 17–19.

Vanderheyden, W.B., Zhang, D.Z., Jayaraman, B., 2011. Modelling wormhole growth and wormhole networks in unconsolidated media using the BP CHOPS model, Paper WHOC11-427 Presented at the World Heavy Oil Congress, Edmonton, Alberta.

Vaziri, H.H., Xiao, Y., Islam, R., Lemoine, E., 2003. Physical modelling study of the influence of shale interbeds and perforation sequence on sand production. J. Pet. Sci. Eng. 37, 11–23.

Vittoratos, E.S., Zhang, L., Turek, E., West, C.C., 2008. Deliberate sand production from heavy oil reservoirs: potent activation of both solution gas and aquifer drives, Paper 2008–501 Presented at the World Heavy Oil Congress, Edmonton, Alberta, March 10–12.

Walton, I.C., Atwood, D.C., Halleck, P.M., Bianco, L.C.B., 2002. Perforating unconsolidated sands: an experimental and theoretical investigation. SPE Drill. Completion September, 141–150.

Ward, C.A., Tikuisis, P., Venter, R.D., 1982. Stability of bubbles in a closed volume of liquid-gas solution. J. Appl. Phys. 53, 6076.

Wong, R.C.K., Smieh, A.M., Kuhlemeyer, R.L., 1994. Oil sand strength parameters at low effective stress: its effect on sand production. J. Can. Pet. Technol. 33 (5), 44–49.

Index

Note: Page number followed by "*f*" and "*t*" refer to figures and tables, respectively.

A

Adsorption isotherm, 124–125
Adsorption of polymers, 67–68
Alipal CD-128 foaming agent, 37–39
Alkali
  generated soap solution, IFT of, 180–181
  interactions between polymer and, 169
  interactions between surfactant and
    adsorption of surfactant, 183–184
    interfacial tensions (IFTs), 180, 183
    optimum salinity and solubilization ratio, 179–180
    phase behavior, 180–183
    salt effect, 179
    synergy between soap and surfactant, 180–183
  synergy between polymer and, 169–171
Alkaline flooding
  alkaline reactions
    alkali–water reactions, 146
    with crude oil, 144–145
    with rock, 145–146
  application conditions of, 149–151
  comparison of alkalis used in, 143–144
  field cases
    Court Bakken Heavy Oil Reservoir, Canada, 164–165
    Hungarian H field, 155–156
    North Gujarat Oil Field, India, 156–157
    Russian Шагщр-Гоаеан Field, 154–155
    Russian Трехозephoe field, 151–154
    Torrance Field, California, 158–159
    Whittier Field, California, 157–158
    Wilmington Field, California, 159–164
  field injection data, 147–149
  recovery mechanisms, 146–147
Alkaline–polymer flooding (AP) project
  enhanced oil recovery (EOR), 345–346
  field applications
    Almy Sands (Isenhour Unit), Wyoming, 171–172, 171*t*
    David Lloydminster "A" Pool, Canada, 174–176
    Etzikom Field, Canada, 176
    Moorcroft West Minnelusa Sand Unit, Wyoming, 172–174, 173*t*
    Thompson Creek Field, Wyoming, 174
    Xing-28 Block, China, 176–177
    Yangsanmu, China, 177–178
  interactions between alkali and polymer, 169
  laboratory test results, 169
  synergy between alkali and polymer, 169–171
Alkaline–surfactant system
  converted fractions of acid into soap, 184, 185*f*
  field cases
    Big Sinking Field, East Kentucky, 186
    White Castle Field, Louisiana, 186–187
  molar fraction of soap in the total amount of surfactants, 185, 185*f*
  sizes and shapes of the aggregates in, 191*t*
Alkaline–surfactant–polymer (ASP) system
  chemical injection concentrations, 191
  condition for synergy, 191–192
  effluent concentration histories, 191, 192*f*
  enhanced oil recovery (EOR), 347
  field applications in China
    between 1990 and 2000, 207
    amounts of chemicals injected, 192–194
    overall performance, 194
  field pilots and applications
    between 1980 and 2000, 207, 218, 219*t*
    Bradford Field, Pennsylvania, 204
    Bridgeport test, 226–227
    Cambridge Minnelusa Field, Wyoming, 196–198
    Daqing Oil Field, 221–225, 222*t*, 223*t*
    Elk Hills project, 226–227
    Etzihom alkaline polymer project, Canada, 226
    Karamay Oil Field, 225–226

Alkaline–surfactant–polymer (ASP) system (*Continued*)
Lagomar LVA-6/9/21 Area, Venezuela, 199–200
Lawrence Field, Illinois, 194–196
Minnelusa Formation, USA, 226
results for tests conducted prior to 1984, 204, 205*t*
Shengli Oil Field, 225
Taber South Mannville B, Canada, 226
Tanner field, Wyoming, 199
US National Petroleum Council (NPC) study, 204
West Kiehl Field, Wyoming, 198–199
White Castle test, 226
field test results
chemical costs, 235*t*
lessons learned in field implementation, 230–232
oil recovery efficiency, 227–229
process application, 229–230
recovery mechanisms, 229
flow diagram of, 238–239, 238*f*
future outlook and focus, 232–234
chemical costs, 234, 235*t*
chemical slugs, 234
offshore reservoirs and applications, 233–234
polymer handling facilities, 232–233
reservoir issues, 234
slug designs, 233
issues, 190–192
chromatographic separation, 191–192
concentration histories in alkaline environment, 192, 193*f*
precipitation, 192
produced emulsion, 190–191
scaling problems, 192
laboratory study, 207–212
aqueous formulation clarity and stability, 208
confinement and heterogeneity of reservoirs, 215–216
dead oil and live fluid experiments, 207–208
displacement efficiencies, 215
effectiveness tests, 215–216
IFT, 208–209, 210*f*
linear coreflood experiment, 210
middle-phase microemulsions, 208
multistage processes, 207
optimal salinity, 207–208
recovery efficiency, 210–212
relative mobility of displacement chemical flooding process, 209–210, 211*f*
single well chemical tracer tests (SWCTT), 215–216
slug and oil phase, 208
solubilization ratios, 208, 209*f*
mechanistic modeling, 212–215
components of, 212
coreflood—Karamay project, 213–214, 213*f*, 214*f*
Daqing B1-FBX ASP pilot test results, 214*f*
history matching of coreflood experiments, 214–215
UTCHEM, 212
recommendations
expansion program, 238
facilities and services, 237
field implementation, 237
initial screening, 235–236
laboratory studies, 237
monitoring and surveillance, 237–238
process design and simulation studies, 237
reservoir studies, 236
screening criteria for, 216–217
acid content, 217
acid number, 217
addition of sequestering or chelating chemicals, 217
lithology, 217
oil viscosity, 216
permeability, 217
requirement of freshwater, 217
reservoir and fluid data for initial process, 236*t*
reservoir temperature, 216–217
steps and estimated timing for, 238–239
synergies and interactions, 189–190
Almy Sands (Isenhour Unit), Wyoming, AP project in, 171–172
"CAT-AN" process, 172
crude oil alkaline-agent screening, 171
inorganic wetting and stabilizing agents used, 172
M-42 reservoir, 171
reservoir and fluid data, 171*t*
soak-and-coat technique, 172
AMPS (2-acrylamide-2-methyl propane-sulfonate), 63–64

Anionics, 117–118. *See also* Surfactants
Aqueous stability test, 120–121
Aqueous stability tests, 124

B

Bei-Yi-Qu-Duan-Xi block, 73
Beverly Hills field, California, 604–613
additional producer treatments, 608
background, 604
BH-15 well, 610–611, 611*f*
discussion of results, 612–613
frontline production wells, 606–608
gas oil ratio (GOR), 612*t*
nutrient treatment, 605–606
OS-8 well, 609–610, 611*f*
OS-9 well, 609*f*
OS-10 well, 609*f*
OS-14 well, 610*f*
OS-1 well test data, 605*f*, 605*t*, 606, 608*t*
pretreatment production, 605
water-injection well treatments, 606–608
Big Sinking Field, East Kentucky, alkaline–surfactant system, 186
BSXX, polymer flooding practice in
individual well injection and production volume design, 109
numerical simulation model, 109
cumulative oil production, 110
financial internal rate of return (FIRR), 111
financial net pay alue (FNPV), 111
history matching of waterflooding before polymer injection, 110
waterflooding performance prediction, 110
performance prediction, 110–111
PI1-4 oil zones of, 108–109
polymer injection formulations design, 108–109
parameters, 109*t*
polymer performance evaluation, 111
water cut and oil production, 111*f*
waterflooding pattern, 110
well pattern and oil strata combination, 107–108, 108*f*
Bubble-point pressure, 9–10
Buckley–Leverett theory, 70–71
Bulk foam rheology, 30

C

C16-18 alpha-olefin sodium sulfonate, 36
Cambridge Minnelusa Field, Wyoming, ASP project in, 196–198
chemical concentrations, 196
corrosion, demulsification, and bactericide program, 197–198
facilities, 197
injection sequence, 196
project economies, 198
chemical cost, 198
incremental maintenance, electrical, and pumping costs, 198
reservoir details, 196
Capillary desaturation curve (CDC), 123
Capillary imbibitions, defined, 286
Capillary number, 122
CAPRI™, 528
Carbonate reservoirs, oil recovery in. *See also* Smart Water (SW) flooding
chemical EOR research, 282
chemicals used in, 291–292
Cottonwood Creek field, Wyoming, 294
Cretaceous Upper Edwards reservoir, Texas, 295–296
Mauddud carbonate reservoir, Bahrain, 292–293
Semoga field, Indonesia, 294–295
Yates field, Texas, 293–294
mechanisms, 289–291
cationics and nonionics, dynamics of, 289–290
IFT between oil and brine, 290
IFT of poly-oxyethylene alcohol (POA), 291
ion-pair complex, 289
problems in, 282–283
upscaling aspects, 286–288
capillary imbibitions, 286
IFT between wetting and nonwetting phase, 286
mobility and capillary pressure parameters, 287–288
oil recovery factor *vs* normalized time, 286–287, 287*f*
using Smart Water (SW) flooding
concentration of NaCl in, 314
effluent concentrations, 307, 308*f*, 309*f*, 311*f*
environmental effects, 315–316
introduction, 306–307
ions and temperature on oil recovery, 309–311
mechanism for wettability modification, 312, 313*f*

Carbonate reservoirs, oil recovery in (*Continued*)
optimization of injected water, 312–314
reactive potential determining ions, 307–312
viscous flood *vs* spontaneous imbibition, 315
wettability alteration using surfactants, 283–286, 289*f*
capillary desaturation curve (CDC), 284
effect of, 285
effect of permeability and porosity, 283
effects of IFT and contact angle on capillary pressure, 286
end-point relative permeability, 285
end-point relative permeability enhancements, 284–285
exponents of relative permeabilities, 285
residual saturation at a low trapping number, 284
residual saturation of conjugate phase, 284–285
trapping number, 284–285
wetting angle effect, 283–284
Carreau equation, 65
Catastrophe theory, 26–27
Cenovus' Christina Lake Thermal Project (CLTP), 439–440
Chaser CD-1040, 41
Chinese polymer flooding projects, 73
CHOPS (cold heavy oil production with sand) process
advantages, 616
in Canada, 616–617
defined, 615–616
field observations, 636–642
BHP well, 640–641, 641*f*
in heterogeneous reservoirs, 645–651
production data, 647*t*, 648*t*
tracer tests, 636–637, 637*f*
field production data, 616
in Loydminster heavy oil fields, 617
mechanisms, 617–618
sand production, 627–645
solution-gas drive process, 618–627
prediction of performance, 652–655
post-CHOPS production, 656–660
sand production, 627–645
background, 627–628
dilated sand field models, 642–643
in dilated zone, 629–630
effective axial stress, 628
field observations, 636–642
forces acting on, 627–628
macroscopic stresses acting on, 628
multiwell CHOPS model, 644
numerical modeling, 642–645
radial effective stress, 628
tangential effective stress, 628
wormhole growth, 630–642
wormhole network field models, 643–645
at Saskatchewan Research Council, 651
solution-gas drive process, 618–627
background, 618–619
Corey-type model, 619
critical gas saturation, 623–625
foamy oil theory, 626–627
at low capillary numbers and pressure depletion rates, 620
mechanical equilibrium of a gas bubble in a liquid, 618
numerical modeling, 626–627
oil chemistry, 625–626
pressure depletion rates, 621–623
recovery factor for light oil ranges, 619–620
relative permeabilities, 619, 627
stratum, 616–617
Chun Huh correlation, 208–209
CO_2-foam, 262–263. *See also* Foams
injection operation
coinjection process, 51–52
diversion from high- to low-permeability layers, 45–46
in East Mallet Unit, Texas, 43–45
in the East Vacuum Grayburg/San Andres Unit, New Mexico, 42–43, 50*f*
laboratory setup for experiments, 32*f*
in McElmo Creek Unit, Utah, 43–45
in North Ward-Estes, Texas, 40–42
in Rangely Weber Sand Unit, Colorado, 39–40
successful SAG processes, 46–51
use of positive-displacement pumps, 31
water WAG process *vs* foam SAG process, 48–49
in Wilmington, California, 37–38
Cold Lake, Canada, CSS project in, 401–402
reservoir conditions, 402
reservoir description, 401–402
soak period, 402

steam generation and fluid handling facilities, 402
steam treatments, 402
Colloidal dispersion gel (CDG), 78–79
Conformance measures in gas flooding, 8
Converted shear rate (equivalent shear rate) of polymer, 66–67
in porous media, 67
COR-180, 35–36
Coreflood tests, 121
Cottonwood Creek field, Wyoming, 294
Court Bakken Heavy Oil Reservoir, Canada, alkaline flooding project in, 164–165
Cretaceous Upper Edwards reservoir, Texas, 295–296
Critical micelle concentration (CMC), 118–119
Crude oil–brine–rock (CBR) system. *See also* Smart Water (SW) flooding
oil recovery by water flooding, 301
wetting properties, 301–302
carbonate, parameter for, 302–304
sandstones, parameter for, 304
smart water flooding, 304–306
Cyclic steam stimulation (CSS) process
Boberg and Lantz model, production response, 391–395
average heat capacity, 392
boundary conditions, 392, 393*f*
heat losses, 392
heated zone, 392
rate of produced heat at downhole condition, 394
field cases
Cold Lake, Canada, 401–402
Gaosheng Field, China, 411
Gudao Field, China, 407–408
Karamay Field, China, 408–410
Liao Shuguang field, China, 404–406
Liaohe Huanxiling Field, China, 406–407
Midway-Sunset field, California, 402–404
mechanism of, 389–391
improvement rate in productivity after steam injection, 391
production rate of steam stimulation, 390
radial flow model, 390*f*
steady-state Darcy flow equation, 389
practice in
completion interval, 400
at deep reservoirs, 399
general producing methods, 396–397
incremental oil recovery and OSR, 400, 401*f*
injection and production parameters, 397–399
monitoring and surveillance, 400–401
number of cycles in actual, 400*f*
production time in actual, 398*f*
soak time in actual, 398*f*
strategy to reduce damages, 398
well total injection rate per cycle (CWE), 398, 399*f*
wellbore heat insulation, 400
screening criteria, 395–396, 395*t*
steam soak and steam huff-and-puff, 389

D

Daqing, polymer flooding practice in
blocks in, 75
concentration histories in alkaline environment, 192
design, 91–102
connectivity factor, 92
effectiveness of, 98, 99*t*
incremental recovery *vs* polymer mass, 99, 100*t*
individual production and injection rate allocation, 101–102
injection formulation options, 95–101
injection rate, 100–101, 101*f*
injection sequence options, 94–95
permeability differential, 92–93
polymer MW, use of, 95–97, 96*t*
polymer solution viscosity and concentration, 97–99, 98*f*
profile modification methods, 94–95
separate layer injection option, 95
swept volume, 99–100
well pattern, 92–93, 94*t*
dynamic performance
issues and solutions during, 105*t*
problems and treatments during different phases, 104
stages and dynamic behavior, 102–104
mechanism
microscopic oil displacement, 86–87
mobility control, 83–84
profile modification, 84–86
relationship of oil fraction and water saturation, 83–84
water cut *vs* average water saturation, 84, 84*t*, 85*f*

Daqing, polymer flooding practice in (*Continued*)
water intake ratios between two permeable layers, 85–86, 86*t*
waterflooding equation, 83
reservoir screening, 87–91
formation water salinity, 90–91
heterogeneity, 89–90
oil viscosity, 90
permeability, 88–89
temperature, 88
type, 87–88
surface facilities
biocide preflush, 105–106
biological degradation of polymer, 105–106
chemical stability of polymer, 105–106
fluid production treatment, 106–107
mechanical stability of polymer, 105–106
mixing and injection work, 105–106, 106*f*
viscosity loss, issue of, 106
Daqing Oil Field, ASP system in, 221–225
Daqing B1-FBX ASP pilot test results, 214*f*
oil recovery efficiency, 227–228
recovery efficiencies, 221
slug designs, 221
slug formulation—test conducted, 223*t*
tests conducted, 222*t*
Darcy velocity, 122
in porous media, 67
Darcy's equation, 54–56, 65–66
David Lloydminster "A" Pool, Canada, AP project in, 174–176
EOR scheme in, 175
incremental oil production, 175–176
injection sequence, 176
reservoir and fluid data, 175*t*
Delaware-Childers M/P Project, 136
mineralogical analysis, 136
oil production, 136
total dissolved solids (TDSs) of formation water, 136
Desulfovibrio hydrocarbonoclasticus, 561–562
Dimensionless time, defined, 287–288
Dispersion, defined, 72
Disproportionate permeability reduction (DPR), 70
Dykstra–Parsons coefficient, 4

E

East Mallet Unit (EMU), Texas, foam field application in, 43–45
East Vacuum Grayburg/San Andres Unit (EVGSAU), New Mexico, foam field application in, 42–43
Effective gas relative permeability, 54–55, 57*f*
Effective gas viscosity, 54–55
El Dorado field of Butler County, Kansas, surfactant–polymer (SP) flooding project in, 130–132
blending mechanisms, 132
design criteria of injection facility, 132
estimated oil saturation, 130
fiberglas and polyvinylchloride piping design, 132
injection sequence patterns, 131
monitoring and controlling bacterial growth, 131
observation well program, 132
overall plant facility, 132
preflush phase, 131
reservoir description and production history, 130–131
sequential M/P flooding project, 130
staged solvent–acid treatment, 131
water-quality monitoring program, 131
Emulsions. *See also* Microemulsions
stability of, 191
types of, 190
water cuts from W/O type to O/W type, 190, 190*t*
Enhanced foam system, 264
Enhanced oil recovery (EOR), 84–85
chemical processes, facility requirements, 339*f*
factory-prefabricated 3000 BPD polymer injection, 339*f*
polymer processing and injection facility, 338, 339*f*
scaling, 338
by gas flooding, field cases, 16–21
handling and processing chemicals on-site, 350–357
alkaline agent handling, processing and metering, 355–357
polymer handling, processing, and metering, 350–354
surfactant handling and metering, 354–355
injection schemes and strategies, 358–359

materials of construction for piping, fittings and valves, 359–360
modes of chemical eor injection, 343–347
alkaline-polymer flooding, 345–346
alkaline-surfactant-polymer flooding, 347
polymer flooding (P), 344
surfactant-polymer flooding, 345
overall project requirements, 339–343
calculations for polymer hydration time, 340, 342*f*
minimum design criteria and operating parameters, 343
spreadsheet setup, 341*f*
physical principle for, 306
in situ combustion (ISC), 454
Smart Water (SW) flooding
in carbonates, 326–329
in limestones, 317–319
in sandstones, 321–326
Statoil Snorre Pilot project, 330–332
water treatment and conditioning, 347–350
Enordet AOS-1618TM surfactant, 36
Enthalpy (H), 363
Equation-of-state (EOS), fluid characterization using, 12
Equilibrium concentration in solution system, 124–125
Etzihom alkaline polymer project, Canada, 226
oil recovery efficiency, 228
Etzikom Field, Canada, AP project in, 176

F

First-contact miscibility (FCM), 10
Flory–Huggins equation, 64–65
Flow resistance factor, defined, 262
Fluid characterization using equation-of-state (EOS), 12
Foam catastrophe theory, 28–30
Foams. *See also* Polymers; Surfactants
application modes
CO_2 foam, 262–263
enhanced foam system, 264
gas coning blocking foam, 264
injection in gas miscible flooding, 264
steam-foam, 262–263
for well stimulation, 264–265
characteristics of, 251–252
CO_2, 8, 23–24
comparison with other gases, 24
coreflood experiments, 31–32
critical pressure (P_{crit}) of, 24
critical temperature (T_{crit}) of, 24
driven EOR processes, 24
impact of emission, 23
limitations resulting from injection of, 24–25
minimum miscibility pressure (MMP) and, 24–25
popularity of, 23–24
supercritical phase, 24
swelling effect, 24
dimensionality-dependent flow characteristics of, 28–30
diversion process, 33–34
diversion process of, 33–34
expression of gas-mobility reduction in, 53–58
calculating, 56–59
field applications
East Mallet Unit (EMU), Texas, 43–45
East Vacuum Grayburg/San Andres Unit (EVGSAU), New Mexico, 42–43
McElmo Creek Unit (MCU), Utah, 43–45
Midway Sunset field, California, 35–36
North Ward-Estes field, Texas, 40–42
Rangely Weber Sand Unit (RWSU), Colorado, 38–39
Rock Creek, Virginia, 38–39
Siggins field, Illinois, 34–35
Wilmington, California, 37–38
flooding applications
injection mode, 266
screening criteria, 265
use of surfactants, 265–266
flow behavior
flow resistance factor, 262
mobility reduction, 261–262
relative permeabilities, 261
rheology, 260–261
viscosity, 260–261
fractional flow curves, 28–30, 29*f*
individual field applications, 268–276
nitrogen foam flooding in a heavy oil reservoir, 271–273
single well polymer-enhanced foam flooding test, 268–271
Snorre foam-assisted-water-alternating-gas (FAWAG) project, 273–276
injection methods, 30–31
lamellae, 24–25
limiting capillary pressure, role of, 32–33

Foams (*Continued*)
mechanisms of flooding to enhance oil recovery, 257–260
coalescence of foam, 259
formation and decay, 258–259
gas phase relative permeability, 260
lamella division, 258, 258*f*
leave-behind process, 259, 259*f*
pore-level generation mechanisms, 258
snap-off process, 258, 258*f*
sweep efficiency, 260
modeling, 28–30
N_2, 8
oil interactions, 34
in porous media
conventional gas–liquid two-phase flow, 25–26, 26*f*
in high-capillary pressure environment, 25–26
high-quality and low-quality regimes, 27–28
lamella creation and coalescence mechanism, 25
local steady-state modeling, 28
mechanistic foam modeling/simulation, 28
in near-horizontal pressure contours, 27–28
in near-vertical pressure contours, 27–28
states and foam generation, 25–27
transition from weak-foam to strong-foam state, 26
results from field application survey
gas used, 268
injection mode, 268
locations of conducted projects, 267
reservoir and process parameters, 267, 267*t*
slug application method, 268
in situ creation, 31–32
stability, 252–257
steam–gas, 9
strong, 25–26
subsurface heterogeneity and, 32–34

G

Gaosheng Field, China, CSS project in, 411
field description, 411
field tests, 411
gas cap breakthrough, 411
heat loss from reservoirs, 411
Gas coning blocking foam, 264
Gas flooding
definition, 1–2
design, 2–3
design steps for a large flood, 2–3
for EOR, field cases, 16–21
Jay field near the Alabama–Florida border, 19–20
Weeks Island, Louisiana field, 17–19
West Texas San Andres dolomite field, 16*f*, 17
injected components in, 1
key to, 2
mixing of oil and gas, 2
mobility control methods, 8
in modern times, 1
primary mechanism for oil recovery, 1–2
screening process, 3–5
single-well method, 8
using carbon dioxide, 7*f*
volumetric sweep efficiency, 2
Gas injection techniques, 5–9
gravity tonguing, issue of, 7
injectivity problems, 8
simultaneous water-gas injection (SWAG), 5–6
tapered WAG (TWAG), 5–6
water-alternating-gas (WAG), 5–6, 6*f*
Gas miscible flooding, foam injection in, 264
Gas–liquid two-phase flow, 54
Gel injection, 79
Gravitational potential energy, 364
Gravity effects on gas flooding, 4–5
Gravity stable process, 19
Greenhouse gases, 23
Gudao Field, China, CSS project in, 407–408
clay swelling, method to prevent, 408
detergents, use of, 408
sand control, 407–408
single well CSS, 407
thin film spreading agents, use of, 408
well completion, 407
Gudong Field, China, urfactant–polymer (SP) flooding project in, 139–141
injection scheme, 141
oil recovery, 139
SP formula design, 140

H

Heat capacity, 361
Heat conduction, 365
Heat convection, 365

Huh equation, 121
Hungarian H field, alkaline flooding project in, 155–156
Hydrodynamic trapping, 125
Hydrolysis rate, 63
Hydrolyzed polyacrylamide (HPAM), 63
  adsorption level of, 68
  alkaline effect on, 66*f*
  MW, polymer concentration and solution viscosity of, 97

I

In situ combustion (ISC), 561–562
  background, 448
  CISC or "burn and turn," 520–525
    in Albania, 521, 523
    for heavy oil production stimulation, 521–524
    at Marguerite Lake, 521
    a priori design, 523–524
    in Romania, 521–523
    sand consolidation, 524–525
    success parameter, 522
    at Suplacu de Barcau, 521–523, 522*f*
    at Videle–Balaria operations, 523
    water injection rates, stimulation of, 524
  co-current, 449
  combustion override split-production horizontal well (COSH) process and, 531–532
  commercial application, 452
  commercially significant applications, 493–497
    active, 500*t*, 503*t*
    Battrum Field, Saskatchewan, Canada, 498
    Bellevue field Louisiana, 498–499, 499*f*, 501*f*
    BSOC Bellevue project, 502
    dry ISC, 494, 502, 504
    Karazhanbas Field, Kazachstan, 497–498
    Moco Zone in Midway Sunset Field, 494
    reservoir properties, 500*t*
    Santhal and Balol projects, 508–510, 509*f*
    Suplacu de Barcau, Romania, 502, 504–508, 506*f*, 507*f*
    very thin layer projects, 494–497, 495*t*
    Videle East Project, 496–497, 496*f*
    wet ISC, 497–511, 509*f*
  commercially unproved applications
    Bottom Water ISC (BW-ISC), 488–491
    at Buffalo Creek reservoir, 492
    in naturally fractured reservoirs (ISC-NFR), 491
    in Russia, 491
    at Zybza Glubokii Yar reservoir, 492, 493*f*
  countercurrent, 449
  design, operation and evaluation
    ignition operation, 454–458
  design and operation
    affected *vs* nonaffected zone, 463–464
    air injection stoppage period, 461
    artificial ignition operation, 455
    conventional peripheral gas displacement, 459–460
    enhanced oil recovery (EOR), 454
    field pilot test, 461–465
    ignition operation, 457*f*
    indications of ignition, 455
    line drive *vs* well pattern application, 458–461
    mathematical modeling, 465–467
    options for expansion of, 458*f*
    performance indices, 462
    peripheral line drive well configuration, 459, 459*f*
    phases, 461
    prediction methods, 465–467
    selection of best location for starting, 462
    South Belridge Field, 465, 466*f*
    spontaneous ignition operation, 455
    Suplacu de Barcau Field, 462–464, 463*f*
    West Balaria Field, 464–465, 464*f*
  dry, 449
    saturation profiles, 449, 450*f*, 451
    temperature profiles, 449, 450*f*, 451, 451*f*
  feasibility of, 452–453
  field applications
    screening guide, 467–469, 469*t*
  fundamentals, 447–467
    front movement, consequences of, 447–448
    heat utilization of, 448
    qualitative description, 447–453
  heat transfer models (HTM), 467
    Chu's model, 467
    Thomas' model, 467
  for heavy oil reservoirs, 452
  in horizontal wells

In situ combustion (ISC) (*Continued*)
at Battrum, Saskatchewan, 525–526
Eyehill Project, Saskatchewan, 525
long-distance oil displacement (LDOD) *vs* short-distance oil displacement (SDOD), 526–528
instrumented pilots
air–oil ratio, 476
Amoco and AOSTRA studies, 479
Bellamy Field, Montana, USA, 484
cold heavy oil production with sand (CHOPS) process, 475–476
commercial projects outside US, 487*t*
dry ISC, 473–474, 476
extra heavy oil reservoirs, 475–476
Gregoire Lake, Athabasca, 476–479
heavy oil reservoirs, 473–475
Joli Fou, Wabasca, Canada, 480–481
Kansas project, 474
KYROCK Project, Kentucky, USA, 481–482, 482*f*, 483*f*
light oil reservoirs, 475
Marguerite Lake, Canada, 480
North Tisdale ISC test, 474
Northwest Asphalt Ridge, Utah, USA, 484–485
oil sand reservoirs, 476–488, 477*t*
reverse ISC, 482–488
Suplacu de Barcau, 473*f*, 474–475
US past commercial, 486*t*
wet ISC, 474
laboratory tests, 453
equipment and procedures used for RTO and CT tests, 453
light oil reservoirs, 512–520
Cedar Creek Anticline, 512–516, 515*f*
data on very, 513*t*
failures, 519–520
Horse Creek reservoir, 516
instrumented pilots, 475
Little Beaver and Pennel projects, 519
Medicine Pole Hills Project, 516, 517*f*
South Buffalo HPAI project, 516, 518*f*
in waterflooded reservoirs, 516–519
at West Hackberry, 519
Williston Basin projects, 512–516
monitoring and evaluation, 469–473
burning efficiency, 470–471
challenges in assessment, 473
of coring wells, 473*f*
fraction of air reacting in reservoir, 471
gas injectivity test, 470
ignition operations, 470
of isolated patterns, 471
location/tracking methods, 471
oxygen utilization efficiency, 470–471
process-conformance factor, 472
remedies for correcting front advancement, 472
operational problems and remedies, 532–534
air injectivity issues, 534–535
critical problems, 533–534
erosion problem, 535
noncritical problems, 534–536
premature arrival of the ISC front, 535–536
pumping problems, 535
sand production and influx issues, 535
uneven global advancement, 535
prediction methods, 465–467
Brigham, Satman, and Soliman method, 465–467
Gates and Ramey method, 465–467
Nelson and McNiel method, 465–467
screening guide, 467–469, 469*t*
easiness of application, 468
high-pressure air injection (HPAI) commercial operations, 469
top-down, 531–532
upgrading considerations, 511
wet, 449, 497–511, 509*f*
brief considerations on application of, 511
saturation profiles, 449, 450*f*, 451
temperature profiles, 449, 450*f*, 451, 451*f*
Inaccessible pore volume (IPV) of polymers, 68–69
Injectivity problems in gas flooding, 8
Instapol Q-41-F, 136
Interfacial tension, 121

J

Joslyn Creek SAGD project, 438

K

Karamay Field, China, CSS project in, 408–410
blocks 97 & 98, 408
field trials, 408–410
first phase, 408–410
performance, 410–411
second phase, 410
tests conducted, 408–410

Karamay Oil Field, ASP system in, 225–226
mechanistic modeling, 214*f*
Kinetic energy, 364
Krafft temperature, 119
KYPAM, 63–64

L

Lagomar LVA-6/9/21 Area, Venezuela, ASP project in, 199–200
Langmuir-type isotherm, 68, 124–125
Latent heat, 361–362
Lawrence Field, Illinois, ASP project in, 194–196
field injection sequence, 195
injection formula, 195
oil saturation level, 195
production rate, 196
reservoir and fluid data, 194, 195*t*
Leverett-J function, 283
Liao Shuguang field, China, CSS project in, 404–406
cyclic steam operations in, 404
Du 66 block, 404
lessons learned, 405–406
oil production, 404–405, 405*f*
prestressed casing, use of, 405
recovery factor, 404
Liaohe Huanxiling Field, China, CSS project in, 406–407
cyclic steam operations in, 406
Jin 45 block, 406
lessons learned, 406–407
reservoir description and condition, 406
LIFTF (LIFT foam), 264
Limiting capillary pressure and foams, 32–33
Loma Novia field, Texas, surfactant–polymer (SP) flooding project in, 127–128
chemicals used for oil recovery, 127–128
reservoir, 127
residual oil saturation, 127
surfactant slug produced, 127–128

M

Mauddud carbonate reservoir, Bahrain, 292–293
McElmo Creek Unit (MCU), Utah, foam field application in, 43–45
Mechanical trapping, 125
Microbial flooding process, 552–554
conclusions and lessons learned, 554
injection facility, 553
pilot tests and results, 553–554
reservoir and fluid data, 553*t*
selection of microbes, 552–553
Microbial-enhanced oil recovery (MEOR)
advantages, 543–544, 545*t*
"Class I Oil Program-Mid Term Activities," 567–568
definition, 543–544, 581
disadvantages, 543–544
essential components, 543–544
characteristics, 549
microbes, 543–544, 548–549
nutrients, 543–544, 548–549
experiments with ZoBell's culture, 561–562
field applications
block Kong-Er-Bei reservoir and fluid data, 556–557, 556*t*
microbial flooding process, 552–554
single-well microbial huff-and-puff process, 551–552
using indigenous microbes, 555–558
well stimulation to remove wellbore or formation damage, 554–555
future studies, 576–577
mechanisms, 544–548
biomass accumulation issues, 548
of biopolymers and biosurfactants, 546
CO_2 generation, 548
heterogeneous reactivity, 547–548
metabolic rate, 546
permeability or microbial reactivity factors, 546–547
microbial permeability profile modification (MPPM) technology, 567–568
at North Blowhorn Creek Oil Unit (NBCU), 567–568, 569*t*
feed and feeding regime, 572*t*
field trials, 569
injection wells, adjustments of, 570
microorganisms, testing of, 574–576
nitrate and phosphate additions, 572–573
nutrient additions, 569*t*, 571, 571*t*
oil production response, 573*t*
performance, 571–572
simulated injection water, testing with, 569
sterile coring devices used, 568
temperature profiles, 573–575
tracer study, 569

Microbial-enhanced oil recovery (MEOR) (*Continued*)
well descriptions, 568
oil degradation at temperatures, 563
origin of, 561–562
patents on, 563–567
aerobic oil-degrading bacteria, 566
Bacillus licheniformis strain JF-2, 564
Bacteriaceae species, 564
denitrifying microorganism, use of, 565
hyperthermophilic anaerobic bacterial strains, 566
indigenous microorganisms, use of, 566
injecting an agent onto the active sites, 565
injecting microorganisms into oil reservoirs, 564
polymeric viscosifier-thickening agent, use of, 564–565
Pseudomonadaceae species, 564
sequentially injecting process, 566
Thauera strain AL 9:8 (ATCC No. PTA 9497), 567
Xanthomonas species, 564
screening criteria, 549–550, 550*t*
formation temperature requirement, 549–550
oil gravity requirement, 549–550
sources of nutrients, 543–544
successful applications, 581
work of, 562–563
Microemulsions
CDC curve, 123
IFT for a middle-phase, 121
lower-phase, 119
middle-phase, 119
types of, 120*f*
upper-phase, 119
viscosity of, 122
water-external phase, 119
Winsor II, 125
Microscopic oil displacement mechanisms, 86–87
Midway Sunset field, California, foam field application in, 35–36
Midway-Sunset field, California, CSS project in, 402–404
cyclic steam operations in, 403–404
producing zones of, 403
reservoir properties, 403
sequential steaming process in, 402, 403*f*
Minas surfactant project, 136–139, 140*t*
interwell tracer tests (ITTs), 139
laboratory coreflood tests, 137–138
presurfactant tracer tests, 138
reservoir and performance description, 137
reservoir surveillance and monitoring program, 138–139
slug sizes and concentrations, 137–138
surfactant field trials, 137–138
Minimum miscibility pressure (MMP) determination, 15–16
Minnelusa Formation, USA, ASP system in, 226
oil recovery efficiency, 228
Miscibility
first-contact miscibility (FCM), 10
for vaporizing, 11
Mixing of polymers, 72
Mobility control methods in gas flooding, 8
Mobility ratio, definition, 6–7
Mobility reduction, foams, 53–58, 261–262
Mobility reduction factor (MRF), 28–30, 45, 52–56, 58*f*
STARS and ECLIPSE model foam behavior, 261–262
Moorcroft West Minnelusa Sand Unit, Wyoming, AP project in, 172–174
biocide application, effects of, 174
Dykstra–Parsons permeability variation coefficient for, 172
oil recovery, 174
primary producing mechanism, 172
produced and injected water analysis in, 173*t*
volumetric sweep improvement program, 173
Movable gel, 78–79
Multicontact test for phase behavior controls, 11

N

Nal-flo B, 136
NEODOL 67-7PO sulfate, 183
NEODOL 25-3S, 181–182, 181*f*
Nitrogen foam flooding in a heavy oil reservoir, analysis, 271–273
application conditions of waterflooding, 271–272
expanded pilot test performance, 272–273
reservoir and well description, 271–272, 272*t*
well patterns, 272

Nonionics, 117–118
North Blowhorn Creek Oil Unit (NBCU), MEOR at, 567–568
  feed and feeding regime, 572*t*
  field trials, 569
  injection wells, adjustments of, 570
  microorganisms, testing of, 574–576
  nitrate and phosphate additions, 572–573
  nutrient additions, 569*t*, 571, 571*t*
  oil production response, 573*t*
  performance, 571–572
  simulated injection water, testing with, 569
  sterile coring devices used, 568
  temperature profiles, 573–575
  tracer study, 569
  well descriptions, 568
North Gujarat Oil Field, India, alkaline flooding project in, 156–157
North Ward-Estes field, Texas, foam field application in, 40–42

O

Oil scan, 120
Oil–gas displacement, 12–16
  for field gas floods, 15
  ternary representation of, 13–15, 13*f*, 14*f*
O.K. Liquid, 35
Optimum salinities, rule for, 179–180
  soap concentration and salinity, 182
  solubilization ratios, 180
Organic oil recovery (OOR). *See also* Microbial-enhanced oil recovery (MEOR)
  in Alton field, Australia, 583
  applications, 581–582, 582*t*
  associated with reservoir plugging, 587–588
  correlations between reservoir parameters and unsuccessful applications, 591–595, 592*t*
  in Devonian reservoir, 585, 586*f*
  in dual-porosity reservoirs, 589–590
  in extreme conditions, application of, 588–589
  field cases
    Beverly Hills field, California, 604–613
    Trial field, Saskatchewan, Canada, 595–604
  Mannville reservoirs, 593, 594*f*
  observations and conclusions, 584
  oil gravities and, 585–587
  oil release mechanism, 582–583, 583*f*
  parameters of reservoirs, 589*t*
  in producing wells, 591–595
  reservoir souring and, 590–591
  screening, 584–585
  in Sparky reservoir, 585–586, 586*f*
  steps in, 581–582
  in tight reservoirs, 591
  in Upper Topanga reservoir, 586–587, 587*f*, 590–591

P

Permeability
  to aqueous phase, 69
  contrast, 32
  disproportionate permeability reduction (DPR), 70
  effective gas relative, 54–55, 57*f*
  gas flooding and, 4, 7–8
  polymer retention and, 68
  reduction, defined, 69
  relative, in polymer flooding, 70
  residual permeability reduction factor, 69
pH effect on hydrolysis, 65
Phase behavior controls, 9–12
  constant-mass expansion tests, 9
  constant-volume depletion tests, 9
  differential liberation tests, 9
  fluid characterization using equation-of-state (EOS), 12
  multicontact test for, 11
  separator tests, 9
  slim-tube experiments for, 10–11
  standard (or basic) pressure, volume, and temperature (PVT) data, 9
  swelling test for, 9–10
Phase behavior tests, 120–121
Phase partitioning, 125
Phase trapping, 125
Pipette tests, 120
Pipettes, 120
Plunger displacement pump, 72
Polyacrylamide (PAM), 63
  hydrolysis of, 63
Polymer flooding projects
  in a Carbonate Reservoir—Vacuum Field, New Mexico, 78
  in the East Bodo Reservoir, Canada, 76
  enhanced oil recovery (EOR), 344
  field performance, 73–74
  in a heterogeneous reservoir, 74–75
  in the Marmul Field, Oman, 77–78
  in the Tambaredjo Field, Suriname, 76–77

Polymer flooding projects (*Continued*)
using high MW and high concentration polymer, 75
wellhead pressures and, 75
Polymers
classification of, 63–64
different forms of, 72
dispersed (concentrated solution), 72
effect on alkaline solution/oil interfacial tension (IFT), 169
in enhanced foam system, 264
flow behavior in porous media, 65–70
adsorption, 67–68
Darcy velocity, 67
equivalent shear rate in, 67
inaccessible pore volume (IPV), 68–69
permeability reduction, 69–70
relative permeability, 70
retention, 67–68
viscosity, 65–67
hydrolysis of, 63
pH effects on, 65
interactions between alkali and, 169
mechanism of flooding, 70–71
economic impact of, 71
fractional flow curves, 70–71, 71*f*
shear stress, 71
in vertical heterogeneous layers, 71
mixing, 72
MW, 69–70
NaCl and, 63
PAM-derived, 63–64
pH effect, 65
pH-sensitive, 63–64
post-polymer conformance control using movable gel, 78–80
power-law model, 65
salinity and concentration effects, 64–65
effective salinity, 64–65
screening criteria for flooding, 72–73
shear effect, 65
solution viscosity, 64–65
solution viscosity and concentration, 97–99, 98*f*
synergy between alkali and, 169–171
sweep efficiency, 170
viscoelastic properties of, 71
xanthan, 63
Power-law equation, 67
Pressure, volume, and temperature (PVT) experiments, 9
Profile modification by polymer flooding, 84–86

R

Rangely Weber Sand Unit (RWSU), Colorado, foam field application in, 38–39
Recovery in gas floods
by condensing gas drive process, 13–14, 13*f*
Jay field near the Alabama–Florida border, 19–20
miscibility, significance of, 12–13
Relative permeabilities, 123–124
effective gas, 54–55, 57*f*
foams, 261
gas phase, 260
polymers, 70
solution-gas drive process, 619
surfactants, 123–124
wettability alteration and, 283, 285
Brooks–Corey model, 285–286
Reservoir screening criteria for polymer flooding, 87–91, 88*t*. *See also* Screening
formation of water salinity, 90–91
heterogeneity, 89–90
oil viscosity, 90
permeability, 88–89, 89*t*
temperature, 88
type, 87–88
Retention of polymers, 67–68
hydrodynamic, 67–68
permeability and, 68
in static bulk tests, 68
Rock Creek, Virginia, foam field application in, 38–39
Russian Шагцр-Гоаеан Field, alkaline flooding project in, 154–155
Russian Трехозерное field, alkaline flooding project in, 151–154
alkaline consumption in, 152
blocks, 151
concentration of tracers, 152
development of the field, 152
field performance, 153–154
laboratory study, 152
monitoring program, 152–153
oil displacement efficiency, 152
pH and carbonate concentrations, 153*f*, 154
reservoir and fluid data, 151*t*
sweep efficiency, 153

S

Salinity scan, 121
Sandstones, Smart Water (SW) flooding. *See also* Smart Water (SW) flooding
adsorption of quinoline onto kaolinite, 322–325, 325*f*
chemical mechanisms, 321
chemical understanding of low salinity EOR mechanism, 321–326
condition for low salinity effects in, 320
crude oil properties, 322, 323*f*
effect of CO_2, 322, 324*f*
low salinity mechanisms, 320–321
pH of effluents, 321–322
salinity effects, 322–325, 324*f*
Satellite Interferometric Synthetic Aperture Radar (InSAR) techniques, 437
Saturation pressure, 363
Saturation temperature, 363
Screening
alkaline–surfactant–polymer (ASP) system, 216–217
acid content, 217
acid number, 217
addition of sequestering or chelating chemicals, 217
initial, 235–236
lithology, 217
oil viscosity, 216
permeability, 217
requirement of freshwater, 217
reservoir and fluid data for initial process, 236*t*
reservoir temperature, 216–217
crude oil alkaline-agent, 171
cyclic steam stimulation (CSS) process, 395–396, 395*t*
in foam flooding applications, 265
gas flooding, 3–5
for gas flooding
gravity effects, 4–5
investment (capital) and operating costs, 5
permeability of reservoirs and, 4
reservoir heterogeneity and conformance, role of, 4
technical factors, 3–4
microbial-enhanced oil recovery (MEOR), 549–550, 550*t*
organic oil recovery (OOR), 584–585
in polymers flooding, criteria for, 72–73
of reservoir in Daqing, 87–91
in situ combustion (ISC), 467–469, 469*t*
for steam flooding, 371–373, 372*t*
Screw pump, 72
Semoga field, Indonesia, 294–295
Sensible heat, 362
Shell's Orion project
start-up process, 422, 422*f*, 428
surveillance employed in, 436*f*
Shengli Oil Field, ASP system in, 225
polymer-enhanced foam flooding, 264
3-slug sequence in, 225
Siggins field, Illinois, foam field application in, 34–35
Single carbon number (SCN) component, 12
Single well polymer-enhanced foam flooding test, 268–271
designed scheme, 269–270
experimental study, 269–270
foam performance, 269
gas rate, 270
gas/liquid ratios, 270–271, 271*f*
oil rate, 270
reservoir and well description, 269
resistance factors, 270–271, 271*f*
test performance, 270–271
water intake profile before foam injection, 270
Single-well microbial huff-and-puff process, 551–552
microbes and effects of microbial products, 551
pilot test, 552
test results, 552
reservoir and fluid description, 551, 551*t*
Slim-tube recoveries, 10–11, 11*f*
pore volumes recovered, 10
Slim-tube test, 10–11
Sloss field, Nebraska, surfactant–polymer (SP) flooding project in, 132–134
design sequence of injection, 132
field operation and pilot performance, 133–134
micellar injection process in, 134
phase-stability tests, 133
reservoir characterization, 133
surface facilities, 134
surfactant adsorption estimation, 133
well arrangement, 132
Smart Water (SW) flooding, 304–306
capillary pressure and relative permeability, impact on, 306
in carbonates
concentration of NaCl in, 314
effluent concentrations, 307, 308*f*, 309*f*, 311*f*

Smart Water (SW) flooding (*Continued*)
environmental effects, 315–316
introduction, 306–307
ions and temperature on oil recovery, 309–311
mechanism for wettability modification, 312, 313*f*
optimization of injected water, 312–314
reactive potential determining ions, 307–312
viscous flood *vs* spontaneous imbibition, 315
chemical understanding of EOR mechanism
in carbonates, 326–328
chalk reservoir Ekofisk in Norwegian sector, 326–327
in sandstones, 328–329
Statoil Snorre Pilot project, 330–332
definition, 306
Ekofisk chalk and, 307, 307*t*
interaction between the cationic surfactant monomers and carboxylic material, 306–307
in limestones, 316–317
condition for low salinity EOR effects in, 317–319
physical principle for EOR, 306
in sandstones
adsorption of quinoline onto kaolinite, 322–325, 325*f*
chemical mechanisms, 321
chemical understanding of low salinity EOR mechanism, 321–326
condition for low salinity effects in, 320
crude oil properties, 322, 323*f*
effect of CO_2, 322, 324*f*
low salinity mechanisms, 320–321
pH of effluents, 321–322
salinity effects, 322–325, 324*f*
surfactant enhanced gravity drainage (SEGD) process, 306–307
validation of, 306
Snorre foam-assisted-water-alternating-gas (FAWAG) project, 273–276
field description, 273
foam injection tests
gas shutoff treatment at well P-18, 273–274
at P-32, 275
at P-39, 275, 276*f*
at P-25A, 274–275
WAG injection at well P-32, 275
Soap-surfactant synergy, 180–183, 183*f*
IFT, effect on, 183
Static mixer, 72
Statoil Snorre Pilot project, 330–332
Steam assisted gravity drainage (SAGD) for oil recovery, 528
Butler's gravity-drainage theory, 413–414
Cold Lake heavy oil field project, 437
definition, 413
drainage rates for three Canadian heavy crudes, 415*f*
evaluation of resources
delineation of wells, 419–420, 420*f*
quality of resource, 416–418
reservoir heterogeneity, effects of, 418
expanding solvent, 440
gravitational force, role of, 414
high-resolution 3D seismic, effect of, 420
integration of subsurface and surface, 440
Joslyn Creek project, 438
McMurray formation geology in Athabasca deposit, 419*f*
production control
Geysering Phenomenon under natural lift, 433–434
steam trap control, 431–432
wellbore lift, 432–433
schematic of, 414*f*
Shell's Orion project
start-up process, 422, 422*f*, 428
surveillance employed in, 436*f*
solvent-enhanced, 440–442
advancement of, 442
oil drainage rate, 441, 442*f*
solvent–steam spectrum, 440–441, 442*f*
start-up process
circulation heating and inter-well communication initialization, 420–423
defined, 420
"hot-pipe" effect, 421
impact on operational parameters and reservoir variables, 422
inter-well formation, 421
pressure gradients effects, 424
pressure-balanced, 422–423
pressure-unbalanced, 422–423
in Shell's Orion project, 422, 422*f*, 428
steam circulation, 421
steam circulation for, 424
steam circulation rate, 422
time to convert to normal, 422

well separation and, 423
wellbore effects, 423–424
steady-state conduction, 413–414
thermal recovery processes, 416
well, reservoir, and facility surveillance (WRFS), 434–438
key aspects, 435
monitoring of reservoir, 435–436
rock deformation evaluation and surface monitoring, 436–438
wellbore pressure and temperature, 435
well completion, design for
carbonate scale, impact of, 427–428
Husky's Lloydminster project, 430*f*
liner plugging issue and treatment, 426–428
reduction and elimination of local steam, 428
sand control liner, 425–426
slotted liners design, 425–426
steam circulation, 424
Surmont project, 431, 440
thermal wellbore insulation, 424–425
using intelligent completion technologies, 429–431
vacuum insulated tubing (VIT), 425
well configuration, 425*f*
wire-wrap screen, 426
well configuration for, 414–416, 415*f*
horizontal wellbores, 414–416
vertical separation of wellbores, 414–416
wind-down processes, 438–440
coinjection of steam and NCG, impact of, 438
numerical evaluation and field experience, 438–439
Steam flooding
estimation of the heated area, 367–370
expression for, 368–369
heat balance equation, 368
steam zone, 368, 370
volumetric heat capacity, 369
field cases
Duri Steam Flood Project, Indonesia, 381–382
Karamay Field, China, 382–383, 383*t*
Kern River, California, 379–381, 380*f*
Qi-40 Block, China, 383–385
WASP in West Coalinga Field, California, 382
heat losses, 366–367
to over- and underburden rocks, 366–367, 369
from produced fluids, 367
from steam zone, 370
from surface pipes, 366
from a wellbore, 366
mechanism of, 371
modes of heat transfer, 364–366
heat conduction, 365
heat convection, 365
thermal radiation, 365–366
oil recovery performance, 370–371
practice in
completion interval, 377–378
formation, 373
injection and production rates, 375–376
injection pattern and well spacing, 374–375
injection schemes, 376
monitoring and surveillance, 379
production facilities, 378
recovery factor and OSR, 377, 377*f*, 378*f*
time to convert steam soak to steam flood, 376–377
water treatment, 378
screening criteria for, 371–373, 372*t*
thermal properties of rock and fluids
enthalpy (H), 363
gravitational potential energy, 364
heat capacity, 361
kinetic energy, 364
latent heat, 361–362
saturation pressure, 363
saturation temperature, 363
sensible heat, 362
steam quality, 363
temperature-dependent oil viscosity, 363–364
thermal or heat diffusivity, 363
total energy of an object, 364
vapor pressure, 363
volumetric heat capacity, 362
Steam quality, 363
Steam-foams, 262–263
field trial, 36
Stefan–Boltzmann law, 365–366
Strong foams, 25–26
Surfactant-alternating gas (SAG) process, 30, 46–51, 266
Surfactant–brine–oil phase behavior, 119–120

Surfactant–gas coinjection processes, 51–52
Surfactant–polymer (SP) flooding
displacement mechanism, 126
enhanced oil recovery (EOR), 345
field performances
Delaware-Childers M/P Project, 136
El Dorado field of Butler County, Kansas, 130–132
Gudong Field, China, 139–141
Loma Novia field, Texas, 127–128
Minas surfactant project, 136–139, 140*t*
Sloss field, Nebraska, 132–134
Torchlight M/P pilot, Wyoming, 134–136
Wichita County Regular field, Texas, 128–130
interactions, 125
screening criteria for, 126
ultralow IFT effect on, 126
United States, field performances in
data, 126
sweep efficiency, 126
Surfactants
adsorption of, 124–125
adsorption on carbonate rock surfaces, 290
cationic and anionic, 289, 291–292
critical micelle concentration (CMC) of, 118–119
defined, 117–118
distribution of molecules, 118*f*
ethoxylated, 290, 290*f*
fluorinated, 266
interactions between alkali and
adsorption of surfactant, 183–184
interfacial tensions (IFTs), 180, 183
oil saturation, 182
optimum salinity and solubilization ratio, 179–180
phase behavior, 180–183
salt effect, 179
synergy between soap and surfactant, 180–183
water–oil ratio (WOR), 180
interfacial tension, 121
ionic nature of, 117–118
Krafft temperature for, 119
parameters to characterize, 118–119
phase trapping, 125
relative permeabilities, 123–124
retention in reservoirs, 124–125
solubilization ratio for, 119
solution phase behavior, 119
in steam-foam applications, 263
used in foam flooding, 265–266
wettability alteration using, 283–286
Surmont project, 431, 440
Swelling test for phase behavior controls, 9–10
bubble-point pressure, 9–10
dew-point behavior, 9–10
for first-contact miscibility (FCM), 10
process, 9–10
purpose, 9

T

Tanner field, Wyoming, ASP project in, 199
Temperature-dependent oil viscosity, 363–364
Tertiary immiscible gas flooding recoveries, 20
THAI™, 528
field testing, 529
in Athabasca oil sands at Conklin, 529
in Kerrobert field, Saskatchewan, 530
fundamental considerations, 530–531
for heavy oil and bitumen recovery, 531
laboratory tests and numerical simulation, 531
for oil sands, 531
laboratory testing, 528–529
Thermal or heat diffusivity, 363
Thermal radiation, 365–366
Thompson Creek Field, Wyoming, AP project in, 174
Three-dimensional fractional flow surface, 28–30
Torchlight M/P pilot, Wyoming, surfactant–polymer (SP) flooding project in, 134–136
injection of high-salinity preflush, 135
laboratory studies conducted, 135–136
micellar fluid injection process in, 135
oil saturation, 135
Tensleep formation, 134–135
Torrance Field, California, alkaline flooding project in, 158–159
Total energy of an object, 364
Trapping number, 122, 284
Trial field, Saskatchewan, Canada, OOR process, 595–604
additional producer applications, 600
background, 595
cumulative oil production, 595
field application process, 596

nutrient treatment, 596–597
pilot project, 597–600
discussion, 604
expanding, 601–603
production response, 598, 599*f*, 600, 601*f*, 602*f*, 603*f*
reservoir screening and laboratory work, 595–596
Upper Shaunavon, 595
TTH displacement processes, 528
TTP-910 modifiers, 74

U

UTCHEM
capillary desaturation curve (CDC), 123
carbonate reservoirs
upscaling aspects of, 287–288
wettability alteration, 283
liquid phase viscosities, 122
mechanistic modeling for ASP process, 212
wettability alteration using surfactants, 283

V

Vapor pressure, 363
VIP-POLYMER™, 109
Viscoelastic properties of polymers, 71
Viscosity
effect of alkali addition, 169
foams, 260–261
of microemulsions, 122
of a polymer solution, 65
water intake ratio and, 87*t*
Volumetric heat capacity, 362
Volumetric sweep efficiency, 2

W

Water-alternating-gas (WAG) technique, 5–6, 37–38, 42–43
benefits, 7
Water-oil ratio (WOR), 34–35
Well stimulation work, 554–555
performance at different wells, 555
in Qianda field, 555, 555*t*
West Kiehl Field, Wyoming, ASP project in, 198–199
White Castle Field, Louisiana, alkaline–surfactant system in, 186–187
carbonate slug (solution), 187
compositions of AS solutions, 187*t*
injectivity sequence, 187
oil recovery, 187
reservoir and fluid properties, 187*t*
sweep efficiency, 187
Whittier Field, California, alkaline flooding project in, 157–158
Wichita County Regular field, Texas, surfactant–polymer (SP) flooding project in, 128–130
falloff and step-rate tests, 128–129
freshwater preflush in, 128
low-tension waterflood (LTWF) process, 128
observation wells, data from, 129
production performance, 130
reservoir engineering studies in, 129–130
surfactant slug produced, 128
Wilmington, California, foam field application in, 37–38
Wilmington Field, California, alkaline flooding project in, 159–164
adsorption loss, 161
area description, 159–160
chemical injection program, 161–162
mini-injection test, 160–161
preflush design, 163
reservoir and fluid data in, 160*t*
scaling problems, 162–164
softened water preflush, effect of, 162
Well B-105, 163
Well B-712, 163
Wormhole, concept of, 631–632

X

Xanthan polymers, 63
Xing-28 Block, China, AP project in, 176–177

Y

Yangsanmu, China, AP project in, 177–178
Yates field, Texas, 293–294

Printed and bound by CPI Group (UK) Ltd, Croydon, CR0 4YY
08/05/2025
01864884-0002